Lecture Notes in Computer Science 2332

Edited by G. Goos, J. Hartmanis, and J. van Leeuwen

Springer
Berlin
Heidelberg
New York
Barcelona
Hong Kong
London
Milan
Paris
Tokyo

Lars Knudsen (Ed.)

Advances in Cryptology – EUROCRYPT 2002

International Conference on the Theory
and Applications of Cryptographic Techniques
Amsterdam, The Netherlands, April 28 – May 2, 2002
Proceedings

Volume Editor

Lars R. Knudsen
Technical University of Denmark, Department of Mathematics
Building 303, 2800 Lyngby, Denmark
E-mail: knudsen@mat.dtu.dk

Cataloging-in-Publication Data applied for

Die Deutsche Bibliothek - CIP-Einheitsaufnahme

Advances in cryptology : proceedings / EUROCRYPT 2002, International Conference on the Theory and Application of Cryptographic Techniques, Amsterdam, The Netherlands, April 28 - May 2, 2002. Lars Knudsen (ed.). - Berlin ; Heidelberg ; New York ; Barcelona ; Hong Kong ; London ; Milan ; Paris ; Tokyo : Springer, 2002
(Lecture notes in computer science ; Vol. 2332)
ISBN 3-540-43553-0

CR Subject Classification (1998): E.3, G.2.1, D.4.6, K.6.5, F.2.1-2, C.2, J.1

ISSN 0302-9743
ISBN 3-540-43553-0 Springer-Verlag Berlin Heidelberg New York

Springer-Verlag Berlin Heidelberg New York
a member of BertelsmannSpringer Science+Business Media GmbH

http://www.springer.de

Typesetting: Camera-ready by author, data conversion by Steingräber Satztechnik GmbH, Heidelberg
Printed on acid-free paper SPIN: 10869749 06/3142 5 4 3 2 1 0

Preface

You are reading the proceedings of EUROCRYPT 2002, the 21st annual Eurocrypt conference. The conference was sponsored by the IACR, the International Association of Cryptologic Research, www.iacr.org, this year in cooperation with the Coding and Crypto group at the Technical University of Eindhoven in The Netherlands. The General Chair, Berry Schoenmakers, was responsible for the local organization, and the conference registration was handled by the IACR Secretariat at the University of California, Santa Barbara, USA. I thank Berry Schoenmakers for all his work and for the pleasant collaboration.

A total of 122 papers were submitted of which 33 were accepted for presentation at the conference. One of the papers is a result of a merger of two submissions. Three additional submissions were withdrawn by the authors shortly after the submission deadline. The program also lists invited talks by Joan Daemen and Vincent Rijmen ("AES and the Wide Trail Strategy") and Stephen Kent ("Rethinking PKI: What's Trust Got To Do with It?"). Also, there was a rump (recent results) session, which Henk van Tilborg kindly agreed to chair.

The reviewing process was a challenging task and many good submissions had to be rejected. Each paper was reviewed by at least three members of the program committee, and papers co-authored by a member of the committee were reviewed by at least five other members. In most cases extensive comments were passed on to the authors. It was a pleasure for me to work with the program committee, whose members all worked very hard over several months. The reviewing process was finalized with a meeting in Copenhagen, on January 13th, 2002.

I am very grateful to the many additional reviewers who contributed with their expertise: Adam Back, Alfred Menezes, Alice Silverberg, Anton Stiglic, Antoon Bosselaers, Ari Juels, Barry Trager, Carlo Blundo, Chan Sup Park, Chong Hee Kim, Christian Paquin, Christophe De Cannière, Craig Gentry, Dae Hyun Yum, Dan Bernstein, Dario Catalano, David Pointcheval, David Wagner, Dong Jin Park, Dorian Goldfeld, Eliane Jaulmes, Emmanuel Bresson, Florian Hess, Frederik Vercauteren, Frédéric Légaré, Frédéric Valette, Glenn Durfee, Guillaume Poupard, Gwenaelle Martinet, Han Pil Kim, Hein Roehrig, Hovav Shacham, Ilya Mironov, Jacques Stern, Jae Eun Kang, Jan Camenisch, Jean-Francois Raymond, Jens Jensen, Jesper Buus Nielsen, Jim Hughes, John Malone-Lee, Jonathan Poritz, Jong Hoon Shin, Katsuyuki Takashima, Kazue Sako, Kenny Paterson, Kyung Weon Kim, Leo Reyzin, Louis Granboulan, Louis Salvail, Markku-Juhani O. Saarinen, Matt Robshaw, Michael Quisquater, Michael Waidner, Michel Mitton, Mike Szydlo, Mike Wiener, Moti Yung, Olivier Baudron, Omer Reingold, Paul Dumais, Paul Kocher, Philippe Chose, Philippe Golle, Pierre-Alain Fouque, Ran Canetti, Richard Jozsa, Ronald Cramer, Sang Gyoo Sim, Sang Jin Lee, Serge Fehr, Shirish Altekar, Simon Blackburn, Stefan Wolf, Steven Galbraith, Svetla Nikova, Tae Gu Kim, Tal Malkin, Tal Rabin, Tetsu Iwata, Toshio Hasegawa, Tsuyoshi Nishioka, Virgil Gligor, Wenbo Mao, Yeon Kyu Park, Yiqun Lisa Yin, Yong Ho Hwang, Yuval Ishai.

My work as program chair was made a lot easier by the electronic submission software written by Chanathip Namprempre for Crypto 2000 with modifications by Andre Adelsbach for Eurocrypt 2001, and by the reviewing software developed and written by Bart Preneel, Wim Moreau, and Joris Claessens for Eurocrypt 2000. I would like to thank Ole da Silva Smith for setting up all this software locally and for the help with the problems I encountered. I am also grateful to Wim Moreau and Chanathip Namprempre for solving some of the problems we had with the software.

On behalf of the general chair I would like to extend my gratitude to the members of the local organizing committee at TU Eindhoven, in particular to Peter Roelse and Gergely Alpár. For financial support of the conference the organizing committee gratefully acknowledges this year's sponsors: Philips Semiconductors Cryptology Competence Center, Mitsubishi Electric Corporation, cv cryptovision, Cryptomathic, ERCIM, CMG, Sectra, EUFORCE, and EIDMA.

Finally, a thank-you goes to all who submitted papers to this conference and last but not least to my family for their love and understanding.

February 2002 Lars Knudsen

EUROCRYPT 2002

April 28–May 2, 2002, Amsterdam, The Netherlands

Sponsored by the

International Association of Cryptologic Research (IACR)

in cooperation with

The Coding and Crypto group at the Technical University of Eindhoven in The Netherlands

General Chair

Berry Schoenmakers, Department of Mathematics and Computing Science, Technical University of Eindhoven, The Netherlands

Program Chair

Lars R. Knudsen, Department of Mathematics, Technical University of Denmark

Program Committee

Dan Boneh .... Stanford University, USA
Stefan Brands .... McGill University School of Computer Science, Montreal, Canada
Christian Cachin .... IBM Research, Zurich, Switzerland
Don Coppersmith .... IBM Research, USA
Ivan Damgård .... Aarhus University, Denmark
Anand Desai .... NTT Multimedia Communications Laboratories, USA
Rosario Gennaro .... IBM Research, USA
Alain Hiltgen .... UBS, Switzerland
Markus Jakobsson .... RSA Laboratories, USA
Thomas Johansson .... University of Lund, Sweden
Antoine Joux .... DCSSI, France
Pil Joong Lee .... Postech, Korea
Arjen Lenstra .... Citibank and Technical University of Eindhoven
Keith Martin .... Royal Holloway, University of London, UK
Mitsuru Matsui .... Mitsubishi Electric, Japan
Phong Q. Nguyen .... CNRS/Ecole Normale Supérieure, France
Kaisa Nyberg .... Nokia Research Center, Finland
Bart Preneel .... Katholieke Universiteit Leuven, Belgium
Reihaneh Safavi-Naini .... University of Wollongong, Australia
Nigel Smart .... University of Bristol, UK
Paul Van Oorschot .... Carleton University, Canada
Rebecca Wright .... DIMACS, USA

Table of Contents

Cryptanalysis I

Cryptanalysis of a Pseudorandom Generator Based on Braid Groups 1
Rosario Gennaro, Daniele Micciancio

Potential Weaknesses of the Commutator Key Agreement Protocol Based on Braid Groups 14
Sang Jin Lee, Eonkyung Lee

Extending the GHS Weil Descent Attack 29
Steven D. Galbraith, Florian Hess, Nigel P. Smart

Public-Key Encryption

Universal Hash Proofs and a Paradigm for Adaptive Chosen Ciphertext Secure Public-Key Encryption 45
Ronald Cramer, Victor Shoup

Key-Insulated Public Key Cryptosystems 65
Yevgeniy Dodis, Jonathan Katz, Shouhuai Xu, Moti Yung

On the Security of Joint Signature and Encryption 83
Jee Hea An, Yevgeniy Dodis, Tal Rabin

Invited Talk

AES and the Wide Trail Design Strategy 108
Joan Daemen, Vincent Rijmen

Information Theory & New Models

Indistinguishability of Random Systems 110
Ueli Maurer

How to Fool an Unbounded Adversary with a Short Key 133
Alexander Russell, Hong Wang

Cryptography in an Unbounded Computational Model 149
David P. Woodruff, Marten van Dijk

Implementational Analysis

Performance Analysis and Parallel Implementation of Dedicated Hash Functions 165
Junko Nakajima, Mitsuru Matsui

Fault Injection and a Timing Channel on an Analysis Technique 181
John A. Clark, Jeremy L. Jacob

Speeding Up Point Multiplication on Hyperelliptic Curves with Efficiently-Computable Endomorphisms 197
Young-Ho Park, Sangtae Jeong, Jongin Lim

Stream Ciphers

Fast Correlation Attacks: An Algorithmic Point of View 209
Philippe Chose, Antoine Joux, Michel Mitton

BDD-Based Cryptanalysis of Keystream Generators 222
Matthias Krause

Linear Cryptanalysis of Bluetooth Stream Cipher 238
Jovan Dj. Golić, Vittorio Bagini, Guglielmo Morgari

Digital Signatures I

Generic Lower Bounds for Root Extraction and Signature Schemes in General Groups 256
Ivan Damgård, Maciej Koprowski

Optimal Security Proofs for PSS and Other Signature Schemes 272
Jean-Sébastien Coron

Cryptanalysis II

Cryptanalysis of SFLASH 288
Henri Gilbert, Marine Minier

Cryptanalysis of the Revised NTRU Signature Scheme 299
Craig Gentry, Mike Szydlo

Key Exchange

Dynamic Group Diffie-Hellman Key Exchange under Standard Assumptions 321
Emmanuel Bresson, Olivier Chevassut, David Pointcheval

Universally Composable Notions of Key Exchange and Secure Channels... 337
Ran Canetti, Hugo Krawczyk

On Deniability in Quantum Key Exchange 352
Donald Beaver

Modes of Operation

A Practice-Oriented Treatment of Pseudorandom Number Generators 368
Anand Desai, Alejandro Hevia, Yiqun Lisa Yin

A Block-Cipher Mode of Operation for Parallelizable Message Authentication 384
John Black, Phillip Rogaway

Invited Talk

Rethinking PKI: What's Trust Got to Do with It? 398
Stephen Kent

Digital Signatures II

Efficient Generic Forward-Secure Signatures with an Unbounded Number of Time Periods 400
Tal Malkin, Daniele Micciancio, Sara Miner

From Identification to Signatures via the Fiat-Shamir Transform: Minimizing Assumptions for Security and Forward-Security 418
Michel Abdalla, Jee Hea An, Mihir Bellare, Chanathip Namprempre

Security Notions for Unconditionally Secure Signature Schemes 434
Junji Shikata, Goichiro Hanaoka, Yuliang Zheng, Hideki Imai

Traitor Tracking & Id-Based Encryption

Traitor Tracing with Constant Transmission Rate 450
Aggelos Kiayias, Moti Yung

Toward Hierarchical Identity-Based Encryption 466
Jeremy Horwitz, Ben Lynn

Multiparty and Multicast

Unconditional Byzantine Agreement and Multi-party Computation
Secure against Dishonest Minorities from Scratch 482
Matthias Fitzi, Nicolas Gisin, Ueli Maurer, Oliver von Rotz

Perfectly Secure Message Transmission Revisited 502
Yvo Desmedt, Yongge Wang

Symmetric Cryptology

Degree of Composition of Highly Nonlinear Functions
and Applications to Higher Order Differential Cryptanalysis 518
Anne Canteaut, Marion Videau

Security Flaws Induced by CBC Padding –
Applications to SSL, IPSEC, WTLS 534
Serge Vaudenay

Author Index 547

Cryptanalysis of a Pseudorandom Generator Based on Braid Groups

Rosario Gennaro[1] and Daniele Micciancio[2,*]

[1] IBM T.J. Watson Research Center, New York, USA,
rosario@watson.ibm.com
[2] University of California, San Diego, La Jolla, California, USA,
daniele@cs.ucsd.edu

Abstract. We show that the decisional version of the Ko-Lee assumption for braid groups put forward by Lee, Lee and Hahn at Crypto 2001 is false, by giving an efficient algorithm that solves (with high probability) the corresponding decisional problem. Our attack immediately applies to the pseudo-random generator and synthesizer proposed by the same authors based on the decisional Ko-Lee assumption, and shows that neither of them is cryptographically secure.

1 Introduction

The search for computationally hard problems to be used as a basis of secure encryption functions is a central problem in cryptography. Recently, braid groups have attracted the attention of many cryptographers as a potential source of computational hardness and many cryptographic protocols have been suggested based on braid groups [1,10,11]. Computational assumptions and cryptographic protocols based on braid groups often resemble similar constructions based on number theoretic groups.

In this paper we point out some fundamental differences between braid groups and number theoretic ones, and show that protocols based on braid groups that are naively designed by analogy with number theoretic groups, can be easily broken. In particular, we show that the decisional version of the Ko-Lee problem put forward by Lee, Lee and Hahn in [11], can be efficiently solved and cannot be used as a basis for the design of secure cryptographic functions. Our attack is extremely efficient: for the values of the security parameters suggested in [11] it only requires a handful of arithmetic operations. Moreover, the attack is asymptotically fast (i.e., polynomial in the input size) and cannot be avoided simply by increasing the value of the security parameter.

Our attack immediately invalidates the security proof of the pseudo-random generator suggested in [11], based on the conjectured (and now disproved) hardness of the decisional Ko-Lee problem. In fact, the scope of our attack extends beyond simply invalidating the computational assumption. Essentially the same techniques used to show that the computational assumption is false, can be used

* Supported in part by NSF Career Award CCR-0093029

L.R. Knudsen (Ed.): EUROCRYPT 2002, LNCS 2332, pp. 1–13, 2002.

to break the "pseudo-random" braid generator proposed in [11] and efficiently distinguish the braids produced by the generator from truly random braids. Our attack applies to the pseudo-random braid generator, as well as the pseudo-random braid synthesizer proposed in [11].

We point out that our attack does not seem to apply to the *computational* version of the Ko-Lee problem, as used in [10] to design a practical encryption scheme in the random oracle model [2]. Without "random oracles", our attack could have been used to extract partial information about the message in the encryption scheme proposed in [10], thereby breaking semantic security. (I.e., the standard notion of security for encryption schemes, see [9].) The use of "random oracles" in [10] protects their encryption scheme from the kind of attacks described in this note, and seems to require a solution to the *computational* version of the Ko-Lee assumption in order to successfully attack the system. A similar "fix" (i.e., applying a hash function modeled as a random oracle to the output of the generator) would clearly apply to the generator of [11] as well, but it would also make the problem studied in [11] completely trivial: in the random oracle model, an extremely efficient pseudo-random generator can be immediately built applying the random oracle directly to the seed, without any computational assumption about braids whatsoever.

It is important to observe that in order to make the result of a Diffie-Hellman or Ko-Lee key exchange look pseudo-random, it is not enough to apply a universal hash function [6]. Indeed universal hash functions produce an almost-random output when starting from an input that has enough entropy to begin with, and this seems to require the *decisional* assumption which we prove to be false for the case of braid groups. In other words, to prove semantic security based only on the *computational* Diffie-Hellman or Ko-Lee assumption the full power of the random oracle model seems to be required.

Our attack does not imply that braid groups cannot be used for cryptographic purposes, and we believe that the use of braid groups as an alternative to number theory in cryptographic applications is an attractive and promising research area. However, extreme care must be used to avoid pitfalls as those exploited in our attack, and cryptographic protocols based on the hardness of computational problems on braid groups must be carefully validated by accurate proofs of security.

Most of the proposed cryptographic schemes based on braid groups, rely on the hardness of the conjugacy problem. Besides [10,11] other schemes include [1] for example. We point out that our attack to the Decisional Ko-Lee assumption, also reveals a more general fact that applies to all the schemes above. The one-way function constructed from the conjugacy problem reveals partial information about its input (it is indeed this partial information that allows us to build our attack). This leakage of information can be avoided by a careful choice of the design parameters (i.e., by working in an appropriate subgroup), which should be incorporated in all schemes based on the braid conjugacy problem. This is however not enough to save the Decisional Ko-Lee Assumption which we show to be insecure for essentially *any* choice of the design parameters.

As a final remark, we point out that the proof of security of the "hard-core" predicate for the conjugacy one-way function described in [11] contains a fundamental flaw (see Sect. 5). Although we do not know of any cryptanalytic attack at the time of this writing, the security of that predicate by no means can be considered proven based on the conjectured intractability of standard problems on braid groups. Until a satisfactory proof of security is found, the only way to get hard-core predicates for the braid conjugacy function is to invoke general results like the Goldreich-Levin predicate [8].

2 Preliminaries

The n-braid group B_n is an infinite non-commutative group defined by the following group presentation:

$$B_n = \left\langle \sigma_1, \ldots, \sigma_{n-1} : \begin{array}{ll} \sigma_i \sigma_j \sigma_i = \sigma_j \sigma_i \sigma_j & \text{if } |i-j| = 1, \text{ and} \\ \sigma_i \sigma_j = \sigma_j \sigma_i & \text{if } |i-j| \geq 2 \end{array} \right\rangle$$

The integer n is called the *braid index*. Elements of B_n are called n-braids. We can give braids the following geometric interpretation. Think of n strands hanging contiguously from the ceiling. Each generator σ_i represents the process of swapping the ith strand with the next one (with the ith strand going under the $(i+1)^{th}$ one).

The multiplication of two braids a, b is then defined as their concatenation (i.e. geometrically putting a on top of b). The identity e is the braid of n straight strands. The inverse of a is the reflection of a across an horizontal line at the bottom.

There is an efficient algorithm to put braids in a normal form. We are not going to use this fact, except for allowing us to define uniquely the *length* of a braid a, which we denote by $|a|$.

PERMUTATIONS VS BRAIDS. It is interesting to note that there is a natural projection from braids to permutations. Geometrically, given a braid a we can define a permutation π_a induced by a as follows. For the ith strand, consider the position j in which this strand ends at the bottom of a, and define $\pi_a(i) = j$. We will make extensive use of this projection from braids to permutations.

CONJUGACY. Given a braid a, we say that another braid y is a *conjugate* of a if there exists a braid x such that $y = xax^{-1}$. It is assumed that solving the conjugacy problem (i.e., retrieving x from a and y) is computationally hard in B_n.

2.1 Key Exchange Based on Braids

At Crypto 2000, [10] suggested a new key exchange protocol based on the conjugacy problem on braid groups. We briefly recall their ideas here.

Notation: For the rest of the paper, we assume that n is even, and let $\ell = \frac{n}{2}$. The key exchange protocol in [10] concerns the braid groups B_n with even n, and the two subgroups $B_L, B_R \subset B_n$ defined as follows:

- B_L is the subgroup generated by $\sigma_1, \ldots, \sigma_{\ell-1}$, i.e., the subgroup of braids that only act on the left strings $1, \ldots, \ell$.
- B_R is the subgroup generated by $\sigma_{\ell+1}, \ldots, \sigma_{n-1}$, i.e., the subgroup of braids that only act on the right strings $\ell+1, \ldots, n$.

The relevant property is that elements of B_L and B_R commute, i.e., for any $x_l \in B_L$ and $x_r \in B_R$ we have $x_l x_r = x_r x_l$.

This property was used in [10] to construct the following key exchange protocol. Let a be a public braid, $a \in_R B_n$, with $|a| = k$ (where k is a security parameter). Alice has a public key y, where $y = x_l a x_l^{-1}$ for $x_l \in_R B_L$, $|x_l| = k$. Similarly Bob has a public key z, where $z = x_r a x_r^{-1}$ for $x_r \in_R B_R$, $|x_r| = k$. The shared key is $s = x_l x_r a x_r^{-1} x_l^{-1}$. Indeed, since x_l and x_r commute, s can be computed given one of the public keys and the other secret key:

$$s = x_l x_r a x_r^{-1} x_l^{-1} = x_r x_l a x_l^{-1} x_r^{-1} = x_r y x_r^{-1} = x_l z x_l^{-1}$$

Notice that it is necessary to assume that conjugacy is hard in B_n for this to be a secure protocol (otherwise an attacker could compute x or w on its own). But the security of the protocol actually relies on a stronger assumption: i.e. that s must be hard to compute given y, z. We call this the *Computational Ko-Lee Assumption.*

Moreover this assumption alone is not sufficient to achieve provable semantic security for the resulting cryptosystem, since if s is used as a shared key it should be random or pseudo-random. In [10] this problem is resolved by setting $key = H(s)$ where H is a suitable hash function and security can be proven in the random oracle model.

There is an analogy with the Diffie-Hellman key exchange, where Alice has a public key $y = a^{x_l}$ and Bob has public key $z = a^{x_r}$, and they share the key $key = H(s)$ where $s = a^{x_l x_r}$. In order to prove security (in the random oracle model) the (Computational) Diffie-Hellman assumption is required, not just the hardness of computing discrete logarithms. Alternatively, one can assume that the Diffie-Hellman problem is hard even in its decisional version: given a, y, z, w it is hard to tell (with probability substantially better than 1/2 if $w = s$ or w is a randomly chosen element in the group generated by a. This Decisional Diffie-Hellman (DDH) assumption has been widely used in cryptography and it is now a relatively established assumption. Under the *DDH* assumption, no random oracle is needed because s is already a pseudo-random group element. (See [3] for further discussion of the DDH problem.)

3 Main Result

In this section we prove the main result of the paper, namely that the decisional version of the Ko-Lee assumption is false. This assumption was suggested in [11] as the basis for some constructions of pseudo-random generators and synthesizers based on braid groups.

The assumption can be considered the equivalent of the Decisional Diffie-Hellman Assumption for the key exchange scheme based on braid groups proposed in [10].

Informally the **Decisional Ko-Lee assumption** says the following.

Given the following public information: a, y, z where
- $a \in_R B_n$, with $|a| = k$,
- $y = x_l a x_l^{-1}$ for $x_l \in_R B_L$, $|x_l| = k$
- $z = x_r a x_r^{-1}$ for $x_r \in_R B_R$, $|x_r| = k$

it is hard to distinguish the "shared key" $s = x_l x_r a x_r^{-1} x_l^{-1}$ from a random conjugate of a of the form waw^{-1}.

The Ko-Lee Decisional Assumption, goes a step further with respect to its computational counterpart. It claims that, not only s is hard to compute, but it's even hard to *distinguish* from a random conjugate of a. In other words, under this assumption the hash function H would not be necessary to prove security since s could be used directly as a random shared key.

More formally the decisional version of the Ko-Lee Assumption can be stated as follows.

Assumption 1. *For every probabilistic polynomial-time Turing machine $\mathcal{D}$, for every polynomial P, for all sufficiently large k*

$$|\Pr[\mathcal{D}(a, x_l a x_l^{-1}, x_r a x_r^{-1}, x_l x_r a x_r^{-1} x_l^{-1}) = 1] - \Pr[\mathcal{D}(a, x_l a x_l^{-1}, x_r a x_r^{-1}, waw^{-1}) = 1]| \leq \frac{1}{P(k)}$$

where the probability is taken over the coin tosses of $\mathcal{D}$ and the following random choices: $a, w \in_R B_n$, $x_l \in_R B_L$, $x_r \in_R B_R$, all braids of length k.

We show that Assumption 1 is false, by exhibiting partial information about s that can be computed from the public information a, y, z. In fact we describe a sequence of attacks. Each attack exploits some specific partial information, and can be avoided by suitably restricting the way a and w are chosen. But then, another attack applies. The sequence of attacks leads to a complete break of the system, showing that for any reasonable choice of probability distribution on the key space a, x_l, x_r, w there is an efficient algorithm that distinguishes $(x_l x_r)a(x_l x_r)^{-1}$ from a random conjugate waw^{-1}.

3.1 The Permutation Attack

The main idea behind the first attack is to focus on the permutation induced by each braid, over the set of strings $\{1, 2, \ldots, n\}$. We are going to use the fact that the braid x_l only acts on the strings on the left, while the braid x_r acts only on the strings on the right.

Our attack shows that the permutation induced by s must satisfy some constraints, which are very unlikely to be satisfied by the permutation induced by a random conjugate. Our attack is actually stronger than needed since it works *for almost any* braid a and not just for a randomly selected one. We start with a basic fact about the permutation induced by a conjugate of a braid.

Fact 2. *Let $a, x, y \in B_n$ be such that $y = xax^{-1}$. Then π_a and π_y have the same cycle structure. Moreover if A is a cycle in π_a, then the corresponding cycle in π_y is $\pi_x^{-1}(A)$.*

The above fact gives some information about the permutation π_x which is hidden in the conjugate y. We are going to use this information to distinguish s from a random conjugate. The only case in which the above fact does not reveal anything about π_x is when π_a is the identity permutation. We say that a is a *pure* braid if π_a is the identity permutation. We are going to show that if a is *not* a pure braid than Assumption 1 is false.

Remark: Before we show that Assumption 1 is false, let us point out that Fact 2 holds regardless of the way we choose the string x used for the conjugation. Thus, we are basically pointing out that the conjugacy problem, although supposed to be hard to solve, does reveal some partial information about x, unless the braid a is chosen to be pure. Since the conjugacy problem is used in all the braid group cryptosystems, this basic fact applies to all of them. To avoid this leakage of partial information, it seems that the all the schemes above should select the braid a as a pure braid.

We now resume the proof that Assumption 1 is false when a is not a pure braid. We distinguish two cases. In the first case, the permutation π_a maps some elements of $\{\ell+1, \ell+2, \ldots, n\}$ to the set $\{1, 2, \ldots, \ell\}$, i.e., from the right half to the left half. (Notice that this is equivalent to the symmetric condition, i.e., the permutation maps some string from the left half to the right half.) The second case covers all the other non-trivial permutations, i.e., the ones that have cycles of at least size 2 and they are all contained in one of the two halves.

Case 1. Let i be an integer in $\{\ell + 1, \ell + 2, \ldots, n\}$ such that $j = \pi_a(i) \in \{1, 2, \ldots, \ell\}$.

We now find the element $j' = \pi_y(i) = \pi_{x_l}^{-1}(\pi_a(\pi_{x_l}(i)))$. Since π_{x_l} only acts on elements on the left, we have that $\pi_{x_l}(i) = i$. Thus $j' = \pi_{x_l}^{-1}(\pi_a(i)) = \pi_{x_l}^{-1}(j)$. In other words we found the mapping of the point j under $\pi_{x_l}^{-1}$.

A similar reasoning tells us that if we take i' such that $\pi_z(i') = j$ we have that $\pi_{x_r}(i') = i$. Indeed by taking the inverse of π_z we get $i' = \pi_{x_r^{-1}}(\pi_{a^{-1}}(\pi_{x_r}(j)))$. Now recall that $j = \pi_a(i)$ is on the left and x_r acts only on the right elements, thus we simplify to $\pi_{x_r}(i') = i$ as desired.

At this point we can check if s was generated from the public keys since if so π_s must map i' into j'. When $s = waw^{-1}$, this is true if and only if π_a maps $\pi_w(i')$ to $\pi_w(j')$. For a randomly generated conjugate, the pair $(\pi_w(i'), \pi_w(j'))$ is distributed (almost) uniformly, so the probability that π_a maps $\pi_w(i')$ to $\pi_w(j')$ is roughly proportional to $1/n$.

Case 2. This case is actually easier than the previous one. We are assuming that π_a does not map any element "across the border" from the two halves. In other words all the cycles of π_a are fully contained in either half and π_a can be written as the product $\pi_a = \pi_{a_l}\pi_{a_r}$ where π_{a_l} acts only on the left elements and π_{a_r} only on the right ones. But then the cycles in π_s will be the union of the cycles on the left half of π_y and the cycles on the right half of π_z. A random conjugate will not have this property with probability close to 1, unless a is a pure braid and π_a is the identity permutation.

We have seen that when the braid a is randomly chosen (or, even more generally, whenever a does not belong to the subgroup of pure braids) it is easy to distinguish triples $x_l a x_l^{-1}, x_r a x_r^{-1}, x_l x_r a (x_l x_r)^{-1}$ (x_l and x_r chosen at random from B_L and B_R) from $x_l a x_l^{-1}, x_r a x_r^{-1}, waw^{-1}$ (where w is just any random braid).

An easy "fix" that comes to mind is to redefine the decisional Ko-Lee assumption, setting w to the product of two random braids $w = w_l w_r$, uniformly chosen from B_L and B_R. In other words, instead of claiming that $s = x_l x_r a (x_l x_r)^{-1}$ is indistinguishable from a random conjugate of the form waw^{-1} with $w \in_R B_n$, one could claim that s is indistinguishable from a conjugate of the form $w_l w_r a w_r^{-1} w_l^{-1}$ where $w_l \in_R B_L$ and $w_r \in_R B_R$. But our distinguisher works with this modified definition as well, as long as a is not a pure braid. Details follows. (1) If permutation π_a maps some element of $\{\ell+1, \ldots, n\}$ to $\{1, \ldots, \ell\}$ we can find i' and j' such that $\pi_s(i') = j'$, exactly the same way we did is Case 1 above. If w is chosen as the product $w_l w_r$ of a left and right half braids, then $\pi_w(i') = \pi_{w_r}(i')$ is distributed uniformly at random in $\{\ell+1, \ldots, n\}$ and $\pi_w(j') = \pi_{w_l}(i')$ is distributed uniformly at random in $\{1, \ldots, \ell\}$. So, the probability that π_a maps $\pi_w(i')$ to $\pi_w(j')$ is at most $1/(n/2) = 2/n$. (2) If permutation π_a is the product $\pi_{a_l}\pi_{a_r}$ of a left and right permutation, then π_s is also the product of a left and right permutation, and we can completely recover π_s as the product of the left half of π_y and the right half of π_z. Also in this case, if $s = waw^{-1}$ is a random conjugate with $w = w_l w_r$, the probability of getting the right permutation π_s is at most $2/n$, unless a is a pure braid and π_a is the identity permutation.

This shows that restricting w to the product $w_l w_r$ of a left and half braid does not make the problem substantially harder. Still we believe that the decisional Ko-Lee problem is more naturally defined with w chosen at random within the subgroup $B_R B_L$, and in the rest of this section we concentrate on this alternative definition.

3.2 The Half-Braid Attack

We now consider the case where a is a pure braid, and show that unless the choice of a is restricted to even a smaller subgroup, it is possible to successfully attack the decisional problem. For a pure braid a, define the left and right projections

$\tau_l(a)$ and $\tau_r(a)$ as the braids (over $n/2$ strings) obtained removing the first or last half of the strings. Now, we are given a, $x_l a x_l^{-1}$, $x_r a x_r^{-1}$ and $s = waw^{-1}$ where either $w = x_l x_r$ or $w = w_l w_r$ for independently and randomly chosen w_l, w_r. It is easy to see that in the first case $\tau_l(x_l a x_l^{-1}) = \tau_l(s)$ and $\tau_l(x_r a x_r^{-1}) = \tau_r(s)$, while in the second case equality does not hold (with high probability) unless $\tau_l(a) = \tau_r(a) = e$ are both the identity braid over $n/2$ strings.

At this point, the only case for which our attacks do not work is when a is chosen as a pure braid with $\tau_l(a) = \tau_r(a) = e$, and w is chosen as the product $w = w_l w_r$ of a left and right braid. This seems an interesting subgroup of braids and an interesting special case of the conjugacy problem. If the conjugacy problem were hard for this special case even in the decisional version, then one could build a pseudo-random generator out of it, fixing the problem of [11]. Unfortunately we will see in the next subsection that even under this restrictions the decisional problem is easy.

3.3 The Single String Attack

Assume we start from a braid a such that $\tau_l(a) = \tau_r(a)$ is the identity. For every $i = n/2 + 1, \ldots, n$, consider the braid $\tau_l^i(a)$ obtained from a removing all right strings except the ith. These are very simple braids: braids obtained running a single string around $n/2$ parallel strings. We claim that unless braids $\tau_l^i(a)$ are the same for all $i = n/2+1, \ldots, n$, we can break the decisional Ko-Lee problem.

Assume they are not all the same and divide the indices $i = n/2, \ldots, n$ according to their equivalence classes, with $i \equiv j$ if and only if $\tau_l^i(a) = \tau_l^j(a)$. Notice that this equivalence relation is efficiently computable because the word problem on braids can be solved in polynomial time. Moreover, since here we are considering a very special class of braids, equivalence can be decided very efficiently.

Notice that the equivalence relation induced by $\tau_l^i(a) = \tau_l^j(a)$ is the same as the one induced by $\tau_l^i(x_l a x_l^{-1}) = \tau_l^j(x_l a x_l^{-1})$. Similarly, $\tau_l^i(x_r a x_r^{-1}) = \tau_l^j(x_r a x_r^{-1})$ induces the same equivalence classes as $\tau_l^i(x_l x_r a (x_l x_r)^{-1}) = \tau_l^j((x_l x_r) a (x_l x_r)^{-1})$. On the other hand, when we compute the conjugate under a right braid x_r the equivalence classes are mapped by the permutation associated to x_r. So, we can decide whether $(w_l w_r) a (w_l w_r)^{-1} = (x_l x_r) a (x_l x_r)^{-1}$ or not by comparing the equivalence relation induced by $\tau_l^i(w_l w_r a (w_l w_r)^{-1}) = \tau_l^j((w_l w_r) a (w_l w_r)^{-1})$ with that of $\tau_l^i(x_r a (x_r)^{-1}) = \tau_l^j((x_r) a (x_r)^{-1})$.

In order to avoid this attach braid a should be chosen in such a way that not only $\tau_r(a) = \tau_l(a) = e$, but also for all $i, j \in \{n/2, \ldots, n\}$, $\tau_l^i(a) = \tau_l^j(a)$.

Using τ_r instead of τ_l we get a symmetric attack that shows that we need an analogous condition $\tau_r^i(a) = \tau_r^j(a)$ for all $i, j = 1, \ldots, n/2$. Now the question is: assume that the braids are chosen in such a way that all above conditions are satisfied; what are the remaining braids? and, is the conjugacy problem for these braids still hard?

3.4 The Trivial Attack

Consider pure braids a satisfying all the conditions stated at the end of the previous subsection:

- $\tau_r(a) = e$
- $\tau_l(a) = e$
- $\tau_l^i(a) = \tau_l^j(a)$ for all $i, j \in \{n/2+1, \ldots, n\}$,
- $\tau_r^i(a) = \tau_r^j(a)$ for all $i, j \in \{1, \ldots, n/2\}$.

It is easy to see that there are only very few braids that satisfy these conditions: the group generated by the braid $a = (\Delta_{n/2}\Delta_n^{-1}\Delta_{n/2})^2$, where Δ_k is the fundamental braid over (the first) k strings. Essentially this is a braid obtained swapping the left and right strings without permuting string in each half, and passing all the right strings over the left strings. The whole operation is performed twice so that the permutation of a is the identity, i.e., a is a pure braid.

But the conjugacy class under left or right braids of a (or any of its powers) is trivial, i.e., $x_l a x_l^{-1} = a$ and $x_r a x_r^{-1} = a$. So, the instances of the decisional Ko-Lee problem have always the form (a, a, a, a) and the answer is always yes.

4 The Attack Extends to the Pseudo-random Constructions

In [11] the Ko-Lee Decisional Assumption is used to construct a pseudo-random generator and a pseudo-random synthesizer. By showing that the underlying assumption is false we have removed the proof of security for the above constructions. However that does not immediately imply an *attack* on the generators.

In this section we show that it is possible to use the above attack to distinguish the output of the generators from random. This is because the constructions are a straightforward application of the conjugacy function. Here we only show the attack on the pseudo-random generator.

The construction is a follows: there are the following public parameters:

- $a \in_R B_n$, with $|a| = k$
- $a_1, \ldots, a_m$, with $a_i = x_i a x_i^{-1}$ for $x_i \in_R B_L$ with $|x_i| = k$.

On input a random seed $w \in_R B_R$ the generator outputs the conjugate of w with respect to all the a_i's. That is, the output is m braids $s_1, \ldots, s_n$ where $s_i = w a_i w^{-1}$. Clearly each component of the output can be distinguished from random using the attack described above.

It should be noted that this attack applies to the "pseudo-random *braid* generator", i.e., a function that on input a short seed, produces a sequence of seemingly random *braids*. In [11] it is suggested that if one wants a "pseudo-random *bit* generator", then universal hashing and the left-over hash lemma can be used to transform a pseudo-random sequence of braids into a pseudo-random sequence of bits. However, the converse is not necessarily true: if the original braid sequence is not pseudo-random, there is no guarantee that applying universal hashing results in a bit-sequence which is computationally indistinguishable from random.

5 On the Bit Security of the Conjugacy Problem

We now offer some remarks on an independent result still contained in [11]. Besides the pseudo-random constructions based on the Ko-Lee Decisional Assumption, [11] claims to construct a hard-core bit for the one-way function induced by the conjugacy problem.

Let $a \in B_n$ and define the following function over B_n, $f(x) = xax^{-1}$. By restricting the input size and assuming the conjugacy problem is hard, it is reasonable to assume that f is a one-way function.

The next question is: does f have any natural hard-core predicate (by natural, we mean something specific to the description of f and not a generic hard-core predicate like the Goldreich-Levin [8] which holds for any one-way function).

We recall that a hard-core predicate π for f is defined as follows. We say that the predicate π over B_n is hard-core if for any efficient (i.e., PPT) Turing machine $\mathcal{A}$ we have that

$$\text{Prob}_{x,a \in B_n}[\mathcal{A}(a, xax^{-1}) = \pi(x)] = \frac{1}{2} + \epsilon \tag{1}$$

where ϵ is a negligible quantity. We stress that the probability of success is taken over *BOTH* the internal coin tosses of $\mathcal{A}$ and the choice of x. In other words if we find an $\mathcal{A}$ that predicts $\pi(x)$ for just 51% of the inputs x, then π is NOT an hard-core bit for f.

Usually proofs for hard-core bits go by contradiction. We assume that there exists an oracle $\mathcal{A}$ that contradicts Equation 1. We then show how to use $\mathcal{A}$ to invert f over a random point. The proof is usually complicated by the fact that $\mathcal{A}$'s responses on any input (a, xax^{-1}) may be wrong and cannot be accepted at face value. Usually the proof contains some random self-reducibility argument (although that's not necessarily the only way to prove hard-core bits). That is instead of querying $\mathcal{A}$ only on (a, xax^{-1}) we also query it on several randomized versions of it and then use some trick to extract the right value of $\pi(x)$ from all these responses (which in the majority are correct).

In [11] such an hard-core predicate and a proof of security is supposedly presented. For the discussion that follows, it is not really important to understand what the hard-core predicate is. For completeness, we briefly recall the definition anyway. Every braid can be uniquely expressed in canonical form as the product $\Delta_n^u \chi_1 \cdots \chi_p$ where Δ_n is the fundamental braid defined in Sect. 3, and χ_p are permutation braids, i.e., braids where each string can be described as a straight line from the initial point to the end point, and at all the crossings the left string goes under the right string. This normal form (called the left canonical form) can be computed in polynomial time. The candidate hard core predicated π proposed in [11] takes a braid x as input, computes the left canonical form $x = \Delta_n^u \chi_1 \cdots \chi_p$ and outputs the parity of u.

The proof of security in [11] misses the whole step of randomizing over the inputs to the predicting oracle. The reduction from predicting π to inverting f assumes that the oracle $\mathcal{A}$ answers correctly *FOR ALL* x with sufficiently high probability (bounded away from a 1/2), but this probability is taken ONLY over the coin tosses of $\mathcal{A}$. If x happen to be one of the bad inputs for which $\mathcal{A}$ gives the wrong answer (possibly for any value of $\mathcal{A}$ internal coin tosses), there is no guarantee that majority voting on $\mathcal{A}$ answers relative to the same input x and independent coin tosses produces the right answer.

Such set of *bad* inputs might constitute 49% of the possible braids, and still π not be a hard-core predicate because (when x is chosen at random) $\mathcal{A}$ has a 1% advantage in predicting $\pi(x)$ over a random guess. (For cryptographic security, even 1% is not a negligible advantage. Formally speaking, the 49% and 1% above should be interpreted as $(50-\epsilon)\%$ and $\epsilon\%$ where $\epsilon(k)$ is a function of the security parameter such that $\epsilon(k) < 1/k^c$ for all $c > 0$ and for all sufficiently large k.)

We were not able to repair the proof of security for the hard-core predicate presented in [11]. Moreover we were not able to construct any other natural hard-core predicate. Thus we conclude that the construction of a hard-core predicate for the conjugacy problem is still an (interesting) open problem.

6 Conclusion

We showed that the decisional version of the Ko-Lee Assumption is false in a very strong sense: for essentially all reasonable input probability distributions, there is a very efficient distinguisher that invalidate the assumption. The attack extends to the pseudo-random generator and synthesizer proposed in [11] which are consequently shown to be insecure.

With regard to the conjugacy problem over braids, which is at the basis of all the cryptographic schemes based on braid groups, we show two facts. (1) The conjugacy problem reveals some partial information about the permutation induced by the input braid, unless the central braid is chosen to be pure; we suggest to incorporate this choice in the design of cryptographic schemes based on the conjugacy problem. (2) The proof of a hard-core bit for the conjugacy problem in [11] is flawed and we were not able to repair it.

In spite of our attacks, braid groups remain an attractive and promising tool for the construction of cryptographic algorithms. However our findings highlight the need for extreme care in analyzing these new protocols and warn about drawing unmotivated conclusions from parallel ones in number-theoretic constructions.

Which leads us, in turn, to many interesting open problems. For example, although we proved that the result of a Ko-Lee exchange is not totally random, it might still be possible to prove that it contains a lot of "computational entropy". If true, this would allow the construction of an efficient secure encryption scheme *without* resorting to the random oracle model, and possibly provide a fix for the pseudorandom constructions presented in [11]. Also, is there a hard-core bit for the conjugacy problem? Another interesting problem is related to the construction of digital signatures. In [13] it is shown that certain groups of points on an supersingular elliptic curves have the property that the Decisional Diffie-Hellman problem is easy, while the Computational Diffie-Hellman problem is presumably hard. This properties can be used to design a digital signature scheme as shown in [5,12,4]. The idea is the following. The secret and public keys of the system are a and $y = g^a$, and the signature of message m is computed as $s = m^a$. It is easy to see that the signature is valid if and only if (g, y, m, s) is a valid DDH sequence. An interesting open question is whether the conjectured hardness of the conjugacy problem for braid groups, together with the techniques to attack the decisional problem presented in this paper, can be used to design a provably secure digital signature scheme based on braid groups.

Acknowledgements

We would like to thank N. Howgrave-Graham for useful discussions and for drawing our attention to the subgroup of pure braids. A preliminary version of this attack was presented at the rump session of Crypto 2001 [7]. We would like to thank E. Lee and S.G. Hahn for useful discussions following the rump session presentation, and the anonymous referees for their comments to the preliminary version of this paper.

References

1. I. Anshel, M. Anshel and D. Goldfeld. An Algebraic Method for Public-Key Cryptography. Mathematical Research Letters, 6 (1999), pp. 287–291.
2. M. Bellare and P. Rogaway, "Random oracles are practical: a paradigm for designing efficient protocols", *1st ACM Conference on Computer and Communications Security,* 1993, 62-73.
3. D. Boneh. The Decision Diffie-Hellman Problem. Third Algorithmic Number Theory Symposium. LNCS 1423, pp.48–63, Springer 1998.
4. D. Boneh, H. Shacham, and B. Lynn. Short signatures from the Weil pairing. Asiacrypt '2001. LNCS 2248, pp. 514–532, Springer-Verlag 2001.
5. S. Brands. An efficient off-line electronic cash system based on the representation problem. Technical Report CS–R9323, CWI (Centre for Mathematics and Computer Science), Amsterdam, 1993.
6. J.L. Carter and M.N. Wegman, Universal classes of hash functions, Journal of Computer and System Sciences 18:143-154, 1979.
7. R. Gennaro, D. Micciancio. Cryptanalysis of a Pseudorandom Generator based on Braid Groups. CRYPTO'2001 rump session, August 2001.
8. O. Goldreich, L. Levin. Hard-core Predicates for any One-way Function. 21st STOC, pp.25-32, 1989.
9. S. Goldwasser, S. Micali. Probabilistic Encryption. Journal of Computer and System Sciences 28:270–299, April 1984.
10. K.H. Ko, S.J. Lee, J.H. Cheon, J.W. Han, J. Kang, C. Park. New Public-Key Cryptosystem Using Braid Groups. CRYPTO'2000, LNCS 1880, pp.166–183, Springer 2000.
11. E. Lee, S.J. Lee, S.G. Hahn. Pseudorandomness from Braid Groups. CRYPTO'2001, Springer 2001.

12. T. Okamoto, D. Pointcheval The Gap problem: a new class of problems for the security of cryptographic primitives Public Key Cryptography, PKC 2001, LNCS 1992, Springer-Verlag 2001.
13. E. R. Verheul Evidence that XTR Is More Secure than Supersingular Elliptic Curve Cryptosystems Eurocrypt'2001. LNCS 2045, p. 195-210

Potential Weaknesses of the Commutator Key Agreement Protocol Based on Braid Groups

Sang Jin Lee[1] and Eonkyung Lee[2]

[1] CMI, Université de Provence, Marseille, France,
sjlee@knot.kaist.ac.kr
[2] Korea Information Security Agency, Seoul, Republic of Korea,
eonkyung@kisa.or.kr

Abstract. The braid group with its conjugacy problem is one of the recent hot issues in cryptography. At CT-RSA 2001, Anshel, Anshel, Fisher, and Goldfeld proposed a commutator key agreement protocol (KAP) based on the braid groups and their colored Burau representation. Its security is based on the multiple simultaneous conjugacy problem (MSCP) plus a newly adopted key extractor. This article shows how to reduce finding the shared key of this KAP to the list-MSCPs in a permutation group and in a matrix group over a finite field. We also develop a mathematical algorithm for the MSCP in braid groups. The former implies that the usage of colored Burau representation in the key extractor causes a new weakness, and the latter can be used as a tool to investigate the security level of their KAP.

Key words: Key agreement protocol, Braid group, Multiple simultaneous conjugacy problem, Colored Burau matrix.

1 Introduction

Current braid cryptographic protocols are based on the intractability of the conjugacy problem: given two conjugate braids a and b, find a conjugator (i.e., find x such that $b = x^{-1}ax$). Because it is hard to find a trapdoor in this problem, some variants have been proposed for the key exchange purpose [2,15].

Anshel *et al.* [2] proposed a key agreement protocol (KAP) assuming the intractability of the following problem: given $a_1, \ldots, a_r, x^{-1}a_1x, \ldots, x^{-1}a_rx \in B_n$, find the conjugator x. We call this problem the *multiple simultaneous conjugacy problem* (MSCP). Loosely speaking, their KAP is as follows: given pairs of n braids $(a_1, x^{-1}a_1x), \ldots, (a_r, x^{-1}a_rx)$, $(b_1, y^{-1}b_1y), \ldots, (b_s, y^{-1}b_sy)$, find the commutator $x^{-1}y^{-1}xy$, where x and y are in the subgroup generated by $\{b_1, \ldots, b_s\}$ and $\{a_1, \ldots, a_r\}$, respectively. The first attack on this KAP is the Length Attack by Hughes and Tannenbaum [14]. They showed that this KAP leaks some information about its private keys x and y for some particular choices of parameters.

At CT-RSA 2001, Anshel *et al.* [1] proposed a new version of their KAP. They adopted a new *key extractor* which transforms a braid into a pair of a

L.R. Knudsen (Ed.): EUROCRYPT 2002, LNCS 2332, pp. 14–28, 2002.

permutation and a matrix over a finite field. Here the matrix is obtained from a multi-variable matrix, called the *colored Burau matrix*, by evaluating the variables at numbers in a finite field. They recommended parameters so as to defeat the Length Attack, the mathematical algorithm for the conjugacy problem, and a potential linear algebraic attack on the key extractor.

Our Results. This article attacks the KAPs in [1,2] from two different angles. Our attacks are partially related to the potential ones already mentioned in their paper.

First, we attack the shared key in [1]. The motivation for this attack is that despite the change of variables in the colored Burau matrix by permutations, the matrix in the final output(i.e., the shared key) is more manageable than braids. We show that the security of the key extractor is based on the problems of listing all solutions to some MSCPs in a permutation group and in a matrix group over a finite field. So if both of the two listing problems are feasible, then we can guess correctly the shared key without solving the MSCP in braid groups.

Second, we attack the private keys in [1,2]. The base problem of these KAPs is different from the standard conjugacy problem in the following two aspects: (i) The conjugation is multiple and simultaneous. That is to say, we have a set of equations $x^{-1}a_i x = c_i$, $i = 1, \ldots, r$, with a single unknown x. On the one hand, the problem is more difficult than the conjugacy problem because we must find a solution which satisfies all the equations simultaneously. On the other hand, the problem is easier because we have multiple equations. (ii) The conjugator x is contained in the subgroup generated by some specific braids $b_1, \ldots, b_s$, which makes the problem easier. We propose a mathematical algorithm for the MSCP in braid groups.

Outline. We review in §2 the braid groups, the canonical form of braid, the colored Burau representation, and the commutator KAP proposed in [1]. We attack the key extractor in §3 and the private key in §4. We close this article with conclusions in §5.

Conventions.

- S_n denotes the n-permutation group. S_n acts on the set $\{1, 2, \ldots, n\}$ from the left so that for $\alpha, \beta \in S_n$ and $1 \le i \le n$, $(\alpha\beta)(i) = \alpha(\beta(i))$. We express a permutation as a product of cycles. A cycle $\alpha = (k_1, k_2, \ldots, k_r)$ means that $\alpha(k_i) = k_{i+1}$ for $i = 1, \ldots, r-1$ and $\alpha(k_r) = k_1$.
- For a prime p, $\mathbf{F}_p$ denotes the field composed of p elements, $\{0, \ldots, p-1\}$.
- $GL_n(R)$, R a ring, denotes the set of all invertible $(n \times n)$-matrices over R.

2 Preliminaries

2.1 Braid Group

Definition 1. *The n-braid group B_n is an infinite non-commutative group defined by the following group presentation*

$$B_n = \left\langle \sigma_1, \ldots, \sigma_{n-1} \;\middle|\; \begin{array}{ll} \sigma_i\sigma_j = \sigma_j\sigma_i, & |i-j| \geq 2 \\ \sigma_i\sigma_{i+1}\sigma_i = \sigma_{i+1}\sigma_i\sigma_{i+1}, & i = 1, \ldots, n-2 \end{array} \right\rangle.$$

The integer n is called the braid index *and the elements in B_n are called* n-braids. *The generators σ_i's are called the* Artin generators.

We can give braids the following geometric interpretation. An n-braid can be thought as a collection of n horizontal strands intertwining themselves. See Figure 1. Each generator σ_i represents the process of swapping the ith strand with the $(i+1)^{\text{st}}$ one, where the strand from upper-left to lower-right is over the other.

We can cut a long geometric braid into simple pieces so that each piece contains only one crossing. This decomposition gives a word $W = \sigma_{i_1}^{\epsilon_1}\sigma_{i_2}^{\epsilon_2}\cdots\sigma_{i_k}^{\epsilon_k}$, $\epsilon = \pm 1$. k is called the *word length* of W.

There is a natural projection $\pi : B_n \to S_n$, sending σ_i to the transposition $(i, i+1)$. Let's denote $\pi(a)$ by π_a, and call it the *induced permutation* of a. The braids whose induced permutation is the identity are called *pure braids*. Conversely, for a permutation $\alpha \in S_n$, we can make a simple braid A_α, called a *permutation braid*, by connecting the ith point on the right to the $\alpha(i)^{\text{th}}$ point on the left by straight lines, where the strand from upper-left to lower-right is over the other at each crossing.

2.2 Canonical Form

We review the canonical form of braids and the related invariants, inf, len, and sup, which can be used in measuring how complicated the given braid is. This section is needed only for §4.

1. A word $\sigma_{i_1}^{\epsilon_1}\cdots\sigma_{i_s}^{\epsilon_s}$ is called a *positive word* if all the exponents ϵ_i's are positive. Because the relations in the group presentation of B_n are equivalences between positive words with the same word-length, the *exponent sum* $\mathrm{e}(\sigma_{i_1}^{\epsilon_1}\cdots\sigma_{i_s}^{\epsilon_s}) = \epsilon_1 + \cdots + \epsilon_s$ is well-defined and invariant under conjugation. If P is a positive word, then $\mathrm{e}(P)$ is equal to the word-length of P. The *positive braid monoid*, denoted by B_n^+, is the set of all braids which can be represented by positive words.
2. The permutation braid corresponding to the permutation $\alpha(i) = (n-i)$ is called the *fundamental braid* and denoted by Δ. It can be written as $\Delta = (\sigma_1)(\sigma_2\sigma_1)\cdots(\sigma_{n-1}\sigma_{n-2}\cdots\sigma_1)$.
3. **Theorem** (See [9,10,4]). Every n-braid a can be decomposed uniquely into

$$a = \Delta^u A_1 A_2 \ldots A_k,$$

where each A_i is a permutation braid and for each $1 \leq i < k$, $A_i A_{i+1}$ is *left-weighted.* For the definition of 'left-weighted', see Appendix A.

4. The expression above is called the *left-canonical form.* The invariants the *infimum*, the *canonical length*, and the *supremum* are defined by $\inf(a) = u$, $\mathrm{len}(a) = k$, and $\sup(a) = u + k$, respectively. Similarly, we can also define the *right-canonical form.* The two canonical forms give the same inf, len, and sup.
5. Let $a = \Delta^u P$ and $b = \Delta^v Q$ be conjugate, where $P, Q \in B_n^+$. Then $\mathrm{e}(a) = u\mathrm{e}(\Delta) + \mathrm{e}(P) = v\mathrm{e}(\Delta) + \mathrm{e}(Q)$. So in a conjugacy class, a braid with greater inf is simpler than the one with smaller inf. Similarly, we can also say that a braid with smaller sup is simpler than the one with greater sup.

2.3 Colored Burau Representation

Morton [20] introduced the colored Burau matrix which is a generalization of the Burau matrix. It would be helpful to see the Burau matrix first. Let $\mathbf{Z}[t^{\pm 1}]$ be the ring of Laurent polynomials $f(t) = a_k t^k + a_{k+1} t^{k+1} + \cdots + a_m t^m$ with integer coefficients (and possibly with negative degree terms). Let $GL_{n-1}(\mathbf{Z}[t^{\pm 1}])$ be the group of $(n-1) \times (n-1)$ invertible matrices over $\mathbf{Z}[t^{\pm 1}]$. Note that a matrix over $\mathbf{Z}[t^{\pm 1}]$ is invertible if and only if its determinant is $\pm t^m$ for some integer m.

For $1 \leq i \leq n-1$, let $C_i(t)$ be the matrix which differs from the identity matrix only at the ith row as shown below. When $i = 1$ or $i = n-1$, the ith row vector is truncated to $(-t, 1, 0, \ldots, 0)$ or $(0, \ldots, 0, t, -t)$.

$$C_i(t) = \begin{pmatrix} 1 & & & & & \\ & \ddots & & & & \\ & & 1 & & & \\ & & t & -t & 1 & \\ & & & & 1 & \\ & & & & & \ddots \\ & & & & & & 1 \end{pmatrix}, \quad C_i^{-1}(t) = \begin{pmatrix} 1 & & & & & \\ & \ddots & & & & \\ & & 1 & & & \\ & & 1 & -\frac{1}{t} & \frac{1}{t} & \\ & & & & 1 & \\ & & & & & \ddots \\ & & & & & & 1 \end{pmatrix}.$$

The *Burau representation* $\rho\colon B_n \to GL_{n-1}(\mathbf{Z}[t^{\pm 1}])$ is defined by $\rho(\sigma_i) = C_i(t)$ [3]. The matrices $C_i(t)$'s satisfy the braid relations, i.e., $C_i(t)C_j(t) = C_j(t)C_i(t)$ for $|i-j| \geq 2$ and $C_i(t)C_{i+1}(t)C_i(t) = C_{i+1}(t)C_i(t)C_{i+1}(t)$ for $i = 1, \ldots, n-2$, and so ρ is a group homomorphism. The elements in $\rho(B_n)$ are called *Burau matrices.* Here is an example of a Burau matrix.

$$\rho(\sigma_1^{-1}\sigma_2\sigma_1^{-1}) = \begin{pmatrix} -\frac{1}{t} & \frac{1}{t} \\ 0 & 1 \end{pmatrix} \begin{pmatrix} 1 & 0 \\ t & -t \end{pmatrix} \begin{pmatrix} -\frac{1}{t} & \frac{1}{t} \\ 0 & 1 \end{pmatrix} = \begin{pmatrix} \frac{1}{t^2} - \frac{1}{t} & -\frac{1}{t^2} + \frac{1}{t} - 1 \\ -1 & 1 - t \end{pmatrix}.$$

Roughly speaking, the *colored Burau matrix* is a refinement of the Burau matrix by assigning σ_i to $C_i(t_{i+1})$ so that the entries of the resulting matrix have several variables. But such a naive construction does not give a group homomorphism. Thus the induced permutations are considered simultaneously.

Let's label the strands of an n-braid by $t_1, \ldots, t_n$, putting the label t_j on the strand which starts from the jth point on the right. Figure 1 shows this labelling for the 3-braid $\sigma_1^{-1}\sigma_2\sigma_1^{-1}\sigma_2$.

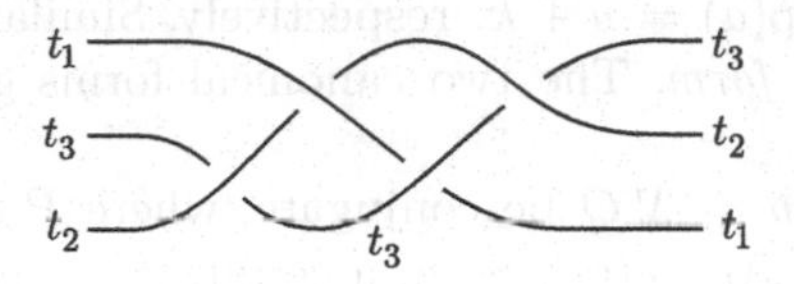

Fig. 1. The labelled braid $\sigma_1^{-1}\sigma_2\sigma_1^{-1}\sigma_2$

Definition 2. *Let $a \in B_n$ be given by a word $\sigma_{i_1}^{\epsilon_1}\sigma_{i_2}^{\epsilon_2}\cdots\sigma_{i_k}^{\epsilon_k}$, $\epsilon_j = \pm 1$. Let t_{j_r} be the label of the under-crossing strand at the rth crossing. Then the* colored Burau matrix *$M_a(t_1, \ldots, t_n)$ of a is defined by*

$$M_a(t_1, \ldots, t_n) = \prod_{r=1}^{k} (C_{i_r}(t_{j_r}))^{\epsilon_r}.$$

We can compute the colored Burau matrix by substituting σ_i^ϵ by $C_i(\cdot)^\epsilon$ and then filling $(\cdot)$ by the variable t_j according to the label of the under-crossing strand. Note that the colored Burau matrix is an invertible matrix over $\mathbf{Z}[t_1^{\pm 1}, \ldots, t_n^{\pm 1}]$, the ring of Laurent polynomials with n variables. In the example $\sigma_1^{-1}\sigma_2\sigma_1^{-1}\sigma_2$ shown in Figure 1, the colored Burau matrix is

$$\begin{aligned} M_{\sigma_1^{-1}\sigma_2\sigma_1^{-1}\sigma_2}(t_1, t_2, t_2) &= C_1(t_3)^{-1}C_2(t_2)C_1(t_1)^{-1}C_2(t_3) \\ &= \begin{pmatrix} -t_3^{-1} & t_3^{-1} \\ 0 & 1 \end{pmatrix}\begin{pmatrix} 1 & 0 \\ t_2 & -t_2 \end{pmatrix}\begin{pmatrix} -t_1^{-1} & t_1^{-1} \\ 0 & 1 \end{pmatrix}\begin{pmatrix} 1 & 0 \\ t_3 & -t_3 \end{pmatrix} \\ &= \begin{pmatrix} -\frac{1}{t_1} - t_2 + \frac{t_2}{t_1} + \frac{1}{t_1 t_3} - \frac{t_2}{t_1 t_3} & \frac{1}{t_1} + t_2 - \frac{t_2}{t_1} \\ -\frac{t_2}{t_1} - t_2 t_3 + \frac{t_2 t_3}{t_1} & t_2 t_3 - \frac{t_2 t_3}{t_1} \end{pmatrix}. \end{aligned}$$

Now we describe the colored Burau group as in [1]. We follow the convention of Morton [20], which is a little different from that in [1]. The permutation group S_n acts on $\mathbf{Z}[t_1^{\pm 1}, \ldots, t_n^{\pm 1}]$ from left by changing variables: for $\alpha \in S_n$, $\alpha(f(t_1, \ldots, t_n)) = f(t_{\alpha(1)}, \ldots, t_{\alpha(n)})$. Then S_n also acts on the matrix group $GL_{n-1}(\mathbf{Z}[t_1^{\pm 1}, \ldots, t_n^{\pm 1}])$ entry-wise: for $\alpha \in S_n$ and $M = (f_{ij})$, $\alpha(M) = (\alpha(f_{ij}))$.

Definition 3. *The* colored Burau group *CB_n is $S_n \times GL_{n-1}(\mathbf{Z}[t_1^{\pm 1}, \ldots, t_n^{\pm 1}])$ with multiplication $(\alpha_1, M_1)\cdot(\alpha_2, M_2) = (\alpha_1\alpha_2, (\alpha_2^{-1}M_1)M_2)$. The* colored Burau representation *$C : B_n \to CB_n$ is defined by $C(\sigma_i) = ((i, i+1), C_i(t_{i+1}))$.*

Then it is easy to see the following: (i) CB_n is a group, where the identity is (e, I_{n-1}) and $(\alpha, M)^{-1} = (\alpha^{-1}, \alpha M^{-1})$, (ii) $C(\sigma_i)$'s satisfy the braid relations and so $C : B_n \to CB_n$ is a group homomorphism, and (iii) for $a \in B_n$, $C(a) = (\pi_a, M_a)$, where π_a is the induced permutation and M_a is the colored Burau matrix in Definition 2.

2.4 Commutator Key Agreement Protocol

Recall the commutator KAP of Anshel *et al.* [1]. Fix a (small) prime number p. Let $K_{n,p}$ be the set of pairs $(\alpha, M) \in S_n \times GL_{n-1}(\mathbf{F}_p)$.

Definition 4. *Let $\tau_1, \ldots, \tau_n$ be distinct invertible integers in $\mathbf{F}_p$. The* key extractor *$E = E_{p,\tau_1,\ldots,\tau_n} : B_n \to K_{n,p}$ is defined by*

$$E(a) = (\pi_a, M_a(\tau_1, \ldots, \tau_n) \bmod p),$$

where reduction 'mod p' means reduction of every entry in the matrix.

Anshel *et al.* gave a very fast algorithm for computing the key extractor in [1]. The running time is $\mathcal{O}(n\ell(\log p)^2)$, where ℓ is the word-length. The idea is that we can compute $E(\sigma_{i_1}^{\epsilon_1} \cdots \sigma_{i_\ell}^{\epsilon_\ell})$ from $\sigma_{i_1}^{\epsilon_1}$ and $E(\sigma_{i_2}^{\epsilon_2} \cdots \sigma_{i_\ell}^{\epsilon_\ell})$. The commutator KAP using the key extractor is constructed as follows.

Public Information
1. An integer $n > 6$. A prime $p > n$.
2. Distinct and invertible integers $\tau_1, \ldots, \tau_n \in \mathbf{F}_p^*$.
3. $(a_1, \ldots, a_r) \in (B_n)^r$ and $(b_1, \ldots, b_s) \in (B_n)^s$.

Private Key
1. Alice's private key is a word $W(b_1, \ldots, b_s)$.
2. Bob's private key is a word $V(a_1, \ldots, a_r)$.

Public Key
1. Alice's public key is $(x^{-1}a_1x, \ldots, x^{-1}a_rx)$, where $x = W(b_1, \ldots, b_s)$.
2. Bob's public key is $(y^{-1}b_1y, \ldots, y^{-1}b_sy)$, where $y = V(a_1, \ldots, a_r)$.

Shared key

$E(x^{-1}y^{-1}xy) = (\pi_{x^{-1}y^{-1}xy}, \ M_{x^{-1}y^{-1}xy}(\tau_1, \ldots, \tau_n) \bmod p)$.

Parameter Recommendation in [1].

- The only restriction on p used in the key extractor is that $p > n$ so that one can choose distinct and invertible elements $\tau_1, \ldots, \tau_n$. One can choose $p < 1000$.
- Take the braid index $n = 80$ or larger and $r = s = 20$. Let each of a_i and b_j be the product of 5 to 10 Artin generators and let each set of public generators involve all the Artin generators of B_n.
- Private keys, x and y, are products of 100 public generators.

3 Linear Algebraic Attack on the Key Extractor

If the KAP [1] is restricted to pure braids, then its key extractor E becomes a group homomorphism. In this case, one can attack the key extractor by linear algebraic methods. To defeat this attack, [1] recommended to choose the private keys x and y such that their induced permutations are sufficiently complex. This section shows that a linear algebraic attack can also be mounted on the KAP even for such parameters. The (list-)MSCP is the following variant of the conjugacy problem.

Definition 5. *Let G be a group. The* MSCP *in G is: given a pair of r-tuples ($r \geq 2$) of elements in G, $(a_1, \ldots, a_r)$ and $(x^{-1}a_1x, \ldots, x^{-1}a_rx)$, find x in polynomial time (in the input length). In this case, the* list-MSCP *in G is to find the list of all such $x \in G$ in polynomial time.*

Their hardness will be discussed later. Henceforth, we will use the notations in §2.4 if there is no confusion from the context.

Theorem 1. *If the induced permutations of the private keys are known, then we can construct four list-MSCPs in $GL_{n-1}(\mathbf{F}_p)$ such that computing the matrix part of the shared key is reduced to solving all these list-MSCPs.*

Proof. By the definition of the colored Burau representation C, we get the following equation

$$\begin{aligned} C(x^{-1}y^{-1}xy) &= (\pi_x, M_x)^{-1}(\pi_y, M_y)^{-1}(\pi_x, M_x)(\pi_y, M_y) \\ &= (\pi_x^{-1}\pi_y^{-1}\pi_x\pi_y, (\pi_y^{-1}\pi_x^{-1}\pi_y\pi_x M_x^{-1})(\pi_y^{-1}\pi_x^{-1}\pi_y M_y^{-1})(\pi_y^{-1}M_x)M_y). \end{aligned}$$

So the matrix part of the shared key is the product of the following four matrices evaluated at $(t_1, \ldots, t_n) = (\tau_1, \ldots, \tau_n)$:

$$(\pi_y^{-1}\pi_x^{-1}\pi_y\pi_x M_x^{-1}), \quad (\pi_y^{-1}\pi_x^{-1}\pi_y M_y^{-1}), \quad (\pi_y^{-1}M_x), \quad \text{and} \quad M_y.$$

Now we propose a method of composing MSCPs in $GL_{n-1}(\mathbf{F}_p)$ for these matrices, assuming that we already know the permutations π_x and π_y. Here we consider only $(\pi_y^{-1}M_x)(\tau_1, \ldots, \tau_n)$. Similar constructions work for the other three matrices. The following technique makes $(\pi_y^{-1}M_x)(\tau_1, \ldots, \tau_n)$ to be a solution to an MSCP with N equations in $GL_{n-1}(\mathbf{F}_p)$ for given N. The basic idea is: if a is a pure braid and $c = x^{-1}ax$, then $(\alpha M_x)(\tau_1, \ldots, \tau_n)$ is a solution to $AX = XC$, where α is an arbitrary permutation, $A = (\alpha\pi_x^{-1}M_a)(\tau_1, \ldots, \tau_n)$, and $C = (\alpha M_c)(\tau_1, \ldots, \tau_n)$.

1. Choose a word $a = U(a_1, \ldots, a_r)$ such that a is a pure braid.
2. Compute $c = U(c_1, \ldots, c_r)$. Then, $c = x^{-1}ax$ and c is also a pure braid.
3. Compute M_a and M_c. Since $ax = xc$ and
$$C(ax) = (\pi_x, (\pi_x^{-1}M_a)M_x), \quad C(xc) = (\pi_x, M_xM_c),$$
we have $(\pi_y^{-1}\pi_x^{-1}M_a)(\pi_y^{-1}M_x) = (\pi_y^{-1}M_x)(\pi_y^{-1}M_c)$.
4. Compute $(\pi_y^{-1}\pi_x^{-1}M_a)(\tau_1, \ldots, \tau_n)$ and $(\pi_y^{-1}M_c)(\tau_1, \ldots, \tau_n)$. Refer to the resulting matrices as A and C, respectively.
5. Repeat the above steps to get a system of equations $A_jX = XC_j$ for $j = 1, \ldots, N$.

It is easy to see that $(\pi_y^{-1}M_x)(\tau_1, \ldots, \tau_n)$ is a solution to the MSCP in $GL_{n-1}(\mathbf{F}_p)$, $\{A_jX = XC_j\}_{1\leq j\leq N}$. □

Now we discuss how to construct the algorithm in Theorem 1 practically, the hardness of the list-MSCPs in S_n and in $GL_{n-1}(\mathbf{F}_p)$, and possible fixes.

How to Generate a Pure Braid $a = U(a_1, \ldots, a_r)$ in the First Step. An easy construction is to choose a word $V(a_1, \ldots, a_r)$ at random and then take a power $U = V^k$, where k is the order of the induced permutation. However, the order of a permutation is the least common multiple of the lengths of cycles, and so it can be too large. For example, for $n = 87$, the maximal order of n-permutations is greater than 10^7. See [19]. One way to avoid this huge order is to choose $V(a_1, \ldots, a_r)$ as a short word. For example, if V is a product of three a_i's, then its induced permutation is a product of 15 to 30 transpositions and so its order is small. (Note that a_i's are products of 5 to 10 Artin generators.) And once we have a pure braid $a = U(a_1, \ldots, a_r)$, then we can also use $W^{-1}aW$ for any word W on a_i's.

Hardness of the List-MSCPs in Permutation Group and in Matrix Group. The MSCP in permutation group is easy. Note that two permutations are conjugate if and only if they have the same cycle decomposition. The MSCP in matrix group is also easy because the equation $AX = XC$ can be considered as a system of homogeneous linear equations in the entries of X. One can use the polynomial time deterministic algorithm by Chistov, Ivanyos, and Karpinski [7].

So the difficulty of the list-MSCP lies only in the number of its solutions. Let G be a group and let $(a_1, \ldots, a_r)$ and $(c_1, \ldots, c_r) \in G^r$ be an instance of a list-MSCP in G. If x_1 and x_2 are two solutions, then $(x_2x_1^{-1})a_i = a_i(x_2x_1^{-1})$ for each i. Hence $x_2 = x_1z$ for some z in $\cap_{i=1}^{r}\mathrm{Cent}(a_i)$, where $\mathrm{Cent}(a_i) = \{g \in G \mid ga_i = a_ig\}$ is the centralizer of a_i. So the number of the solutions is exactly the cardinality of the subgroup $\cap_{i=1}^{r}\mathrm{Cent}(a_i)$. We don't have the average cardinality of this subgroup when G is either S_n or $GL_{n-1}(\mathbf{F}_p)$. But it does not seem large for generic a_i's in S_n or in $GL_{n-1}(\mathbf{F}_p)$ from the following observation.

For $a_1, \ldots, a_r \in S_n$, let $z \in \cap_{i=1}^{r}\mathrm{Cent}(a_i)$. Then for each $1 \leq i \leq r$ and for each $1 \leq k \leq n$, if k lies in an m-cycle of a_i, then so does $z(k)$. Moreover, if $(k_1, \ldots, k_m)$ and $(l_1, \ldots, l_m)$ are m-cycles in a_i such that $z(k_1) = l_1$, then $z(k_j) = l_j$ for all $2 \leq j \leq m$. For example, let $a_1 = (1,2)(4,5)(8,9)$, $a_2 = (3,4)(5,7,8)$, and $a_3 = (1,5,6)(9,10)$ be the cycle decompositions of permutations in S_{10}. Let $z \in \cap_{i=1}^{3}\mathrm{Cent}(a_i)$. Then $z(1) = 1$ because $z(1)$ must lie in a 2-cycle of a_1, in an 1-cycle of a_2, and in a 3-cycle of a_3 simultaneously. In addition, $z(k) = k$ for $k = 2, 5, 6$ because z must fix all the numbers in a cycle of a_i containing 1, for any $i = 1, 2, 3$. By continuing this argument we can see that z is the identity permutation. The list-MSCP in $GL_{n-1}(\mathbf{F}_p)$ can be discussed similarly using the Jordan canonical form of matrices. See §5.6 in [8].

Possible Fix. To defeat our linear algebraic attack, at least one of the list-MSCPs in permutation group and in matrix group must be infeasible. Here, we discuss how to make the induced list-MSCP in permutation group infeasible. One way to do so is to use pure braids. But if all the braids $(a_1, \ldots, a_r)$ and $(b_1, \ldots, b_s)$ are pure, then the induced permutation is nothing more than the identity. Hence, we can consider the following simple cases.

1. Choose $(a_1, \ldots, a_r)$ in B_n and $(b_1, \ldots, b_s)$ in the pure braid group. Then the induced permutation of $x = W(b_1, \ldots, b_s)$ is the identity but it is impossible

to list all $y = V(a_1, \ldots, a_r)$ because the equation $y^{-1}b_i y = d_i$ gives no information about π_y.
2. Choose $(a_1, \ldots, a_r)$ and $(b_1, \ldots, b_s)$ so that the induced permutations of a_i's fix $\{1, \ldots, \lfloor \frac{n}{2} \rfloor\}$ and those of b_j's fix $\{\lfloor \frac{n}{2} \rfloor + 1, \ldots, n\}$.

In both cases, the list-MSCP in permutation group is infeasible. But these fixes give disadvantage that the braids a_i's or b_j's become complicated, and so the KAP becomes less secure against the Length Attack.

4 Attack on the Private Key

This section proposes an attack on the private keys of the commutator KAPs in [1,2] by solving the MSCPs in braid groups; given $(a_1, \ldots, a_r)$ and $(c_1, \ldots, c_r)$ in $(B_n)^r$, find $x \in B_n$ such that $c_i = x^{-1}a_i x$ for all i simultaneously. We start with some discussions.

Uniqueness of the Solution to the MSCP in Braid Groups. For generic choice of $a_1, \ldots, a_r$, the solution x is unique up to a power of Δ^2 (See Appendix B). That is, if x' is another solution such that $x'^{-1}a_i x' = c_i$ for all i, then $x' = \Delta^{2k}x$ for some integer k. Note that $x'^{-1}y^{-1}x'y = x^{-1}y^{-1}xy$ for any y, because Δ^2 is a central element. Therefore, it suffices to find $\Delta^{2k}x$ for any k.

Length Attack. The commutator KAPs in [1,2] have the following condition in addition to the standard MSCP: x is contained in the subgroup generated by some publicly known braids $b_1, \ldots, b_s$. This fact is crucial to the Length Attack of J. Hughes *et al.* [14]. They showed that the KAP is vulnerable to the Length Attack when b_j's are complicated and x is a product of a small number of $b_j^{\pm 1}$'s. And same for a_i's and y. To defeat the Length Attack, Anshel *et al.* [1] recommended using simple a_i's and b_j's and complicated x and y as mentioned in §2.4. Our attack of this section is strong when a_i's and b_j's are simple and it does not depend on how complicated x and y are.

Which Braid Is Simpler in the Conjugacy Class? Recall the discussion in §2.2. Let $a = \Delta^u A_1 \ldots A_k$ and $c = \Delta^v C_1 \ldots C_\ell$ be the left canonical forms of conjugate braids. It is natural to say that a is *simpler* than c if (i) the word-length of $A_1 \ldots A_k$ is smaller than that of $C_1 \ldots C_\ell$, or (ii) they have same word-length but k is smaller than ℓ. The former is equivalent to $u = \inf(a) > v = \inf(c)$ and the latter is equivalent to $\inf(a) = \inf(c)$ and $\sup(a) < \sup(b)$.

Mathematical Algorithm for the Conjugacy Problem. The conjugacy problem in braid groups is: given $(a, c) \in (B_n)^2$, decide whether they are conjugate and if so, find $x \in B_n$ such that $x^{-1}ax = c$. The algorithm for the conjugacy problem, first proposed by Garside [13] and still being improved [10,9,4,5,11], works as follows:

1. For each element in B_n, the *super summit set* is defined as the set of all conjugates which have the minimal canonical length. Then it is a finite set and two braids are conjugate if and only if the corresponding super summit sets coincide.
2. Given a braid $a \in B_n$, one can compute an element in the super summit set easily.
3. Given two elements u and v in the same super summit set, there is a chain leading from u to v, where successive elements are conjugated by a permutation braid.

How to Develop an Algorithm for the MSCP in Braid Groups. There are several directions in designing an algorithm for the MSCP, depending on the characteristics of the instances. This section focuses on the fact that $(a_1, \ldots, a_r)$ is simple because a_i's are products of a few Artin generators but $(c_1, \ldots, c_r)$ are usually complicated because $c_i = x^{-1}a_i x$ for a complicated braid x. See [1].

Let $\tau : B_n \to B_n$ be the isomorphism defined by $\tau(\sigma_i) = \sigma_{n-i}$. Hence τ^k, k-composition of τ, is the identity for k even, and τ for k odd.

Proposition 1. *Let $a, c, x \in B_n$ and $c = x^{-1}ax$. Let $c = \Delta^w C_1 \ldots C_l$ be the left-canonical form. If $\inf(a) > \inf(c)$, then $x = x_0 H$ for some $x_0 \in B_n$ with $\inf(x) = \inf(x_0)$ and for a permutation braid H determined by $H = \Delta\tau^w(C_1^{-1})$.*

Proof. It is a restatement of the Cycling Theorem in Appendix C (Theorem 4.1 of [9] and Theorem 5.1 of [4]). The statements in [4,9] look different from the above, but the argument of their proofs is exactly Proposition 1. □

We can understand this proposition in the following way.

- $x = x_0 H$ and $\inf(x) = \inf(x_0)$ means that x_0 is simpler than x: if $x_0 = \Delta^u P$ and $x = \Delta^u Q$ are the left canonical forms, then the word-length of P is smaller than that of Q.
- Let $c' = HcH^{-1}$. Then $x^{-1}ax = c$ implies that $x_0^{-1}ax_0 = c'$. Thus we get a conjugacy problem with simpler solution.
- The condition $\inf(a) > \inf(c)$ means that c is more complicated than a. And H is determined not by x but by $c(= x^{-1}ax)$.
- Consequently, we can interpret Proposition 1 as follows: *if c is more complicated than a, then we can find H such that the solution to the conjugacy problem for (a, c') is simpler than that for (a, c), where $c' = HcH^{-1}$.*

Definition 6. *For $a_1, \ldots, a_r \in B_n$, define $C^{\inf}(a_1, \ldots, a_r)$ as the set of all $(u_1, \ldots, u_r)$ such that $\inf(u_i) \geq \inf(a_i)$ for all i and there exists some $w \in B_n$ satisfying $u_i = w^{-1}a_i w$ for all i simultaneously.*

Theorem 2. *Let $(a_1, \ldots, a_r)$ and $(c_1, \ldots, c_r)$ be an instance of an MSCP in B_n and x a positive braid such that $x^{-1}a_i x = c_i$ for all i. Assume that a_i's and c_i's are already in the left canonical form. Then we can compute positive braid x_0*

and $(c'_1, \ldots, c'_r)$ *such that* $(c'_1, \ldots, c'_r) \in C^{\inf}(a_1, \ldots, a_r)$ *and* $c'_i = x_0 c_i x_0^{-1}$ *for all* i*, in time proportional to*

$$n(\log n)|x| \left(|x| + \sum_{i=1}^{r} (|a_i| + |c_i|) \right), \quad (1)$$

where $|\cdot|$ *denotes the word-length in generators. Moreover* $x = x_1 x_0$ *for some positive braid* x_1*, in particular the word-length of* x_1 *is less than that of* x*.*

Proof. We exhibit an algorithm that computes x_0 and hence $(c'_1, \ldots, c'_r)$.

Input: $(a_1, \ldots, a_r), (c_1, \ldots, c_r) \in (B_n)^r$.
Initialization: $x_0 = e$(identity braid), $c'_i = c_i$ for all i.
Loop:
STEP 1: If $\inf(c_i) \geq \inf(a_i)$ for all i, then STOP
STEP 2: Choose k such that $\inf(c_k) < \inf(a_k)$.
Compute the permutation braid H by applying Proposition 1 to (a_k, c_k).
STEP 3: $x_0 \leftarrow H x_0$, $c'_i \leftarrow H c'_i H^{-1}$ for all i. GO TO STEP 1
Output: x_0 and $(c'_1, \ldots, c'_r)$.

Because H in Proposition 1 is a suffix of x, so is x_0 at each step in the above algorithm. Whenever Proposition 1 is applied in the loop, the word-length of x_0 strictly increases and its final length is bounded above by $|x|$. So the algorithm stops in at most $|x|$ repetitions of the loop.

All the computations involved is to compute simple conjugations such as HaH^{-1}, $a \in B_n$ and H a permutation braid, which can be done in time $\mathcal{O}(n(\log n)|a|)$ and simple multiplications of the form Hx_0, which can be done in time $\mathcal{O}(n(\log n)|x_0|)$. So the whole complexity is (1). □

Note that the a_i's are much simpler than c_i's [1] and that the newly obtained braids c'_i's are at least as simple as a_i's in terms of 'inf'.

Now we have simple instance $(a_1, \ldots, a_r)$ and $(c'_1, \ldots, c'_r)$. The natural question is how to solve the MSCP for this new instance. It uses a variant of the Convexity Theorem [4,9]. See Appendix C.

Theorem 3. *Given* $(c'_1, \ldots, c'_r) \in C^{\inf}(a_1, \ldots, a_r)$*, there exists a chain of elements in* $C^{\inf}(a_1, \ldots, a_r)$ *from* $(a_1, \ldots, a_r)$ *to* $(c'_1, \ldots, c'_r)$*, where successive elements are simultaneously conjugated by a permutation braid. In other words, there is a sequence* $(a_1, \ldots, a_r) \to (a'_1, \ldots, a'_r) \to \cdots \to (a_1^{(k)}, \ldots, a_r^{(k)}) = (c'_1, \ldots, c'_r)$ *such that for each* j*, there is a permutation braid* H_j *satisfying* $a_i^{(j+1)} = H_j^{-1} a_i^{(j)} H_j$ *for all* i *simultaneously.*

Proof. This is a restatement of the Convexity Theorem in Appendix C. □

By Theorem 2 and Theorem 3, we can solve any MSCP in finite time. But the computational complexity of a naive implementation is exponential with respect to the braid index n and involves the cardinality of the set $C^{\inf}(a_1, \ldots, a_r)$. There seems to be no previous result concerning the cardinality of $C^{\inf}(a_1, \ldots, a_r)$.

If the instances are extremely simple, for example when all a_i's are positive braids, then the set $C^{\inf}(a_1, \ldots, a_r)$ will be very small, so that the MSCP is feasible. But the MSCP for generic instances needs more work.

Possible Improvement of the Algorithm for the MSCP in Braid Groups. Recently N. Franco and J. González-Meneses [11] improved the algorithm for the conjugacy problem in braid groups. The complexity of their algorithm to compute the super summit set is $\mathcal{O}(N\ell^2 n^4 \log n)$, where N is the cardinality of the super summit set, n is the braid index, and ℓ is the word-length of the given braid. The complexity of the old algorithm was $\mathcal{O}(N\ell^2 (n!) n \log n)$. With respect to the braid index n, the complexity was reduced from exponential function to polynomial. We expect that their idea can also be applied to the MSCP and our algorithm can be improved so that the computational complexity is a polynomial in (n, r, ℓ, N), where ℓ is the maximal word-length of a_i's and N is the cardinality of the set $C^{\inf}(a_1, \ldots, a_r)$.

5 Concluding Remarks

For the commutator key agreement protocols of Anshel *et al.* [2,1], we have proposed two kinds of attacks: a linear algebraic attack on the key extractor and a mathematical algorithm solving the MSCP in braid groups.

Our linear algebraic attack has shown that given the induced permutations of the private keys, computing the matrix part of the shared key $E(x^{-1}y^{-1}xy)$ is reduced to some list-MSCPs in $GL_{n-1}(\mathbf{F}_p)$. So one can compute the entire shared key very efficiently if the list-MSCPs in S_n and in $GL_{n-1}(\mathbf{F}_p)$ are feasible.

On the other hand, we have proposed an algorithm for the MSCP in braid groups that is suitable for the instance, $(a_1, \ldots, a_r)$ and $(c_1, \ldots, c_r)$, where a_i's are simple and c_i's are complicated. It consists of two steps. We first transform $(c_1, \ldots, c_r)$ into $(c'_1, \ldots, c'_r)$ where each c'_i is at least as simple as a_i, and then find the conjugator. The first step is really efficient. However, there is no polynomial time algorithm for the second step.

It is interesting to study the (in-)feasibility of the list-MSCPs in permutation groups and in matrix groups, and to improve the mathematical algorithm for the MSCP in braid groups.

Acknowledgement

The authors are grateful to Rosario Genaro, Dorian Goldfeld, Hi-joon Chae, Gabor Ivanyos, the anonymous referees, and Eurocrypt 2002 program committee for their helpful comments or suggestions. The first author was supported by postdoctoral fellowship program of KOSEF. And the second author was supported by 2002 R&D project, Development of models & schemes for the security analysis of cryptographic protocols, of MIC.

References

1. I. Anshel, M. Anshel, B. Fisher, and D. Goldfeld, *New Key Agreement Protocols in Braid Group Cryptography*, Topics in Cryptology—CT-RSA 2001 (San Francisco, CA), 13–27, Lecture Notes in Comput. Sci., 2020, Springer, Berlin, 2001.

2. I. Anshel, M. Anshel, and D. Goldfeld, *An algebraic method for public-key cryptography*, Math. Res. Lett. **6**(1999), no. 3-4, 287–291.
3. J.S. Birman, *Braids, links and the mapping class group*, Ann. Math. Studies 82, Princeton Univ. Press, 1974.
4. J.S. Birman, K.H. Ko, and S.J. Lee, *A new approaches to the world and conjugacy problem in the braid groups*, Adv. Math. **139**(1998), no. 2, 322–353.
5. J.S. Birman, K.H. Ko, and S.J. Lee, *The infimum, supremum and geodesic length of a braid conjugacy class*, Adv. Math. **164**(2001), no. 1, 41–56.
6. J.C. Cha, K.H. Ko, S.J. Lee, J.W. Han, and J.H. Cheon, *An Efficient Implementation of Braid Groups*, Advances in Cryptology—ASIACRYPT 2001, (Gold Coast, Queensland, Australia), 144–156, Lecture Notes in Comput. Sci., 2248, Springer, Berlin, 2001.
7. A. Chistov, G. Ivanyos, and M. Karpinski, *Polynimial time algorithms for modules over finite dimensional algebras*, Proc. Int. Symp. on Symbolic and Algebraic Computation (ISSAC) '97, ACM. 68–74, 1997.
8. C.G. Cullen, *Matrices and Linear Transformations*, Addison-Wesley, 1972.
9. E.A. Elrifai and H.R. Morton, *Algorithms for positive braids*, Quart. J. Math. Oxford Ser. (2) **45**(1994), no. 180, 479–497.
10. D. Epstein, J. Cannon, D. Holt, S. Levy, M. Paterson, and W. Thurston, *Word processing in groups*, Jones and Bartlett Publishers, Boston, MA, 1992.
11. N. Franco and J. González-Meneses, *Conjugacy problem for braid groups and Garside groups*, arXiv:math.GT/0112310, preprint 2001.
12. A. Fathi, F. Laudenbach, and V. Poénaru, *Travaux de Thurston sur les surfaces*, Astérisque, 66–67, 1979.
13. F.A. Garside, *The braid group and other groups*, Quart. J. Math. Oxford Ser. (2) **20**(1969), 235–254.
14. J. Hughes and A. Tannenbaum, *Length-based attacks for certain group based encryption rewriting systems*, Institute for Mathematics and Its Applications, April, 2000, Minneapolis, MN, Preprint number 1696, URL `http://www.ima.umn.edu/preprints/apr2000/1696.pdf`.
15. K.H. Ko, S.J. Lee, J.H. Cheon, J.H. Han, J.S. Kang, and C. Park, *New public key cryptosystem using braid groups*, Advances in cryptology—CRYPTO 2000 (Santa Barbara, CA), 166–183, Lecture Notes in Comput. Sci., 1880, Springer, Berlin, 2000.
16. E. Lee, S.J. Lee, and S.G. Hahn, *Pseudorandomness from Braid Groups*, Advances in cryptology—CRYPTO 2001 (Santa Barbara, CA), 486–502, Lecture Notes in Comput. Sci., 2139, Springer, Berlin, 2001.
17. J. Los, *Pseudo-Anosov maps and invariant train track in the disc: a finite algorithm*, Proc. Lond. Math. Soc. **66**, 400-430, 1993.
18. J. McCarthy, *Normalizers and centralizers of psuedo-Anosov mapping clases*, Available at `http://www.mth.msu.edu/~mccarthy/research/`.
19. W. Miller, *The maximum order of an element of a finite symmetric group*, Amer. Math. Monthly **94**(1987), no. 6, 497–506.
20. H.R. Morton, *The multivariable Alexander polynomial for a closed braid*, Low-dimensional topology (Funchal, 1998), 167–172, Contemp. Math., 233, Amer. Math. Soc., Providence, RI, 1999.

A Left-Weighted

For a positive braid P, the *starting set* $S(P)$ and the *finishing set* $F(P)$ is defined as follows.

$$S(P) = \{i \mid P = \sigma_i Q \quad \text{for some } Q \in B_n^+\},$$
$$F(P) = \{i \mid P = Q\sigma_i \quad \text{for some } Q \in B_n^+\}$$

For positive braids P and Q, we say that PQ is *left-weighted* if $F(P) \supset S(Q)$ and *right-weighted* if $F(P) \subset S(Q)$.

B Uniqueness of the Solution to the MSCP in Braid Groups

Proposition 2. *For generic instances, the MSCP* $x^{-1}a_i x = c_i$ *for* $i = 1, \dots, r$, *has unique solution up to power of* Δ^2.

Proof. Let x_1 and x_2 be two solutions, i.e., $x_1^{-1}a_i x_1 = x_2^{-1}a_i x_2$ for each i. Therefore $x_1 x_2^{-1}$ commutes with each a_i and so it commutes with any element in the subgroup generated by $a_1, \dots, a_r$.

If we choose an n-braid at random, then it is pseudo-Anosov [12,16,17]. So we may assume that we can choose two pseudo-Anosov braids, a and a', from the subgroup generated by $\{a_1, \dots, a_r\}$ such that they have different invariant measured foliations and that there is no symmetry on the foliations. Then a braid commuting with both of a and a' is of the form Δ^{2k} for some integer k [18]. □

C Algorithm for the Conjugacy Problem

1. The *super summit set* of a is defined by the set of all braids a' such that a and a' are conjugate and $\inf(a')$ is maximal and $\sup(a')$ is minimal in the conjugacy class of a. We can compute the super summit set by two theorems, the Cycling Theorem and the Convexity Theorem.
2. Let $a = \Delta^u A_1 \cdots A_k$ be the left-canonical form of a. The *cycling* $\mathbf{c}(a)$ and the *decycling* $\mathbf{d}(a)$ of a is defined by

$$\mathbf{c}(a) = \Delta^u A_2 \cdots A_k \tau^u(A_1),$$
$$\mathbf{d}(a) = \Delta^u \tau^{-u}(A_k) A_1 \cdots A_{k-1}.$$

   In general, the braids on the right hand sides are not in canonical forms, and so must be rearranged into canonical forms before the operations are repeated.
3. **Cycling Theorem** (See [4,9,10])
   (i) If $\inf(a)$ is not maximal in the conjugacy class, then $\inf \mathbf{c}^l(a) > \inf(a)$ for some l.

(ii) If $\sup(a)$ is not minimal in the conjugacy class, then $\sup \mathbf{d}^l(a) < \sup(a)$ for some l.

(iii) So the maximal value of inf and the minimum value of sup can be achieved simultaneously. In particular, the super summit set is not empty for any braid.

4. **Convexity Theorem** (See [4,9,10])
Let $c = x^{-1}ax$, $\inf(a) = \inf(c)$, and $\sup(a) = \sup(c)$. Let $H_1H_2 \cdots H_k$ be the left-canonical form of x. Then
$$\inf(H_1^{-1}aH_1) \geq \inf(a) \quad \text{and} \quad \sup(H_1^{-1}aH_1) \leq \sup(a),$$
that is, $H_1^{-1}aH_1$ is as simple as a with respect to both of inf and sup.

5. It is clear that $\inf(a) = \inf(\tau(a))$ and $\sup(a) = \sup(\tau(a))$. So by the Convexity Theorem, we know that if both a and c are contained in the same super summit set, then there is a finite sequence $a = a_0 \to a_1 \to \cdots \to a_k = c$ such that for each $i = 1, \ldots, k$, a_i is contained in the super summit set and $a_i = H_i^{-1}a_{i-1}H_i$ for some permutation braid H_i.
So the conjugacy problem can be solved by the following steps: (i) given a and c, compute elements a' and c' contained in the super summit set, by using the Cycling Theorem, and (ii) for all permutation braids A_α, $\alpha \in S_n$, compute $A_\alpha^{-1}a'A_\alpha$ and collect the ones with the same inf and sup as a'. Do the same thing for all the other elements in the super summit set until any new element cannot be obtained.
The first step can be done in polynomial time and the second step is done in exponential time because the number of n-permutations is $n!$.

Extending the GHS Weil Descent Attack

Steven D. Galbraith[1], Florian Hess[2], and Nigel P. Smart[2]

[1] Mathematics Department, Royal Holloway University of London,
Egham, Surrey TW20 0EX, United Kingdom
Steven.Galbraith@rhul.ac.uk

[2] Department of Computer Science, University of Bristol,
Merchant Venturers Building, Woodland Road, Bristol, BS8 1UB, United Kingdom
{florian,nigel}@cs.bris.ac.uk

Abstract. In this paper we extend the Weil descent attack due to Gaudry, Hess and Smart (GHS) to a much larger class of elliptic curves. This extended attack applies to fields of composite degree over $\mathbb{F}_2$. The principle behind the extended attack is to use isogenies to find an elliptic curve for which the GHS attack is effective. The discrete logarithm problem on the target curve can be transformed into a discrete logarithm problem on the isogenous curve.
A further contribution of the paper is to give an improvement to an algorithm of Galbraith for constructing isogenies between elliptic curves, and this is of independent interest in elliptic curve cryptography.
We show that a larger proportion than previously thought of elliptic curves over $\mathbb{F}_{2^{155}}$ should be considered weak.

1 Introduction

The technique of Weil descent to solve the elliptic curve discrete logarithm problem (ECDLP) was first proposed by Frey [6]. This strategy was elaborated on further by Galbraith and Smart [9]. The work of Gaudry, Hess and Smart [10] gave a very efficient algorithm to reduce the ECDLP to the discrete logarithm in a Jacobian of a hyperelliptic curve over $\mathbb{F}_q$. Since subexponential algorithms exist for the discrete logarithm problem in high genus curves, this gives a possible method of attack against the ECDLP. We refer to the method of [10] as the GHS attack.

Menezes and Qu [15] analysed the GHS attack in some detail and demonstrated that it did not apply to the case when $q = 2$ and n is prime. Since this is the common case in real world applications, the work of Menezes and Qu means that the GHS attack does not apply to most deployed systems. However, there are a few deployed elliptic curve systems which use the fields $\mathbb{F}_{2^{155}}$ and $\mathbb{F}_{2^{185}}$. Hence there is considerable interest as to whether the GHS attack makes all curves over these fields vulnerable. In [18] Smart examined the GHS attack for elliptic curves with respect to the field extension $\mathbb{F}_{2^{155}}/\mathbb{F}_{2^{31}}$ and concluded that such a technique was unlikely to work for any curve defined over $\mathbb{F}_{2^{155}}$.

Jacobson, Menezes and Stein [11] also examined the field $\mathbb{F}_{2^{155}}$, this time using the GHS attack down to the subfield $\mathbb{F}_{2^5}$. They concluded that such a

L.R. Knudsen (Ed.): EUROCRYPT 2002, LNCS 2332, pp. 29–44, 2002.

strategy could be used in practice to attack around 2^{33} isomorphism classes of elliptic curves defined over $\mathbb{F}_{2^{155}}$. Since there are about 2^{156} isomorphism classes of elliptic curves defined over $\mathbb{F}_{2^{155}}$, the probability of finding one where the GHS attack is applicable is negligible.

In this paper we extend the GHS attack to a much larger number of elliptic curves over certain composite fields of even characteristic.

The main principle behind the paper is the following. Let E_1 be an elliptic curve over a finite field $\mathbb{F}_{q^n}$ and suppose that the GHS attack transforms the discrete logarithm problem in $E(\mathbb{F}_{q^n})$ into one on a curve of genus g over $\mathbb{F}_q$. Now let E_2 be an elliptic curve over $\mathbb{F}_{q^n}$ which is isogenous to E_1 (i.e., $\#E_1(\mathbb{F}_{q^n}) = \#E_2(\mathbb{F}_{q^n})$). The GHS method is not usually invariant under isogeny, so the genus which arises from the GHS attack on E_2 can be different to the one for E_1. There are two ways this property might be exploited:

- To solve a discrete logarithm problem on an elliptic curve E_1 over $\mathbb{F}_{q^n}$ for which the GHS attack is not effective, one could try to find an isogenous curve E_2 for which the GHS attack is effective.
- It is often possible to construct a 'weak' elliptic curve E_2 over $\mathbb{F}_{q^n}$ for which the GHS attack is particularly successful (this is essentially what was done by [11]). One might 'hide' such a curve by taking an isogeny to a curve E_1 for which the GHS attack is not effective. Knowledge of the 'trapdoor' (i.e., the isogeny) would enable one to solve the discrete logarithm. This approach might have both malicious and beneficial applications.

We achieve the first point as follows. Given an elliptic curve E_1 over $\mathbb{F}_{q^n}$ with $N = \#E_1(\mathbb{F}_{q^n})$ the strategy is to search over all elliptic curves which are vulnerable to the GHS attack (using the method of Section 4) until one is found which has N points (this is checked by 'exponentiating' a random point). Once such an 'easy' curve is found one can construct an isogeny explicitly using the method of Section 3, which is an improved version of the algorithm of Galbraith [8].

This process extends the power of the GHS attack considerably. For instance, with $K = \mathbb{F}_{2^{155}}$ and $k = \mathbb{F}_{2^5}$, Jacobson, Menezes and Stein [11] found that there are only 2^{33} curves for which the GHS attack is feasible. Using our techniques the number of isomorphism classes of curves which are vulnerable to attack is increased to around 2^{104}. This is a significant breakthrough in the power of the GHS attack.

Regarding the second point, we show that it is possible in principle to construct a trapdoor discrete logarithm problem using this approach. But such systems would not be practical.

As an aside, we note that the methods of this paper give a way to unify the treatment of subfield curves (sometimes called Koblitz curves) with the general case. Given an elliptic curve E_1 defined over $\mathbb{F}_q$ then special techniques are required to perform Weil descent with respect to $\mathbb{F}_{q^n}/\mathbb{F}_q$. By taking an isogeny $\phi : E_1 \to E_2$ such that E_2 is defined over $\mathbb{F}_{q^n}$ one can use the GHS method. However, we emphasise that subfield curves are only used when the extension degree n is a large prime, and Weil descent is not successful in this case.

We only describe the extended GHS strategy in the case of fields of characteristic two, though the principles can of course be easily adapted to the general case. We stress that, in the case of characteristic two, our results only apply to extension fields of composite degree.

The remainder of the paper is organised as follows. In Section 2 we explain the GHS attack and the analysis of Menezes and Qu. In Section 3 we discuss the method of Galbraith for finding isogenies between elliptic curves, in addition we sketch a new version of Galbraith's algorithm which requires much less memory. In Section 4 we describe how to obtain an explicit list of isomorphism classes of elliptic curves for which the GHS attack can be successfully applied. In Section 5 we examine the implications of using isogenies to extend the GHS attack.

2 The GHS Attack

Let us first set up some notation. Throughout this paper we let E denote an elliptic curve over the field $K = \mathbb{F}_{q^n}$ where $q = 2^r$. Let k denote the subfield $\mathbb{F}_q$. To simplify the discussion, and since those cases are the most important, we always assume that r and n are odd. We also assume that n is a prime. We stress that it is easy to obtain analogous results in the more general case.

Define $\sigma : K \to K$ to be the q-power Frobenius automorphism, and let $\pi : K \to K$ denote the absolute Frobenius automorphism $\pi : \alpha \to \alpha^2$. Therefore, $\sigma = \pi^r$.

The elliptic curve discrete logarithm problem (ECDLP) is the following: given $P \in E(K)$ and $Q \in \langle P \rangle$ find an integer λ such that $Q = [\lambda]P$. The apparent intractability of the ECDLP forms the basis for the security of cryptographic schemes based on elliptic curves.

Let l denote the order of the point P. To avoid various well known attacks, namely those described in [16], [17], [14] and [7], one chooses the curve such that l is a prime of size $l \approx q^n$. One also ensures that l does not divide $q^{ni} - 1$, for all "small" values of i.

The GHS attack is as follows. One takes an elliptic curve defined, as above, over $\mathbb{F}_{q^n}$, with a large subgroup of prime order l. We assume the curve is given by an equation of the form

$$E : Y^2 + XY = X^3 + aX^2 + b \text{ where } a \in \{0,1\}, b \in K.$$

We may assume that $a \in \{0,1\}$ since r and n are odd. Then one constructs the Weil restriction of scalars $W_{E/k}$ of E over k, this is an n-dimensional abelian variety over k. The variety $W_{E/k}$ is then intersected with $n-1$ carefully chosen hyperplanes so as to obtain a hyperelliptic curve C over the field k. Let g denote the genus of C.

In addition, the GHS attack gives an explicit and efficient group homomorphism from $E(K)$ to the Jacobian $J_C(k)$ of the curve C. Assuming some mild conditions, $J_C(k)$ will contain a subgroup of order l and the image of the subgroup of order l in $E(K)$ will be a non-trivial subgroup of order l in $J_C(k)$.

The genus of C is equal to either 2^{m-1} or $2^{m-1}-1$, where m is determined as follows.

Theorem 1 ([10]). *Let $b_i = \sigma^i(b)$, then m is given by*

$$m = m(b) = \dim_{\mathbb{F}_2}\left(\mathrm{Span}_{\mathbb{F}_2}\left\{(1, b_0^{1/2}), \ldots, (1, b_{n-1}^{1/2})\right\}\right).$$

In particular we have $1 \leq m \leq n$. If m is too small then the size of $J_C(k)$, which is $\approx q^g$, will be too small to contain a subgroup of size l. If m is too large then, although we can translate discrete logarithm problems to the hyperelliptic setting, this does not help us to solve the original ECDLP in practice.

Menezes and Qu proved the following theorem which characterises the smallest value of $m > 1$ and the elliptic curves which give rise to such m.

Theorem 2 ([15]). *Keeping the notation as above, and considering the GHS technique for Weil restriction of E from K down to k. Suppose n is an odd prime. Let t denote the multiplicative order of two modulo n and let $s = (n-1)/t$. Then*

1. *The polynomial $x^n - 1$ factors over $\mathbb{F}_2$ as $(x-1)f_1 f_2 \cdots f_s$ where the f_i's are distinct irreducible polynomials of degree t. For $1 \leq i \leq s$ define*

$$B_i = \{b \in \mathbb{F}_{q^n} : (\sigma - 1) f_i(\sigma) b = 0\}.$$

2. *For all $1 \leq i \leq s$ and all $b \in B_i$ the elliptic curves*

$$Y^2 + XY = X^3 + b,$$
$$Y^2 + XY = X^3 + \alpha X^2 + b$$

have $m(b) \leq t+1$, where α is a fixed element of K of trace one with respect to $K/\mathbb{F}_2$ (when r and n are odd we may take $\alpha = 1$).

3. *If $m(b) = t+1$ then E must be one of the previous curves for some i and some $b \in B_i$.*
4. *The cardinality of the set $B = \cup_{i=1}^{s} B_i$ is $qs(q^t - 1) + q$.*

In particular, $m(b) = t+1$ is the smallest attainable value of $m(b)$ (apart from the trivial value $m = 1$) using the GHS technique for Weil restriction down to $\mathbb{F}_q$.

Menezes and Qu use the above theorem to show that if n is a prime in the range $160 \leq n \leq 600$ and $q = 2$ then the GHS attack will be infeasible.

If we consider smaller prime values of n we see that $n = 31$ is particularly interesting, since we obtain the particularly low value of $t = 5$ and $s = 6$. Thus for the field $\mathbb{F}_{2^{155}}$ there are around 2^{33} elliptic curves whose Weil restriction down to $\mathbb{F}_{2^5}$ contains a hyperelliptic curve of genus 31 or 32. However, the next admissible size of t is 10, which would correspond to hyperelliptic curves of genus 2^{11} or $2^{11} - 1$. The algorithms for solving the discrete logarithm problem on curves of genus 2^{11} over $\mathbb{F}_{2^5}$ do have subexponential complexity, but the problem has grown to such a size (10000 bits) that this is not an efficient way to solve a 155 bit elliptic curve discrete logarithm problem.

We also need to take into account the Weil restriction from $\mathbb{F}_{2^{155}}$ down to $\mathbb{F}_{2^{31}}$. This will always lead to values of m equal to 1 or 5, thus the most useful hyperelliptic curves have genus 15 or 16. It was shown in [18] that solving a discrete logarithm problem on a curve of genus 16 over the field $\mathbb{F}_{2^{31}}$ is infeasible using current technology.

It would therefore appear that, for the field $\mathbb{F}_{2^{155}}$, by avoiding the $\approx 2^{33}$ curves which gives rise to $t = 5$ means one need not worry about the GHS attack. However, as we have explained in the introduction, our new results show that this argument is not true.

3 Constructing Isogenies

Let $K = \mathbb{F}_{q^n}$ be a finite field. Let $\Sigma = \sigma^n$ be the q^n-th power Frobenius. Let E_1 and E_2 be two non-supersingular elliptic curves over K which are isogenous over K (i.e., $\#E_1(K) = \#E_2(K)$). We wish to find an explicit representation for an isogeny

$$\phi : E_1 \to E_2.$$

In [8] the following result was proved.

Theorem 3. *There is an algorithm to compute ϕ which in the worst case takes $O(q^{3n/2+\epsilon})$ operations in $\mathbb{F}_{q^n}$ and requires at worst $O(q^{n+\epsilon})$ space. The average case complexity is $O(q^{n/4+\epsilon})$ operations.*

Galbraith in [8] gives the algorithm only in the case of prime fields of large characteristic. However, he also discusses how to extend the algorithm to arbitrary finite fields using the techniques of Couveignes [5] and Lercier [13]. It is clear that the complexity estimates are the same for the different types of finite fields.

In this section we sketch a version of Galbraith's algorithm which has polynomial storage requirement. The technique to obtain an algorithm with reduced storage requirement is inspired by ideas of Pollard [17]. This new version has expected average case running time $O(q^{n/4+\epsilon})$. The sketched algorithm applies over any base field.

Let t denote the common trace of Frobenius of E_1 and E_2. We have $\Sigma^2 - t\Sigma + q^n = 0$. Set $\Delta = t^2 - 4q^n$. The endomorphism rings $\mathrm{End}(E_i)$ are orders in the imaginary quadratic field $\mathbb{Q}(\sqrt{\Delta})$. The maximal order of $\mathbb{Q}(\sqrt{\Delta})$ we shall denote by $\mathcal{O}$, and its class number by h_Δ. We have $h_\Delta < \sqrt{|\Delta|} \ln |\Delta|$ (see [4] Ex. 5.27). Since the Frobenius lies in $\mathrm{End}(E_i)$ we have $\mathbb{Z}[\Sigma] \subseteq \mathrm{End}(E_i) \subseteq \mathcal{O}$.

The algorithm for finding an isogeny $\phi : E_1 \to E_2$ consists of a number of stages.

Stage 0: Reduce to finding an isogeny between two curves whose endomorphism ring is the maximal order.

Stage 1: Use a random walk to determine an ideal of $\mathcal{O}$ corresponding to an isogeny between the elliptic curves.

Stage 2: Smooth the ideal (using ideas from index calculus algorithms for ideal class groups in quadratic fields).

Stage 3: Extract an isogeny corresponding to the smooth ideal output by the previous stage.

In the next subsections we outline the various stages. We note that the main operations in Stages 1 and 2 can be parallelised.

In the course of our algorithm we will need to pass from isogenies to ideals and vice-versa. We shall now explain how to perform this subtask. The ideas in this section are based on subprocedures of the Schoof-Elkies-Atkin (SEA) algorithm. For an overview of this we refer to Chapter VII of [3].

Let l be a prime. An l-isogeny between two elliptic curves E_1 and E_2 with endomorphism ring $\mathcal{O}$ corresponds to an $\mathcal{O}$-ideal $\mathfrak{l}$ of norm l. We shall concentrate on the more complicated case where l splits in $\mathcal{O}$, leaving the ramified case to the reader.

Let j denote the j-invariant of an elliptic curve E over K such that $\mathrm{End}(E) \cong \mathcal{O}$. Let l denote a prime which splits in $\mathcal{O}$ (the maximal order). The characteristic polynomial of Frobenius factorises as

$$X^2 - tX + q^n \equiv (X - \mu)(X - \lambda) \pmod{l},$$

where $\mu, \lambda \in K$. By Dedekind's Theorem the prime l splits in $\mathcal{O}$ into the product of the ideals

$$\mathfrak{l}_1 = (l, \Sigma - \mu), \mathfrak{l}_2 = (l, \Sigma - \lambda).$$

In addition, the modular polynomial $\Phi_l(j, X)$ has two roots in K. These roots correspond to two j-invariants j_1 and j_2, and these are the j-invariants of the elliptic curves E_1 and E_2 that are l-isogenous to E.

We wish to determine the correct association between $\{j_1, j_2\}$ and $\{\mathfrak{l}_1, \mathfrak{l}_2\}$. To do this we use techniques from the Elkies variant of Schoof's algorithm. Fix a j-invariant, say j_1, and determine the subgroup C_1 of $E[l]$ which lies in the kernel of the isogeny from E to E_1. We then determine whether μ or λ is an eigenvalue for this isogeny by checking whether

$$\Sigma(P) = [\mu]P \text{ or } [\lambda]P \text{ for } P \in C_1.$$

If μ is an eigenvalue then j_1 corresponds to $\mathfrak{l}_1$ and j_2 corresponds to $\mathfrak{l}_2$, otherwise the correspondence is the opposite.

Using the above techniques one can also solve the following inverse problem. Given j and a prime ideal $\mathfrak{l}_1$, determine the j-invariant of the isogenous curve corresponding to the isogeny determined by $\mathfrak{l}_1$.

In either direction the method requires operations on polynomials of degree $O(l)$. Hence, the total complexity will be $O((\log q^n)l^2)$ field operations, since the main bottleneck is computing x^{q^n} modulo the a polynomial in x of degree $(l-1)/2$.

Stage 0:

Using Kohel's algorithm [12] we find, for each i, a chain of isogenies from E_i to an elliptic curve E_i' whose endomorphism ring is the maximal order $\mathcal{O}$.

This is the part of the procedure which gives us the worst case running time. Let c be the largest integer such that $c^2|\Delta$ and $\Delta/c^2 \equiv 0,1 \pmod 4$. If c contains a large prime factor then this stage will not be efficient. This will lead to a large worst case complexity for our algorithm. However, on average c turns out to be both small and smooth, and so this stage is particularly simple. In fact if $c = 1$ (i.e., the order $\mathbb{Z}[\Sigma]$ is the maximal order) then **Stage 0** can be eliminated completely.

By abuse of notation for the rest of the description we shall set $E_i = E_i'$ and $j_i = j(E_i)$.

Stage 1:

We define a random walk on the j-invariants of elliptic curves. More specifically, we will consider pairs of the form $(j, \mathfrak{a})$, where j is the j-invariant of some elliptic curve and $\mathfrak{a}$ is an element of the ideal class group of $\mathcal{O}$. The random walk only depends on the value of j.

The steps of the random walk will be l-isogenies for primes l in a set $\mathcal{F}$ of small primes. The set $\mathcal{F}$ must satisfy two important properties. First, the primes $\mathfrak{l}$ corresponding to the primes $l \in \mathcal{F}$ should generate the ideal class group of $\mathcal{O}$ (otherwise it may not be possible to get a collision). Second, there should be enough primes in $\mathcal{F}$ that the walk "looks random". The set $\mathcal{F}$ is chosen as the set of primes which split in $\mathcal{O}$ (some ramified primes can also be used) which are less than some bound L. In theory we should take $L = 6(\log \Delta)^2$. In practice, the set $\mathcal{F}$ can be taken to be rather small; it is usually enough that $\mathcal{F}$ contain about 16 distinct split primes.

We require a function

$$f : K \to \mathcal{F} \times \{0, 1\}$$

which should be deterministic but have a distribution close to uniform. The function f will be used to define the random walk. We usually construct this function using bits in the representation of the element of K.

Recall we have two j-invariants j_1 and j_2 of two isogenous elliptic curves and we wish to determine the isogeny between them, using as small amount of memory as possible. For this we use the ideas of Pollard.

We define a step of our random walk given a j-invariant $j_k^{(i)}$ as follows: First compute $(l, b) = f(j_k^{(i)})$. Then factor $\Phi_l(j_k^{(i)}, X)$ to obtain one or two new j-invariants. Using the bit b select one of the j-invariants in a deterministic manner and call it $j_k^{(i+1)}$. Use the technique from earlier to determine the prime ideal $\mathfrak{l}$ corresponding to the isogeny from $j_k^{(i)}$ to $j_k^{(i+1)}$. Finally, update the original pair $(j_k^{(i)}, \mathfrak{a}_k^{(i)})$ to $(j_k^{(i+1)}, \mathfrak{a}_k^{(i+1)})$ where

$$\mathfrak{a}_k^{(i+1)} = \text{Reduce}\left(\mathfrak{a}_k^{(i)} \cdot \mathfrak{l}\right).$$

A simplified presentation of the algorithm is as follows. We take a random walk of $T = O(\sqrt{h_\Delta}) = O(q^{n/4})$ steps, starting with the initial value $(j_1^{(0)} = j_1, \mathfrak{a}_1^{(0)} = (1))$. Only the final position $(j_1^{(T)}, \mathfrak{a}_1^{(T)})$ is stored. Then start a second random walk from the initial value ($j_2^{(0)} = j_2, \mathfrak{a}_2^{(0)} = (1)$). Eventually, after an expected T steps, we will find a value of S such that

$$j_1^{(T)} = j_2^{(S)}.$$

If such a collision is not found then the initial j-invariants may be 'randomised', by taking known isogenies and computing the corresponding j-invariants.

In practice one uses a set of distinguished j-invariants and has many processors running in parallel (starting on differently randomised j-invariants).

Once a collision is found we know that the isogeny from j_1 to j_2 is represented by the ideal

$$\mathfrak{a} = \mathfrak{a}_1^{(T)}/\mathfrak{a}_2^{(S)}.$$

It is possible to construct a chain of isogenies from E_1 to E_2 by following the paths in the random walk, but this is much longer than necessary. Instead, as we will show in the discussion of **Stage 2** , one can obtain an isogeny which can be easily represented in a short and compact format.

To analyse the complexity of **Stage 1** we notice that since the random walk is on a set of size h_Δ then we expect a collision to occur after $\sqrt{\pi h_\Delta/2}$ steps, by the birthday paradox. In the unlikely event that a collision does not occur after this many steps, we start again with related initial j-invariants, or repeat the process using a different function f. Since each step of the walk requires at most $O((\log q^n)L^2)$ field operations we obtain a final complexity for **Stage 1** of

$$O((\log q^n)L^2\sqrt{h_\Delta}) = O\left((\log q^n)^6 q^{n/4}\right) = O(q^{n/4+\epsilon}).$$

Stage 2:

Now we have two j-invariants j_1 and j_2 and an ideal $\mathfrak{a}$ representing an isogeny between j_1 and j_2. We can assume that $\mathfrak{a}$ is a reduced ideal. In this stage we will replace $\mathfrak{a}$ by a smooth ideal. Of course, the ideal $\mathfrak{a}$ was originally constructed as a smooth product of ideals, but this representation has enormous (exponential) length. Hence we desire a representation which is more suitable for computation. This is accomplished using techniques from index calculus algorithms for imaginary quadratic fields.

We choose a factor base $\mathcal{F}'$ as a set of prime ideals of $\mathcal{O}$ which are split or ramified in $\mathcal{O}$ and of size less than some bound L', which should be chosen to optimise the performance (which depends on smoothness probabilities).

We repeatedly compute the following reduced (this can be distributed) ideal

$$\mathfrak{b} = \text{Reduce}\left(\mathfrak{a} \prod_{\mathfrak{l}_i \in \mathcal{F}'} \mathfrak{l}_i^{a_i}\right),$$

where the integers a_i are chosen randomly, until the ideal $\mathfrak{b}$ factorises over the factor base $\mathcal{F}'$ as

$$\mathfrak{b} = \prod_{\mathfrak{l}_i \in \mathcal{F}'} \mathfrak{l}_i^{b_i}.$$

We then have

$$\mathfrak{a} \equiv \prod_{\mathfrak{l}_i \in \mathcal{F}'} \mathfrak{l}_i^{b_i - a_i}. \quad (1)$$

The size of the b_i's are bounded since the ideal $\mathfrak{b}$ is reduced. We choose L' (see below) so that we heuristically expect to require at most $q^{n/4}$ choices of the a_i before we obtain a value of $\mathfrak{b}$ which is sufficiently smooth. Hence, if we assume $|a_i| \leq t$ then we will require

$$t^{\#\mathcal{F}'} \geq q^{n/4}.$$

In other words the value of t can be taken to be of polynomial size in $n\log(q)$. Hence, we require polynomial storage to hold the isogeny as a smooth ideal.

To estimate the running time we need to examine the probability of obtaining a smooth number. We are essentially testing whether an integer of size $\sqrt{\Delta}$, i.e. the norm of a reduced ideal, factors over a factor base of integers less than L'. There is an optimal choice for L', but to obtain our result it is enough to take $L' = (\log(q^n))^2$. Standard estimates give an asymptotic smoothness probability of approximately u^{-u} where $u = \log(\Delta)/\log(L)$. In our case the probability is

$$u^{-u} \approx q^{n(-1+c/(\log\log q))/4}$$

for some constant c. Therefore the complexity of **Stage 2** is $q^{n/4+\epsilon}$.

For real life applications (since we need to use the modular polynomials of degree less than L') we actually select $L' = 1000$. The probability is of finding a factorisation of this form is $\Psi(x, 1000)/x$ where $\Psi(N, b)$ is the number of b-smooth integers less than N.

For completeness we give the following table on approximate values of

$$\Psi(x, 1000)$$

for values of x of interest in our situation. These values have been computed by Dan Bernstein [2].

$x = q^{n/2}$	$\approx$ $\log_2 \Psi(x, 1000)$	$\approx$ $\log_2(\Psi(x, 1000)/x)$
2^{50}	39	−11
2^{60}	45	−15
2^{70}	51	−19
2^{80}	56	−24
2^{90}	61	−29
2^{100}	66	−34
2^{110}	71	−39
2^{120}	75	−45

For the typical case, we must factor an integer of size 2^{80} over a factor base of primes less than 1000. From the table we find the probability of success is 2^{-24}, which gives a total complexity less than $q^{n/4} = 2^{160/4} = 2^{40}$. Similarly, from the table we see that for all values of n in the range $100 \leq n \leq 240$, **Stage 2** of our algorithm will run in time $O(q^{n/4+\epsilon})$.

Stage 3:

Finally we can write down the isogeny between E_1 and E_2. This is done by taking each prime ideal in equation (1) and applying the method previously given to determine the associated j-invariant. We comment that negative powers of a prime $\mathfrak{l}$ correspond to positive powers of the complex conjugate ideal $\bar{\mathfrak{l}}$. The actual isogeny is determined by Vélu's formulae [19].

Chaining these isogenies together we obtain the desired map from E_1 to E_2. In practice we do not actually write down the isogeny but simply evaluate the isogeny on the points of interest.

The ideal $\mathfrak{b}$ in **Stage 2** will have norm at most $O(\sqrt{\Delta})$. The smooth representation of the ideal equivalent to $\mathfrak{a}$ will have at most $O(\log \Delta)$ not necessarily distinct factors in it, each factor corresponding to an isogeny of degree at most L. Hence, the mapping of points from E_1 to E_2, given the smooth representation of the ideal equivalent to $\mathfrak{a}$, can be performed in time polynomial in $\log q^n$.

4 Finding Vulnerable Curves

Theorem 2 shows that there is a set of possible values for b such that the elliptic curves $Y^2 + XY = X^3 + b$ and $Y^2 + XY = X^3 + X^2 + b$ have a specific small value for m. For our application it is necessary to be able to generate at least one representative for each isogeny class of elliptic curves with this small m. In this section we discuss methods to achieve this.

Note that another approach would be to list all members of the given isogeny class, but this almost always requires more than $O(q^{n/2})$ operations.

By Theorem 2, the set of possible values b is equal to the union of the sets

$$B_i = \{b \in \mathbb{F}_{q^n} : (\sigma - 1)f_i(\sigma)b = 0\}.$$

for $1 \leq i \leq s$.

An element b lies in two of these sets if there are indices i and j such that $(\sigma - 1)f_i(\sigma)b = (\sigma - 1)f_j(\sigma)b = 0$. Since $\gcd(f_i(x), f_j(x)) = 1$ it follows that $(\sigma - 1)b = 0$. Therefore $B_i \cap B_j = \mathbb{F}_q$ for all $i \neq j$.

Each of these sets will be handled in turn, so from now on we fix an index i and define $f(x) = (x-1)f_i(x)$ and $B = \{b \in \mathbb{F}_{q^n} : f(\sigma)b = 0\}$.

Let α be a normal basis generator for $K = \mathbb{F}_{2^{rn}}$ over $\mathbb{F}_2$. In other words $\{\alpha, \alpha^2, \ldots, \alpha^{2^{nr-1}}\}$ is a vector space basis for K over $\mathbb{F}_2$. It is a fundamental fact that such an element α exists and that $K = \{g(\pi)\alpha : g(x) \in \mathbb{F}_2[x]\} = \{g(\sigma)\alpha : g(x) \in \mathbb{F}_q[x]\}$.

The following result is an easy exercise.

Lemma 1. *Let the notation be as above. Write $h(x) = (x^n - 1)/f(x)$ and define $\alpha' = h(\sigma)\alpha$. Then*

$$B = \{g(\sigma)\alpha' : g(x) \in \mathbb{F}_q[x]\}. \tag{2}$$

Indeed, in the above it is enough to let the $g(x)$ run over a set of representatives for the quotient ring $\mathbb{F}_q[x]/(f(x))$ (i.e., over all elements of $\mathbb{F}_q[x]$ of degree less than $\deg(f(x))$).

Lemma 1 gives an efficient algorithm to compute representatives for each set B which has running time $O(\#B)$. Clearly $\#B = q^{\deg(f(x))}$.

By taking the union of these sets we obtain all the values for b which we require. In the notation of Theorem 2 we have s values for i and $\deg(f_i(x)) = t$, therefore $\#B_i = q^{t+1}$ for each index i and, since the intersection of any two B_i has size q it follows that $\#(\cup_i B_i) = qs(q^t - 1) + q$ as claimed in Theorem 2.

In the appendix we describe a more efficient search strategy to find whether there are any isogenous curves with a small value of m. The refined method is designed to give at least one representative for each isogeny class. The full list of candidates has size $sq^{t+1}/(nr)$ and the complexity of generating this list is $O(sq^{t+1}/(nr))$ operations in K.

5 Implications

We stress that for curves over large prime fields the techniques of Weil restriction do not apply, and for curves over fields of the form $\mathbb{F}_{2^p}$, where p is prime, Menezes and Qu [15] showed that for curves of cryptographic interest the GHS attack does not apply. Hence, for the rest of this section we will concentrate on the case where $K = \mathbb{F}_{q^n}$ where $q = 2^r$ is a non-trivial power of two and n is a prime such that $5 \leq n < 43$.

Theorem 2 states that an upper bound on the number of isomorphism classes with the smallest non-trivial value of m is given roughly by $2sq^{t+1}$. Hence the probability that a random curve is vulnerable to the GHS attack is roughly sq^{t+1-n}.

We now consider the number of isogeny classes which contain a curve with the smallest non-trivial value of m. The total number of isogeny classes of curves is around $2q^{n/2}$, due to the Hasse-Weil bounds and the fact that we are in characteristic two. We make the heuristic assumption that the elliptic curves with small m distribute over the isogeny classes similar to arbitrary elliptic curves. Note that the 2-power Frobenius preserves the value of m. We make the heuristic assumption that this is the only isogeny with this property. Hence, the number of non-isomorphic curves with the smallest non-trivial value of m (up to 2-power Frobenius action) is

$$\frac{2sq^{t+1}}{nr}.$$

Thus we deduce that the probability that a random elliptic curve over K lies in an isogeny class which contains a curve with the smallest non-trivial value of m is approximated by $p = \min(1, \mathfrak{p})$ where

$$\mathfrak{p} = \frac{sq^{t+1-n/2}}{nr}.$$

We now consider some special cases of small prime n. Recall that $st = n-1$. For the cases with $s = 1$ the GHS attack does not reduce the problem to one which is significantly easier. Our extension is not interesting in that case.

When $s = 2$ we obtain an interesting case. The original GHS method applies to a random curve with probability about $2/q^{(n-1)/2}$ in this case. The extended approach should apply with probability about $2q^{1/2}/(nr)$ which is larger than one when n is fixed and $q \to \infty$. In other words, we should eventually be able to consider all curves using the new method.

However, our method is not feasible in this case since the (reduced) set B of Theorem 2 and the end of section 4 has size $2q^{(n+1)/2}/(nr)$ and so the cost of finding all the curves with small m is greater than the Pollard methods for solving the original ECDLP.

The only prime value of n with $s \geq 3$ in the range $5 \leq n < 43$ is $n = 31$ where $s = 6$ and $t = 5$. This is particularly interesting given that the field $\mathbb{F}_{2^{155}}$ is used for an elliptic curve group in the IPSEC set of protocols for key agreement. See [1] for a description of the curve used and [18] for a previous analysis of this curve using the GHS attack with $n = 5$.

When $n = 31$ the proportion of all curves which succumb to the basic GHS attack is approximately $6q^{-25}$. For the extended attack, assuming the values of m are distributed evenly over the isogeny classes, the proportion of all vulnerable curves is approximately

$$\frac{6q^{-9.5}}{31r}.$$

The complexity of the extended method is as follows. Given an elliptic curve E with N points we search all the curves which have small m (using the method of the Appendix) until we find one with N points (this is checked by exponentiating a random point). This search takes less than $6q^6/(31r)$ steps. If such a curve is found then construct an isogeny using Section 3 in $O(q^{7.75+\epsilon})$ operations in most cases. For $r = 5$ we obtain a total of about $O(2^{26} + 2^{39})$ operations. Jacobson, Menezes and Stein [11] stated that solving the discrete logarithm problem on the hyperelliptic curve is feasible for curves of cryptographic interest when $m = t + 1 = 6$ and $r = 5$. The total complexity is expected to be dominated by $O(2^{39})$, which is quite feasible.

For the IPSEC curve over $\mathbb{F}_{2^{155}}$ this means there is roughly a 2^{-52} chance that the curve can be attacked using the extended GHS attack, as opposed to 2^{-122} for the standard GHS attack. Using the methods in Section 4 we searched all isogeny classes to see if there was a curve with $m = 6$ which was isogenous to the IPSEC curve. This search took 31 days on a 500 Mhz Pentium III using the Magma package. Not surprisingly we did not find an isogenous curve and so we can conclude that the IPSEC curve is not susceptible to the extended GHS attack. This shows that further research into Weil descent is required before it can be shown that all elliptic curves over composite extension fields are weak.

We comment that for most small primes $n \geq 43$ with $s \geq 3$ the value of t is so large that the GHS method is not effective, except for the well-known case of $n = 127$ which has $t = 7$ (the next smallest values are $n = 73$ with $t = 9$ and

$n = 89$ with $t = 11$). If we consider elliptic curves over $\mathbb{F}_{q^{127}}$ then the extended GHS method applies with probability roughly $18q^{-55.5}/(127r)$ compared with probability $18q^{-119}$ for the usual GHS attack.

We briefly mention one possible extension of the ideas of this paper. Given an ECDLP in $E(\mathbb{F}_{q^n})$ one could enlarge the field to $\mathbb{F}_{q^{nl}}$ for a small value of l. One could then perform the GHS attack with respect to the extension $\mathbb{F}_{q^{nl}}/\mathbb{F}_q$ and the number of curves with small m might be increased. The drawback of this approach is that the final discrete logarithm problem which must be solved has grown in size. We expect that this approach would not be useful in practice.

Finally we comment on the possibility of using Weil descent and isogenies to construct a trapdoor for the ECDLP. Suppose $E/\mathbb{F}_{q^n}$ is vulnerable to the GHS attack and suppose one has an isogeny $\phi : E' \to E$ such that E' is not vulnerable to the GHS attack. Then one could publish E' and yet solve the ECDLP using the trapdoor ϕ.

The methods of this paper allow an attacker to find ϕ whenever the (reduced) set B of Theorem 2 is small enough. Since B has size $sq^{t+1}/(nr)$ it follows that t should be large for the trapdoor discrete logarithm scheme. On the other hand, to solve the DLP using the GHS attack it is necessary that t be small.

The most suitable case seems to be $n = 7$. The set B has size $O(q^4)$ while the GHS attack reduces to a DLP on a genus 8 curve (and Gaudry's algorithm requires $O(q^{2+\epsilon})$ operations). However, these parameters do not appear to result in a truly practical system.

Acknowledgement

The authors thank Pierrick Gaudry for useful conversations. The authors also thank Edlyn Teske, Alfred Menezes, Michael Müller and Kim Nguyen for comments.

References

1. IETF. The Oakley Key Determination Protocol. *IETF RFC 2412*, Nov 1998.
2. D.J. Bernstein. Bounds on $\Psi(x, y)$. **http://cr.yp.to/psibound.html**.
3. I.F. Blake, G. Seroussi and N.P. Smart. *Elliptic Curves in Cryptography.* Cambridge University Press, 1999.
4. H. Cohen, *A course in computational number theory.* Springer GTM 138 1993.
5. J.-M. Couveignes. Computing l-isogenies using the p-torsion. *Algorithmic Number Theory Symposium- ANTS II*, Springer-Verlag LNCS 1122, 59–65, 1996.
6. G. Frey. How to disguise an elliptic curve. Talk at ECC' 98, Waterloo.
7. G. Frey and H. Rück. A remark concerning m-divisibility and the discrete logarithm in the divisor class group of curves. *Math. Comp.*, **62**, 865–874, 1994.
8. S.D. Galbraith. Constructing isogenies between elliptic curves over finite fields. *LMS J. Comput. Math.*, **2**, 118–138, 1999.
9. S.D. Galbraith and N.P. Smart. A Cryptographic application of Weil descent. *Codes and Cryptography*, Springer-Verlag LNCS 1746, 191–200, 1999.
10. P. Gaudry, F. Hess and N.P. Smart. Constructive and destructive facets of Weil descent on elliptic curves. *J. Cryptology*, **15**, 19–46, 2002.

11. M. Jacobson, A. Menezes and A. Stein. Solving elliptic curve discrete logarithm problems using Weil descent. *J. Ramanujan Math. Soc.*, **16**, No. 3, 231–260, 2001.
12. D. Kohel. *Endormorphism rings of elliptic curves over finite fields.* Phd Thesis, Berkeley, 1996.
13. R. Lercier. Computing isogenies in $\mathbb{F}_{2^n}$. *Algorithmic Number Theory Symposium-ANTS II*, Springer-Verlag LNCS 1122, 197–212, 1996.
14. A. Menezes, T. Okamoto and S. Vanstone. Reducing elliptic curve logarithms to logarithms in finite fields. *IEEE Trans. on Infor. Th.*, **39**, 1639–1646, 1993.
15. A. Menezes and M. Qu. Analysis of the Weil descent attack of Gaudry, Hess and Smart. *Topics in Cryptology - CT-RSA 2001*, Springer-Verlag LNCS 2020, 308–318, 2001.
16. S. Pohlig and M. Hellman. An improved algorithm for computing logarithms over $GF(p)$ and its cryptographic significance. *IEEE Trans. on Infor. Th.*, **24**, 106–110, 1978.
17. J. Pollard. Monte Carlo methods for index computations mod p. *Math. Comp.*, **32**, 918–924, 1978.
18. N.P. Smart. How secure are elliptic curves over composite extension fields? *EUROCRYPT '01*, Springer-Verlag LNCS 2045, 30–39, 2001.
19. J. Vélu. Isogénies entre courbes elliptiques. *Comptes Rendus l'Acad. Sci. Paris, Ser. A*, **273**, 238-241 1971.

Appendix: A More Refined Search Strategy

Recall that the goal is to produce a representative of each isogeny class of elliptic curves with small values for m. Since r and n are odd we may assume that all elliptic curves have the form $E : y^2 + xy = x^3 + ax^2 + b$ with $a \in \{0, 1\}$. If E is an elliptic curve which has a small value for m then $E^\pi : y^2 + xy = x^3 + ax^2 + \pi(b)$ is an isogenous elliptic curve. Since $f(\sigma)\pi b = \pi f(\sigma) b = 0$ it follows that E^π necessarily has the same small value for m. Rather than listing all $b \in B$ it would be better to find a set $B' \subset B$ such that $B = \{\pi^j(b) : b \in B', 1 \leq j \leq rn\}$.

From Lemma 1 we have $B = \{g(\sigma)\alpha' : g(x) \in \mathbb{F}_q[x]\}$. We want to work in terms of π rather than $\sigma = \pi^r$. An analogous argument to that used to prove Lemma 1 gives

$$B = \{g(\pi)\alpha' : g(x) \in \mathbb{F}_2[x]\}. \tag{3}$$

In this case the polynomials $g(x) \in \mathbb{F}_2[x]$ may be taken to have degree less than $\deg(f(x^r))$. Given any two elements $b_j = g_j(\pi)\alpha'$ for $j \in \{1, 2\}$ and any two elements $c_j \in \mathbb{F}_2$ we clearly have $c_1 b_1 + c_2 b_2 = (c_1 g_1 + c_2 g_2)(\pi)\alpha' \in B$. It is easy to show that B is a module over $\mathbb{F}_2[\pi]$ and that the following result holds.

Lemma 2. *Let notation be as above. There is an isomorphism of rings from $\mathbb{F}_2[\pi]$ to $\mathbb{F}_2[x]$ and there is a corresponding isomorphism of modules from the $\mathbb{F}_2[\pi]$-module B to the $\mathbb{F}_2[x]$-module $\mathbb{F}_2[x]/(f(x^r))$. An isomorphism is given by*

$$g(\pi)\alpha' \longmapsto g(x).$$

Consider the factorisation $f(x^r) = \prod_{j=1}^{l} f_j(x)$ into irreducibles. The polynomials $f_j(x)$ are all distinct since $x^{rn} - 1$ has no repeated roots (when r and n are both odd then $\gcd(x^{rn}-1, rnx^{rn-1}) = 1$). The Chinese remainder theorem for polynomials implies that we have the following isomorphism of $\mathbb{F}_2[x]$-modules

$$\mathbb{F}_2[x]/(f(x^r)) \cong \bigoplus_{j=1}^{l} \mathbb{F}_2[x]/(f_j(x)). \tag{4}$$

The terms on the right hand side are finite fields $K_j = \mathbb{F}_2[x]/(f_j(x))$.

Combining Lemma 2 and equation (4) we see that the $\mathbb{F}_2[\pi]$-module B is isomorphic to the $\mathbb{F}_2[x]$-module $(\oplus_j K_j)$. We need to understand the action of π on elements of B, and this corresponds to multiplication of elements of $(\oplus_j K_j)$ by x.

Hence, write N_j for the normal subgroup of K_j^* generated by x, so that $x(gN_j) = gN_j$ for any $g \in K_j^*$. It follows that the cosets of K_j^*/N_j give representatives for the x-orbits of elements of K_j^* (the zero element of K_j must be handled separately). We write $c_j = [K_j^* : N_j]$ for the index (i.e., the number of distinct cosets). Let ζ_j be a generator for the cyclic group K_j^*, then $N_j = \langle \zeta_j^{c_j} \rangle$ (i.e., $x = \zeta_j^{dc_j}$ for some d which depends on ζ_j).

However, it is not possible to study the fields K_j individually. Instead, we have to consider the product $\oplus_j K_j$ and have to consider that the action of x on this product is multiplication by x on every coordinate.

The following result gives an explicit set of representatives for $(\oplus_j K_j)/\langle x \rangle$. This is essentially achieved by forming a diagonalisation (under the action of x) of the set $\oplus_j K_j$.

Lemma 3. *The set*

$$\{g_1 \oplus \cdots \oplus g_{a-1} \oplus \zeta_a^b : 1 \le a \le l, 0 \le b < c_a, g_j \in K_j\} \cup \{0\}$$

is a set of representatives for $(\oplus_j K_j)/\langle x \rangle$.

Proof. The result follows from two facts. First, every element of $\oplus_j K_j$ is of the given form if we disregard the condition $0 \le b < c_a$. Second, x acts as

$$x^i(g_1 \oplus \cdots \oplus g_{a-1} \oplus \zeta_a^b) = x^i(g_1) \oplus \cdots \oplus x^i(g_{a-1}) \oplus x^i(\zeta_a^b),$$

and $x^i(\zeta_a^b) = \zeta_a^{b+idc_a}$. □

This set of representatives can be easily mapped to B as follows. For any $g_1 \oplus \cdots \oplus g_l$ we can represent each $g_j \in K_j = \mathbb{F}_2[x]/(f_j(x))$ as a polynomial $g_j(x) \in \mathbb{F}_2[x]$. The Chinese remainder algorithm gives $g(x) \in \mathbb{F}_2[x]/(f(x))$ which reduces to each $g_j(x)$ modulo $(f_j(x))$. We then obtain the corresponding value $b = g(\pi)\alpha' \in B$.

Thus we obtain an algorithm to compute a set B' of elements whose π-orbits generate B.

Example

We give an example of the refined approach in a simplified way. Let us consider $K = \mathbb{F}_{2^{155}}$, $n = 31$, $r = 5$, $q = 2^5$ and $m = 6$. The factorisation of $x^{31} - 1$ over $\mathbb{F}_2$ is

$$(x-1)\prod_{i=1}^{6} f_i(x)$$

with $f_i(x)$ of degree 5. For brevity we skip the actual values of the $f_i(x)$ and subsequent polynomials. We obtain $s = 6$, $t = 5$ and $B_i \cap B_j = \mathbb{F}_q$ for $i \neq j$.

We now write $f(x)$ and B instead of $f_i(x)$ and B_i for all i in turn (note the difference in defining $f(x)$ compared to before). Over $\mathbb{F}_2$ we have the factorisation

$$f(x^r) = g_1(x)g_2(x)$$

with $\deg(g_1(x)) = 5$ and $\deg(g_2(x)) = 20$. Using the Chinese remainder theorem again yields

$$\begin{aligned} B &\cong \mathbb{F}_2[x]/((x^r-1)f(x^r)) \\ &\cong \mathbb{F}_2[x]/((x^r-1)g_1(x)) \oplus \mathbb{F}_2[x]/(g_2(x)). \end{aligned}$$

The first quotient ring above contains $2^{\deg((x^r-1)g_1(x))} = 1024$ elements. The second quotient ring $\mathbb{F}_2[x]/(g_2(x))$ is isomorphic to $\mathbb{F}_{2^{20}}$ and the index of the group generated by the element $x + (g_2(x))$ in the full multiplicative group equals $(2^{20} - 1)/nr = 6765$. Let $\zeta \in \mathbb{F}_2[x]/(g_2(x))$ be a generator of the full multiplicative group. Mapping all elements of $\mathbb{F}_2[x]/((x^r - 1)g_1(x))$ and the elements $\{\zeta^j \mid 0 \leq j < 6765\} \cup \{0\}$ under the above isomorphisms to K gives sets of elements $B_1, B_2 \subseteq K$ such that

$$B = \bigcup_{j=0}^{rn-1} \pi^j(B_1 + B_2).$$

In $B_1 + B_2$ clearly only the 1024 elements of B_1 can possibly be conjugated under π which makes up only a very small fraction of all the $1024 \cdot 6676$ elements of $B_1 + B_2$.

We conclude that altogether for $1 \leq i \leq 6$ we obtain a set of $6 \cdot 1024 \cdot 6766$ representatives of classes under the action of powers of π in $\cup_i B_i$ with only small redundancy due to double occurrences or the action of powers of π.

Universal Hash Proofs and a Paradigm for Adaptive Chosen Ciphertext Secure Public-Key Encryption

Ronald Cramer and Victor Shoup

[1] BRICS & Dept. of Computer Science, Aarhus University
cramer@brics.dk
[2] IBM Zurich Research Laboratory
sho@zurich.ibm.com

Abstract. We present several new and fairly practical public-key encryption schemes and prove them secure against adaptive chosen ciphertext attack. One scheme is based on Paillier's Decision Composite Residuosity assumption, while another is based in the classical Quadratic Residuosity assumption. The analysis is in the standard cryptographic model, i.e., the security of our schemes does not rely on the Random Oracle model. Moreover, we introduce a general framework that allows one to construct secure encryption schemes in a generic fashion from language membership problems that satisfy certain technical requirements. Our new schemes fit into this framework, as does the Cramer-Shoup scheme based on the Decision Diffie-Hellman assumption.

1 Introduction

It is generally considered that the "right" notion of security for security for a general-purpose public-key encryption scheme is that of *security against adaptive chosen ciphertext attack*, as defined by Rackoff and Simon [RS].

Rackoff and Simon present a scheme that can be proven secure against adaptive chosen ciphertext attack under a reasonable intractability assumption; however, their scheme requires the involvement of a trusted third party that plays a special role in registering users (both senders and receivers). Dolev, Dwork, and Naor [DDN] present a scheme that can be proven secure against adaptive chosen ciphertext attack under a reasonable intractability assumption, and which does not require a trusted third party.

Although these schemes run in polynomial time, they are horrendously impractical. Up until now, the only *practical* scheme that has been proposed that can be proven secure against adaptive chosen ciphertext attack under a reasonable intractability assumption is that of Cramer and Shoup [CS1, CS3]. This scheme is based on the Decision Diffie-Hellman (DDH) assumption, and is not much less efficient than traditional ElGamal encryption.

Other practical schemes have been proposed and *heuristically* proved secure against adaptive chosen ciphertext. More precisely, these schemes are proven secure under reasonable intractability assumptions in the *Random Oracle model*

L.R. Knudsen (Ed.): EUROCRYPT 2002, LNCS 2332, pp. 45–64, 2002.

[BR]. While the Random Oracle model is a useful heuristic, a proof in the Random Oracle model does not rule out all possible attacks (see [CGH]).

1.1 Our Contributions

We present several new and fairly practical public-key encryption schemes and prove them secure against adaptive chosen ciphertext attack. One scheme is based on Paillier's Decision Composite Residuosity (DCR) assumption [P], while another is based in the classical Quadratic Residuosity (QR) assumption. The analysis is in the standard cryptographic model, i.e., the security of our schemes does not rely on the Random Oracle model. Also, our schemes do not rely on the involvement of a trusted third party.

We also introduce the notion of a *universal hash proof system.* Essentially, this is a special kind of non-interactive zero-knowledge proof system for a language. We do not show that universal hash proof systems exist for all NP languages, but we do show how to construct *very efficient* universal hash proof systems for a general class of group-theoretic language membership problems.

Given an efficient universal hash proof system for a language with certain natural cryptographic indistinguishability properties, we show how to construct an efficient public-key encryption scheme secure against adaptive chosen ciphertext attack in the standard model. Our construction only uses the universal hash proof system as a primitive: no other primitives are required, although even more efficient encryption schemes can be obtained by using hash functions with appropriate collision-resistance properties.

We show how to construct efficient universal hash proof systems for languages related to the DCR and QR assumptions. From these we get corresponding public-key encryption schemes that are secure under these assumptions.

The DCR-based scheme is very practical. It uses an n-bit RSA modulus N (with, say, $n = 1024$). The public and private keys, as well as the ciphertexts, require storage for $O(n)$ bits. Encryption and decryption require $O(n)$ multiplications modulo N^2.

The QR-based scheme is somewhat less practical. It uses an n-bit RSA modulus N as above, as well as an auxiliary parameter t (with, say, $t = 128$). The public and private keys require $O(nt)$ bits of storage, although ciphertexts require just $O(n + t)$ bits of storage. Encryption and decryption require $O(nt)$ multiplications modulo N.

We also show that the original Cramer-Shoup scheme follows from of our general construction, when applied to a universal hash proof system related to the DDH assumption.

For lack of space, some details have been omitted from this extended abstract. We refer the reader to the full length version of this paper [CS2] for these details.

2 Universal Projective Hashing

Let X and Π be finite, non-empty sets. Let $H = (H_k)_{k \in K}$ be a collection of functions indexed by K, so that for every $k \in K$, H_k is a function from X into

Π. Note that we may have $H_k = H_{k'}$ for $k \neq k'$. We call $\mathbf{F} = (H, K, X, \Pi)$ a *hash family*, and each H_k a *hash function.*

We now introduce the concept of *universal projective hashing.* Let $\mathbf{F} = (H, K, X, \Pi)$ be a hash family. Let L be a non-empty, proper subset of X. Let S be a finite, non-empty set, and let $\alpha : K \to S$ be a function. Set $\mathbf{H} = (H, K, X, L, \Pi, S, \alpha)$.

Definition 1. $\mathbf{H} = (H, K, X, L, \Pi, S, \alpha)$, *as above, is called a* projective hash family (for (X, L)) *if for all* $k \in K$, *the action of* H_k *on* L *is determined by* $\alpha(k)$.

Definition 2. *Let* $\mathbf{H} = (H, K, X, L, \Pi, S, \alpha)$ *be a projective hash family, and let* $\epsilon \geq 0$ *be a real number. Consider the probability space defined by choosing* $k \in K$ *at random.*

We say that $\mathbf{H}$ *is* ϵ-universal *if for all* $s \in S$, $x \in X \setminus L$, *and* $\pi \in \Pi$, *it holds that*

$$\Pr[H_k(x) = \pi \ \wedge \ \alpha(k) = s] \leq \epsilon \Pr[\alpha(k) = s].$$

We say that $\mathbf{H}$ *is* ϵ-universal$_2$ *if for all* $s \in S$, $x, x^* \in X$, *and* $\pi, \pi^* \in \Pi$ *with* $x \notin L \cup \{x^*\}$, *it holds that*

$$\Pr[H_k(x) = \pi \wedge H_k(x^*) = \pi^* \ \wedge \alpha(k) = s] \leq \epsilon \Pr[H_k(x^*) = \pi^* \ \wedge \alpha(k) = s].$$

We will sometimes refer to the value of ϵ in the above definition as the *error rate* of $\mathbf{H}$.

Note that if $\mathbf{H}$ is ϵ-universal$_2$, then it is also ϵ-universal (note that $|X| \geq 2$).

We can reformulate the above definition as follows. Let $\mathbf{H}=(H,K,X,L,\Pi,S,\alpha)$ be a projective hash family, and consider the probability space defined by choosing $k \in K$ at random. $\mathbf{H}$ is ϵ-universal means that conditioned on a fixed value of $\alpha(k)$, even though the value of H_k is completely determined on L, for any $x \in X \setminus L$, the value of $H_k(x)$ can be guessed with probability at most ϵ. $\mathbf{H}$ is ϵ-universal$_2$ means that in addition, for any $x^* \in X \setminus L$, conditioned on fixed values of $\alpha(k)$ and $H_k(x^*)$, for any $x \in X \setminus L$ with $x \neq x^*$, the value of $H_k(x)$ can be guessed with probability at most ϵ.

We will need a variation of universal projective hashing, which we call *smooth projective hashing.*

Let $\mathbf{H} = (H, K, X, L, \Pi, S, \alpha)$ be a projective hash family. We define two random variables, $U(\mathbf{H})$ and $V(\mathbf{H})$, as follows. Consider the probability space defined by choosing $k \in K$ at random, $x \in X \setminus L$ at random, and $\pi' \in \Pi$ at random. We set $U(\mathbf{H}) = (x, s, \pi')$ and $V(\mathbf{H}) = (x, s, \pi)$, where $s = \alpha(k)$ and $\pi = H_k(x)$.

Definition 3. *Let* $\epsilon \geq 0$ *be a real number. A projective hash family* $\mathbf{H}$ *is* ϵ-smooth *if* $U(\mathbf{H})$ *and* $V(\mathbf{H})$ *are* ϵ*-close (i.e., the statistical distance between them is at most* ϵ*).*

Our definition of universal and universal$_2$ projective hash families are quite strong: so strong, in fact, that in many instances it is impossible to efficiently implement them. However, in all our applications, it is sufficient to efficiently implement a projective hash family that effectively *approximates* a universal or universal$_2$ projective hash family. To this end, we define an appropriate notion of *distance* between projective hash families.

Definition 4. *Let $\delta \geq 0$ be a real number. Let $\mathbf{H} = (H, K, X, L, \Pi, S, \alpha)$ and $\mathbf{H}^* = (H^*, K^*, X, L, \Pi, S, \alpha^*)$ be projective hash families. We say that* $\mathbf{H}$ *and* $\mathbf{H}^*$ *are* δ-close *if the distributions $(H_k, \alpha(k))$ (for random $k \in K$) and $(H^*_{k^*}, \alpha^*(k^*))$ (for random $k^* \in K^*$) are δ-close.*

2.1 Some Elementary Reductions

We mention very briefly here some reductions between the above notions. Details are presented in [CS2]. First, via a trivial "t-fold parallelization," we can reduce the error rate of a universal or universal$_2$ family of projective hash functions from ϵ to ϵ^t. Second, we can efficiently convert an ϵ-universal family of projective hash functions into an ϵ-universal$_2$ family of projective hash functions. Third, using a pair-wise independent family of hash functions, and applying the Leftover Hash Lemma (a.k.a., Entropy Smoothing Lemma; see, e.g., [L, p. 86]), we can efficiently convert an ϵ-universal family of projective hash functions into a δ-smooth family of projective hash functions whose outputs are a-bit strings, provided ϵ and a are not too large and δ is not too small. These last two constructions are useful from a theoretical perspective, but we will not actually need them to obtain any of our concrete encryption schemes.

3 Subset Membership Problems

In this section we define a class of languages with some natural cryptographic indistinguishability properties. The definitions below capture the natural properties of well-known cryptographic problems such as the Quadratic Residuosity and Decision Diffie-Hellman problems, as well as others.

A *subset membership problem* $\mathbf{M}$ specifies a collection $(I_\ell)_{\ell \geq 0}$ of distributions. For every value of a security parameter $\ell \geq 0$, I_ℓ is a probability distribution of *instance descriptions.*

An instance description Λ specifies finite, non-empty sets X, L, and W, such that L is a proper subset of X, as well as a binary relation $R \subset X \times W$. For all $\ell \geq 0$, $[I_\ell]$ denotes the instance descriptions that are assigned non-zero probability in the distribution I_ℓ. We write $\Lambda[X, L, W, R]$ to indicate that the instance Λ specifies X, L, W and R as above. For $x \in X$ and $w \in W$ with $(x, w) \in R$, we say that w is a *witness* for x. Note that it would be quite natural to require that for all $x \in X$, we have $(x, w) \in R$ for some $w \in W$ if and only if $x \in L$, and that the relation R is efficiently computable; however, we will not make these requirements here, as they are not necessary for our purposes. The actual role of a witness will become apparent in the next section.

A subset membership problem also provides several algorithms. For this purpose, we require that instance descriptions, as well as elements of the sets X and

W, can be uniquely encoded as bit strings of length polynomially bounded in ℓ. The following algorithms are provided:

- an efficient *instance sampling algorithm* that samples the distribution I_ℓ. We only require that the output distribution of this algorithm is statistically close to I_ℓ. In particular, with negligible probability, it may output something that is not even an element of $[I_\ell]$.
- an efficient *subset sampling algorithm* that given an instance $\Lambda[X, L, W, R] \in [I_\ell]$, outputs a random $x \in L$, together with a witness $w \in W$ for x. We only require that the distribution of the output value x is statistically close to the uniform distribution on L. However, we do require that the output x is always in L.
- an efficient algorithm that given an instance $\Lambda[X, L, W, R] \in [I_\ell]$ and a bit string ζ, checks whether ζ is a valid binary encoding of an element of X.

This completes the definition of a subset membership problem.

We say a subset membership problem is *hard* if it is computationally hard to distinguish (Λ, x) from (Λ, y), where $\Lambda[X, L, W, R]$ is randomly sampled from I_ℓ, x is randomly sampled from L, and y is randomly sampled from $X \setminus L$.

4 Universal Hash Proof Systems

4.1 Hash Proof Systems

Let **M** be a subset membership problem, as defined in §3, specifying a sequence $(I_\ell)_{\ell\geq 0}$ of instance distributions.

A *hash proof system (HPS)* **P** for **M** associates with each instance $\Lambda[X,L,W,R]$ of **M** a projective hash family $\mathbf{H} = (H, K, X, L, \Pi, S, \alpha)$ for (X, L).

Additionally, **P** provides several efficient algorithms to carry out basic operations we have defined for an associated projective hash family; namely, sampling $k \in K$ at random, computing $\alpha(k) \in S$ given $k \in K$, and computing $H_k(x) \in \Pi$ given $k \in K$ and $x \in X$. We call this latter algorithm the *private evaluation algorithm for* **P**. Moreover, a crucial property is that the system provides an efficient algorithm to compute $H_k(x) \in \Pi$, given $\alpha(k) \in S$, $x \in L$, and $w \in W$, where w is a witness for x. We call this algorithm the *public evaluation algorithm for* **P**. The system should also provide an algorithm that recognizes elements of Π.

4.2 Universal Hash Proof Systems

Definition 5. *Let $\epsilon(\ell)$ be a function mapping non-negative integers to non-negative reals. Let* **M** *be a subset membership problem specifying a sequence $(I_\ell)_{\ell\geq 0}$ of instance distributions. Let* **P** *be an HPS for* **M**.

We say that **P** *is $\epsilon(\ell)$-*universal (respectively, -universal$_2$, -smooth) *if there exists a negligible function $\delta(\ell)$ such that for all $\ell \geq 0$ and for all $\Lambda[X, L, W, R] \in [I_\ell]$, the projective hash family $\mathbf{H} = (H, K, X, L, \Pi, S, \alpha)$ that* **P** *associates with*

Λ is $\delta(\ell)$-close to an $\epsilon(\ell)$-universal (respectively, -universal$_2$, -smooth) projective hash family $\mathbf{H}^* = (H^*, K^*, X, L, \Pi, S, \alpha^*)$.

Moreover, if this is the case, and $\epsilon(\ell)$ is a negligible function, then we say that **P** *is* strongly universal (respectively, universal$_2$, smooth).

It is perhaps worth remarking that if a hash proof system is strongly universal, and the underlying subset membership problem is hard, then the problem of evaluating $H_k(x)$ for random $k \in K$ and arbitrary $x \in X$, given only x and $\alpha(k)$, must be hard.

We also need an extension of this notion.

The definition of an *extended* HPS **P** for **M** is the same as that of ordinary HPS for **M**, except that for each $\ell \geq 0$ and for each $\Lambda = \Lambda[X, L, W, R] \in [I_\ell]$, the proof system **P** associates with Λ a finite set E along with a projective hash family $\mathbf{H} = (H, K, X \times E, L \times E, \Pi, S, \alpha)$ for $(X \times E, L \times E)$. Note that in this setting, to compute $H_k(x, e)$ for $x \in L$ and $e \in E$, the public evaluation algorithm takes as input $\alpha(k) \in S$, $x \in L$, $e \in E$, and a witness $w \in W$ for x, and the private evaluation algorithm takes as input $k \in K$, $x \in X$, and $e \in E$. We shall also require that elements of E are uniquely encoded as bit strings of length bounded by a polynomial in ℓ, and that **P** provides an algorithm that efficiently determines whether a bit string is a valid encoding of an element of E.

Definition 5 can be modified in the obvious way to define *extended* $\epsilon(\ell)$-*universal$_2$ HPS's* (we do not need any of the other notions, nor are they particularly interesting).

Note that based on the constructions mentioned in §2.1, given an HPS that is (say) 1/2-universal, we can construct a strongly universal HPS, a (possibly extended) strongly universal$_2$ HPS, and a strongly smooth HPS. However, in most special cases of practical interest, there are much more efficient constructions.

5 A General Framework for Secure Public-Key Encryption

In this section, we present a general technique for building secure public-key encryption schemes using appropriate hash proof systems for a hard subset membership problem.

Let **M** be a subset membership problem specifying a sequence $(I_\ell)_{\ell \geq 0}$ of instance distributions. We also need a strongly smooth hash proof system **P** for **M**, as well as a strongly universal$_2$ extended hash proof system $\hat{\mathbf{P}}$ for **M**. We discuss **P** and $\hat{\mathbf{P}}$ below in greater detail.

To simplify the notation, we will describe the scheme with respect to a fixed value $\ell \geq 0$ of the security parameter, and a fixed instance description $\Lambda[X, L, W, R] \in [I_\ell]$. Thus, it is to be understood that the key generation algorithm for the scheme generates this instance description, using the instance sampling algorithm provided by **M**, and that this instance description is a part of the public key as well; alternatively, in an appropriately defined "multi-user setting," different users could work with the same instance description.

With Λ fixed as above, let $\mathbf{H} = (H, K, X, L, \Pi, S, \alpha)$ be the projective hash family that $\mathbf{P}$ associates with Λ, and let $\hat{\mathbf{H}} = (\hat{H}, \hat{K}, X \times \Pi, L \times \Pi, \hat{\Pi}, \hat{S}, \hat{\alpha})$ be the projective hash family that $\hat{\mathbf{P}}$ associates with Λ. We require that Π is an abelian group, for which we use additive notation, and that elements of Π can be efficiently added and subtracted.

We now describe the key generation, encryption, and decryption algorithms for the scheme, as they behave for a fixed instance description Λ, with corresponding projective hash families $\mathbf{H}$ and $\hat{\mathbf{H}}$, as above. The message space is Π.

Key Generation: Choose $k \in K$ and $\hat{k} \in \hat{K}$ at random, and compute $s = \alpha(k) \in S$ and $\hat{s} = \hat{\alpha}(\hat{k}) \in \hat{S}$. Note that all of these operations can be efficiently performed using the algorithms provided by $\mathbf{P}$ and $\hat{\mathbf{P}}$. The public key is $(s, \hat{s})$. The private key is $(k, \hat{k})$.

Encryption: To encrypt a message $m \in \Pi$ under a public key as above, one does the following. Generate a random $x \in L$, together with a corresponding witness $w \in W$, using the subset sampling algorithm provided by $\mathbf{M}$. Compute $\pi = H_k(x) \in \Pi$, using the public evaluation algorithm for $\mathbf{P}$ on inputs s, x, and w. Compute $e = m + \pi \in \Pi$. Compute $\hat{\pi} = \hat{H}_{\hat{k}}(x, e) \in \hat{\Pi}$, using the public evaluation algorithm for $\hat{\mathbf{P}}$ on inputs $\hat{s}$, x, e, and w. The ciphertext is $(x, e, \hat{\pi})$.

Decryption: To decrypt a ciphertext $(x, e, \hat{\pi}) \in X \times \Pi \times \hat{\Pi}$ under a secret key as above, one does the following. Compute $\hat{\pi}' = \hat{H}_{\hat{k}}(x, e) \in \hat{\Pi}$, using the private evaluation algorithm for $\hat{\mathbf{P}}$ on inputs $\hat{k}$, x, and e. Check whether $\hat{\pi} = \hat{\pi}'$; if not, then output reject and halt. Compute $\pi = H_k(x) \in \Pi$, using the private evaluation algorithm for $\mathbf{P}$ on inputs k and x. Compute $m = e - \pi \in \Pi$, and output the message m.

It is to be implicitly understood that when the decryption algorithm is presented with a ciphertext, this ciphertext is actually just a bit string, and that the decryption algorithm must parse this string to ensure that it properly encodes some $(x, e, \hat{\pi}) \in X \times \Pi \times \hat{\Pi}$; if not, the decryption algorithm outputs reject and halts.

We remark that to implement this scheme, all we really need is a 1/2-universal HPS, since we can convert this into appropriate strongly smooth and strongly universal$_2$ HPS's using the general constructions discussed in §2.1. Indeed, the Leftover Hash construction mentioned in §2.1 gives us a strongly smooth HPS whose hash outputs are bit strings of a given length a, and so we can take the group Π in the above construction to be the group of a-bit strings with "exclusive or" as the group operation.

Theorem 1. *The above scheme is secure against adaptive chosen ciphertext attack, assuming* $\mathbf{M}$ *is a hard subset membership problem.*

We very briefly sketch here the main ideas of the proof. Complete details may be found in [CS2].

First, we recall the definition of security. We consider an adversary A that sees the public key and also has access to a *decryption oracle*. A may also query (only once) an *encryption oracle*: A submits two messages m_0, m_1 to the oracle, which chooses $\beta \in \{0,1\}$ at random, and returns an encryption σ^* of m_β to A. The only restriction on A is that subsequent to the invocation of the encryption oracle, he may not submit σ^* to the decryption oracle. At the end of the game, A outputs a bit $\hat{\beta}$. Security means that the probability that $\beta = \hat{\beta}$ is negligibly close to $1/2$, for any polynomially bounded A.

To prove the security of the above scheme, suppose that an adversary A can guess the bit β with probability that is bounded away from $1/2$ by a non-negligible amount. We show how to use this adversary to distinguish x^* randomly chosen from $X \setminus L$ from x^* randomly chosen from L. On input x^*, the distinguishing algorithm D interacts with A as in the above attack game, using the key generation and decryption algorithms of the above scheme; however, to implement the encryption oracle, it uses the given value of x^*, along with the *private* evaluation algorithms for $\mathbf{P}$ and $\hat{\mathbf{P}}$, to construct a ciphertext $\sigma^* = (x^*, e^*, \hat{\pi}^*)$. At the end of A's attack, D outputs 1 if $\beta = \hat{\beta}$, and 0 otherwise.

If x^* is randomly chosen from L, the interaction between A and D is essentially equivalent to the behavior of A in the above attack game, and so D outputs 1 with probability bounded away from $1/2$ by a non-negligible amount.

However, if x^* is randomly chosen from $X \setminus L$, then it is easy to see that the strongly universal$_2$ property for $\hat{\mathbf{P}}$ implies that with overwhelming probability, D rejects all ciphertexts $(x, e, \hat{\pi})$ with $x \in X \setminus L$ submitted to the decryption oracle, and if this is the case, the strongly smooth property for $\mathbf{P}$ implies that the target ciphertext σ^* hides almost all information about m_β. From this it follows that D outputs 1 with probability negligibly close to $1/2$.

6 Universal Projective Hash Families: Constructions

We now present group-theoretic constructions of universal projective hash families.

6.1 Diverse Group Systems and Derived Projective Hash Families

Let X, L and Π be finite abelian groups, where L is a proper subgroup of X. We will use additive notation for these groups.

Let $\mathrm{Hom}(X, \Pi)$ denote the group of all homomorphisms $\phi : X \to \Pi$. This is also a finite abelian group for which we use additive notation as well. For $\phi, \phi' \in \mathrm{Hom}(X, \Pi)$, $x \in X$, and $a \in \mathbb{Z}$, we have $(\phi + \phi')(x) = \phi(x) + \phi'(x)$, $(\phi - \phi')(x) = \phi(x) - \phi'(x)$, and $(a\phi)(x) = a\phi(x) = \phi(ax)$. The zero element of $\mathrm{Hom}(X, \Pi)$ sends all elements of X to $0 \in \Pi$.

Definition 6. *Let X, L, Π be as above. Let $\mathcal{H}$ be a subgroup of $\mathrm{Hom}(X, \Pi)$. We call $\mathbf{G} = (\mathcal{H}, X, L, \Pi)$ a* group system.

Let $\mathbf{G} = (\mathcal{H}, X, L, \Pi)$ be a group system, and let $g_1, \ldots, g_d \in L$ be a set of generators for L. Let $\mathbf{H} = (H, K, X, L, \Pi, S, \alpha)$, where (1) for randomly chosen

$k \in K$, H_k is uniformly distributed over $\mathcal{H}$, (2) $S = \Pi^d$, and (3) the map $\alpha : K \to S$ sends $k \in K$ to $(\phi(g_1), \ldots, \phi(g_d)) \in S$, where $\phi = H_k$.

It is easily seen that $\mathbf{H}$ is a projective hash family. To see this, note that if $x \in L$, then there exist $w_1, \ldots, w_d \in \mathbb{Z}$ such that $x = \sum_{i=1}^{d} w_i g_i$; now, for $k \in K$ with $H_k = \phi$ and $\alpha(k) = (\mu_1, \ldots, \mu_d)$, we have $H_k(x) = \sum_{i=1}^{d} w_i \mu_i$. Thus, the action of H_k on L is determined by $\alpha(k)$, as required.

Definition 7. *Let* $\mathbf{G}$ *be a group system as above and let* $\mathbf{H}$ *be a projective hash family as above. Then we say that* $\mathbf{H}$ *is a projective hash family* derived from $\mathbf{G}$.

Looking ahead, we remark that the reason for defining α in this way is to facilitate efficient implementation of the public evaluation algorithm for a hash proof system with which $\mathbf{H}$ may be associated. In this context, if a "witness" for x is $(w_1, \ldots, w_d)$ as above, then $H_k(x)$ can be efficiently computed from $\alpha(k)$ and $(w_1, \ldots, w_d)$, assuming arithmetic in Π is efficiently implemented.

Our first goal is to investigate the conditions under which a projective hash family derived from a group system is ϵ-universal for some $\epsilon < 1$. Some notation: for an element g of a group G, $\langle g \rangle$ denotes the subgroup of G generated by g; likewise, for a subset U of G, $\langle U \rangle$ denotes the subgroup of G generated by U.

Definition 8. *Let* $\mathbf{G} = (\mathcal{H}, X, L, \Pi)$ *be a group system. We say that* $\mathbf{G}$ *is* diverse *if for all* $x \in X \setminus L$*, there exists* $\phi \in \mathcal{H}$ *such that* $\phi(L) = \langle 0 \rangle$*, but* $\phi(x) \neq 0$.

It is not difficult to see that diversity is a necessary condition for a group system if any derived projective hash family is to be ϵ-universal for some $\epsilon < 1$. We will show in Theorem 2 below that any projective hash family derived from a diverse group system is ϵ-universal, where $\epsilon = 1/\tilde{p}$, and $\tilde{p}$ is the smallest prime dividing $|X/L|$.

6.2 A Universal Projective Hash Family

Throughout this section, $\mathbf{G} = (\mathcal{H}, X, L, \Pi)$ denotes a group system, $\mathbf{H} = (H, K, X, L, \Pi, S, \alpha)$ denotes a projective hash family derived from $\mathbf{G}$, and $\tilde{p}$ denotes the smallest prime dividing $|X/L|$.

Definition 9. *For a set* $Y \subset X$*, let us define* $\mathcal{A}(Y)$ *to be the set of* $\phi \in \mathcal{H}$ *such that* $\phi(x) = 0$ *for all* $x \in Y$*; that is,* $\mathcal{A}(Y)$ *is the collection of homomorphisms in* $\mathcal{H}$ *that annihilate* Y.

It is clear that $\mathcal{A}(Y)$ is a subgroup of $\mathcal{H}$, and that $\mathcal{A}(Y) = \mathcal{A}(\langle Y \rangle)$.

Definition 10. *For* $x \in X$*, let* $\mathcal{E}_x : \mathcal{H} \to \Pi$ *be the map that sends* $\phi \in \mathcal{H}$ *to* $\phi(x) \in \Pi$*. Let us also define* $\mathcal{I}(x) = \mathcal{E}_x(\mathcal{A}(L))$.

Clearly, $\mathcal{E}_x$ is a group homomorphism, and $\mathcal{I}(x)$ is a subgroup of Π.

Lemma 1 below is a straightforward re-statement of Definition 8. The proofs of Lemmas 2, 3, and 4 below may be found in [CS2].

Lemma 1. $\mathbf{G}$ *is diverse if and only if for all* $x \in X \setminus L$*,* $\mathcal{A}(L \cup \{x\})$ *is a proper subgroup of* $\mathcal{A}(L)$.

Lemma 2. *If p is a prime dividing $|\mathcal{A}(L)|$, then p divides $|X/L|$.*

Lemma 3. *If $\mathbf{G}$ is diverse, then for all $x \in X \setminus L$, $|\mathcal{I}(x)|$ is at least $\tilde{p}$.*

Lemma 4. *Let $s \in \alpha(K)$ be fixed. Consider the probability space defined by choosing $k \in \alpha^{-1}(s)$ at random, and let $\rho = H_k$. Then ρ is uniformly distributed over a coset $\psi_s + \mathcal{A}(L)$ of $\mathcal{A}(L)$ in $\mathcal{H}$, the precise coset depending on s.*

In Lemma 4, there are many choices for the "coset leader" $\psi_s \in \mathcal{H}$; however, let us fix one such choice arbitrarily, so that for the for the rest of this section ψ_s denotes this coset leader.

Theorem 2. *Let $s \in \alpha(K)$ and $x \in X$ be fixed. Consider the probability space defined by choosing $k \in \alpha^{-1}(s)$ at random, and let $\pi = H_k(x)$. Then π is uniformly distributed over a coset of $\mathcal{I}(x)$ in Π (the precise coset depending on s and x). In particular, if $\mathbf{G}$ is diverse, then $\mathbf{H}$ is $1/\tilde{p}$-universal.*

Proof. Let $\rho = H_k$. By Lemma 4, ρ is uniformly distributed over $\psi_s + \mathcal{A}(L)$. Since $\pi = \rho(x)$, it follows that π is uniformly distributed over $\mathcal{E}_x(\psi_s + \mathcal{A}(L)) = \psi_s(x) + \mathcal{I}(x)$. That proves the first statement of the theorem. The second statement follows immediately from Lemma 3, and the fact that $|\psi_s(x) + \mathcal{I}(x)| = |\mathcal{I}(x)|$.

□

6.3 A Universal$_2$ Projective Hash Family

We continue with the notation established in §6.2; in particular, $\mathbf{G} = (\mathcal{H}, X, L, \Pi)$ denotes a group system, $\mathbf{H} = (H, K, X, L, \Pi, S, \alpha)$ denotes a projective hash family derived from $\mathbf{G}$, and $\tilde{p}$ denotes the smallest prime dividing $|X/L|$.

Starting with $\mathbf{H}$, and applying the constructions mentioned in §2.1, we can obtain a universal$_2$ projective hash family. However, by exploiting the group structure underlying $\mathbf{H}$, we can construct a more efficient universal$_2$ projective hash family $\hat{\mathbf{H}}$.

Let E be an arbitrary finite set. $\hat{\mathbf{H}}$ is to be a projective hash family for $(X \times E, L \times E)$. Fix an injective encoding function $\Gamma : X \times E \to \{0, \ldots, \tilde{p}-1\}^n$, where n is sufficiently large.

Let $\hat{\mathbf{H}} = (\hat{H}, K^{n+1}, X \times E, L \times E, \Pi, S^{n+1}, \hat{\alpha})$, where $\hat{H}$ and $\hat{\alpha}$ are defined as follows. For $\boldsymbol{k} = (k', k_1, \ldots, k_n) \in K^{n+1}$, $x \in X$, and $e \in E$, we define $\hat{H}_{\boldsymbol{k}}(x, e) = H_{k'}(x) + \sum_{i=1}^{n} \gamma_i H_{k_i}(x)$, where $(\gamma_1, \ldots, \gamma_n) = \Gamma(x, e)$, and we define $\hat{\alpha}(\boldsymbol{k}) = (\alpha(k'), \alpha(k_1), \ldots, \alpha(k_n))$. It is clear that $\hat{\mathbf{H}}$ is a projective hash family. We shall prove:

Theorem 3. *Let $\hat{\mathbf{H}}$ be as above. Let $\boldsymbol{s} \in \alpha(K)^{n+1}$, $x, x^* \in X$, and $e, e^* \in E$ be fixed, where $(x, e) \neq (x^*, e^*)$. Consider the probability space defined by choosing $\boldsymbol{k} \in \hat{\alpha}^{-1}(\boldsymbol{s})$ at random, and let $\pi = \hat{H}_{\boldsymbol{k}}(x, e)$ and $\pi^* = \hat{H}_{\boldsymbol{k}}(x^*, e^*)$. Then π is uniformly distributed over a coset of $\mathcal{I}(x)$ in Π (the precise coset depending on s, x, and e), and π^* is uniformly and independently distributed over a coset of $\mathcal{I}(x^*)$ in Π (the precise coset depending on s, x^*, and e^*). In particular, if the underlying group system $\mathbf{G}$ is diverse, then $\hat{\mathbf{H}}$ is $1/\tilde{p}$-universal$_2$.*

Before proving this theorem, we state another elementary lemma. Let $M \in \mathbb{Z}^{a \times b}$ be an integer matrix with a rows and b columns. Let $\mathcal{G}$ be a finite abelian group. Let $\mathbf{T}(M, \mathcal{G}) : \mathcal{G}^b \to \mathcal{G}^a$ be the map that sends $\boldsymbol{u} \in \mathcal{G}^b$ to $\boldsymbol{v} \in \mathcal{G}^a$, where $\boldsymbol{v}^\top = M\boldsymbol{u}^\top$; here, $(\cdots)^\top$ denotes transposition. Clearly, $\mathbf{T}(M, \mathcal{G})$ is a group homomorphism.

Lemma 5. *Let M and $\mathcal{G}$ be as above. If for all primes p dividing $|\mathcal{G}|$, the rows of M are linearly independent modulo p, then $\mathbf{T}(M, \mathcal{G})$ is surjective.*

See [CS2] for a proof of this lemma.

Proof of Theorem 3. Let $\boldsymbol{s} = (s', s_1, \ldots, s_n)$, $(\gamma_1, \ldots, \gamma_n) = \Gamma(x, e)$, and $(\gamma_1^*, \ldots, \gamma_n^*) = \Gamma(x^*, e^*)$. Let $(\rho', \rho_1, \ldots, \rho_n) = (H_{k'}, H_{k_1}, \ldots, H_{k_n})$.

Now define the matrix $M \in \mathbb{Z}^{2 \times (n+1)}$ as

$$M = \begin{pmatrix} 1 & \gamma_1 & \gamma_2 & \cdots & \gamma_n \\ 1 & \gamma_1^* & \gamma_2^* & \cdots & \gamma_n^* \end{pmatrix},$$

so that if $(\tilde{\rho}, \tilde{\rho}^*)^\top = M(\rho', \rho_1, \ldots, \rho_n)^\top$, then we have $(\pi, \pi^*) = (\rho(x), \rho^*(x^*))$.

By the definition of Γ, and by Lemma 2, we see that $(\gamma_1, \ldots, \gamma_n)$ and $(\gamma_1^*, \ldots, \gamma_n^*)$ are distinct modulo any prime p that divides $\mathcal{A}(L)$. Therefore, Lemma 5 implies that the map $\mathbf{T}(M, \mathcal{A}(L))$ is surjective. By Lemma 4, $(\rho', \rho_1, \ldots, \rho_n)$ is uniformly distributed over $(\psi_{s'} + \mathcal{A}(L), \psi_{s_1} + \mathcal{A}(L), \ldots, \psi_{s_n} + \mathcal{A}(L))$. Thus, $(\tilde{\rho}, \tilde{\rho}^*)$ is uniformly distributed over $(\tilde{\psi} + \mathcal{A}(I), \tilde{\psi}^* + \mathcal{A}(I))$, where $(\tilde{\psi}, \tilde{\psi}^*)^\top = M(\psi_{s'}, \psi_{s_1}, \ldots, \psi_{s_n})^\top$. It follows that (π, π^*) is uniformly distributed over $(\tilde{\psi}(x) + \mathcal{I}(x), \tilde{\psi}^*(x^*) + \mathcal{I}(x^*))$.

That proves the first statement of the theorem. The second statement now follows from Lemma 3. □

If $\tilde{p}$ is small, then the t-fold parallelization mentioned in §2.1 can be used to reduce the error to at most $1/\tilde{p}^t$ for a suitable value of t. However, this comes at the cost of a multiplicative factor t in efficiency. We now describe another construction that achieves an error rate of $1/\tilde{p}^t$ that comes at the cost of just an *additive* factor of $O(t)$ in efficiency.

Let $t \geq 1$ be fixed, and let E be an arbitrary finite set. Our construction yields a projective hash family $\hat{\mathbf{H}}$ for $(X \times E, L \times E)$. We use the same name $\hat{\mathbf{H}}$ for this projective hash family as in the construction of Theorem 3, because when $t = 1$, the constructions are identical. Fix an injective encoding function $\Gamma : X \times E \to \{0, \ldots, \tilde{p} - 1\}^n$, where n is sufficiently large.

Let $\hat{\mathbf{H}} = (\hat{H}, K^{n+2t-1}, X \times E, L \times E, \Pi, S^{n+2t-1}, \hat{\alpha})$, where $\hat{H}$ and $\hat{\alpha}$ are defined as follows. For $\boldsymbol{k} = (k_1', \ldots, k_t', k_1, \ldots, k_{n+t-1}) \in K^{n+2t-1}$, $x \in X$, and $e \in E$, we define $\hat{H}_{\boldsymbol{k}}(x, e) = (\pi_1, \ldots, \pi_t)$, where $\pi_j = H_{k_j'}(x) + \sum_{i=1}^n \gamma_i H_{k_{i+j-1}}(x)$ for $j = 1, \ldots, t$, and $(\gamma_1, \ldots, \gamma_n) = \Gamma(x, e)$. We also define $\hat{\alpha}(\boldsymbol{k}) = (\alpha(k_1'), \ldots, \alpha(k_t'), \alpha(k_1), \ldots, \alpha(k_{n+t-1}))$. Again, it is clear that $\hat{\mathbf{H}}$ is a projective hash family.

Theorem 4. *Let $\hat{\mathbf{H}}$ be as above. Let $\boldsymbol{s} \in \alpha(K)^{n+2t-1}$, $x, x^* \in X$, and $e, e^* \in E$ be fixed, where $(x, e) \neq (x^*, e^*)$. Consider the probability space defined by choosing $\boldsymbol{k} \in \hat{\alpha}^{-1}(\boldsymbol{s})$ at random, and let $\boldsymbol{\pi} = \hat{H}_{\boldsymbol{k}}(x, e)$ and $\boldsymbol{\pi}^* = \hat{H}_{\boldsymbol{k}}(x^*, e^*)$.*

Then $\boldsymbol{\pi}$ is uniformly distributed over a coset of $\mathcal{I}(x)^t$ in Π^t (the precise coset depending on s, x, and e), and $\boldsymbol{\pi}^*$ is uniformly and independently distributed over a coset of $\mathcal{I}(x^*)^t$ in Π^t (the precise coset depending on s, x^*, and e^*). In particular, if the underlying group system $\mathbf{G}$ is diverse, then $\hat{\mathbf{H}}$ is $1/\tilde{p}^t$-universal$_2$.

Proof. Let $(\gamma_1,\ldots,\gamma_n) = \Gamma(x,e)$, and $(\gamma_1^*,\ldots,\gamma_n^*) = \Gamma(x^*,e^*)$. Let $\boldsymbol{\rho} = (H_{k_1'},\ldots, H_{k_t'}, H_{k_1},\ldots,H_{k_{n+t-1}}) \in \mathcal{H}^{n+2t-1}$. Now define the matrix $M \in \mathbb{Z}^{2t\times(n+2t-1)}$ as

$$M = \left(\begin{array}{cccc|cccccccc} 1 & & & & \gamma_1 & \gamma_2 & \cdots & \gamma_n & & & & \\ & 1 & & & & \gamma_1 & \gamma_2 & \cdots & \gamma_n & & & \\ & & \ddots & & & & \ddots & \ddots & & \ddots & & \\ & & & 1 & & & & \gamma_1 & \gamma_2 & \cdots & \gamma_n & \\ 1 & & & & \gamma_1^* & \gamma_2^* & \cdots & \gamma_n^* & & & & \\ & 1 & & & & \gamma_1^* & \gamma_2^* & \cdots & \gamma_n^* & & & \\ & & \ddots & & & & \ddots & \ddots & & \ddots & & \\ & & & 1 & & & & \gamma_1^* & \gamma_2^* & \cdots & \gamma_n^* & \end{array} \right)$$

$$\underbrace{\qquad\qquad}_{t \text{ columns}} \quad \underbrace{\qquad\qquad\qquad\qquad}_{n+t-1 \text{ columns}}$$

so that if $(\tilde{\rho}_1,\ldots,\tilde{\rho}_t,\tilde{\rho}_1^*,\ldots,\tilde{\rho}_t^*)^\top = M\boldsymbol{\rho}^\top$, then $\boldsymbol{\pi} = (\tilde{\rho}_1(x),\ldots,\tilde{\rho}_t(x))$ and $\boldsymbol{\pi}^* = (\tilde{\rho}_1^*(x),\ldots,\tilde{\rho}_t^*(x))$.

Claim. The rows of M are linearly independent modulo p for any prime p dividing $|\mathcal{A}(L)|$.

The theorem is implied by the claim, as we now argue. By Lemma 5, the map $\mathbf{T}(M,\mathcal{A}(L))$ is surjective. By Lemma 4, $\boldsymbol{\rho}$ is uniformly distributed over a coset of $\mathcal{A}(L)^{n+2t-1}$ in $\mathcal{H}^{n+2t-1}$. It follows that $(\tilde{\rho}_1,\ldots,\tilde{\rho}_t, \tilde{\rho}_1^*,\ldots,\tilde{\rho}_t^*)$ is uniformly distributed over a coset of $\mathcal{A}(L)^{2t}$ in $\mathcal{H}^{2t}$, and therefore, $\boldsymbol{\pi}$ and $\boldsymbol{\pi}^*$ are uniformly and independently distributed over cosets of $\mathcal{I}(x)^t$ and $\mathcal{I}(x^*)^t$, respectively, in Π^t.

That proves the first statement of the theorem. The second statement of the theorem now follows from Lemma 3.

The proof of the claim is omitted for lack of space. See [CS2] for details. □

6.4 Examples of Diverse Group Systems

In this section, we discuss two examples of diverse group systems that have cryptographic importance.

Example 1

Let G be a group of prime of prime order q, and let $X = G^r$, i.e., X is the direct product of r copies of G. Let L be any proper subgroup of X, and let $\mathcal{H} = \mathrm{Hom}(X,G)$. Consider the group system $\mathbf{G} = (\mathcal{H},X,L,G)$.

It is easy to show that $\mathbf{G}$ is diverse, and that in fact, a projective hash family derived from $\mathbf{G}$ is $1/q$-universal, or equivalently, 0-smooth. See [CS2] for details.

Example 2

Let X be a cyclic group of order $a = bb'$, where $b' > 1$ and $\gcd(b, b') = 1$, and let L be the unique subgroup of X of order b. Let $\mathcal{H} = \mathrm{Hom}(X, X)$, and consider the group system $\mathbf{G} = (\mathcal{H}, X, L, X)$.

The group X is isomorphic to $\mathbb{Z}_a$. If we identify X with $\mathbb{Z}_a$, then $\mathcal{H}$ can be identified with $\mathbb{Z}_a$ as follows: for every $\nu \in \mathbb{Z}_a$, define $\phi_\nu \in \mathcal{H}$ to be the map that sends $x \in \mathbb{Z}_a$ to $x \cdot \nu \in \mathbb{Z}_a$.

The group X is of course also isomorphic to $\mathbb{Z}_b \times \mathbb{Z}_{b'}$. If we identify X with $\mathbb{Z}_b \times \mathbb{Z}_{b'}$, then L corresponds to $\mathbb{Z}_b \times \langle 0 \rangle$. Moreover, we can identify $\mathcal{H}$ with $\mathbb{Z}_b \times \mathbb{Z}_{b'}$ as follows: for $(\nu, \nu') \in \mathbb{Z}_b \times \mathbb{Z}_{b'}$, let $\psi_{\nu,\nu'} \in \mathcal{H}$ be the map that sends $(x, x') \in \mathbb{Z}_b \times \mathbb{Z}_{b'}$ to $(x \cdot \nu, x' \cdot \nu') \in \mathbb{Z}_b \times \mathbb{Z}_{b'}$.

Under the identification in the previous paragraph, it is evident that $\mathcal{A}(L)$ is the subgroup of $\mathcal{H}$ generated by $\psi_{0,1}$. If we take any $(x, x') \in X \setminus L$, so that $x' \neq 0$, we see that $\psi_{0,1}(x, x') = (0, x')$. Thus, $\psi_{0,1} \notin \mathcal{A}(L \cup \{(x, x')\})$, which shows that $\mathbf{G}$ is diverse. Therefore, a projective hash family derived from $\mathbf{G}$ is $1/\tilde{p}$-universal, where $\tilde{p}$ is the smallest prime dividing b'.

It is also useful to characterize the group $\mathcal{I}(x, x') = \mathcal{E}_{x,x'}(\mathcal{A}(L))$. Evidently, since $\mathcal{A}(L) = \langle \psi_{0,1} \rangle$, we must have $\mathcal{I}(x, x') = \langle 0 \rangle \times \langle x' \rangle$.

7 Concrete Encryption Schemes

We present two new public-key encryption schemes secure against adaptive chosen ciphertext attack. The first scheme is based on Paillier's Decision Composite Residuosity assumption, and the second is based on the classical Quadratic Residuosity assumption. Both are derived from the general construction in §5.

One can also show that the public-key encryption scheme from [CS1] can be viewed as a special case of our general construction, based on Example 1 in §6.4. However, for lack of space, we refer the reader to [CS2] for the details.

7.1 Schemes Based on the Decision Composite Residuosity Assumption

Derivation

Let p, q, p', q' be distinct odd primes with $p = 2p' + 1$ and $q = 2q' + 1$, and where p' and q' are both λ bits in length. Let $N = pq$ and $N' = p'q'$. Consider the group $\mathbb{Z}_{N^2}^*$ and the subgroup P of $\mathbb{Z}_{N^2}^*$ consisting of all Nth powers of elements in $\mathbb{Z}_{N^2}^*$. Note that $\lambda = \lambda(\ell)$ is a function of the security parameter ℓ.

Paillier's Decision Composite Residuosity (DCR) assumption is that given only N, it is hard to distinguish random elements of $\mathbb{Z}_{N^2}^*$ from random elements of P. We shall assume that "strong" primes, such as p and q above, are sufficiently dense (as is widely conjectured and supported by empirical evidence). This implies that such primes can be efficiently generated, and that the DCR assumption with the restriction to strong primes is implied by the DCR assumption without this restriction.

We can decompose $\mathbb{Z}_{N^2}^*$ as an internal direct product $\mathbb{Z}_{N^2}^* = G_N \cdot G_{N'} \cdot G_2 \cdot T$, where each group G_τ is a cyclic group of order τ, and T is the subgroup of $\mathbb{Z}_{N^2}^*$ generated by $(-1 \bmod N^2)$. This decomposition is unique, except for the choice of G_2 (there are two possible choices). For any $x \in \mathbb{Z}_{N^2}^*$, we can express x uniquely as $x = x(G_N)x(G_{N'})x(G_2)x(T)$, where for each G_τ, $x(G_\tau) \in G_\tau$, and $x(T) \in T$. Note that the element $\xi = (1 + N \bmod N^2) \in \mathbb{Z}_{N^2}^*$ has order N, i.e., it generates G_N, and that $\xi^a = (1 + aN \bmod N^2)$ for $0 \le a < N$.

Let $X = \{(a \bmod N^2) \in \mathbb{Z}_{N^2}^* : (a \mid N) = 1\}$, where $(\cdot \mid \cdot)$ is the Jacobi symbol. It is easy to see that $X = G_N G_{N'} T$. Let L be the subgroup of Nth powers of X, i.e., $L = G_{N'} T$. These groups X and L will define our subset membership problem.

Our instance description Λ will contain N, along with a random generator g for L. It is easy to generate such a g: choose a random $\mu \in \mathbb{Z}_{N^2}^*$, and set $g = -\mu^{2N}$. With overwhelming probability, such a g will generate L; indeed, the output distribution of this sampling algorithm is $O(2^{-\lambda})$-close the uniform distribution over all generators.

Let us define the set of witnesses as $W = \{0, \ldots, \lfloor N/2 \rfloor\}$. We say $w \in W$ is a witness for $x \in X$ if $x = g^w$. To generate $x \in L$ at random together with a corresponding witness, we simply generate $w \in W$ at random, and compute $x = g^w$. The output distribution of this algorithm is not the uniform distribution over L, but one that is $O(2^{-\lambda})$-close to it.

This completes the description of our subset membership problem. It is easy to see that it satisfies all the basic requirements specified in §3.

Next, we argue that the DCR assumption implies that this subset membership problem is hard. Suppose we are given x sampled at random from $\mathbb{Z}_{N^2}^*$ (respectively, P). If we choose $b \in \{0, 1\}$ at random, then $x^2(-1)^b$ is uniformly distributed over X (respectively, L). This implies that distinguishing X from L is at least as hard as distinguishing $\mathbb{Z}_{N^2}^*$ from P, and so under the DCR assumption, it is hard to distinguish X from L. It is easy to see that this implies that it is hard to distinguish $X \setminus L$ from L as well.

Now it remains to construct appropriate strongly smooth and strongly universal$_2$ HPS's for the construction in §5. To do this, we first construct a diverse group system (see Definition 8), from which we can then derive the required HPS's.

Fix an instance description Λ, where Λ specifies an integer N – defining groups X and L as above – along with a generator g for L. Let $\mathcal{H} = \mathrm{Hom}(X, X)$ and consider the group system $\mathbf{G} = (\mathcal{H}, X, L, X)$. As discussed in Example 2 in §6.4, $\mathbf{G}$ is a diverse group system; moreover, for $x \in X$, we have $\mathcal{I}(x) = \langle x(G_N) \rangle$; thus, for $x \in X \setminus L$, $\mathcal{I}(x)$ has order p, q, or N, according to whether $x(G_N)$ has order p, q, or N.

For $k \in \mathbb{Z}$, let $H_k \in \mathrm{Hom}(X, X)$ be the kth power map; that is, H_k sends $x \in X$ to $x^k \in X$. Let $K_* = \{0, \ldots, 2NN' - 1\}$. As discussed in Example 2 in §6.4, the correspondence $k \mapsto H_k$ yields a bijection between K_* and $\mathrm{Hom}(X, X)$.

Consider the projective hash family $\mathbf{H}_* = (H, K_*, X, L, X, L, \alpha)$, where H and K_* are as in the previous paragraph, and α maps $k \in \mathbb{Z}$ to $H_k(g) \in L$.

Clearly, $\mathbf{H}_*$ is a projective hash family derived from $\mathbf{G}$, and so by Theorem 2, it is $2^{-\lambda}$-universal. From this, we can obtain a corresponding HPS $\mathbf{P}$; however, as we cannot readily sample elements from K_*, the projective hash family $\mathbf{H}$ that $\mathbf{P}$ associates with the instance description Λ is slightly different than $\mathbf{H}_*$; namely, we use the set $K = \{0, \ldots, \lfloor N^2/2 \rfloor\}$ in place of the set K_*, but otherwise, $\mathbf{H}$ and $\mathbf{H}_*$ are the same. It is readily seen that the uniform distribution on K_* is $O(2^{-\lambda})$-close to the uniform distribution on K, and so $\mathbf{H}$ and $\mathbf{H}_*$ are also $O(2^{-\lambda})$-close (see Definition 4). It is also easy to verify that all of the algorithms that $\mathbf{P}$ should provide are available.

So we now have a $2^{-\lambda(\ell)}$-universal HPS $\mathbf{P}$. We could easily convert $\mathbf{P}$ into a strongly smooth HPS by applying the Leftover Hash Lemma construction mentioned in §2.1 to the underlying universal projective hash family $\mathbf{H}_*$. However, there is a much more direct and practical way to proceed, as we now describe.

According to Theorem 2, for any $s, x \in X$, if k is chosen at random from K_*, subject to $\alpha(k) = s$, then $H_k(x)$ is uniformly distributed over a coset of $\mathcal{I}(x)$ in X. As discussed above, $\mathcal{I}(x) = \langle x(G_N) \rangle$, and so is a subgroup of G_N. Moreover, for random $x \in X \setminus L$, we have $\mathcal{I}(x) \neq G_N$ with probability at most $2^{-\lambda+1}$.

Now define the map $\chi : \mathbb{Z}_{N^2} \to \mathbb{Z}_N$ that sends $(a + bN \bmod N^2)$, where $0 \leq a, b < N$, to $(b \bmod N)$. This map does not preserve any algebraic structure; however, it is easy to see that the restriction of χ to any coset of G_N in X is a one-to-one map from that coset onto $\mathbb{Z}_N$ (see [CS2] for details).

Let us define $\mathbf{H}_*^\times = (H^\times, K_*, X, L, \mathbb{Z}_N, L, \alpha)$, where for $k \in \mathbb{Z}$, $H_k^\times = \chi \circ H_k$. That is, $\mathbf{H}_*^\times$ is the same as $\mathbf{H}_*$, except that in $\mathbf{H}_*^\times$, we pass the output of the hash function for $\mathbf{H}_*$ through χ. From the observations in the previous two paragraphs, it is clear that $\mathbf{H}_*^\times$ is a $2^{-\lambda+1}$-smooth projective hash family. From $\mathbf{H}_*^\times$ we get a corresponding approximation $\mathbf{H}^\times$ (using K in place of K_*), and from this we get corresponding $2^{-\lambda(\ell)+1}$-smooth HPS $\mathbf{P}^\times$.

We can apply the construction in Theorem 3 to $\mathbf{H}_*$, obtaining a $2^{-\lambda}$-universal$_2$ projective hash family $\hat{\mathbf{H}}_*$ for $(X \times \mathbb{Z}_N, L \times \mathbb{Z}_N)$. From $\hat{\mathbf{H}}_*$ we get a corresponding approximation $\hat{\mathbf{H}}$ (using K in place of K_*), and from this we get a corresponding $2^{-\lambda(\ell)}$-universal$_2$ extended HPS $\hat{\mathbf{P}}$.

We could build our encryption scheme directly using $\hat{\mathbf{P}}$; however, we get more compact ciphertexts if we modify $\hat{\mathbf{H}}_*$ by passing its hash outputs through χ, just as we did in building $\mathbf{H}_*^\times$, obtaining the analogous projective hash family $\hat{\mathbf{H}}_*^\times$ for $(X \times \mathbb{Z}_N, L \times \mathbb{Z}_N)$. From Theorem 4, and the above discussion, it is clear that $\hat{\mathbf{H}}_*^\times$ is also $2^{-\lambda}$-universal$_2$. From $\hat{\mathbf{H}}_*^\times$ we get a corresponding approximation $\hat{\mathbf{H}}^\times$ (using K in place of K_*), and from this we get a corresponding $2^{-\lambda(\ell)}$-universal$_2$ extended HPS $\hat{\mathbf{P}}^\times$.

The Encryption Scheme

We now present in detail the encryption scheme obtained from the HPS's $\mathbf{P}^\times$ and $\hat{\mathbf{P}}^\times$ above.

We describe the scheme for a fixed value of N that is the product of two $(\lambda + 1)$-bit strong primes. The message space for this scheme is $\mathbb{Z}_N$.

Let X, L, and χ be as defined above. Also, let $W = \{0, \ldots, \lfloor N/2 \rfloor\}$ and $K = \{0, \ldots, \lfloor N^2/2 \rfloor\}$, as above. Let $R = \{0, \ldots, 2^\lambda - 1\}$, and let $\Gamma : \mathbb{Z}_{N^2} \times \mathbb{Z}_N \to R^n$ be an efficiently computable injective map for an appropriate $n \geq 1$. For sufficiently large λ, $n = 7$ suffices.

Key Generation: Choose $\mu \in \mathbb{Z}_{N^2}^*$ at random and set $g = -\mu^{2N} \in L$. Choose $k, \tilde{k}, \hat{k}_1, \ldots, \hat{k}_n \in K$ at random, and compute $s = g^k \in L$, $\tilde{s} = g^{\tilde{k}} \in L$, and $\hat{s}_i = g^{\hat{k}_i} \in L$ for $i = 1, \ldots, n$. The public key is $(g;\ s;\ \tilde{s};\ \hat{s}_1, \ldots, \hat{s}_n)$. The private key is $(k;\ \tilde{k};\ \hat{k}_1, \ldots, \hat{k}_n)$.

Encryption: To encrypt a message $m \in \mathbb{Z}_N$ under a public key as above, one does the following. Choose $w \in W$ at random, and compute $x = g^w \in L$, $y = s^w \in L$, $\pi = \chi(y) \in \mathbb{Z}_N$, and $e = m + \pi \in \mathbb{Z}_N$. Compute $\hat{y} = \tilde{s}^w \prod_{i=1}^n \hat{s}_i^{\gamma_i w} \in L$ and $\hat{\pi} = \chi(\hat{y}) \in \mathbb{Z}_N$, where $(\gamma_1, \ldots, \gamma_n) = \Gamma(x, e) \in R^n$. The ciphertext is $(x, e, \hat{\pi})$.

Decryption: To decrypt a ciphertext $(x, e, \hat{\pi}) \in X \times \mathbb{Z}_N \times \mathbb{Z}_N$ under a secret key as above, one does the following. Compute $\hat{y} = x^{\tilde{k} + \sum_{i=1}^n \gamma_i \hat{k}_i} \in X$ and $\hat{\pi}' = \chi(\hat{y}) \in \mathbb{Z}_N$, where $(\gamma_1, \ldots, \gamma_n) = \Gamma(x, e) \in R^n$. Check whether $\hat{\pi} = \hat{\pi}'$; if not, then output reject and halt. Compute $y = x^k \in X$, $\pi = \chi(y) \in \mathbb{Z}_N$, and $m = e - \pi \in \mathbb{Z}_N$, and then output m.

Note that in the decryption algorithm, we are assuming that $x \in X$, which implicitly means that the decryption algorithm should check that $x = (a \bmod N^2)$ with $(a \mid N) = 1$, and reject the ciphertext if this does not hold.

This is *precisely* the scheme that our general construction in §5 yields. Thus, the scheme is secure against adaptive chosen ciphertext attack, provided the DCR assumption holds.

Minor variations. To get a more efficient scheme, we could replace Γ by a collision resistant hash function (CRHF), obtaining an even more efficient scheme with a smaller value of n, possibly even $n = 1$. It is straightforward to adapt our general theory to show that the resulting scheme is still secure against adaptive chosen ciphertext attack, assuming Γ is a CRHF. In fact, with a more refined analysis, it suffices to assume that Γ is a universal one-way hash function (UOWHF) [NY1]. In [CS2], we present a number of further variations on this scheme.

7.2 Schemes Based on the Quadratic Residuosity Assumption

Derivation

Let p, q, p', q' be distinct odd primes with $p = 2p' + 1$ and $q = 2q' + 1$, and where p' and q' are both λ bits in length. Let $N = pq$ and let $N' = p'q'$. Consider the group $\mathbb{Z}_N^*$, and let X be the subgroup of elements $(a \bmod N) \in \mathbb{Z}_N^*$ with Jacobi symbol $(a \mid N) = 1$, and let L be the subgroup of squares (a.k.a., quadratic residues) of $\mathbb{Z}_N^*$. Note that L is a subgroup of X of index 2. Also, note that $\lambda = \lambda(\ell)$ is a function of the security parameter ℓ.

The Quadratic Residuosity (QR) assumption is that given only N, it is hard to distinguish random elements of X from random elements of L. This implies that it is hard to distinguish random elements of $X \setminus L$ from random elements of L.

As in §7.1, we shall assume that strong primes (such as p and q) are sufficiently dense.

The groups X and L above will define our subset membership problem.

We can decompose $\mathbb{Z}_N^*$ as an internal direct product $\mathbb{Z}_N^* = G_{N'} \cdot G_2 \cdot T$, where each group G_τ is a cyclic group of order τ, and T is the subgroup of $\mathbb{Z}_N^*$ generated by $(-1 \bmod N)$. This decomposition is unique, except for the choice of G_2 (there are two possible choices).

It is easy to see that $X = G_{N'}T$, so it is a cyclic group, and that $L = G_{N'}$.

Our instance description Λ will contain N, along with a random generator g for L. It is easy to generate such a g: choose a random $\mu \in \mathbb{Z}_N^*$, and set $g = \mu^2$. With overwhelming probability, such a g will generate L; indeed, the output distribution of this sampling algorithm is $O(2^{-\lambda})$-close the uniform distribution over all generators.

Let us define the set of witnesses as $W = \{0, \ldots, \lfloor N/4 \rfloor\}$. We say $w \in W$ is a witness for $x \in X$ if $x = g^w$. To generate $x \in L$ at random together with a corresponding witness, we simply generate $w \in W$ at random, and compute $x = g^w$. The output distribution of this algorithm is not the uniform distribution over L, but is $O(2^{-\lambda})$-close to it.

This completes the description of our subset membership problem. It is easy to see that it satisfies all the basic requirements specified in §3. As already mentioned, the QR assumption implies that this is a hard subset membership problem.

Now it remains to construct appropriate strongly smooth and strongly universal$_2$ HPS's for the construction in §5. To do this, we first construct a diverse group system (see Definition 8), from which we can then derive the required HPS's.

Fix an instance description Λ, where Λ specifies an integer N – defining groups X and L as above – along with a generator g for L. Let $\mathcal{H} = \mathrm{Hom}(X, X)$ and consider the group system $\mathbf{G} = (\mathcal{H}, X, L, X)$.

As discussed in Example 2 in §6.4, $\mathbf{G}$ is a diverse group system; moreover, for $x \in X$, if we decompose x as $x = x(L) \cdot x(T)$, where $x(L) \in L$ and $x(T) \in T$, then we have $\mathcal{I}(x) = \langle x(T) \rangle$; thus, for $x \in X \setminus L$, $\mathcal{I}(x) = T$.

For $k \in \mathbb{Z}$, let $H_k \in \mathrm{Hom}(X, X)$ be the kth power map; that is, H_k sends $x \in X$ to $x^k \in X$. Let $K_* = \{0, \ldots, 2N' - 1\}$. As discussed Example 2 in in §6.4, the correspondence $k \mapsto H_k$ yields a bijection between K_* and $\mathrm{Hom}(X, X)$.

Consider the projective hash family $\mathbf{H}_* = (H, K_*, X, L, X, L, \alpha)$, where H and K_* are as in the previous paragraph, and α maps $k \in \mathbb{Z}$ to $H_k(g) \in L$. Clearly, $\mathbf{H}_*$ is a projective hash family derived from $\mathbf{G}$, and so by Theorem 2, it is 1/2-universal. From this, we can obtain a corresponding HPS $\mathbf{P}$; however, as we cannot readily sample elements from K_*, the projective hash family $\mathbf{H}$ that $\mathbf{P}$ associates with the instance description Λ is slightly different than $\mathbf{H}_*$;

namely, we use the set $K = \{0, \ldots, \lfloor N/2 \rfloor\}$ in place of the set K_*, but otherwise, $\mathbf{H}$ and $\mathbf{H}_*$ are the same. It is readily seen that the uniform distribution on K_* is $O(2^{-\lambda})$-close to the uniform distribution on K, and so $\mathbf{H}$ and $\mathbf{H}_*$ are also $O(2^{-\lambda})$-close. It is also easy to verify that all of the algorithms that $\mathbf{P}$ should provide are available.

So we now have a 1/2-universal HPS $\mathbf{P}$. We can apply the t-fold parallelization mentioned in §2.1 to $\mathbf{H}_*$, using a parameter $t = t(\ell)$, to get a 2^{-t}-universal projective hash family $\bar{\mathbf{H}}_*$. From $\bar{\mathbf{H}}_*$ we get a corresponding approximation $\bar{\mathbf{H}}$ (using K in place of K_*), and from this we get corresponding 2^{-t}-universal HPS $\bar{\mathbf{P}}$.

Now, we could easily convert $\bar{\mathbf{P}}$ into a strongly smooth HPS by applying the Leftover Hash Lemma construction mentioned in §2.1 to the underlying projective hash family $\bar{\mathbf{H}}_*$. However, there is a much more direct and practical way to proceed, as we now describe.

According to Theorem 2, for any $s, x \in X$, if k is chosen at random from K_*, subject to $\alpha(k) = s$, then $H_k(x)$ is uniformly distributed over a coset of $\mathcal{I}(x)$ in X. As discussed above, for $x \in X \setminus L$, $\mathcal{I}(x) = T$.

Now define the map $\chi : \mathbb{Z}_N \to \mathbb{Z}_2$ as follows: for $x = (a \bmod N) \in \mathbb{Z}_N^*$, with $0 \le a < N$, let $\chi(x) = 1$ if $a > N/2$, and $\chi(x) = 0$ otherwise. It is easy to verify that the restriction of χ to any coset of T in X (which is a set of the form $\{\pm x\}$ for some $x \in X$) is a one-to-one map from that coset onto $\mathbb{Z}_2$.

Let us define $\mathbf{H}_*^\times = (H^\times, K_*, X, L, \mathbb{Z}_N, L, \alpha)$, where for $k \in \mathbb{Z}$, $H_k^\times = \chi \circ H_k$. That is, $\mathbf{H}_*^\times$ is the same as $\mathbf{H}_*$, except that in $\mathbf{H}_*^\times$, we pass the output of the hash function for $\mathbf{H}_*$ through χ. From the observations in the previous two paragraphs, it is clear that $\mathbf{H}_*^\times$ is a 1/2-universal, and so 0-smooth, projective hash family.

We can apply the t-fold parallelization mentioned in §2.1 to $\mathbf{H}_*^\times$ with the parameter $t = t(\ell)$ to get a 0-smooth projective hash family $\bar{\mathbf{H}}_*^\times$ whose hash output space is $\mathbb{Z}_2^t$. From $\bar{\mathbf{H}}_*^\times$ we get a corresponding approximation $\bar{\mathbf{H}}^\times$ (using K in place of K_*), and from this we get corresponding 0-smooth HPS $\bar{\mathbf{P}}^\times$.

We can apply the construction in Theorem 4 to $\mathbf{H}_*$, using a parameter $\hat{t} = \hat{t}(\ell)$, obtaining a $2^{-\hat{t}}$-universal$_2$ projective hash family $\hat{\mathbf{H}}_*$ for $(X \times \mathbb{Z}_2^t, L \times \mathbb{Z}_2^t)$. From $\hat{\mathbf{H}}_*$ we get a corresponding approximation $\hat{\mathbf{H}}$ (using K in place of K_*), and from this we get a corresponding $2^{-\hat{t}(\ell)}$-universal$_2$ extended HPS $\hat{\mathbf{P}}$.

We could build our encryption scheme directly using $\hat{\mathbf{P}}$; however, we get more compact ciphertexts if we modify $\hat{\mathbf{H}}_*$ by passing its hash outputs through χ, just as we did in building $\mathbf{H}_*^\times$, obtaining the analogous projective hash family $\hat{\mathbf{H}}_*^\times$ for $(X \times \mathbb{Z}_2^t, L \times \mathbb{Z}_2^t)$. From Theorem 4, and the above discussion, it is clear that $\hat{\mathbf{H}}_*^\times$ is also $2^{-\hat{t}}$-universal$_2$. From $\hat{\mathbf{H}}_*^\times$ we get a corresponding approximation $\hat{\mathbf{H}}^\times$ (using K in place of K_*), and from this we get a corresponding $2^{-\hat{t}(\ell)}$-universal$_2$ extended HPS $\hat{\mathbf{P}}^\times$.

The Encryption Scheme

We now present in detail the encryption obtained using the HPS's $\bar{\mathbf{P}}^\times$ and $\hat{\mathbf{P}}^\times$ above.

We describe the scheme for a fixed value of N that is product of two $(\lambda+1)$-bit strong primes. The message space for this scheme is $\mathbb{Z}_2^t$, where $t = t(\ell)$ is an auxiliary parameter. Note that t may be any size – it need not be particularly large. We also need an auxiliary parameter $\hat{t} = \hat{t}(\ell)$. The value of $\hat{t}$ should be large; more precisely, $2^{-\hat{t}(\ell)}$ should be a negligible function in ℓ.

Let X, L, and χ be as defined above. Also as above, let $K = \{0, \ldots, \lfloor N/2 \rfloor\}$, and $W = \{0, \ldots, \lfloor N/4 \rfloor\}$. Let $\Gamma : \mathbb{Z}_N \times \mathbb{Z}_2^t \to \{0,1\}^n$ be an efficiently computable injective map for an appropriate $n \geq 1$.

Key Generation: Choose $\mu \in \mathbb{Z}_N^*$ at random and set $g = \mu^2 \in L$. Randomly choose $k_1, \ldots, k_t$, $\tilde{k}_1, \ldots, \tilde{k}_{\hat{t}}$, $\hat{k}_1, \ldots, \hat{k}_{n+\hat{t}-1} \in K$. Compute $s_i = g^{k_i} \in L$ for $i = 1, \ldots, t$, $\tilde{s}_i = g^{\tilde{k}_i} \in L$ for $i = 1, \ldots, \hat{t}$, and $\hat{s}_i = g^{\hat{k}_i} \in L$ for $i = 1, \ldots, n + \hat{t} - 1$. The public key is $(g;\ s_1, \ldots, s_t;\ \tilde{s}_1, \ldots, \tilde{s}_{\hat{t}};\ \hat{s}_1, \ldots, \hat{s}_{n+\hat{t}-1})$. The private key is $(k_1, \ldots, k_t;\ \tilde{k}_1, \ldots, \tilde{k}_{\hat{t}};\ \hat{k}_1, \ldots, \hat{k}_{n+\hat{t}-1})$.

Encryption: To encrypt a message $m \in \mathbb{Z}_2^t$ under a public key as above, one does the following. Choose $w \in W$ at random, and compute $x = g^w \in L$, and $y_i = s_i^w \in L$ for $i = 1, \ldots, t$. Compute $\pi = (\chi(y_1), \ldots, \chi(y_t)) \in \mathbb{Z}_2^t$ and $e = m + \pi \in \mathbb{Z}_2^t$. Compute $\tilde{z}_i = \tilde{s}_i^w \in L$ for $i = 1, \ldots, t$, $\hat{z}_i = \hat{s}_i^w \in L$ for $i = 1, \ldots, n + \hat{t} - 1$, and $\hat{y}_i = \tilde{z}_i \prod_{j=1}^{n} (\hat{z}_{i+j-1})^{\gamma_j} \in L$ for $i = 1, \ldots, \hat{t}$, where $(\gamma_1, \ldots, \gamma_n) = \Gamma(x, e) \in \{0,1\}^n$. Compute $\hat{\pi} = (\chi(\hat{y}_1), \ldots, \chi(\hat{y}_{\hat{t}})) \in \mathbb{Z}_2^{\hat{t}}$. The ciphertext is $(x, e, \hat{\pi})$.

Decryption: To decrypt a ciphertext $(x, e, \hat{\pi}) \in X \times \mathbb{Z}_2^t \times \mathbb{Z}_2^{\hat{t}}$ under a private key as above, one does the following. Compute $\hat{y}_i = x^{\tilde{k}_i + \sum_{j=1}^{n} \gamma_j \hat{k}_{i+j-1}} \in X$ for $i = 1, \ldots, \hat{t}$, where $(\gamma_1, \ldots, \gamma_n) = \Gamma(x, e) \in \{0,1\}^n$. Compute $\hat{\pi}' = (\chi(\hat{y}_1), \ldots, \chi(\hat{y}_{\hat{t}})) \in \mathbb{Z}_2^{\hat{t}}$. Check whether $\hat{\pi} = \hat{\pi}'$; if not, then output reject and halt. Compute $y_i = x^{k_i} \in X$ for $i = 1, \ldots, t$, $\pi = (\chi(y_1), \ldots, \chi(y_t)) \in \mathbb{Z}_2^t$, and $m = e - \pi \in \mathbb{Z}_2^t$, and then output m.

Note that in the decryption algorithm, we are assuming that $x \in X$, which implicitly means that the decryption algorithm should check that $x = (a \bmod N)$ with $(a \mid N) = 1$.

This is *precisely* the scheme that our general construction in §5 yields. Thus, the scheme is secure against adaptive chosen ciphertext attack, provided the QR assumption holds.

As in §7.1, if we replace Γ by a CRHF we get an even more efficient scheme with a smaller value of n. In fact, just a UOWHF suffices. In [CS2], we describe further variations on this scheme.

While this scheme is not nearly as efficient as our schemes based on the DDH and DCR assumptions, it is based on an assumption that is perhaps qualitatively weaker than either of these assumptions. Nevertheless, it is perhaps not so impractical. Consider some concrete security parameters. Let N be a 1024-bit number. If we use this scheme just to encrypt a symmetric encryption key, then we might let $t = 128$. We also let $\hat{t} = 128$. Let Γ be a hash function like SHA-1, so that $n = 160$. With these choices of parameters, the size of a public or private key will be less than 70KB. Ciphertexts are quite compact, requiring 160 bytes. An encryption takes less than 600 1024-bit exponentiations modulo N, a decryption will require about half as many exponentiations modulo N, and there are a number of further optimizations that are applicable as well.

Acknowledgments

Thanks to Ivan Damgaard for noting an improvement in the $1/p$-bound stated in Theorem 2, and thanks to Amit Sahai and Yehuda Lindell for useful discussions.

References

[BR] M. Bellare and P. Rogaway. Random oracles are practical: a paradigm for designing efficient protocols. In *Proc. ACM Computer and Communication Security '93,* ACM Press, 1993.

[CGH] R. Canetti, O. Goldreich, and S. Halevi. The random oracle model, revisited. In *Proc. STOC '98,* ACM Press, 1998.

[CS1] R. Cramer and V. Shoup. A practical public key cryptosystem secure against adaptive chosen cipher text attacks. In *Proc. CRYPTO '98,* Springer Verlag LNCS, 1998.

[CS2] R. Cramer and V. Shoup. Universal hash proofs and a paradigm for adaptive chosen ciphertext secure public key encryption. Cryptology ePrint Archive, Report 2001/085, 2001. `http://eprint.iacr.org`.

[CS3] R. Cramer and V. Shoup. Design and analysis of practical public-key encryption schemes secure against adaptive chosen ciphertext attack. Cryptology ePrint Archive, Report 2001/108, 2001. `http://eprint.iacr.org`.

[DDN] D. Dolev, C. Dwork, and M. Naor. Non-malleable cryptography. *SIAM Journal on Computing,* 30:391–437, 2000. Extended abstract in *Proc. STOC '91,* ACM Press, 1991.

[L] M. Luby. *Pseudorandomness and Cryptographic Applications.* Princeton University Press, 1996.

[NY1] M. Naor and M. Yung. Universal one-way hash functions and their cryptographic applications. In *Proc. STOC '89,* ACM Press, 1989.

[P] P. Paillier. Public-key cryptosystems based on composite degree residue classes. In *Proc. EUROCRYPT '99,* Springer Verlag LNCS, 1999.

[RS] C. Rackoff and D. Simon. Non-interactive zero knowledge proof of knowledge and chosen ciphertext attacks. In *Proc. CRYPTO '91,* Springer Verlag LNCS, 1991.

Key-Insulated Public Key Cryptosystems

Yevgeniy Dodis[1], Jonathan Katz[2], Shouhuai Xu[3], and Moti Yung[4]

[1] Department of Computer Science, New York University
dodis@cs.nyu.edu
[2] Department of Computer Science, Columbia University
jkatz@cs.columbia.edu
[3] Department of Information and Software Engineering, George Mason University
sxu1@gmu.edu
[4] CertCo, Inc.
moti@cs.columbia.edu

Abstract. Cryptographic computations (decryption, signature generation, etc.) are often performed on a relatively insecure device (e.g., a mobile device or an Internet-connected host) which cannot be trusted to maintain secrecy of the private key. We propose and investigate the notion of *key-insulated security* whose goal is to minimize the damage caused by secret-key exposures. In our model, the secret key(s) stored on the insecure device are refreshed at discrete time periods via interaction with a physically-secure – but computationally-limited – device which stores a "master key". All cryptographic computations are still done on the insecure device, and the public key remains unchanged. In a (t, N)-key-insulated scheme, an adversary who compromises the insecure device and obtains secret keys for up to t periods of his choice is unable to violate the security of the cryptosystem for *any* of the remaining $N-t$ periods. Furthermore, the scheme remains secure (for *all* time periods) against an adversary who compromises *only* the physically-secure device. We focus primarily on key-insulated public-key encryption. We construct a (t, N)-key-insulated encryption scheme based on any (standard) public-key encryption scheme, and give a more efficient construction based on the DDH assumption. The latter construction is then extended to achieve chosen-ciphertext security.

1 Introduction

Motivation. Exposure of secret keys is perhaps the most devastating attack on a cryptosystem since it typically means that security is entirely lost. This problem is probably the greatest threat to cryptography in the real world: in practice, it is typically easier for an adversary to obtain a secret key from a naive user than to break the computational assumption on which the system is based. The threat is increasing nowadays with users carrying mobile devices which allow remote access from public or foreign domains.

Two classes of methods exist to deal with this problem. The first tries to prevent key exposure altogether. While this is an important goal, it is not always practical. For example, when using portable devices to perform cryptographic

L.R. Knudsen (Ed.): EUROCRYPT 2002, LNCS 2332, pp. 65–82, 2002.

operations (e.g., decrypting transmissions using a mobile phone) one must expect that the device itself may be physically compromised in some way (e.g., lost or stolen) and thus key exposure is inevitable. Furthermore, complete prevention of key exposure – even for non-mobile devices – will usually require some degree of physical security which can be expensive and inconvenient. The second approach assumes that key exposure will inevitably occur and seeks instead to minimize the damage which results when keys are obtained by an adversary. Secret sharing [35], threshold cryptography [13,12], proactive cryptography [32], exposure-resilient cryptography [9] and forward-secure signatures [3,5] may all be viewed as different means of taking this approach.

The most successful solution will involve a combination of the above approaches. Physical security may be ensured for a *single* device and thus we may assume that data stored on this device will remain secret. On the other hand, this device may be computationally limited or else not suitable for a particular application and thus we are again faced with the problem that some keys will need to be stored on insecure devices which are likely to be compromised during the lifetime of the system. Therefore, techniques to minimize the damage caused by such compromises must also be implemented.

Our Model. We focus here on a notion we term *key-insulated security.* Our model is the following (the discussion here focuses on public-key encryption, yet the term applies equally-well to the case of digital signatures). The user begins by registering a single public key PK. A "master" secret key SK^* is stored on a device which is physically secure and hence resistant to compromise. All decryption, however, is done on an insecure device for which key exposure is expected to be a problem. The lifetime of the protocol is divided into distinct periods $1, \ldots, N$ (for simplicity, one may think of these time periods as being of equal length; e.g., one day). At the beginning of each period, the user interacts with the secure device to derive a temporary secret key which will be used to decrypt messages sent during that period; we denote by SK_i the temporary key for period i. On the other hand, the public key PK used to encrypt messages does *not* change at each period; instead, ciphertexts are now labeled with the time period during which they were encrypted. Thus, encrypting M in period i results in ciphertext $\langle i, C\rangle$.

The insecure device, which does all actual decryption, is vulnerable to repeated key exposures; specifically, we assume that up to $t < N$ periods can be compromised (where t is a parameter). Our goal is to minimize the effect such compromises will have. Of course, when a key SK_i is exposed, an adversary will be able to decrypt messages sent during time period i. Our notion of security (informally) is that this is all an adversary can do. In particular, the adversary will be unable to determine any information about messages sent during *all* time periods other than those in which a compromise occurred. This is the strongest level of security one can expect in such a model. We call a scheme satisfying the above notion (t, N)-key-insulated.

If the physically-secure device is completely trusted, we may have this device generate (PK, SK^*) itself, keep SK^*, and publish PK. When a user requests a key for period i, the device may compute SK_i and send it to the user. More involved methods are needed when the physically-secure device is *not* trusted by the user. In this, more difficult case (which we consider here), the user may generate (PK, SK) himself, publish PK, and then derive keys SK^*, SK_0. The user then sends SK^* to the device and stores SK_0 himself. When the user requests a key for period i, the device sends "partial" key SK'_i to the user, who may then compute the "actual" key SK_i using SK_{i-1} and SK'_i. In this way, the user's security is guaranteed during *all* time periods with respect to the device itself, provided that the knowledge of SK^* *alone* is not sufficient to derive any of the actual keys SK_i. We note that this strong security guarantee is essential when a single device serves many different users, offering them protection against key exposure. In this scenario, users may trust this device to update their keys, but may not want the device to be able to read their encrypted traffic. Thus, there is no reason this device should have complete (or *any*!) knowledge of their "actual" keys. Finally we note that assuring that the devices are synchronized to the same period (so that only one secret key per period is given by the physically secure device) and that they handle proper authenticated interaction is taken care of by an underlying protocol (which is outside our model).

Other Applications. Besides the obvious application to minimizing the risk of key exposures across multiple time periods, key-insulated security may also be used to protect against key exposures across multiple locations, or users. For example, a company may establish a single public key and distribute (different) secret keys to its various employees; each employee is differentiated by his "non-cryptographic ID" i (e.g., a social security number or last name), and can use his own secret key SK_i to perform the desired cryptographic operation. This approach could dramatically save on the public key size, and has the property that the system remains secure (for example, encrypted messages remain hidden) for all employees whose keys are not exposed.

A key-insulated scheme may also be used for purposes of delegation [22]; here, a user (who has previously established a public key) delegates his rights in some specified, limited way to a second party. In this way, even if up to t of the delegated parties' keys are lost, the remaining keys – and, in particular, the user's secret key – are secure.

Finally, we mention the application of key escrow by legal authorities. For example, consider the situation in which the FBI wants to read email sent to a particular user on a certain date. If a key-insulated scheme (updated daily) is used, the appropriate key for up to t desired days can be given to the FBI without fear that this will enable the FBI to read email sent on other days. A similar application (with weaker security guarantees) was considered by [2].

Our Contributions. We introduce the notion of key-insulated security and construct efficient schemes secure under this notion. Although our definition may be applied to a variety of cryptographic primitives, we focus here on public-key encryption. In Section 3, we give a generic construction of a (t, N)-key-insulated encryption scheme based on any (standard) public-key encryption scheme. Section 4 gives a more efficient construction which is secure under the DDH assumption.

Both of these schemes achieve semantic security; however, we show in Section 5 how the second scheme can be improved to achieve chosen-ciphertext security. In a companion paper [15], we consider key-insulated security of signature schemes.

Related Work. Arriving at the right definitions and models for the notion we put forth here has been somewhat elusive. For example, Girault [21] considers a notion similar to key-insulated security of signature schemes. However, [21] does not present any formal definitions, nor does it present schemes which are provably secure. Recently and concurrently with our work, other attempts at formalizing key-insulated public-key encryption have been made [36,30]. However, these works consider only a *non-adaptive* adversary who chooses which time periods to expose at the outset of the protocol, whereas we consider the more natural and realistic case of an *adaptive* adversary who may choose which time periods to expose at any point during protocol execution. Furthermore, the solution of [36] for achieving chosen-ciphertext security is proven secure in the random oracle model; our construction of Section 5 is proven secure against chosen-ciphertext attacks in the standard model ([30] does not address chosen-ciphertext security at all). Finally, our definition of security is stronger than that considered in [36,30]. Neither work considers the case of an untrusted, physically-secure device. Additionally, [30] require only that an adversary cannot fully determine an un-exposed key SK_i; we make the much stronger requirement that an adversary cannot break the underlying cryptographic scheme for any (set of) un-exposed periods.

Our notion of security complements the notion of forward security for digital signatures.[1] In this model [3,5], an adversary who compromises the system during a particular time period obtains *all* the secret information which exists at that point in time. Clearly, in such a setting one cannot hope to prevent the adversary from signing messages associated with future time periods (since the adversary has all relevant information), even though no explicit key exposures happen during those periods. Forward-secure signatures, however, prevent the adversary from signing messages associated with *prior* time periods. Many improved constructions of forward-secure signatures have subsequently appeared [1,28,25,31].

Our model uses a stronger assumption in that we allow for (a limited amount of) physically-secure storage which is used exclusively for key updates and is not used for the actual cryptographic computations. As a consequence, we are able to obtain a much stronger level of security in that the adversary is unable to sign/decrypt messages at *any* non-compromised time period, both in the future and in the past.

An identity-based encryption scheme may be converted to an $(N-1, N)$-key-insulated encryption scheme by viewing the period number as an "identity" and having the physically-secure device implement the trusted third party. In

[1] Although forward-security also applies to public-key encryption, forward-secure encryption schemes are not yet known. The related notion of "perfect forward secrecy" [14], where the parties exchange ephemeral keys on a per-session basis, is incomparable to our notion here.

fact, our notion of (t, N)-key-insulated encryption *with a fully trusted device* can be viewed as a relaxation of identity-based encryption, where we do not insist on $t = N - 1$. Only recently Boneh and Franklin [7] have proposed a practical, identity-based encryption scheme; they also mention the above connection. It should be noted, however, that the security of their scheme is proven in the random oracle model under a very specific, number-theoretic assumption. By focusing on key-insulated security for $t \ll N$, as we do here, schemes based on alternate assumptions and/or with improved efficiency and functionality may be designed.

2 Definitions

2.1 The Model

We now provide a formal model for key-insulated security, focusing on the case of public-key encryption (other key-insulated primitives can be defined similarly; e.g., signature schemes are treated in [15]). Our definition of a key-updating encryption scheme parallels the definition of a key-evolving signature scheme which appears in [5], with one key difference: in a key-updating scheme there is some data (in particular, SK^*) which is never erased since it is stored on a physically-secure device. However, since the physically-secure device may not be fully trusted, new security concerns arise.

Definition 1. *A* key-updating (public-key) encryption scheme *is a 5-tuple of poly-time algorithms* $(\mathcal{G}, \mathcal{U}^*, \mathcal{U}, \mathcal{E}, \mathcal{D})$ *such that:*

- $\mathcal{G}$, *the* key generation algorithm, *is a probabilistic algorithm which takes as input a security parameter* 1^k *and the total number of time periods* N. *It returns a public key* PK, *a master key* SK^*, *and an initial key* SK_0.
- $\mathcal{U}^*$, *the* device key-update algorithm, *is a deterministic algorithm which takes as input an index* i *for a time period (throughout, we assume* $1 \leq i \leq N$*) and the master key* SK^*. *It returns the partial secret key* SK'_i *for time period* i.
- $\mathcal{U}$, *the* user key-update algorithm, *is a deterministic algorithm which takes as input an index* i, *secret key* SK_{i-1}, *and a partial secret key* SK'_i. *It returns secret key* SK_i *for time period* i *(and erases* SK_{i-1}, SK'_i*).*
- $\mathcal{E}$, *the* encryption algorithm, *is a probabilistic algorithm which takes as input a public-key* PK, *a time period* i, *and a message* M. *It returns a ciphertext* $\langle i, C \rangle$.
- $\mathcal{D}$, *the* decryption algorithm, *is a deterministic algorithm which takes as input a secret key* SK_i *and a ciphertext* $\langle i, C \rangle$. *It returns a message* M *or the special symbol* $\perp$.

We require that for all messages M, $\mathcal{D}_{SK_i}(\mathcal{E}_{PK}(i, M)) = M$.

A key-updating encryption scheme is used as one might expect. A user begins by generating $(PK, SK^*, SK_0) \leftarrow \mathcal{G}(1^k, N)$, registering PK in a central location (just as he would for a standard public-key scheme), storing SK^* on a physically-secure device, and storing SK_0 himself. At the beginning of time period i, the user requests $SK'_i = \mathcal{U}^*(i, SK^*)$ from the secure device. Using SK'_i and SK_{i-1}, the user may compute $SK_i = \mathcal{U}(i, SK_{i-1}, SK'_i)$. This key may be used to decrypt messages sent during time period i without further access to the device. After computation of SK_i, the user must erase SK'_i and SK_{i-1}. Note that encryption is always performed with respect to a fixed public key PK which need not be changed. Also note that the case when the device is fully trusted corresponds to $SK_0 = \perp$ and $SK_i = SK'_i$.

Random-Access Key Updates. All the schemes we construct will have a useful property we call *random-access key updates.* For any current period j and any desired period i, it is possible to update the secret key from SK_j to SK_i in "one shot". Namely, we can generalize the key updating algorithms $\mathcal{U}^*$ and $\mathcal{U}$ to take a pair of periods i and j such that $\mathcal{U}^*((i,j), SK^*)$ outputs the partial key SK'_{ij} and $\mathcal{U}((i,j), SK_j, SK'_{ij})$ outputs SK_i. Our definition above implicitly fixes $j = i - 1$. We remark that random-access key updates are impossible to achieve in the forward-security model.

2.2 Security

The are three types of exposures we protect against: (1) ordinary *key exposure*, which models (repeated) compromise of the insecure storage (i.e., leakage of SK_i); (2) *key-update* exposure, which models (repeated) compromise of the insecure device during the key-updating step (i.e., leakage of SK_{i-1} and SK'_i); and (3) *master key exposure*, which models compromise of the physically-secure device (i.e., leakage of SK^*; this includes the case when the device itself is untrusted).

To formally model key exposure attacks, we give the adversary access to two (possibly three) types of oracles. The first is a *key exposure oracle* $\mathsf{Exp}_{SK^*,SK_0}(\cdot)$ which, on input i, returns the temporary secret key SK_i (note that SK_i is uniquely defined by SK^* and SK_0). The second is a *left-or-right encryption oracle* [4], $\mathsf{LR}_{PK,\boldsymbol{b}}(\cdot,\cdot,\cdot)$, where $\boldsymbol{b} = b_1 \ldots b_N \in \{0,1\}^N$, defined as: $\mathsf{LR}_{PK,\boldsymbol{b}}(i,M_0,M_1) \stackrel{\text{def}}{=} \mathcal{E}_{PK}(i, M_{b_i})$. This models encryption requests by the adversary for time periods and message pairs of his choice. We allow the adversary to interleave encryption requests and key exposure requests, and in particular the key exposure requests of the adversary may be made adaptively and in any order. Finally, we may also allow the adversary access to a *decryption oracle* $\mathcal{D}^*_{SK^*,SK_0}(\cdot)$ that, on input $\langle i, C\rangle$, computes $\mathcal{D}_{SK_i}(\langle i, C\rangle)$. This models a chosen-ciphertext attack by the adversary.

The vector $\boldsymbol{b}$ for the left-or-right oracle will be chosen randomly, and the adversary succeeds by guessing the value of b_i for any un-exposed time period i. Informally, a scheme is secure if any probabilistic polynomial time (PPT) adversary has success negligibly close to $1/2$. More formally:

Definition 2. *Let $\Pi = (\mathcal{G}, \mathcal{U}^*, \mathcal{U}, \mathcal{E}, \mathcal{D})$ be a key-updating encryption scheme. For adversary A, define the following:*

$$\mathsf{Succ}_{A,\Pi}(k) \stackrel{\text{def}}{=} \Pr\Big[(PK, SK^*, SK_0) \leftarrow \mathcal{G}(1^k, N); \boldsymbol{b} \leftarrow \{0,1\}^N; \\ (i, b') \leftarrow A^{\mathsf{LR}_{PK,\boldsymbol{b}}(\cdot,\cdot,\cdot), \mathsf{Exp}_{SK^*,SK_0}(\cdot), \mathcal{O}(\cdot)}(PK) : b' = b_i\Big],$$

*where i was never submitted to $\mathsf{Exp}_{SK^*,SK_0}(\cdot)$, and $\mathcal{O}(\cdot) = \perp$ for a plaintext-only attack and $\mathcal{O}(\cdot) = \mathcal{D}^*_{SK^*,SK_0}(\cdot)$ for a chosen-ciphertext attack (in the latter case the adversary is not allowed to query $\mathcal{D}^*(\langle i, C\rangle)$ if $\langle i, C\rangle$ was returned by $\mathsf{LR}(i,\cdot,\cdot)$). Π is (t, N)-*key-insulated *if, for any* PPT A *who submits at most t requests to the key-exposure oracle, $|\mathsf{Succ}_{A,\Pi}(k) - 1/2|$ is negligible.*

As mentioned above, we may also consider attacks in which an adversary breaks in to the user's storage while a key update is taking place (i.e., the exposure occurs between two periods $i-1$ and i); we call this a *key-update exposure* at period i. In this case, the adversary receives SK_{i-1}, SK'_i, and (can compute) SK_i. Informally, we say a scheme has *secure key updates* if a key-update exposure at period i is equivalent to key exposures at periods $i-1$ and i and no more. More formally:

Definition 3. *Key-updating encryption scheme Π has* secure key updates *if the view of any adversary A making a key-update exposure request at period i can be perfectly simulated by an adversary A' who makes key exposure requests at periods $i-1$ and i.*

This property is desirable in real-world implementations of a key-updating encryption scheme since an adversary who gains access to the user's storage is likely to have access for several consecutive time periods (i.e., until the user detects or re-boots), including the key updating steps.

We also consider attacks which compromise the physically-secure device (this includes attacks in which this device is untrusted). Here, our definition requires that the encryption scheme be secure against an adversary which is given SK^* as input. Note that we do *not* require security against an adversary who compromises both the user's storage and the secure device – in our model this is impossible since, given SK^* and SK_i, an adversary can compute SK_j (at least for $j > i$) by himself.

Definition 4. *Let Π be a key-updating scheme which is (t, N)-key-insulated. For any adversary B, define the following:*

$$\mathsf{Succ}_{B,\Pi}(k) \stackrel{\text{def}}{=} \Pr\Big[(PK, SK^*, SK_0) \leftarrow \mathcal{G}(1^k, N); \boldsymbol{b} \leftarrow \{0,1\}^N; \\ (i, b') \leftarrow B^{\mathsf{LR}_{PK,\boldsymbol{b}}(\cdot,\cdot,\cdot), \mathcal{O}(\cdot)}(PK, SK^*) : b' = b_i\Big],$$

where $\mathcal{O}(\cdot) = \perp$ for a plaintext-only attack and $\mathcal{O}(\cdot) = \mathcal{D}^*_{SK^*,SK_0}(\cdot)$ for a chosen-ciphertext attack (in the latter case the adversary is not allowed to query $\mathcal{D}^*(\langle i, C\rangle)$ if $\langle i, C\rangle$ was returned by $\mathsf{LR}(i,\cdot,\cdot)$). Π is strongly (t, N)-key-insulated *if, for any* PPT B, *$|\mathsf{Succ}_{B,\Pi}(k) - 1/2|$ is negligible.*

3 Generic Semantically-Secure Construction

Let (G, E, D) be any semantically secure encryption scheme. Rather than giving a separate (by now, standard) definition, we may view it simply as a $(0,1)$-key-insulated scheme. Namely, only one secret key SK is present, and any PPT adversary, given PK and the left-or-right-oracle $\mathsf{LR}_{PK,b}$, cannot predict b with success non-negligibly different from $1/2$. Hence, our construction below can be viewed as an amplification of a $(0,1)$-key-insulated scheme into a general (t, N)-key-insulated scheme.

We will assume below that $t, \log N = O(\mathsf{poly}(k))$, where k is our security parameter. Thus, we allow exponentially-many periods, and can tolerate exposure of any polynomial number of keys. We assume that E operates on messages of length $\ell = \ell(k)$, and construct a (t, N)-key-insulated scheme operating on messages of length $L = L(k)$.

Auxiliary Definitions. We need two auxiliary definitions: that of an *all-or-nothing transform* [34,8] (AONT) and a *cover-free family* [18,16]. Informally, an AONT splits the message M into n secret shares $x_1, \ldots, x_n$ (and possibly one public share z), and has the property that (1) the message M can be efficiently recovered from *all* the shares $x_1, \ldots, x_n, z$, but (2) missing even a single share x_j gives "no information" about M. As such, it is a generalization of $(n-1, n)$-secret sharing. We formalize this, modifying the conventional definitions [8,9] to a form more compatible with our prior notation.

Definition 5. *An efficient randomized transformation $\mathcal{T}$ is called an (L, ℓ, n)-AONT if: (1) on input $M \in \{0,1\}^L$, $\mathcal{T}$ outputs $(X, z) \stackrel{\text{def}}{=} (x_1, \ldots, x_n, z)$, where $x_j \in \{0,1\}^\ell$; (2) there exists an efficient inverse function $\mathcal{I}$ such that $\mathcal{I}(X, z) = M$; (3) $\mathcal{T}$ satisfies the indistinguishability property described below.*
Let $X_{-j} = (x_1, \ldots, x_{j-1}, x_{j+1}, \ldots, x_n)$ and $\mathcal{T}_{-j}(M) = (X_{-j}, z)$, where $(X, z) \leftarrow \mathcal{T}(M)$. Define the left-or-right oracle $\mathsf{LR}_b(j, M_0, M_1) \stackrel{\text{def}}{=} \mathcal{T}_{-j}(M_b)$, where $b \in \{0,1\}$. For any PPT A, we let $\mathsf{Succ}_{A,\mathcal{T}}(k) \stackrel{\text{def}}{=} \Pr[b \leftarrow \{0,1\}; b' \leftarrow A^{\mathsf{LR}_b(\cdot,\cdot,\cdot)}(1^k) : b' = b]$. We require that $|\mathsf{Succ}_{A,\mathcal{T}}(k) - 1/2|$ is negligible.

A family of subsets $S_1 \ldots S_N$ over some universe U is said to be *t-cover-free* if no t subsets $S_{i_1}, \ldots, S_{i_t}$ contain a (different) subset S_{i_0}; in other words, for all $\{i_0, \ldots, i_t\}$ with $i_0 \notin \{i_1, \ldots, i_t\}$, we have $S_{i_0} \not\subseteq \cup_{j=1}^t S_{i_j}$. A family is said to be *(t, α)-cover-free*, where $0 < \alpha < 1$, if, for all $\{i_0, \ldots, i_t\}$ with $i_0 \notin \{i_1, \ldots, i_t\}$, we have $|S_{i_0} \setminus \cup_{j=1}^t S_{i_j}| \geq \alpha |S_{i_0}|$. Such families are well known and have been used several times in cryptographic applications [10,29,20]. In what follows, we fix $\alpha = 1/2$ for simplicity, and will use the following (essentially optimal) result, non-constructively proven by [18] and subsequently made efficient by [29,24].

Theorem 1 ([18,29,24]). *For any N and t, one can efficiently construct a $(t, \frac{1}{2})$-cover-free collection of N subsets $S_1, \ldots, S_N$ of $U = \{1, \ldots, u\}$, with $|S_i| = n$ for all i, satisfying: $u = \Theta(t^2 \log N)$ and $n = \Theta(t \log N)$.*

Since we assumed that $t, \log N = O(\mathsf{poly}(k))$, we have $u, n = O(\mathsf{poly}(k))$ as well.

Construction. For simplicity, we first describe the scheme which is not strongly secure (see Definition 4), and then show a modification making it strongly secure. Let $S_1, \ldots, S_N \subset [u] \stackrel{\text{def}}{=} \{1 \ldots u\}$ be the $(t, \frac{1}{2})$-cover-free family of n-element sets, as given by Theorem 1. Also, let $\mathcal{T}$ be a secure (L, ℓ, n)-AONT. Our (t, N)-key-insulated scheme will have a set of u independent encryption/decryption keys (sk_r, pk_r) for our basic encryption E, of which only the subset S_i will be used at time period i. Specifically, the public key of the scheme will be $PK = \{pk_1, \ldots, pk_u\}$, the secret key at time i will be $SK_i = \{sk_r \,:\, r \in S_i\}$, and the master key (for now) will be $SK^* = \{sk_1, \ldots, sk_u\}$. We define the encryption of $M \in \{0,1\}^L$ at time period i as:

$$\mathcal{E}_{PK}(i, M) = \langle\ i,\ (E_{pk_{r_1}}(x_1), \ldots, E_{pk_{r_n}}(x_n),\ z)\ \rangle,$$

where $(x_1, \ldots, x_n, z) \leftarrow \mathcal{T}(M)$ and $S_i = \{r_1, \ldots, r_n\}$. To decrypt $\langle i, (y_1, \ldots, y_n, z) \rangle$ using $SK_i = \{sk_r \,:\, r \in S_i\}$, the user first recovers the x_j's from the y_j's using D, and then recovers the message $M = \mathcal{I}(x_1, \ldots, x_n, z)$. Key updates are trivial: the device sends the new key SK_i and the user erases the old key SK_{i-1}. Obviously, the scheme supports secure key updates as well as random-access key updates.

Security. We informally argue the (t, N)-key-insulated security of this scheme (omitting the formal proof due to space limitations). The definition of the AONT implies that the system is secure at time period i provided the adversary misses *at least one* key sk_r, where $r \in S_i$. Indeed, semantic security of E implies that the adversary completely misses the shares encrypted with sk_r in this case, and hence has no information about the message M. On the other hand, if the adversary learn any t keys $SK_{i_1}, \ldots, SK_{i_t}$, he learns the auxiliary keys $\{sk_r \,:\, r \in S_{i_1} \cup S_{i_2} \ldots \cup S_{i_t}\}$. Hence, the necessary and sufficient condition for (t, N)-key-insulated security is exactly the t-cover freeness of the S_i's! The parameter $\alpha = \frac{1}{2}$ is used to improve the exact security of our reduction.

Theorem 2. *The generic scheme Π described above is (t, N)-key-insulated with secure key updates, provided (G, E, D) is semantically-secure, $\mathcal{T}$ is a secure (L, ℓ, n)-AONT, and the family $S_1, \ldots, S_N$ is $(t, \frac{1}{2})$-cover-free. Specifically, breaking the security of Π with advantage ε implies the same for either (G, E, D) or $\mathcal{T}$ with advantage at least $\Omega(\varepsilon/t)$.*

Strong Key-Insulated Security. The above scheme is not *strongly* (t, N)-key-insulated since the device stores all the secret keys $(sk_1, \ldots, sk_u)$. However, we can easily fix this problem. The user generates one extra key pair (sk_0, pk_0). It publishes pk_0 together with the other public keys, but keeps sk_0 for itself (never erasing it). Assuming now that $\mathcal{T}$ produces $n+1$ secret shares $x_0, \ldots, x_n$ rather than n, we just encrypt the first share x_0 with pk_0 (and the others, as before, with the corresponding keys in S_i). Formally, let $S'_i = S_i \cup \{0\}$, the master key is still $SK^* = \{sk_1, \ldots, sk_u\}$, but now $PK = \{pk_0, pk_1, \ldots, pk_u\}$ and the i-th secret key is $SK_i = \{sk_r \,:\, r \in S'_i\}$. Strong key-insulated security of this scheme follows a similar argument as in Theorem 2.

Efficiency. The main parameters of the scheme are: (1) the size of PK and SK^* are both $u = O(t^2 \log N)$; and (2) the user's storage and the number of local encryptions per global encryption are both $n = O(t \log N)$. In particular, the surprising aspect of our construction is that it supports an *exponential* number of periods N and the main parameters depend mainly on t, the number of exposures we allow. Since t is usually quite small (say, $t = O(1)$ and certainly $t \ll N$), we obtain good parameters considering the generality of the scheme. (In Section 4 we use a specific encryption scheme and achieve $|PK|, |SK^*| = O(t)$ and $|SK_i| = O(1)$.)

Additionally, the choice of a secure (L, ℓ, n)-AONT defines the tradeoff between the number of encrypted bits L compared to the total encryption size, which is $(\beta n\ell + |z|)$, where β is the expansion of E, and $|z|$ is the size of the public share. In particular, if $L = \ell$, we can use any traditional $(n-1, n)$-secret sharing scheme (e.g., Shamir's scheme [35], or even XOR-sharing: pick random x_j's subject to $M = \bigoplus x_j$). This way we have no public part, but the ciphertext increases by a factor of βn as compared to the plaintext. Computationally-secure AONT's allow for better tradeoffs. For example, using either the computational secret sharing scheme of [27], or the AONT constructions of [9], we can achieve $|z| = L$, while ℓ can be as small as the security parameter k (in particular, $\ell \ll L$). Thus, we get *additive* increase $\beta n\ell$, which is essentially independent of L. Finally, in the random oracle model, we could use the construction of [8] achieving $|z| = 0$, $L = \ell(n-1)$, so the ciphertext size is $\beta\ell n \approx \beta L$. Finally, in practice one would use the above scheme to encrypt a random key K (which is much shorter than M) for a symmetric-key encryption scheme, and concatenate to this the symmetric-key encryption of M using K.

Adaptive vs. Non-adaptive Adversaries. Theorem 2 holds for an *adaptive* adversary who makes key exposure requests based on all information collected so far. We notice, however, that both the security and the efficiency of our construction could be somewhat improved for non-adaptive adversaries, who choose the key-exposure periods $i_1, \ldots, i_t$ at the outset of the protocol (which is the model of [36,30,2]). For example, it is easy to see that we no longer lose the factor t in the security of our reduction in Theorem 2. As for the efficiency, instead of using an AONT (which is essentially an $(n-1, n)$-secret sharing scheme), we can now use any $(n/2, n)$-"ramp" secret sharing scheme [6]. This means that n shares reconstruct the secret, but any $n/2$ shares yield no information about the secret. Indeed, since our family is $(t, \frac{1}{2})$-cover-free, any non-exposed period will have the adversary miss more than half of the relevant secret keys. For non-adaptive adversaries, we know at the outset which secret keys are non-exposed, and can use a simple hybrid argument over these keys to prove the security of the modified scheme. For example, we can use the "ramp" generalization of Shamir's secret sharing scheme[2] proposed by Franklin and Yung [19], and achieve $L = \ell n/2$ instead of $L = \ell$ resulting from regular Shamir's $(n-1, n)$-scheme.

[2] Here the message length $L = \ell n/2$, and the ℓ-bit parts $m_1, \ldots, m_{n/2}$ of M are viewed as the $n/2$ lower order coefficients of an otherwise random polynomial of degree $(n-1)$ over $GF[2^\ell]$. This polynomial is then evaluated at n points of $GF[2^\ell]$ to give the final n shares.

4 Semantic Security Based on DDH

In this section, we present an efficient strongly (t, N)-key-insulated scheme, whose semantic security can be proved under the DDH assumption.

We first describe the basic encryption scheme we build upon. The key generation algorithm $\mathsf{Gen}(1^k)$ selects a random prime q with $|q| = k$ such that $p = 2q+1$ is prime. This defines a unique subgroup $\mathbb{G} \subset \mathbb{Z}_p^*$ of size q in which the DDH assumption is assumed to hold; namely, it is hard to disinguish a random tuple (g, h, u, v) of four independent elements in $\mathbb{G}$ from a random tuple satisfying $\log_g u = \log_h v$. Given group $\mathbb{G}$, key generation proceeds by selecting random elements $g, h \in \mathbb{G}$ and random $x, y \in \mathbb{Z}_q$. The public key consists of g, h, and the Pedersen commitment [33] to x and y: $z = g^x h^y$. The secret key contains both x and y. To encrypt $M \in \mathbb{G}$, choose random $r \in \mathbb{Z}_q$ and compute $(g^r, h^r, z^r M)$. To decrypt (u, v, w), compute $M = w/u^x v^y$. This scheme is very similar to El Gamal encryption [17], except it uses two generators. It has been recently used by [26] in a different context.

Our Scheme. Our (t, N)-key-insulated scheme builds on the above basic encryption scheme and is presented in Figure 1. The key difference is that, after choosing $\mathbb{G}, g, h$, as above, we select two random polynomials $f_x(\tau) \stackrel{\text{def}}{=} \sum_{j=0}^t x_j^* \tau^j$ and $f_y(\tau) \stackrel{\text{def}}{=} \sum_{j=0}^t y_j^* \tau^j$ over $\mathbb{Z}_q$ of degree t. The public key consists of g, h and Pedersen commitments $\{z_0^*, \ldots, z_t^*\}$ to the coefficients of the two polynomials (see Figure 1). The user stores the constant terms of the two polynomials (i.e., x_0^* and y_0^*) and the remaining coefficients are stored by the physically-secure device. To encrypt during period i, first z_i is computed from the public key as $z_i \stackrel{\text{def}}{=} \Pi_{j=0}^t (z_j^*)^{i^j}$. Then (similar to the basic scheme), encryption of message M is done by choosing $r \in \mathbb{Z}_q$ at random and computing $(g^r, h^r, z_i^r M)$. Using our notation from above, it is clear that $z_i = g^{f_x(i)} h^{f_y(i)}$. Thus, as long as the user has secret key $SK_i = (f_x(i), f_y(i))$ during period i, decryption during that period may be done just as in the basic scheme. As for key evolution, the user begins with $SK_0 = (x_0^*, y_0^*) = (f_x(0), f_y(0))$. At the start of any period i, the device transmits partial key $SK_i' = (x_i', y_i')$ to the user. Note that (cf. Figure 1) $x_i' = f_x(i) - f_x(i-1)$ and $y_i' = f_y(i) - f_y(i-1)$. Thus, since the user already has SK_{i-1}, the user may easily compute SK_i from these values. At this point, the user erases SK_{i-1}, and uses SK_i to decrypt for the remainder of the time period.

Theorem 3. *Under the DDH assumption, the encryption scheme of Figure 1 is strongly (t, N)-key-insulated under plaintext-only attacks. Furthermore, it has secure key updates and supports random-access key updates.*

Proof. Showing secure key updates is trivial, since an adversary who exposes keys SK_{i-1} and SK_i can compute the value SK_i' by itself (and thereby perfectly simulate a key-update exposure at period i). Similarly, random-access key updates can be done using partial keys $SK_{ij}' = (x_{ij}', y_{ij}')$, where $x_{ij}' = f_x(i) - f_x(j)$, $y_{ij}' = f_y(i) - f_y(j)$. The user can then compute $x_i = x_j + x_{ij}'$ and $y_i = y_j + y_{ij}'$.

$\mathcal{G}(1^k)$: $(g,h,q) \leftarrow \mathsf{Gen}(1^k)$; $x_0^*, y_0^*, \ldots, x_t^*, y_t^* \leftarrow \mathbb{Z}_q$ $z_0^* := g^{x_0^*} h^{y_0^*}, \ldots, z_t^* := g^{x_t^*} h^{y_t^*}$; $PK := (g,h,z_0^*,\ldots,z_t^*)$ $SK^* := (x_1^*, y_1^*, \ldots, x_t^*, y_t^*)$; $SK_0 := (x_0^*, y_0^*)$ return PK, SK^*, SK_0	
$\mathcal{U}^*(i, SK^* = (x_1^*, y_1^*, \ldots, x_t^*, y_t^*))$: $x_i' := \sum_{j=1}^t x_j^* \left(i^j - (i-1)^j\right)$ $y_i' := \sum_{j=1}^t y_j^* \left(i^j - (i-1)^j\right)$ return $SK_i' = (x_i', y_i')$	$\mathcal{U}(i, SK_{i-1} = (x_{i-1}, y_{i-1}), SK_i' = (x_i', y_i'))$: $x_i := x_{i-1} + x_i'$ $y_i := y_{i-1} + y_i'$ return $SK_i = (x_i, y_i)$
$\mathcal{E}_{(g,h,z_0^*,\ldots,z_y^*)}(i, M)$: $z_i := \Pi_{j=0}^t (z_j^*)^{i^j}$ $r \leftarrow \mathbb{Z}_q$ $C := (g^r, h^r, z_i^r M)$ return $\langle i, C \rangle$	$\mathcal{D}_{(x_i,y_i)}(\langle i, C = (u,v,w) \rangle)$: $M := w / u^{x_i} v^{y_i}$ return M

Fig. 1. Semantically-secure key-updating encryption scheme based on DDH.

We now show that the scheme satisfies Definition 2. By a standard hybrid argument [4], it is sufficient to consider an adversary A who asks a single query to its left-or-right oracle (for some time period i of A's choice) and must guess the value b_i. So we assume A makes only a single query to the LR oracle during period i for which it did not make a key exposure request. In the original experiment (cf. Figure 1), the output of $\mathsf{LR}_{PK,\boldsymbol{b}}(i, M_0, M_1)$ is defined as follows: choose $r \in \mathbb{Z}_q$ at random and output $(g^r, h^r, z_i^r M_{b_i})$. Given a tuple (g,h,u,v) which is either a DDH tuple or a random tuple, modify the original experiment as follows: the output of $\mathsf{LR}_{PK,\boldsymbol{b}}(i, M_0, M_1)$ will be $(u, v, u^{x_i} v^{y_i} M_b)$. Note that if (g,h,u,v) is a DDH tuple, then this is a perfect simulation of the original experiment. On the other hand, if (g,h,u,v) is a random tuple then, under the DDH assumption, the success of any PPT adversary in this modified experiment cannot differ by more than a negligible amount from its success in the original experiment. It is important to note that, in running the experiment, we can answer all of A's key exposure requests correctly since all secret keys are known. Thus, in contrast to [36,30], we may handle an adaptive adversary who chooses when to make key exposure requests based on all information seen during the experiment.

Assume now that (g,h,u,v) is a random tuple and $\log_g h \neq \log_u v$ (this will occur with all but negligible probability). We claim that the adversary's view in the modified experiment is independent of $\boldsymbol{b}$. Indeed, the adversary knows only t values of $f_x(\cdot)$ and $f_y(\cdot)$ (at points other than i), and since both $f_x(\cdot)$ and $f_y(\cdot)$ are random polynomials of degree t, the values x_i, y_i $(= f_x(i), f_y(i))$ are *information-theoretically* uniformly distributed, subject only to:

$$\log_g z_i = x_i + y_i \log_g h. \qquad (1)$$

Consider the output of the encryption oracle $(u, v, u^{x_i} v^{y_i} M_b)$. Since:

$$\log_u (u^{x_i} v^{y_i}) = x_i + y_i \log_u v, \qquad (2)$$

and (1) and (2) are linearly independent, the conditional distribution of $u^{x_i}v^{y_i}$ (conditioned on b_i and the adversary's view) is uniform. Thus, the adversary's view is independent of b_i (and hence $\boldsymbol{b}$). This implies that the success probability of A in this modified experiment is $1/2$, and hence the success probability of A in the original experiment is at most negligibly different from $1/2$.

We now consider security against the physically-secure device; in this case, there are no key exposure requests but the adversary learns SK^*. Again, it is sufficient to consider an adversary who asks a single query to its left-or-right oracle (for time period i of its choice) and must guess the value b_i. Since SK^* only contains the t highest-order coefficients of t-degree polynomials, the pair (x_i, y_i) is information-theoretically uniformly distributed (for all i) subject to $x_i + y_i \log_g h = \log_g z_i$. An argument similar to that given previously shows that the success probability of the adversary is at most negligibly better than $1/2$, and hence the scheme satisfies Definition 4.

5 Chosen-Ciphertext Security Based on DDH

We may modify the scheme given in the previous section so as to be resistant to chosen-ciphertext attacks. In doing so, we build upon the chosen-ciphertext-secure (standard) public-key encryption scheme of Cramer and Shoup [11].

<table>
<tr><td colspan="2">$\mathcal{G}(1^k)$: $(g, h, q) \leftarrow \mathsf{Gen}(1^k)$; $H \leftarrow \mathsf{UOWH}(1^k)$
for $i = 0$ to t and $n = 0$ to 2:
$x^*_{i,n}, y^*_{i,n} \leftarrow \mathbb{Z}_q$
for $i = 0$ to t:
$z^*_i := g^{x^*_{i,0}} h^{y^*_{i,0}}$; $c^*_i := g^{x^*_{i,1}} h^{y^*_{i,1}}$; $d^*_i := g^{x^*_{i,2}} h^{y^*_{i,2}}$
$PK := (g, h, H, \{z^*_i, c^*_i, d^*_i\}_{0 \le i \le t})$
$SK^* := (\{x^*_{i,n}, y^*_{i,n}\}_{2 \le i \le t,\, 0 \le n \le 2})$; $SK_0 := (\{x^*_{i,n}, y^*_{i,n}\}_{0 \le i \le 1,\, 0 \le n \le 2})$
return PK, SK^*, SK_0</td></tr>
<tr><td>$\mathcal{U}^*(i, SK^*)$:
for $n = 0$ to 2:
$x'_{i,n} := \sum_{j=2}^{t} x^*_{j,n} \left(i^j - (i-1)^j\right)$
$y'_{i,n} := \sum_{j=2}^{t} y^*_{j,n} \left(i^j - (i-1)^j\right)$
return $SK'_i = (\{x'_{i,n}, y'_{i,n}\}_{0 \le n \le 2})$</td><td>$\mathcal{U}(i, SK_{i-1}, SK'_i)$:
for $n = 0$ to 2:
$x_{i,n} = x_{i-1,n} + x'_{i,n} + x_{1,n}$
$y_{i,n} = y_{i-1,n} + y'_{i,n} + y_{1,n}$
return $SK_i = (\{x_{i,n}, y_{i,n}, x_{1,n}, y_{1,n}\}_{0 \le n \le 2})$</td></tr>
<tr><td>$\mathcal{E}_{PK}(i, M)$:
$z_i := \Pi_{j=0}^{t} (z^*_j)^{i^j}$; $c_i := \Pi_{j=0}^{t} (c^*_j)^{i^j}$
$d_i := \Pi_{j=0}^{t} (d^*_j)^{i^j}$
$r \leftarrow \mathbb{Z}_q$
$C := (g^r, h^r, z_i^r M, (c_i d_i^\alpha)^r)$,
where $\alpha \stackrel{\text{def}}{=} H(g^r, h^r, z_i^r M)$
return $\langle i, C \rangle$</td><td>$\mathcal{D}_{SK_i}(\langle i, C = (u, v, w, e) \rangle)$:
$\alpha := H(u, v, w)$
if $u^{x_{i,1} + x_{i,2}\alpha} v^{y_{i,1} + y_{i,2}\alpha} \neq e$
return $\perp$
else $M := w / u^{x_{i,0}} v^{y_{i,0}}$
return M</td></tr>
</table>

Fig. 2. Chosen-ciphertext-secure key-updating encryption scheme based on DDH.

We briefly review the "basic" Cramer-Shoup scheme (in part to conform to the notation used in Figure 2). Given generators g, h of group $\mathbb{G}$ (as described in the previous section), secret keys $\{x_n, y_n\}_{0 \le n \le 2}$ are chosen randomly from $\mathbb{Z}_q$. Then, public-key components $z = g^{x_0}h^{y_0}$, $c = g^{x_1}h^{y_1}$, and $d = g^{x_2}h^{y_2}$ are computed. In addition, a function H is randomly chosen from a family of universal one-way hash functions. The public key is (g, h, z, c, d, H).

To encrypt a message $M \in \mathbb{G}$, a random element $r \in \mathbb{Z}_q$ is chosen and the ciphertext is: $(g^r, h^r, z^r M, (cd^\alpha)^r)$, where $\alpha = H(g^r, h^r, z^r M)$. To decrypt a ciphertext (u, v, w, e), we first check whether $u^{x_1+x_2\alpha}v^{y_1+y_2\alpha} = e$. If not, we output $\perp$. Otherwise, we output $M = w/u^{x_0}v^{y_0}$.

In our extended scheme (cf. Figure 2), we choose six random, degree-t polynomials (over $\mathbb{Z}_q$) $f_{x_0}, f_{y_0}, f_{x_1}, f_{y_1}, f_{x_2}, f_{y_2}$, where $f_{x_n}(\tau) \stackrel{\text{def}}{=} \sum_{j=0}^{t} x_{j,n}^* \tau^j$ and $f_{y_n}(\tau) \stackrel{\text{def}}{=} \sum_{j=0}^{t} y_{j,n}^* \tau^j$ for $0 \le n \le 2$. The public key consists of g, h, H, and Pedersen commitments to the coefficients of these polynomials. The user stores the constant term *and* the coefficient of the linear term for each of these polynomials, and the remaining coefficients are stored by the physically-secure device.

To encrypt during period i, a user first computes z_i, c_i, and d_i by evaluating the polynomials "in the exponent" (see Figure 2). Then, just as in the basic scheme, encryption of M is performed by choosing random $r \in \mathbb{Z}_q$ and computing $(g^r, h^r, z_i^r M, (c_i d_i^\alpha)^r)$, where $\alpha \stackrel{\text{def}}{=} H(g^r, h^r, z_i^r M)$. Notice that $z_i = g^{f_{x_0}(i)}h^{f_{y_0}(i)}$, $c_i = g^{f_{x_1}(i)}h^{f_{y_1}(i)}$, and $d_i = g^{f_{x_2}(i)}h^{f_{y_2}(i)}$. Thus, the user can decrypt (just as in the basic scheme) as long as he has $f_{x_n}(i), f_{y_n}(i)$ for $0 \le n \le 2$. In fact, the secret key SK_i includes these values; in addition, the secret key at all times includes the linear coefficients $x_{1,0}^*, y_{1,0}^*, \ldots, x_{1,2}^*, y_{1,2}^*$. These values are used to help update SK_i.

Theorem 4. *Under the DDH assumption, the encryption scheme of Figure 2 is strongly $(t-2, N)$-key-insulated under chosen-ciphertext attacks. Furthermore, the scheme has secure key updates and supports random-access key updates.*

Proof. That the scheme has secure key updates is trivial, since SK_i' may be computed from SK_{i-1} and SK_i. Random-access key updates are done analogously to the scheme of the previous section. We now show the key-insulated security of the scheme (cf. Definition 2). A standard hybrid argument [4] shows that it is sufficient to consider an adversary A who makes only a single request to its left-or-right oracle (for time period i of the adversary's choice) and must guess the value b_i. We stress that polynomially-many calls to the decryption oracle are allowed.

Assume A makes a single query to the LR oracle during period i for which it did not make a key exposure request. In the original experiment (cf. Figure 2), the output of $\mathsf{LR}_{PK,\mathbf{b}}(i, M_0, M_1)$ is as follows: choose $r \leftarrow \mathbb{Z}_q$ and output $(g^r, h^r, z_i^r M_{b_i}, (c_i d_i^\alpha)^r)$, where α is as above. As in the proof of Theorem 3, we now modify the experiment. Given a tuple (g, h, u, v) which is either a DDH tuple or a random tuple, we define the output of $\mathsf{LR}_{PK,\mathbf{b}}(i, M_0, M_1)$ to be $(u, v, \tilde{w} = u^{x_{i,0}}v^{y_{i,0}}M_{b_i}, \tilde{e} = u^{x_{i,1}+x_{i,2}\alpha}v^{y_{i,1}+y_{i,2}\alpha})$, where $\alpha \stackrel{\text{def}}{=} H(u, v, \tilde{w})$. Note that if (g, h, u, v) is a DDH tuple, then this results in a perfect simulation of the original experiment. On the other hand, if (g, h, u, v) is a random tuple, then, under the DDH assumption, the success of any PPT adversary cannot differ by a non-negligible amount from its success in the original experiment. As in the proof of Theorem 3, note that, in running the experiment, we can answer all of A's key exposure queries. Thus, the proof handles an adaptive adversary whose key exposure requests may be made based on all information seen up to that point.

Assume now that (g, h, u, v) is a random tuple and $\log_g h \neq \log_u v$ (this happens with all but negligible probability). We show that, with all but negligible probability, the adversary's view in the modified experiment is independent of $\boldsymbol{b}$. The proof parallels [11, Lemma 2]. Say a ciphertext $\langle i, (u', v', w', e')\rangle$ is *invalid* if $\log_g u' \neq \log_h v'$. Then:

Claim. If the decryption oracle outputs $\perp$ for all invalid ciphertexts during the adversary's attack, then the value of b_i (and hence $\boldsymbol{b}$) is independent of the adversary's view.

The adversary knows at most $t-2$ values of $f_{x_0}(\cdot)$ and $f_{y_0}(\cdot)$ (at points other than i) and additionally knows the values $x^*_{1,0}$ and $y^*_{1,0}$ (the linear terms of these polynomials). Since $f_{x_0}(\cdot)$ and $f_{y_0}(\cdot)$ are random polynomials of degree t, the values $x_{i,0}, y_{i,0}$ $(= f_{x_0}(i), f_{y_0}(i))$ are uniformly distributed subject to:

$$\log_g z_i = x_{i,0} + y_{i,0} \log_g h. \tag{3}$$

Furthermore, when the decryption oracle decrypts valid ciphertexts $\langle i,(u',v',w',e')\rangle$, the adversary only obtains linearly-dependent relations $r' \log_g z_i = r' x_{i,0} + r' y_{i,0} \log_g h$ (where $r' \stackrel{\text{def}}{=} \log_g u'$). Similarly, decryptions of valid ciphertexts at other time periods do not further constrain $x_{i,0}, y_{i,0}$. Now consider the third component $u^{x_{i,0}} v^{y_{i,0}} M_{b_i}$ of the encryption oracle (the only one which depends on b_i). Since:

$$\log_u (u^{x_{i,0}} v^{y_{i,0}}) = x_{i,0} + y_{i,0} \log_u v, \tag{4}$$

and (3) and (4) are linearly independent, the conditional distribution of $u^{x_{i,0}} v^{y_{i,0}}$ (conditioned on b_i and the adversary's view) is uniform. Thus, the adversary's view is independent of b_i. The following claim now completes the proof of key-insulated security:

Claim. With all but negligible probability, the decryption oracle will output $\perp$ for all invalid ciphertexts.

Consider a ciphertext $\langle j, (u', v', w', e')\rangle$, where j represents a period during which a key exposure request was not made. We show that, with all but negligible probability, this ciphertext is rejected if it is invalid. There are two cases to consider: (1) $j = i$ (recall that i is the period during which the call to the LR oracle is made) and (2) $j \neq i$.

When $j = i$, the proof of the claim follows the proof of [11, Claim 2] exactly. The adversary knows at most $t-2$ values of $f_{x_1}(\cdot), f_{y_1}(\cdot), f_{x_2}(\cdot)$, and $f_{y_2}(\cdot)$ (at points other than i) and additionally knows the linear coefficients of these polynomials. Since these are all random polynomials of degree t, the values $(x_{i,1}, y_{i,1}, x_{i,2}, y_{i,2})$ are uniformly distributed subject to:

$$\log_g c_i = x_{i,1} + y_{i,1} \log_g h \quad (5)$$
$$\log_g d_i = x_{i,2} + y_{i,2} \log_g h \quad (6)$$
$$\log_u \tilde{e} = x_{i,1} + \alpha x_{i,2} + (\log_u v)\, y_{i,1} + (\log_u v)\, \alpha\, y_{i,2}, \quad (7)$$

where (7) comes from the output of the encryption oracle. If the submitted ciphertext $\langle i, (u', v', w', e') \rangle$ is invalid and $(u', v', w', e') \neq (u, v, \tilde{w}, \tilde{e})$, there are three possibilities:

Case 1. $(u', v', w') = (u, v, \tilde{w})$. In this case, $v' \neq \tilde{v}$ ensures that the decryption oracle will reject.

Case 2. $(u', v', w') \neq (u, v, \tilde{w})$ but $H(u', v', w') = H(u, v, \tilde{w})$. This violates the security of the universal one-way hash family and hence cannot occur with non-negligible probability. See [11].

Case 3. $H(u', v', w') \neq H(u, v, \tilde{w})$. The decryption oracle will reject unless:

$$\log_{u'} e' = x_{i,1} + \alpha' x_{i,2} + (\log_{u'} v')\, y_{i,1} + (\log_{u'} v')\, \alpha'\, y_{i,2}. \quad (8)$$

But (5)–(8) are all linearly independent, from which it follows that the decryption oracle rejects except with probability $1/q$. (As in [11], each rejection further constrains the values $(x_{i,1}, y_{i,1}, x_{i,2}, y_{i,2})$; however, the kth query will be rejected except with probability at most $1/(q - k + 1)$.)

When $j \neq i$, the values $(x_{i,1}, y_{i,1}, x_{i,2}, y_{i,2}, x_{j,1}, y_{j,1}, x_{j,2}, y_{j,2})$ are uniformly distributed subject only to (5)–(7) and:

$$\log_g c_j = x_{j,1} + y_{j,1} \log_g h \quad (9)$$
$$\log_g d_j = x_{j,2} + y_{j,2} \log_g h. \quad (10)$$

Here, we make crucial use of the fact that the adversary has made at most $t - 2$ key exposure requests – had the adversary learned $t - 1$ points on the polynomials, this (along with knowledge of the linear coefficients) would yield additional linear relations (e.g., between $x_{i,1}$ and $x_{j,1}$), and the proof of security would not go through.

If the ciphertext $\langle j, (u', v', w', e') \rangle$ submitted by the adversary is invalid, the decryption oracle will reject unless:

$$\log_{u'} e' = x_{j,1} + \alpha' x_{j,2} + (\log_{u'} v')\, y_{j,1} + (\log_{u'} v')\, \alpha'\, y_{j,2}. \quad (11)$$

Clearly, however, (5)–(7) and (9)–(11) are all linearly independent, from which it follows that the decryption oracle rejects except with probability $1/q$. This completes the proof of $(t - 2, N)$-key-insulated security.

The key to the proof above (informally) is that the adversary learns only $t - 1$ "pieces of information" about the polynomials $f_{x_1}(\cdot), f_{y_1}(\cdot), f_{x_2}(\cdot)$, and

$f_{y_2}(\cdot)$ (i.e., their values at $t-2$ points and their linear coefficients). Hence, before any calls to the decryption oracle have been made, the pair $(x_{i,1}, x_{j,1})$ (for example) is uniformly distributed. The proof of *strong* key-insulated security follows exactly the same arguments given above once we notice that SK^* gives only $t-1$ "pieces of information" as well (i.e., the $t-1$ leading coefficients). We omit further details.

We note that a trivial modification to the scheme achieves $(t-1, N)$-key-insulated security with minimal added complexity: choose random elements $\{\tilde{x}_{1,n}, \tilde{y}_{1,n}\}_{0 \le n \le 2}$, then set $\hat{x}_{1,n} = x_{1,n} + \tilde{x}_{1,n}$ and $\hat{y}_{1,n} = y_{1,n} + \tilde{y}_{1,n}$ for $0 \le n \le 2$. Now, include $\{\tilde{x}_{1,n}, \tilde{y}_{1,n}\}_{0 \le n \le 2}$ with SK^* and store $\{\hat{x}_{1,n}, \hat{y}_{1,n}\}_{0 \le n \le 2}$ as part of SK_0 (and have these values be part of SK_i at all time periods). Key updates are done in the obvious way. Note that SK^* only stores $t-1$ "pieces of information" about the random, degree-t polynomials; furthermore, $t-1$ key exposures only reveal $t-1$ "pieces of information" as well. Thus, a proof of security follows the proof of the above theorem.

Acknowledgment

Shouhuai Xu was partially supported by an NSF grant to the Laboratory for Information Security Technology at George Mason University.

References

1. M. Abdalla and L. Reyzin. A New Forward-Secure Digital Signature Scheme. Asiacrypt'00.
2. M. Abe and M. Kanda. A Key Escrow Scheme with Time-Limited Monitoring for One-Way Communication. ACISP '00.
3. R. Anderson. Invited lecture. ACM CCCS '97.
4. M. Bellare, A. Desai, E. Jokipii, and P. Rogaway. A Concrete Security Treatment of Symmetric Encryption: Analysis of the DES Modes of Operation. FOCS '97.
5. M. Bellare and S.K. Miner. A Forward-Secure Digital Signature Scheme. Crypto '99.
6. G. Blakley and C. Meadows. Security of Ramp Schemes. Crypto '84.
7. D. Boneh and M. Franklin. Identity-Based Encryption from the Weil Pairing. Crypto '01.
8. V. Boyko. On the Security Properties of the OAEP as an All-or-Nothing Transform. Crypto '99.
9. R. Canetti, Y. Dodis, S. Halevi, E. Kushilevitz, and A. Sahai. Exposure-Resilient Functions and All-Or-Nothing-Transforms. Eurocrypt '00.
10. B. Chor, A. Fiat, and M. Naor. Tracing Traitors. Crypto '94.
11. R. Cramer and V. Shoup. A Practical Public-Key Cryptosystem Provably Secure against Adaptive Chosen-Ciphertext Attacks. Crypto '98.
12. A. De Santis, Y. Desmedt, Y. Frankel, and M. Yung. How to Share a Function Securely. STOC 94.
13. Y. Desmedt and Y. Frankel. Threshold cryptosystems. Crypto'89.
14. W. Diffie, P. van Oorschot and M. Wiener. Authentication and Authenticated Key Exchanges. *Designs, Codes and Cryptography*, 2:107–125, 1992.

15. Y. Dodis, J. Katz, S. Xu and M. Yung. Key-Insulated Signature Schemes. Manuscript, 2002.
16. A. Dyachkov and V. Rykov. A Survey of Superimposed Code Theory. In *Problems of Control and Information Theory*, vol. 12, no. 4, 1983.
17. T. El Gamal. A Public-Key Cryptosystem and a Signature Scheme Based on the Discrete Logarithm. *IEEE Transactions of Information Theory*, 31(4): 469–472, 1985.
18. P. Erdos, P. Frankl, and Z. Furedi. Families of Finite Sets in which no Set is Covered by the Union of r Others. In *Israel J. Math.*, 51(1-2): 79–89, 1985.
19. M. Franklin, M. Yung. Communication Complexity of Secure Computation. STOC '92.
20. E. Gafni, J. Staddon, and Y. L. Yin. Efficient Methods for Integrating Traceability and Broadcast Encryption. Crypto '99.
21. M. Girault. Relaxing Tamper-Resistance Requirements for Smart Cards Using (Auto)-Proxy Signatures. CARDIS '98.
22. O. Goldreich, B. Pfitzmann, and R.L. Rivest. Self-Delegation with Controlled Propagation – or – What if You Lose Your Laptop? Crypto '98.
23. S. Goldwasser, S. Micali, and R.L. Rivest. A Digital Signature Scheme Secure Against Adaptive Chosen-Message Attacks. SIAM J. Computing 17(2): 281–308 (1988).
24. P. Indyk. Personal communication.
25. G. Itkis and L. Reyzin. Forward-Secure Signatures with Optimal Signing and Verifying. Crypto '01.
26. S. Jarecki and A. Lysyanskaya. Concurrent and Erasure-Free Models in Adaptively-Secure Threshold Cryptography. Eurocrypt '00.
27. H. Krawczyk. Secret Sharing Made Short. Crypto '93.
28. H. Krawczyk. Simple Forward-Secure Signatures From any Signature Scheme. ACM CCCS '00.
29. R. Kumar, S. Rajagopalan, and A. Sahai. Coding Constructions for Blacklisting Problems without Computational Assumptions. Crypto '99.
30. C.-F. Lu and S.W. Shieh. Secure Key-Evolving Protocols for Discrete Logarithm Schemes. RSA 2002, to appear.
31. T. Malkin, D. Micciancio, and S. Miner. Efficient Generic Forward-Secure Signatures With an Unbounded Number of Time Periods. These proceedings.
32. R. Ostrovsky and M. Yung. How to Withstand Mobile Virus Attacks. PODC '91.
33. T. Pedersen. Non-Interactive and Information-Theoretic Secure Verifiable Secret Sharing. Crypto '91.
34. R. Rivest. All-or-Nothing Encryption and the Package Transform. FSE '97.
35. A. Shamir. How to share a secret. *Comm. ACM*, 22(11):612–613, 1979.
36. W.-G. Tzeng and Z.-J. Tzeng. Robust Key-Evolving Public-Key Encryption Schemes. Available at `http://eprint.iacr.org`.

On the Security of Joint Signature and Encryption

Jee Hea An[1], Yevgeniy Dodis[2], and Tal Rabin[3]

[1] SoftMax Inc., San Diego, USA (Work done while at UCSD)
jeehea@cs.ucsd.edu

[2] Department of Computer Science, New York University, USA
dodis@cs.nyu.edu

[3] IBM T.J. Watson Research Center, USA
talr@watson.ibm.com

Abstract. We formally study the notion of a joint signature and encryption in the public-key setting. We refer to this primitive as *signcryption*, adapting the terminology of [35]. We present two definitions for the security of signcryption depending on whether the adversary is an outsider or a legal user of the system. We then examine generic sequential composition methods of building signcryption from a signature and encryption scheme. Contrary to what recent results in the symmetric setting [5, 22] might lead one to expect, we show that classical "encrypt-then-sign" ($\mathcal{E}t\mathcal{S}$) and "sign-then-encrypt" ($\mathcal{S}t\mathcal{E}$) methods are both *secure* composition methods in the public-key setting.

We also present a new composition method which we call "commit-then-encrypt-and-sign" ($\mathcal{C}t\mathcal{E}\&\mathcal{S}$). Unlike the generic sequential composition methods, $\mathcal{C}t\mathcal{E}\&\mathcal{S}$ applies the expensive signature and encryption operations *in parallel*, which could imply a gain in efficiency over the $\mathcal{S}t\mathcal{E}$ and $\mathcal{E}t\mathcal{S}$ schemes. We also show that the new $\mathcal{C}t\mathcal{E}\&\mathcal{S}$ method elegantly combines with the recent "hash-sign-switch" technique of [30], leading to efficient *on-line/off-line* signcryption.

Finally and of independent interest, we discuss the *definitional* inadequacy of the standard notion of chosen ciphertext (CCA2) security. We suggest a natural and very slight relaxation of CCA2-security, which we call generalized CCA2-security (gCCA2). We show that gCCA2-security suffices for all known uses of CCA2-secure encryption, while no longer suffering from the definitional shortcomings of the latter.

1 Introduction

Signcryption. Encryption and signature schemes are fundamental cryptographic tools for providing privacy and authenticity, respectively, in the public-key setting. Until very recently, they have been viewed as important but *distinct* basic building blocks of various cryptographic systems, and have been designed and analyzed separately. The separation between the two operations can be seen as a natural one as encryption is aimed at providing privacy while signatures are used to enable authentication, and these are two fundamentally different security

L.R. Knudsen (Ed.): EUROCRYPT 2002, LNCS 2332, pp. 83–107, 2002.

goals. Yet clearly, there are many settings where both are needed, perhaps the most basic one is in secure e-mailing, where each message should be authenticated and encrypted. A straightforward solution to offering simultaneously both privacy and authenticity might be to compose the known solutions of each of the two components. But given that the combination of the two security goals is so common, and in fact a basic task, it stands to reason that a tailored solution for the combination should be given. Indeed, a cryptographic tool providing both authenticity and privacy has usually been called an *authenticated encryption*, but was mainly studied in the symmetric setting [6,5,22]. This paper will concentrate on the corresponding study in the public key setting, and will use the term *signcryption* to refer to a "joint signature and encryption". We remark that this term was originally introduced and studied by Zheng in [35] with the primary goal of reaching greater efficiency than when carrying out the signature and encryption operations separately. As we will argue shortly, efficiency is only one (albeit important) concern when designing a secure joint signature and encryption. Therefore, we will use the term "signcryption" for *any* scheme achieving both privacy and authenticity in the public key setting, irrespective of its performance, as long as it satisfies a formal definition of security we develop in this paper. Indeed, despite presenting some security arguments, most of the initial work on signcryption [35,36,26,19] lacked formal definitions and analysis. This paper will provide such a formal treatment, as well as give new general constructions of signcryption.

Signcryption as a Primitive? Before devoting time to the definition and design of (additional) signcryption schemes one must ask if there is a need for defining signcryption as a separate primitive. Indeed, maybe one should forgo this notion and always use a simple composition of a signature and encryption? Though we show in the following that these compositions, in many instances, yield the desired properties, we still claim that a separate notion of signcryption is extremely useful. This is due to several reasons. First, under certain definitions of security (i.e., so called CCA2-security as explained in Section 8), the straightforward composition of a secure signature and encryption does *not* necessarily yield a secure signcryption. Second, as we show in Section 3, there are quite subtle issues with respect to signcryption – especially in the public-key setting – which need to be captured in a formal definition. Third, there are other interesting constructions for signcryption which do not follow the paradigm of sequentially composing signatures and encryption. Fourth, designing tailored solutions might yield efficiency (which was the original motivation of Zheng [35]). Finally, the usage of signcryption as a primitive might conceptually simplify the design of complex protocols which require both privacy and authenticity.

Summarizing the above discussion, we believe that the study of signcryption *as a primitive* is important and can lead to very useful, general as well as specific, paradigms for achieving privacy and authenticity at the same time.

Our Results. This paper provides a formal treatment of signcryption and analyzes several general constructions for this primitive. In particular, we note that the problem of defining signcryption in the public key setting is more involved than the corresponding task in the symmetric setting studied by [5,22], due to the asymmetric nature of the former. For example, full-fledged signcryption needs to be defined in the *multi-user* setting, where some issues with user's identities need to be addressed. In contrast, authenticated encryption in the symmetric setting can be fully defined in a simpler *two-user* setting. Luckily, we show that it suffices to design and analyze signcryption schemes in the two-user setting as well, by giving a generic transformation to the multi-user setting.

We give two definitions for security of signcryption depending on whether the adversary is an outsider or a legal user of the network (i.e., either the sender or the receiver). In both of these settings, we show that the common "encrypt-then-sign" ($\mathcal{E}t\mathcal{S}$) and "sign-then-encrypt" ($\mathcal{S}t\mathcal{E}$) methods in fact yield a *secure* signcryption, provided an *appropriate* definition of security is used. Moreover, when the adversary is an outsider, these composition methods can actually provide *stronger* privacy or authenticity properties for the resulting signcryption scheme than the assumed security properties on the base encryption or signature scheme. Specifically, the security of the base signature scheme can help amplify the privacy of $\mathcal{E}t\mathcal{S}$, while the security of the base encryption scheme can do the same to the authenticity of $\mathcal{S}t\mathcal{E}$. We remark that these possibly "expected" results are nevertheless somewhat surprising in light of recent "negative" indications from the symmetric setting [5,22], and illustrate the need for rigorous definitions for security of signcryption.

In addition, we present a novel construction of signcryption, which we call "commit-then-encrypt-and-sign" ($\mathcal{C}t\mathcal{E}\&\mathcal{S}$). Our scheme is a general way to construct signcryption from any signature and encryption schemes, while utilizing in addition a commitment scheme. This method is quite different from the obvious sequential composition paradigm. Moreover, unlike the previous sequential methods, the $\mathcal{C}t\mathcal{E}\&\mathcal{S}$ method applies the expensive signature and encryption operations *in parallel*, which could imply a gain in efficiency. We also show that our construction naturally leads to a very efficient way to implement *off-line signcryption*, where the sender can prepare most of the authenticated ciphertext in advance and perform very little on-line computation.

Finally and of independent interest, we discuss the *definitional* inadequacy of the standard notion of chosen ciphertext (CCA2) security [13,4]. Motivated by our applications to signcryption, we show that the notion of CCA2-security is syntactically ill-defined, and leads to artificial examples of "intuitively CCA2-secure" schemes which do not meet the formal definition (such observations were also made by [8,9]). We suggest a natural and very slight relaxation of CCA2-security, which we call *generalized* CCA2-*security* (gCCA2). We show that gCCA2-security suffices for all known uses of CCA2-secure encryption, while no longer suffering from the definitional shortcomings of the latter.

Related Work. The initial works on signcryption [35,36,26,19] designed several signcryption schemes, whose "security" was informally based on various number-theoretic assumptions. Only recently (and independently of our work) Baek et al. [3] showed that the original scheme of Zheng [35] (based on shortened ElGamal signatures) can be shown secure in the random oracle model under the gap Diffie-Hellman assumption.

We also mention the works of [34,29], which used Schnorr signature to amplify the security of ElGamal encryption to withstand a chosen ciphertext attack. However, the above works concentrate on providing privacy, and do not provide authenticity, as required by our notion of signcryption.

Recently, much work has been done about authenticated encryption in the symmetric (private-key) setting. The first formalizations of authenticated encryption in the symmetric setting were done by [21,6,5]. The works of [5,22] discuss the security of generic composition methods of a (symmetric) encryption and a message authentication code (MAC). In particular, a lot of emphasis in these works is given to the study of sufficient conditions under which a given composition method can *amplify* (rather than merely preserve) the privacy property of a given composition method from the chosen plaintext (CPA) to the chosen ciphertext (CCA2) level. From this perspective, the "encrypt-then-mac" method – which always achieves such an amplification due to a "strongly unforgeable" MAC – was found generically preferable to the "mac-then-encrypt" method, which does so only in specific (albeit very useful) cases [22]. In contrast, An and Bellare [1] study a symmetric question of under which conditions a "good" privacy property on the base encryption scheme can help amplify the authenticity property in the "mac-then-encrypt" (or "encrypt-with-redundancy") method. On a positive side, they found that chosen ciphertext security on the base encryption scheme is indeed sufficient for that purpose. As we shall see in Section 4, all these results are very related to our results about "sign-then-encrypt" and "encrypt-then-sign" methods for signcryption when the adversary is an "outsider".

Another related paradigm for building authenticated encryption is the "encode-then-encipher" method of [6]: add randomness and redundancy, and then encipher (i.e., apply a pseudorandom permutation) rather than encrypt. Even though a strong pseudorandom permutation is often more expensive than encryption, [6] shows that very simple *public* redundancy functions are sufficient – in contrast to the "encrypt-with-redundancy" method, where no public redundancy can work [1].

Finally, we mention recently designed modes of operations for block ciphers that achieve both privacy and authenticity in the symmetric setting: RFC mode of [21], IACBC and IAPM modes of [20], OCB mode of [28], and SNCBC mode of [1].

2 Definitions

In this section we briefly review the (public-key) notions of encryption, signature and commitment schemes. In addition, we present our extended definition for CCA2.

2.1 Encryption

Syntax. An encryption scheme consists of three algorithms: $\mathcal{E} = (\mathsf{Enc\text{-}Gen}, \mathsf{Enc}, \mathsf{Dec})$. $\mathsf{Enc\text{-}Gen}(1^k)$, where k is the security parameter, outputs a pair of keys $(\mathsf{EK}, \mathsf{DK})$. EK is the encryption key, which is made public, and DK is the decryption key which is kept secret. The randomized encryption algorithm Enc takes as input a key EK and a message m from the associated message space $\mathcal{M}$, and internally flips some coins and outputs a ciphertext e; we write $e \leftarrow \mathsf{Enc}_{\mathsf{EK}}(m)$. For brevity, we will usually omit EK and write $e \leftarrow \mathsf{Enc}(m)$. The deterministic decryption algorithm Dec takes as input the ciphertext e, the secret key DK, and outputs some message $m \in \mathcal{M}$, or $\perp$ in case e was "invalid". We write $m \leftarrow \mathsf{Dec}(e)$ (again, omitting DK). We require that $\mathsf{Dec}(\mathsf{Enc}(m)) = m$, for any $m \in \mathcal{M}$.

Security of Encryption. When addressing the security of the schemes, we deal with two issues: what we want to achieve (security goal) and what are the capabilities of the adversary (attack model). In this paper we will talk about the most common security goal: indistinguishability of ciphertexts [16], which we will denote by IND. A related notion of *non-malleability* will be briefly discussed in Section 8.

Intuitively, indistinguishability means that given a randomly selected public key, no PPT (probabilistic polynomial time) adversary $\mathcal{A}$ can distinguish encryptions of any two messages m_0, m_1 chosen by $\mathcal{A}$: $\mathsf{Enc}(m_0) \approx \mathsf{Enc}(m_1)$. Formally, we require that for any PPT $\mathcal{A}$, which runs in two stages, find and guess, we have

$$\Pr\left[\, b=\tilde{b} \;\middle|\; \begin{array}{c} (\mathsf{EK},\mathsf{DK}) \leftarrow \mathsf{Enc\text{-}Gen}(1^k),\ (m_0, m_1, \alpha) \leftarrow \mathcal{A}(\mathsf{EK}, \mathsf{find}), \\ b \stackrel{R}{\leftarrow} \{0,1\},\ e \leftarrow \mathsf{Enc}_{\mathsf{EK}}(m_b),\ \tilde{b} \leftarrow \mathcal{A}(e, \alpha, \mathsf{guess}) \end{array} \right] \le \frac{1}{2} + \mathsf{negl}(k)$$

Here and elsewhere $\mathsf{negl}(k)$ is some negligible function in the security parameter k, and α is some internal state information $\mathcal{A}$ saves and uses in the two stages.

We now turn to the second issue of security of encryption – the attack model. We consider three types of attack: CPA, $\mathsf{CCA1}$ and $\mathsf{CCA2}$. Under the *chosen plaintext* (or CPA) attack, the adversary is not given any extra capabilities other than encrypting messages using the public encryption key. A more powerful type of *chosen ciphertext* attack gives $\mathcal{A}$ access to the decryption oracle, namely the ability to decrypt arbitrary ciphertexts of its choice. The first of this type of attack is the *lunch-time* ($\mathsf{CCA1}$) attack [27], which gives access only in the find stage (i.e., before the challenge ciphertext e is given). The second is $\mathsf{CCA2}$ on which we elaborate in the following.

$\mathsf{CCA2}$ *Attacks.* The *adaptive chosen ciphertext attack* [13] ($\mathsf{CCA2}$) gives access to the decryption oracle in the guess stage as well. As stated, the $\mathsf{CCA2}$ attack does not make sense since $\mathcal{A}$ can simply ask to decrypt the challenge e. Therefore, we need to restrict the class of ciphertexts e' that $\mathcal{A}$ can give to the decryption oracle in the guess stage. The minimal restriction is to have $e' \neq e$, which is the way the $\mathsf{CCA2}$ attack is usually defined. As we will argue in Section 8, stopping at this minimal (and needed) restriction in turn restricts the class of encryption schemes that we intuitively view as being "secure". In particular, it is not robust to syntactic changes in the encryption (e.g., appending a harmless random bit to a secure encryption suddenly makes it "insecure" against CCA2). Leaving further discussion to Section 8, we now define a special case of the CCA2 attack which does not suffer from the above syntactic limitations and suffices for all the uses of the CCA2-secure encryption we are aware of.

We first generalize the CCA2 attack with respect to some equivalence relation $\mathcal{R}(\cdot,\cdot)$ on the ciphertexts. $\mathcal{R}$ is defined as part of the encryption scheme, it can depend on the public key EK, but must have the following property: if $\mathcal{R}(e_1, e_2) = \mathit{true} \Rightarrow \mathsf{Dec}(e_1) = \mathsf{Dec}(e_2)$. We call such $\mathcal{R}$ *decryption-respecting.* Now $\mathcal{A}$ is forbidden to ask any e' equivalent to e, i.e. $\mathcal{R}(e, e') = \mathit{true}$. Since $\mathcal{R}$ is reflexive, this at least rules out e, and since $\mathcal{R}$ is decryption-respecting, it only restricts ciphertexts that decrypt to the same value as the decryption of e (i.e. m_b). We note that the usual CCA2 attack corresponds to the equality relation. Now we say that the encryption scheme is secure against *generalized* CCA2 (or gCCA2) if there *exists* some efficient decryption-respecting relation $\mathcal{R}$ w.r.t. which it is CCA2-secure. For example, appending a harmless bit to gCCA2-secure encryption or doing other easily recognizable manipulation still leaves it gCCA2-secure.

We remark that the notion of gCCA2-security was recently proposed in [32] (under the name *benign malleability*) for the ISO public key encryption standard. In the private-key setting, [22] uses equivalences relations to define "loose ciphertext unforgeability".

2.2 Signatures

Syntax. A signature scheme consists of three algorithms: $\mathcal{S} = (\mathsf{Sig\text{-}Gen}, \mathsf{Sig}, \mathsf{Ver})$. $\mathsf{Sig\text{-}Gen}(1^k)$, where k is the security parameter, outputs a pair of keys (SK, VK). SK is the signing key, which is kept secret, and VK is the verification key which is made public. The randomized signing algorithm Sig takes as input a key SK and a message m from the associated message space $\mathcal{M}$, internally flips some coins and outputs a signature s; we write $s \leftarrow \mathsf{Sig}_{\mathsf{SK}}(m)$. We will usually omit SK and write $s \leftarrow \mathsf{Sig}(m)$. Wlog, we will assume that the message m can be determined from the signature s (e.g., is part of it), and write $m = \mathsf{Msg}(s)$ to denote the message whose signature is s. The deterministic verification algorithm Ver takes as input the signature s, the public key VK, and outputs the answer a which is either succeed (signature is valid) or fail (signature is invalid). We write $a \leftarrow \mathsf{Ver}(s)$ (again, omitting VK). We require that $\mathsf{Ver}(\mathsf{Sig}(m)) = \mathsf{succeed}$, for any $m \in \mathcal{M}$.

Security of Signatures. As with the encryption, the security of signatures addresses two issues: what we want to achieve (security goal) and what are the capabilities of the adversary (attack model). In this paper we will talk about the the most common security goal: *existential unforgeability* [17], denoted by UF. This means that any PPT adversary $\mathcal{A}$ should have a negligible probability of generating a valid signature of a "new" message. To clarify the meaning of "new", we will consider the following two attack models. In the *no message attack* (NMA), $\mathcal{A}$ gets no help besides VK. In the *chosen message attack* (CMA), in addition to VK, the adversary $\mathcal{A}$ gets full access to the signing oracle Sig, i.e. $\mathcal{A}$ is allowed to query the signing oracle to obtain valid signatures $s_1, \ldots, s_n$ of arbitrary messages $m_1, \ldots, m_n$ adaptively chosen by $\mathcal{A}$ (notice, NMA corresponds to $n = 0$). Naturally, $\mathcal{A}$ is considered successful only if it forges a valid signature s of a message m not queried to signing oracle: $m \notin \{m_1 \ldots m_n\}$. We denote the resulting security notions by UF-NMA and UF-CMA, respectively.

We also mention a slightly stronger type of unforgeability called *strong unforgeability*, denoted sUF. Here $\mathcal{A}$ should not only be unable to generate a signature of a "new" message, but also be unable to generate even a different signature of an already signed message, i.e. $s \notin \{s_1, \ldots, s_n\}$. This only makes sense for the CMA attack, and results in a security notion we denote by sUF-CMA.

2.3 Commitment

Syntax. A (non-interactive) commitment scheme consists of three algorithms: $\mathcal{C} = (\mathsf{Setup}, \mathsf{Commit}, \mathsf{Open})$. The setup algorithm $\mathsf{Setup}(1^k)$, where k is the security parameter, outputs a public commitment key CK (possibly empty, but usually consisting of public parameters for the commitment scheme). Given a message m from the associated message space $\mathcal{M}$ (e.g., $\{0,1\}^k$), $\mathsf{Commit}_{\mathsf{CK}}(m; r)$ (computed using the public key CK and additional randomness r) produces a commitment pair (c, d), where c is the *commitment* to m and d is the *decommitment*. We will usually omit CK and write $(c, d) \leftarrow \mathsf{Commit}(m)$. Sometimes we will write $c(m)$ (resp. $d(m)$) to denote the commitment (resp. decommitment) part of a randomly generated (c, d). The last (deterministic) algorithm $\mathsf{Open}_{\mathsf{CK}}(c, d)$ outputs m if (c, d) is a *valid* pair for m (i.e. could have been generated by $\mathsf{Commit}(m)$), or $\perp$ otherwise. We require that $\mathsf{Open}(\mathsf{Commit}(m)) = m$ for any $m \in \mathcal{M}$.

Security of Commitment. Regular commitment schemes have two security properties:

Hiding. No PPT adversary can distinguish the commitments to any two message of its choice: $c(m_1) \approx c(m_2)$. That is, $c(m)$ reveals "no information" about m. Formally, for any PPT $\mathcal{A}$ which runs in two stages, find and guess, we have

$$\Pr\left[b = \tilde{b} \,\middle|\, \begin{array}{c} \mathsf{CK} \leftarrow \mathsf{Setup}(1^k), (m_0, m_1, \alpha) \leftarrow \mathcal{A}(\mathsf{CK}, \mathsf{find}), \\ b \overset{R}{\leftarrow} \{0,1\}, (c, d) \leftarrow \mathsf{Commit}_{\mathsf{CK}}(m_b), \tilde{b} \leftarrow \mathcal{A}(c;\ \alpha, \mathsf{guess}) \end{array}\right] \leq \frac{1}{2} + \mathsf{negl}(k)$$

Binding. Having the knowledge of CK, it is computationally hard for the adversary $\mathcal{A}$ to come up with c, d, d' such that (c, d) and (c, d') are valid commitment pairs for m and m', but $m \neq m'$ (such a triple c, d, d' is said to cause a *collision*). That is, $\mathcal{A}$ cannot find a value c which it can open in two different ways.

Relaxed Commitments. We will also consider *relaxed* commitment schemes, where the (strict) binding property above is replaced by the **Relaxed Binding** property: for any PPT adversary $\mathcal{A}$, having the knowledge of CK, it is computationally hard for $\mathcal{A}$ to come up with a message m, such that when $(c, d) \leftarrow \mathsf{Commit}(m)$ is generated, $\mathcal{A}(c, d, \mathsf{CK})$ produces, with non-negligible probability, a value d' such that (c, d') is a valid commitment to some $m' \neq m$. Namely, $\mathcal{A}$ cannot find a collision using a *randomly generated* $c(m)$, even for m of its choice.

To justify this distinction, first recall the concepts of collision-resistant hash function (CRHF) families and universal one-way hash function (UOWHF) families. For both concepts, it is hard to find a colliding pair $x \neq x'$ such that $H(x) = H(x')$, where H is a function randomly chosen from the corresponding family. However, with CRHF, we first select the function H, and for UOWHF the adversary has to select x before H is given to it. By the result of Simon [33], UOWHF's are strictly weaker primitive than CRHF (in particular, they can be built from regular one-way functions [24]). We note two classical results about (regular) commitment schemes: the construction of such a scheme by [11,18], and the folklore "hash-then-commit" paradigm (used for committing to long messages by hashing them first). Both of these results require the use of CRHF's, and it is easy to see that UOWHF's are not sufficient to ensure (strict) binding for either one of them. On the other hand, it is not very hard to see that UOWHF's suffice to ensure relaxed binding in both cases. Hence, basing some construction on relaxed commitments (as we will do in Section 5) has its merits over regular commitments.

Trapdoor Commitments. We also define a very useful class of commitment schemes, known as (non-interactive) *trapdoor commitments* [7] or *chameleon hash functions* [23]. In these schemes the setup algorithm $\mathsf{Setup}(1^k)$ outputs a pair of keys $(\mathsf{CK}, \mathsf{TK})$. That is, in addition to the public commitment key CK, it also produces a *trapdoor* key TK. Like regular commitments, trapdoor commitments satisfy the hiding property and (possibly relaxed) binding properties. Additionally, they have an efficient switching algorithm Switch, which allows one to find arbitrary collisions using the trapdoor key TK.

Given any commitment (c, d) to some message m and *any* message m', $\mathsf{Switch}_{\mathsf{TK}}((c, d), m')$ outputs a valid commitment pair (c, d') to m' (note, c is the same!). Moreover, having the knowledge of CK, it is computationally hard to come up with two messages m, m' such that the adversary can distinguish $\mathsf{Commit}_{\mathsf{CK}}(m')$ (random commitment pair for m') from $\mathsf{Switch}_{\mathsf{TK}}(\mathsf{Commit}_{\mathsf{CK}}(m), m')$ (faked commitment pair for m' obtained from a random pair for m).

We note that the trapdoor collisions property is much stronger (and easily implies) the hiding property (since the switching algorithm does not change $c(m)$). Moreover, the hiding property is *information-theoretic.* We also note that very efficient trapdoor commitment schemes exist based on factoring [23,30] or discrete log [23,7]. In particular, the switching function requires just one modulo addition and one modulo multiplication for the discrete log based solution. Less efficient constructions based on more general assumptions are known as well [23].

3 Definition of Signcryption in the Two-User Setting

The definition of signcryption is a little bit more involved than the corresponding definition of authenticated encryption in the symmetric setting. Indeed, in the *symmetric* setting, we only have one specific pair of users who (1) share a single key; (2) trust each other; (3) "know who they are"; and (4) care about being protected from "the rest of the world". In contrast, in the *public* key setting each user independently publishes its public keys, after which it can send/receive messages to/from any other user. In particular, (1) each user should have an explicit identity (i.e., its public key); (2) each signcryption has to explicitly contain the (presumed) identities of the sender S and the receiver R; (3) each user should be protected from every other user. This suggests that signcryption should be defined in the *multi-user* setting. Luckily, we show that we can first define and study the crucial properties of signcryption in the stand-alone two-user setting, and then add identities to our definitions and constructions to achieve the full-fledged multi-user security. Thus, in this section we start with a simple two-user setting, postponing the extension to multi-user setting to Section 7.

Syntax. A signcryption scheme $\mathcal{SC}$ consists of three algorithms: $\mathcal{SC} = (\mathsf{Gen}, \mathsf{SigEnc}, \mathsf{VerDec})$. The algorithm $\mathsf{Gen}(1^k)$, where k is the security parameter, outputs a pair of keys $(\mathsf{SDK}, \mathsf{VEK})$. SDK is the user's sign/decrypt key, which is kept secret, and VEK the user's verify/encrypt key, which is made public. Note, that in the signcryption setting *all* participating parties need to invoke Gen. For a user P, denote its keys by SDK_P and VEK_P. The randomized *signcryption* (sign/encrypt) algorithm SigEnc takes as input the sender S's secret key SDK_S and the receiver R's public key VEK_R and a message m from the associated message space $\mathcal{M}$, and internally flips some coins and outputs a signcryption (ciphertext) u; we write $u \leftarrow \mathsf{SigEnc}(m)$ (omitting SDK_S, VEK_R). The deterministic *de-signcryption* (verify/decrypt) algorithm VerDec takes as input the signcryption (ciphertext) e, the receiver R's secret key SDK_R and the sender S's public key VEK_S, and outputs $m \in \mathcal{M} \cup \{\bot\}$, where $\bot$ indicates that the message was not encrypted or signed properly. We write $m \leftarrow \mathsf{VerDec}(u)$ (again, omitting the keys). We require that $\mathsf{VerDec}(\mathsf{SigEnc}(m)) = m$, for any $m \in \mathcal{M}$.

Security of Signcryption. Fix the sender S and the receiver R. Intuitively, we would like to say that S's authenticity is protected, and R's privacy is protected. We will give two formalizations of this intuition. The first one assumes that the adversary $\mathcal{A}$ is an outsider who only knows the public information $pub = (\mathsf{VEK}_R, \mathsf{VEK}_S)$. We call such security *Outsider security*. The second, stronger notion, protects S's authenticity even against R, and R's privacy even against S. Put in other words, it assumes that the adversary $\mathcal{A}$ is a legal user of the system. We call such security *Insider security.*

Outsider Security. We define it against the strongest security notions on the signature (analogs of UF-CMA or sUF-CMA) and encryption (analogs of IND-gCCA2 or IND-CCA2), and weaker notions could easily be defined as well. We assume that the adversary $\mathcal{A}$ has the public information $pub = (\mathsf{VEK}_S, \mathsf{VEK}_R)$. It also has oracle access to the functionalities of both S and R. Specifically, it can mount a chosen message attack on S by asking S to produce signcryption u of an arbitrary message m. In other words, $\mathcal{A}$ has access to the *signcryption oracle*. Similarly, it can mount a chosen ciphertext attack on R by giving R any candidate signcryption u and receiving back the message m (where m could be $\perp$), i.e. $\mathcal{A}$ has access to the *de-signcryption oracle*. Notice, $\mathcal{A}$ cannot by itself run either the signcryption or the de-signcryption oracles due to the lack of corresponding secret keys SDK_S and SDK_R.

To break the UF-CMA security of the signcryption scheme, $\mathcal{A}$ has to come up with a *valid* signcryption u of a "new" message m, which it did not ask S to signcrypt earlier (notice, $\mathcal{A}$ is not required to "know" m when producing u). The scheme is *Outsider-secure* in the UF-CMA sense if any PPT $\mathcal{A}$ has a negligible chance of succeeding. (For sUF-CMA, $\mathcal{A}$ only has to produce u which was not returned by S earlier.)

To break the indistinguishability of the signcryption scheme, $\mathcal{A}$ has to come up with two messages m_0 and m_1. One of these will be signcrypted at random, the corresponding signcryption u will be given to $\mathcal{A}$, and $\mathcal{A}$ has to guess which message was signcrypted. To succeed in the CCA2 attack, $\mathcal{A}$ is only disallowed to ask R to de-signcrypt the challenge u. For gCCA2 attack, similarly to the encryption scenario, we first define CCA2 attack against a given efficient decryption-respecting relation $\mathcal{R}$ (which could depend on $pub = (\mathsf{VEK}_R, \mathsf{VEK}_S)$ but not on any of the secret keys). As before, decryption-respecting means that $\mathcal{R}(u, u') = \mathit{true} \Rightarrow \mathsf{VerDec}(u) = \mathsf{VerDec}(u')$. Thus, CCA2 attack w.r.t. $\mathcal{R}$ disallows $\mathcal{A}$ to de-signcrypt any u' equivalent to the challenge u. Now, for Outsider-security against CCA2 w.r.t. $\mathcal{R}$, we require $\Pr[\mathcal{A} \text{ succeeds}] \leq \frac{1}{2} + \mathsf{negl}(k)$. Finally, the scheme is *Outsider-secure* in the IND-gCCA2 sense if it is Outsider-secure against CCA2 w.r.t. *some* efficient decryption-respecting $\mathcal{R}$.

Insider Security. We could define Insider security in a similar manner by defining the capabilities of $\mathcal{A}$ and its goals. However, it is much easier to use *already existing* security notions for signature and encryption schemes. Moreover, this will capture the intuition that "signcryption = signature + encryption". More precisely, given any signcryption scheme $\mathcal{SC} = (\mathsf{Gen}, \mathsf{SigEnc}, \mathsf{VerDec})$, we define the corresponding *induced* signature scheme $\mathcal{S} = (\mathsf{Sig\text{-}Gen}, \mathsf{Sig}, \mathsf{Ver})$ and encryption scheme $\mathcal{E} = (\mathsf{Enc\text{-}Gen}, \mathsf{Enc}, \mathsf{Dec})$.

- **Signature $\mathcal{S}$.** The generation algorithm Sig-Gen runs $\mathsf{Gen}(1^k)$ twice to produce two key pairs $(\mathsf{SDK}_S, \mathsf{VEK}_S)$ and $(\mathsf{SDK}_R, \mathsf{VEK}_R)$. Let $pub = \{\mathsf{VEK}_S, \mathsf{VEK}_R\}$ be the public information. We set the signing key to $\mathsf{SK} = \{\mathsf{SDK}_S, pub\}$, and the verification key to $\mathsf{VK} = \{\mathsf{SDK}_R, pub\}$. Namely, the public verification key (available to the adversary) *contains the secret key of the receiver R*. To sign a message m, $\mathsf{Sig}(m)$ outputs $u = \mathsf{SigEnc}(m)$, while the verification algorithm $\mathsf{Ver}(u)$ runs $m \leftarrow \mathsf{VerDec}(u)$ and outputs succeed iff $m \neq \perp$. We note that the verification is indeed polynomial time since VK includes SDK_R.
- **Encryption $\mathcal{E}$.** The generation algorithm Enc-Gen runs $\mathsf{Gen}(1^k)$ twice to produce two key pairs $(\mathsf{SDK}_S, \mathsf{VEK}_S)$ and $(\mathsf{SDK}_R, \mathsf{VEK}_R)$. Let $pub = \{\mathsf{VEK}_S,$

$\mathsf{VEK}_R\}$ be the public information. We set the encryption key to $\mathsf{EK} = \{\mathsf{SDK}_S, pub\}$, and the decryption key to $\mathsf{DK} = \{\mathsf{SDK}_R, pub\}$. Namely, the public encryption key (available to the adversary) *contains the secret key of the sender* S. To encrypt a message m, $\mathsf{Enc}(m)$ outputs $u = \mathsf{SigEnc}(m)$, while the decryption algorithm $\mathsf{Dec}(u)$ simply outputs $\mathsf{VerDec}(u)$. We note that the encryption is indeed polynomial time since EK includes SDK_S.

We say that the signcryption is *Insider-secure* against the corresponding attack (e.g. gCCA2/CMA) on the privacy/authenticity property, if the corresponding induced encryption/signature is secure against the same attack.[1] We will aim to satisfy IND-gCCA2-security for encryption, and UF-CMA-security for signatures.

Should We Require Non-repudiation? We note that the conventional notion of digital signatures supports *non-repudiation*. Namely, the receiver R of a correctly generated signature s of the message m can hold the sender S responsible to the contents of m. Put differently, s is unforgeable and publicly verifiable. On the other hand, non-repudiation does not *automatically* follow from the definition of signcryption. Signcryption only allows the *receiver* to be convinced that m was sent by S, but does not necessarily enable a third party to verify this fact.

We believe that non-repudiation should not be part of the *definition* of signcryption security, but we will point out which of our schemes achieves it. Indeed, non-repudiation might be needed in some applications, while explicitly undesirable in others (e.g., this issue is the essence of undeniable [10] and chameleon [23] signature schemes).

Insider vs. Outsider Security. We illustrate some of the differences between Insider and Outsider security. For example, Insider-security for authenticity implies non-repudiation "in principle". Namely, non-repudiation is certain at least when the receiver R is willing to reveal its secret key SDK_R (since this induces a regular signature scheme), or may be possible by other means (like an appropriate zero-knowledge proof). In contrast, Outsider-security leaves open the possibility that R can generate – using its secret key – valid signcryptions of messages that were not actually sent by S. In such a case, non-repudiation cannot be achieved no matter what R does.

Despite the above issues, however, it might still seem that the distinction between Insider- and Outsider-security is a bit contrived, especially for privacy. Intuitively, the Outsider-security protects the privacy of R when talking to S from outside intruders, who do not know the secret key of S. On the other hand, Insider-security assumes that the sender S *is* the intruder attacking the privacy of R. But since S is the *only* party that can send valid signcryptions from S to R, this seems to make little sense. Similarly for authenticity, if non-repudiation is *not* an issue, then Insider-security seems to make little sense; as it assumes that R

[1] One small technicality for the gCCA2-security. Recall, the equivalence relation $\mathcal{R}$ can depend on the public encryption key – in this case $\{\mathsf{SDK}_S, pub\}$. We strengthen this and allow it to depend only on pub (i.e. disallow the dependence on sender's secret key SDK_S).

is the intruder attacking the authenticity of S, and simultaneously the *only* party that needs to be convinced of the authenticity of the (received) data. And, indeed, in many settings *Outsider-security might be all one needs* for privacy and/or authenticity. Still, there are some cases where the extra strength of the Insider-security might be important. We give just one example. Assume an adversary $\mathcal{A}$ happens to steal the key of S. Even though now $\mathcal{A}$ can send fake messages "from S to R", we still might not want $\mathcal{A}$ to understand previous (or even future) recorded signcryptions sent from honest S to R. Insider-security will guarantee this fact, while the Outsider-security might not.

Finally, we note that *achieving* Outsider-security could be significantly easier than Insider-security. One such example will be seen in Theorems 2 and 3. Other examples are given in [2], who show that authenticated encryption in the *symmetric setting* could be used to build Outsider-secure signcryption which is not Insider-secure. To summarize, one should carefully examine if one really needs the extra guarantees of Insider-security.

4 Two Sequential Compositions of Encryption and Signature

In this section, we will discuss two methods of constructing signcryption schemes that are based on sequential generic composition of encryption and signature: encrypt-then-sign ($\mathcal{E}t\mathcal{S}$) and sign-then-encrypt ($\mathcal{S}t\mathcal{E}$).

Syntax. Let $\mathcal{E} = (\mathsf{Enc\text{-}Gen}, \mathsf{Enc}, \mathsf{Dec})$ be an encryption scheme and $\mathcal{S} = (\mathsf{Sig\text{-}Gen}, \mathsf{Sig}, \mathsf{Ver})$ be a signature scheme. Both $\mathcal{E}t\mathcal{S}$ and $\mathcal{S}t\mathcal{E}$ have the same generation algorithm $\mathsf{Gen}(1^k)$. It runs $(\mathsf{EK}, \mathsf{DK}) \leftarrow \mathsf{Enc\text{-}Gen}(1^k)$, $(\mathsf{SK}, \mathsf{VK}) \leftarrow \mathsf{Sig\text{-}Gen}(1^k)$ and sets $\mathsf{VEK} = (\mathsf{VK}, \mathsf{EK})$, $\mathsf{SDK} = (\mathsf{SK}, \mathsf{DK})$. To describe the signcryptions from sender S to receiver R more compactly, we use the shorthands $\mathsf{Sig}_S(\cdot)$, $\mathsf{Enc}_R(\cdot)$, $\mathsf{Ver}_S(\cdot)$ and $\mathsf{Dec}_R(\cdot)$ indicating whose keys are used but omitting which specific keys are used, since the latter is obvious (indeed, Sig_S always uses SK_S, Enc_R – EK_R, Ver_S – VK_S and Dec_R – DK_R).

Now, we define "encrypt-then-sign" scheme $\mathcal{E}t\mathcal{S}$ by $u \leftarrow \mathsf{SigEnc}(m;(\mathsf{SK}_S,\mathsf{EK}_R)) = \mathsf{Sig}_S(\mathsf{Enc}_R(m))$. To de-signcrypt u, we let $\tilde{m} = \mathsf{Dec}_R(\mathsf{Msg}(u))$ provided $\mathsf{Ver}_S(u) = \mathsf{succeed}$, and $\tilde{m} = \perp$ otherwise. We then define $\mathsf{VerDec}(u; (\mathsf{DK}_R, \mathsf{VK}_S)) = \tilde{m}$. Notice, we do not mention $(\mathsf{EK}_S, \mathsf{DK}_S)$ and $(\mathsf{SK}_R, \mathsf{VK}_R)$, since they are not used to send the message from S to R. Similarly, we define "sign-then-encrypt" scheme $\mathcal{S}t\mathcal{E}$ by $u \leftarrow \mathsf{SigEnc}(m; (\mathsf{SK}_S, \mathsf{EK}_R)) = \mathsf{Enc}_R(\mathsf{Sig}_S(m))$. To de-signcrypt u, we let $s = \mathsf{Dec}_R(u)$, and set $\tilde{m} = \mathsf{Msg}(s)$ provided $\mathsf{Ver}_S(s) = \mathsf{succeed}$, and $\tilde{m} = \perp$ otherwise. We then define $\mathsf{VerDec}(u; (\mathsf{DK}_R, \mathsf{VK}_S)) = \tilde{m}$.

Insider-Security. We now show that both $\mathcal{E}t\mathcal{S}$ and $\mathcal{S}t\mathcal{E}$ are secure composition paradigms. That is, they *preserve* (in terms of Insider-security) or even *improve* (in terms of Outsider-security) the security properties of $\mathcal{E}$ and $\mathcal{S}$. We start with Insider-security.

Theorem 1. *If $\mathcal{E}$ is* IND-gCCA2*-secure, and $\mathcal{S}$ is* UF-CMA*-secure, then $\mathcal{E}t\mathcal{S}$ and $\mathcal{S}t\mathcal{E}$ are both* IND-gCCA2*-secure and* UF-CMA*-secure in the Insider-security model.*

The proof of this result is quite simple (and is omitted due to space limitations). However, we remark the crucial use of gCCA2-security when proving the security of $\mathcal{E}t\mathcal{S}$. Indeed, we can call two signcryptions u_1 and u_2 equivalent for $\mathcal{E}t\mathcal{S}$, if each u_i is a valid signature (w.r.t. $\mathcal{S}$) of $e_i = \mathsf{Msg}(u_i)$, and e_1 and e_2 are equivalent (e.g., equal) w.r.t. to the equivalence relation of $\mathcal{E}$. In other words, a *different* signature of the *same* encryption clearly corresponds to the same message, and we should not reward the adversary for achieving such a trivial[2] task.

Remark 1. We note that $\mathcal{S}t\mathcal{E}$ achieves non-repudiation. On the other hand, $\mathcal{E}t\mathcal{S}$ might not achieve obvious non-repudiation, except for some special cases. One such important case concerns encryption schemes, where the decryptor can reconstruct the randomness r used by the encryptor. In this case, presenting r such that $\mathsf{Enc}_R(m;\ r) = e$, and u is a valid signature of e yields non-repudiation. We note that, for the *Insider-security* in the public-key setting, we cannot hope to *amplify* the security of the "base" signature or encryption, unlike the symmetric setting, where a proper use of a MAC allows one to increase the privacy from CPA to CCA2-security (see [5,22]). For example, in the Insider-security for encryption, the adversary is acting as the sender and holds the signing key. Thus, it is obvious that the use of this signing key cannot protect the receiver and increase the quality of the encryption. Similar argument holds for signatures. Thus, the result of Theorem 1 is the most optimistic we can hope for in that it at least *preserves* the security of the base signature and encryption, while simultaneously achieving *both* functionalities.

Outsider-Security. On the other hand, we show that in the weaker Outsider-security model, it is possible to amplify the security of encryption using signatures, as well as the security of signatures using encryption, *exactly* like in the symmetric setting [5,22,1]. This shows that Outsider-security model is quite similar to the symmetric setting: namely, from the adversarial point of view the sender and the receiver "share" the secret key $(\mathsf{SDK}_S, \mathsf{SDK}_R)$.

Theorem 2. *If $\mathcal{E}$ is* IND-CPA*-secure, and $\mathcal{S}$ is* UF-CMA*-secure, then $\mathcal{E}t\mathcal{S}$ is* IND-gCCA2*-secure in the Outsider- and* UF-CMA*-secure in the Insider-security models.*

We omit the proof due to space limitations. Intuitively, either the de-signcryption oracle always returns $\perp$ to the gCCA2-adversary, in which case it is "useless" and IND-CPA-security of $\mathcal{E}$ is enough, or the adversary can submit a valid signcryption $u = \mathsf{Sig}(\mathsf{Enc}(\cdot))$ to this oracle, in which case it breaks the UF-CMA-security of the "outside" signature $\mathcal{S}$.

[2] The task is indeed trivial in the Insider-security model, since the adversary has the signing key.

Theorem 3. *If $\mathcal{E}$ is* IND-gCCA2*-secure, and $\mathcal{S}$ is* UF-NMA*-secure, then $\mathcal{S}t\mathcal{E}$ is* IND-gCCA2*-secure in the Insider- and* UF-CMA*-secure in the Outsider-security models.*

We omit the proof due to space limitations. Intuitively, the IND-gCCA2-security of the "outside" encryption $\mathcal{E}$ makes the CMA attack of UF-CMA-adversary $\mathcal{A}$ "useless", by effectively hiding the signatures corresponding to $\mathcal{A}$'s queried messages, hence making the attack reduced to NMA.

5 Parallel Encrypt and Sign

So far we concentrated on two basic sequential composition methods, "encrypt-then-sign" and "sign-then-encrypt". Another natural generic composition method would be to both encrypt the message and sign the message, denoted $\mathcal{E}\&\mathcal{S}$. This operation simply outputs a pair (s, e), where $s \leftarrow \mathsf{Sig}_S(m)$ and $e \leftarrow \mathsf{Enc}_R(m)$. One should observe that $\mathcal{E}\&\mathcal{S}$ preserves the authenticity property but obviously does *not* preserve the privacy of the message as the signature s might reveal information about the message m. Moreover, if the adversary knows that $m \in \{m_0, m_1\}$ (as is the case for IND-security), it can see if s is a signature of m_0 or m_1, thus breaking IND-security. This simple observation was also made by [5,22]. However, we would like to stress that this scheme has a great advantage: it allows one to parallelize the expensive public key operations, which could imply significant efficiency gains.

Thus, the question which arises is under which conditions can we design a secure signcryption scheme which would also yield itself to efficiency improvements such as parallelization of operations. More concretely, there is no reason why we should apply Enc_R and Sig_S to m itself. What if we apply some efficient "pre-processing" transformation T to the message m, which produces a pair (c, d), and then sign c and encrypt d in parallel? Under which conditions on T will this yield a secure signcryption? Somewhat surprisingly, we show a very general result: instantiating T as a commitment scheme would enable us to both achieve a signcryption scheme and parallelize the expensive public key operations. More precisely, *relaxed commitment is necessary and sufficient*! In the following we explain this result in more detail.

Syntax. Clearly, the values (c, d) produced by $T(m)$ should be such that m is recoverable from (c, d), But which exactly the *syntax* (but not yet the *security*) of a commitment scheme, as defined in Section 2.3. Namely, T could be viewed as the message commitment algorithm Commit, while the message recovery algorithm is the opening algorithm Open, and we want $\mathsf{Open}(\mathsf{Commit}(m)) = m$. For a technical reason, we will also assume there exists at most one valid c for every value of d. This is done without loss of generally when commitment schemes are used. Indeed, essentially all commitment schemes have, and can always be assumed to have, $d = (m, r)$, where r is the randomness of $\mathsf{Commit}(m)$, and $\mathsf{Open}(c, (m, r))$ just checks if $\mathsf{Commit}(m; r) = (c, (m, r))$ before outputting m.

Now, given any such (possibly insecure) $\mathcal{C} = (\mathsf{Setup}, \mathsf{Commit}, \mathsf{Open})$, an encryption scheme $\mathcal{E} = (\mathsf{Enc\text{-}Gen}, \mathsf{Enc}, \mathsf{Dec})$ and a signature scheme $\mathcal{S} = (\mathsf{Sig\text{-}Gen}, \mathsf{Sig}, \mathsf{Ver})$, we define a new composition paradigm, which we call "commit-then-encrypt-and-sign": shortly, $\mathcal{C}t\mathcal{E}\&\mathcal{S} = (\mathsf{Gen}, \mathsf{SigEnc}, \mathsf{VerDec})$. For simplicity, we assume for now that all the participants share the same common commitment key CK (e.g., generated by a trusted party). $\mathsf{Gen}(1^k)$ is the same as for $\mathcal{E}t\mathcal{S}$ and $\mathcal{S}t\mathcal{E}$ compositions: set $\mathsf{VEK} = (\mathsf{VK}, \mathsf{EK})$, $\mathsf{SDK} = (\mathsf{SK}, \mathsf{DK})$. Now, to signcrypt a message m from S to R, the sender S first runs $(c, d) \leftarrow \mathsf{Commit}(m)$, and outputs signcryption $u = (s, e)$, where $s \leftarrow \mathsf{Sig}_S(c)$ and $e \leftarrow \mathsf{Enc}_R(d)$. Namely, we sign the commitment c and encrypt the decommitment d. To de-signcrypt, the receiver R validates $c = \mathsf{Msg}(s)$ using $\mathsf{Ver}_S(s)$ and decrypts $d = \mathsf{Dec}_R(e)$ (outputting $\perp$ if either fails). The final output is $\tilde{m} = \mathsf{Open}(c, d)$. Obviously, $\tilde{m} = m$ if everybody is honest.

Main Result. We have defined the new composition paradigm $\mathcal{C}t\mathcal{E}\&\mathcal{S}$ based purely on the syntactic properties of $\mathcal{C}$, $\mathcal{E}$ and $\mathcal{S}$. Now we formulate which security properties of $\mathcal{C}$ are necessary and sufficient so that our signcryption $\mathcal{C}t\mathcal{E}\&\mathcal{S}$ preserves the security of $\mathcal{E}$ and $\mathcal{S}$. As in Section 4, we concentrate on UF-CMA and IND-gCCA2 security. Our main result is as follows:

Theorem 4. *Assume that $\mathcal{E}$ is IND-gCCA2-secure, $\mathcal{S}$ is UF-CMA-secure and $\mathcal{C}$ satisfies the syntactic properties of a commitment scheme. Then, in the Insider-security model, we have:*

- *$\mathcal{C}t\mathcal{E}\&\mathcal{S}$ is IND-gCCA2-secure $\iff$ $\mathcal{C}$ satisfies the hiding property.*
- *$\mathcal{C}t\mathcal{E}\&\mathcal{S}$ is UF-CMA-secure $\iff$ $\mathcal{C}$ satisfies the relaxed binding property.*

Thus, $\mathcal{C}t\mathcal{E}\&\mathcal{S}$ preserves security of $\mathcal{E}$ and $\mathcal{S}$ iff $\mathcal{C}$ is a secure relaxed commitment. In particular, any secure regular commitment $\mathcal{C}$ yields secure signcryption $\mathcal{C}t\mathcal{E}\&\mathcal{S}$.

We prove our theorem by proving two related lemmas of independent interest. Define auxiliary encryption scheme $\mathcal{E}' = (\mathsf{Enc\text{-}Gen}', \mathsf{Enc}', \mathsf{Dec}')$ where (1) $\mathsf{Enc\text{-}Gen}' = \mathsf{Enc\text{-}Gen}$, (2) $\mathsf{Enc}'(m) = (c, \mathsf{Enc}(d))$, where $(c, d) \leftarrow \mathsf{Commit}(m)$, and (3) $\mathsf{Dec}'(c, e) = \mathsf{Open}(c, \mathsf{Dec}(d))$.

Lemma 1. *Assume $\mathcal{E}$ is IND-gCCA2-secure encryption. Then $\mathcal{E}'$ is IND-gCCA2-secure encryption iff $\mathcal{C}$ satisfies the hiding property.*

Proof. For one direction, we show that if $\mathcal{C}$ does not satisfy the hiding property, then $\mathcal{E}$ cannot even be IND-CPA-secure, let alone IND-gCCA2-secure. Indeed, if some adversary $\mathcal{A}$ can find m_0, m_1 s.t. $c(m_0) \not\approx c(m_1)$, then obviously $\mathsf{Enc}'(m_0) \equiv (c(m_0), \mathsf{Enc}(d(m_0))) \not\approx (c(m_1), \mathsf{Enc}(d(m_1))) \equiv \mathsf{Enc}'(m_1)$, contradicting IND-CPA-security.

Conversely, assume $\mathcal{C}$ satisfies the hiding property, and let $\mathcal{R}$ be the decryption-respecting equivalence relation w.r.t. which $\mathcal{E}$ is IND-CCA2-secure. We let the equivalence relation $\mathcal{R}'$ for $\mathcal{E}'$ be $\mathcal{R}'((c_1, e_1), (c_2, e_2)) = true$ iff $\mathcal{R}(e_1, e_2) = true$ and $c_1 = c_2$. It is easy to see that $\mathcal{R}'$ is decryption-respecting, since if $d_i =$

$\mathsf{Dec}(e_i)$, then $\mathcal{R}'((c_1, e_1), (c_2, e_2)) = true$ implies that $(c_1, d_1) = (c_2, d_2)$, which implies that $m_1 = \mathsf{Open}(c_1, d_1) = \mathsf{Open}(c_2, d_2) = m_2$.

We now show IND-CCA2-security of $\mathcal{E}'$ w.r.t. $\mathcal{R}'$. For that, let Env_1 denote the usual environment where we place any adversary $\mathcal{A}'$ for $\mathcal{E}'$. Namely, (1) in find Env_1 honestly answers the decryption queries of $\mathcal{A}'$; (2) after m_0 and m_1 are selected, Env_1 picks a random b, sets $(c_b, d_b) \leftarrow \mathsf{Commit}(m_b)$, $e_b \leftarrow \mathsf{Enc}(d_b)$ and returns $\tilde{e} = \mathsf{Enc}'(m_b) = (c_b, e_b)$; (3) in guess, Env_1 honestly answers decryption query $e' = (c, e)$ provided $\mathcal{R}'(e', \tilde{e}) = false$. We can assume that $\mathcal{A}'$ never asks a query (c, e) where $\mathcal{R}(e, e_b) = true$ but $c \neq c_b$. Indeed, by our assumption only the value $c = c_b$ will check with d_b, so the answer to queries with $c \neq c_b$ is $\perp$ (and $\mathcal{A}'$ knows it). Hence, we can assume that $\mathcal{R}'(e', \tilde{e}) = false$ implies that $\mathcal{R}'(e, e_b) = false$. We let $\mathsf{Succ}_1(\mathcal{A}')$ denote the probability $\mathcal{A}'$ succeeds in predicting b. Then, we define the following "fake" environment Env_2. It is identical to Env_1 above, except for one aspect: in step (2) it would return bogus encryption $\tilde{e} = (c(0), e_b)$, i.e. puts the commitment to the zero string 0 instead of the expected c_b. In particular, step (3) is the same as before with the understanding that $\mathcal{R}'(e', \tilde{e})$ is evaluated with the fake challenge $\tilde{e}$. We let $\mathsf{Succ}_2(\mathcal{A}')$ be the success probability of $\mathcal{A}$ in Env_2.

We make two claims: (a) using the hiding property of $\mathcal{C}$, no PPT adversary $\mathcal{A}'$ can distinguish Env_1 from Env_2, i.e. $|\mathsf{Succ}_1(\mathcal{A}') - \mathsf{Succ}_2(\mathcal{A}')| \leq \mathsf{negl}(k)$; (b) using IND-gCCA2-security of $\mathcal{E}$, $\mathsf{Succ}_2(\mathcal{A}') < \frac{1}{2} + \mathsf{negl}(k)$, for any PPT $\mathcal{A}'$. Combined, claims (a) and (b) imply the lemma.

Proof of Claim (a). If for some $\mathcal{A}'$, $\mathsf{Succ}_1(\mathcal{A}') - \mathsf{Succ}_2(\mathcal{A}') > \varepsilon$ for non-negligible ε, we create $\mathcal{A}_1$ that will break the hiding property of $\mathcal{C}$. $\mathcal{A}_1$ picks $(\mathsf{EK}, \mathsf{DK}) \leftarrow \mathsf{Enc\text{-}Gen}(1^k)$ by itself, and runs $\mathcal{A}'$ (answering its decryption queries using DK) until $\mathcal{A}'$ outputs m_0 and m_1. At this stage $\mathcal{A}_1$ picks a random $b \leftarrow \{0, 1\}$, and claims to be able to distinguish $c(0)$ from $c_b = c(m_b)$. When presented with $\tilde{c}$ – a commitment to either 0 or m_b – $\mathcal{A}_1$ will return to $\mathcal{A}'$ the "ciphertext" $\tilde{e} = (\tilde{c}, e_b)$. $\mathcal{A}_1$ will then again run $\mathcal{A}'$ to completion refusing to decrypt e' such that $\mathcal{R}'(e', \tilde{e}) = true$. When $\mathcal{A}'$ outputs $\tilde{b}$, $\mathcal{A}_1$ says that the message was m_b if $\mathcal{A}'$ succeeds ($\tilde{b} = b$), and says 0 otherwise. It is easy to check that in case $\tilde{c} = c(m_b) = c_b$, $\mathcal{A}'$ was run exactly in Env_1, otherwise – in Env_2, which easily implies that $\Pr(\mathcal{A}_1 \text{ succeeds}) \geq \frac{1}{2} + \frac{\varepsilon}{2}$, a contradiction.

Proof of Claim (b). If for some $\mathcal{A}'$, $\mathsf{Succ}_2(\mathcal{A}') > \frac{1}{2} + \varepsilon$, we create $\mathcal{A}_2$ which will break IND-gCCA2-security of $\mathcal{E}$. Specifically, $\mathcal{A}_2$ can simulate the decryption query $e' = (c, e)$ of $\mathcal{A}'$ by asking its own decryption oracle to decrypt $d = \mathsf{Dec}(e)$, and returning $\mathsf{Open}(c, d)$. When $\mathcal{A}'$ outputs m_0 and m_1, $\mathcal{A}_2$ sets $(c_i, d_i) \leftarrow \mathsf{Commit}(m_i)$ and claims to distinguish d_0 and d_1. When given challenge $e_b \leftarrow \mathsf{Enc}(d_b)$ for unknown b, $\mathcal{A}_2$ gives $\mathcal{A}'$ the challenge $\tilde{e} = (c(0), e_b)$. Then, again, $\mathcal{A}_2$ uses its own decryption oracle to answer all queries $e' = (c, e)$ as long as $\mathcal{R}'(e', \tilde{e}) = false$. From the definition of $\mathcal{R}'$ and our assumption earlier, we see that $\mathcal{R}(e, e_b) = false$ as well, so all such queries are legal. Since $\mathcal{A}_2$ exactly recreates the environment Env_2 for $\mathcal{A}'$, $\mathcal{A}_2$ succeeds with probability $\mathsf{Succ}_2(\mathcal{A}') > \frac{1}{2} + \varepsilon$.

We note that the first part of Theorem 4 follows using exactly the same proof as Lemma 1. Only few small changes (omitted) are needed due to the fact that the commitment is now signed. We remark only that IND-gCCA2 security is again important here. Informally, IND-gCCA2-security is robust to easily recognizable and invertible changes of the ciphertext. Thus, signing the commitment part – which is polynomially verifiable – does not spoil IND-gCCA2-security.

We now move to the second lemma. We define auxiliary signature scheme $\mathcal{S}' = (\mathsf{Sig\text{-}Gen}', \mathsf{Sig}', \mathsf{Ver}')$ as follows: (1) $\mathsf{Sig\text{-}Gen}' = \mathsf{Sig\text{-}Gen}$, (2) $\mathsf{Sig}'(m) = (\mathsf{Sig}(c), d)$, where, $(c, d) \leftarrow \mathsf{Commit}(m))$, (3) $\mathsf{Ver}'(s, d) = \mathsf{succeed}$ iff $\mathsf{Ver}(s) = \mathsf{succeed}$ and $\mathsf{Open}(\mathsf{Msg}(s), d) \neq \perp$.

Lemma 2. *Assume $\mathcal{S}$ is* UF-CMA*-secure signature. Then $\mathcal{S}'$ is* UF-CMA*-secure signature iff $\mathcal{C}$ satisfies the relaxed binding property.*

Proof. For one direction, we show that if $\mathcal{C}$ does not satisfy the relaxed binding property, then $\mathcal{S}'$ cannot be UF-CMA-secure. Indeed, assume for some adversary $\mathcal{A}$ can produce m such that when $(c, d) \leftarrow \mathsf{Commit}(m)$ is generated and given to $\mathcal{A}$, $\mathcal{A}$ can find (with non-negligible probability ε) a value d' such that $\mathsf{Open}(c, d') = m'$ and $m' \neq m$. We build a forger $\mathcal{A}'$ for $\mathcal{S}'$ using $\mathcal{A}$. $\mathcal{A}'$ gets m from $\mathcal{A}$, and asks its signing oracle to sign m. $\mathcal{A}'$ gets back (s, d), where s is a valid signature of c, and (c, d) is a random commitment pair for m. $\mathcal{A}'$ gives (c, d) to $\mathcal{A}$, and gets back (with probability ε) the value d' such that $\mathsf{Open}(c, d') = m'$ different from m. But then (s, d') is a valid signature (w.r.t. $\mathcal{S}'$) of a "new" message m', contradicting the UF-CMA-security of $\mathcal{S}$.

Conversely, assume some forger $\mathcal{A}'$ breaks the UF-CMA-security of $\mathcal{S}'$ with non-negligible probability ε. Assume $\mathcal{A}'$ made (wlog exactly) $t = t(k)$ oracle queries to Sig' for some polynomial $t(k)$. For $1 \leq i \leq t$, we let m_i be the i-th message $\mathcal{A}'$ asked to sign, and (s_i, d_i) be its signature (where $(c_i, d_i) \leftarrow \mathsf{Commit}(m_i)$ and $s_i \leftarrow \mathsf{Sig}(c_i)$). We also let m, s, d, c have similar meaning for the message that $\mathcal{A}'$ forged. Finally, let Forged denote the event that $c \notin \{c_1, \ldots, c_t\}$. Notice,

$$\varepsilon < \Pr(\mathcal{A}' \text{ succeeds}) = \Pr(\mathcal{A}' \text{ succeeds } \wedge \mathsf{Forged}) + \Pr(\mathcal{A}' \text{ succeeds } \wedge \overline{\mathsf{Forged}})$$

Thus, at least one of the probabilities above is $\geq \varepsilon/2$. We show that the first case contradicts the UF-CMA-security of $\mathcal{S}$, while the second case contradicts the relaxed binding property of $\mathcal{C}$.

Case 1: $\Pr(\mathcal{A}' \textit{ Succeeds } \wedge \mathsf{Forged}) \geq \varepsilon/2$. We construct a forger $\mathcal{A}_1$ for $\mathcal{S}$. It simulates the run of $\mathcal{A}'$ by generating a commitment key CK by itself, and using its own signing oracle to answer the signing queries of $\mathcal{A}'$: set $(c_i, d_i) \leftarrow \mathsf{Commit}(m_i)$, get $s_i \leftarrow \mathsf{Sig}'(c_i)$ from the oracle, and return (s_i, d_i). When $\mathcal{A}'$ forges a signature (s, d) of m w.r.t. $\mathcal{S}'$, $\mathcal{A}_1$ forges a signature s of c w.r.t. $\mathcal{S}$. Notice, c is a "new forgery" in $\mathcal{S}$ iff Forged happens. Hence, $\mathcal{A}_1$ succeeds with probability at least $\varepsilon/2$, a contradiction to UF-CMA-security of $\mathcal{S}$.

Case 2: $\Pr(\mathcal{A}' \textit{ Succeeds } \wedge \overline{\mathsf{Forged}}) \geq \varepsilon/2$. We construct an adversary $\mathcal{A}_2$ contradicting the relaxed binding property of $\mathcal{C}$. $\mathcal{A}_2$ will generate its own key pair

$(\mathsf{SK},\mathsf{VK}) \leftarrow \mathsf{Sig\text{-}Gen}(1^k)$, and will also pick a random index $1 \le i \le t$. It simulates the run of $\mathcal{A}'$ in a standard manner (same way as $\mathcal{A}_1$ above) up to the point where $\mathcal{A}'$ asks its i-th query m_i. At this stage $\mathcal{A}_2$ outputs m_i as its output to the find stage. When receiving back random $(c_i, d_i) \leftarrow \mathsf{Commit}(m_i)$, it uses them to sign m_i as before (i.e., returns $(\mathsf{Sig}(c_i), d_i)$ to $\mathcal{A}'$), and keeps simulating the run of $\mathcal{A}'$ in the usual manner. When $\mathcal{A}$ outputs the forgery (s, d) of a message m, $\mathcal{A}_2$ checks if $c_i = c$ ($\mathsf{Msg}(s)$) and $m_i \ne m$. If this fails, it fails as well. Otherwise, it outputs d as its final output to the collide stage. We note that when Forged does not happen, i.e. $c \in \{c_1 \dots c_t\}$, we have $c = c_i$ with probability at least $1/t$. Thus, with overall non-negligible probability $\varepsilon/(2t)$ we have that: (1) $m \ne m_i$ ($\mathcal{A}'$ outputs a new message m); (2) $c_i = c$ (Forged did not happen and $\mathcal{A}_2$ correctly guessed i such that $c_i = c$); (3) $\mathsf{Open}(c, d) = m$ and $\mathsf{Open}(c, d_i) = m_i$. But this exactly means that $\mathcal{A}_2$ broke the relaxed binding property of $\mathcal{C}$, a contradiction.

We note that the second part of Theorem 4 follows using exactly the same proof as Lemma 2. Only few small changes are needed due to the fact that the decommitment is now encrypted (e.g., the adversary chooses its own encryption keys and performs decryptions on its own). This completes the proof of Theorem 4.

Remark 2. We note that $\mathcal{C}t\mathcal{E}\&\mathcal{S}$ achieves non-repudiation by Lemma 2. Also note that the necessity of relaxed commitments holds in the weaker Outsider-security model as well. Finally, we note that $\mathcal{C}t\mathcal{E}\&\mathcal{S}$ paradigm successfully applies to the symmetric setting as well.

Remark 3. We remark that in practice, $\mathcal{C}t\mathcal{E}\&\mathcal{S}$ could be faster or slower than the sequential $\mathcal{E}t\mathcal{S}$ and $\mathcal{S}t\mathcal{E}$ compositions, depending on the specifics $\mathcal{C}$, $\mathcal{E}$ and $\mathcal{S}$. For most efficiency on the commitment side, however, one can use the simple commitment $c = H(m, r)$, $d = (m, r)$, where r is a short random string and H is a cryptographic hash function (analyzed as a random oracle). For provable security, one can use an almost equally efficient commitment scheme of [11,18] based on CRHF's.

6 On-Line/Off-Line Signcryption

Public-key operations are expensive. Therefore, we examine the possibility of designing signcryption schemes which could be run in two phases: (1) the *off-line* phase, performed before the messages to be signcrypted is known; and (2) the *on-line* phase, which uses the message and the pre-computation of the off-line stage, to efficiently produce the required signcryption. We show that the $\mathcal{C}t\mathcal{E}\&\mathcal{S}$ paradigm is ideally suited for such a task, but first we recall a similar notion for ordinary signatures.

On-Line/Off-Line Signatures. On-line/Off-line signatures where introduced by Even et al. [14] who presented a general methodology to transform any signature scheme into a more efficient on-line/off-line signature (by using so called "one-time" signatures). Their construction, however, is mainly of theoretical interest. Recently, Shamir and Tauman [30] introduced the following much more efficient method to generate on-line/off-line signatures, which they called "hash-sign-switch". The idea is to use *trapdoor commitments* (see Section 2.3) in the following way. The signer S chooses two pairs of keys: regular signing keys $(\mathsf{SK}, \mathsf{VK}) \leftarrow \mathsf{Sig\text{-}Gen}(1^k)$, and trapdoor commitment keys $(\mathsf{TK}, \mathsf{CK}) \leftarrow \mathsf{Setup}(1^k)$. S keeps $(\mathsf{SK}, \mathsf{TK})$ secret, and publishes $(\mathsf{VK}, \mathsf{CK})$. In the off-line phase, S prepares $(c, d_0) \leftarrow \mathsf{Commit}_{\mathsf{CK}}(0)$, and $s \leftarrow \mathsf{Sig}_{\mathsf{SK}}(c)$. In the on-line phase, when the message m arrives, S creates "fake" decommitment $(c, d) \leftarrow \mathsf{Switch}_{\mathsf{TK}}((c, d_0), m)$ to m, and outputs (s, d) as the signature. To verify, the receiver R checks that s is a valid signature of $c = \mathsf{Msg}(s)$, and $\mathsf{Open}_{\mathsf{CK}}(c, d) = m$.

Notice, this is very similar to the auxiliary signature scheme $\mathcal{S}'$ we used in Lemma 2. The only difference is that the "fake" pair (c, d) is used instead of $\mathsf{Commit}(m)$. However, by the trapdoor collisions property of trapdoor commitments, we get that $(c, d) \approx \mathsf{Commit}(m)$, and hence Lemma 2 – true for any commitment scheme – implies that this modified signature scheme is indeed secure (more detailed proof is given in [30]). Thus, the resulting signature $\mathcal{S}''$ essentially returns the same $(\mathsf{Sig}(c), d)$ as $\mathcal{S}'$, except that the expensive signature Sig is computed in the off-line phase.

"Hash-Sign-Switch" for Signcryption. Now, we could use the on-line/off-line signature $\mathcal{S}''$ above with any of our composition paradigms: $\mathcal{E}t\mathcal{S}$, $\mathcal{S}t\mathcal{E}$ or $\mathcal{C}t\mathcal{E}\&\mathcal{S}$. In all cases this would move the actual signing operation into the off-line phase. For example, $\mathcal{E}t\mathcal{S}$ will (essentially) return $(\mathsf{Sig}(c(e)), d(e))$, where $e \leftarrow \mathsf{Enc}(m)$; while $\mathcal{S}t\mathcal{E}$ will return $\mathsf{Enc}(\mathsf{Sig}(c(m)), d(m))$. We could also apply it "directly" to the $\mathcal{C}t\mathcal{E}\&\mathcal{S}$ scheme. However, $\mathcal{C}t\mathcal{E}\&\mathcal{S}$ scheme *already uses commitments*! So let us see what happens when we use a trapdoor commitment $\mathcal{C}$ instead of any general commitment. We see that we still return $(\mathsf{Sig}(c), \mathsf{Enc}(d))$ (where $(c, d) \leftarrow \mathsf{Switch}(\mathsf{Commit}(0), m) \approx \mathsf{Commit}(m)$), except the expensive signature part is performed off-line, exactly as we wish. Thus, $\mathcal{C}t\mathcal{E}\&\mathcal{S}$ yields a more efficient (and provably secure by Theorem 4) on-line/off-line implementation than the one we get by blindly applying the "hash-sign-switch" technique to the $\mathcal{E}t\mathcal{S}$ or $\mathcal{S}t\mathcal{E}$ schemes.

We remark that in this scheme the trapdoor key TK has to be known to the sender, but not to the receiver. Hence, each user P has to generate its own pair $(\mathsf{TK}, \mathsf{CK})$ during key generation, keeping TK as part of SDK_P. Also, P should use its own CK_P when sending messages, and the sender's CK when receiving messages. Notice, since trapdoor commitments are *information-theoretically* hiding, there is no danger for the receiver that the sender chooses a "bad" commitment key (the hiding property is satisfied for all CK's, and it is in sender's interest to choose CK so that the binding is satisfied as well).

Adding On-Line/Off-Line Encryption. We have successfully moved the expensive public-key signature to the off-line phase. What about public-key encryption? We can use the folklore technique of integrating public- and secret-key encryptions: $\mathsf{Enc}'_{\mathsf{EK}}(m) = (\mathsf{Enc}_{\mathsf{EK}}(r), E_r(m))$. Namely, we encrypt a random secret-key r for symmetric encryption E, and then encrypt the actual message m using E with the key r. Clearly, we can do the (much more expensive) public-key encryption $\mathsf{Enc}_{\mathsf{EK}}(r)$ in the off-line stage. Surprisingly, this folklore technique, which is being extensively used in practice, has only recently been formally analyzed in the CCA2-setting by [12]. Translated to our terminology, IND-gCCA2-secure Enc and E yield IND-gCCA2-secure Enc′ above ([12] showed this for regular IND-CCA2-security). As a side remark, in the random oracle model, clever integration of public- and secret-key encryption allows us to get IND-CCA2-secure Enc′ starting from much less secure base encryption Enc (e.g., see [15,25]). Thus, making encryption off-line can also amplify its security in this setting.

Final Scheme. To summarize, we get the following very efficient on-line/off-line signcryption scheme from any signature $\mathcal{S}$, public-key encryption $\mathcal{E}$, trapdoor commitment $\mathcal{C}$, and symmetric encryption E: (1) in the off-line stage generate $(c, d_0) \leftarrow \mathsf{Commit}_{\mathsf{CK}_S}(0)$, and prepare $e_1 \leftarrow \mathsf{Enc}_{\mathsf{EK}_R}(r)$, and $s \leftarrow \mathsf{Sig}_{\mathsf{SK}_S}(c)$; (2) in the on-line stage, create $(c, d) \leftarrow \mathsf{Switch}_{\mathsf{TK}_S}((c, d_0), m)$, $e_2 \leftarrow E_r(d)$, and return $(s, (e_1, e_2))$. In essence, we efficiently compute and return $(\mathsf{Sig}(c), (\mathsf{Enc}(r), E_r(d)))$, where $(c, d) \approx \mathsf{Commit}(m)$. Since the switching operation and the symmetric encryption are usually very fast, we get significant efficiency gain. Decryption and verification are obvious.

7 Multi-user Setting

Syntax. So far we have concentrated on the network of two users: the sender S and the receiver R. Once we move to the full-fledged multi-user network, several new concerns arise. First, users must now have identities. We denote by ID_P the identity of user P. We do not impose any constraints on the identities, other than they should be easily recognizable by everyone in the network, and that users can easily obtain the public key VEK_P from ID_P (e.g., ID_P could be VEK_P). Next, we change the syntax of the signcryption algorithm SigEnc to both take and output the identity of the sender and the receiver. Specifically, (1) the signcryption for user S, on input, $(m, \mathsf{ID}_{S'}, \mathsf{ID}_{R'})$, uses $\mathsf{VEK}_{R'}$ and generates $(u, \mathsf{ID}_S, \mathsf{ID}_{R'})$ provided $\mathsf{ID}_S = \mathsf{ID}_{S'}$; (2) the de-signcryption for user R, on input $(u, \mathsf{ID}_{S'}, \mathsf{ID}_{R'})$, uses $\mathsf{VEK}_{S'}$ and outputs $\tilde{m}$ provided $\mathsf{ID}_R = \mathsf{ID}_{R'}$. It must be clear from which S' the message $\tilde{m}$ came from. Otherwise this will not be able to satisfy the security property described below.

Security. To break the Outsider-security between a pair of designated users S and R, $\mathcal{A}$ is assumed to have all the secret keys beside SDK_S and SDK_R, and has access to the signcryption oracle of S (which it can call with *any* $\mathsf{ID}_{R'}$ and not just ID_R) and the de-signcryption oracle for R (which it can call with *any* $\mathsf{ID}_{S'}$ and not just ID_S). Naturally, to break the UF-CMA-security, $\mathcal{A}$ has to come up with a valid signcryption $(u, \mathsf{ID}_S, \mathsf{ID}_R)$ of the message m such that $(m, \mathsf{ID}_S, \mathsf{ID}_R)$ was not queried earlier to the signcryption oracle of S. Similarly, to break IND-gCCA2-security of encryption, $\mathcal{A}$ has to come up with m_0 and m_1 such that it can distinguish $\mathsf{SigEnc}(m_0, \mathsf{ID}_S, \mathsf{ID}_R)$ from $\mathsf{SigEnc}(m_1, \mathsf{ID}_S, \mathsf{ID}_R)$. Of course, given a challenge $(u, \mathsf{ID}_S, \mathsf{ID}_R)$, $\mathcal{A}$ is disallowed to ask the de-signcryption oracle for R a query $(u', \mathsf{ID}_S, \mathsf{ID}_R)$ where $\mathcal{R}(u, u') = true$.

We define Insider-security in an analogous manner, except now the adversary has all the secret keys except SDK_S when attacking authenticity or SDK_R when attacking privacy. Also, for UF-CMA-security, a forgery $(u, \mathsf{ID}_S, \mathsf{ID}_{R'})$ of a message m is "new" as long as $(m, \mathsf{ID}_S, \mathsf{ID}_{R'})$ was not queried (even though $(m, \mathsf{ID}_S, \mathsf{ID}_{R''})$ could be queried). Similarly, $\mathcal{A}$ could choose to distinguish signcryptions $(m_0, \mathsf{ID}_{S'}, \mathsf{ID}_R)$ from $(m_1, \mathsf{ID}_{S'}, \mathsf{ID}_R)$ (for any S'), and only has the natural restriction on asking de-signcryption queries of the form $(u, \mathsf{ID}_{S'}, \mathsf{ID}_R)$, but has no restrictions on using $\mathsf{ID}_{S''} \neq \mathsf{ID}_{S'}$.

Extending Signcryption. We can see that the signcryption algorithms that we use so far have to be upgraded, so that they use the new inputs ID_S and ID_R in non-trivial manner. For example, if the $\mathcal{E}t\mathcal{S}$ method is used in the multi-user setting, the adversary $\mathcal{A}$ can easily break the gCCA2-security, even in the Outsider-model. Indeed, given the challenge $u = (\mathsf{Sig}_S(e), \mathsf{ID}_S, \mathsf{ID}_R)$, where $e = \mathsf{Enc}_R(m_b)$, $\mathcal{A}$ can replace the sender's signature with its own by computing $u' = (\mathsf{Sig}_{\mathcal{A}}(e), \mathsf{ID}_{\mathcal{A}}, \mathsf{ID}_R)$ and ask R to de-signcrypt it. Since $\mathcal{A}$ has no restrictions on using $\mathsf{ID}_{\mathcal{A}} \neq \mathsf{ID}_S$ in its de-signcryption oracle queries, $\mathcal{A}$ can effectively obtain the decryption of e (i.e. m_b). Similar attack on encryption holds for the $\mathcal{S}t\mathcal{E}$ scheme, while in $\mathcal{C}t\mathcal{E}\&\mathcal{S}$ both the encryption and the signature suffer from these trivial attacks.

It turns out there is a general simple solution to this problem. For any signcryption scheme $\mathcal{SC} = (\mathsf{Gen}, \mathsf{SigEnc}, \mathsf{VerDec})$ designed for the two-user setting (like $\mathcal{E}t\mathcal{S}$, $\mathcal{S}t\mathcal{E}$, $\mathcal{C}t\mathcal{E}\&\mathcal{S}$), we can transform it into a multi-user signcryption scheme $\mathcal{SC}' = (\mathsf{Gen}, \mathsf{SigEnc}', \mathsf{VerDec}')$ as follows: $\mathsf{SigEnc}'_S(m, \mathsf{ID}_S, \mathsf{ID}_R) = (\mathsf{SigEnc}_S(m, \mathsf{ID}_S, \mathsf{ID}_R), \mathsf{ID}_S, \mathsf{ID}_R)$, and $\mathsf{VerDec}'_R(u, \mathsf{ID}_S, \mathsf{ID}_R)$ gets $(m, \alpha, \beta) = \mathsf{VerDec}_R(u)$ and outputs m only if $\alpha = \mathsf{ID}_S$ and $\beta = \mathsf{ID}_R$. It is easy to see that the security properties of the two-user signcryption scheme is preserved in the multi-user setting by the transformation; namely, whatever properties $\mathcal{SC}$ has in the two-user setting, $\mathcal{SC}'$ will have in the multi-user setting. (The proof is simple and is omitted. Intuitively, however, the transformation effectively binds the signcryption output to the users, by computing signcryption as a function of the users' identities.)

Moreover, one can check quite easily that we can be a little more efficient in our compositions schemes. Namely, whatever security was proven in the two-user setting remains unchanged for the multi-user setting as long as we follow these simple changes:

1. Whenever *encrypting* something, include the identity of the *sender* ID_S together with the encrypted message.
2. Whenever *signing* something, include the identity of the *receiver* ID_R together with the signed message .
3. On the receiving side, whenever either the identity of the sender or of the receiver do not match what is expected, output $\perp$.

Hence, we get the following new analogs for $\mathcal{E}t\mathcal{S}$, $\mathcal{S}t\mathcal{E}$ and $\mathcal{C}t\mathcal{E}\&\mathcal{S}$:

- $\mathcal{E}t\mathcal{S}$ returns $(\mathsf{Sig}_S(\mathsf{Enc}_R(m, \mathsf{ID}_S), \mathsf{ID}_R), \mathsf{ID}_S, \mathsf{ID}_R)$.
- $\mathcal{S}t\mathcal{E}$ returns $(\mathsf{Enc}_R(\mathsf{Sig}_S(m, \mathsf{ID}_R), \mathsf{ID}_S), \mathsf{ID}_S, \mathsf{ID}_R)$.
- $\mathcal{C}t\mathcal{E}\&\mathcal{S}$ returns $(\mathsf{Sig}_S(c, \mathsf{ID}_R), \mathsf{Enc}_R(d, \mathsf{ID}_S), \mathsf{ID}_S, \mathsf{ID}_R)$, where $(c, d) \leftarrow \mathsf{Commit}(m)$.

8 On CCA2 Security and Strong Unforgeability

This section will be mainly dedicated to the conventional notion of CCA2-attack for encryption. Much of the discussion also applies to a related notion of strong unforgeability, sUF, for signatures. Despite the fact that one specifies the attack model, and the other – the adversary's goal, we will see that the relation between gCCA2/CCA2, and UF/sUF notions is quite similar. We will argue that: (1) gCCA2-attack and UF-security are better suited for a "good" *definition* than their stronger but syntactically ill CCA2 and sUF counterparts; (2) it is unlikely that the extra strength of CCA2 w.r.t. gCCA2 and sUF w.r.t. UF will find any useful applications.

Of course, what is stated above is a subjective opinion. Therefore, we briefly remark which of our previous results for signcryption (stated for gCCA2/UF notions) extend to the CCA2/sUF notions. Roughly, half of the implications still hold, while the other half fails to do so. As one representative example, $\mathcal{E}t\mathcal{S}$ is no longer CCA2-secure even if $\mathcal{E}$ is CCA2-secure. A "counter-example" comes when we use a perfectly possible UF-CMA-secure signature scheme $\mathcal{S}$ which always appends a useless bit during signing. By simply flipping this bit on the challenge ciphertext, CCA2-adversary is now "allowed" to use the decryption oracle and recover the plaintext. The artificial nature of this "counter-example" is perfectly highlighted by Theorem 1, which shows that the IND-gCCA2-security of $\mathcal{E}t\mathcal{S}$ is preserved.

Definitional Necessity. Even more explicitly, appending a useless (but harmless) bit to a CCA2-secure encryption no longer leaves it CCA2-secure. It seems a little disturbing that this clearly harmless (albeit useless) modification does not satisfy the *definition* of "secure encryption". The common answer to the above criticism is that there is nothing wrong if we became overly strict with our definitions, as long as (1) the definitions do not allow for "insecure" schemes; and (2) we can meet them. In other words, the fact that some secure, but "useless" constructions are ruled out can be tolerated. However, as we illustrated for the first time, the conventional CCA2 notion *does* rule out some secure "useful" constructions as well. For example, it might have led one to believe that the $\mathcal{E}t\mathcal{S}$ scheme is generically insecure and should be avoided, while we showed that this is not the case.

Relation to Non-malleability. We recall that the concept of indistinguishability is very useful in terms of *proving* schemes secure, but it is not really "natural". It is generally believed that a more useful security notion – and the one really important in applications – is that of *non-malleability* [13] (denoted NM), which we explain in a second. Luckily, it is known [13,4] that IND-CCA2 is equivalent to NM-CCA2, which "justifies" the use of IND-CCA2 as a simpler notion to work with. And now that we relaxed IND-CCA2 to IND-gCCA2, a valid concern arises that we loose the above equivalence, and therefore the justification for using indistinguishability as our security notion. A closer look, however, reveals that this concern is merely a syntactic triviality. Let us explain.

In essence, NM-security roughly states the following: upon seeing some unknown ciphertext e, the only thing the adversary can extract – which bears any relevance to the corresponding plaintext m – is the encryption of this plaintext (which the adversary has anyway). The current formalization of non-malleability additionally requires that the only such encryption e' that $\mathcal{A}$ can get is e itself. However, unlike the first property, the last requirement does not seem crucial, provided that *anybody can tell that the ciphertext e' encrypts the same message as e, by only looking at e and e'.* In other words, there could possibly be no harm even if $\mathcal{A}$ can generate $e' \neq e$: anyone can tell that $\mathsf{Dec}(e) = \mathsf{Dec}(e')$, so there is no point to even change e to e'. Indeed, we can relax the formalization of non-malleability (call if gNM) by using a decryption-respecting relation $\mathcal{R}$, just like we did for the CCA2 attack: namely, $\mathcal{A}$ is not considered successful if it outputs e' s.t. $\mathcal{R}(e, e') = \mathit{true}$. Once this is done, the equivalence between "gNM-CCA2" and IND-gCCA2 holds again.

Applicational Necessity. The above argument also indicates that gCCA2-security is sufficient for all applications where chosen ciphertext security matters (e.g., those in [31,9,8]). Moreover, it is probably still a slight overkill in terms of a necessary and sufficient formalization of "secure encryption" from the applicational point of view. Indeed, we tried to relax the notion of CCA2-security to the minimum extent possible, just to avoid the syntactic problems of CCA2-security. In particular, we are not aware of any "natural" encryption scheme in the gap between gCCA2 and CCA2-security. The only thing we are saying is that the *notion* of gCCA2 security is more robust to syntactic issues, seems more applicable for studying generic properties of "secure encryption", while also being sufficient for its applications.

Strong Unforgeability. Finally, we briefly remark on the concept of sUF-security for signatures. To the best of our knowledge, the extra guarantees of this concept have no realistic applications (while suffering similar syntactic problems as CCA2-security does). Indeed, once the message m is signed, there is no use to produce a different signature of the same message: the adversary already has a valid signature of m. The only "application" we are aware of is building CCA2-secure encryption from a CPA-secure encryption, via the $\mathcal{E}t\mathcal{S}$ method. As we demonstrated in Theorem 2, sUF-security is no longer necessarily once we accept the concept of gCCA2-security.

References

1. J. An and M. Bellare, "Does encryption with redundancy provide authenticity?," In *Eurocrypt '01*, pp. 512–528, LNCS Vol. 2045.
2. J. An and Y. Dodis, "Secure integration of symmetric- and public-key authenticated encryption." Manuscript, 2002.
3. J. Baek, R. Steinfeld, and Y. Zheng, "Formal proofs for the security of signcryption," In *PKC '02*, 2002.
4. M. Bellare, A. Desai, D. Pointcheval and P. Rogaway, "Relations among notions of security for public-key encryption schemes," In *Crypto '98*, LNCS Vol. 1462.
5. M. Bellare and C. Namprempre, "Authenticated Encryption: Relations among Notions and Analysis of the Generic Composition Paradigm," In *Asiacrypt '00*, LNCS Vol. 1976.
6. M. Bellare, P. Rogaway, "Encode-Then-Encipher Encryption: How to Exploit Nonces or Redundancy in Plaintexts for Efficient Cryptography," In *Asiacrypt '00*, LNCS Vol 1976.
7. G. Brassard, D. Chaum, and C. Crépeau, "Minimum disclosure proofs of knowledge," *JCSS*, 37(2):156–189, 1988.
8. R. Canetti, "Universally Composable Security: A New Paradigm for Cryptographic Protocols," In *Proc. 42st FOCS*, pp. 136–145. IEEE, 2001.
9. R. Canetti and H. Krawczyk, "Analysis of Key-Exchange Protocols and Their Use for Building Secure Channels," In *Eurocrypt '01*, pp. 453–474, LNCS Vol. 2045.
10. D. Chaum and H. Van Antwerpen, "Undeniable signatures," In *Crypto '89*, pp. 212–217, LNCS Vol. 435.
11. I. Damgård, T. Pedersen, and B. Pfitzmann, "On the existence of statistically hiding bit commitment schemes and fail-stop signatures," In *Crypto '93*, LNCS Vol. 773.
12. G. Di Crescenzo, J. Katz, R. Ostrovsky, and A. Smith, "Efficient and Non-interactive Non-malleable Commitment," In *Eurocrypt '01*, pp. 40–59, LNCS Vol. 2045.
13. D. Dolev, C. Dwork and M. Naor, "Non-malleable cryptography," In *Proc. 23rd STOC*, ACM, 1991.
14. S. Even, O. Goldreich, and S. Micali, "On-Line/Off-Line Digital Schemes," In *Crypto '89*, pp. 263–275, LNCS Vol. 435.
15. E. Fujisaki and T. Okamoto, "Secure integration of asymmetric and symmetric encryption schemes," In *Crypto '99*, pp. 537–554, 1999, LNCS Vol. 1666.
16. S. Goldwasser and S. Micali, "Probabilistic encryption," *JCSS*, 28(2):270–299, April 1984.
17. S. Goldwasser, S. Micali, and R. Rivest, "A digital signature scheme secure against adaptive chosen-message attacks," *SIAM J. Computing*, 17(2):281–308, April 1988.
18. S. Halevi and S. Micali, "Practical and provably-secure commitment schemes from collision-free hashing," In *Crypto '96*, pp. 201–215, 1996, LNCS Vol. 1109.
19. W. He and T. Wu, "Cryptanalysis and Improvement of Petersen-Michels Signcryption Schemes," *IEE Computers and Digital Communications*, 146(2):123–124, 1999.
20. C. Jutla, "Encryption modes with almost free message integrity," In *Eurocrypt '01*, pp. 529–544, LNCS Vol. 2045.
21. J. Katz and M. Yung, "Unforgeable Encryption and Chosen Ciphertext Secure Modes of Operation," In *FSE '00*, pp. 284–299, LNCS Vol. 1978.

22. H. Krawczyk, "The Order of Encryption and Authentication for Protecting Communications (or: How Secure Is SSL?)," In *Crypto '01*, pp. 310–331, LNCS Vol. 2139.
23. H. Krawczyk and T. Rabin, "Chameleon Signatures," In *NDSS '00*, pp. 143–154, 2000.
24. M. Naor and M. Yung, "Universal One-Way Hash Functions and their Cryptographic Applications," In *Proc. 21st STOC*, pp. 33–43, ACM, 1989.
25. T. Okamoto and D. Pointcheval, "React: Rapid enhanced-security asymmetric cryptosystem transform," In *CT-RSA '01*, pp. 159–175, 2001, LNCS Vol. 2020.
26. H. Petersen and M. Michels, "Cryptanalysis and Improvement of Signcryption Schemes," *IEE Computers and Digital Communications*, 145(2):149–151, 1998.
27. C. Rackoff and D. Simon, "Non-Interactive zero-knowledge proof of knowledge and chosen ciphertext attack," In *Crypto '91*, LNCS Vol. 576.
28. P. Rogaway, M. Bellare, J. Black, and T. Krovetz, "OCB: A Block-Cipher Mode of Operation for Efficient Authenticated Encryption," In *Proc. 8th CCS*, ACM, 2001.
29. C. Schnorr and M. Jakobsson, "Security of Signed ElGamal Encryption," In *Asiacrypt '00*, pp. 73–89, LNCS Vol. 1976.
30. A. Shamir and Y. Tauman, "Improved Online/Offline Signature Schemes," In *Crypto '01*, pp. 355–367, LNCS Vol. 2139.
31. V. Shoup, "On Formal Models for Secure Key Exchange," Technical Report RZ 3120, IBM Research, 1999.
32. V. Shoup, "A proposal for an ISO standard for public key encryption (version 2.1)," Manuscript, Dec. 20, 2001.
33. D. Simon, "Finding Collisions on a One-Way Street: Can Secure Hash Functions Be Based on General Assumptions?," In *Eurocrypt '98*, pp. 334–345, LNCS Vol. 1403.
34. Y. Tsiounis and M. Yung, "On the Security of ElGamal Based Encryption," In *PKC '98*, pp. 117–134, LNCS Vol. 1431.
35. Y. Zheng, "Digital Signcryption or How to Achieve Cost(Signature & Encryption) $\ll$ Cost(Signature) + Cost(Encryption)," In *Crypto '97*, pp. 165–179, 1997, LNCS Vol. 1294.
36. Y. Zheng and H. Imai, "Efficient Signcryption Schemes on Elliptic Curves," *Information Processing Letters*, 68(5):227–233, December 1998.

AES and the Wide Trail Design Strategy

Joan Daemen[1] and Vincent Rijmen[2,3]

[1] ProtonWorld, Zweefvliegtuigstraat 10, B-1130 Brussel, Belgium
Joan.Daemen@protonworld.com
[2] CRYPTOMAThIC, Lei 8A, B-3001 Leuven, Belgium
Vincent.Rijmen@cryptomathic.com
[3] IAIK, Graz University of Technology, Inffeldgasse 16a/1, A-8010 Graz, Austria

Rijndael is an iterated block cipher that supports key and block lengths of 128 to 256 bits in steps of 32 bits. It transforms a plaintext block into a ciphertext block by iteratively applying a single round function alternated by the addition (XOR) of a round keys. The round keys are derived from the cipher key by means of a key schedule. As a result of the wide trail strategy, the round function of Rijndael consists of three dedicated steps that each have a particular role. Rijndael versions with a block length of 128 bits, and key lengths of 128, 192 and 256 bits have been adopted as the Advanced Encryption Standard (AES).

The main cryptographic criterion in the design of Rijndael has been its resistance against differential and linear cryptanalysis. Differential cryptanalysis exploits differential trails ("characteristics") with high probability, Linear cryptanalysis exploits linear trails ("linear approximation") with high correlations ("bias"). Differential and linear trails have in common that they are both structures that propagate over multiple rounds. We show that in key-alternating ciphers such as Rijndael, their probability/correlation is independent of the value of the key. This greatly simplifies the estimation of the resistance against linear and differential cryptanalysis.

The structure of the Rijndael round function imposes strict upper limits to the correlation and probability of multiple-round trails. By combining diffusion operations based on MDS codes with byte transpositions in the round function, we obtain a provable lower bound of 25 active S-boxes in any four-round trail.

Linear trails and differential trails may combine to give rise to correlation and difference propagation probability values that are significantly higher than those of individual trails. We show the effect of the combination of trails for the case of a fixed key for any iterated block cipher and averaged over all round keys for key-alternating ciphers.

The most powerful attacks against Rijndael are saturation attacks. These attacks exploit the byte-oriented structure of the cipher and can break round-reduced variants of Rijndael up to 6 (128-bit key and state) or 7 rounds.

Rijndael can be completely specified with operations in GF(2^8). How the elements of GF(2^8) are represented in bytes can be seen as a detail of the specification, important for interoperability only. We can make abstraction from the representation of the elements of GF(2^8) and consider a block cipher that operates on strings of elements of GF(2^8). We call this generalization RIJNDAEL-GF.

L.R. Knudsen (Ed.): EUROCRYPT 2002, LNCS 2332, pp. 108–109, 2002.

Rijndael is an instance of RIJNDAEL-GF, where the representation of the elements has been specified.

Intuitively, it seems obvious that if Rijndael has a cryptographic weakness, this is inherited by RIJNDAEL-GF and any instance of it. Traditionally, linear and differential cryptanalysis are done at the bit level and hence require to choose one specific representation of $GF(2^8)$. We demonstrate how to conduct differential and linear propagation analysis at the level of elements of $GF(2^8)$, without having to deal with representation issues.

Indistinguishability of Random Systems

Ueli Maurer*

ETH Zurich, Department of Computer Science,
maurer@inf.ethz.ch

Abstract. An $(\mathcal{X}, \mathcal{Y})$-random system takes inputs $X_1, X_2, \ldots \in \mathcal{X}$ and generates, for each new input X_i, an output $Y_i \in \mathcal{Y}$, depending probabilistically on $X_1, \ldots, X_i$ and $Y_1, \ldots, Y_{i-1}$. Many cryptographic systems like block ciphers, MAC-schemes, pseudo-random functions, etc., can be modeled as random systems, where in fact Y_i often depends only on X_i, i.e., the system is stateless. The security proof of such a system (e.g. a block cipher) amounts to showing that it is indistinguishable from a certain perfect system (e.g. a random permutation).
We propose a general framework for proving the indistinguishability of two random systems, based on the concept of the equivalence of two systems, conditioned on certain events. This abstraction demonstrates the common denominator among many security proofs in the literature, allows to unify, simplify, generalize, and in some cases strengthen them, and opens the door to proving new indistinguishability results.
We also propose the previously implicit concept of quasi-randomness and give an efficient construction of a quasi-random function which can be used as a building block in cryptographic systems based on pseudo-random functions.

Key words. Indistinguishability, random systems, pseudo-random functions, pseudo-random permutations, quasi-randomness, CBC-MAC.

1 Introduction

1.1 Indistinguishability

Indistinguishability of two systems, introduced by Blum and Micali [7] for defining pseudo-random bit generators, is a central concept in cryptographic security definitions and proofs. The simplest distinguisher problem is that for two random variables: The success probability (or advantage) of the optimal distinguisher is just the distance of the two probability distributions. As a slight generalization, one can define indistinguishability for infinite sequences of random variables, e.g. of a pseudo-random bit generator from a true random bit generator [7].

It is substantially more difficult to investigate the indistinguishability of two *interactive* random systems **F** and **G** because the distinguisher can *adaptively* choose its inputs (also called queries) to the system, depending on the outputs seen for previous inputs. Every distinguisher **D** defines a pair of generally very

* Supported in part by the Swiss National Science Foundation, grant 2000-055466.98

L.R. Knudsen (Ed.): EUROCRYPT 2002, LNCS 2332, pp. 110–133, 2002.

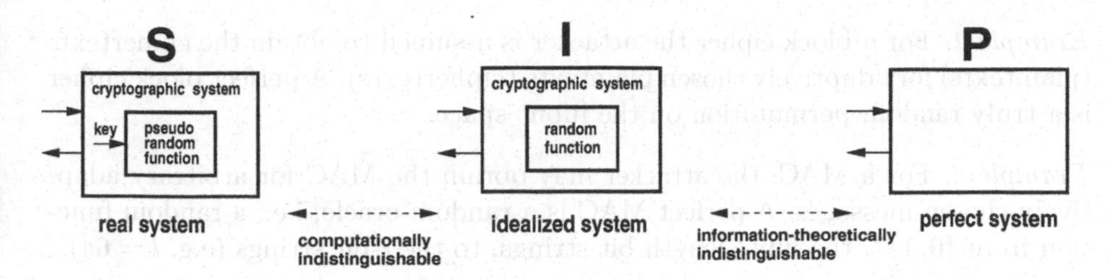

Fig. 1. Real system **S**, idealized system **I**, and perfect system **P**.

complex random experiments, one when **D** queries **F** and the other one when **D** queries **G**. A security proof requires to prove an upper bound, holding for every **D**, on the difference of the probability of some event in the corresponding two experiments. In general, this is a hard probability-theoretic problem.

1.2 Security Proofs Based on Pseudo-random Functions

The security of many cryptographic systems (e.g., block ciphers, message authentication codes, challenge-response protocols) is based on the assumption that a certain component (e.g. DES, IDEA, or Rijndael) used in the construction is a pseudo-random function (PRF) [8]. Such systems are proven secure, relative to this assumption, by showing that any algorithm for breaking the system can be transformed into a distinguisher for the PRF. For example, in a classic paper, Luby and Rackoff [10] showed how to construct a secure block cipher from any pseudo-random function, and Bellare et al. [2] proved the security of the CBC-MAC. The following general steps can be used to prove the security of a cryptographic system based on a pseudo-random function (cf. Fig. 1):

1. The attacker's capabilities, i.e., the types and number of allowed queries to **S** are defined. Moreover, security of **S** is defined by specifying what it means for the attacker to break **S**, and a purely theoretical *perfect system* **P** is defined which is trivially secure (see examples below).
2. One considers an *idealized system* **I** obtained from **S** by replacing the PRF by a truly random function and proves that **I** and **P** are information-theoretically indistinguishable: no adaptive computationally unbounded distinguisher algorithm **D** has a non-negligible advantage unless it queries the system for an infeasibly large (e.g. super-polynomial) number of queries.[1]
3. Hence, because **S** is computationally indistinguishable from **I** if the underlying function is pseudo-random, **S** is also computationally indistinguishable from **P**. Because **P** is unbreakable, there exists no breaking algorithm for **S** since it could directly be used as a distinguisher for **S** and **P**.

[1] This is the only technical step in such a proof. It is purely information-theoretic, not involving complexity theory, and is the subject of this paper.

Example 1. For a block cipher the attacker is assumed to obtain the ciphertexts (plaintexts) for adaptively chosen plaintexts (ciphertexts). A perfect block cipher is a truly random permutation on the input space.

Example 2. For a MAC, the attacker may obtain the MAC for arbitrary adaptively chosen messages. A perfect MAC is a random oracle, i.e., a random function from $\{0,1\}^*$, the finite-length bit strings, to the l-bit strings (e.g. $l = 64$).

1.3 Previous Work

Many authors were intrigued by the complexity of certain security proofs in the literature, most notably [10], and have given shorter proofs for these and more general results. It is beyond the scope of this paper to discuss all of these results, but a few are mentioned below. Patarin [14,15] developed a technique called "coefficient H method" and used it to analyze Feistel ciphers, even with more than four rounds [16]. To the best of our knowledge, the concept of conditioning events in security proofs was first made explicit in [11] and [12] where, using appropriate conditioning events, the proof for the Luby-Rackoff construction and generalizations thereof was shown to boil down to simple collision arguments (but the proof was stated only for non-adaptive distinguishers). Naor and Reingold [18] generalized the Luby-Rackoff constructions. In a sequence of papers (e.g., see [21, 22]), Vaudenay developed decorrelation theory and applied it to the design of block ciphers and the analysis of constructions like the CBC-MAC. Petrank and Rackoff [17] gave a generalized treatment of the CBC-MAC.

1.4 Contributions of the Paper and Sketch of the Framework

This paper defines the natural concept of a random system and proposes a general framework for proving the indistinguishability of two random systems **F** and **G** by identifying internal events such that, conditioned on these events, **F** and **G** are equivalent, i.e., have the identical input-output behavior.

The advantage in distinguishing **F** and **G** with k queries and unbounded computing power is shown to be at most the probability of success in provoking one of these events *not* to occur (Theorem 1). Under a certain condition, adaptive strategies can be shown to be not more powerful than non-adaptive strategies, thus allowing to eliminate the distinguisher from the analysis (Theorem 2 and Corollary 1).

The framework is illustrated for a few application areas and by giving simple and intuitive analyses and generalizations of some classical results. Due to the high level of abstraction, one can apply the basic techniques in settings where previous proof techniques appeared to be too complex or where changing a small detail in the construction requires a complete rehash of the proof.

Moreover, in some cases one can prove stronger bounds. For instance, under certain conditions one can prove that if a construction involves several components, each indistinguishable from a certain perfect system, then the overall system is distinguishable from its perfect counterpart with probability only the *product* (rather than the sum or the maximum) of the maximal distinguishing probabilities of the component systems (Theorem 3).

1.5 A Motivating Example

The security proof [2] for the CBC-MAC (cf. Fig. 6), and several generalizations thereof, will follow as a simple consequence of our framework (see Section 6). Roughly speaking, the proof consists of the following simple steps. First, conditioned on the event that all inputs to the internal random function **R** (modeling the PRF used in an actual implementation), corresponding to a final block of a message, are distinct, the CBC-MAC behaves like a random oracle, i.e., a perfect MAC. Second, one can hence restrict attention to algorithms trying to prevent this event from occurring by any adaptive choice of the inputs. Third, since the outputs are independent of the inputs, given this event, one can restrict the analysis to non-adaptive strategies, which turn out to be easy to analyze.

1.6 Quasi-randomness

The general idea behind such cryptographic constructions is to "package" a given amount of randomness such that it appears to any observer as a random system **S** which behaves essentially like a (in some sense) perfect random system **P** containing a much larger amount of randomness. If **S** is computationally indistinguishable from **P**, it is generally called pseudo-random (with respect to **P**). Informally, we call **S** *quasi-random* (with respect to **P**) if it is indistinguishable from **P**, provided only that the *amount of interaction* (e.g. the number of queries) is bounded, but with otherwise unbounded computational resources.

An important question, addressed in this paper, is how an efficient quasi-random system **S** of a certain type can be constructed, using as few random bits as possible, and indistinguishable from the corresponding perfect system **P** for as many queries as possible.

1.7 Outline of the Paper

In Section 3 we introduce the concepts of a random automaton and of a random system as well as the equivalence of such systems. We also define monotone conditions and event sequences, the conditional equivalence of random systems, cascades of random systems, and the invocation of a random system by another random system. In Section 4 we define the indistinguishability of random systems, prove a few general results on indistinguishability, and discuss the framework for indistinguishability proofs based on conditional equivalence as well as consequences thereof. In Section 5 we apply the framework to the construction of quasi-random functions, and in Sections 6 and 7 to the analysis and security proofs of MAC's and of pseudo-random permutations, respectively.

The treatment is more general than necessary just for proving the results in Sections 5–7. Due to space limitations, many proofs are omitted (but see [13]).

2 Notation and Preliminaries

Random variables and concrete values they can take on are usually denoted by capital and small letters, respectively. For a set $\mathcal{S}$, an *$\mathcal{S}$-sequence* is an infinite (or possibly finite) sequence $s = s_1, s_2, \ldots$ of elements of $\mathcal{S}$. Prefixes of sequences (of values or random variables) are denoted by a superscript, e.g. s^k denotes the finite sequence $[s_1, s_2, \ldots, s_k]$. For a list L of random variables over the same alphabet, $\mathrm{dist}(L)$ denotes the event that all values in L are distinct. Let $p_{\mathrm{coll}}(n, k)$ denote the probability that k independent random variables with uniform distribution over a set of size n contain a collision, i.e., that they are not all distinct. Of course, $p_{\mathrm{coll}}(n, k) = 1 - \prod_{i=1}^{k-1}\left(1 - \frac{i}{n}\right) < \frac{k^2}{2n}$.

In the context of this paper one considers different random experiments, and when analyzing probabilities it is crucial to be precise about which random experiment is considered. The random experiment is usually defined by one or several defining, usually independent, random variables. We will use these defining random variables as superscripts when denoting probabilities. For example, if $\mathbf{F}$ denotes the system under investigation and $\mathbf{D}$ the distinguisher, then $P^{\mathbf{DF}}$ denotes probabilities in the combined random experiment where $\mathbf{D}$ queries $\mathbf{F}$. In contrast $P^{\mathbf{F}}$ denotes probabilities in the simpler random experiment involving only the selection of $\mathbf{F}$, without even considering a distinguisher. If no superscript is used, the random experiment is clear from the context.

We use the following notation for probability distributions. If $\mathcal{A}$ and $\mathcal{B}$ are events and U and V are random variables with ranges $\mathcal{U}$ and $\mathcal{V}$, respectively, then $P_{U\mathcal{A}|V\mathcal{B}}$ denotes the corresponding conditional probability distribution, a function $\mathcal{U} \times \mathcal{V} \to \mathbf{R}^+$. Thus $P_{U\mathcal{A}|V\mathcal{B}}(u, v)$ for $u \in \mathcal{U}$ and $v \in \mathcal{V}$ is well-defined (except if $P_{V\mathcal{B}}(v) = 0$ in which case it is undefined). Note that $P_{\mathcal{A}}$ is equivalent to $P(\mathcal{A})$. For an event E, $\overline{E}$ denotes the complement of E. Equality of probability distributions means equality as functions, i.e., for all arguments. This extends to the equality of conditional probability distributions, even if one of them contains additional random variables in the conditioning set, meaning that equality holds for all possible values. For example, $P_{Y^i|X^k} = P_{Y^i|X^i}$ for $k > i$ means that for all x^k and y^i, $P_{Y^i|X^k}(y^i, x^k) = P_{Y^i|X^i}(y^i, x^i)$.

3 Random Systems and Monotone Event Sequences

3.1 Sources, Random Automata, and Random Systems

Definition 1. *An $\mathcal{X}$-*source $\mathbf{S}$ *is an infinite sequence $\mathbf{S} = S_1, S_2, \ldots$ of random variables $S_i \in \mathcal{X}$, characterized by the sequence $P^{\mathbf{S}}_{S_i|S^{i-1}}$ of conditional probability distributions. This also defines the distributions $P^{\mathbf{S}}_{S^i} := \prod_{j=1}^{i} P^{\mathbf{S}}_{S_i|S^{i-1}}$.*

In the following we consider systems which take inputs (or queries) $X_1, X_2, \ldots \in \mathcal{X}$ and generate, for each new input X_i, an output $Y_i \in \mathcal{Y}$. Such a system can be deterministic or probabilistic, and it can be stateless or contain internal memory. A stateless deterministic system is simply a function $\mathcal{X} \to \mathcal{Y}$.

Fig. 2. Left: An $(\mathcal{X}, \mathcal{Y})$-random system $\mathbf{F}$ takes inputs $X_1, X_2, X_3, \ldots \in \mathcal{X}$ and outputs $Y_1, Y_2, Y_3, \ldots \in \mathcal{Y}$, where Y_i is generated after receiving input X_i. It is characterized by the sequence of conditional probability distributions $P^{\mathbf{F}}_{Y_i|X^iY^{i-1}}$ for $i \geq 1$. Right: Random system $\mathbf{F}$ with a monotone event sequence $\mathcal{A} = A_0, A_1, A_2, \ldots$, denoted $\mathbf{F}^{\mathcal{A}}$.

Definition 2. A *random function* $\mathcal{X} \to \mathcal{Y}$ is a random variable which takes as values functions $\mathcal{X} \to \mathcal{Y}$. A deterministic system with state space Σ is called an $(\mathcal{X}, \mathcal{Y})$-*automaton* and is described by an infinite sequence $f_1, f_2, \ldots$ of functions, with $f_i : \mathcal{X} \times \Sigma \to \mathcal{Y} \times \Sigma$, where $(Y_i, S_i) = f_i(X_i, S_{i-1})$, S_i is the state at time i, and an initial state S_0 is fixed. An $(\mathcal{X}, \mathcal{Y})$-*random automaton* $\mathbf{F}$ is like an automaton but $f_i : \mathcal{X} \times \Sigma \times \mathcal{R} \to \mathcal{Y} \times \Sigma$ (where $\mathcal{R}$ is the space of the internal randomness), together with a probability distribution over $\mathcal{R} \times \Sigma$ specifying the internal randomness and the initial state.[2]

A large variety of constructions and definitions in the cryptographic literature can be interpreted as random functions, including pseudo-random functions, pseudo-random permutations, and MAC schemes. We consider the more general concept of a (stateful) random system because this is just as simple and because distinguishers can also be modeled as random systems.

The observable input-output behavior of a random automaton $\mathbf{F}$ is referred to as a random system. In the following we use the terms random automaton and random system interchangeably when no confusion is possible.

Definition 3. An $(\mathcal{X}, \mathcal{Y})$-*random system* $\mathbf{F}$ is an infinite[3] sequence of conditional probability distributions $P^{\mathbf{F}}_{Y_i|X^iY^{i-1}}$ for $i \geq 1$.[4] Two random automata $\mathbf{F}$ and $\mathbf{G}$ are *equivalent*, denoted $\mathbf{F} \equiv \mathbf{G}$, if they correspond to the same random system, i.e., if $P^{\mathbf{F}}_{Y_i|X^iY^{i-1}} = P^{\mathbf{G}}_{Y_i|X^iY^{i-1}}$ for $i \geq 1$.[5]

The above definition is very general and captures systems that answer several types of queries (in which case the input set $\mathcal{X}$ is the union of the query sets) and for which the behavior depends on the index i. Note that a source can be interpreted as a special type of random system for which the input is ignored, i.e., the outputs are independent of the inputs. We will often assume that the input and output alphabets of a random system are clear from the context.

Let us discuss a few special examples of random systems. Throughout, the symbols $\mathbf{B}$, $\mathbf{R}$, $\mathbf{P}$, and $\mathbf{O}$ are used exclusively for the systems defined below.

[2] $\mathbf{F}$ can also be considered as a random variable taking on as values $(\mathcal{X}, \mathcal{Y})$-automata.

[3] Random systems with finite-length input sequences could also be defined.

[4] $P^{\mathbf{F}}_{Y_i|X^iY^{i-1}}$ is a function $\mathcal{Y} \times \mathcal{X}^i \times \mathcal{Y}^{i-1} \to \mathbf{R}^+$ such that, for all $x^i \in \mathcal{X}^i$ and $y^{i-1} \in \mathcal{Y}^{i-1}$, $\sum_{y_i \in \mathcal{Y}} P^{\mathbf{F}}_{Y_i|X^iY^{i-1}}(y_i, x^i, y^{i-1}) = 1$.

[5] The distribution $P^{\mathbf{F}}_{Y^i|X^i} = \prod_{j=1}^{i} P^{\mathbf{F}}_{Y_i|X^iY^{i-1}}$ is also defined. $P^{\mathbf{F}}_{Y_i|X^iY^{i-1}}(y_i, x^i, y^{i-1})$ can be undefined for values x^i and y^{i-1} with $P^{\mathbf{F}}_{Y^{i-1}|X^i}(y^{i-1}, x^i) = 0$.

Definition 4. An $(\mathcal{X}, \mathcal{Y})$-*beacon* [19] $\mathbf{B}$ is a random system (actually a source) for which $Y_1, Y_2, \ldots$ are independent and uniformly distributed over $\mathcal{Y}$, independent of the inputs $X_1, X_2, \ldots$. A *uniform random function (URF)* $\mathbf{R} : \mathcal{X} \to \mathcal{Y}$ (a *uniform random permutation (URP)* $\mathbf{P}$ on $\mathcal{X}$) is a random function with uniform distribution over all functions from $\mathcal{X}$ to $\mathcal{Y}$ (permutations on $\mathcal{X}$). A $\mathcal{Y}$-*random oracle* $\mathbf{O}$ is a random function with input alphabet $\mathcal{X} = \{0,1\}^*$ with $P^{\mathbf{O}}_{Y_i|X_i}(y, x) = 1/|\mathcal{Y}|$ for all $i \geq 1$, $x \in \mathcal{X}$ and $y \in \mathcal{Y}$.

3.2 Monotone Conditions and Event Sequences

For a given $(\mathcal{X}, \mathcal{Y})$-random function or automaton $\mathbf{F}$, the evaluation of Y_i usually requires the evaluation of some internal random variables.[6] Consider the internal sequence of random variables $U_1, U_2, \ldots$. In the sequel it is very useful to consider an internal condition defined, for each i, after input X_i is entered. As a simple example, the condition could be $\text{dist}(U^i)$, i.e., that $U_1, \ldots, U_i$ are all distinct.

Such an internal condition can be modeled as a binary random variable, say Z_i, indicating whether the condition is satisfied ($Z_i = 1$) or not ($Z_i = 0$) after input X_i has been given. If Z_i is taken as part of the ith output of $\mathbf{F}$, i.e., the ith output is the pair (Y_i, Z_i) instead of just Y_i, then this corresponds to a $(\mathcal{X}, \mathcal{Y} \times \{0,1\})$-random system.[7] One can also define several such conditions for $\mathbf{F}$, each corresponding to a binary random variable.

We will only consider *monotone* conditions, meaning that once it fails to be satisfied it remains so for all future inputs. For example, the condition $\text{dist}(U^i)$ is obviously monotone. If U_i is a vector in some vector space, another monotone condition is that $U_1, \ldots, U_i$ are linearly independent.

For a random automaton $\mathbf{F}$ and a given monotone internal condition we will often be interested in $\mathbf{F}$'s behavior only as long as the condition is satisfied. For example, a URF behaves like a beacon as long as the inputs are distinct. We therefore consider the monotone sequence $\mathcal{A} = A_0, A_1, A_2, \ldots$ of events, where A_i is the event that the condition is satisfied (and $\overline{A_i}$ is the complementary event) and where A_0 is for convenience defined to be the certain event (cf. Fig. 2).

We will also consider two or more monotone conditions simultaneously. For two monotone event sequences (MES) $\mathcal{A}$ and $\mathcal{B}$ defined for $\mathbf{F}$, $\mathcal{A} \wedge \mathcal{B}$ denotes the MES defined by $(\mathcal{A} \wedge \mathcal{B})_i = A_i \wedge B_i$ for $i \geq 1$, and $\mathcal{A} \vee \mathcal{B}$ is defined analogously.

Definition 5. For MESs $\mathcal{A}$ and $\mathcal{C}$ defined for random automata $\mathbf{F}$ and $\mathbf{G}$, respectively, $\mathbf{F}$ *with* $\mathcal{A}$ *is equivalent to* $\mathbf{G}$ *with* $\mathcal{C}$, denoted $\mathbf{F}^{\mathcal{A}} \equiv \mathbf{G}^{\mathcal{C}}$, if $P^{\mathbf{F}}_{Y_i A_i | X^i Y^{i-1} A_{i-1}} = P^{\mathbf{G}}_{Y_i C_i | X^i Y^{i-1} C_{i-1}}$ for $i \geq 1$.[8]

We refer to later sections for examples.

---

[6] For example, in the CBC-MAC U_i could be the input to the internal random function corresponding to the last block of the ith message.

[7] One can also think of an internal device (or genie) in $\mathbf{F}$ which beeps when the condition fails to be satisfied ($Z_i = 0$).

[8] Note that $\mathbf{F}^{\mathcal{A}} \equiv \mathbf{G}^{\mathcal{C}}$ does not imply $\mathbf{F} \equiv \mathbf{G}$.

Definition 6. For a random system $\mathbf{F}$ with MES $\mathcal{A} = A_0, A_1, A_2, \ldots$, $\mathbf{F}$ *conditioned on* $\mathcal{A}$ *is equivalent to* $\mathbf{G}$, denoted $\mathbf{F}|\mathcal{A} \equiv \mathbf{G}$, if $P^{\mathbf{F}}_{Y_i|X^iY^{i-1}A_i} = P^{\mathbf{G}}_{Y_i|X^iY^{i-1}}$ for $i \geq 1$, for all arguments for which $P^{\mathbf{F}}_{Y_i|X^iY^{i-1}A_i}$ is defined. More generally, if $\mathcal{A}$ and $\mathcal{B}$ are defined for $\mathbf{F}$, then we write $\mathbf{F}^{\mathcal{B}}|\mathcal{A} \equiv \mathbf{G}^{\mathcal{C}}$ if $P^{\mathbf{G}}_{Y_iC_i|X^iY^{i-1}C_{i-1}} = P^{\mathbf{F}}_{Y_iB_i|X^iY^{i-1}B_{i-1}A_i}$ for $i \geq 1$.

Definition 7. One can *adjoin* an MES $\mathcal{C}$ to a random system $\mathbf{G}$ by defining C_i as depending probabilistically on X^i and Y^i, i.e., by a sequence of distributions $P^{\mathbf{G}}_{C_i|X^iY^iC_{i-1}}$. If an MES $\mathcal{C}$ is already defined for $\mathbf{G}$, then one can adjoin a further MES $\mathcal{D}$ according to a sequence $P^{\mathbf{G}}_{D_i|X^iY^iC_iD_{i-1}}$ of distributions.[9]

Lemma 1. **(i)** *If* $\mathbf{F}^{\mathcal{A}} \equiv \mathbf{G}^{\mathcal{C}}$, *then* $\mathbf{F}|\mathcal{A} \equiv \mathbf{G}|\mathcal{C}$[10] *(but not vice versa).*
(ii) *If* $\mathbf{F}|\mathcal{A} \equiv \mathbf{G}$, *then* $\mathbf{F}^{\mathcal{A}} \equiv \mathbf{G}^{\mathcal{C}}$ *for some MES* $\mathcal{C}$ *adjoined to* $\mathbf{G}$.
(iii) *More generally, if* $\mathbf{F}^{\mathcal{B}}|\mathcal{A} \equiv \mathbf{G}^{\mathcal{C}}$, *then* $\mathbf{F}^{\mathcal{A}\wedge\mathcal{B}} \equiv \mathbf{G}^{\mathcal{C}\wedge\mathcal{D}}$ *for some MES* $\mathcal{D}$.
(iv) *If* $\mathbf{F}|\mathcal{A} \equiv \mathbf{G}|\mathcal{C}$ *and* $P^{\mathbf{F}}_{A_i|X^iY^{i-1}A_{i-1}} \leq P^{\mathbf{G}}_{C_i|X^iY^{i-1}C_{i-1}}$ *for* $i \geq 1$ *(and for all* x^i *and* y^{i-1}*), then one can adjoin an MES* $\mathcal{D}$ *to* $\mathbf{G}$ *such that* $\mathbf{F}^{\mathcal{A}} \equiv \mathbf{G}^{\mathcal{C}\wedge\mathcal{D}}$.

Proof. Claim (i) is obvious. Claim (ii) follows from (iii), which follows by defining the MES $\mathcal{D}$ via $P^{\mathbf{G}}_{D_i|X^iY^iC_iD_{i-1}} = P^{\mathbf{F}}_{A_i|X^iY^{i-1}A_{i-1}B_{i-1}}$. The proof uses $P^{\mathbf{G}}_{Y_iC_i|X^iY^{i-1}C_{i-1}D_{i-1}} = P^{\mathbf{G}}_{Y_iC_i|X^iY^{i-1}C_{i-1}}$ (since $P^{\mathbf{G}}_{D_{i-1}|X^iY^iC_i} = P^{\mathbf{G}}_{D_{i-1}|X^iY^{i-1}C_{i-1}}$) and $P^{\mathbf{G}}_{Y_iC_i|X^iY^{i-1}C_{i-1}} = P^{\mathbf{F}}_{Y_iB_i|X^iY^{i-1}B_{i-1}A_i}$ (from $\mathbf{F}^{\mathcal{A}} \equiv \mathbf{G}^{\mathcal{C}}$). The proof of (iv) is omitted. □

The following lemma states the trivial fact that given that all inputs are distinct, a random function behaves like a beacon. The proof is obvious.

Lemma 2. *Let* $\mathcal{C}$ *(*$\mathcal{D}$*) be an MES defined on the inputs (outputs) of a system.*
(i) $\mathbf{F}|\mathcal{C} \equiv \mathbf{F}$ *for every random system* $\mathbf{F}$.
(ii) *If* $\mathbf{F}^{\mathcal{A}} \equiv \mathbf{G}^{\mathcal{B}}$, *then* $\mathbf{F}^{\mathcal{A}\wedge\mathcal{C}} \equiv \mathbf{G}^{\mathcal{B}\wedge\mathcal{C}}$ *and* $\mathbf{F}^{\mathcal{A}\wedge\mathcal{D}} \equiv \mathbf{G}^{\mathcal{B}\wedge\mathcal{D}}$.
(iii) *If* C_i *implies that the first* i *inputs are distinct, then* $\mathbf{R}^{\mathcal{C}} \equiv \mathbf{B}^{\mathcal{C}}$ *and* $\mathbf{R}|\mathcal{C} \equiv \mathbf{B}$.

3.3 Cascades and Invocations of Random Systems

Definition 8. The *cascade* of an $(\mathcal{X}, \mathcal{Y})$-random system $\mathbf{F}$ and a $(\mathcal{Y}, \mathcal{Z})$-random system $\mathbf{G}$, denoted $\mathbf{FG}$, is the $(\mathcal{X}, \mathcal{Z})$-random system defined as applying $\mathbf{F}$ to the input sequence and $\mathbf{G}$ to the output of $\mathbf{F}$ (cf. Fig. 3). For MESs $\mathcal{A}$ and $\mathcal{B}$ defined for $\mathbf{F}$ and $\mathbf{G}$, respectively, $\mathcal{A}$, $\mathcal{B}$, and $\mathcal{A}\wedge\mathcal{B}$ are defined naturally for $\mathbf{FG}$.

Lemma 3. **(i)** *For any source* $\mathbf{S}$ *and any (compatible)* $\mathbf{E}$ *we have* $\mathbf{ES} \equiv \mathbf{S}$.
(ii) *If* $\mathbf{F}|\mathcal{A} \equiv \mathbf{G}$, *then* $\mathbf{EF}|\mathcal{A} \equiv \mathbf{EG}$ *for any compatible* $\mathbf{E}$.

---

[9] Informally, one connects an *independent* component, characterized by $P^{\mathbf{G}}_{D_i|X^iY^iC_iD_{i-1}}$, to the input and output of $\mathbf{G}$ and to the indicator random variable of $\mathcal{C}$ which generates the indicator random variable for $\mathcal{D}$.

[10] $\mathbf{F}|\mathcal{A} \equiv \mathbf{G}|\mathcal{C}$ should be read as: there exists $\mathbf{H}$ such that $\mathbf{F}|\mathcal{A} \equiv \mathbf{H}$ and $\mathbf{G}|\mathcal{C} \equiv \mathbf{H}$.

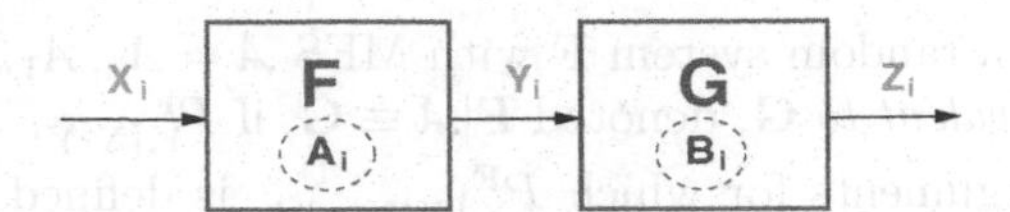

Fig. 3. The cascade of an $(\mathcal{X}, \mathcal{Y})$-random system $\mathbf{F}$ and a $(\mathcal{Y}, \mathcal{Z})$-random system $\mathbf{G}$, denoted $\mathbf{FG}$. For $\mathbf{F}^{\mathcal{A}}$ and $\mathbf{G}^{\mathcal{B}}$, $\mathbf{FG}^{\mathcal{A} \wedge \mathcal{B}}$ is defined naturally.

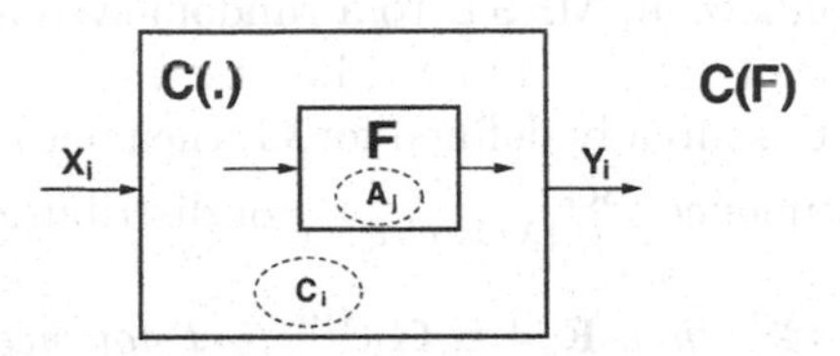

Fig. 4. A random system $\mathbf{C}(.)$ invoking an internal random system $\mathbf{F}$, then the combined random system is $\mathbf{C}(\mathbf{F})$.

We denote by $\mathbf{C}(.)$ a random system that *invokes* an internal random system (with specified input and output alphabets). If the internal system is $\mathbf{F}$, then the combined random system is $\mathbf{C}(\mathbf{F})$ (cf. Fig. 4). For the evaluation of the output Y_i for a given input X_i to $\mathbf{C}(\mathbf{F})$, $\mathbf{F}$ is called zero, one, or several times, where the inputs to $\mathbf{F}$ and even the number of such inputs may depend on the state of $\mathbf{C}(.)$, hence on $X_1, \ldots, X_i$.[11]

An MES, say $\mathcal{C} = C_0, C_1, C_2, \ldots$, can be defined also for such a system $\mathbf{C}(.)$. If $\mathcal{A}$ is an MES defined for the invoked $\mathbf{F}$, one can associate a natural corresponding MES $\tilde{\mathcal{A}} = \tilde{A}_0, \tilde{A}_1, \tilde{A}_2, \ldots$ with $\mathbf{C}(\mathbf{F})$, where $\tilde{A}_i$ is the event that the A-event occurs for $\mathbf{F}$ up to the evaluation of the ith input to $\mathbf{C}(\mathbf{F})$. If $\mathbf{F}$ is called t times for each input to $\mathbf{C}(\mathbf{F})$, then $\tilde{A}_i = A_{ti}$. Let $m_{\mathbf{C}(.)}(k)$ be the maximal number of evaluations of any internal system $\mathbf{F}$ for any sequence of k inputs to $\mathbf{C}(\mathbf{F})$, if it is defined.

The following lemma states the simple fact that by replacing a random system by an equivalent random system, the overall behavior of a system does not change. Let $\mathbf{C}(.)$ be any random system and let $\mathbf{F}$ and $\mathbf{G}$ be input/output compatible with $\mathbf{C}(.)$. Let $\mathcal{A}, \mathcal{B}$, and $\mathcal{C}$ be defined for $\mathbf{C}(.)$, $\mathbf{F}$ and $\mathbf{G}$, respectively.

Lemma 4. **(i)** *If* $\mathbf{F} \equiv \mathbf{G}$, *then* $\mathbf{C}(\mathbf{F}) \equiv \mathbf{C}(\mathbf{G})$ *and* $\mathbf{C}(\mathbf{F})^{\mathcal{C}} \equiv \mathbf{C}(\mathbf{G})^{\mathcal{C}}$.
(ii) *If* $\mathbf{F}^{\mathcal{A}} \equiv \mathbf{G}^{\mathcal{B}}$, *then* $\mathbf{C}(\mathbf{F})^{\tilde{\mathcal{A}}} \equiv \mathbf{C}(\mathbf{G})^{\tilde{\mathcal{B}}}$ *and* $\mathbf{C}(\mathbf{F})^{\tilde{\mathcal{A}} \wedge \mathcal{C}} \equiv \mathbf{C}(\mathbf{G})^{\tilde{\mathcal{B}} \wedge \mathcal{C}}$.[12]

Proof. The lemma follows directly from the fact that the probability distribution of all random variables and events occurring in $\mathbf{C}(.)$, when including $\mathcal{A} = A_0, A_1, A_2, \ldots$ (or $\mathcal{B} = B_0, B_1, B_2, \ldots$), is the product of conditional distributions defined by the random system and by $\mathbf{C}(.)$. The conditional distributions

[11] Formally, $\mathbf{C}(.)$ is not a random system without specifying an argument $\mathbf{F}$.
[12] Note, however, that $\mathbf{F}|\mathcal{A} \equiv \mathbf{G}$ does not imply $\mathbf{C}(\mathbf{F})|\tilde{\mathcal{A}} \equiv \mathbf{C}(\mathbf{G})$.

Fig. 5. Distinguishing two $(\mathcal{X}, \mathcal{Y})$-random systems **F** and **G** by means of a distinguisher **D**. The figure shows the two random experiments under consideration.

defined by $\mathbf{C}(.)$ are trivially identical and those defined by **F** (or **G**) are identical in both cases because of $\mathbf{F}^{\mathcal{A}} \equiv \mathbf{G}^{\mathcal{B}}$. □

4 Indistinguishability Proofs for Random Systems

4.1 Distinguishers for Random Systems

We consider the problem of distinguishing two $(\mathcal{X}, \mathcal{Y})$-random systems **F** and **G** by means of a computationally unbounded, possibly probabilistic adaptive distinguisher algorithm (or simply distinguisher) **D** asking at most k queries, for some k (cf. Fig. 5). The distinguisher generates X_1 as an input to **F** (or **G**), receives the output Y_1, then generates X_2, receives Y_2, etc. Finally, after receiving Y_k, it outputs a binary decision bit. More formally:

Definition 9. A *distinguisher for* $(\mathcal{X}, \mathcal{Y})$*-random systems* is a $(\mathcal{Y}, \mathcal{X})$-random system **D** together with an initial value $X_1 \in \mathcal{X}$ which outputs a binary decision value after some specified number k of queries to the system. Without loss of generality we can assume that **D** outputs a binary value after every query and that this sequence is monotone (0 never followed by 1), i.e., we can define the MES $\mathcal{E} = E_0, E_1, E_2, \ldots$ where E_i is the event that **D** outputs 1 after the i-th query. Application of **D** to a random system **F** (cf. Fig. 5) means that X_1 is the first input to **F**, the i-th input and output of **D** are Y_i and $\tilde{X}_i$, respectively, and $X_i := \tilde{X}_{i-1}$ for $i \geq 2$ is the i-th input to **F**.

Definition 10. The maximal advantage, of any distinguisher issuing k queries, for distinguishing **F** and **G**, is

$$\Delta_k(\mathbf{F}, \mathbf{G}) \;:=\; \max_{\mathbf{D}} \left| P^{\mathbf{DF}}(E_k) - P^{\mathbf{DG}}(E_k) \right|.$$

We summarize a few simple facts used in many security proofs. The inequalities hold for any compatible random automata or random systems.

Lemma 5. (i) $\Delta_k(\mathbf{F}, \mathbf{H}) \leq \Delta_k(\mathbf{F}, \mathbf{G}) + \Delta_k(\mathbf{G}, \mathbf{H})$.
(ii) $\Delta_k(\mathbf{C}(\mathbf{F}), \mathbf{C}(\mathbf{G})) \leq \Delta_{k'}(\mathbf{F}, \mathbf{G})$, *where* $k' = m_{\mathbf{C}(.)}(k)$.
(iii) $\Delta_k(\mathbf{FF'}, \mathbf{GG'}) \leq \Delta_k(\mathbf{F}, \mathbf{G}) + \Delta_k(\mathbf{F'}, \mathbf{G'})$.
(iv) (Informal.) *If* $\Delta_k(\mathbf{F}, \mathbf{G})$ *is negligible in* k *and* **G** *is computationally indistinguishable from* **H**, *then* **F** *is also computationally indistinguishable from* **H**.

Proof. (i) follows by a simple application of the triangle inequality $|c-a| \leq |b-a|+|c-b|$ for any real a, b, and c, applied to $a = P^{\mathbf{DF}}(E_k)$, $b = P^{\mathbf{DG}}(E_k)$, and $c = P^{\mathbf{DH}}(E_k)$ for any distinguisher $\mathbf{D}$. To prove (ii), suppose for the sake of contradiction that there exists a distinguisher for $\mathbf{C}(\mathbf{F})$ and $\mathbf{C}(\mathbf{G})$, asking at most k queries, with advantage greater than $\Delta_{k'}(\mathbf{F}, \mathbf{G})$. By simulating $\mathbf{C}(.)$ one can construct a distinguisher for $\mathbf{F}$ and $\mathbf{G}$ with the same advantage, asking at most k' queries. This is a contradiction. Now we prove (iii). From (ii) we have $\Delta_k(\mathbf{FF}', \mathbf{GF}') \leq \Delta_k(\mathbf{F}, \mathbf{G})$ and $\Delta_k(\mathbf{GF}', \mathbf{GG}') \leq \Delta_k(\mathbf{F}', \mathbf{G}')$. Now we apply (i) to the random systems $\mathbf{FF}'$, $\mathbf{GF}'$, and $\mathbf{GG}'$. The proof of (iv) is omitted. □

It is easy to see that the described view of a distinguisher $\mathbf{D}$ is equivalent to an alternative view where $\mathbf{D}$ is given access to a blackbox containing $\mathbf{F}$ or $\mathbf{G}$ with probability $\frac{1}{2}$ each, where $\mathbf{D}$ must guess which of the two is the case. The best success probability with k queries is $\frac{1}{2} + \frac{1}{2}\Delta_k(\mathbf{F}, \mathbf{G})$.

4.2 Indistinguishability Proofs Based on Conditional Equivalence

In this section we prove that if $\mathbf{F}|\mathcal{A} \equiv \mathbf{G}$ for some MES $\mathcal{A}$ (or if $\mathbf{F}^{\mathcal{A}} \equiv \mathbf{G}^{\mathcal{B}}$), then a distinguisher $\mathbf{D}$ for distinguishing $\mathbf{F}$ from $\mathbf{G}$ with k queries (according to the view described above) *must* provoke the event $\overline{A_k}$ in $\mathbf{F}$ in order to have a non-zero advantage. Informally this could be proved by assuming a genie sitting inside $\mathbf{F}$ and beeping when it sees that $\overline{A_i}$ occurs for some i. The genie's help can only help since it could always be ignored, and given the genie's help, the optimal strategy would be to guess "$\mathbf{F}$" if the genie beeps and to flip a fair coin between $\mathbf{F}$ and $\mathbf{G}$ otherwise. Therefore we consider distinguishers $\mathbf{D}$ that try to provoke the event $\overline{A_k}$.

Definition 11. For a random system $\mathbf{F}$ with MES $\mathcal{A}$, let

$$\nu(\mathbf{F}, \overline{A_k}) := \max_{\mathbf{D}} P^{\mathbf{DF}}(\overline{A_k})$$

be the maximal probability, for any adaptive strategy $\mathbf{D}$, of provoking $\overline{A_k}$ in $\mathbf{F}$. Moreover, let

$$\mu(\mathbf{F}, \overline{A_k}) := \max_{x^k} P^{\mathbf{F}}_{\overline{A_k}|X^k}(x^k)$$

be the maximal probability of $\overline{A_k}$ for non-adaptive algorithms querying $\mathbf{F}$.

Lemma 6. (i) $\mu(\mathbf{F}, \overline{A_k}) \leq \nu(\mathbf{F}, \overline{A_k})$.
(ii) *If* $\mathbf{F}^{\mathcal{A}} \equiv \mathbf{G}^{\mathcal{B}}$, *then* $\nu(\mathbf{F}, \overline{A_k}) = \nu(\mathbf{G}, \overline{B_k})$.
(iii) $\nu(\mathbf{F}, \overline{A_k \vee B_k}) \leq \nu(\mathbf{F}, \overline{A_k}) + \nu(\mathbf{F}, \overline{B_k})$ *if* $\mathcal{A}$ *and* $\mathcal{B}$ *are defined for* $\mathbf{F}$.
(iv) *For any system* $\mathbf{C}(.)$ *with MES* $\mathcal{C}$, *invoking* $\mathbf{F}$, $\nu(\mathbf{C}(\mathbf{F}), \overline{C_k}) \leq \nu(\mathbf{C}(.), \overline{C_k})$[13] *and* $\nu(\mathbf{C}(\mathbf{F}), \overline{\tilde{A}_k}) \leq \nu(\mathbf{F}, \overline{A_{k'}})$, *where* $k' = m_{\mathbf{C}(.)}(k)$.
(v) *If* $\mathcal{A}$ *is defined on the inputs of* $\mathbf{F}$, *then* $\mu(\mathbf{EF}, \overline{A_k}) = \mu(\mathbf{E}, \overline{A_k})$ *for any* $\mathbf{E}$.

[13] $\nu(\mathbf{C}(.), \overline{C_k})$ is defined as the maximal probability of provoking event $\overline{C_k}$ in $\mathbf{C}(.)$ for algorithms with full control of the input to $\mathbf{C}(.)$ and the internal interface.

Proof. (i) holds because the set of adaptive strategies includes the non-adaptive ones. Claim (ii) follows from $\nu(\mathbf{F},\overline{A_k}) = 1 - \nu(\mathbf{F},A_k)$ and $\nu(\mathbf{G},\overline{B_k}) = 1 - \nu(\mathbf{G},B_k)$, using $\nu(\mathbf{F},A_k) = \nu(\mathbf{G},B_k)$ which follows from Lemma 4. Claim (iii) is a simple application of the union bound together with the fact that if different systems $\mathbf{D}$ can be used to provoke $\overline{A_k}$ and $\overline{B_k}$, this can only improve the success probability. Claim (iv) follows from the fact that $\mathbf{C}(.)$ can be used as a possible algorithm for provoking $\overline{A_k}$ in $\mathbf{F}$, and similarly $\mathbf{F}$ can be used as the random system in an algorithm for provoking $\overline{B_k}$ in $\mathbf{C}(.)$. Claim (v) is trivial. □

Lemma 7. *If $\mathbf{F}^{\mathcal{A}} \equiv \mathbf{G}^{\mathcal{B}}$, then for any (compatible) distinguisher $\mathbf{D}$ and any event E_k defined in $\mathbf{D}$ after k queries,*

$$\left|P^{\mathbf{DF}}(E_k) - P^{\mathbf{DG}}(E_k)\right| \le P^{\mathbf{DF}}(\overline{A_k}) = P^{\mathbf{DG}}(\overline{B_k}).$$

Proof. Lemma 4 gives $P^{\mathbf{DF}}(E_k \wedge A_k) = P^{\mathbf{DG}}(E_k \wedge B_k) \le P^{\mathbf{DG}}(E_k)$. Thus

$$P^{\mathbf{DF}}(E_k) = P^{\mathbf{DF}}(E_k \wedge A_k) + P^{\mathbf{DF}}(E_k \wedge \overline{A_k}) \le P^{\mathbf{DG}}(E_k) + P^{\mathbf{DF}}(\overline{A_k}).$$

$P^{\mathbf{DG}}(E_k) \le P^{\mathbf{DF}}(E_k) + P^{\mathbf{DG}}(\overline{B_k})$ follows by symmetry, and $P^{\mathbf{DF}}(\overline{A_k}) = P^{\mathbf{DG}}(\overline{B_k})$ follows from Lemma 4. □

Theorem 1. (i) *If $\mathbf{F}^{\mathcal{A}} \equiv \mathbf{G}^{\mathcal{B}}$ or $\mathbf{F}|\mathcal{A} \equiv \mathbf{G}$, then $\Delta_k(\mathbf{F},\mathbf{G}) \le \nu(\mathbf{F},\overline{A_k})$.*
(ii) *If $\mathbf{F}^{\mathcal{B}}|\mathcal{A} \equiv \mathbf{G}^{\mathcal{C}}$, then $\Delta_k(\mathbf{F},\mathbf{G}) \le \nu(\mathbf{F},\overline{A_k} \vee \overline{B_k}) \le \nu(\mathbf{F},\overline{A_k}) + \nu(\mathbf{G},\overline{C_k})$.*
(iii) *If $\mathbf{F}|\mathcal{A} \equiv \mathbf{G}|\mathcal{C}$ and $P^{\mathbf{F}}_{A_i|X^iY^{i-1}A_{i-1}} \le P^{\mathbf{G}}_{C_i|X^iY^{i-1}C_{i-1}}$ for $i \ge 1$, then $\Delta_k(\mathbf{F},\mathbf{G}) \le \nu(\mathbf{F},\overline{A_k})$.*

Proof. The first claim of (i) is a special case of Lemma 7, where $\mathbf{D}$ is the distinguisher with MES $\mathcal{E}$. The second claim of (i) is a special case of (ii), which is proved as follows. According to Lemma 1 (iii) we have $\mathbf{F}^{\mathcal{A}\wedge\mathcal{B}} \equiv \mathbf{G}^{\mathcal{C}\wedge\mathcal{D}}$ for some MES $\mathcal{D}$ defined for $\mathbf{G}$. Thus we can apply (i). The last inequality of (ii) follows because for any $\mathbf{D}$, $P^{\mathbf{DF}}(\overline{A_k} \vee \overline{B_k}) \le P^{\mathbf{DF}}(\overline{A_k}) + P^{\mathbf{DF}}(\overline{B_k}|A_k)$, and since if $P^{\mathbf{DF}}(\overline{A_k})$ and $P^{\mathbf{DF}}(\overline{B_k}|A_k)$ can be maximized separately by choices of $\mathbf{D}$, this is an upper bound on $\max_{\mathbf{D}} P^{\mathbf{DF}}(\overline{A_k} \vee \overline{B_k})$. Moreover, $\max_{\mathbf{D}} P^{\mathbf{DF}}(\overline{B_k}|A_k) = \max_{\mathbf{D}} P^{\mathbf{DG}}(\overline{C_k}) = \nu(\mathbf{G},\overline{C_k})$. To prove (iii), adjoin the MES $\mathcal{D}$ to $\mathbf{G}$ as in Lemma 1 (iv) and apply (i) of this theorem. □

4.3 Adaptive versus Non-adaptive Strategies

It is generally substantially easier to analyze non-adaptive as opposed to adaptive strategies, e.g. for distinguishing two random systems. The following theorem states simple and easily checkable conditions for a random system $\mathbf{F}$ with MES $\mathcal{A}$ which implies that no adaptive strategy for provoking $\overline{A_k}$ is better than the best non-adaptive strategy. The optimal strategy hence selects (one of) the fixed input sequence(s) x^k that minimizes $P^{\mathbf{F}}_{A_k|X^k}(x^k)$ (or equivalently, maximizes $P^{\mathbf{F}}_{\overline{A_k}|X^k}(x^k)$). Hence the system $\mathbf{D}$ (over choices of which the definition of $\nu(\mathbf{F},\overline{A_k})$ maximizes) can be eliminated from the analysis.

Theorem 2. *If a random system* $\mathbf{F}$ *with MES* $\mathcal{A}$ *satisfies*

$$P^{\mathbf{F}}_{A_i|X^iY^{i-1}A_{i-1}} = P^{\mathbf{F}}_{A_i|X^iA_{i-1}} \tag{1}$$

for $i \geq 1$, *which holds if*

$$P^{\mathbf{F}}_{Y^i|X^iA_i} = P^{\mathbf{G}}_{Y^i|X^i} \tag{2}$$

for $i \geq 1$, *for some system* $\mathbf{G}$ *(actually,* $\mathbf{G} \equiv \mathbf{F}|\mathcal{A}$*), then* $\nu(\mathbf{F}, \overline{A_k}) = \mu(\mathbf{F}, \overline{A_k})$.

Corollary 1. **(i)** *If* $\mathcal{A}$ *is defined on the inputs of* $\mathbf{F}$*, then* $\mathbf{F}$ *satisfies (1).*
(ii) *If* $\mathbf{F}$ *with* $\mathcal{A}$ *satisfy (1), then so does* $\mathbf{FG}$ *with* $\mathcal{A}$ *for any (compatible)* $\mathbf{G}$.
(iii) *If* $\nu(\mathbf{F}, \overline{A_k}) = \mu(\mathbf{F}, \overline{A_k})$*, then* $\nu(\mathbf{FG}, \overline{A_k}) = \mu(\mathbf{F}, \overline{A_k})$ *for any* $\mathbf{G}$.
(iv) *If* $\mathcal{A}$ *is defined on the inputs of* $\mathbf{F}$ *and* $\mathbf{F}|\mathcal{A} \equiv \mathbf{U}$ *for a source* $\mathbf{U}$*, then* $\nu(\mathbf{EF}, \overline{A_k}) = \mu(\mathbf{E}, \overline{A_k})$ *for any* $\mathbf{E}$.
(v) *If* A_i *(*B_i*) is defined on the inputs (outputs) of* $\mathbf{F}$ *and* $\mathbf{F}^{\mathcal{B}}|\mathcal{A} \equiv \mathbf{U}^{\mathcal{B}}$ *for a source* $\mathbf{U}$*, then* $\nu(\mathbf{EF}, \overline{A_k} \vee \overline{B_k}) \leq \mu(\mathbf{E}, \overline{A_k}) + \mu(\mathbf{U}, \overline{B_k})$ *for any* $\mathbf{E}$.
(vi) *If* $\mathcal{A}$ *is defined on the inputs of* $\mathbf{F}$ *and* $\mathbf{F}|\mathcal{A} \equiv \mathbf{B}$*, then for any random system* $\mathbf{C}(.)$ *such that* $\mathbf{C}(\mathbf{B}) \equiv \mathbf{B}$*,* $\nu(\mathbf{C}(\mathbf{F}), \overline{A_k}) = \mu(\mathbf{C}(\mathbf{F}), \overline{A_k})$.

4.4 Exploiting Independent Events

Consider a random system $\mathbf{C}(.,.)$ invoking two independent random systems $\mathbf{F}$ and $\mathbf{G}$ with MESs $\mathcal{A}$ and $\mathcal{B}$, respectively. For each input to $\mathbf{C}(\mathbf{F},\mathbf{G})$, $\mathbf{F}$ and $\mathbf{G}$ can be called several times. For a given k, let k' and k'' be the maximal number of invocations of $\mathbf{F}$ and $\mathbf{G}$, respectively, for any input sequence to $\mathbf{C}(\mathbf{F},\mathbf{G})$ of length k.

Theorem 3. *If* $\mathbf{C}(\mathbf{F},\mathbf{G})|(\tilde{\mathcal{A}} \vee \tilde{\mathcal{B}}) \equiv \mathbf{H}$*, then*

$$\Delta_k(\mathbf{C}(\mathbf{F},\mathbf{G}),\mathbf{H}) \leq \nu(\mathbf{F}, \overline{\tilde{A}_{k'}}) \cdot \nu(\mathbf{G}, \overline{\tilde{B}_{k''}}).$$

Proof. We have

$$\begin{aligned}
\Delta_k(\mathbf{C}(\mathbf{F},\mathbf{G}),\mathbf{H}) &\leq \nu(\mathbf{C}(\mathbf{F},\mathbf{G}), \overline{\tilde{A}_{k'}} \wedge \overline{\tilde{B}_{k''}}) = \max_{\mathbf{D}} P^{\mathbf{DCFG}}(\overline{\tilde{A}_{k'}} \wedge \overline{\tilde{B}_{k''}}) \\
&= \max_{\mathbf{D}} \left(P^{\mathbf{DCFG}}(\overline{\tilde{A}_{k'}}) \cdot P^{\mathbf{DCFG}}(\overline{\tilde{B}_{k''}}|\overline{\tilde{A}_{k'}}) \right) \\
&\leq \underbrace{\max_{\mathbf{D}} P^{\mathbf{DCFG}}(\overline{\tilde{A}_{k'}})}_{=\nu(\mathbf{C}(\mathbf{F},\mathbf{G}),\overline{\tilde{A}_{k'}}) \leq \nu(\mathbf{F},\overline{\tilde{A}_{k'}})} \cdot \underbrace{\max_{\mathbf{D}} P^{\mathbf{DCFG}}(\overline{\tilde{B}_{k''}}|\overline{\tilde{A}_{k'}})}_{\leq \nu(\mathbf{G},\overline{\tilde{B}_{k''}})}.
\end{aligned}$$

The last inequality holds because in the expression on the last line the two maximizations over choices of $\mathbf{D}$ are independent, as opposed to the previous line. We have $\nu(\mathbf{C}(\mathbf{F},\mathbf{G}), \overline{\tilde{A}_{k'}}) \leq \nu(\mathbf{F}, \overline{\tilde{A}_{k'}})$ by Lemma 6 (iv) and $\max_{\mathbf{D}} P^{\mathbf{DCFG}}(\overline{\tilde{B}_{k''}}|\overline{\tilde{A}_{k'}}) \leq \nu(\mathbf{G}, \overline{\tilde{B}_{k''}})$ because for every particular choices for $\mathbf{D}$, $\mathbf{C}$, and $\mathbf{F}$, the probability of $\overline{\tilde{B}_{k''}}$ is at most $\nu(\mathbf{G}, \overline{\tilde{B}_{k''}})$, whether or not $\overline{\tilde{A}_{k'}}$ occurs for these choices. Thus the bound on $\nu(\mathbf{G}, \overline{\tilde{B}_{k''}})$ also holds on average. □

Corollary 2. *Let* $\mathbf{F}$ *with MES* $\mathcal{A}$ *and* $\mathbf{G}$ *with MES* $\mathcal{B}$ *be random permutations such that* $\mathbf{F}|\mathcal{A} \equiv \mathbf{P}$ *and* $\mathbf{G}|\mathcal{B} \equiv \mathbf{P}$. *Then* $\Delta_k(\mathbf{FG}, \mathbf{P}) \le \nu(\mathbf{F}, \overline{\tilde{A}_{k'}}) \cdot \nu(\mathbf{G}, \overline{\tilde{B}_{k''}})$.

Proof. We have $\mathbf{FG}|(\mathcal{A} \vee \mathcal{B}) \equiv \mathbf{P}$, hence Theorem 3 can be applied.[14] □

For two $(\mathcal{X}, \mathcal{Y})$-random automata $\mathbf{F}$ and $\mathbf{G}$ and a group operation $\star$ on $\mathcal{Y}$, let $\mathbf{F} \star \mathbf{G}$ denote the random automaton obtained by using $\mathbf{F}$ and $\mathbf{G}$ in parallel (with the same input) and combining the two outputs using $\star$.

Corollary 3. *If* $\mathbf{F}|\mathcal{A} \equiv \mathbf{G}|\mathcal{B} \equiv \mathbf{R}$, *then* $\Delta_k(\mathbf{F} \star \mathbf{G}, \mathbf{R}) \le \nu(\mathbf{F}, \overline{A_k}) \cdot \nu(\mathbf{G}, \overline{B_k})$.

Proof. We have $(\mathbf{F} \star \mathbf{G})|(\mathcal{A} \vee \mathcal{B}) \equiv \mathbf{R}$, hence Theorem 3 can be applied. □

5 Applications to Quasi-random Functions

5.1 Quasi-random Functions

Definition 12. For a function $d : \mathbf{N} \to \mathbf{R}^+$, a random function or random system $\mathbf{F}$ is called a *$d(k)$-quasi-random function* ($d(k)$-QRF for short) if $\Delta_k(\mathbf{F}, \mathbf{R}) \le d(k)$ for $k \ge 1$. Quasi-random permutations, beacons and oracles are defined analogously, replacing $\mathbf{R}$ by $\mathbf{P}$, $\mathbf{B}$, and $\mathbf{O}$, respectively.

By concatenating, for any w, 2^w outputs of a $d(k)$-QRF $\{0,1\}^l \to \{0,1\}^m$ one obtains a $\tilde{d}(k)$-QRF $\{0,1\}^{l-w} \to \{0,1\}^{2^w m}$ for $\tilde{d}(k) = d(2^w k)$, thus increasing the output size by a factor 2^w at the expense of reducing the input size by w bits.

The problem considered in this section is to *expand* the input size substantially at the sole expense of increasing $d(k)$ moderately, i.e., to expand a given supply of random bits into a much larger supply of apparently random bits.

This general problem is important because the core of a cryptographic system based on a PRF corresponds to the construction of a quasi-random system of the same type from a URF $\mathbf{R}$. In any such construction, $\mathbf{R}$ can be replaced by a QRF, possibly constructed recursively from smaller QRF's, where at the lowest level the randomness is replaced by the PRF. This can for instance be used to avoid the birthday problem when collisions are a security issue (see below).

For any $d(k)$-QRF $\mathbf{G} : \{0,1\}^L \to \{0,1\}^M$ constructed from a URF $\mathbf{R} : \{0,1\}^l \to \{0,1\}^m$ it is obvious that $d(k)$ cannot be negligible for $k > 2^l m/M$, i.e., when the internal randomness is exhausted. One could achieve $d(k) = 0$ for up to $k \approx 2^l m/M$ by defining $\mathbf{G}$ as the evaluation of a polynomial whose coefficients are taken from the function table of $\mathbf{R}$, but this construction would be exponentially inefficient since the entire table of $\mathbf{R}$ must be read for each evaluation of $\mathbf{G}$. Efficiency, i.e., the number of evaluations of $\mathbf{R}$ required for one evaluation of $\mathbf{G}$, is an important parameter of a construction. There is a trade-off between the efficiency and the degree $d(k)$ of indistinguishability.

[14] The corollary also follows from Vaudenay's nice proof [22] (stated in our terminology) that $\Delta_k(\mathbf{FG}, \mathbf{P}) \le \Delta_k(\mathbf{F}, \mathbf{P}) \cdot \Delta_k(\mathbf{G}, \mathbf{P})$ for two random permutations $\mathbf{F}$ and $\mathbf{G}$.

5.2 An Efficient Construction of a Quasi-random Function

We now propose the construction of an efficient QRF $\mathbf{C}(\mathbf{F}) : \{0,1\}^L \rightarrow \{0,1\}^m$ from a QRF $\mathbf{F} : \{0,1\}^l \rightarrow \{0,1\}^m$, for $L \gg l$. The basic idea for the definition of $\mathbf{C}(.)$ is to map an argument to $\mathbf{C}(.)$ to a list of t arguments for $\mathbf{F}$ and to XOR the corresponding values of $\mathbf{F}$. In fact, we can (but need not) use the convention that if a list contains a value more than once, these values are ignored, resulting in fewer than t values being XORed.

One can associate, in a natural manner, with each such set of t values a characteristic vector, with at most t 1-entries, in the vector space $\{0,1\}^{2^l}$. The described XORing operation corresponds to computing the scalar product of the characteristic vector with the function table of $\mathbf{F}$ (interpreted as a vector in $(\{0,1\}^m)2^l$).

Hence Lemma 11 in the Appendix implies that, given the event that these k vectors (for the k arguments to $\mathbf{C}(.)$) are linearly independent, the construction is equivalent to a URF (and also a beacon). Therefore Theorem 1 (i) can be applied.

It only remains to find a mapping $\mathbf{H} : \{0,1\}^L \rightarrow S$, where S is the subset of the vector space $\{0,1\}^{2^l}$ consisting of the vectors of weight at most t. The internal randomness of $\mathbf{H}$ can actually be taken from the function table of $\mathbf{F}$ (say for the z highest values, where z is an appropriate small number). For this to be secure, the mapping $\mathbf{H}$ must be restricted slightly to generate vectors with no 1-entry in the last z coordinates.

Lemma 12 in the Appendix shows that $\mathbf{H}$ can be implemented by using a $2t$-wise random function $\mathbf{E} : \{0,1\}^L \times \{1,\ldots,t\} \rightarrow \{0,\ldots,2^l - z - 1\}$. For an argument $x \in \{0,1\}^L$ of $\mathbf{H}$, $\mathbf{E}(x,i)$ for $1 \leq i \leq t$ is evaluated and the corresponding characteristic vector is formed.[15] Note that the z unit vectors with 1-entries in one of the top z positions must also be taken into account in Lemma 12, but they are of course linearly independent of the k vectors discussed above.

Hence we have outlined the proof of the following theorem.

Theorem 4. *For a $d(k)$-QRF $\mathbf{F}$, $\mathbf{C}(\mathbf{F})$ is a $\tilde{d}(k)$-QRF for $\tilde{d}(k) = k\left(\frac{kt}{2^l}\right)^t + d(tk+z)$.*

The term $k(kt/2^l)^t$ is very small, even for $k \gg 2^{l/2}$ for which collisions among random values in the input space of $\mathbf{F}$ would be very probable. This was called "security beyond the birthday barrier" in [1].[16] Already for moderate values of t, the described construction achieves a negligible $\tilde{d}(k)$ for $k \approx 2^{lt/(t+1)}$, i.e., far beyond the birthday barrier.

The above construction ideas apply in other contexts as well, for instance the use of some values of a PRF as the key of another component in a manner that

---

[15] Such a function $\mathbf{E}$ can for instance be obtained by evaluation of a polynomial of degree $2t$ over an appropriate finite field of size at least $t2^L$.

[16] This fact was pointed out already in [12], Theorem 2, where the basic idea of XORing several values of a function to go beyond the birthday bound was proposed.

does not compromise security. Note that the security of the XOR-MAC [3] and of other constructions based on linearly independent inputs (e.g. [1]) follow directly from Lemma 11 as well as a (non-adaptive) analysis of the linear independence event. For the XOR-MAC the analysis of this event is trivial.

6 Applications to MAC's

A secure MAC-scheme is a PRF $\mathcal{M} \to \{0,1\}^l$ for $\mathcal{M} = \cup_{i=1}^{L}\{0,1\}^i$ for some maximal message length L and an appropriate security parameter l. If $L = \infty$, then this corresponds to a pseudo-random oracle.

A very natural construction originating in [23] and used in many later papers (e.g. see [5, 20] and the discussion and references therein) is to apply an ϵ-almost universal hash function[17] $\mathbf{U} : \mathcal{M} \to \mathcal{X}$ for some $\mathcal{X}$ to the message and to apply a PRF $\mathbf{F} : \mathcal{X} \to \{0,1\}^l$ to the result. Such a scheme has two keys, those of $\mathbf{U}$ and $\mathbf{F}$, but in fact the $\mathbf{U}$-key can be obtained by evaluating $\mathbf{F}$ for an appropriate number z of fixed arguments, as follows easily from our framework. More precisely, $\mathbf{U}(.)$ is a random system[18] invoking $\mathbf{F}$ some z times to set up the key of $\mathbf{U}$ and then applies it to the input.[19] Of course, the key can be cached so that only one evaluation of $\mathbf{F}$ is needed for each input.

The security proof of such a scheme is trivial in our framework. The following theorem implies that $\mathbf{U}(\mathbf{F})$ is a computationally secure MAC for any PRF $\mathbf{F}$.

Theorem 5. *For a $d(k)$-QRF $\mathbf{F}$, $\mathbf{U}(\mathbf{F})$ is a $\tilde{d}(k)$-QRO for $\tilde{d}(k) = \epsilon(k+z)^2/2 + d(k+z)$.*

Proof. Define A_i as the event that all inputs to $\mathbf{F}$ are distinct, including the z fixed values needed for the key setup for $\mathbf{U}$. Lemma 5 (i) implies $\Delta_k(\mathbf{U}(\mathbf{F}), \mathbf{R}) \leq \Delta_k(\mathbf{U}(\mathbf{F}), \mathbf{U}(\mathbf{R})) + \Delta_k(\mathbf{U}(\mathbf{R}), \mathbf{R})$. Lemma 5 (ii) implies $\Delta_k(\mathbf{U}(\mathbf{F}), \mathbf{U}(\mathbf{R})) \leq d(k+z)$. Moreover, $\mathbf{U}(\mathbf{R})|\mathcal{A} \equiv \mathbf{R}$ and hence, using Theorem 1 (i), $\Delta_k(\mathbf{U}(\mathbf{R}), \mathbf{R}) \leq \nu(\mathbf{U}(\mathbf{R}), \overline{A_k})$. Using Corollary 1 (vi) together with $\mathbf{R}|\mathcal{A} \equiv \mathbf{B}$ and $\mathbf{U}(\mathbf{B}) \equiv \mathbf{B}$ gives $\nu(\mathbf{U}(\mathbf{R}), \overline{A_k}) = \mu(\mathbf{U}(\mathbf{R}), \overline{A_k})$, hence one can restrict attention to non-adaptive strategies. Now, for any fixed input sequence to $\mathbf{U}(\mathbf{R})$, A_k is the union of $\binom{k+z}{2} < (k+z)^2/2$ collision events, each with probability at most ϵ. Application of the union bound concludes the proof. □

As a further demonstration of the general applicability of the framework, we give a simple security proof of a generalized version of the CBC-MAC (e.g., see Fig. 6 and [2]), with which we assume the reader is familiar. We do not wish to make an *a priori* assumption about the maximal message length, hence we need a prefix-free encoding $\sigma : \{0,1\}^* \to \{0,1\}^*$ of the binary strings which does not significantly expand the length. A good choice is to prepend a block encoding

[17] $P(\mathbf{U}(x) = \mathbf{U}(x')) \leq \epsilon$ for any $x \neq x'$. Actually, $\mathbf{U}$ must satisfy $P(\mathbf{U}(x) = y) \leq \epsilon$ for any x and y (which is usually the case).

[18] This is a cascade $\mathbf{UF}$, but this notation is incorrect because $\mathbf{U}$ depends on $\mathbf{F}$.

[19] As an alternative, a fixed value of $\mathbf{F}$ could be used as the key to generate the key of $\mathbf{U}$ pseudo-randomly. The security of such a scheme follows also from our analysis.

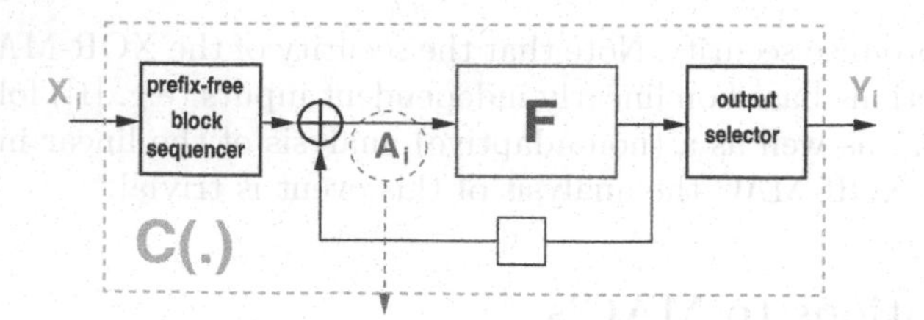

Fig. 6. The CBC-MAC. The $(\{0,1\}^*, \{0,1\}^l)$-random system $\mathbf{C}(\mathbf{F})$ is defined by applying some prefix-free encoding σ to the message, then padding the result with 0's to complete the last block, then applying the CBC feedback construction with a random function (or more generally a random automaton) $\mathbf{F}$, and taking the last output (for a given message) as the MAC-value for that message.

the length of the message, but from a theoretical viewpoint this restricts the message length and hence does not yield a true quasi-random oracle.[20]

Let $\mathbf{C}(\mathbf{F})$ be the $(\{0,1\}^*, \{0,1\}^l)$-random system defined by applying σ to the message, then padding with 0's to fill the last block, and then applying the CBC-MAC with a random function (or more generally a random system) $\mathbf{F}$ (cf. Fig. 6). A result similar in spirit to the following theorem was stated (without proof) independently by Petrank and Rackoff [17].

Theorem 6. *If* $\mathbf{F}$ *is a* $d(k)$*-QRF, then* $\mathbf{C}(\mathbf{F})$ *is a* $\tilde{d}(k)$*-quasi-random oracle for* $\tilde{d}(k) = n^2 2^{-(l+1)} + d(n)$*, where* n *is the total number of blocks of all* k *messages issued by the distinguisher.*

Proof. Lemma 5 (i) implies $\Delta_k(\mathbf{C}(\mathbf{F}), \mathbf{O}) \leq \Delta_k(\mathbf{C}(\mathbf{F}), \mathbf{C}(\mathbf{R})) + \Delta_k(\mathbf{C}(\mathbf{R}), \mathbf{O})$. Lemma 5 (ii) implies $\Delta_k(\mathbf{C}(\mathbf{F}), \mathbf{C}(\mathbf{R})) \leq d(n)$. Consider the event A_i that all inputs to $\mathbf{F}$ are distinct, up to and including the processing of the i-th message, except those inputs to $\mathbf{F}$ that are trivially equal because the prefix of the actual message processed so far is also a prefix of a previous message. Because due to σ no (encoded) message is a prefix of another message, A_i implies that for a given message x_i the last input to $\mathbf{F}$ (for x_i) is distinct from all previous inputs to $\mathbf{F}$ (for $x_1, \ldots, x_{i-1}$). Hence $\mathbf{C}(\mathbf{R})|\mathcal{A} \equiv \mathbf{O}$ and by Theorem 1 (i) we have $\Delta_k(\mathbf{C}(\mathbf{R}), \mathbf{O}) \leq \nu(\mathbf{C}(\mathbf{R}), \overline{A_k})$. Equation (2) is satisfied (for $\mathbf{G} = \mathbf{B}$) for all i since $P^{\mathbf{C}(\mathbf{R})}_{Y^i|X^i A_i}$ is the uniform distribution over $\{\{0,1\}^l\}^i$ for all input values (resulting in A_i being satisfied). Hence $\nu(\mathbf{C}(\mathbf{R}), \overline{A_k}) = \mu(\mathbf{C}(\mathbf{R}), \overline{A_k})$ and one can restrict attention to non-adaptive strategies, which are easy to analyse.

For any given k input messages $x_1, \ldots, x_k$ of arbitrary lengths, but consisting of a total of n blocks, $\overline{A_k}$ corresponds to the event that a collision occurs among

[20] A true prefix-free encoding $\sigma : \{0,1\}^* \to \{0,1\}^*$ can be obtained as follows. Let $\overline{n}$ be the standard binary representation of the integer n, and let $l(x)$ be the length of the binary string x. It is not difficult to see that the mapping $\sigma : \{0,1\}^* \to \{0,1\}^*$ defined by $r = l(\overline{l(x)}) - 1$ and $\sigma(x) := 0^r 1 || \overline{l(x)} || x$ is prefix-free. For instance, $\sigma(1100010111001) = 0001110111000101110 01$. This encoding is efficient: $l(\sigma(x)) \approx l(x) + 2 \log l(x)$. It can be improved to $l(\sigma(x)) \approx l(x) + \log l(x)$ by using the encoding $x \mapsto \sigma(\overline{l(x)})||x$.

$n - w(x^k)$ independent and uniformly random values, where $w(x^k)$ is the total number of blocks in the messages $x_1, \ldots, x_k \in (\{0,1\}^l)^*$ which belong to a prefix (say of x_i) that was also the prefix of a previous message $x_1, \ldots, x_{i-1}$ (see above), i.e., $P^{\mathbf{C}(\mathbf{R})}_{A_k|X^k}(x^k) = p_{\text{coll}}(2^l, n - w(x^k)) \leq p_{\text{coll}}(2^l, n) \leq n^2 2^{-(l+1)}$.[21] □

7 Applications to the Analysis of Random Permutations

7.1 Random Permutations

For a random permutation[22] $\mathbf{Q}$, the inverse is also a random permutation and is denoted by $\mathbf{Q}^{-1}$. Remember that $\mathbf{P}$ denotes a uniform random permutation. Let $(\mathbf{E}, \mathbf{G})$ be any pair of (possibly dependent[23]) random permutations.

Lemma 8. **(i)** $\mathbf{EPG} \equiv \mathbf{P}$. *Moreover, if* $\mathbf{Q}|\mathcal{A} \equiv \mathbf{P}$, *then* $\mathbf{EQG}|\mathcal{A} \equiv \mathbf{P}$.
(ii) *For a MES* $\mathcal{C}$ *defined on the outputs of* $(\mathcal{X}, \mathcal{Y})$*-random systems such that* C_i *implies that the first* i *outputs are distinct, we have* $\mathbf{R}|\mathcal{C} \equiv \mathbf{P}|\mathcal{C}$ *and* $\mathbf{R}^{\mathcal{C}} \equiv \mathbf{P}^{\mathcal{C} \wedge \mathcal{D}}$ *for some MES* $\mathcal{D}$ *adjoined to* $\mathbf{P}$.

Proof. $\mathbf{EPG} \equiv \mathbf{P}$ is a special case of the second statement when A_i is the certain event for all i. We have $EQG|\mathcal{A} \equiv EPG$ for any two fixed permutations E and G because E and G simply correspond to relabelings of the input and output alphabets of $\mathbf{Q}$. Hence this equivalence also holds if the pair (E, G) is a random variable. Now we prove (ii). We have $\mathbf{R}|\mathcal{C} \equiv \mathbf{P}|\mathcal{C}$ since conditioned on the output being distinct, both $\mathbf{R}$ and $\mathbf{P}$ generate completely new random outputs. Moreover, $P^{\mathbf{R}}_{C_i|X^iY^{i-1}C_{i-1}} \leq P^{\mathbf{P}}_{C_i|X^iY^{i-1}C_{i-1}}$ is a simple consequence of the fact that for a given X^i with distinct values (i.e., $\text{dist}(X_1, \ldots, X_i)$), only Y^i with distinct values are consistent with $\mathbf{P}$, whereas other values for Y^i are consistent with $\mathbf{R}$, but C_i cannot hold for these Y^i. Now apply Lemma 1 (iv). □

Definition 13. *A pairwise independent permutation (PIP) [18]* $\mathbf{Q}$ *is a random permutation such that for any two inputs* x *and* x', $\mathbf{Q}(x)$ *and* $\mathbf{Q}(x')$ *are a completely random pair of (distinct) values.*[24]

7.2 Two Feistel Rounds with Random Functions

Let $\mathcal{R}$ be a set and let $\star$ be a group operation on $\mathcal{R}$. Typically $\mathcal{R} = \{0,1\}^l$ for some l and $\star$ is bitwise XOR. We now consider permutations on $\mathcal{R}^2$, i.e., on

[21] The proof goes through for more general versions of the CBC-MAC. For example, in addition to letting the input to $\mathbf{F}$ be the current message block XORed with the previous output of $\mathbf{F}$, as in the CBC-MAC, one could XOR in any further function of all the previous message blocks and all the previous outputs of $\mathbf{F}$ (except the last). Such a modification could make sense if one considers the risk that $\mathbf{F}$ might not be a PRF and hence wants to build in extra complexity for heuristic security.

[22] Much of this section can be generalized to the more general concept of a permutation random system, i.e., a $(\mathcal{X}, \mathcal{X})$-random system $\mathbf{Q}$ which for all i is a random permutation on $\mathcal{X}^i$.

[23] However, the pair $(\mathbf{E}, \mathbf{G})$ is, as always, assumed to be independent of $\mathbf{Q}$.

[24] A PIP can for instance be implemented by interpreting all quantities as elements of a finite field F and setting $\mathbf{Q}(x) = ax + b$ for random $a, b \in F$ with $a \neq 0$.

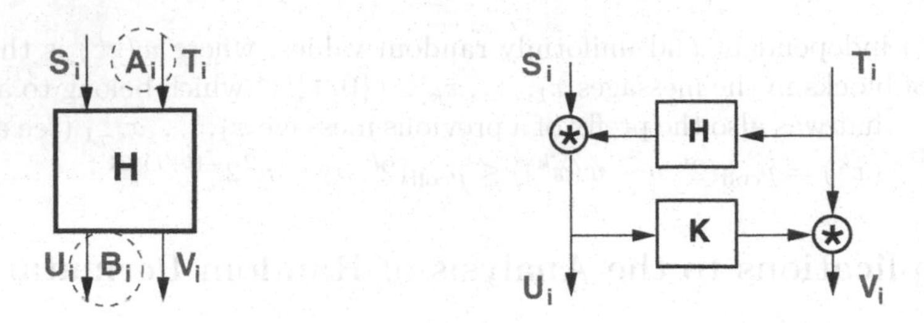

Fig. 7. Left side: Notation for random systems whose inputs and outputs are pairs. $A_i := \text{dist}(T^i)$ and $B_i := \text{dist}(U^i)$. Right side: Special case; two Feistel rounds with random systems $\mathbf{H}$ and $\mathbf{K}$, denoted $\mathbf{M}(\mathbf{H},\mathbf{K})$.

pairs which can be considered as "left" and "right" halves, or as high and low part when the pair is interpreted as a single element of, say, a field. For any random function $\mathbf{F} : \mathcal{R}^2 \to \mathcal{R}^2$ we can define the following random variables (see Figure 7, left): (S_i, T_i) is the i-th input and (U_i, V_i) are the i-th output. We define two MES, $A_i := \text{dist}(T^i)$ and $B_i := \text{dist}(U^i)$, used throughout Section 7.

For two random functions $\mathcal{R} \to \mathcal{R}$, $\mathbf{H}$ and $\mathbf{K}$, let $\mathbf{M}(\mathbf{H},\mathbf{K})$ be the $\mathcal{R}^2$-random permutation defined by two Feistel rounds with $\mathbf{H}$ and $\mathbf{K}$ (see Figure 7, right).[25] More precisely, $U_i = S_i \star \mathbf{H}(T_i)$ and $V_i = T_i \star \mathbf{K}(U_i)$. Let $\mathbf{R} : \mathcal{R}^2 \to \mathcal{R}^2$ be a URF, and let $\mathbf{R}'$ and $\mathbf{R}''$ be URF's $\mathcal{R} \to \mathcal{R}$. We have

Lemma 9. $\mathbf{M}(\mathbf{R}',\mathbf{R}'')^{\mathcal{A}\wedge\mathcal{B}} \equiv \mathbf{B}^{\mathcal{A}\wedge\mathcal{B}} \equiv \mathbf{R}^{\mathcal{A}\wedge\mathcal{B}} \equiv \mathbf{P}^{\mathcal{A}\wedge\mathcal{B}\wedge\mathcal{D}}$ *for some MES* $\mathcal{D}$.

Proof. Given A_i, the joint distribution of (U_i,V_i) and B_i is identical for $\mathbf{M}(\mathbf{R}',\mathbf{R}'')$, for $\mathbf{B}$, and for $\mathbf{R}$, independent of the input: U_i and V_i are independent new random values and B_i is determined by U^i. Hence $\mathbf{M}(\mathbf{R}',\mathbf{R}'')^{\mathcal{A}\wedge\mathcal{B}} \equiv \mathbf{B}^{\mathcal{A}\wedge\mathcal{B}} \equiv \mathbf{R}^{\mathcal{A}\wedge\mathcal{B}}$. The last equivalence follows from $\mathbf{R}^{\mathcal{B}} \equiv \mathbf{P}^{\mathcal{B}\wedge\mathcal{D}}$ (Lemma 8 (ii)) and because $\mathcal{A}$ is defined on the inputs and thus Lemma 2 (ii) can be applied. □

7.3 Mono-directional Luby-Rackoff and Naor-Reingold

The following theorem generalizes the one-directional Luby-Rackoff [10] and Naor-Reingold [18] results (cf. Fig. 8 left) and follows easily from our framework.

Theorem 7. *Let* $\mathbf{L} := \mathbf{E}\mathbf{M}(\mathbf{R}',\mathbf{R}'')$ *for some random permutation* $\mathbf{E}$. *Then* $\Delta_k(\mathbf{L},\mathbf{P}) \leq \mu(\mathbf{E},\overline{A_k}) + p_{\text{coll}}(|\mathcal{R}|,k)$. *If* $\mathbf{E}$ *is a PIP (Naor-Reingold) or if* $\mathbf{E}$ *is a Feistel round with another random function* $\mathbf{R}'''$ *(Luby-Rackoff), then* $\Delta_k(\mathbf{L},\mathbf{P}) \leq 2 \cdot p_{\text{coll}}(|\mathcal{R}|,k) < k^2/|\mathcal{R}|$.

Proof. Using Lemma 9 and Lemma 4 we obtain

$$\mathbf{L}^{\mathcal{A}\wedge\mathcal{B}} \equiv \mathbf{E}\mathbf{B}^{\mathcal{A}\wedge\mathcal{B}} \equiv \mathbf{E}\mathbf{P}^{\mathcal{A}\wedge\mathcal{B}\wedge\mathcal{D}} \tag{3}$$

[25] This can easily be generalized from random functions to random automata.

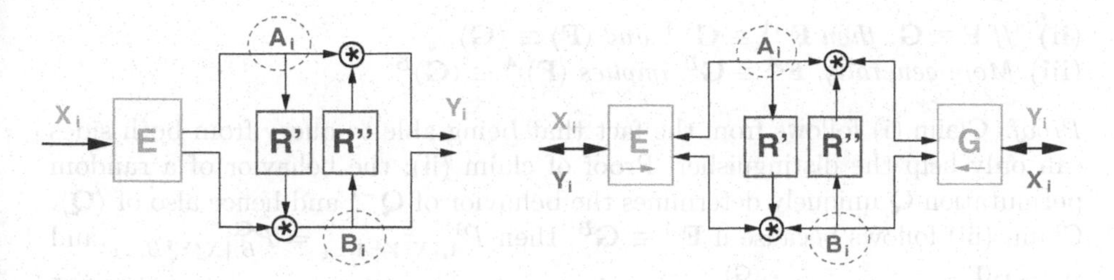

Fig. 8. Illustration for the one-directional (left) and bidirectional (right) Luby-Rackoff and Naor-Reingold results and generalizations thereof.

(with the events A_i defined internally). Lemma 8 (i) yields the first step of

$$\Delta_k(\mathbf{L},\mathbf{P}) = \Delta_k(\mathbf{L},\mathbf{EP}) \le \nu(\mathbf{L},\overline{A_k}\vee\overline{B_k}) = \nu(\mathbf{EB},\overline{A_k}\vee\overline{B_k})$$

and the next two steps follow from (3) and Theorem 1 (i), and from (3) and Lemma 6 (ii), respectively. Now obviously (and by Corollary 1 (v)), $\nu(\mathbf{EB},\overline{A_k}\vee\overline{B_k}) \le \mu(\mathbf{E},\overline{A_k}) + \mu(\mathbf{B},\overline{B_k})$ where $\mu(\mathbf{B},\overline{B_k}) = p_{\rm coll}(|\mathcal{R}|,k)$. The second claim follows by a trivial analysis of a collision event among k random values. □

Remark. Theorem 7, besides being more general, is also slightly stronger than that of [18] and [10] (see also [9]) where an additional term $k^2/(|\mathcal{R}|)^2$ appears on the right side. This weaker bound would in our context be obtained by proving $\Delta_k(\mathbf{L},\mathbf{R}) < k^2/|\mathcal{R}|$ and then using $\Delta_k(\mathbf{R},\mathbf{P}) \le k^2/|\mathcal{R}|^2$. One could also append an additional random permutation $\mathbf{G}$, as follows directly from Corollary 1 (iii).

7.4 Bidirectional Permutations

Definition 14. For an $\mathcal{X}$-random permutation $\mathbf{Q}$, let $\langle\mathbf{Q}\rangle$ be the *bidirectional permutation*[26] $\mathbf{Q}$ with access from both sides (i.e., one can query both $\mathbf{Q}$ and $\mathbf{Q}^{-1}$). More precisely, $\langle\mathbf{Q}\rangle$ is the random function $\mathcal{X}\times\{0,1\}\to\mathcal{X}$ defined as follows:

$$\langle\mathbf{Q}\rangle(U_i,D_i) = \begin{cases} \mathbf{Q}(U_i) & \text{if } D_i = 0 \\ \mathbf{Q}^{-1}(U_i) & \text{if } D_i = 1\,. \end{cases}$$

If $\mathcal{A}$ is defined for $\mathbf{Q}$, $\mathcal{A}$ can also be defined naturally for $\langle\mathbf{Q}\rangle$: Let $V_i := \langle\mathbf{Q}\rangle(U_i,D_i)$, and let X_i and Y_i be the i-th input and output of $\mathbf{Q}$ (i.e., if $D_i=0$, then $X_i=U_i$ and $Y_i=V_i$, and if $D_i=1$, then $Y_i=U_i$ and $X_i=V_i$). Recall that $P^{\mathbf{Q}}_{Y_iA_i|X^iY^{i-1}A_{i-1}} = P^{\mathbf{Q}}_{Y_i|X^iY^{i-1}A_{i-1}}\cdot P^{\mathbf{Q}}_{A_i|X^iY^iA_{i-1}}$. Now we let $P^{\langle\mathbf{Q}\rangle}_{A_i|X^iY^iA_{i-1}} := P^{\mathbf{Q}}_{A_i|X^iY^iA_{i-1}}$.

Lemma 10. *For any random permutatio* $\mathbf{F}$ *and* $\mathbf{G}$*,*
(i) $\Delta_k(\mathbf{F},\mathbf{G}) \le \Delta_k(\langle\mathbf{F}\rangle,\langle\mathbf{G}\rangle)$.[27]

[26] This definition is motivated by considering a block cipher which in a mixed chosen-plaintext and chosen-ciphertext attack can be queried from both sides.

[27] $\Delta_k(\langle\mathbf{F}\rangle,\langle\mathbf{G}\rangle)$ can be much larger than $\Delta_k(\mathbf{F},\mathbf{G})$ because inverse queries may help the distinguisher significantly.

(ii) *If* $\mathbf{F} \equiv \mathbf{G}$, *then* $\mathbf{F}^{-1} \equiv \mathbf{G}^{-1}$ *and* $\langle \mathbf{F} \rangle \equiv \langle \mathbf{G} \rangle$.
(iii) *More generally,* $\mathbf{F}^{\mathcal{A}} \equiv \mathbf{G}^{\mathcal{B}}$ *implies* $\langle \mathbf{F} \rangle^{\mathcal{A}} \equiv \langle \mathbf{G} \rangle^{\mathcal{B}}$.

Proof. Claim (i) follows from the fact that being able to query from both sides can only help the distinguisher. Proof of claim (ii): the behavior of a random permutation $\mathbf{Q}$ uniquely determines the behavior of $\mathbf{Q}^{-1}$ and hence also of $\langle \mathbf{Q} \rangle$. Claim (iii) follows because if $\mathbf{F}^{\mathcal{A}} \equiv \mathbf{G}^{\mathcal{B}}$, then $P^{\mathbf{F}}_{A_i|X^iY^iA_{i-1}} = P^{\mathbf{G}}_{B_i|X^iY^iB_{i-1}}$ and thus $P^{\langle \mathbf{F} \rangle}_{A_i|U^iD^iV^iA_{i-1}} = P^{\langle \mathbf{G} \rangle}_{B_i|U^iD^iV^iB_{i-1}}$. □

The following theorem generalizes Theorem 3.2 of [18] in several ways. The proof is omitted.

Theorem 8. *Let* $\mathbf{L}$ *be defined as* $\mathbf{L} := \mathbf{EM}(\mathbf{R}', \mathbf{R}')\mathbf{G}^{-1}$ *(cf. Fig. 8 right).*
(i) *If* $\mathbf{E}$ *and* $\mathbf{G}^{-1}$ *are independent PIP's, then* $\Delta_k(\langle \mathbf{L} \rangle, \langle \mathbf{P} \rangle) < k^2/|\mathcal{R}|$.
(ii) *If* $\mathbf{E}$ *is a PIP and* $\mathbf{G} = \mathbf{E}^{-1}$, *then* $\Delta_k(\langle \mathbf{L} \rangle, \langle \mathbf{P} \rangle) < 4k^2/|\mathcal{R}|$.
(iii) *If* $\mathbf{R}' = \mathbf{R}''$, *i.e.,* $\mathbf{L} := \mathbf{EM}(\mathbf{R}', \mathbf{R}')\mathbf{E}^{-1}$, *then* $\Delta_k(\langle \mathbf{L} \rangle, \langle \mathbf{P} \rangle) < 8k^2/|\mathcal{R}|$.
(iv) *Moreover, if* $\mathcal{R} = GF(q)$ *is a field and* $\mathbf{E}$ *is also derived from* $\mathbf{R}'$ *by a linear polynomial* $ax + b$ *over* $GF(q^2)$ *with* a *and* b *defined by* $a = (\mathbf{R}(\xi_1)||\mathbf{R}(\xi_2))$ *and* $b = (\mathbf{R}(\xi_3)||\mathbf{R}(\xi_4))$ *for some fixed* $\xi_1, \xi_2, \xi_3, \xi_4 \in GF(q)$, *then* $\Delta_k(\langle \mathbf{L} \rangle, \langle \mathbf{P} \rangle) < 8(k+1)^2/|\mathcal{R}| + 1/|\mathcal{R}|^2$.

8 Conclusions

We have described a general framework for indistinguishability proofs of the most general form of random systems. The purpose of the framework is to prove results at the most general and abstract level, and this leads to substantial simplifications in actual security proof (making them for example tractable for a textbook) and to new security proofs that before may have appeared unrealistic. It would be a pleasure to see the framework at work in future security proofs.

We suggest as an open problem to find constructions of QRF's from QRF's better than that of Section 5, i.e., with either higher security (degree of indistinguishability) or lower complexity (number of evaluations of $\mathbf{F}$), or both. However, it is possible that this construction is quite close to optimal.

Acknowledgments

I would like to thank Thomas Holenstein, Olaf Keller, Krzysztof Pietrzak, and Renato Renner for many very helpful comments and for a careful proofreading, and Markus Stadler for discussions at an early stage of this work.

References

1. M. Bellare, O. Goldreich, and H. Krawczyk, Stateless evaluation of pseudorandom functions: security beyond the birthday barrier, *Advances in Cryptology – CRYPTO '99*, Lecture Notes in Computer Sc., vol. 1666, pp. 270–287, Springer-Verlag, 1999.

2. M. Bellare, J. Kilian, and P. Rogaway, The security of the cipher block chaining message authentication code, *Advances in Cryptology – CRYPTO '94*, Lecture Notes in Computer Science, vol. 839, pp. 341–358, Springer-Verlag, 1995.
3. M. Bellare, J. Guérin, and P. Rogaway, XOR MACs: New methods for message authentication using finite pseudorandom functions, *Advances in Cryptology – CRYPTO '95*, Lecture Notes in Computer Science, vol. 963, Springer-Verlag, 1994.
4. D. J. Bernstein, How to stretch random functions: The security of protected counter sums, *Journal of Cryptology*, vol. 12, pp. 185–192, Springer-Verlag, 1999.
5. J. Black, S. Halevi, H. Krawczyk, T. Krovetz, and P. Rogaway, UMAC: Fast and secure message authentication, *Advances in Cryptology – CRYPTO '99*, Lecture Notes in Computer Science, vol. 1666 pp. 216–233, Springer-Verlag, 1999.
6. R. E. Blahut, *Principles and practice of information theory*, Addison-Wesley Publishing Company, 1988.
7. M. Blum and S. Micali, How to generate cryptographically strong sequences of pseudo-random bits, *SIAM J. on Computing*, vol. 13, no. 4, pp. 850–864, 1984.
8. O. Goldreich, S. Goldwasser, and S. Micali, How to construct random functions, *Journal of the ACM*, vol. 33, no. 4, pp. 210–217, 1986.
9. M. Luby, *Pseudorandomness and Cryptographic Applications*, Princeton University Press, 1996.
10. M. Luby and C. Rackoff, How to construct pseudo-random permutations from pseudo-random functions, *SIAM J. on Computing*, vol. 17, no. 2, pp. 373–386, 1988.
11. U. M. Maurer, Conditionally-perfect secrecy and a provably-secure randomized cipher, *Journal of Cryptology*, vol. 5, pp. 53–66, Springer-Verlag, 1992.
12. ——, A simplified and generalized treatment of Luby-Rackoff pseudo-random permutation generators, *Advances in Cryptology – EUROCRYPT '92*, Lecture Notes in Computer Science, vol. 658, pp. 239–255, Springer-Verlag, 1992.
13. ——, Extended version of this paper, see `www.crypto.ethz.ch/publications/`.
14. J. Patarin, Etude des générateurs de permutations basés sur le Schéma du D.E.S., Ph. D. Thesis, INRIA, Le Chesnay, France, 1991. An extract appeared in: J. Patarin, New results on pseudorandom permutation generators based on the DES scheme, *Advances in Cryptology – CRYPTO'91*, J. Feigenbaum (ed.), Lecture Notes in Computer Science, Vol. 576, Springer-Verlag, pp. 301–312, 1992.
15. ——, How to construct pseudorandom permutations from a single pseudorandom function, *Advances in Cryptology – EUROCRYPT '92*, R. Rueppel (ed.), Lecture Notes in Computer Science, vol. 658, pp. 256–266, Springer-Verlag, 1992.
16. ——, About Feistel schemes with six (or more) rounds, *Fast Software Encryption*, Lecture Notes in Computer Science, vol. 1372, pp. 103–121, Springer-Verlag, 1998.
17. E. Petrank and C. Rackoff, CBC MAC for real-time data sources, *Journal of Cryptology*, vol. 13, no. 3, pp. 315–338, 2000.
18. M. Naor and O. Reingold, On the construction of pseudorandom permutations: Luby-Rackoff revisited, *Journal of Cryptology*, vol. 12, no. 1, pp. 29–66, 1999.
19. M. O. Rabin, Transaction protection by beacons, *J. Comp. Sys. Sci.*, vol. 27, pp. 256–267, 1983.
20. V. Shoup, On fast and provably secure message authentication based on universal hashing, *Advances in Cryptology – CRYPTO '96*, Lecture Notes in Computer Science, vol. 1109, pp. 313–328, Springer-Verlag, 1996.
21. S. Vaudenay, Provable security for block ciphers by decorrelation, *Proceedings of STACS'98*, Lecture Notes in Computer Science, vol. 1373, Springer-Verlag, pp. 249–275, 1998.
22. ——, On provable security for conventional ciphers, in *Proc. of ICISC'99*, Lecture Notes in Computer Science, Springer-Verlag, 1999.

23. M. N. Wegman and J. L. Carter, New hash functions and their use in authentication and set equality, *J. of Computer and System Sciences*, vol. 22, pp. 265–279, 1981.

Appendix

Lemma 11. *Let* $\mathbf{U} = [U_1, \ldots, U_n]$ *with* $U_i \in GF(q)$ *be a vector of random variables with uniform distribution* $GF(q)^n$*, and define the random function* $\mathbf{K} : GF(q)^n \to GF(q)$ *as the scalar product of the input vector* $\mathbf{x} = [x_1, \ldots, x_n] \in GF(q)^n$ *and* $\mathbf{U}$,

$$\mathbf{K}(\mathbf{x}) = \langle \mathbf{x}, \mathbf{U} \rangle = \sum_{j=1}^{n} x_j U_j.$$

Then$\mathbf{K}^{\mathcal{A}} \equiv \mathbf{R}^{\mathcal{A}} \equiv \mathbf{B}^{\mathcal{A}}$ *with* A_i *as the event that* $\mathbf{x}_1, \ldots, \mathbf{x}_i$ *are linearly independent.*

Proof. For a list $\mathbf{v}^k = [\mathbf{v}_1, \ldots, \mathbf{v}_k]$ of vectors in a finite-dimensional vector space, let $span(\mathbf{v}^k)$ denote the subspace spanned by $\mathbf{v}_1, \ldots, \mathbf{v}_k$ and let $dim(\mathbf{v}^k)$ denote its dimension. If $\mathbf{v}_1, \ldots, \mathbf{v}_k$ are linearly independent, then $dim(\mathbf{v}^k) = k$.

Let $T \subseteq GF(q)^n$ be a set of input vectors to $\mathbf{K}$, and let $\mathbf{K}(T)$ denote the corresponding list of values of $\mathbf{K}$. We prove[28] that $H(\mathbf{K}(T)) = dim(T)r$, where $r = \log q$. This clearly implies that for any set of linearly independent vectors the corresponding function values have maximal entropy, as is to be proved. Linear dependence implies functional dependence, hence $H(\mathbf{K}(T)) = H(\mathbf{K}(span(T))) = H(\mathbf{K}(span(B)))$, where B is any basis of $span(T)$ and has cardinality $B = dim(T)$. Thus $H(\mathbf{K}(T)) \leq dim(T)r$. On the other hand, it follows from linear algebra that T can be complemented by a set T' of size $n - dim(T)$ such that $T \cup T'$ spans the entire space $GF(q)^n$. Hence $H(\mathbf{K}|\mathbf{K}(T)) \leq (n - dim(T))r$. Because $H(\mathbf{K}) = H(\mathbf{K}(T)) + H(\mathbf{K}|\mathbf{K}(T)) = nr$ we must have equality in the two previous inequalities. □

Let $S_n := \{1, \ldots, n\}$. The characteristic vector in $\{0,1\}^n$ of a subset S' of S_n has a 1 at position i if and only if $i \subseteq S'$. For multi-sets or lists of elements of S_n, we define the characteristic vector to have a 1-entry only for those elements of S_n that occur *exactly once*.

The proof of following lemma is straight-forward.

Lemma 12. *If* kt *elements of* S_n *are selected* b*-wise independently (for* $b \geq 2t$*) and interpreted as* k *lists of* t *elements,* $V_i = [V_{i1}, \ldots, V_{it}]$ *for* $1 \leq i \leq k$*, then their characteristic vectors* $W_1, \ldots, W_k$ *are linearly independent with probability at least* $1 - k\left(\frac{kt}{n}\right)^t$.

[28] See [6] for definitions of the entropy $H(X)$ and the conditional entropy $H(X|Y)$.

How to Fool an Unbounded Adversary with a Short Key

Alexander Russell and Hong Wang

Department of Computer Science and Engineering,
University of Connecticut, Storrs, Connecticut 06269, USA,
acr@engr.uconn.edu,hongmuw@engr.uconn.edu

Abstract. We consider the symmetric encryption problem which manifests when two parties must securely transmit a message m with a short shared secret key. As we permit arbitrarily powerful adversaries, any encryption scheme must leak information about m – the mutual information between m and its ciphertext cannot be zero. Despite this, we present a family of encryption schemes which guarantee that for any message space in $\{0,1\}^n$ with minimum entropy $n - \ell$ and for any Boolean function $h : \{0,1\}^n \to \{0,1\}$, no adversary can predict $h(m)$ from the ciphertext of m with more than $1/n^{\omega(1)}$ advantage; this is achieved with keys of length $\ell + \omega(\log n)$. In general, keys of length $\ell + s$ yield a bound of $2^{-\Theta(s)}$ on the advantage. These encryption schemes rely on no unproven assumptions and can be implemented efficiently.

1 Introduction

One of the simplest and most secure encryption systems is the *one time pad*: two parties who have agreed on a uniformly selected secret key $s \in \{0,1\}^n$ can exchange a single message $m \in \{0,1\}^n$ by transmitting $m \oplus s$, this parity being taken componentwise. If we think of the message m and the secret key s as independent random variables, then it is easy to see that the message m and the ciphertext $m \oplus s$ are uncorrelated: we say that this encryption system offers *perfect secrecy*.

One unfortunate consequence of this absolute security guarantee is that any such system must use a fresh secret key $s \in \{0,1\}^n$ for each new message of length n. Indeed, regardless of the system employed, if a uniformly selected message $m \in \{0,1\}^n$ is encrypted with a key of length $k < n$, then at least $n - k$ bits of "information" about m have leaked into the ciphertext. (See, e.g., [25] for a formal discussion of message equivocation.) Despite this, we construct a family of encryption systems utilizing short keys which guarantee that for any message space with sufficient min-entropy, no adversary can predict any Boolean function of the message m with non-negligible advantage; specifically, if the message space has min-entropy $n - \ell$, and secret keys of length $\ell + s$ are utilized, no Boolean function can be predicted with advantage $2^{-\Theta(s)}$. These systems rely on no unproven assumptions, and encryption (and decryption) can be computed efficiently. The precise notion of security is described below.

L.R. Knudsen (Ed.): EUROCRYPT 2002, LNCS 2332, pp. 133–148, 2002.

Of course, if a pseudorandom generator exists, then it is possible to construct encryption systems with satisfactory security guarantees against resource-bounded adversaries, even when the length of the message m exceeds the length of the key. A traditionally accepted notion of security in this resource-bounded case is that of *semantic security* [11], though a number of stronger (and important) notions exist (see, e.g., [4,9,17,19]). A system with semantic security guarantees that observation of $E(m)$, the encryption of a message m, offers essentially no advantage to a bounded adversary in predicting any Boolean function of the message m. (This Boolean function may be some specific bit of m, or, perhaps, a complicated function capturing some global property of m.) Furthermore, this guarantee is offered *regardless of the a priori distribution of the message* m. In the last section of the paper, we discuss some potential applications of the information-theoretic encryption systems of Sects. 3 and 4 to this complexity-theoretic framework. Specifically, we observe that a hybrid approach can reduce the complexity of the resulting system at the expense of weakening (in a controlled fashion) the notion of semantic security. Finally, we mention that if the adversary is space-limited and the parties have access to a long public random string, strong privacy guarantees can be obtained with short keys [15,2].

Returning to the case for unbounded adversaries, we say that an encryption system offers *entropically bounded security* if for all message distributions with sufficient min-entropy, and all pieces of partial information $h : \{0,1\}^* \to \{0,1\}$, observation of the ciphertext of m offers no adversary non-negligible advantage in prediction of $h(m)$. If the definition is strengthened so that it applies for all message spaces and the error terms in the advantage are removed, then we exactly recover the definition of perfect secrecy. (See the next section for precise definitions.) Initially, we give a simple encryption system offering entropic security in the case when the adversary has *no* a priori information about the message (i.e., the message distribution is uniform); the scheme can be realized with keys of length $\omega(\log n)$. We then show that for message spaces with min-entropy $n - \ell$, an encryption system offering entropically bounded security can be realized with keys of length $\ell + \omega(\log n)$.

The two main theorems in the article, Theorem 2 and Theorem 3, are both instantiations of common paradigms in cryptography. The first is an information-theoretic variant of the standard practice of encrypting a short seed which is then used for a pseudorandom generator (in our case, this will be an ϵ-biased space). The second is a variant of the "simple embedding schemes" often used in practice, where a message is encrypted by applying a one-way permutation after a suitable (bijective) hash function. The system of Bellare and Rogaway [5] is also theoretical evidence for the quality of such schemes.

In Sect. 2 we give basic definitions, including a brief discussion of ϵ-biased spaces, universal hash functions, and the Fourier transform over $\mathbb{Z}_2^n$, which will be used in the main results, presented in Sects. 3 and 4. In Sect. 5, we discuss some applications of these theorems to resource-bounded encryption systems.

2 Definitions

Definition 1. *A pair* $(\mathcal{E}, \mathcal{D})$ *is a* symmetric encryption system with parameters (ℓ_s, ℓ_e) *if*

1. $\ell_s : \mathbb{N} \to \mathbb{N}$ *and* $\ell_e : \mathbb{N} \to \mathbb{N}$ *are functions, determining the length of the secret key and the length of the encryption for messages of length* n,
2. $\mathcal{E} = \{E_n : \{0,1\}^n \times \{0,1\}^{\ell_s(n)} \to \{0,1\}^{\ell_e(n)} \mid n \geq 1\}$ *is the family of encryption functions, and*
3. $\mathcal{D} = \{D_n : \{0,1\}^{\ell_e(n)} \times \{0,1\}^{\ell_s(n)} \to \{0,1\}^n \mid n \geq 1\}$ *is the family of decryption functions,*

so that for all $n \geq 1$, $m \in \{0,1\}^n$, *and* $s \in \{0,1\}^{\ell_s(n)}$, $D_n(E_n(m,s),s) = m$. *When the length* n *of the message can be inferred from context, we write* $E(m,s)$ $(D(m,s))$ *rather than* $E_n(m,s)$ $(D_n(m,s))$. *When an encryption system is clear from context, we let* S_n *denote the random variable uniform on the set* $\{0,1\}^{\ell_s(n)}$.

As defined above, encryption and decryption are deterministic; in Sect. 4 we shall consider the case when the encryption algorithm may depend on some private randomness.

Definition 2. *A* message space $\mathcal{M}$ *is a sequence of random variables* $\mathcal{M} = \{M_n \mid n \geq 1\}$ *so that* M_n *takes values in* $\{0,1\}^n$. *(When we couple* $\mathcal{M}$ *with an encryption system, we always assume that* M_n *and* S_n *are independent.)*

A symmetric encryption system $(\mathcal{E}, \mathcal{D})$ with parameters (ℓ_s, ℓ_e) is said to possess *perfect secrecy* if for all message spaces $\mathcal{M}$, all $n > 0$, and all $e \in \mathbf{im}\, E_n$, $\Pr[M_n = m] = \Pr[M_n = m \mid E(M_n, S_n) = e]$. An equivalent definition of perfect secrecy is the following:

Definition 3 (Perfect Secrecy). *A symmetric encryption system* $(\mathcal{E}, \mathcal{D})$ *with parameters* (ℓ_s, ℓ_e) *is said to possess* perfect secrecy *if for all* $\mathcal{M}$, *all* $n \geq 1$, *and all functions* $f : \{0,1\}^{\ell_e(n)} \to \{0,1\}$, *there is a random variable* G_f, *independent of* M_n, *so that for every* $h : \{0,1\}^n \to \{0,1\}$,

$$\Pr[f(E(M_n, S_n)) = h(M_n)] = \Pr[G_f = h(M_n)] \ .$$

Intuitively, this asserts that if there is an adversary f which can predict some Boolean function of m based on the ciphertext of m, then there is another adversary G_f which can predict this same Boolean function of m *without* even witnessing the ciphertext. (If one suitably changes this definition so that the function f and the random variable G_f are polynomial time computable and allows for negligible error, then one obtains the notion of semantic security.)

A random variable M_n taking values in $\{0,1\}^n$ has *min-entropy* $n - \ell$ when $\forall m_o \in \{0,1\}^n$, $\Pr[M_n = m_o] \leq 2^{-n+\ell}$. A message space $\mathcal{M}$, is said to have min-entropy $n - \ell(n)$ when the random variable M_n possesses this property for each n.

Definition 4. *We say that an encryption system possesses* $\ell(n)$-entropic security *if for every message space* $\mathcal{M}$ *with min-entropy* $n - \ell(n)$, *every* $n > 0$, *and all functions* $f : \{0,1\}^{\ell_e(n)} \to \{0,1\}$, *there is a random variable* G_f, *independent of* M_n, *so that for every* $h : \{0,1\}^n \to \{0,1\}$,

$$|\Pr[f(E(M_n)) = h(M_n)] - \Pr[G_f = h(M_n)]| = n^{-\omega(1)} .$$

Observe that if no constraint is placed on the min-entropy in $\mathcal{M}$ and the $n^{-\omega(1)}$ error term is removed, we recover the definition of perfect secrecy. We will construct two encryption systems, $(\mathcal{E}^u, \mathcal{D}^u)$ and $(\mathcal{E}^k, \mathcal{D}^k)$, so that

- E^u possesses 0-entropic security (i.e., provides security when the message space is uniform) and uses keys of length $w(n)\log n$, where $w(n)$ is any function tending to infinity. E^u (and D^u) can be computed in time $O(w(n)n \log^{1+c} n)$ for any $c > 0$.
- E^k possesses ℓ-entropic security (i.e., provides security when the message space has min-entropy $n - \ell$) so long as $\ell(n) \leq k(n) - \omega(\log n)$ and uses keys of length $k(n)$. E^k (and D^k) can be computed in time $O\left(n \log^2 n \log\log n\right)$.

These constructions make use of ϵ-biased sample spaces and universal hash functions, defined below.

2.1 ϵ-Biased Sample Spaces

Definition 5. *A set* $S \subseteq \{0,1\}^n$ *is called* ϵ-biased *(or an* ϵ-biased sample space*) if for all nonempty* $\alpha \subset [n] = \{1, \ldots, n\}$, $\left|\mathsf{Exp}_{s \in S}\left[\prod_{a \in \alpha}(-1)^{s_a}\right]\right| \leq \epsilon$.

Small sets with these properties were initially constructed by Naor and Naor [16] and Peralta [18]. We will use a construction, due to Alon, Goldreich, Håstad and Peralta [1], which gives an ϵ-biased sample space in $\{0,1\}^n$ of size about $(\frac{n}{\epsilon})^2$. The sample space is given as the image of a certain function $\sigma_{n,m} : \mathbb{F}_{2^m} \times \mathbb{F}_{2^m} \to \{0,1\}^n$. (Here $\mathbb{F}_{2^n}$ denotes the finite field with 2^n elements.) To define σ, let bin : $\mathbb{F}_{2^m} \to \{0,1\}^m$ be a bijection satisfying $\text{bin}(0) = 0^m$ and $\text{bin}(x + y) = bin(x) \oplus bin(y)$, where $\alpha \oplus \beta$ denotes the componentwise exclusive or of α and β. Then $\sigma(x, y) = r = (r_0, \ldots, r_{n-1})$, where $r_i = \langle bin(x^i), bin(y)\rangle_2$, the inner product, modulo two, of x^i and y. The size of the sample space is 2^{2m}. Let $S_{n,m} \subset \{0,1\}^n$ be the collection of points so defined. They show that

Theorem 1 ([1]). $S_{n,m} = \boldsymbol{im}\,\sigma_{m,n}$ *is* $\frac{n-1}{2^m}$*-biased.*

Observe that when $m = \lceil \log n\epsilon^{-1} \rceil$, $\frac{n-1}{2^m} \leq \epsilon$. As elements of $S_{m,n}$ are constructed during the encryption (and decryption) phase of the 0-entropic encryption system, we analyze the complexity of computing the function above. First, we need to find an irreducible polynomial p of degree m over the finite field $\mathbb{F}_2$. As the degree of the polynomial will correspond to the quantity ϵ, we can be somewhat flexible concerning the degree of the irreducible polynomial and use an explicit family: for each $c \in \mathbb{N}$, the polynomial $p_c(x) = x^{2m} + x^m + 1$, where $m = 3^c$, is irreducible over $\mathbb{F}_2$. (See [14, Exercise 3.96].) Computation of $\sigma = \sigma_{m,n}$ for a pair (x, y) is performed on a component by component basis: given x^i, computation of x^{i+1} requires a single multiplication in $\mathbb{F}_{2^m} \cong \mathbb{F}_2[x]/(p_c)$. Using fast polynomial multiplication, computing this product takes $O(m \log m \log\log m)$ time (see [22], or the discussion in [3, p. 232]). As p_c is sparse (it has only 3 nonzero terms), reducing this result modulo p_c requires $O(m)$ time. Hence computation of $\sigma(x, y)$ requires $O(nm \log m \log\log m))$ time. In order for $S = \mathbf{im}\,\sigma$ to be ϵ-biased, we take m to be the smallest integer of form $2 \cdot 3^c$ larger than $\lceil \log(n/\epsilon) \rceil$; in this case the above running time is $O(n \log(n/\epsilon) \log\log(n/\epsilon) \log\log\log(n/\epsilon))$. To simplify notation, we let $\sigma_{n,\epsilon}$ denote $\sigma_{n,m}$ for this value of m.

2.2 k-Wise Independent Permutations

Definition 6. *A family of permutations $\mathcal{P} \subset \{f : X \to X\}$ is a family of k-*wise independent permutations *[24] if for all distinct $s_1, \ldots, s_k \in X$ and all distinct $t_1, \ldots, t_k \in X$, $\Pr_{\phi \in \mathcal{P}}[\forall i, \phi(s_i) = t_i] = \prod_{i=0}^{t-1} \frac{1}{|X|-i}$.*

We will use a family of 3-wise independent permutations, described below. See Rees [20] for a more detailed description.

Let V be a two-dimensional vector space over $\mathbb{F}$, a finite field. For two nonzero vectors $\boldsymbol{v}$ and $\boldsymbol{w}$ in this space, we write $\boldsymbol{v} \sim \boldsymbol{w}$ when $\boldsymbol{v} = c\boldsymbol{w}$ for some $c \in \mathbb{F}$ (so that the two vectors span the same one-dimensional subspace). This is an equivalence relation; we write $[\boldsymbol{v}]$ for the equivalence class containing $\boldsymbol{v}$. Projective 2-space over $\mathbb{F}$ is then $P_2(\mathbb{F}) = \{[\boldsymbol{v}] \mid \boldsymbol{v} \neq \mathbf{0}\}$. We let $\mathrm{GL}_2(\mathbb{F})$ denote the set of non-singular 2×2 matrices over $\mathbb{F}$, and $\mathrm{PGL}_2(\mathbb{F}) = \mathrm{GL}_2(\mathbb{F})/\{cI | c \in \mathbb{F}\}$, where I is the identity matrix. An element ϕ of $\mathrm{PGL}_2(\mathbb{F})$ acts on $P_2(\mathbb{F})$ in a natural (and well-defined) way, mapping $[\boldsymbol{v}]$ to $[\phi(\boldsymbol{v})]$. It is not difficult to show that for any distinct $[\boldsymbol{u_1}], [\boldsymbol{u_2}], [\boldsymbol{u_2}] \in P_2(\mathbb{F})$ and any distinct $[\boldsymbol{v_1}], [\boldsymbol{v_2}], [\boldsymbol{v_2}] \in P_2(\mathbb{F})$, there is in fact a unique $\phi \in \mathrm{PGL}_2(\mathbb{F})$ so that $\phi([\boldsymbol{u_i}]) = [\boldsymbol{v_i}]$ for each i. In particular, $\mathrm{PGL}_2(\mathbb{F})$ is a 3-wise independent family of permutations. As multiplication and inversion in a finite field $\mathbb{F}_p$, for a prime p, may be accomplished in time $O(\log p (\log\log p)^2 \log\log\log p)$ time [23,21], evaluation of an element $\phi \in \mathrm{PGL}_2(\mathbb{F}_p)$ also has this complexity.

Proposition 1. *$PGL_2(\mathbb{F}_p)$ is a 3-wise independent set of permutations of $P_2(\mathbb{F}_p)$.*

2.3 Fourier Analysis of Boolean Functions

Let $L(\mathbb{Z}_2^n) = \{f : \mathbb{Z}_2^n \to \mathbb{R}\}$ denote the set of real valued functions on $\mathbb{Z}_2^n = \{0,1\}^n$. Though our interest shall be in Boolean functions, it will be temporarily convenient to consider this richer space. $L(\mathbb{Z}_2^n)$ is a vector space over $\mathbb{R}$ of dimension 2^n, and has a natural inner product: for $f, g \in L(\mathbb{Z}_2^n)$, define $\langle f, g \rangle = 2^{-n} \sum_{x \in \{0,1\}^n} f(x)g(x)$. For a subset $\alpha \subset \{1, \ldots, n\}$, define the function $\chi_\alpha : \{0,1\}^n \to \mathbb{R}$ so that $\chi_\alpha(x) = \prod_{a \in \alpha} (-1)^{x_a}$. These functions χ_α are the *characters* of $\mathbb{Z}_2^n = \{0,1\}^n$. Among their many wonderful properties is the fact that *the characters form an orthonormal basis for* $L(\mathbb{Z}_2^n)$. To see this, observe that $\forall \alpha \subset [n]$, $\sum_{x \in \{0,1\}^n} \chi_\alpha(x) = 2^n$ when $\alpha = \emptyset$, and 0 otherwise. Furthermore, for $\alpha, \beta \subset [n]$, $\chi_\alpha(x)\chi_\beta(x) = \chi_{\alpha \oplus \beta}(x)$, where $\alpha \oplus \beta$ denotes the symmetric difference of α and β, so that $\langle \chi_\alpha, \chi_\beta \rangle = 1$ when $\alpha = \beta$, and 0 otherwise. Considering that there are 2^n characters, pairwise orthogonal, they span $L(\mathbb{Z}_2^n)$, as promised. Any function $f : \{0,1\}^n \to \mathbb{R}$ may then be written in terms of this basis: $f = \sum_\alpha \widehat{f}_\alpha \chi_\alpha$, where $\widehat{f}_\alpha = \langle f, \chi_\alpha \rangle$ is the projection of f onto χ_α. These coefficients $\widehat{f}_\alpha$, $\alpha \subset [n]$, are the *Fourier coefficients* of f, and, as we have above observed, uniquely determine the function f.

Given the above, it is easy to establish the *Plancherel* equality:

Proposition 2. *Let $f \in L(\mathbb{Z}_2^n)$. Then $\sum_\alpha \widehat{f}_\alpha^2 = \frac{1}{2^n} \sum_x f(x)^2$.*

As always, $\widehat{f}_\emptyset = \mathsf{Exp}[f]$ and, when the range of f is $\{\pm 1\}$, $\sum_\alpha \widehat{f}_\alpha^2 = \|f\|_2^2 = 1$.

3 Security for Uniformly Distributed Message Spaces

We begin by constructing a simple encryption system offering security in the case when the adversary has no a priori knowledge concerning the message (i.e., the message space is uniform). The next section will develop a more flexible encryption system which can tolerate general message distributions, so long as they have sufficient min-entropy.

Theorem 2. *Let $(\mathcal{E}^u, \mathcal{D}^u)$ be the encryption system given by $D_n^u(m,s) = E_n^u(m,s) = m \oplus \sigma_{n,\epsilon}(s)$ where $|m| = n$ and s is selected randomly in the domain of $\sigma_{n,\epsilon}$ (so $|s| = O(\log n\epsilon^{-1})$). Then for $\epsilon = n^{-\omega(1)}$, this encryption system offers 0-entropic security. E_n^u and D_n^u can be computed in time $O(n \log(n/\epsilon) \log\log(n/\epsilon)$ $\log\log\log(n/\epsilon))$.*

Proof. For $n \geq 1$, let M_n be uniform on $\{0,1\}^n$; when n is understood we drop the subscripts. For simplicity, we treat h as a function with range $\{\pm 1\}$ rather than $\{0,1\}$. We must show that for every $f : \{0,1\}^n \to \{\pm 1\}$, there is a random variable G_f, independent of M, so that for all functions $h : \{0,1\}^n \to \{\pm 1\}$

$$|\mathsf{Pr}[f(M \oplus \sigma(S)) = h(M)] - \mathsf{Pr}[G_f = h(M)]| \leq n^{-\omega(1)} \ .$$

(Here S is uniform on the domain of $\sigma_{n,\epsilon}$.) The random variable G_f is defined in terms of the function f; though G_f is independent of M, G_f will predict $h(M)$ nearly as well as does f. Let M' be uniform on $\{0,1\}^n$ and independent of M (and S). Define $G_f = f(M')$; then $\mathsf{Pr}[G_f = h(M)] = \mathsf{Pr}[f(M') = h(M)]$. We begin with a lemma providing an upper bound on this prediction probability which is independent of f:

Lemma 1. *Let $T(m) = \mathsf{Exp}_S[h(m \oplus \sigma(S))] - \mathsf{Exp}_{M'}[h(M')]$; then*

$$|\mathsf{Pr}\left[f(M') = h(M)\right] - \mathsf{Pr}\left[f(M) = h(M \oplus \sigma(S))\right]| \leq \frac{1}{2} \mathop{\mathsf{Exp}}_{M}\left[|T(M)|\right] \ .$$

Proof. Define $c(m, m')$ so that $c(m, m') = 1$ when $f(m) = h(m')$ and 0 if $f(m) \neq h(m')$. As $h(\cdot)$ takes values in the set $\{\pm 1\}$, we can rewrite $c(m, m') = \frac{1}{2}[1 + f(m)h(m')]$ and

$$\begin{aligned}
&|\mathsf{Pr}\,[f(M') = h(M)] - \mathsf{Pr}\,[f(M) = h(M \oplus \sigma(S))]| \\
&= \left| \mathop{\mathsf{Exp}}_{M,M'}[c(M, M')] - \mathop{\mathsf{Exp}}_{M,S}[c(M, M \oplus \sigma(S))] \right| \\
&= \left| \mathop{\mathsf{Exp}}_{M}\left[\frac{f(M)}{2}\left(\mathop{\mathsf{Exp}}_{M'}[h(M')] - \mathop{\mathsf{Exp}}_{S}[h(M \oplus \sigma(S))]\right)\right] \right| \\
&\leq \frac{1}{2} \mathop{\mathsf{Exp}}_{M}\left[\left|\mathop{\mathsf{Exp}}_{M'}[h(M')] - \mathop{\mathsf{Exp}}_{S}[h(M \oplus \sigma(S))]\right|\right] = \frac{1}{2} \mathop{\mathsf{Exp}}_{M}[|T(M)|]\,.
\end{aligned}$$

□

We apply the second moment method to control $\mathsf{Exp}_M[|T(M)|]$. Observe that $\mathsf{Exp}_M[h(M)] = \widehat{h}_\emptyset$, so

$$\begin{aligned}
T(m) &= \mathop{\mathsf{Exp}}_{S}\left[\sum_{\alpha \neq \emptyset} \widehat{h}_\alpha \chi_\alpha(m \oplus \sigma(S))\right] \\
&= \sum_{\alpha \neq \emptyset} \widehat{h}_\alpha \mathop{\mathsf{Exp}}_{S}[\chi_\alpha(m \oplus \sigma(S))] = \sum_{\alpha \neq \emptyset} \widehat{h}_\alpha \chi_\alpha(m) \mathop{\mathsf{Exp}}_{S}[\chi_\alpha(\sigma(S))]\ .
\end{aligned}$$

Then $\mathsf{Exp}_M\,[T(M)] = \sum_{\alpha \neq \emptyset} \widehat{h}_\alpha \mathsf{Exp}_S\,[\chi_\alpha(\sigma(S))]\, \mathsf{Exp}_M\,[\chi_\alpha(M)] = 0$. Now, the random variables $\widehat{h}_\alpha \chi_\alpha(M \oplus \sigma(S))$ and $\widehat{h}_\beta \chi_\beta(M \oplus \sigma(S))$ are pairwise independent (recall that M is uniform) so that

$$\begin{aligned}
\mathop{\mathsf{Var}}_{M}[T(M)] &= \mathop{\mathsf{Var}}_{M}\left[\sum_{\alpha \neq \emptyset} \widehat{h}_\alpha \chi_\alpha(M) \mathop{\mathsf{Exp}}_{S}[\chi_\alpha(\sigma(S))]\right] \\
&= \sum_{\alpha \neq \emptyset} \widehat{h}_\alpha^2 \mathop{\mathsf{Exp}}_{S}[\chi_\alpha(\sigma(S))]^2 \mathop{\mathsf{Var}}_{M}[\chi_\alpha(M))] \leq \epsilon^2 \sum_{\alpha \neq \emptyset} \widehat{h}_\alpha^2 \leq \epsilon^2
\end{aligned}$$

by the Plancherel equality (see Section 2.3) and the fact that $\mathsf{Var}_M\,[\chi_\alpha(M)] = 1$. Now, applying Chebyshev's inequality, we have $\mathsf{Pr}\,[|T(M)| > \lambda] < \epsilon^2 \lambda^{-2}$.

Selecting $\lambda = \epsilon^{\frac{2}{3}}$, we have

$$\begin{aligned}
\mathop{\mathsf{Exp}}_{M}[|T(M)|] &\leq \mathsf{Pr}\,[|T(M)| > \lambda] \cdot \max_{m} |T(m)| + \mathsf{Pr}\,[|T(M)| \leq \lambda] \cdot \lambda \\
&\leq \frac{\epsilon^2}{\lambda^2} \cdot 2 + (1 - \frac{\epsilon^2}{\lambda^2}) \cdot \lambda \leq 3\epsilon^{\frac{2}{3}}\,.
\end{aligned}$$

Hence $|\mathsf{Pr}[f(M \oplus \sigma(S)) = h(M)] - \mathsf{Pr}[G_f = h(M)]| < \frac{3}{2}\epsilon^{\frac{2}{3}}$. As $\epsilon = n^{-\omega(1)}$, this completes the proof. The bound on $|s|$ and the running time of E^u and D^u follow from Section 2.1. □

4 Security for Entropically Rich Message Spaces

In this section we describe a symmetric encryption system offering ℓ-entropic security; keys of length $\ell + \omega(\log n)$ suffice. In preparation, we will slightly enrich our notion of symmetric encryption system by allowing the encryption function(s) to be stochastic: for each n, E_n may depend on m, the message, s, the secret key, and ϕ, some private random coins of the encryption function. To keep the notation uniform, we let Φ_n be the random variable on which E_n may depend. Φ_n is independent of M_n and S_n.

For convenience, we will assume that the message space is $\mathbb{Z}_{p+1}$ for a prime p. (So we treat $\mathcal{M} = \{M_p \mid p \text{ prime}\}$, where M_p is a random variable taking values in $\mathbb{Z}_{p+1}$.) To keep our notation uniform, we let $n = \log(p+1)$ and then say that $M_p \in \mathbb{Z}_{p+1}$ has min-entropy $n - \ell$ if $\Pr[M_p = m_0] \leq 2^{-n+\ell}$. Now, we select an artificial bijection $L : \mathbb{Z}_{p+1} \to P_2(\mathbb{F}_p)$, so that

$$L(z) = \left[\begin{pmatrix} 1 \\ z \end{pmatrix}\right], \text{ for } 0 \leq z \leq p-1, \text{ and } \qquad L(p) = \left[\begin{pmatrix} 0 \\ 1 \end{pmatrix}\right] .$$

L can be computed in linear time; L^{-1} can be computed by single inversion modulo p. Having fixed this bijection, we will treat the 3-wise independent functions $\mathrm{PGL}_2(\mathbb{F}_p)$, described in Sect. 2.2, as if they act on $\mathbb{Z}_{p+1}$.

Theorem 3. *Let $(\mathcal{E}^k, \mathcal{D}^k)$ be the encryption system with $E_p^k(m, s; \phi) = (\phi(m) + s, \phi)$, where the message $m \in \mathbb{Z}_{p+1}$, s is the secret key, chosen in the set $\{0, 1, \ldots, 2^k - 1\} \subset \mathbb{Z}_{p+1}$, and ϕ is an element of $PGL_2(\mathbb{F}_p)$ selected at random by E_n^k. (Here $+$ is modulo $p+1$.) Decryption is defined analogously. Then if $k = \ell + \omega(\log n)$, this encryption system offers ℓ-entropic security. Furthermore, E_n^k and D_n^k can be computed in time $O(n \log^2 n \log\log n)$.*

Proof. We need to show that for every message space $\mathcal{M}$ of sufficient min-entropy and every $f : \mathbf{im}\, E_p^k \to \{0,1\}$, there is a random variable G_f, independent of M_p, such that for all functions $h : \mathbb{Z}_{p+1} \to \{0,1\}$

$$|\Pr[f(\Phi(M) + S, \Phi) = h(M)] - \Pr[G_f = h(M)]| \leq \frac{1}{n^{\omega(1)}} .$$

Here S is uniform on $K = \{0, \ldots, 2^k - 1\} \subset \mathbb{Z}_{p+1}$ and Φ is uniform on $\mathrm{PGL}_2(\mathbb{F}_p)$; Φ, S, and M are independent.

We define the random variable G_f, which can predict $h(M)$ nearly as well as can f even without observing $E(M, S; \Phi)$. As in the proof above, let M', S', and Φ' be a random variables with the same distributions as M, S, and Φ; all independent. Define the random variable $G_f = f(\Phi'(M') + S', \Phi')$. Observe that

$$\Pr[G_f = h(M)] = \Pr[f(\Phi'(M') + S', \Phi') = h(M)] = \Pr[f(\Phi(M') + S, \Phi) = h(M)] .$$

We begin by recording an analogue of Lemma 1 for this cryptosystem which allows us to remove the dependence on f.

Lemma 2. *For any* $h : \mathbb{Z}_{p+1} \to \{0,1\}$ *and* $f : \mathbb{Z}_{p+1} \times PGL_2(F_p) \to \{0,1\}$,

$$|\Pr[f(\Phi(M)+S,\Phi) = h(M')] - \Pr[f(\Phi(M)+S,\Phi) = h(M)]| \le \mathop{\mathsf{Exp}}_{M,\Phi,S}\left[\left|\mathop{\mathsf{Exp}}_{M'}[h(M')] - \mathop{\mathsf{Exp}}_{M',S'}[h(M') \,|\, \Phi(M') + S' = \Phi(M) + S]\right|\right].$$

Proof. For simplicity, use two new functions $\tilde{f}$ and $\tilde{h}$ which take values in $\{\pm 1\}$ rather than $\{0,1\}$. Let $\tilde{f}(x) = 2f(x) - 1$ and $\tilde{h}(x) = 2h(x) - 1$. Then

$$\begin{aligned}
&|\Pr[f(\Phi(M)+S,\Phi) = h(M')] - \Pr[f(\Phi(M)+S,\Phi) = h(M)]| \\
&= \frac{1}{2}\left|\mathop{\mathsf{Exp}}_{M,M',\Phi,S}\left[\tilde{f}(\Phi(M)+S,\Phi)\tilde{h}(M')\right] - \mathop{\mathsf{Exp}}_{M,\Phi,S}\left[\tilde{f}(\Phi(M)+S,\Phi)\tilde{h}(M)\right]\right| \\
&= \frac{1}{2}\left|\mathop{\mathsf{Exp}}_{\Phi,M,S}\left[\tilde{f}(\Phi(M)+S,\Phi)\left(\mathop{\mathsf{Exp}}_{M'}[\tilde{h}(M')] - \tilde{h}(M)\right)\right]\right| \\
&= \frac{1}{2}\left|\mathop{\mathsf{Exp}}_{\Phi}\left[\sum_{e\in\mathbb{Z}_{p+1}} \Pr[\Phi(M)+S = e]\cdot \tilde{f}(e,\Phi)\cdot\right.\right. \\
&\qquad\qquad \left.\left.\mathop{\mathsf{Exp}}_{M,S}\left[\left(\mathop{\mathsf{Exp}}_{M'}[\tilde{h}(M')] - \tilde{h}(M)\right) \;\middle|\; \Phi(M)+S = e\right]\right]\right| \\
&\le \frac{1}{2}\mathop{\mathsf{Exp}}_{\Phi}\left[\sum_{e} \Pr[\Phi(M)+S = e]\cdot\left|\mathop{\mathsf{Exp}}_{M}[\tilde{h}(M)] - \mathop{\mathsf{Exp}}_{M,S}\left[\tilde{h}(M)\,\middle|\,\Phi(M)+S = e\right]\right|\right]
\end{aligned}$$

Observe now that for any functions $g_1 : \mathbb{Z}_{p+1} \to \mathbb{R}$ and $g_2 : \mathbb{Z}_{p+1} \to \mathbb{Z}_{p+1}$,

$$\sum_e \Pr_x[g_2(x) = e]\, \mathop{\mathsf{Exp}}_x[g_1(g_2(x)) \mid g_2(x) = e] = \mathop{\mathsf{Exp}}_x[g_1(g_2(x))] \;;$$

the statement of the lemma follows. □

Now, for an element $s_0 \in K$, let $K_{s_0} = \{s - s_0 \bmod (p+1) \mid s \in K\}$. Then define

$$G^{s_0}_{m,\phi} = \mathop{\mathsf{Exp}}_{M}[h(M)] - \mathop{\mathsf{Exp}}_{M,S'}[h(M) \,|\, \phi(M) + (S' - s_0) = \phi(m)] \;,$$

where $S' - s_0$ is uniform on K_{s_0}. From the lemma above, it suffices to show that for every $s_0 \in K$, $\mathsf{Exp}_{M,\Phi}[|G^{s_0}_{M,\Phi}|]$ is small. So fix $s_0 \in K$. We let $p_m = \Pr[M = m]$. For a message $m_0 \in \mathbb{Z}_{p+1}$ and an element $w_0 \in \mathbb{Z}_{p+1}$, let $\mathcal{P}_{w_0\to m_0} = \{\phi \in \mathrm{PGL}_2(\mathbb{F}_p) \mid \phi(m_0) = w_0\}$. We will handle each of these "slices" of $\mathrm{PGL}_2(\mathbb{F}_p)$ separately. For any permutation ϕ in $\mathcal{P}_{w_0\to m_0}$ we can consider the sums

$$A^\phi_0 = \sum_{m\in\phi^{-1}(w_0+K_{s_0})} p_m \qquad \text{and} \qquad B^\phi_0 = \sum_{m\in\phi^{-1}(w_0+K_{s_0})} p_m \cdot h(m) \;;$$

then

$$\frac{B^\phi_0}{A^\phi_0} = \mathop{\mathsf{Exp}}_{M}[h(M) \mid \phi(M) \in w_0 + K_{s_0}] \,. \tag{1}$$

(Here $w_0 + K_{s_0}$ denotes the set $\{w_0 + s \bmod (p+1) \mid s \in K_{s_0}\}$.) We would like to see that this quotient is well-behaved (i.e., near the expected value of $h(M)$,) for most ϕ.

Let Φ_0 be a uniform random variable in $\mathcal{P}_{m_0 \to w_0}$ and let X_m be the random variable taking the value p_m if $\Phi_0(m) \in w_0 + K_{s_0}$, and 0 otherwise. Then $\sum_{m \in \mathbb{Z}_{p+1}} X_m = A_0^{\Phi_0}$ and $\sum_{m \in h^{-1}(1)} X_m = B_0^{\Phi_0}$ so that

$$\begin{aligned}\mathsf{Exp}[A_0^{\Phi_0}] &= \mathsf{Exp}\left[p_{m_0} + \sum_{m \neq m_0} X_m\right] = p_{m_0} + \sum_{m \neq m_0} (\mathsf{Pr}[\Phi_0(m) \in w_0 + K_{s_0}] \cdot p_m) \\ &= p_{m_0}\left(1 - \frac{2^k - 1}{p+1}\right) + \frac{2^k - 1}{p+1}.\end{aligned}$$

Hence $\frac{2^k-1}{p+1} \leq \mathsf{Exp}[A_0^{\Phi_0}] \leq p_{m_0}\left(1 - \frac{2^k-1}{p+1}\right) + \frac{2^k-1}{p+1} \leq 2^{-n+k} + p_{m_0}$. Let $\overline{h} = \mathsf{Exp}_M[h(M)]$; then, similarly,

$$\overline{h} \cdot \frac{2^k - 1}{p+1} \leq \mathop{\mathsf{Exp}}_{\Phi_0}[B_0^{\Phi_0}] \leq \overline{h} \cdot \frac{2^k - 1}{p+1} + p_{m_0}\left(1 - \frac{2^k - 1}{p+1}\right).$$

Recalling that the distribution of M has min-entropy $n - \ell$,

$$\frac{\mathsf{Exp}_{\Phi_0}[B_0^{\Phi_0}]}{\mathsf{Exp}_{\Phi_0}[A_0^{\Phi_0}]} - \overline{h} \leq \frac{2^\ell}{2^k - 1},$$

and similarly,

$$\frac{\mathsf{Exp}_{\Phi_0}[B_0^{\Phi_0}]}{\mathsf{Exp}_{\Phi_0}[A_0^{\Phi_0}]} - \overline{h} \geq \frac{\overline{h} \cdot (2^k - 1)}{2^k + (p+1) \cdot p_{m_0} - 1} - \overline{h} \geq -2^{-k+\ell}.$$

Hence,

$$\left| \frac{\mathsf{Exp}_{\Phi_0}[B_0^{\Phi_0}]}{\mathsf{Exp}_{\Phi_0}[A_0^{\Phi_0}]} - \overline{h} \right| \leq 2 \cdot 2^{-k+\ell}.$$

We wish to insure that $A_0^{\Phi_0}$ and $B_0^{\Phi_0}$ are close to their expected vales. In preparation for applying Chebyshev's inequality, we compute their variances. We have

$$\mathop{\mathsf{Var}}_{\Phi_0}[A_0^{\Phi_0}] = \mathsf{Var}[\sum_{m \in \mathbb{Z}_p} X_m] = \sum_{m \neq m_0} \mathop{\mathsf{Var}}_{\Phi_0}[X_m] + \sum_{m_1 \neq m_2 \neq m_0} \mathop{\mathsf{Cov}}_{\Phi_0}[X_{m_1}, X_{m_2}].$$

Now, $\sum_{m \neq m_0} \mathsf{Var}[X_m] \leq \sum_{m \neq m_0} \mathsf{Exp}[X_m^2] \leq 2^{-2n+k+\ell}$, and these variables are pairwise negatively correlated (so that $\mathsf{Cov}[X_{m_1}, X_{m_2}] < 0$ for $m_1 \neq m_2$, both distinct from m_0). Then, $\mathsf{Var}_{\Phi_0}[A_0^{\Phi_0}] < 2^{-2n+k+\ell}$. Similarly, $\mathsf{Var}_{\Phi_0}[B_0^{\Phi_0}] < \overline{h} \cdot 2^{-2n+k+\ell}$. Observe that 3-wise independence is required here.

By Chebyshev's inequality we have

$$\mathop{\mathsf{Pr}}_{\Phi_0}\left[\left|A_0^{\Phi_0} - \mathsf{Exp}[A_0^{\Phi_0}]\right| \geq \delta_a\right] \leq \frac{\mathsf{Var}[A_0^{\Phi_0}]}{\delta_a^2} < \frac{2^k}{2^{2n-\ell} \cdot \delta_a^2}, \tag{2}$$

and

$$\Pr_{\Phi_0}\left[\left|B_0^{\Phi_0} - \mathsf{Exp}[B_0^{\Phi_0}]\right| \geq \delta_b\right] \leq \frac{\mathsf{Var}[B_0^{\Phi_0}]}{\delta_b^2} < \frac{\overline{h} \cdot 2^k}{2^{2n-\ell} \cdot \delta_b^2} \ . \tag{3}$$

If ϕ is an element of $\mathcal{P}_0$ for which both

$$\left|A_0^\phi - \mathop{\mathsf{Exp}}_{\Phi_0}[A_0^\phi]\right| \leq \delta_a \qquad \text{and} \qquad \left|B_0^\phi - \mathop{\mathsf{Exp}}_{\phi}[B_0^{\Phi_0}]\right| \leq \delta_b \ ,$$

we have in particular, from equation (1),

$$\left|\frac{\mathsf{Exp}_{\Phi_0}[B_0^{\Phi_0}]}{\mathsf{Exp}_{\Phi_0}[A_0^{\Phi_0}]} - \mathop{\mathsf{Exp}}_M[h(M) \mid \phi(M) \in w_0 + K_{s_0}]\right| \leq \frac{2(\delta_a + \delta_b)}{\mathsf{Exp}_{\Phi_0}[A_0^{\Phi_0}]} \leq 2(\delta_a+\delta_b)\cdot 2^{n-k} \ ,$$

(assuming that $\delta_a, \delta_b < 1/2$). When $\delta_a = \delta_b = \frac{\epsilon_1}{4\cdot 2^{n-k}}$, this is the statement that

$$\left|\mathop{\mathsf{Exp}}_M[h(M) \mid \phi(m) \in w_0 + K_{s_0}] - \mathop{\mathsf{Exp}}_m[h(m)]\right| < \epsilon_1 + 2 \cdot 2^{-k+\ell} \ .$$

For this fixed s_0, we say that $(\phi, w) \in \mathrm{PGL}_2(p) \times \mathbb{Z}_{p+1}$ is *ϵ-concealing* if

$$\left|\mathop{\mathsf{Exp}}_M[h(M) \mid \phi(M) \in w + K_{s_0}] - \overline{h}\right| < \epsilon \ .$$

From inequalities (2) and (3), for any fixed m_0, w_0 and $\epsilon_1 > 0$ we see that

$$\Pr\left[(\Phi_0, w_0) \text{ is not } (\epsilon_1 + 2 \cdot 2^{-k+\ell})\text{-concealing}\right] \leq \frac{2 \cdot 2^k}{2^{2n-\ell} \cdot \delta_a^2} < \frac{2^5}{\epsilon_1^2 \cdot 2^{k-\ell}} \ .$$

Now, for any fixed m_0 and $\epsilon > 2 \cdot 2^{-k+\ell}$,

$$\begin{aligned} &\Pr[(\Phi, \Phi(m_0)) \text{ is } \epsilon\text{-concealing}] \\ &= \sum_w \Pr[\Phi \in \mathcal{P}_{m_0 \to w}] \cdot \Pr[(\Phi, w) \text{ is } \epsilon\text{-concealing} \mid \Phi \in \mathcal{P}_{m_0 \to w}] \\ &\geq 1 - \frac{2^5}{\epsilon^2 \cdot 2^{k-\ell} - 4\epsilon + 4 \cdot 2^{-k+\ell}}. \end{aligned}$$

For a random pair (m, ϕ), we will (lower) bound the probability that $(\phi, \phi(m))$ is ϵ-concealing pair for s_0, since $\mathsf{Exp}_{M,\Phi}[|G_{M,\Phi}^{s_0}|]$ is no greater than

$$\epsilon \Pr[(\Phi, \Phi(M)) \text{ is } \epsilon\text{-concealing}] + (1 - \Pr[(\Phi, \Phi(M)) \text{ is } \epsilon\text{-concealing}]) \ .$$

For specific m, we define Y_m to be the random variable taking the value 1 if $(\Phi, \Phi(m))$ is ϵ-concealing, and 0 otherwise. Now,

$$\begin{aligned} \Pr[(\Phi, \Phi(M)) \text{ is } \epsilon\text{-concealing}] &= \mathop{\mathsf{Exp}}_{M,\Phi}[Y_M] = \sum_m p_m \cdot \mathop{\mathsf{Exp}}_{\Phi}[Y_m] \\ &= \sum_m p_m \Pr[(\Phi, \Phi(m)) \text{ is } \epsilon\text{-concealing}] \geq 1 - \frac{2^5}{\epsilon^2 \cdot 2^{k-\ell} - 4\epsilon + 4 \cdot 2^{-k+\ell}}, \end{aligned}$$

so that for all $\epsilon > 2 \cdot 2^{-k+\ell}$, $\mathsf{Exp}_{M,\Phi}\left[\left|G^{s_0}_{M,\Phi}\right|\right] \leq \epsilon(1-\delta)+\delta < \epsilon+\delta$, where $\delta^{-1} = 2^{-5}\cdot(\epsilon^2 \cdot 2^{k-\ell} - 4\epsilon + 4\cdot 2^{-k+\ell})$. Select $\epsilon = 4\cdot 2^{\frac{-k+\ell}{3}}$. As $k = \ell + \omega(\log n)$, we can be guaranteed that $\epsilon = n^{-\omega(1)}$ and so for all s, $\mathsf{Exp}_{M,\Phi}\left[\left|G^{s}_{M,\Phi}\right|\right] = n^{-\omega(1)}$. Hence, $\mathsf{Exp}_{M,\Phi,S}\left[\left|G^{S}_{m,\phi}\right|\right] = n^{-\omega(1)}$, which, considering the above lemma, completes the proof. Note that if, in general, the keys have length $\ell + s$, the advantage is bounded by $2^{-\Theta(s)}$. □

5 Applications to Asymmetric Encryption

In many practical situations requiring public-key cryptography, encryption and decryption (with, e.g., RSA) are so expensive that they are used only to exchange a session key, which is then fed into some cheaper symmetric system. A similar approach is possible with the systems above.

As mentioned in the introduction, a natural analogue of the notion of perfect secrecy for resource-bounded adversaries is the notion of *semantic security*. This is the guarantee that no bounded adversary can predict any piece of partial information about the message with non-negligible advantage. There are a range of complexity-theoretic assumptions which can give rise to such systems. In general, stronger assumptions allow implementations with improved efficiency. Firstly, constructions are possible under "generic" assumptions, e.g., existence of a one-way trapdoor permutation [7]. Under the stronger assumption that factoring is hard, a system of Blum and Goldwasser [6] based on the Rabin functions ($x \mapsto x^2 \bmod pq$) encrypts (in a semantically secure fashion) an n-bit message in time $O(nk\,\mathrm{poly}(\log k))$. Here k is the security parameter of the system. (It is interesting to note that under assumptions of a presumably stronger flavor, Cramer and Shoup [8] show that a constant number of exponentiations over a group suffice to encrypt a group element, in such a way that the resulting system is secure against even (adaptive) chosen ciphertext attack. In particular, hardness of the Diffie-Hellman decision problem is sufficient.)

As the encryption schemes of Theorem 3 are quite efficient, it is interesting to consider the (public-key) system obtained by applying an extant public key system to securely transmit the (short) shared key of E^b. Specifically, if E_{pub} is the encryption algorithm for a public key system offering semantic security, one can study the behavior of the scheme which encrypts a message m as $(\phi(m) + z, \phi, E_{\mathrm{pub}}(z; P, R))$; here P denotes the public key, R the random string required for E, and z the and ϕ the variables of Theorem 3. If, for example, the public key system is taken to be that of Blum and Goldwasser mentioned above, the hybrid system has running time $O([\ell + w(k)\log k]k\,\mathrm{poly}(\log k) + n\,\mathrm{poly}(\log n))$, where w is any function which tends to infinity. (Here k is the security parameter of the system.) *Note, however, that the resulting security guarantee will be weaker than that of semantic security: it requires message spaces of min-entropy $n - \ell$.*

The proof of security for this system follows the proof of Theorem 3 except that one needs to initially argue that availability of the semantically secure encryption of the secret key does not interfere with the security guarantee. This fact relies on a variant of an "elision" lemma originally proved in [11], for which we give a new, streamlined proof.

For definitions of public-key cryptosystem, semantic security, and indistinguishability of encryptions we refer the reader to [11].

We shift notation in this section to agree with [12]: when x is a variable and S a random variable, $x \leftarrow S$ denotes the assignment of x according to S. If S is simply a set, we abuse the notation by allowing S to represent the random variable uniform on S. In the sequel, we will use the term "algorithm" to refer to a probabilistic polynomial time Turing machine. Furthermore, a "message generator" is an algorithm which, given 1^k, produces a output in the set $\{0,1\}^n$ (determined by the random coins of M), where n is polynomially bounded in k. Whenever a probability is expressed it is understood that the random coins of any algorithm appearing inside the brackets are to be included in the probability space. When the underlying probability space of a variable x is clear from context, we may simply write $\mathsf{Pr}_x[P(x)]$, or elide x altogether. A public-key encryption scheme is described by a triple (G, E, D): here G is a key generator algorithm which, given 1^k, generates a pair (P, S); $E(m, P, R)$ is the encryption algorithm, operating on a message m, the public key P, and a random string R; and $D(c, S)$ is the decryption algorithm, operating on a ciphertext c and the private key S.

The following lemma, which generalizes the original elision lemma of [11], is due to [10]. We give a streamlined proof which improves upon previous proofs in the sense that it *requires no sampling* on the part of the constructed algorithm (F, in the proof below). It gives an error bound which depends only on a natural 2-norm of the message distribution. Roughly, the lemma asserts that a cryptosystem offering indistinguishability of encryptions possesses the property that any efficient computation performed with observation of $E(m)$, an encryption, (and, perhaps, some related information) may as well have been performed without it. In the system mentioned above (coupling E^b with E_{pub}), this allows us to disregard the fact that the bounded adversary has witnessed a semantically secure encryption of the shared secret key s; the result is a security guarantee of the form appearing in Theorem 3 but for efficient adversaries.

Lemma 3. *Let (G, E, D) denote an encryption system possessing indistinguishability of encryptions. Then for every message space M and algorithm A, there is an algorithm B so that for all polynomials Q_1, all efficiently computable $f : \{0,1\}^* \to \{0,1\}^*$, and every polynomial Q_2, $\exists k_0, \forall k > k_0$ and $\forall h : \{0,1\}^* \to \{0,1\}^*$,*

$$\mathsf{Pr}[A(1^k, P, f(s,m), E(m; P, R)) = h(s,m)] \leq \mathsf{Pr}[B(1^k, f(s,m)) = h(s,m)] + \frac{1}{Q_2(k)}.$$

The first probability is taken over $m \leftarrow M(1^k)$, $(P, S) \leftarrow G(1^k)$, $s \leftarrow \{0,1\}^{Q_1(k)}$, and R. The second probability is taken over $m \leftarrow M(1^k)$ and $s \leftarrow \{0,1\}^{P(k)}$.

Proof. The algorithm B uses A as a black box: given 1^k and $f(s,m)$, B proceeds as follows: generate $m' \leftarrow M(1^k)$, $(P,S) \leftarrow G(1^k)$, and a random string R of appropriate length; return $v = A(1^k, P, f(s,m), E(m'; P, R))$.

Observe that $\Pr[B(1^k, f(s,m)) = h(s,m)]$ is exactly the probability $\Pr[A(1^k, P, f(s,m), E(m'; P, R)) = h(s,m)]$. In this case, the lemma is a consequence of the following claim:

Claim. For every message space M, efficient algorithm A, every polynomial Q_1, efficiently computable $f : \{0,1\}^* \to \{0,1\}^*$, and every polynomial Q_2, $\exists k_0, \forall k > k_0$ and $\forall h : \{0,1\}^* \to \{0,1\}^*$,

$$\Pr[A(1^k, P, f(s,m), E(m; P, R)) = h(s,m)] \leq \Pr[A(1^k, P, f(s,m), E(m'; P, R)) = h(s,m)] + \frac{1}{Q_2(k)},$$

where each probability is taken over $m \leftarrow M(1^k)$, $m' \leftarrow M(1^k)$, $(P,S) \leftarrow G(1^k)$, $s \leftarrow \{0,1\}^{Q_1(k)}$, and R.

Proof (of Claim). Suppose not. Then there is a polynomial Q_2, a message space M, and an algorithm A, a polynomial Q_1 and a function f so that $\forall k_0, \exists k > k_0$,

$$\Pr_{s,m,R,P}\left[A(1^k, P, f(s,m), E(m; P, R)) = h(s,m)\right] > \Pr_{s,m,m',R,P}\left[A(1^k, P, f(s,m), E(m'; P, R)) = h(s,m)\right] + \epsilon$$

where $\epsilon = \epsilon(k) = \frac{1}{Q_2(k)}$. For a pair of messages m, m', define $P_{m,m'}$ to be the probability $\Pr_{s,R,P}[A(1^k, P, f(s,m), E(m'; P, R)) = h(s,m)]$ and $P_{m,*} = \mathrm{Exp}_{m'}\left[P_{m,m'}\right]$. Observe, then, that

$$\Pr_{s,m,R,P}\left[A(1^k, P, f(s,m), E(m; P, R)) = h(s,m)\right] = \mathop{\mathrm{Exp}}_{m}[P_{m,m}], \text{ and}$$
$$\Pr_{s,m,m',R,P}\left[A(1^k, P, f(s,m), E(m'; P, R)) = h(s,m)\right] = \mathop{\mathrm{Exp}}_{m}[P_{m,*}].$$

In particular, $\mathrm{Exp}_m[P_{m,m}] - \mathrm{Exp}_m[P_{m,*}] > \epsilon$.

Now, we build an algorithm F which, given random m_0 and m_1, can distinguish an encryption of m_0 from one of m_1. The algorithm F proceeds as follows: given m_0, m_1 and $\alpha = E(m_i; P, R)$, *(i.)* j is chosen uniformly in $\{0,1\}$, s is chosen uniformly in $\{0,1\}^{P(n)}$, and R is chosen uniformly among strings of appropriate length, *(ii.)* $E(m_j; P, R)$ and $f(s, m_0)$ are computed, *(iii.)* $A(1^k, f(s, m_0), E(m_j; P, R))$ is simulated, resulting in the value v_j, and $A(1^k, f(s, m_0), \alpha)$ is simulated, resulting in the value v, and, finally, *(iv.)* if $v = v_j$, the j is output; otherwise $1 - j$ is output.

Let $I_n = \{A(1^k, P, f(s,m'), E(m; P, R)) \mid m, m' \in \{0,1\}^n, s \in \{0,1\}^{Q_1(n)}, R\}$ be values that algorithm A can take, when restricted to those inputs possible when $|m| = n$. Then, for $v \in I_n$, let

$$D^s_{m',m}(v) = \Pr_{R,P}[A(1^k, P, f(s,m'), E(m; P, R)) = v] ,$$

so that $P_{m',m} = \mathsf{Exp}_s[D^s_{m',m}(h(s,m'))]$. Now, for a particular pair m_0, m_1, the probability $\mathsf{Pr}[F(m_0,m_1,\alpha)=i]$ is

$$\sum_{i'=0}^{1}\sum_{j'=0}^{1} \mathsf{Pr}[i=i' \wedge j=j'] \cdot \mathsf{Pr}[F(m_0,m_1,\alpha)=i' \mid i=i', j=j']$$

$$= \mathop{\mathsf{Exp}}_s \left[\frac{1}{4} \sum_v D^s_{m_0,m_0}(v)^2 + 2(1 - \sum_v D^s_{m_0,m_0}(v) \cdot D^s_{m_0,m_1}(v)) + \sum_v D^s_{m_0,m_1}(v)^2 \right]$$

$$\overset{*}{\geq} \frac{1}{2} + \frac{1}{4}(\mathop{\mathsf{Exp}}_s \left[D^s_{m_0,m_0}(h(s,m_0)) - D^s_{m_0,m_1}(h(s,m_0)) \right])^2.$$

which is $\frac{1}{2} + \frac{1}{4}(P_{m_0,m_0} - P_{m_0,m_1})^2$. Here inequality $\overset{*}{\geq}$ follows because $\mathsf{Exp}[X]^2$ never exceeds $\mathsf{Exp}[X^2]$ for any random variable. Then

$$\mathop{\mathsf{Pr}}_{m_0,m_1}[F(m_0,m_1,\alpha)=i] \geq \mathop{\mathsf{Exp}}_{m_0,m_1}\left[\frac{1}{2} + \frac{1}{4} \cdot (P_{m_0,m_0} - P_{m_0,m_1})^2\right]$$

$$\geq \frac{1}{2} + \frac{1}{4} \cdot (\mathop{\mathsf{Exp}}_{m_0,m_1}[P_{m_0,m_0} - P_{m_0,m_1}])^2 = \frac{1}{2} + \frac{1}{4} \cdot (\mathop{\mathsf{Exp}}_{m_0}[P_{m_0,m_0} - \mathop{\mathsf{Exp}}_{m_1}[P_{m_0,m_1}]])^2$$

$$= = \frac{1}{2} + \frac{1}{4} \cdot (\mathop{\mathsf{Exp}}_m[P_{m,m}] - \mathop{\mathsf{Exp}}_m[P_{m,*}])^2 \geq \frac{1}{2} + \frac{\epsilon^2}{4}.$$

Hence (E,D) does not offer indistinguishability of encryptions. □

As mentioned above, the Lemma follows immediately from the Claim. □

References

[1] Noga Alon, Oded Goldreich, Johan Håstad, and René Peralta. Simple constructions of almost k-wise independent random variables. In *31st Annual Symposium on Foundations of Computer Science*, volume II, pages 544–553, St. Louis, Missouri, 22–24 October 1990. IEEE.

[2] Yonatan Aumann and Michael O. Rabin. Information theorctically secure communication in the limited storage space model. In Michael Wiener, editor, *Advances in Cryptology – CRYPTO '99*, volume 1666 of *Lecture Notes in Computer Science*, pages 65–79. Springer-Verlag, 1999.

[3] Eric Bach and Jeffrey Shallit. *Algorithmic number theory. Vol. 1.* MIT Press, Cambridge, MA, 1996. Efficient algorithms.

[4] M. Bellare, A. Desai, A. Pointcheval, and P. Rogaway. Relations among notions of public-key cryptosystems. In Krawczyk [13], page 540.

[5] Mihir Bellare and Phillip Rogaway. Optimal asymmetric encryption. In Alfredo De Santis, editor, *Advances in Cryptology – EUROCRYPT 94*, volume 950 of *Lecture Notes in Computer Science*, pages 92–111. Springer-Verlag, 1995, 9–12 May 1994.

[6] Manuel Blum and Shafi Goldwasser. An *efficient* probabilistic public-key encryption scheme which hides all partial information. In G. R. Blakley and David Chaum, editors, *Advances in Cryptology: Proceedings of CRYPTO 84*, volume 196 of *Lecture Notes in Computer Science*, pages 289–299. Springer-Verlag, 1985, 19–22 August 1984.

[7] Manuel Blum and Silvio Micali. How to generate cryptographically strong sequences of pseudo-random bits. *SIAM Journal on Computing*, 13(4):850–864, November 1984.
[8] Ronald Cramer and Victor Shoup. A practical public key cryptosystem provably secure against adaptive chosen ciphertext attack. In Krawczyk [13], pages 13–25.
[9] Danny Dolev, Cynthia Dwork, and Moni Naor. Non-malleable cryptography (extended abstract). In *Proceedings of the Twenty Third Annual ACM Symposium on Theory of Computing*, pages 542–552, New Orleans, Louisiana, 6–8 May 1991.
[10] Oded Goldreich. A uniform-complexity treatment of encryption and zero-knowledge. *Journal of Cryptology*, 6(1):21–53, 1993.
[11] Shafi Goldwasser and Silvio Micali. Probabilistic encryption. *Journal of Computer and System Sciences*, 28(2):270–299, April 1984.
[12] Shafi Goldwasser, Silvio Micali, and Ronald L. Rivest. A digital signature scheme secure against adaptive chosen-message attacks. *SIAM Journal on Computing*, 17(2):281–308, April 1988.
[13] Hugo Krawczyk, editor. *Advances in Cryptology - CRYPTO '98*, volume 1462 of *Lecture Notes in Computer Science*. Springer-Verlag, 23–27 August 1998.
[14] Rudolf Lidl and Harald Niederreiter. *Finite Fields*, volume 20 of *Encyclopedia of Mathematics and its Applications*. Addison-Wesley Publishing Company, Reading, Massachusetts, 1983.
[15] Ueli M. Maurer. Conditionally-perfect secrecy and a provably-secure randomized cipher. *Journal of Cryptology*, 5(1):53–66, 1992.
[16] Joseph Naor and Moni Naor. Small-bias probability spaces: Efficient constructions and applications. *SIAM Journal on Computing*, 22(4):838–856, August 1993.
[17] Moni Naor and Moti Yung. Public-key cryptosystems provably secure against chosen ciphertext attacks. In *Proceedings of the Twenty Second Annual ACM Symposium on Theory of Computing*, pages 427–437, Baltimore, Maryland, 14–16 May 1990.
[18] Rene Peralta. On the distribution of quadratic residues and nonresidues modulo a prime number. *Mathematics of Computation*, 58(197):433–440, 1992.
[19] Charles Rackoff and Daniel R. Simon. Non-interactive zero-knowledge proof of knowledge and chosen ciphertext attack. In J. Feigenbaum, editor, *Advances in Cryptology - CRYPTO '91*, volume 576 of *Lecture Notes in Computer Science*, pages 433–444. Springer-Verlag, 1992, 11–15 August 1991.
[20] E. G. Rees. *Notes on Geometry*. Springer-Verlag, 1983.
[21] A. Schönhage. Schnelle berechnung von kettenbruchentwicklungen. *Acta Informatica*, 1:139–144, 1971.
[22] A. Schönhage. Schnelle Multiplikation von Polynomen über Körpern der Charakteristik 2. *Acta Informat.*, 7(4):395–398, 1976/77.
[23] A. Schönhage and V. Strassen. Schnelle multiplikation gro*ß*er zahlen. *Computing*, 7:281–292, 1971.
[24] Mark N. Wegman and J. Lawrence Carter. New classes and applications of hash functions. In *20th Annual Symposium on Foundations of Computer Science*, pages 175–182, San Juan, Puerto Rico, 29–31 October 1979. IEEE.
[25] Dominic Welsh. *Codes and cryptography*. The Clarendon Press Oxford University Press, New York, 1988.

Cryptography in an Unbounded Computational Model

David P. Woodruff[1] and Marten van Dijk[1,2]

[1] MIT Laboratories for Computer Science, Cambridge, USA,
dpwood@mit.edu, marten@caa.lcs.mit.edu
[2] Philips Research Laboratories, Eindhoven, The Netherlands

Abstract. We investigate the possibility of cryptographic primitives over nonclassical computational models. We replace the traditional finite field F_n^* with the infinite field $\mathbb{Q}$ of rational numbers, and we give all parties unbounded computational power. We also give parties the ability to sample random real numbers. We determine that secure signature schemes and secure encryption schemes do not exist. We then prove more generally that it is impossible for two parties to agree upon a shared secret in this model. This rules out many other cryptographic primitives, such as Diffie-Hellman key exchange, oblivious transfer and interactive encryption.

1 Introduction

In the classical model of cryptography, parties represent data as a sequence of bits, and have a small set of bit operations to work with. Usually some parties are restricted to a polynomial number of operations in the size of a security parameter. In our model, all parties start with the field of rational numbers $\mathbb{Q}$, and have a certain set of operations to work with. We will consider the standard sets of field operations {+, -, *, /}. We give all parties the ability to sample a uniform distribution of real numbers over a bounded interval. Furthermore, all parties have unbounded computational power; that is to say, all parties can perform any finite number of field operations.

It is critical that we give all parties the ability to sample from a uniform distribution of real numbers so that they can generate random secrets that are unpredictable to an adversary. Otherwise any "secret" used by one party could be generated by another party. Indeed, the rational numbers are countable. Therefore, any adversary could simply enumerate elements of his field until he encounters another party's secret since he has unbounded computational time. When we allow sampling from the reals, we are allowing parties to sample from an uncountable domain. Therefore, an adversary cannot simply enumerate the elements of his field to find another party's secret.

The existence of many cryptographic primitives, such as signature schemes, encryption schemes, and identification protocols, depends on the existence of one-way functions and trapdoor functions. In [5] Rompel shows that one-way functions are necessary and sufficient for secure signature schemes to exist in

L.R. Knudsen (Ed.): EUROCRYPT 2002, LNCS 2332, pp. 149–164, 2002.

the standard computational model. The proof relies on the bit-representation of numbers in the number field the parties are working in. In our model, bit-representations play no role. Parties are equipped with infinite-precision registers with the ability to perform any field operation on irrational numbers in constant time.

We do not need to speculate about the existence of one-way functions in this model. Over the rational numbers, a party can sample a random real number r and publish its square r^2. It is impossible to deduce r to infinite precision from r^2 and $\mathbb{Q}$ using only the operations $\{+, -, *, /\}$. Even if one were to sample real numbers, there are only a countable number of real numbers that could help one deduce r from $\mathbb{Q}(r^2)$, but we're drawing from an uncountable set. Hence, there is zero probability of deducing r from r^2 and $\mathbb{Q}$, so the function $f(r) \rightarrow r^2$, where r is a real number, is a one-way function in this model.

Given that we have one-way functions, it is only natural to ask which cryptographic primitives are possible. In [2], an elegant proof of knowledge was presented over the ruler-compass constructible points, and then extended to an authentication protocol. What, if any other, primitives are possible over the ruler-compass constructible points? In our model, we will see that authentication protocols exist but secure signatures schemes and public-key encryption schemes do not. We conjecture the same to be true of the ruler-compass constructible points.

Secion 2 covers some standard techniques in modern algebra, focusing mainly on the theory of field extensions. The theorems presented in this section are crucial to understanding the impossibility proofs in the remaining sections. Section 3 presents an authentication protocol in this model. Section 4 shows that secure encryption schemes do not exist and section 5 shows the same for secure signature schemes. Finally, Section 6 shows more generally that it is impossible to share a secret in this model.

2 Algebraic Preliminaries

The proof of knowledge presented in [2] over the ruler-compass constructible points is based on the idea that trisecting an arbitrary angle is impossible with only a ruler and a compass. Although it is well-known that one cannot trisect an arbitrary angle, the proof is not well-known. Proving that signature schemes and encryption schemes are not possible over the rationals requires field-theoretic techniques similar to those used in [1] where angle trisection is shown to be impossible. We state and develop some of these techniques here. We assume familiarity with the definition of a field. We shall restrict our attention to infinite subfields of the real numbers.

A real number x is said to be *algebraic* over a field F if x is a root of a polynomial $p(t)$, with coefficients in F in the indeterminate t. If no such polynomial exists, x is said to be *transcendental* over F. For example, $\sqrt{2}$ is algebraic over $\mathbb{Q}$ because it satisfies the polynomial $p(t) = t^2 - 2$. We can think of a transcendental element over a field F as a "variable" over that field. For example, the symbol "y" and the number π are transcendental over $\mathbb{Q}$ because they do not satisfy a polynomial $p(t)$ with rational coefficients [4]. A new field can be obtained by taking the set-theoretic union of the elements of F with x, then closing up under all of the field operations {+, -, *, /}. This new field, denoted $F(x)$, is the minimal field containing F and x, i.e., the intersection of all fields containing F and x.

The new field $F(x)$ can be thought of as a vector space over F. A *basis* for this vector space is a set of elements $\{v_\alpha\}$ such that every element of $F(x)$ can be written as a unique finite linear combination of the form $f_1 v_{\alpha_1} + f_2 v_{\alpha_2} + \ldots + f_n v_{\alpha_n}$, where $f_i \in F$ for all i. The dimension of this vector space is defined as the number of elements in any basis. If x is algebraic over F, then there exists a polynomial $q(t)$ of minimal degree such that $q(x) = 0$. It is a theorem of algebra [3] that, if x is algebraic over F and $q(t)$ denotes its minimal polynomial over F, then the set $\{1, x, x^2, x^3, \ldots, x^{(n-1)}\}$ forms a basis for the extension field $F(x)$ viewed as a vector space over F, where n is the degree of $q(t)$. Hence, the dimension of this vector space is equal to the degree of $q(t)$. Call this degree the degree of the *field extension* $F(x)/F$ and denote it by $[F(x) : F]$. If x is transcendental over F, then there is no finite basis of $F(x)$ over F. In this case $[F(x) : F] = \infty$. Furthermore, the elements of $F(x)$ constitute the set of all elements of the form $p(x)/q(x)$, $q(x) \neq 0$, where p and q are polynomials with coefficients in F in the indeterminate x.

More generally, any field extension K/F can be viewed as a vector space over F. The degree $[K : F]$ of this extension denotes the (possibly infinite) number of elements in any basis of K/F. It is a well-known fact that if we have the field inclusions $F \subset L \subset K$, then the degree $[K : F]$ of the extension K/F is equal to the product of the degrees $[K : L]$ and $[L : F]$. We will use this fact frequently and refer to it as the *Tower Law.* For example, since $\sqrt{2}$ is irrational, it does not lie in $\mathbb{Q}$. It satisfies the polynomial $p(t) = t^2 - 2$. Clearly, $p(t)$ is the polynomial of minimal degree of $\sqrt{2}$ over $\mathbb{Q}$, as otherwise there would be a polynomial $q(t) = q_1 t + q_2$, such that $q(\sqrt{2}) = q_1\sqrt{2} + q_2 = 0$, implying $\sqrt{2} = -q_2/q_1$ and therefore that $\sqrt{2}$ is rational. Hence, $[\mathbb{Q}(\sqrt{2}) : \mathbb{Q}] = 2$. It is not hard to see that $\sqrt{3}$ is not in the field $\mathbb{Q}(\sqrt{2})$ (indeed, $\sqrt{3} \neq q_1 + q_2\sqrt{2}$ for any q_1, q_2 in $\mathbb{Q}$). Since $\sqrt{3}$ satisfies the polynomial $p(t) = t^2 - 3$ over $\mathbb{Q}$, it also satisfies this polynomial over $\mathbb{Q}(\sqrt{2})$, and since it is not contained in $\mathbb{Q}(\sqrt{2})$, this polynomial has minimal degree. Hence, we have the field inclusions $\mathbb{Q} \subset \mathbb{Q}(\sqrt{2}) \subset \mathbb{Q}(\sqrt{2}, \sqrt{3})$, where $[\mathbb{Q}(\sqrt{2}) : \mathbb{Q}] = 2$ and $[\mathbb{Q}(\sqrt{2}, \sqrt{3}) : \mathbb{Q}(\sqrt{2})] = 2$, so by the Tower Law $[\mathbb{Q}(\sqrt{2}, \sqrt{3}) : \mathbb{Q}] = 4$. A basis of $\mathbb{Q}(\sqrt{2}, \sqrt{3})$ as a vector space over $\mathbb{Q}$ is $\{1, \sqrt{2}, \sqrt{3}, \sqrt{6}\}$.

Also note that if $[K : F] = 1$, then $K = F$. Indeed, $[K : F] = 1$ implies that there is only one element in any basis of K over F. Consider the set $\{1\}$, where 1 is the identity element of F. Trivially, this is a linearly independent set, and since we know the size of any basis is one, $\{1\}$ also spans K over F, so $\{1\}$ is a basis. Any element of K can be written as $f \cdot 1$, for $f \in F$. This implies $K = F$.

Given a field extension $F(x)/F$, we can adjoin another element y to the field $F(x)$, obtaining the field $F(x)(y)$. It is a standard fact that $F(x)(y) = F(y)(x)$. We will let $F(x, y) = F(x)(y) = F(y)(x)$.

We now define the *algebraic closure* of a field. Consider an infinite field F. Consider the set S of all polynomials $p(t)$ with coefficients in F in the indeterminate t. Suppose we take the minimal field containing F and adjoin all the roots of all the polynomials of S. This new field will be called the algebraic closure of F. For the countably infinite fields we shall be dealing with, it is known [4] that the cardinality of the algebraic closure of F is also countable.

We will need some specific facts concerning transcendental field extensions. Let x be transcendental over a field F and let $K = F(x)$. Then any element $u \in K$ can be written as $p(x)/q(x)$, where p and q are relatively prime polynomials with coefficients in F in the indeterminate x. We have that $F(u) \subset K$. In [4] it is shown that the degree $[K : F(u)]$ equals $max\{deg(p(x)), deg(q(x))\}$.

We will also need some specific results concerning the intermediate fields of a field extension. A field L such that $F \subset L \subset K$ is called an intermediate field of the field extension K/F. If x is transcendental over K, it is clearly transcendental over F since $F \subset K$. Conversely, if every element $k \in K$ is algebraic over F and if x is transcendental over F, it is also transcendental over K. This follows from the transitivity property of being algebraic, namely, if x is algebraic over K, and K is algebraic over F, then x is algebraic over F [4].

Suppose x is transcendental over F, and K is an algebraic extension of F, then the intermediate fields L of K/F are in bijective correspondence with the intermediate fields of $K(x)/F(x)$. The bijection sends an intermediate field L of K/F to the intermediate field $L(x)$ of $K(x)/F(x)$. The inverse sends an intermediate field G of $K(x)/F(x)$ to $G \cap K$. The intuition behind this fact is that x, being transcendental over F, plays no role in factoring the minimal polynomials of elements of K over F. Since the intermediate fields of $K(x)/F(x)$ are determined by these polynomials, the intermediate fields of $K(x)/F(x)$ are exactly those of K/F with the additional element x adjoined. This result also holds if K is a transcendental extension of F and x is transcendental over K. See [4] for more details.

The final theorem that we will need, due to Lüroth [4], states that if x is transcendental over a field F, then the intermediate fields L of the field extension $F(x)/F$ all have the form $F(u)$, where u has the form $p(x)/q(x)$, where p and q are polynomials with coefficients in F and $q \neq 0$.

3 An Identification Protocol

Here is a simple zero-knowledge proof of knowledge similar to that in [2]. Suppose Alice wishes to identify herself to Bob. She samples a random real number r and publishes $p = r^2$. Because finding the exact square root of r^2 over $\mathbb{Q}$ with only the operations {+, -, *, /} is impossible, Alice knows she is the only one who knows r. Also, even if parties are allowed to sample random real numbers, the probability is zero that any number sampled will help an adversary compute r from $\mathbb{Q}$ and r^2. Here's the protocol:

1. Alice samples a real number s. She gives Bob $t = s^2$.
2. Bob flips a coin and tells Alice the result.
3. – If Bob said "heads", then Alice gives Bob s, and Bob checks that $s^2 = t$.
   – If Bob said "tails", then Alice gives Bob $u = rs$, and Bob checks that $u^2 = pt$.

We sketch a proof of the three properties of zero-knowledge: completeness, soundness, and zero-knowledge. For completeness, note that if Alice and Bob follow the protocol, then Bob always accepts Alice's proof of identity. For soundness, note that anyone impersonating Alice cannot respond to both of Bob's challenges because he cannot know both s and rs, as otherwise he could compute $(rs)/s = r$, contradicting the fact that it is not possible to compute r given only $\mathbb{Q}$ and r^2 in our model of computation. Hence, with each iteration of the above protocol an impersonator can succeed with probability at most 1/2. After k iterations, the probability that Bob will be fooled by the impersonator is at most 2^{-k}.

To show the protocol is zero-knowledge, we construct a simulator to produce transcripts of Bob's view in the protocol. Bob's views are of the form $(t, \text{"heads"}, s)$ or $(t, \text{"tails"}, u)$. The first can be simulated by choosing s at random and setting t to be s^2. The second can be simulated by choosing u at random and taking t to be u^2/p. As stated in [2], even if Bob is to use a nonuniform distribution, his view can be simulated by probing and resetting him. If k rounds are executed in series, the expected number of trials of the simulator is $2k$. If k flips are sent in parallel, then the expected number of trials is 2^k; this is not a problem since there are no complexity assumptions in our computational model.

4 The Impossibility of Secure Public-Key Encryption Schemes

We now address the possibility of secure encryption schemes in this model. We would like an encrypter to be able to encrypt an arbitrary real number of his choice, even after the public and secret keys have been generated. Intuitively, such encryption schemes cannot exist because both the message space and the ciphertext space are uncountably infinite, whereas the set of numbers that are "algebraically dependent" on any finite set of secret keys is only countably infinite. Since all parties are restricted to finite time, only a finite set of secret keys can be generated. Hence, the trapdoor information that comes with knowledge of the set of secret keys can only help decrypt a countable number of messages. We now formalize this intuition.

We first consider a special scenario. Suppose Alice starts with the field $\mathbb{Q}$. She then samples a random real number SK to be her secret key. She now has the field $\mathbb{Q}(SK)$. Suppose she then performs some finite number of field operations in the field $\mathbb{Q}(SK)$ to compute her public key PK, another element of $\mathbb{Q}(SK)$. She then publishes PK. We first consider the case when the degree $[\mathbb{Q}(SK) : \mathbb{Q}(PK)]$ is finite.

We would like Bob to be able to encrypt an arbitrary real number m using Alice's public key PK, generating a ciphertext c. Given the ciphertext c and PK, we do not want an adversary to be able to decrypt c to obtain the original message m. However, we do want Alice to be able to use her secret key SK, together with c, to decrypt c and recover the original message m. Collecting this information, we have the following tower of fields:

$$\mathbb{Q}(PK, c) \subset \mathbb{Q}(PK, m) \subset \mathbb{Q}(SK, c).$$

Indeed, the inclusion $\mathbb{Q}(PK, c) \subset \mathbb{Q}(PK, m)$ holds because, given PK and m, the encrypter can compute c with only field operations, and hence $c \in \mathbb{Q}(PK, m)$. The inclusion $\mathbb{Q}(PK, m) \subset \mathbb{Q}(SK, c)$ holds because, given SK and c, the legitimate decrypter Alice can recover m with only field operations.

Let's now inspect the degrees of these field extensions. Set $n = [\mathbb{Q}(SK) : \mathbb{Q}(PK)]$. Then $[\mathbb{Q}(SK, c) : \mathbb{Q}(PK, c)]$ is at most n, since adjoining c to both fields can only reduce the degree of the minimal polynomial of SK over $\mathbb{Q}(PK)$. We show that $\mathbb{Q}(SK, c) = \mathbb{Q}(SK, m)$. We know $\mathbb{Q}(SK, c) \supset \mathbb{Q}(SK, m)$ from the tower of fields above. Furthermore, given SK anyone can recompute PK since $\mathbb{Q}(PK) \subset \mathbb{Q}(SK)$, and given PK and m anyone can recompute c. Therefore, $\mathbb{Q}(SK, m) \supset \mathbb{Q}(SK, c)$. We deduce that $[\mathbb{Q}(SK, c) : \mathbb{Q}(PK, m)] = [\mathbb{Q}(SK, m) : \mathbb{Q}(PK, m)]$. Since m is a general real number, $[\mathbb{Q}(SK, m) : \mathbb{Q}(PK, m)]$ also equals n. Applying the Tower Law, we have that $[\mathbb{Q}(PK, m) : \mathbb{Q}(PK, c)][\mathbb{Q}(SK, c) : \mathbb{Q}(PK, m)] = [\mathbb{Q}(SK, c) : \mathbb{Q}(PK, c)]$. Since $[\mathbb{Q}(SK, c) : \mathbb{Q}(PK, c)]$ is at most n, and since $[\mathbb{Q}(SK, c) : \mathbb{Q}(PK, m)]$ is exactly n, we see that $[\mathbb{Q}(PK, m) : \mathbb{Q}(PK, c)]$ must equal 1. Hence, $\mathbb{Q}(PK, m) = \mathbb{Q}(PK, c)$. Therefore the message m lies in the adversary's field. Since the adversary has unbounded computational time, and since his field $\mathbb{Q}(PK, c)$ is countable, he can enumerate each of the elements of his field and run the public encryption algorithm on each of them until he finds the unique message m which encrypts to c.

Hence, for the above scenario, any encryption scheme is not secure. So we modify the scenario in a couple of ways. Suppose instead of a single secret key SK and a single public key PK, Alice uses n secret keys $SK_1, ..., SK_n$, and m public keys $PK_1, ..., PK_m$. If each SK_i is algebraic over $\mathbb{Q}(PK_1, ..., PK_m)$, $[\mathbb{Q}(SK_1, ..., SK_n) : \mathbb{Q}(PK_1, ..., PK_m)]$ will still be finite. Replacing SK with $SK_1, ..., SK_n$ and PK with $PK_1, ..., PK_m$ in the above argument, we conclude that even in this case secure encryption is not possible. Note that since all parties are restricted to a finite number of operations, there can be at most a finite number of public and secret keys generated.

For now, we will continue to assume that the degree of the legitimate decrypter's field over the adversary's field is finite. For convenience, we will assume that there is one secret key SK and one public key PK. From the results in the previous paragraph, the following arguments easily generalize to the case of multiple public-secret keys in so long as each secret key is algebraic over the field $\mathbb{Q}(PK_1, ...PK_m)$ where m is the number of public keys. Whereas before we restricted the encrypter to field operations when encrypting a message m, we now allow the encrypter to sample real numbers as he encrypts and we allow the adversary to sample real number as well.

We first show that giving the adversary the power to sample real numbers will not help him. The encrypter will have the field $\mathbb{Q}(PK, m, r_1, ..., r_m)$ where r_i is a sampled real number for all i. Note that the number of real numbers sampled is necessarily finite. Now, the adversary has the field $\mathbb{Q}(PK, c) \subset \mathbb{Q}(PK, m, r_1, ...r_m)$. For the adversary to gain anything by sampling real numbers, he must be able to generate, via sampling and field operations, an element of $\mathbb{Q}(PK, m, r_1, r_2, ..., r_m) \setminus \mathbb{Q}(PK, c)$. Suppose he draws m random reals $s_1, ..., s_m$. He now has the field $\mathbb{Q}(PK, c, s_1, ..., s_m)$. Every element of his field has the form $p(PK, c, s_1, ..., s_m)/q(PK, c, s_1, ..., s_m)$, for p and q polynomials with rational coefficients in the indeterminates $PK, c, s_1, ..., s_m$. To generate an element y in $\mathbb{Q}(PK, m, r_1, ..., r_m) \setminus \mathbb{Q}(PK, c)$, we must have some expression $p(PK, c, s_1, ..., s_m)/q(PK, c, s_1, ..., s_m) = y$. We know that p/q is not in $\mathbb{Q}(PK, c)$ since y is assumed not to lie in $\mathbb{Q}(PK, c)$. Note that not all of the coefficients of the s_i in the expression p/q can be zero and not all of the s_i in p can cancel with those in q; for example, we cannot have the cancellation $(s_1 + s_2)/(2(s_1 + s_2)) = 1/2$, for then p/q would actually be an element of $\mathbb{Q}(PK, c)$. But then we have found a nontrivial relation among the s_i over the field $\mathbb{Q}(PK, c)$. If the s_i are random real numbers, this occurs with probability zero since the field $\mathbb{Q}(PK, c)$ is countable, whereas the real numbers are uncountable. Hence, sampling does not help the adversary.

We now allow the encrypter to probabilistically encrypt; that is to say, we give him the ability to sample real numbers. We still necessarily have the tower of fields $\mathbb{Q}(PK, c) \subset \mathbb{Q}(PK, m) \subset \mathbb{Q}(SK, c)$, where, if the encryption scheme is to be secure, then each of the above inclusions must be a proper inclusion. However, the encrypter's field is no longer $\mathbb{Q}(PK, m)$, but rather $\mathbb{Q}(PK, m, r_1, ...r_m)$, where each r_i is a sampled real number. However, the argument given above still implies that the inclusions in this tower cannot be proper. That is to say, $\mathbb{Q}(PK, c) = \mathbb{Q}(PK, m)$. Hence, even if the adversary is not able to recover the original field $\mathbb{Q}(PK, m, r_1, ...r_m)$ of the encrypter, he can still recover $\mathbb{Q}(PK, m)$ and hence recover m.

We now consider the case where $[\mathbb{Q}(SK) : \mathbb{Q}(PK)]$ is infinite. We still want the inclusions in the tower of fields

$$\mathbb{Q}(PK, c) \subset \mathbb{Q}(PK, m) \subset \mathbb{Q}(SK, c),$$

to be proper.

We want both to be able to encrypt an arbitrary real number m and to have a ciphertext c decrypt to a unique message m. Hence, the number of distinct ciphertexts is at least as large as the number of distinct messages. These observations imply that the number of possible ciphertexts is uncountably infinite. Since any element y which is algebraic over $\mathbb{Q}(SK)$ is in the algebraic closure of $\mathbb{Q}(SK)$, and since the algebraic closure of $\mathbb{Q}(SK)$ is countable, there is zero probability that the ciphertext c will be algebraic over $\mathbb{Q}(SK)$. Hence, c is transcendental over $\mathbb{Q}(SK)$ with probability 1.

Since c is transcendental over $\mathbb{Q}(SK)$, and hence over $\mathbb{Q}(PK)$, the intermediate fields of $\mathbb{Q}(SK, c)/\mathbb{Q}(PK, c)$ are of the form $L(c)$, where L is an intermediate field of $\mathbb{Q}(SK)/\mathbb{Q}(PK)$. By Lüroth's theorem, all intermediate fields of $\mathbb{Q}(SK)/\mathbb{Q}(PK)$ have the form $\mathbb{Q}(u)$, where u has the form $p(SK)/q(SK)$, for p and q are polynomials with coefficients in $\mathbb{Q}$ in the indeterminate SK and $q \neq 0$. For the inclusions in the above tower of fields to be proper, $\mathbb{Q}(PK, m)$ must be of the form $\mathbb{Q}(u, c)$. But u has the form $p(SK)/q(SK)$ with $u \notin \mathbb{Q}(PK, c)$, and such a u is impossible for the encrypter to generate since all he has are PK and m, which are each algebraically independent of SK. Even if he were to sample real numbers, he has zero probability of generating an element u of the form $p(SK)/q(SK)$. Therefore the field $\mathbb{Q}(PK, m)$ cannot contain an element of the form $p(SK)/q(SK)$, and therefore $\mathbb{Q}(PK, m)$ is forced to equal $\mathbb{Q}(PK, c)$.

5 The Impossibility of Secure Signature Schemes

We now shift our attention to the possibility of secure signature schemes in this model. We will show the strongest possible result, that even one-time signature schemes cannot exist.

We first need to define exactly what we mean by a signature scheme. We would like the signer to be able to sign an arbitrary real number m that is not fixed at the time of key generation. If we were to remove this constraint and instead allow the signer to specify a finite sequence of messages $m_1, ..., m_N$ which he would like to be able to sign with a given keypair, secure signature schemes would in fact be possible. A secure signature scheme can be built on the fact that finding a square root of an arbitary real number r is impossible in the field $\mathbb{Q}(r^2)$. Let Alice be the signer, Bob the verifier. Here's the protocol:

1. Initialization: Alice decides upon a finite sequence of messages $(m_1, m_2, ..., m_N)$ she would like to be able to sign with the public-secret keypair she is about to create. She then samples N real numbers $r_1, r_2, ..., r_N$. The ordered set $(r_1, r_2, ..., r_N)$ forms Alice's secret key. Alice publishes the two ordered sets $(r_1^2, r_2^2, ..., r_N^2)$ and $(m_1, m_2, ..., m_N)$.
2. Signing: To sign the message m_i for $1 \leq i \leq N$), Alice sends the pair (m_i, r_i).
3. Verifying: Bob verifies the pair (m_i, s) by computing i from m_i and checking that $s^2 = r_i^2$.

It is easy to verify the security of the above signature scheme. Also, since all parties are given unbounded computational time, N can be chosen to be arbitrarily large.

We can improve this signature scheme by reducing the number of real numbers sampled to exactly one. This more efficient protocol is based on the fact that finding an nth root of an arbitrary real number r is impossible in the field $\mathbb{Q}(r)$. Here's the protocol:

1. Initialization: Alice decides upon a finite ordered set of messages $(m_1, m_2, ..., m_N)$ she would like to sign with the key pair she is about to generate. She

then calls the subroutine $primeConvolve(N)$, described below to get the ordered set $(n_1, n_2, ..., n_N)$ and the integer P. She samples a real number r, which is her secret key. She publishes the ordered sets $(m_1, m_2, ..., m_N)$ and $(n_1, n_2, ..., n_N)$ along with the real number r^P and the integer P.
2. Signing: To sign the message m_i for $1 \leq i \leq N$, Alice sends the pair (m_i, u^{n_i}).
3. Verifying: Bob verifies the pair (m_i, s) by computing i from m_i and checking that $s^{(P/n_i)} = r^P$.

We describe the subroutine $primeConvolve(N)$ in English. For every nonempty subset S of $\{1, 2, ..., N\}$, the subroutine chooses a unique prime p_S. It then defines t_i for $1 \leq i \leq N$ to be the product $\Pi_{i \in S} p_S$. Finally, it returns the ordered set $(t_1, ..., t_n)$ and the product $\Pi_{S \subset \{1,...,N\}} p_S$.

$PrimeConvolve(N)$ is used to thwart a gcd attack by an adversary who uses an adaptive chosen-message attack. $PrimeConvolve(N)$ generates a set T of N elements with the property that $\forall T' \subset T$, the gcd$(y \in T')$ does not divide x for $x \in T \setminus T'$. For example, calling $primeConvolve(3)$ could return the set

$$(2 \cdot 7 \cdot 11 \cdot 17, 3 \cdot 7 \cdot 13 \cdot 17, 5 \cdot 11 \cdot 13 \cdot 17)$$

and the product $2 \cdot 3 \cdot 5 \cdot 7 \cdot 13 \cdot 17$. If, for example, messages m_1 and m_2 have been signed, then an adversary will learn $r^{(2 \cdot 7 \cdot 11 \cdot 17)}$ and $r^{(3 \cdot 7 \cdot 13 \cdot 17)}$, from which he can compute $gcd(r^{(2 \cdot 7 \cdot 11 \cdot 17)}, r^{(3 \cdot 7 \cdot 13 \cdot 17)}) = r^{(7 \cdot 17)}$. This is the smallest power of r that can be obtained by the adversary if r was chosen to be a random real number. Now, 7 does not divide $5 \cdot 11 \cdot 13 \cdot 17$; so an adversary cannot compute $r^{(5 \cdot 11 \cdot 13 \cdot 17)}$ so it is not possible for him to forge message m_3.

The above two signature schemes suffer because the message space is fixed to a finite subset of the real numbers at the time of key generation. We now show that, if we remove this constraint and instead allow Alice the ability to sign arbitrary real numbers after the time of key generation, then even one-time schemes are not secure.

Suppose Alice starts with the field $\mathbb{Q}$. She then samples a random real number SK that will be her secret key. She is left with the field $\mathbb{Q}(SK)$. Suppose she then performs some finite number of field operations in the field $\mathbb{Q}(SK)$ to compute her public key PK, another element of $\mathbb{Q}(SK)$. She then publishes PK. We consider the case where the degree $[\mathbb{Q}(SK) : \mathbb{Q}(PK)]$ is finite. Let m be the message to be signed, $\sigma(m)$ its signature. For the moment, suppose that $\sigma(m)$ can be generated from $\mathbb{Q}(SK, m)$ with field operations alone. We would like the inclusions in the following tower of fields to be proper:

$$\mathbb{Q}(PK, m) \subset \mathbb{Q}(PK, m, \sigma(m)) \subset \mathbb{Q}(SK, m)$$

The leftmost field is known by an adversary trying to forge the signature $\sigma(m)$. The rightmost field is known by the legitimate signer Alice. The field in between is known by all after m has been signed. If the inclusion $\mathbb{Q}(PK, m) \subset \mathbb{Q}(PK, m, \sigma(m))$ were not proper, the adversary could run the public verification algorithm on each element of his field to determine if it is in fact a valid signature for m. Since his field is enumerable, he will find $\sigma(m)$ in finite time.

We also want the inclusion $\mathbb{Q}(PK, m, \sigma(m)) \subset \mathbb{Q}(SK, m)$ to be proper. Otherwise, after viewing one signature $\sigma(m)$, anyone could enumerate through the field $\mathbb{Q}(PK, m, \sigma(m))$ to discover Alice's secret key SK.

Since m is a general real, m is transcendental over $\mathbb{Q}(SK)$, and hence over $\mathbb{Q}(PK)$. Therefore, the intermediate fields of $\mathbb{Q}(SK, m)/\mathbb{Q}(PK, m)$ all have the form $L(m)$, where L is an intermediate field of $\mathbb{Q}(SK)/\mathbb{Q}(PK)$. Now, in the case $[\mathbb{Q}(SK) : \mathbb{Q}(PK)]$ finite, there are only a finite number of intermediate fields of $\mathbb{Q}(SK)/\mathbb{Q}(PK)$. After one message m has been signed, the public learns the field $\mathbb{Q}(PK, m, \sigma(m)) = L(m)$ for some L. Hence, given any future message m' to be signed, the probability that the signature $\sigma(m')$ is in the field $L(m')$ is nonzero. Since an adversary has learned the field L, he can simply adjoin the message m' to the field L, obtaining the field $L(m')$. He can then enumerate elements of this field until he finds $\sigma(m')$, which can be verified using the public verification algorithm. The probability that he forges an arbitrary future message m' is nonzero. Therefore, after obtaining the signature for only one message, there is a nonnegligible probability that the signature of any future message can be forged. Hence, even one-time signature schemes are not possible in this model.

We now allow parties to sample real numbers. Intuitively, sampling real numbers cannot help the adversary since only a countable subset of the real numbers helps, and he is drawing from an uncountable set; see Section 3.2 for more detail. Now, if the signer is allowed to sample real numbers, the tower of fields changes to

$$\mathbb{Q}(PK, m) \subset \mathbb{Q}(PK, m, \sigma(m)) \subset \mathbb{Q}(SK, m, r).$$

For the signature scheme to be secure, each of the above inclusions must be a proper inclusion. We may not have the inclusion $\mathbb{Q}(PK, m, \sigma(m)) \subset \mathbb{Q}(SK, m)$, so the argument given above does not apply. Note, however, if there exists a message m whose signature $\sigma(m)$ does not lie in the field $\mathbb{Q}(SK, m)$, then it is necessarily transcendental over $\mathbb{Q}(PK, m)$. This assertion follows from Lüroth's theorem; see Section 2. But then there can be no public verification algorithm involving PK, m, and $\sigma(m)$ over $\mathbb{Q}$ since $\sigma(m)$ does not satisfy any algebraic relation over $\mathbb{Q}(PK, m)$.

Finally, we consider the case where $[\mathbb{Q}(SK) : \mathbb{Q}(PK)]$ is infinite. The analysis in this case is similar to that given in the previous case where we allowed the signer to use randomness. As always, we want the inclusions in the following tower of fields to be proper:

$$\mathbb{Q}(PK, m) \subset \mathbb{Q}(PK, m, \sigma(m)) \subset \mathbb{Q}(SK, m)$$

Since SK is transcendental over $\mathbb{Q}(SK)$, Lüroth's theorem tells us that the only intermediate fields of $\mathbb{Q}(SK)/\mathbb{Q}(PK)$ are transcendental extensions of $\mathbb{Q}(PK)$. Therefore, all intermediate fields of $\mathbb{Q}(SK, m)/\mathbb{Q}(PK, m)$ are transcendental extensions of $\mathbb{Q}(PK, m)$. If the signature scheme is to be secure, $\sigma(m)$ cannot be in $\mathbb{Q}(PK, m)$. Then $\mathbb{Q}(PK, m, \sigma(m))$ would necessarily be a transcendental extension of $\mathbb{Q}(PK, m)$, and hence, $\sigma(m)$ would be transcendental over $\mathbb{Q}(PK, m)$. As we argued previously, in this case, there can be no public verification algorithm of $\sigma(m)$ over $\mathbb{Q}(PK, m)$ because $\sigma(m)$ does not satisfy any algebraic relation over $\mathbb{Q}(PK, m)$.

6 The Impossibility of Secret Sharing

We now generalize the impossibility of public-key encryption in this model to the impossibility of sharing a secret. The impossibility of sharing a secret will immediately rule out public-key encryption, interactive encryption, Diffie-Hellman key exchange, and oblivious transfer. We will consider an arbitrary two-party protocol and show that no such protocol establishes a shared secret.

A protocol between Alice and Bob consists of a sequence of steps. Let F_A be the field generated by Alice and let F_B be the field generated by Bob. During each step information may be revealed to the public. Let F_P be the field generated by the public information. There are two types of steps, either Alice (Bob) selects a random element thereby extending her associated field or Alice (Bob) transmits an element from her field to Bob (Alice). Due to the transmission a transmitted element is revealed to the public.

Step 1. *A trancedental element x over $\mathbb{Q}(F_A, F_B)$ is selected by Alice:*

$$(F_A, F_B, F_P) \to (F_A(x), F_B, F_P),$$

or a trancedental element x over $\mathbb{Q}(F_A, F_B)$ is selected by Bob:

$$(F_A, F_B, F_P) \to (F_A, F_B(x), F_P).$$

Step 2. *Alice selects an element x in F_A and transmits it to Bob:*

$$(F_A, F_B, F_P) \to (F_A, F_B(x), F_P(x)),$$

or Bob selects an element x in F_B and transmits it to Alice:

$$(F_A, F_B, F_P) \to (F_A(x), F_B, F_P(x)).$$

To show the impossibility of secret sharing over rational numbers we need to prove

$$F_A \cap F_B = F_P \tag{1}$$

after each step of the protocol. In other words all shared information can be computed by the public by means of field operations. In the unbounded computing model there does not exist a secret since the field F_P is countable. We need to prove that (1) is invariant under steps 1 and 2.

In the remainder we assume w.l.o.g. that Bob selects x in both steps. Steps 1 and 2 are invariant under:

Invariant 1. *F_A, F_B, and F_P are fields such that $F_P \subseteq F_A \cap F_B$. Furthermore $F_A = \mathbb{Q}(A)$ and $F_B = \mathbb{Q}(B)$ for finite sets of real numbers A and B.*

Proof. From the invariant we infer that $F_P \subseteq F_A \cap F_B \subseteq F_A \cap F_B(x)$ and secondly $F_P(x) \subseteq (F_A \cap F_B)(x) \subseteq F_A(x) \cap F_B(x) = F_A(x) \cap F_B$ for $x \in F_B$. □

In general however, we can not prove that $F_A \cap F_B = F_P$ is invariant under step 2. For example, take $F_A = \mathbb{Q}(\sqrt{6}, \sqrt{15})$, $F_B = \mathbb{Q}(\sqrt{2}, \sqrt{5})$, $F_P = \mathbb{Q}$, and $x = \sqrt{2} \in F_B$. Clearly, $F_A(x) = \mathbb{Q}(\sqrt{2}, \sqrt{3}, \sqrt{5})$ implying that $F_A(x) \cap F_B = \mathbb{Q}(\sqrt{2}, \sqrt{5})$ while $F_P(x) = \mathbb{Q}(\sqrt{2})$.

Thus in order to prove that (1) is invariant under steps 1 and 2 we need to introduce a stronger invariant.

Lemma 1. *Let $G \subseteq F$ be fields such that $[G(v) : G] = [F(v) : F]$. Then either v is trancedental over F or there exists a basis $\mathcal{X} = \{1, v, v^2, \ldots, v^{n-1}\}$ of $F(v)$ over F which is also a basis of $G(v)$ over G.*

Proof. The basis $\mathcal{X}$ of $F(v)$ over F is linear independent over $G \subseteq F$. Since $[G(v) : G] = [F(v) : F]$ and $\mathcal{X}$ does not depend on F, $\mathcal{X}$ is a basis of $G(v)$ over G. □

Now we are ready to formulate the stronger invariant.

Invariant 2. *There exist real numbers a_i, $1 \leq i \leq n$, such that*

$$F_B = F_P(a_1, a_2, \ldots, a_n)$$

and

$$\mathbb{Q}(F_A, F_B) = F_A(a_1, a_2, \ldots, a_n)$$

with

$$[F_A(a_1, \ldots, a_{i+1}) : F_A(a_1, \ldots, a_i)] = [F_P(a_1, \ldots, a_{i+1}) : F_P(a_1, \ldots, a_i)]$$

for all $0 \leq i \leq n-1$.

Initially, $F_A = F_B = F_P = \mathbb{Q}$ and invariant 2 holds for $n = 0$. The next lemmas will be used to prove that invariant 2 implies (1).

Lemma 2. *Let $G \subseteq F$ and let v be trancedental over F. Then $F \cap G(v) = G$.*

Proof. Let $x \in G(v)$. Then there exist polynomials $f(.)$ and $g(.)$ with coefficients in $G \subseteq F$ and $g(v) \neq 0$ such that $x = f(v)/g(v)$. If x is also in F then either v is algebraic over F or $f(v)/g(v)$ does not depend on v, that is $x \in G$. □

We define the vector space

$$G[\mathcal{X}] = \left\{ x = \sum_{\gamma \in \mathcal{X}} x_\gamma \cdot \gamma : x_\gamma \in G \right\}.$$

If $\mathcal{X}$ is a basis of $G(v)$ over G then $G(v) = G[\mathcal{X}]$.

Lemma 3. *Let $G \subseteq F$ and let $\mathcal{X}$ be a finite linear independent set over F with $1 \in \mathcal{X}$. Then $F \cap G[\mathcal{X}] = G$.*

Proof. Let $x \in G[\mathcal{X}]$. Then there exist coefficients $x_\gamma \in G \subseteq F$ such that $x = \sum_{\gamma \in \mathcal{X}} x_\gamma \cdot \gamma$. If x is also in F then $x = x_1 \in G$ since $\mathcal{X}$ is linearly independent over F. □

Theorem 1. *Invariant 2 implies $F_A \cap F_B = F_P$.*

Proof. Let $F_i = F_A(a_1, \dots, a_i)$ and $G_i = F_P(a_1, \dots, a_i)$. By lemma 1, invariant 2 implies either a_{i+1} is trancedental over F_i or there exists a basis $\mathcal{X}$ of $F_i(a_{i+1})$ over F_i which is also a basis of $G_i(a_{i+1})$ over G_i. According to lemmas 2 and 3 respectively, $F_i \cap G_i(a_{i+1}) = G_i$. Since $F_A \subseteq F_i$, $F_A \cap G_i(a_{i+1}) \subseteq G_i$, that is

$$F_A \cap G_{i+1} = F_A \cap G_i(a_{i+1}) \subseteq F_A \cap G_i.$$

Hence,

$$F_P \subseteq F_A \cap F_B = F_A \cap G_n \subseteq \dots \subseteq F_A \cap G_0 = F_A \cap F_P = F_P.$$

□

The next invariant is like invariant 2 where the A's and a's are interchanged with the B's and b's. Because of the symmetry theorem 1 also holds for this invariant.

Invariant 3. *There exist real numbers b_i, $1 \leq i \leq m$, such that*

$$F_A = F_P(b_1, b_2, \dots, b_n)$$

and

$$\mathbb{Q}(F_A, F_B) = F_B(b_1, b_2, \dots, b_m)$$

with

$$[F_B(b_1, \dots, b_{i+1}) : F_B(b_1, \dots, b_i)] = [F_P(b_1, \dots, b_{i+1}) : F_P(b_1, \dots, b_i)]$$

for all $0 \leq i \leq m-1$.

The next lemmas are used to show that invariants 2 and 3 are equivalent. The proof of Lemma 5 is left to the appendix.

Lemma 4. *Consider the chain of fields $G \subseteq H \subseteq F$ and suppose that $[F(v) : F] = [G(v) : G]$. Then $[F(v) : F] = [H(v) : H] = [G(v) : G]$.*

Proof. According to lemma 1 either v is trancedental over F or there exists a basis $\mathcal{X}$ of $F(v)$ over F which is also a basis of $G(v)$ over G. If v is trancedental over F then it is also trancedental over its subfields G and H in which case $[F(v) : F] = [H(v) : H] = [G[v] : G] = \infty$. If $\mathcal{X}$ is a basis over F then it is linear independent over H, hence $[H(v) : H] \geq [F(v) : F]$. If $\mathcal{Y}$ is a basis over H then it is linear independent over G, hence $[G(v) : G] \geq [H(v) : H]$. Since $[F(v) : F] = [G(v) : G]$, equalities hold everywhere. □

Lemma 5. *Let G be a field. If $[G(u,v):G(u)] = [G(v):G]$ then also $[G(v,u):G(v)] = [G(u):G]$.*

Theorem 2. *Invariants 2 and 3 are equivalent.*

Proof. Suppose that invariant 2 holds. By invariant 1 there exist real numbers b_i, $1 \leq i \leq m$, such that $F_A = F_P(b_1, b_2, \ldots, b_m)$. Let

$$H_{i,j} = F_P(a_1, \ldots, a_i)(b_1, \ldots, b_j),$$

$F_i = F_A(a_1, \ldots, a_i)$, and $G_i = F_P(a_1, \ldots, a_i)$. Notice that $H_{n,j} = F_B(b_1, \ldots, b_j)$, $H_{0,j} = F_P(b_1, \ldots, b_j)$, and $H_{n,m} = \mathbb{Q}(F_A, F_B)$. Clearly, $G_i \subseteq H_{i,j} \subseteq H_{i,j}(b_{j+1}) \subseteq F_i)$. By using invariant 2 and twice applying lemma 4 we obtain

$$\begin{aligned}[F_i(a_{i+1}) : F_i] &= [H_{i,j}(b_{j+1}, a_{i+1}) : H_{i,j}(b_{j+1})] \\ &= [H_{i,j}(a_{i+1}) : H_{i,j}] = [G_i(a_{i+1}) : G_i].\end{aligned}$$

By lemma 5 we conclude $[H_{i,j}(b_{j+1}, a_{i+1}) : H_{i,j}(a_{i+1})] = [H_{i,j}(b_{j+1}) : H_{i,j}]$, that is

$$[H_{i+1,j+1} : H_{i+1,j}] = [H_{i,j+1} : H_{i,j}].$$

Repeating this process gives

$$[H_{n,j+1} : H_{n,j}] = [H_{0,j+1} : H_{0,j}],$$

which is equivalent to invariant 3. □

Notice that the above proof holds for all real numbers b_i, $1 \leq i \leq m$, such that $F_A = F_P(b_1, b_2, \ldots, b_m)$. We may reformulate both invariants accordingly.

Now we are ready to prove the correctness of both invariants under steps 1 and 2. Consider step 1. Bob selects a trancedental element x over $\mathbb{Q}(F_A, F_B)$. Take $a_{n+1} = x$. Notice that

$$[\mathbb{Q}(F_A, F_B)(x) : \mathbb{Q}(F_A, F_B)] = [\mathbb{Q}(F_B)(x) : \mathbb{Q}(F_B)]. \tag{2}$$

Hence, invariant 2 holds again:

$$F_B(x) = F_P(a_1, \ldots, a_{n+1})$$

and

$$\mathbb{Q}(F_A, F_B(x)) = \mathbb{Q}(F_A, F_B)(x) = F_A(a_1, \ldots, a_{n+1})$$

together with the corresponding degree requirements.

Consider step 2. Bob selects an element $x \in F_B$ which he transmits to Alice. Invariant 3 holds prior to this step: $F_A = F_P(b_1, \ldots, b_m)$ and $\mathbb{Q}(F_A, F_B) = F_B(b_1, \ldots, b_m)$ with

$$[F_B(b_1, \ldots, b_{i+1}) : F_B(b_1, \ldots, b_i)] = [F_P(b_1, \ldots, b_{i+1}) : F_P(b_1, \ldots, b_i)]$$

for $0 \leq i \leq m-1$. Notice that $F_P \subseteq F_P(x) \subseteq F_B$. By repeadetly applying lemma 4 we obtain

$$[F_B(b_1,\ldots,b_{i+1}) : F_B(b_1,\ldots,b_i)] = [F_P(x)(b_1,\ldots,b_{i+1}) : F_P(x)(b_1,\ldots,b_i)]$$

for $0 \leq i \leq m-1$. Since $F_A(x) = F_P(x)(b_1,\ldots,b_m)$ and $\mathbb{Q}(F_A(x), F_B) = \mathbb{Q}(F_A, F_B) = F_B(b_1,\ldots,b_m)$, invariant 3 holds again. By theorem 2 both invariants hold again after each step.

Theorem 3. *Invariants 2 and 3 are invariant under steps 1 and 2.*

The proof of the invariants being invariant under step 1 only requires the condition (2), which is satisfied for step 1 because x is trancedental over $\mathbb{Q}(F_A, F_B)$.

7 Conclusion

In summary, we have shown that although authentication protocols and one-way functions exist in the this model, secure signature schemes, secure encryption schemes, and secret sharing schemes do not. If we replace the operations {+, -, *, /} with the operations {+, -, *, /, x^y}, where x^y denotes the operation of raising an arbitrary number x to an arbitrary power y, we are able to recover many cryptographic primitives, such as Diffie-Hellman Key Exchange, secure signature schemes, and secure encryption schemes. Of course we still allow all parties the ability to sample real numbers. We would like to determine a set of necessary and sufficient conditions for a set of operations to admit certain cryptographic primitives.

Acknowledgments

Special thanks to Ron Rivest for helping frame these problems as field-theoretic problems. Also, thanks to Xiaowen Xin for helping prepare this document and for motivation.

References

1. Artin, M., "Algebra," Prentice-Hall, 1991.
2. Burmester, M., Rivest, R., Shamir, A., *Geometric Cryptography*, http://theory.lcs.mit.edu/~rivest/publications.html, 1997.
3. Kaplansky, I., "Fields and Rings," Second Edition, University of Chicago Press, 1972.
4. Morandi, P., "Field and Galois Theory, Graduate Texts in Mathematics," Volume 167, Springer-Verlag, 1996.
5. Rompel, J., *One-way Functions are Necessary and Sufficient for Secure Signatures*, ACM Symp. on Theory of Computing **22** (1990), 387–394.

A Proof of Lemma 5

If $[G(u,v):G(u)] = [G(v):G] < \infty$ then the proof follows from

$$[G(u,v):G(u)][G(u):G] = [G(u,v):G] = [G(v,u):G(v)][G(v):G].$$

If $[G(u,v):G(u)] = [G(v):G] = \infty$ then v is trancedental over $G(u)$. We distinguish two cases. Firstly, if u is trancedental over $G(v)$ then it is also trancedental over G, hence, $[G(v,u):G(v)] = [G(u):G] = \infty$.

Secondly, suppose that u is algebraic over $G(v)$. We will show that u is algebraic over G and that a basis of $G(u)$ over G is also linearly independent over $G(v)$, which implies $[G(u):G] \leq [G(v,u):G(v)]$. A basis $\mathcal{X} = \{1, u, u^2, \ldots, u^{n-1}\}$ of $G(v,u)$ over $G(v)$ exists and is also linearly independent over G and part of $G(u)$, which implies $[G(u):G] \geq [G(v,u):G(v)]$ and equality must hold.

We are in the case that v is trancedental over $G(u)$ and u is algebraic over $G(v)$. Then there exist a finite and strictly positive number of non-zero coefficients $u_i \in G(v)$ such that $0 = \sum_i u_i \cdot u^i$. Each coefficient u_i is in $G(v)$ and can be expressed as $u_i = f_i(v)/g_i(v)$, where $f_i(.)$ and $g_i(.)$ are polynomials with coefficients in G. Define $h_i(v) = f_i(v)\prod_{j\neq i} g_j(v)$. Then $\sum_i h_i(v) \cdot u^i = 0$. Polynomial $h_i(.)$ has coefficients in G, therefore $h_i(v) = \sum_j h_{i,j} \cdot v^j$ for finitely many non-zero coefficients $h_{i,j} \in G$. We obtain

$$0 = \sum_j \left\{ \sum_i h_{i,j} \cdot u^i \right\} \cdot v^j.$$

The inner sums are in $G(u)$. Since v is trancedental over $G(u)$, these inner sums are equal to 0. If u is trancedental over G then all coefficients $h_{i,j} = 0$. This implies that $h_i(v) = 0$. All $f_j(v) \neq 0$, therefore $g_i(v) = 0$, hence, $u_j = 0$. However, there is a strictly positive number of non-zero coefficients u_j. Concluding, u is not trancedental but algebraic over G.

Since u is algebraic over G there exists a finite basis $\mathcal{X}$ of $G(u)$ over G with $G(u) = G[\mathcal{X}]$. We want to show that $\mathcal{X}$ is linearly independent over $G(v)$. Suppose that $\sum_{\gamma\in\mathcal{X}} x_\gamma \cdot \gamma = 0$ for some $x_\gamma \in G(v)$. For the coefficients x_γ there exist polynomials $f_\gamma(.)$ and $g_\gamma(.)$ with coefficients in G with $g_\gamma(v) \neq 0$ such that $x_\gamma = f_\gamma(v)/g_\gamma(v)$. Define $h_\gamma(v) = f_\gamma(v)\prod_{\sigma\neq\gamma} g_\sigma(v)$. Then $\sum_{\gamma\in\mathcal{X}} h_\gamma(v) \cdot \gamma = 0$. Polynomial $h_\gamma(.)$ has coefficients in G, therefore $h_\gamma(v) = \sum_j h_{\gamma,j} \cdot v^i$ for finitely many non-zero coefficients $h_{\gamma,j} \in G$. We obtain

$$0 = \sum_j \left\{ \sum_{\gamma\in\mathcal{X}} h_{\gamma,j} \cdot \gamma \right\} \cdot v^j.$$

The inner sums are in $G[\mathcal{X}] = G(u)$. Since v is trancedental over $G(u)$, these inner sums are equal to 0. Set $\mathcal{X}$ is linearly independent over G, hence, all coefficients $h_{\gamma,j} = 0$. This implies that $h_\gamma(v) = 0$. All $f_\sigma(v) \neq 0$, therefore $g_\gamma(v) = 0$, hence, $x_\gamma = 0$.

Performance Analysis and Parallel Implementation of Dedicated Hash Functions

Junko Nakajima and Mitsuru Matsui

Mitsubishi Electric Corporation,
5-1-1 Ofuna, Kamakura, Kanagawa 247-8501, Japan,
{june15,matsui}@iss.isl.melco.co.jp

Abstract. This paper shows an extensive software performance analysis of dedicated hash functions, particularly concentrating on Pentium III, which is a current dominant processor. The targeted hash functions are MD5, RIPEMD-128 -160, SHA-1 -256 -512 and Whirlpool, which fully cover currently used and future promising hashing algorithms. We try to optimize hashing speed not only by carefully arranging pipeline scheduling but also by processing two or even three message blocks in parallel using MMX registers for 32-bit oriented hash functions. Moreover we thoroughly utilize 64-bit MMX instructions for maximizing performance of 64-bit oriented hash functions, SHA-512 and Whirlpool. To our best knowledge, this paper gives the first detailed measured performance analysis of SHA-256, SHA-512 and Whirlpool.

Keywords. dedicated hash functions, parallel implementations, Pentium III

1 Introduction

A one-way and collision resistant hash function is one of the most important cryptographic primitives that require utmost speed particularly in software due to its heavy use for creating a digital finger printing of a long message. Historically the first constructions of hash functions were based on strong block ciphers and many efforts have since been done for their design and proof of security. However since this design approach does not necessarily result in fast hash functions in practice and often their hashing speed is much slower than underlying block ciphers, many "dedicated hash functions" suitable for software implementation on modern processors have been proposed and are now widely used in real world applications.

Therefore, performance analysis of hash functions in real environments is recognized as an important research topic and many studies have been done on this topic. Among them, a paper presented at CRYPTO'96 by Bosselaers et al. [BGV96] showed an excellent fast implementation and performance evaluation of dedicated hash functions (of that time) on the Pentium processor, which was a dominant processor at the time of the publication. They also gave a thorough critical path analysis of the MD-family, particularly concentrated on SHA-1 in the paper presented at Eurocrypt'97 [BGV97].

L.R. Knudsen (Ed.): EUROCRYPT 2002, LNCS 2332, pp. 165–180, 2002.

Recently NIST published three new dedicated hash functions with a larger hash size; SHA-256, SHA-384 and SHA-512 [FIP01], of which SHA-386 is essentially a truncation of the hashed value of SHA-512. These new hash functions have much more complex structure than SHA-1. Also in the European NESSIE project, a new 512-bit hash function Whirlpool was proposed [BR00]. The structure of Whirlpool is very similar to Rijndael, where the block size of its underlying block cipher is 512. All these new hash algorithms are under discussion for an inclusion in the next version of the ISO/IEC 10118 standard.

The purpose of this paper is to include these new generations as well as currently used dedicated hash functions and give an extensive performance analysis with actual implementations and measured cycle counts in a real processor platform. We particularly concentrate on the Pentium III processor, but most of our programs also run on Pentium II and Celeron in the same efficiency since these processors largely share their internal architecture. Note that Pentium 4 has a new architecture with richer SIMD instructions (but some instructions take a larger number of cycles now), which we do not deal with in this paper.

For the 32-bit oriented hash functions (MD5, RIPEMD-128, RIPEMD-160, SHA-1 and SHA-256), we perform not only a straightforward coding using 32-bit x86 registers, but also give an implementation method that enables fast parallel hashing of two or even three independent message blocks in parallel using 64-bit MMX registers (and x86 registers simultaneously) on Pentium III. Specifically, the two-block parallel method assigns the two blocks to upper and lower 32-bit halves of the MMX registers, and the three-block parallel method moreover assigns the third block to the x86 registers. For the 64-bit oriented hash functions (SHA-512 and Whirlpool), we fully utilize the MMX registers and instructions to extract maximal hashing performance. For another example of an optimization of a cryptographic algorithm using the MMX technology, see [Lip98].

The internal architecture of Pentium II/III is totally different from that of Pentium. Coding on Pentium was a programmers' paradise; estimating a cycle count of a given piece of code is not very difficult and, consequently, serious efforts to optimize a program by carefully arranging instructions were always rewarded. Unfortunately this is not the case for Pentium II/III. Intel documented the hardware architecture of Pentium II/III well [Int01][Int02][Int03], but still it is no longer possible to correctly predict how many cycles a given code takes without an actual measurement. This is partly due to an out-of-order execution nature of the processor, and also probably due to undocumented and unknown pipeline stall factors that only the hardware designers of Pentium II/III know.

In our experiences, even a well tuned-up code on Pentium II/III usually runs 10%-15% slower than what we expect. Filling this gap is a programmers' nightmare; it is a groping task with endless trial and error and very often such efforts are not rewarded at all. So in our implementations of hash functions, we first tried to write a code so that its data dependency chain could be as short as possible, and then re-scheduled the code to remove possible pipeline stall factors until the measured performance became up to 10%-15% slower than our best (fastest) estimation. This means that if a long dependency chain dominates

Table 1. Feature of dedicated hash functions

Algorithms	Endianess	Message Block Size (bits)	Digest Size (bits)	Word Size (bits)	The Number of Steps	Message Scheduling
MD5	Little	512	128	32	64	NO
RIPEMD-128	Little	512	128	32	2×64	NO
RIPEMD-160	Little	512	160	32	2×80	NO
SHA-1	Big	512	160	32	80	YES
SHA-256	Big	512	256	32	64	YES
SHA-512	Big	1024	512	64	80	YES
Whirlpool	Neutral	512	512	-	10*	YES

(*) Since Whirlpool has an architecture based on a block cipher, we will use the term "round" instead of "step" in this paper.

speed of an algorithm, Pentium may run in a smaller number of cycles than Pentium II/III.

Our implementation and performance measurement results show that, for MD5 and the RIPEMD family, the three-block parallel hashing on Pentium III reduces a cycle count of one block operation significantly as compared with the straightforward implementation. Also our instruction scheduling of SHA-1 and SHA-256 works excellently on Pentium III. In particular, the pipeline efficiency of SHA-1 reaches 2.52 μops/cycle. Since at most two integer/logical μops can be executed simultaneously on Pentium III, this shows that memory access instructions in the message scheduling part that was introduced in the SHA family are "hidden" in other logical and arithmetic instructions, which contributes to an effective use of the pipelines.

Another interesting result is that the two 512-bit hash functions SHA-512 and Whirlpool run almost at the same speed on Pentium III, while these algorithms have totally different architectures and design philosophies (and Whirlpool looks a much simpler algorithm). This does not seem to be a part of design principles of Whirlpool as far as we know. It should be also noted that since SHA-512 is designed for a pure 64-bit environment, it suffers performance penalties from some missing 64-bit instructions on Pentium III. Hence on a genuine 64-bit processor, SHA-512 might outperform Whirlpool.

2 Dedicated Hash Functions

Table 1 shows features of seven dedicated hash functions we deal with in this paper, and Table 2 summarizes the definitions of operations in these hash functions. So far there have been a lot of concrete proposals for efficient hash functions. The first constructions for hash functions were based on an efficient conventional encryption scheme such as DES. Although some trust has been built up in the security of these proposals, their software performance was not very well for the practical use, since they are typically a couple of times slower than the corresponding block cipher.

Table 2. Definitions of the operations for dedicated hash functions

Algorithm: Operations in one step	Nonlinear round functions at bit level
MD5:	$f_i(x,y,z) = (x \wedge y) \vee (\neg x \wedge z)$ $\quad 0 \le i \le 15$
Main stream	$f_i(x,y,z) = (x \wedge z) \vee (y \wedge \neg z)$ $\quad 16 \le i \le 31$
$A := B + (A + f_i(B,C,D) + X_{t[i]} + K_i)^{\lll s_i}$	$f_i(x,y,z) = x \oplus y \oplus z$ $\quad 32 \le i \le 47$
	$f_i(x,y,z) = y \oplus (x \vee \neg z)$ $\quad 48 \le i \le 63$
RIPEMD-128:	$f_i(x,y,z) = x \oplus y \oplus z$ $\quad 0 \le i \le 15$
Main streams 1 and 2	$f_i(x,y,z) = (x \wedge y) \vee (\neg x \wedge z)$ $\quad 16 \le i \le 31$
$A := B + (A + f_i(B,C,D) + X_{t[i]} + K_i)^{\lll s_i}$	$f_i(x,y,z) = (x \vee \neg y) \oplus z$ $\quad 32 \le i \le 47$
	$f_i(x,y,z) = (x \wedge z) \vee (y \wedge \neg z)$ $\quad 48 \le i \le 63$
	In their second stream,
	f_i is applied in the reversed order.
RIPEMD-160:	$f_i(x,y,z) = x \oplus y \oplus z$ $\quad 0 \le i \le 15$
Main streams 1 and 2	$f_i(x,y,z) = (x \wedge y) \vee (\neg x \wedge z)$ $\quad 16 \le i \le 31$
$A := (A + f_i(B,C,D) + X_{t[i]} + K_i)^{\lll s_i} + E$	$f_i(x,y,z) = (x \vee \neg y) \oplus z$ $\quad 32 \le i \le 47$
$C := C^{\lll 10}$	$f_i(x,y,z) = (x \wedge z) \vee (y \wedge \neg z)$ $\quad 48 \le i \le 63$
	$f_i(x,y,z) = x \oplus (y \vee \neg z)$ $\quad 64 \le i \le 79$
	In their second stream,
	f_i is applied in the reversed order.
SHA-1:	
Message scheduling	$f_i(x,y,z) = (x \wedge y) \vee (\neg x \wedge z)$ $\quad 0 \le i \le 19$
$X_i := (X_{i-3} \oplus X_{i-8} \oplus X_{i-14} \oplus X_{i-16})^{\lll 1}$	$f_i(x,y,z) = x \oplus y \oplus z$ $\quad 20 \le i \le 39$
Main stream	$f_i(x,y,z) = (x \wedge y) \oplus (x \wedge z) \oplus (y \wedge z)$ $\quad 40 \le i \le 59$
$E := A^{\lll 5} + f_i(B,C,D) + X_i + K_i + E$	$f_i(x,y,z) = x \oplus y \oplus z$ $\quad 60 \le i \le 79$
$B := B^{\lll 30}$	
SHA-256:	
Message scheduling	$Ch(x,y,z) = (x \wedge y) \vee (\neg x \wedge z)$
$X_i := \sigma_1(X_{i-2}) + X_{i-7} + \sigma_0(X_{i-15}) + X_{i-16}$	$Maj(x,y,z)$
$16 \le i \le 63$	$= (x \wedge y) \oplus (x \wedge z) \oplus (y \wedge z)$
Main stream	$\Sigma_0(x) = x^{\ggg 2} \oplus x^{\ggg 13} \oplus x^{\ggg 22}$
$T_1 := H + \Sigma_1(E) + Ch(E,F,G) + K_i + X_i$	$\Sigma_1(x) = x^{\ggg 6} \oplus x^{\ggg 11} \oplus x^{\ggg 25}$
$T_2 := \Sigma_0(A) + Maj(A,B,C)$	$\sigma_0(x) = x^{\ggg 7} \oplus x^{\ggg 18} \oplus x^{\gg 3}$
$D := D + T_1$	$\sigma_1(x) = x^{\ggg 17} \oplus x^{\ggg 19} \oplus x^{\gg 10}$
$H := T_1 + T_2$	
SHA-512:	
Message scheduling	$Ch(x,y,z) = (x \wedge y) \vee (\neg x \wedge z)$
$X_i := \sigma_1(X_{i-2}) + X_{i-7} + \sigma_0(X_{i-15}) + X_{i-16}$	$Maj(x,y,z)$
$16 \le i \le 79$	$= (x \wedge y) \oplus (x \wedge z) \oplus (y \wedge z)$
Main stream	$\Sigma_0(x) = x^{\ggg 28} \oplus x^{\ggg 34} \oplus x^{\ggg 39}$
$T_1 := H + \Sigma_1(E) + Ch(E,F,G) + K_i + X_i$	$\Sigma_1(x) = x^{\ggg 14} \oplus x^{\ggg 18} \oplus x^{\ggg 41}$
$T_2 := \Sigma_0(A) + Maj(A,B,C)$	$\sigma_0(x) = x^{\ggg 1} \oplus x^{\ggg 8} \oplus x^{\gg 7}$
$D := D + T_1$	$\sigma_1(x) = x^{\ggg 19} \oplus x^{\ggg 61} \oplus x^{\gg 6}$
$H := T_1 + T_2$	
Whirlpool:	
Message scheduling	
$E_i^r = \sum_{j=0}^{7} \text{Table}_j[(E^{r-1} \oplus K^{r-1})_{(i-j)mod8,j}]$	$a = \begin{pmatrix} a_0 \\ a_1 \\ \vdots \\ a_7 \end{pmatrix} = \begin{pmatrix} a_{00} & a_{01} & \cdots & a_{07} \\ a_{10} & a_{11} & \cdots & a_{17} \\ \vdots & \vdots & \ddots & \vdots \\ a_{70} & a_{71} & \cdots & a_{77} \end{pmatrix}$
Main stream	
$X_i^r = \sum_{j=0}^{7} \text{Table}_j[(X^{r-1} \oplus E^r)_{(i-j)mod8,j}]$	
$(1 \le r \le 10)$	$a = X, E$ or K

A, B, C, D, E, F, G, H	: Intermediate variables
K	: Fixed constant value
X	: Message block (and its derivatives)
$x^{\gg n}$, $x^{\ll n}$	: Right/Left shift of x by n bits
$x^{\ggg n}$, $x^{\lll n}$	: Right/Left rotate shift of x by n bits

MD4, proposed by R. Rivest in 1990 [R90,R492], is the first dedicated hash function that targeted at *speed* in software (MD2 has a totally different architecture and is slow). It was particularly designed to archive high performance on 32-bit processors that were the architecture of the future at the time. The algorithm is based on a simple set of primitive operations such as `add`, `xor`, `or` etc. on 32-bit words. Additionally, MD4 was designed to be favorable to a "little-endian" architecture, which obviously fits the Intel processor architecture perfectly. We do not deal with MD4 in this paper because it was broken several years ago [Dob98]. Almost all hash functions discussed hereafter are direct descendants of MD4 and inherit its structural characteristics. The **MD5** algorithm is an extension of the MD4 algorithm, increasing the number of steps from 48 to 64 [R592], and it is one of the most widely used dedicated hash algorithms in real applications. Some weaknesses of MD5 are reported in [Dob96][Rob96], although a collision of MD5 has not been found.

RIPEMD-128 and **RIPEMD-160** [DBP96] evolved out of RIPEMD, which was developed as a strengthened version of MD4 by the RIPE consortium. RIPEMD-128 was developed out of RIPEMD-160 as a plug-in substitute for RIPEMD. They also have a little-endian architecture. One of the new characteristics of these hash functions is that they have essentially two parallel instances of MD4. Each of these instances includes 64 and 80 steps for RIPE MD-128 and RIPEMD-160, respectively. This structure has a potential ability to realize high performance when implemented on pipelined or superscalar execution mechanisms. Additionaly, each step of RIPEMD-160 has a rotate shift operation by 10 bits. Both of RIPEMD-128 and RIPEMD-160 are standardized as dedicated hash functions in ISO/IEC 10118 [ISO97].

SHA-1 [FIP95] designed by NSA and published by NIST was also based on the design of MD4. The round functions of SHA-1 are exactly the same as those of MD4 while the total number of steps is 80. However, the SHA family went over to "big-endian", throwing away suitability for Intel processors. Moreover, it newly introduced a "message scheduling part". The design criteria of these new characteristics have not been made public. Although the role of the message schedule in the SHA family is similar to that of the key schedule in block ciphers, the message schedule must be done for each block like the on-the-fly implementation in block ciphers. The message schedule and main stream can be done independently, which means that it has a high capability for parallel execution, considering its rather high complexity. SHA-1 is also standardized as one of the dedicated hash functions in ISO/IEC 10118 [ISO97].

Recently **SHA-256** and **SHA-512** [FIP01] were proposed to keep up with possible future applications where longer message digests are required . They have almost the same basic components, except that SHA-256 operates on eight 32-bit words, while SHA-512 operates on eight 64-bit words. SHA-512 is the first dedicated hash function designed for a genuine 64-bit processor. SHA-256 and SHA-512 also have a message schedule part, which has a much more complex structure than that of SHA-1, including many shift operations in both message scheduling part and main stream. Another feature of these algorithms is that exactly the same operations are used in all steps.

Whirlpool does not belong to the MD-family nor to the SHA-family. It directly inherits its structure from Rijndael. The state of Whirlpool consists of a 8×8 matrix over $GF(2^8)$, while that of Rijndael was designed on a 4×4 matrix over $GF(2^8)$. A transformation from a state to a next state is given by eight 8-bit to 64-bit look-up tables. Hence Whirlpool demonstrates best performance in true 64-bit environments, but can also realize good performance on any processor. Whirlpool has a message scheduling part, which is exactly the same as its main stream. Whirlpool is endian-neutral. Recently the designer of Whirlpool announced a tweak of the algorithm to improve its hardware efficiency. SHA-256, -512 and this tweak of Whirlpool are under discussion for an inclusion in the next version of the ISO/IEC 10118 standard.

3 A Brief Overview of Pentium III Processor

Pentium III supports all x86 instructions with eight 32-bit registers, and additional MMX instructions with eight 64-bit registers (MMX registers). The main motivation of the MMX instructions is to enable 16×4 / 32×2 parallel operations for multimedia applications. Although Pentium III is not a full 64-bit processor such as the Alpha processor, — a 64-bit addition instruction is missing, for instance —, the MMX instructions are attractive to wider applications since most of them work in one cycle including a 64-bit memory load instruction. Pentium III also provides XMM instructions (SSE) with eight 128-bit registers, but we are not able to use them for fast hashing because they operate on floating-point data elements.

The following shows some of the essential topics for optimizing on Pentium II/III, which greatly owes to an excellent guidebook for optimizing Pentium family written by Agner Fog [Fog00].

Instruction Decoding. In the decoding stage, instruction codes are broken down into simpler micro-operations (μops), where one instruction usually consists of one to four μops. Pentium II/III has three decoders that can work in parallel and theoretically can generate six μops per cycle. However due to limitations coming from instruction fetch rules and decoder capabilities, the code length and order of instructions heavily affect efficiency of the decoding speed. A typical decoding rate of real applications is two to three μops per cycle on an average.
The decoding speed is not an important issue for MD5 and the RIPEMD family since the performance bottleneck of these hash functions is actually a long data dependency chain, but it can be a critical factor for algorithms with high parallel computation capabilities such as the SHA family. Generally, optimizing decoding performance without causing other pipeline stalls is an extremely difficult puzzle to solve.

Register Renaming. After the decoding stage, all permanent registers (x86 and MMX registers) are renamed into internal registers. This mechanism

solves fake data dependency chains caused by register starvation. The renaming is controlled by the register alias table (RAT), which can handle only three μops per cycle. This means that overall performance of Pentium II/III can not exceed three μops per cycle. Moreover the RAT can read two permanent registers in a cycle, and hence depending on the situation we should use an absolute addressing mode instead of a register indirect addressing mode.
Some hash functions repeatedly use a fixed constant value. If a register is free, it is common to assign the constant value to the register, which is best in terms of instruction length. However in order to avoid the register read stalls, we often handled the value using an immediate addressing mode if possible, or an absolute addressing mode via memory, particularly in the case of MMX instructions.

Execution Units. Pentium II/III has five independent execution ports p0 to p4 to carry out a sequence of μops in an out-of-order manner. p0 and p1 are mainly for arithmetic and logical operations, p2 for load and p3 and p4 for store operations. An important consequence of this architecture is that if arithmetic and logical operations are a dominant factor of performance, load and store operations can be "hidden" behind them. In other words, we should store a temporary value not in a free register but in memory in such a case.
The SHA family requires memory operations more frequently than MD5 and the RIPEMD family due to the existence of the message scheduling part, and also due to a large hash size for SHA-256 and SHA-512. If properly implemented these hash functions can take full advantage of hiding load/store operations behind other operations.

Other Topics. In Pentium II/III, a partial register/memory access (reading from a register/memory after writing to a part of it) causes a heavy performance penalty; i.e. a register/memory must be basically read/written in the same size at the same memory address. In this paper, the partial memory access can be a problem only in the 64-bit oriented hash functions SHA-512 and Whirlpool as will be shown in a later section.
According to [Fog00], the retirement stage of μops can be a bottleneck of performance. We however can do little about this stall factor, while the partial register/memory stalls can be avoided by programmers.

4 API and How to Measure Performance

We developed our assembly language programs on the following hardware and software environments. The size of RAM memory is not an important issue because we designed the programs so that everything could be on the first level cache (code 16KB + data 16KB).

Hardware: IBM Compatible PC with Pentium III 800MHz and 256MB RAM
Software: Windows 98, Visual C++ 6.0, MASM 6.15

We described a hashing logic in a subroutine form callable from C language and measured its execution time from outside the subroutine, which reflects hashing performance in real applications. The subroutine API is that an input message to be compressed and a resultant hashed value are represented as a byte sequence passed by pointers, which is also common as an interface of hash functions. Note that this means that big-endian algorithms, specifically the SHA family, must perform a byte-order swap operation inside the subroutine and it is counted as a part of hashing time. We assume a padding operation is done outside this routine.

In the next section, we will give a method of coding for hashing two or three independent messages of the same block length in parallel. For simplicity, we adopted the same API for this method as that for a straightforward implementation, appending the second/third message to the first/second message for each block. Hence a procedure for combining two separate 4-byte words into one 8-byte MMX register is also counted as a part of hashing time. If we interleave different messages in every word (4-byte), we can skip this procedure, but did not adopt this approach because assuming such interface looks uncommon in real applications.

We actually measured time for hashing a total of 8KB message bytes, which is half of the first level data cache size. To maximize speed, we unrolled an inner loop as long as the code is fully covered by the first level code cache. The cycle counting was done using the `rdtsc` (read time stamp counter) instruction, as shown in [Fog00]. The timing was measured several times to remove possible negative effects on performance due to interrupts by an operating system. Also note that our programs are neither data-dependent nor self-modified.

5 Implementations of Hash Functions on Pentium III

5.1 Implementation Methods

We designed three types of assembly language codes for 32-bit oriented hash functions (MD5, RIPEMD-128 -160, SHA-1 -256); namely, straightforward, two-block parallel and three-block parallel as follows. The second and third methods fully extract the power of parallel execution capabilities of MMX. For 64-bit oriented hash functions (SHA-512 and Whirlpool), we implemented the straightforward version only (no way to parallelize).

Method 1: Straightforward

MD5, RIPEMD-128, RIPEMD-160, SHA-1
All words of the main stream are always held in four or five x86 registers and the remaining registers are used as temporary variables. This is the most common implementation of dedicated hash functions in software.

SHA-256, SHA-512
All eight 32-bit/64-bit words of the main stream are stored in memory because of the register starvation of the Pentium family, and x86/MMX regis-

ters are basically used only as temporary variables. Additionally, for SHA-512, 32-bit x86 registers are essentially used to realize a 64-bit addition operation, which is a missing instruction in Pentium III.

Whirlpool
A 64-byte state matrix is always held in all eight MMX registers. However, since the state information of the preceding round is necessary to generate the state matrix of the next round, it must be, after all, stored in memory at the end of each round.

Method 2: Two-Block Parallel

This method is applied only to 32-bit oriented hash functions (**MD5, RIPEMD -128, RIPEMD-160, SHA-1, SHA-256**), where a word of one message is loaded on the upper half of 64-bit register/memory and a word of another message is loaded on the lower half. This enables fast parallel hashing computation of two independent messages, but the following penalties peculiar to Pentium III should be taken into consideration.

1. The code length of an MMX instruction is usually longer (typically by one byte) than that of an equivalent x86 instruction, which may lead to a performance penalty due to an inefficient instruction decoding.
2. Since MMX instructions do not have an immediate addressing mode, a 64-bit immediate value must be processed via memory, which increases the number of memory access μops (p2, p3 and p4). But this penalty is often able to be hidden in integer μops (p0 and p1).
3. A parallel rotate shift instruction is missing (if a shift count is not a multiple of 16). To do this on Pentium III, four instructions are needed; that is, `movq` (copy), `psrld` (right shift), `pslld` (left shift) and `pxor` (xor).

Our implementation will show that gains by parallel computation exceed the penalties caused by these factors.

Method 3: Three-Block Parallel

This method is a combination of the two methods above, which is hence applicable to 32-bit oriented dedicated hash functions. Since methods 1 and 2 use different types of registers, these two programs can "coexist", that is, can be executed in parallel without depriving each other of hardware resources. Although this is not an essential methodological improvement of implementation, it is expected that a better instruction scheduling leads to further improvement of hashing speed, particularly if a long data dependency chain dominates the speed of a target algorithm. Possible penalties that should be noted in this case are:

1. Code size: The code size becomes big because the entire code is simply a merged combination of the two implementations. But this does not lead to an actual big penalty issue as long as the entire code is within the first level cache.

2. Register read stall: The heavy simultaneous use of x86 and MMX registers easily create register read stalls. It is often very difficult to completely remove the possibility of this stall without causing other penalties.

Our implementation will show that this three-block method actually gives a performance improvement for MD5, RIPEMD-128, RIPEMD-160.

5.2 Implementation Results and Discussions

Table 7 shows our implementation results of the seven dedicated hash functions. We manually counted the number of μops of one block operation, referring to [Fog00]. "p01" denotes logical and integer μops that use pipes p0 and/or p1. "p2" and "p34" denote memory read/write μops that use pipes p2 and p3, p4, respectively. Note again that the cycle counts of the SHA family include time for endian conversion and that the cycle counts of methods 2 and 3 include time for merging two 32-bit words into one 64-bit register.

MD5

The performance of our straightforward implementation of MD5 on Pentium III is approximately the same as that shown in [Bos97] on Pentium. This reflects the fact that the frequency of memory access operations is very low and a long data dependency chain is an actual dominant factor of its hashing speed. In fact, the μops/cycle value of our program is only 1.64 while the maximal performance that Pentium III can achieve is 3 μops/cycle.

The two-block and three-block parallel implementations significantly improve the efficiency of hashing. In particular, it can be seen that the three-block version is almost perfectly scheduled (1.83 p01 μops/cycle) and 30% faster than the straightforward version.

RIPEMD-128, RIPEMD-160

The same tendency can be seen for RIPEMD-128 and RIPEMD-160 as for MD5. One possible reason that RIPEMD-160 has a better μops/cycle value for two-block and three-block versions is the existence of the operation "$B^{\lll 10}$" that was introduced in RIPEMD-160. This operation takes four μops on MMX, which can be executed independently with other operations.

[Note] Since the RIPEMD family has two parallel instances, it is possible to assign each of the two instances to each of two types of registers, that is, one to 32-bit registers and the other to 64-bit MMX registers. This makes a parallel execution *inside* one message block possible, but unfortunately this did not run very fast, probably because assigning the full 64-bit registers for only one instance execution was too inefficient reducing overall performance. Another possibility to utilize the parallelism of the RIPEMD family might be interleaving the first instance and the second instance (our current code executes the second one after finishing the first one). Although this method leads to an increase in a total number of μops due to the register starvation, the parallelism of the resultant code will be improved. However we have

Table 3. Coding example of one step of round 2 of RIPEMD-128 on a Pentium III processor. The chaining variable A, B, C, D is stored in x86 registers **eax** through **edx** or MMX registers mm0 through mm3. In the three-block parallel implementation, these codes are interleaved.

Straightforward code using x86	Two-block code using MMX
`mov esi, edx`	`paddd mm0, [eax+8*X]`
`add eax, (X+4)[edi]`	`movq mm5, mm3`
`xor esi, ecx`	`pxor mm5, mm2`
`and esi, ebx`	`paddd mm0, K`
`xor esi, edx`	`pand mm5, mm1`
`lea eax, [eax+esi+K]`	`pxor mm5, mm3`
`rol eax, s`	`paddd mm0, mm5`
	`movq mm5, mm0` ; left rotate
	`pslld mm0, s` ; shift of mm0
	`psrld mm5, 32-s` ; by s bits
	`pxor mm0, mm5` ;

not succeeded, so far, in a speed-up of a straightforward implementation of RIPEMD-128 or RIPEMD-160 using this technique.

SHA-1

One big difference between the SHA family and the (RIPE)MD family is the existence of a message scheduling part. The message scheduling part of the SHA family is independent of the main stream and contains many memory access operations enabling the straightforward SHA-1 implementation to optimally exploit the increased hardware parallelism of Pentium III. The value 2.52 μops/cycles of our straightforward version is the highest of our programs.

Because of this highly parallel feature of SHA-1, the performance improvement of the two-block version is not so big as that of the (RIPE)MD family. Moreover the three-block version is rather slower than the two-block version. This suggests that, as far as an instruction scheduling efficiency of SHA-1 is concerned, our implementation has reached almost an optimal level on Pentium III (1.82 p01 μops/cycle).

SHA-256

SHA-256 (and SHA-512) uses eight words and it is no longer possible to keep all internal values on the permanent registers. They must be stored in memory and be read from/written to memory in each step, which inevitably increases the frequency of memory access. However, since these memory access μops can be mostly carried out in parallel with logical and integer uops, the pipeline scheduling of SHA-256 works excellently (2.32 μops/cycle).

The two-block parallel implementation of SHA-256 runs 15% faster than the straightforward implementation, but this improvement is smaller than that in other hash algorithms. This is because SHA-256 (and SHA-512) has

Table 4. Coding example of one step of round 2 of SHA-1 on a Pentium III processor. The chaining variable A, B, C, D, E is stored in x86 registers `eax,ebx,ecx,edx,ebp` or MMX registers `mm0` through `mm4`. In the three-block parallel implementation, these codes are interleaved. Instructions marked "m.s." are for message scheduling part.

Straightforward code using x86	Two-block code using MMX
`mov esi,W+(s*4)` ; m.s.	`movq mm6,WW+(s*8)` ; m.s.
`mov edi,edx`	`movq mm5,mm1`
`xor esi,W+(s1*4)` ; m.s.	`pxor mm6,WW+(s1*8)` ; m.s.
`xor edi,ecx`	`pxor mm5,mm2`
`xor esi,W+(s2*4)` ; m.s.	`pxor mm6,WW+(s2*8)` ; m.s.
`xor edi,ebx`	`pxor mm5,mm3`
`xor esi,W+(s3*4)` ; m.s.	`pxor mm6,WW+(s3*8)` ; m.s.
`add ebp,edi`	`paddd mm4,mm5`
`rol esi,1` ; m.s.	`movq mm7,mm6` ; m.s.
`mov edi,eax`	`paddd mm4,CONSTonMEM`
`add ebp,esi`	`paddd mm6,mm6` ; m.s.
`rol edi,5`	`psrld mm7,31` ; m.s.
`rol ebx,30`	`movq mm5,mm1`
`lea ebp,[ebp+edi+CONST]`	`pxor mm6,mm7` ; m.s.
`mov W+(s*4),esi` ; m.s.	`psrld mm1, 2`
	`paddd mm4, mm6`
	`movq WW+(s*8),mm6` ; m.s.
	`pslld mm5,30`
	`movq mm6,mm0`
	`movq mm7,mm0`
	`pslld mm6,5`
	`paddd mm4, mm6`
	`psrld mm7,27`
	`pxor mm1,mm5`
	`paddd mm4,mm7`

many rotate shift operations in both the message scheduling part and the main stream, which causes a non-negligible increase of the number of μops when realized in 64-bit MMX registers. The performance of the three-block parallel version was not good for the same reason as for SHA-1.

[**Note**] The architecture of SHA-256 is quite different from that of SHA-1. But interestingly, the rate of the number of memory access μops to all μops is almost the same (approximately 30%), which is ideal in terms of Pentium III scheduling. Is this a hidden design criteria of the SHA family??

SHA-512

SHA-512 is a 64-bit oriented hash function and hence it is essential to utilize MMX registers and instructions. However, since Pentium III does not have a 64-bit addition instruction (but fortunately it does have a 64-bit shift operation! Pentium 4 supports both instructions), we realized it by simply combining two 32-bit additions. Although this naive method obviously suffers a non-negligible performance penalty from 32-bit from/to 64-bit register transfer, we do not know a faster method to do this. Another penalty that we should note is a partial memory access. Below left is a straightforward method for coding the last part of one step, but this causes a partial memory access at the beginning of the next step. Below right is the corrected code we adopted, which is free from the penalty and additionally saves one μop.

The pipeline scheduling of our implementation is good (2.24 μops/cycle), but we feel that there is still room for performance improvement since the hash algorithm itself allows higher parallel execution capability.

Table 5. 64-bit addition and partial memory stall on a Pentium III processor.

	μops		μops
`add [mem+0],eax`	; 4	`add eax,[mem+0]`	; 2
`adc [mem+4],ebx`	; 6	`adc ebx,[mem+4]`	; 3
		`movd mm0,eax`	; 1
		`movd mm1,ebx`	; 1
(next step)		`punpckldq mm0,mm1`	; 1
`movq mm0,[mem]`	; 1 ← **Stall**	`movq [mem],mm0`	; 2

Whirlpool

The simplest realization of Whirlpool is to keep the state matrix in eight MMX registers and eight 8-bit to 64-bit look-up tables on memory, but these tables completely cover all the 16KB data cache of Pentium III, which leads to data cache miss penalties. We hence had only four tables on memory and generated the remaining from them when necessary using the `pshufw` (packed shuffle word) instruction.

This instruction works only in Pentium III (not in Pentium II and Celeron); all other instructions throughout our codes run on Pentium II and Celeron. Also taking into consideration reducing the frequency of memory access and

Table 6. An essential part of Whirlpool on a Pentium III processor.

```
mov   edx, [Matrix Address] ; preceding state matrix
movzx esi, dl               ; address generation
pxor  mm0, Table1[esi*8]    ; current state matrix
movzx esi, dh
pxor  mm1, Table2[esi*8]
shr   esi, 16
movzx esi, dl
pxor  mm2, Table3[esi*8]
movzx esi, dh
pxor  mm3, Table4[esi*8]
```

Table 7. Performance Figure

Algorithm	method	μops/block [p01]	[p2]	p[34]	[total]	cycles /block	cycles /byte	μops/cycle [p01]	[total]	size (bytes) code	data	Pentium[Bos97] #inst.	cycles	i/c
MD5	1	503	69	10	582	354	5.53	1.42	1.64	1750	32	577	337	1.71
MD5	2	783	148	40	971	276	4.31	1.42	1.76	3331	704			
MD5	3	1283	217	50	1550	234	3.66	1.83	2.21	5259	736			
RIPEMD-128	1	875	141	18	1034	602	9.41	1.45	1.72	2707	48	1024	592	1.73
RIPEMD-128	2	1379	252	48	1679	477	7.45	1.45	1.76	5703	280			
RIPEMD-128	3	2251	393	66	2710	425	6.64	1.77	2.13	8379	320			
RIPEMD-160	1	1421	176	22	1619	911	14.23	1.56	1.78	4227	56	1639	1013	1.62
RIPEMD-160	2	2533	383	52	2968	738	11.53	1.72	2.01	10110	320			
RIPEMD-160	3	3951	559	74	4584	726	11.34	1.81	2.10	14305	376			
SHA-1	1	1100	295	174	1569	623	9.73	1.77	2.52	4664	356	1469	837	1.76
SHA-1	2	1928	389	170	2487	531	8.30	1.82	2.34	8567	248			
SHA-1	3	2977	684	376	4037	559	8.73	1.78	2.41	13175	340			
SHA-256	1	2491	609	418	3518	1519	23.73	1.64	2.32	10202	144			
SHA-256(*)	2	4385	689	418	5492	1318	20.59	1.66	2.08	6705	1232			
SHA-256(*)	3	6921	1352	852	9125	1417	22.14	1.63	2.15	14142	1816			
SHA-512(*)	1	8828	1522	1186	11536	5143	40.18	1.72	2.24	7203	1496			
Whirlpool(*)	1	3328	1634	384	5346	2337	36.52	1.42	2.29	2456	8496			

(*) only partially loop-unrolled in a message block to reduce code size within the first level cache (16KB). All other implementations are fully loop-unrolled.

removing the possibility of partial register/memory stalls, we wrote the entire algorithm by repeating the piece of code shown in Table 6.

The pipeline of this program works very well, achieving 2.29 μops/cycle. The resultant hashing speed (cycles/byte) is more than 3.5 times faster than that of the designers' C implementation [BR00]. Very interestingly, this performance is almost the same as that of SHA-512; Whirlpool is slightly (within 10%) faster than SHA-512, but this difference looks within "a margin of implementation".

6 Conclusions

This paper showed a performance analysis and speeding-up method of dedicated hash functions on the Pentium III processor. A further improvement of software performance is ongoing.

An overall performance of a processor can be achieved by (1) high instruction parallelism, (2) high SIMD parallelism and (3) high clock frequency. The index "cycle/byte" is the most common performance measure of cryptographic algorithms, which direct reflects (1) and (2), but not (3). So we should note that this measure is appropriate to compare software performance of given cryptographic algorithms on a fixed target processor, but not necessarily appropriate to evaluate hardware performance of given processors on a fixed cryptographic algorithm.

High processor performance has been achieved by improving (1) and (3) in the past, but in recent processors this seems to be rapidly shifting to (2) and (3). This means that use of parallel execution of multiple blocks will be much more important and have a practical impact in near future.

Acknowledgments

We would like to thank Antoon Bosselaers for his careful reading and many valuable remarks.

References

[BR00] P. Barreto, V. Rijmen, "The Whirlpool hashing function," *First open NESSIE Workshop record,* Leuven, 13-14 November 2000. The document is available at http://www.cryptonessie.org/workshop/submissions/whirlpool.zip.

[BGV96] A. Bosselaers, R. Govaerts, J. Vandewalle, "Fast hashing on Pentium," *Advances in Cryptology, Proceedings Crypto '96, LNCS 1109,* N. Koblitz, Ed., Springer-Verlag, 1996, pp. 298-312.

[BGV97] A. Bosselaers, R. Govaerts and J. Vandewalle, "SHA: A design for parallel architectures?," *Advances in Cryptology, Proceedings Eurocrypt'97, LNCS 1233,* W. Fumy, Ed., Springer-Verlag, 1997, pp. 348-362.

[Bos97] A. Bosselaers, "Even faster hashing on the Pentium," presented at the rump session of Eurocrypt'97. Available at http://www.esat.kuleuven.ac.be/cosicart /pdf/AB-9701.pdf.

[Dob98] H. Dobbertin, "Cryptanalysis of MD4," *J. Cryptology,* Vol. 11, pp. 253-271, 1998.

[Dob96] H. Dobbertin, "The status of MD5 after a recent attack," *Cryptobytes,* Vol. 2, No. 2, pp. 1-6, 1996. Available at ftp://ftp.rsasecurity.com/pub/cryptobytes/ crypto2n2.pdf

[DBP96] H. Dobbertin, A. Bosselaers, B. Preneel, "RIPEMD-160, a strengthened version of RIPEMD," *Fast Software Encryption, LNCS 1039,* D. Gollmann, Ed., Springer-Verlag, 1996, pp. 71-82. The final version is available at http:// www.esat.kuleuven.ac.be/cosicart/pdf/AB-9601.pdf

[FIP95] Federal Information Processing Standards (FIPS) Publication 180-1, *Secure Hash Standard (SHS),* U.S. DoC/NIST, April 17, 1995.

[FIP01] Draft Federal Information Processing Standards (FIPS) Publication 180-2, *Secure Hash Standard (SHS),* U.S. DoC/NIST, May 30, 2001.

[Fog00] Agner Fog, How to Optimize for the Pentium Microprocessors, 03 July 2000. Available at http://www.agner.org/assem/

[Int01] Intel, *Intel Architecture Optimization. Reference Manual,* 1999. Order Number 245127-001. Available at http://www.intel.com/design/pentiumIII/manuals/

[Int02] Intel, *Intel Architecture Optimization Manual,* 1997. Order Number 242816-003. Available at http://www.intel.com/design/pentium/manuals/

[Int03] Intel, *Intel Architecture Software Developer's Manual,* 2001.
Volume 1 Basic Architecture (Order Number 245470)
Volume 2 Instruction Set Reference (Order Number 245471)
Volume 3 System Programming Guide (Order Number 245472)
Available at http://www.intel.com/design/pentiumIII/manuals/

[ISO97] ISO/IEC 10118-3, *"Information technology - Security techniques - Hash-functions -Part 3: Dedicated hash-functions,"* IS 10118, 1997.

[Lip98] H. Lipmaa, "IDEA, A Cipher for Multimedia Architectures?," *Selected Areas in Cryptography '98, LNCS 1556,* Henk Meijer, Eds., Springer-Verlag, 1998, pages 248–263. Available at http://www.tcs.hut.fi/ helger/papers/lip98/.

[PRB98] B. Preneel, V. Rijmen, A. Bosselaers, "Recent developments in the design of conventional cryptographic algorithms," *Computer Security and Industrial Cryptography, State of the Art and Evolution, LNCS 1528,* B. Preneel, V. Rijmen, Eds., Springer-Verlag, 1998, pp. 106-131.

[R90] R.L. Rivest, "The MD4 message digest algorithm," *Advances in Cryptology, Proceedings Crypto '90, LNCS 537,* S. Vanstone, Ed., Springer-Verlag, 1991, pp. 303-311.

[R492] R.L. Rivest, "The MD4 message-digest algorithm," *Request for comments (RFC) 1320,* Internet Activities Board, Internet Privacy Task Force, April 1992.

[R592] R.L. Rivest, "The MD5 message-digest algorithm," *Request for comments (RFC) 1321,* Internet Activities Board, Internet Privacy Task Force, April 1992.

[Rob96] M. Robshaw, "On recent results for MD2, MD4 and MD5," *RSA laboratories' Bulletin,* No. 4, November 1996. Available at ftp://ftp.rsasecurity.com/pub/pdfs/bulletn4.pdf

Fault Injection and a Timing Channel on an Analysis Technique

John A. Clark and Jeremy L. Jacob

Dept. of Computer Science, University of York, York YO10 5DD, England, UK,
{jac,jeremy}@cs.york.ac.uk

Abstract. Attacks on cryptosystem *implementations* (e.g. security fault injection, timing analysis and differential power analysis) are amongst the most exciting developments in cryptanalysis of the past decade. Altering the internal state of a cryptosystem or profiling the system's computational dynamics can be used to gain a huge amount of information. This paper shows how fault injection and timing analysis can be interpreted for a simulated annealing attack on Pointcheval's Permuted Perceptron Problem (PPP) identification schemes. The work is unusual in that it concerns fault injection and timing analysis on an *analysis technique*. All recommended sizes of the PPP schemes are shown to be unsafe.

Keywords: Heuristic Optimisation, timing channels, identification schemes

1 Introduction: Zero Knowledge and NP-Hard Solution Techniques

Since the introduction of zero-knowledge proofs by Goldwasser et al. in 1985 [5] several schemes have been proposed. Some make use of number theoretic results [3]. Others have sought to make use of the computational intractability of known NP-complete problems. Since Shamir exemplified the concept using the Permuted Kernel Problem (PKP)[12], others have offered systems based on Syndrome Decoding [14], Constrained Linear Equations (CLEs) [15] and the Permuted Perceptron Problem (PPP) [11].

Heuristic optimisation techniques such as genetic algorithms [4] and simulated annealing [7] have shown their worth over a huge number of engineering disciplines. It comes as no surprise that they have been investigated as cryptanalysis tools. Most work has been concerned with elementary ciphers [16,6,10,13] but recent work has included attacks on full-strength zero knowledge schemes. Pointcheval [11] gives results of attacks using simulated annealing on his PPP-based schemes. Knudsen and Meier [8] have recently improved on those results using an unusual and sophisticated attack. Their work (based on properties of sets of solutions obtained from multiple runs) is a direct challenge to the 'standard' way of applying heuristic optimisation techniques and indicates that there is considerable room for more sophistication in their application to cryptanalysis.

L.R. Knudsen (Ed.): EUROCRYPT 2002, LNCS 2332, pp. 181–196, 2002.

This paper describes two further very non-standard approaches based loosely around extant cryptanalysis notions of security fault injection [1] and timing analysis [9]. For comparison, Pointcheval's schemes are the subject of attack. In Section 3 Problem Warping (an interpretation of security fault injection) is used to attack the Perceptron Problem (PP). Cost functions whose minimisation is highly unlikely to lead to an actual solution are used. In a sense, the search is used to find a solution to a warped (different but related) problem. It turns out that the solutions obtained in this way are highly correlated with the solution to the actual problem (much more so than solutions obtained by attempting to solve the original problem directly). Furthermore, combining solutions from differently warped problem searches also provides an efficient way of obtaining secret information. In Section 4 the same technique is used to attack the more difficult Permuted Perceptron Problem (PPP). In addition, a form of timing side-channel with huge power is also demonstrated. The search process is monitored as it moves to its final solution. Some elements of the solution take particular values early in the search and then never change. Observing *when* particular solution elements get 'stuck' in this way can reveal over half the secret vector in a single run. The authors believe that this is the first time *the computational dynamics of an analysis technique* have been used to reveal information. Section 5 draws conclusions and indicates future work. First, the particular schemes analysed are described together with an outline of simulated annealing.

2 Preliminaries

2.1 The Perceptron Problem and Permuted Perceptron Problem

In 1995 Pointcheval [11] suggested what seems a promising scheme based on the *Perceptron Problem* (PP). In fact, he chose a variant of this problem that is much harder to solve known as the *Permuted Perceptron Problem* (PPP). If instances of these problems can be solved the identification schemes are broken. The protocols used to implement the identification schemes are not described here (the reader is referred to [11] for details). This paper concentrates on attacking the underlying NP-complete problems. The notation of [11] and [8] will be used. A column vector whose entries have value +1 or -1 is termed an ϵ -vector. Similarly, a matrix whose entries have value +1 or -1 is termed an ϵ -matrix.

- **Perceptron Problem:** :
  Input: An m by n ϵ-matrix A.
  Problem: Find an ϵ-vector V of size n such that
  $(AV)_i \geq 0$ for all $i = 1, ..., m$.

- **Permuted Perceptron Problem:**
  Input: An m by n ϵ-matrix A and a multiset S of non-negative numbers of size m.
  Problem: Find an ϵ-vector V of size n such that
  $\{\{(AV)_i | i = \{1, ..., m\}\}\} = S$.

In the PP we require that image elements $(AV)_i$ be non-negative, in the PPP we require that these elements have a particular distribution (histogram). If n is odd (even) then the $(AV)_i$ must all take odd (even) numbered values. Pointcheval's PPP schemes used only odd values for n (see below). It is always possible to generate feasible instances of these problems. The matrix A and column vector V are generated randomly. If $(AV)_i < 0$ then the elements a_{ij} of the ith row are negated. This method of generation introduces significant structure into the problem. In particular, the majority vectors of the entries for columns of A are correlated with the corresponding elements of V (as indicated in [11] and [8]). The security of the scheme relies on the computational intractability of exploiting this structure.

Any PPP solution is obviously a solution to the corresponding PP since the PPP simply imposes an extra histogram (multiset) constraint. A solution to the PP is not necessarily a solution to a related PPP. Pointcheval investigated the complexity of generating PPP solutions by the repeated generation of PP solutions. He indicated that matrices of the form $(m, n) = (m, m+16)$ gave best practical security and offered three particular sizes: $(101, 117)$, $(131, 147)$ and $(151, 167)$.

2.2 Simulated Annealing

Simulated annealing is a combinatorial optimisation technique based loosely on the physical annealing process of molten metals. An informal description is given below followed by a detailed one.

States and State Cost. Candidate solutions to the problem at hand form the states over which the search will range. With each state V is associated a cost, $cost(V)$, that gives some measure of how undesirable that state is (in physical annealing a high energy state is undesirable). In attacking the Perceptron and Permuted Perceptron Probems, the current state (solution) will be some ϵ-vector V_{curr} of size n. The choice of cost function is a crucial issue (discussed in Section 3).

The Neighborhood. The search moves from state to state in an attempt to find a state V_{best} with minimum cost over all states. The search may move only to another state that is 'close to' or 'in the neighborhood of' the current one, i.e. it is a *local* search. If V_{curr} is the current state vector, then the local neighborhood $Neighborhood(V_{curr})$ is the set of ϵ-vectors of size n obtained from V_{curr} by negating a single element (i.e. changing a 1 to a -1 or vice versa).

Accepting and Rejecting Moves. Simulated annealing combines hill-climbing with an ability to accept worsening state moves to provide for escape from local optima. From the current state V_{curr} a neighbouring state V_{neigh} is generated randomly. If $cost(V_{neigh}) < cost(V_{curr})$ then V_{neigh} becomes the current state (this is the 'hill-climbing', though perhaps 'valley diving' would be a better term for minimisation problems.) If not, then the state may be accepted probabilistically in a way that depends on the *temperature* T of the search (see below) and the extent to which the target state is worse (in terms of cost). The worse a target state is, the less likely it is a move to that state will be taken.

Cooling It All Down. In analogy with the physical annealing process, simulated annealing has a control parameter T, known as the temperature. Initially the temperature is high and virtually any move is accepted. Gradually the temperature is cooled and it becomes ever harder to accept worsening moves. Eventually the process 'freezes' and only improving moves are accepted at all. If no move has been accepted for some time then the search halts. We now describe the algorithm in detail.

The technique has the following principal parameters:

- the temperature T
- the cooling rate $\alpha \in (0,1)$
- the number of moves N considered at each temperature cycle
- the number $MaxFailedCycles$ of consecutive failed temperature cycles (where no move is accepted) before the search aborts
- the maximum number IC_{Max} of temperature cycles considered before the search aborts

The initial temperature T_0 is obtained by the technique itself. The other values are typically supplied by the user. In the work described here they remain fixed during a run. More advanced approaches allow these parameters to vary dynamically during the search. The simulated annealing algorithm is as follows:

1. Let T_0 be the start temperature. Increase this temperature until the percentage of moves accepted within an inner loop of N trials exceeds some threshold (e.g. 95%).
2. Set $IC = 0$ (iteration count), $finished = false$ and $ILSinceLastAccept = 0$ (number of inner loops since a move was accepted) and randomly generate an initial current solution V_{curr}.
3. while(not $finished$) do 3a-3d
   (a) **Inner Loop:** repeat N Times
      i. $V_{\text{new}} = \text{generateMoveFrom}(V_{\text{curr}})$
      ii. calculate change in cost
      $\Delta_{\text{cost}} = \text{cost}(V_{\text{new}}) - \text{cost}(V_{\text{curr}})$
      iii. If $\Delta_{\text{cost}} < 0$ then accept the move, i.e. $V_{\text{curr}} = V_{\text{new}}$
      iv. Otherwise generate a value u from a uniform(0,1) random variable. If $\exp^{-\Delta_{\text{cost}}/T} > u$ then accept the move, otherwise reject it.
   (b) if no move has been accepted in most recent inner loop then
   $ILSinceLastAccept = ILSinceLastAccept + 1$
   else $ILSinceLastAccept = 0$
   (c) $T = T * \alpha$, $IC = IC + 1$
   (d) if (ILSinceLastAccept > MaxFailedCycles) or ($IC > IC_{max}$) then $finished = true$
4. The state V_{best} giving the lowest cost over the whole search is taken as the final 'solution'.

Note that as T decreases to 0 then $\exp^{-\Delta_{\text{cost}}/T}$ also tends to 0 if $\Delta_{\text{cost}} > 0$ and so the chances of accepting a worsening move become vanishingly small as the temperature is lowered. In all the work reported here, the authors have used a value 0.95 for the geometric cooling parameter α and a value of 400 for N.

In attacking the PPP it will be useful also to record *when* each solution element (i.e. $V_0, \ldots, V_{n-1}$) changed for the last time. For current purposes, the time of last change to an element V_k is deemed to be the index IC of the inner loop in which the last change was made to that element (i.e. when a neighboring solution was obtained by flipping V_k and a move to that solution was taken). These times form the basis of the timing channel indicated earlier.

3 Attacking the Perceptron Problem by Problem Warping

Both Pointcheval [11] and subsequently Knudsen and Meier [8] have attacked the PP using a cost function of the form.

$$Cost(V') = \sum_{i=1}^{m} max\{-(AV')_i, 0\}$$

Pointcheval uses the annealing process directly to obtain solutions to the PP and reports 'We have carried out many tests on square matrices ($m = n$) and on some other sizes, and during a day, we can find a solution to any instance of PP with size less than about 200.' Knudsen and Meier use an iterative procedure, each stage using multiple runs of the annealing algorithm. At each stage commonality between solutions is determined and then fixed for subsequent stages. They report obtaining solutions for various sizes including $(m, n) = (151, 167)$ far quicker (a factor of 180) than those reported earlier.

In both cases the cost function used is very *direct*. It is an obvious characterisation of what the search process is required to achieve. Direct cost functions are, however, not always the most effective. Examination of the way problem instances are generated reveals that small values of $w_i = (AV)_i$ are more likely than larger values. The initial distribution of $(AV)_i$ is (essentially) binomial, with values ranging (potentially) from $-n$ to n. The negation of particular matrix rows simply folds the distribution at 0. This causes difficulties for the search process since attempting to cause negative w_i to become positive by flipping the value of some element V_j is likely to cause various small but positive w_k to become negative. It is just too easy for the search to get stuck in such local optima. One solution is to encourage the w_i to assume values far from 0. This is easily effected: rather than punish when w_i is negative, punish when $w_i < K$ for some positive value K, i.e. use a cost function of the form

$$Cost(V') = g \sum_{i=1}^{m} (max\{K - (AV')_i, 0\})^R$$

The cost function is *a means to an end.* A 'good' cost function is one that 'works', i.e. one that guides the search to obtain desired results. Choice of cost function is a subtle matter and experience with many domains suggests that experimentation is essential. This explains the inclusion of R as a parameter. There would seem no a priori reason to restrict R to 1.0 (the value used by previous researchers), and the work reported below shows that higher values give very good results. Here g magnifies the effect of changes when the current solution is changed and is really intended as a weighting factor when the cost function is extended for the PPP problem (see Section 4).

By varying the parameters of this cost function quite radical improvements can be brought in effectiveness. Experiments were carried out for problem instances of sizes $(201, 217)$, $(401, 417)$, $(501, 517)$ and $(601, 617)$. For the $(201, 217)$ problems three values of K were used: $20, 15, 10$. For the $(401, 417)$ problems four values of K were used: $30, 25, 20, 15$. For the rest $K = 25$ was used. In the $(201, 217)$ and $(401, 417)$ cases $R = 2.0$ and in the others $R = 3.0$. A weighting value $g = 20$ was used throughout. For each configuration of parameters ten runs were carried out for each problem instance.

The results are shown in Tables 1 and 2. It was found that there were few direct simulated annealing solutions of the largest PP instances. However, it was often found that flipping a small number of annealing solution bits (e.g. 1, 2 or 3) provided a solution to the PP instance.

All $(201, 217)$ problem instances gave rise to some solution with Problem 0 being the most resilient (only three out of thirty annealing solutions gave rise to a PP solution and then only after three-bit enumerative search). Four of the ten $(401, 417)$ problems produced direct (0-bit search) simulated annealing solutions.

Table 1. Number of successes after simulated annealing plus {0,1,2,3}-bit hill-climbing for (201,217) and (401,417) instances

PP(201,217):30 Runs

	0	1	2	3		0	1	2	3
Pr 0	0	0	0	3	Pr 5	0	4	6	5
Pr 1	3	6	2	11	Pr 6	3	6	12	5
Pr 2	1	11	6	8	Pr 7	4	7	14	2
Pr 3	8	12	6	3	Pr 8	3	14	2	9
Pr 4	0	4	5	4	Pr 9	1	1	5	4

PP(401,417):40 Runs

	0	1	2		0	1	2
Pr 0	0	0	1	Pr 5	0	1	0
Pr 1	0	0	2	Pr 6	1	2	6
Pr 2	0	0	1	Pr 7	0	11	6
Pr 3	1	4	14	Pr 8	0	2	9
Pr 4	1	3	6	Pr 9	3	12	11

Table 2. Number of successes after simulated annealing plus {0,1,2,3}-bit hill-climbing for (501,517) and (601,617) instances

PP(501,517):10 Runs

	0	1	2	3		0	1	2	3
Pr 0	0	0	0	0	Pr 5	0	0	0	2
Pr 1	0	0	1	1	Pr 6	0	0	0	1
Pr 2	0	2	2	4	Pr 7	0	0	0	1
Pr 3	0	1	1	3	Pr 8	0	0	0	0
Pr 4	0	0	0	0	Pr 9	0	1	3	4

PP(601,617):10 Runs

	0	1	2	3		0	1	2	3
Pr 0	0	0	0	1	Pr 5	0	0	2	2
Pr 1	0	0	0	1	Pr 6	0	2	1	1
Pr 2	0	0	0	0	Pr 7	0	0	0	0
Pr 3	0	0	0	2	Pr 8	0	0	0	0
Pr 4	0	0	0	0	Pr 9	0	0	0	0

All problems were solved by some run followed by at most an enumerative 2-bit search. For the $(501, 517)$ problems seven produced a solution (with up to 3-bit search used). Half the $(601, 617)$ problems gave rise to a solution. No claim to optimality is made here. For the larger problem sizes only one cost function has been used and then with only ten runs for each problem. The results serve as a simple demonstration of how small changes may matter greatly. The use of cost functions whose minimisation does not lead to the required solution we term *Problem Warping.* The solutions obtained by the warping are highly correlated with the actual defining solution of the problem. For the $(201, 217)$ problems, the best solution over the 30 runs for each problem ranged from 79.2% – 87.1 % correct. For the $(401, 417)$, $(501, 517)$ and $(601, 617)$ problems, the ranges were 83.4 % – 87.5%, 80.6% – 86.4% and 77.5% – 86.1%. This is generally much better than solutions obtained using the standard cost function ($K = 0$ and $R = 1$).

That an enumerative search should be required to obtain PP solutions is not surprising. The cost functions used do not define what it means to be a solution to the PP and the annealing has attempted to solve the problem it was posed. However, the results show that the cost functions used do characterise in some way what it means to be 'close' to a PP solution. The enumerative search can be considered as a second stage optimisation with respect to the traditional cost function (i.e. $K = 0$). The authors would suggest a playful guideline for optimisation researchers in cryptography – if you cannot solve a problem, solve a different one. It might just help!

Application of Problem Warping has allowed instances of the Perceptron Problem with secret vectors **three times longer** than hitherto handled in the literature (an increase in secret state space from around 2^{200} to 2^{617}). This is a huge increase in power and stresses how fragile is current understanding of the power of heuristic optimisation for cryptanalysis (including our own). However, the real power of Problem Warping will be seen in the next section. Its real power lies in its application to the Permuted Perceptron Problem.

4 Attacking the Permuted Perceptron Problem

In 1999 Knudsen and Meier [8] showed that the $(m, n) = (101, 117)$ schemes recommended by Pointcheval were susceptible to a sophisticated attack based on an understanding of patterns in the results obtained during repeated runs of an annealing process. Essentially, their initial simulated annealing process is the standard one (with the number of trials at each temperature cycle equal to n) but with a modified cost function given by

$$Cost(V') = g \sum_{i=1}^{m} (max\{K - (AV')_i, 0\})^R + \sum_{i=1}^{n} (|H(i) - H'(i)|)^R$$

$H(k) = \#\{j : (AV)_j = k\}$, i.e. the number of the $w_i = (AV)_i$ that have value k. H is the reference histogram for the target solution V (i.e. the histogram of the values in AV). Similarly H' is the histogram for the current solution V'.

The histograms apply only to positive $(AV)_i$ elements. In all the experiments reported in [8] $R = 1.0$ and $K = 0$. There, repeated runs are carried out and commonality of the outputs from these runs is noted. Loosely speaking if all runs of the technique agree on certain secret element values there is a good chance that the agreed values are the correct ones. Agreed bits are fixed and the process carried out repeatedly until all bits agree. Unfortunately some (small number of) bits unanimously agreed in this way are actually wrong, and an enumerative search is made for these bits.

4.1 ClearBox Cryptanalysis – Looking inside the Box

Virtually all applications of optimisation techniques in cryptography view optimisation as a black box technique. A problem is served as input, the optimisation algorithm is applied, and some output is obtained (a candidate secret in the PP and PPP examples). However, in moving from starting solution to eventual solution the heuristic algorithm will have evaluated a cost function at many (possibly hundreds of) thousands of points. Each such evaluation is a source of information for the guidance process. In the black-box approach this information is simply thrown away. For the PPP, the information loss is huge.

As the temperature cools in an application of simulated annealing it becomes more difficult to accept worsening moves. At some stage an element will assume the value of 1 (or -1) and then never change for the rest of the search, i.e. it gets stuck at that value. It is found that some bits have a considerable tendency to get stuck earlier than others when annealing is applied. (Indeed this observation is at the root of Chardaire et al.'s variant of annealing known as thermo-statistical persistency [2].) One could ask 'Why?'. The answer is that the structure of the problem instance defined by the matrix and reference histogram exerts such influence as to cause this. The bits that get stuck early **tend to get stuck at the correct values.** Once a bit has got stuck at the wrong value it is inevitable that other bits will subsequently get stuck at wrong values too. However, it is unclear how many bits will get stuck at the right value before a wrong value is fixed. This has been investigated for various problem sizes and cost functions. Three problem sizes were considered as shown in Table 3. For each problem size a cost function is defined by a value of g, a value of K and a value of R. Thirty problem instances were created for each problem size. For each problem and each cost function ten runs of the annealing process were carried out. The runs were assessed on two criteria: number of bits set correctly in the final solution and number of bits initially stuck correctly before a bit became stuck at an incorrect value.

Table 3. Cost function parameter values. All combinations of g, K and R were used.

(m,n)	Values of g_1	Values of K	Values of R
(101,117)	20,10,5	1,3,5,7,9,11,13,15	2,1.5,1
(131,147)	20,10	7, 10, 13, 16	2,1
(151,167)	20,15,10,5	5, 10, 15, 20	2,1

Thus, for (101, 117) instances there were $3 \times 8 \times 3 = 72$ cost functions and so 720 runs in total for each problem. The results are shown in Table 4. For each problem the maximum number of correctly set bits in a final solution (i.e. the final result of an annealing run) is recorded together with the maximum number bits fixed correctly in a solution before a bit was set incorrectly (usually these will not be simultaneously achieved by one single solution).

Table 4. Maximum final bits correct (FBC) and maximum initial bits correct (IBC) over all runs. Total number of runs shown for each problem size. Thirty problem instances were attacked for each problem size.

Prob	FBC	IBC	Prob	FBC	IBC	Prob	FBC	IBC
Pr 0	102	50	Pr 0	126	42	Pr 0	148	72
Pr 1	100	45	Pr 1	135	68	Pr 1	142	64
Pr 2	103	45	Pr 2	128	64	Pr 2	145	66
Pr 3	99	53	Pr 3	126	672	Pr 3	157	88
Pr 4	101	46	Pr 4	130	39	Pr 4	147	58
Pr 5	108	72	Pr 5	131	70	Pr 5	140	67
Pr 6	99	39	Pr 6	126	47	Pr 6	151	86
Pr 7	101	56	Pr 7	128	56	Pr 7	135	48
Pr 8	104	55	Pr 8	123	52	Pr 8	143	55
Pr 9	106	56	Pr 9	139	75	Pr 9	150	95
Pr 10	102	56	Pr 10	129	51	Pr 10	149	61
Pr 11	107	56	Pr 11	123	48	Pr 11	145	70
Pr 12	101	58	Pr 12	134	57	Pr 12	143	49
Pr 13	104	42	Pr 13	132	62	Pr 13	138	63
Pr 14	102	47	Pr 14	124	37	Pr 14	147	58
Pr 15	102	56	Pr 15	122	59	Pr 15	141	63
Pr 16	101	39	Pr 16	124	41	Pr 16	151	56
Pr 17	103	51	Pr 17	121	42	Pr 17	144	82
Pr 18	103	40	Pr 18	130	62	Pr 18	147	98
Pr 19	103	50	Pr 19	129	53	Pr 19	137	47
Pr 20	105	62	Pr 20	132	67	Pr 20	136	69
Pr 21	107	68	Pr 21	128	59	Pr 21	140	59
Pr 22	106	58	Pr 22	129	97	Pr 22	142	55
Pr 23	103	62	Pr 23	127	61	Pr 23	146	67
Pr 24	103	53	Pr 24	126	43	Pr 24	138	69
Pr 25	100	56	Pr 25	127	72	Pr 25	147	69
Pr 26	104	51	Pr 26	132	44	Pr 26	145	61
Pr 27	98	53	Pr 27	125	68	Pr 27	146	68
Pr 28	105	57	Pr 28	126	38	Pr 28	141	64
Pr 29	103	56	Pr 29	123	50	Pr 29	143	80
Size (101,117) 720 runs			Size (131,147) 160 runs			Size (151,167) 320 runs		

4.2 Making Best Use of Available Information

Consider Ax for any solution vector x. Flipping any single element of x causes the components $(Ax)_i$ to change by ± 2. Similarly, flipping any two bits of x causes the components to change by ± 4 or else stay the same. Flipping three bits causes the components to change by ± 2 or ± 6. Generalising, if x may be transformed into the secret generating solution V by changing an even number of bits, then $(Ax)_i = (AV)_i \pm 4k$ for some integer k. Similarly, if an odd number of bit changes are needed then $(Ax)_i = (AV)_i \pm 4k + 2$. For any x let

$$SUMA(x) = \#\{i : (Ax)_i = 4k + 1, for\ some\ k\}$$

$$SUMB(x) = \#\{i : (Ax)_i = 4k + 3, for\ some\ k\}$$

$SUMA(V) = H(1) + H(5) + \ldots$ and $SUMB(V) = H(3) + H(7) + \ldots$ where H is the publicly available reference histogram. If V is obtained from x by an even number of bit changes, then we have $SUMA(V) = SUMA(x)$ and also $SUMB(V) = SUMB(x)$. If V is obtained from x by an odd number of bit changes, then $SUMA(V) = SUMB(x)$ and $SUMB(V) = SUMA(x)$. Only one of $SUMA(V)$ and $SUMB(V)$ can be odd (since their sum, n, is odd). Thus, for any vector x it is possible to determine whether it differs from V by an even or odd number of bits using the respective values of SUMA(x) and SUMA(V).

Suppose V is the actual secret and x is a solution obtained by annealing. If x is a high performing solution (with few bits wrong) then $(Ax)_i$ will typically be very close to $(AV)_i$. For the (101,117) problem instances, if $(Ax)_i = 1$ then the average actual value of $(AV)_i$ was 6.02. For (131,147) and (151,167) instances the averages were 6.23 and 6.46.

Suppose that $(Ax)_i = 1$ and ten bits are wrong. Typically it will be the case that $(AV)_i \in \{1, 5, 9, 13\}$. This observation has a big impact on enumerative search. For the sake of argument suppose that $(Ax)_i = (AV)_i = 1$. Then flipping the ten wrong bit values to obtain the actual secret must have no effect on the resulting value of $(Ax)_i$. This means that for five wrong bits we must have $a_{ij}x_j = 1$ and for the other five we must have $a_{ij}x_j = -1$. This reduces any enumerative search. For example, searching over 117 bits would usually require C^{117}_{10} (around 4.4×10^{15}) but now requires a search of order around $C^{58}_5 \times C^{57}_5$ (around 2.1×10^{13}). This assumes that for solution x $\#\{x_j : a_{ij}x_j = 1\} = 58$ and $\#\{x_j : a_{ij}x_j = -1\} = 57$ (or vice versa). In practice, this may not be the case but any skew actually reduces the complexity of the search. In this respect, it may be computationally advantageous to consider some $(Ax)_i < 0$. For example, if $(Ax)_i = -7$ and there are 10 bits wrong then $(AV)_i$ must be in the range 1..13 with the smaller values much more likely. If $(AV)_i = 1$ then there must be seven wrong bits currently with $a_{ij}x_j = -1$ and three with $a_{ij}x_j = 1$. This is a powerful mechanism that will be used repeatedly.

One has to guess the relationship of $(Ax)_i$ to $(AV)_i$. This will generally add only a factor of about four to the search (and often less). One has also to determine how many bits are actually wrong too. One can start by assuming that the solution vector has the minimum number of bits wrong yet witnessed and engage in enumerative searches. If these fail, simply increment the number of bits assumed incorrect by 2 and repeat the search processes (only even numbers or odd numbers of wrong bits need be considered). The complexity of the search is dominated by the actual number of wrong bits (searches assuming fewer numbers of wrong bits are trivial by comparison). The complexities reported in this paper therefore assume knowledge of the number of wrong bits in the current solution.

4.3 The Direct Attack

It is obvious that 'warping' the cost function produces results that are indeed better than those obtained under the natural cost function. Thus, in the $(101, 117)$ problems three (5, 11 and 22) have given rise to solutions with 10 bits or fewer wrong (from the FBC column of Table 4). Once the highest performing solution has been selected (a factor of 720) an enumerative search of order $C_5^{58} \times C_5^{57}$ (which is less than 2^{45}) will find the solution in these cases. For the $(131, 147)$ and $(151, 167)$ instances extreme results are also occasionally produced. (131,147) Problem 9 gave rise to one solution with only 8 bits wrong. (151,167) Problem 3 similarly gave rise to a solution with only 10 bits wrong. This would require a total search of approximately $320 \times C_5^{84} \times C_5^{83}$ which is less than 2^{60}. Thus, even a fairly brutal search will suffice on occasion, even for the biggest sizes. This is not the most efficient way of solving the problem however.

4.4 Timing Supported Attack

The largest number of initially settled bits can clearly leak a huge amount of information. For (151,167) problems 18, 9 and 3 some solution was obtained whose first 98, 95,88 initially stuck bits were correct. The respective complexities of brute force search over the remaining entries would be of order $2^{69}, 2^{72}, 2^{79}$. Although not within the traditional 2^{64} distance they are sufficiently close to render use of the PPP scheme impossible. For (131,147) there would appear to be an outlier problem 22 with 97 initial bits correct. This leaves a search of order 2^{50}.

Another approach would be to consider in turn all possible pairs of solutions obtained. One pair contains a solution VMAX with the maximum number of bits correct and a solution VINIT with the maximum number of initial bits correct. This pair could form the basis for the subsequent search and we can calculate the computational complexity of finding the exact solution. Obtaining this pair requires a search factor equal to the number of runs squared.

Assume that at least the first I bits initially fixed in VINIT are correct. Change the corresponding bits in VMAX to agree with those in VINIT. These bits are now excluded from the subsequent search – the search will be over the remaining $n - I$ bits of the modified VMAX. For example, suppose in the $(101, 117)$ case that the best initial solution provides us with at least 37 bits (from Table 4 this applies to all 30 problems). This leaves us with 80 bits over which to conduct the remaining search. Suppose ten wrong bits remain. The total complexity of the whole search is now approximately

$$720 \times 720 \times C_5^{40} \times C_5^{40} = 2.24 \times 2^{57}$$

Enumerative searches can be performed under optimistic assumptions and these can be progressively relaxed leading to more complex searches. Assuming the number of initially set bits is known, and the number of bits wrong in the best final solution is known, the complexities of the searches are given in Table 5. We have given some very conservative attacks. Some cost functions are clearly more effective than others and it would be possible to restrict attention to a subset of the set consider so far. For the (151,167) problem size Tables 6 and 7 give summary results for each problem instance (over all cost functions) and for each cost function (over all problem instances).

Table 5. Search complexities (log 2) of Timing Supported Attacks on (101,117)(upper), (131,147)(middle) and (151,167)(lower) size schemes. Thirty problem instances at each size.

66	71	65	69	69	47	75	65	60	56
63	54	64	64	67	63	71	64	67	64
56	51	56	59	63	66	62	70	58	62
85	59	74	76	77	67	84	77	87	47
76	88	64	67	91	85	89	94	71	76
65	76	57	77	85	72	71	77	86	88
80	94	88	56	86	95	70	111	95	68
81	86	97	100	86	96	78	83	72	108
99	99	97	86	97	83	89	85	95	85

4.5 Other Attacks

Other attacks are possible. For example taking the majority vector over all solution runs (whatever the cost function) can on occasion leak a great deal of information. Commonality of solution elements of repeated runs is at the heart of Knudsen and Meier's technique. This strategy can be adopted here. If runs agree under widespread problem deformation (i.e. using multiple cost functions) then there is often good cause to believe they agree correctly. Rather than insist on absolute agreement, we can rank the secret bits according to the degree of agreement. Frequently the top ranked bits are correct though this method is somewhat erratic. Table 8 shows the number of top ranked bits correct for each (151,167) problem. Thus, for problem 1 the 41 bits that gave rise to most agreement over all runs (the 320 runs indicated in Table 4) were actually correct. We can see that for problems 3, 9 and 18 the most agreed 78, 87 and 88 bits were correct (in the sense that the majority vector is right). This is very significant since around half of all bits are revealed without any kind of enumerative search being deployed.

Table 6. Summary results over all cost functions for the (151,167) problem instances. For each problem instance and cost function the minimum final bits correct, the average final bits correct and the maximum final bits correct over the ten annealing runs were calculated. For each problem instance columns 2–4 record the averages of such results over all cost functions. Columns 5–7 record similar information for the initial bits correct.

Problem	Final Bits Correct			Initial Bits Correct		
	Av.Min	Over.Av	Av.Max	Av.Min	Over.Av	Av.Max
Prob 0	130.84	135.82	140.34	6.84	22.06	43.81
Prob 1	125.66	130.25	136.12	7.41	23.68	43.28
Prob 2	129.31	135.24	140.94	8.78	25.78	47.59
Prob 3	133.16	140.88	147.19	23.72	44.34	66.81
Prob 4	127.53	132.96	138.44	9	24.3	41.25
Prob 5	128.22	132	135.72	9.66	27.13	47.22
Prob 6	136.06	140.78	145.22	12.91	31.31	53.69
Prob 7	116.94	123.18	129.22	8.28	21.33	35.56
Prob 8	129.56	133.32	137.22	5.16	16.9	33.09
Prob 9	135.34	139.48	143.84	25.03	49.25	73
Prob 10	127.34	132.09	136.94	3.84	16.82	34.78
Prob 11	127.91	135.78	141.53	23.66	39.82	57.62
Prob 12	124.69	131.02	137.34	6.19	20.69	35.25
Prob 13	122.69	127.63	132.44	10.66	25.44	42.16
Prob 14	127.91	132.17	137.12	9.91	25.47	43.5
Prob 15	125.25	130.67	134.91	5.59	18.6	35.62
Prob 16	132.75	139.54	145.41	3.84	17.9	38.12
Prob 17	125.91	131.04	136.38	24.12	42.57	61.72
Prob 18	133.78	138.31	143.12	30.06	53.65	72.81
Prob 19	122.94	127.63	132.25	1.09	11.93	28.03
Prob 20	122.16	126.53	131.56	10.25	26.03	43.94
Prob 21	126.62	131.68	136.69	8.72	24.37	43.34
Prob 22	123.28	129.99	135.91	4.06	17.48	36.09
Prob 23	127.34	133.53	139.31	7.25	21.68	40.88
Prob 24	120.28	126.09	131.94	16.78	35.12	54.59
Prob 25	130.16	136.05	140.38	8	25.65	46.06
Prob 26	134.09	138.25	141.94	6.84	23.44	45.97
Prob 27	131.22	136.85	142.19	5.97	23.45	46.22
Prob 28	119.03	125.63	133.38	6.19	20.47	37.62
Prob 29	129.12	134.59	139.47	18.25	37.94	59.44

It is also possible to add up the sticking times of components over all runs. When these are ranked (the highest being the one that took least aggregate time to get stuck) the results can also leak information. In some cases only 1 bit of the first ranked 100 for the (151,167) gave rise to a majority vector component that was incorrect.

Table 7. Summary results for each cost function over all (151,167) problem instances. For each problem instance and cost function the minimum final bits correct, the average final bits correct and the maximum final bits correct over the ten annealing runs were calculated. For each cost function columns 2–4 record the averages of such results over all problem instances. Columns 5–7 record similar information for the initial bits correct.

Parameters	Final Bits Correct			Initial Bits Correct		
(K,g,R)	Av.Min	Over.Av	Av.Max	Av.Min	Over.Av	Av.Max
(20,20,2)	132.47	136.02	139.23	12.9	29.25	49.97
(20,20,1)	129.93	132.71	135.07	8.47	24.13	40.73
(20,15,2)	132.17	135.82	138.9	14.8	30.57	49.27
(20,15,1)	130.03	132.49	135	9.33	24.86	43.27
(20,10,2)	132.6	135.74	138.57	12.77	29.54	49.1
(20,10,1)	129.97	132.37	134.53	10.3	24.3	41.8
(20,5,2)	132.13	135.8	138.9	13.7	30.71	49.33
(20,5,1)	129.73	132.29	135.07	8.47	24.44	41.97
(15,20,2)	129.7	135.13	139.63	12.23	29.67	48.1
(15,20,1)	129.57	133.67	137.6	10.57	26.64	45.63
(15,15,2)	130.87	135.32	140.1	12.37	30.43	50.7
(15,15,1)	130.1	133.76	137.33	12.1	28.09	46.3
(15,10,2)	129.63	135.15	140.1	12.53	30.85	49.83
(15,10,1)	129.23	133.57	137.3	12.83	29.77	47.9
(15,5,2)	130.5	135.49	140.13	14.9	31.16	51.67
(15,5,1)	129.63	133.67	137.3	10.6	27.61	47.07
(10,20,2)	126.67	133.01	140.23	10.53	28.56	48.87
(10,20,1)	128	134.06	139.47	11.27	28.89	50.6
(10,15,2)	126.6	133.39	140.1	11.53	29.25	47.23
(10,15,1)	128.87	134.17	138.9	12.4	29.28	49.47
(10,10,2)	126.93	133.54	139.73	10.67	28.06	49.3
(10,10,1)	128.57	133.91	139.13	14.27	29.95	47.03
(10,5,2)	126.6	133.43	140.53	12.63	28.5	47.73
(10,5,1)	128.8	134.2	139.73	12.77	29.48	49.27
(5,20,2)	120.4	128.16	135.97	7.23	21.24	38.93
(5,20,1)	122	130.1	138.77	8.87	23.74	43.23
(5,15,2)	120.27	129.18	137	7.97	23.81	44
(5,15,1)	122.5	130.04	137.53	7.47	23.73	43.27
(5,10,2)	121.37	129.17	137.03	7.57	22.54	42.1
(5,10,1)	122.8	130.13	137.2	8.9	24.15	44.6
(5,5,2)	121.23	129.19	136.87	8.13	21.95	40.67
(5,5,1)	122.37	130.22	137.77	8.87	23.75	42.77

Table 8. Top N agreed correct for problem instances of size (151,167)

Probs 0–9	0	41	0	78	0	30	0	0	0	87
Probs 10–19	0	41	0	0	0	39	0	62	88	0
Probs 20–29	0	36	0	0	36	32	33	36	28	46

5 Summary and Conclusions

We have demonstrated that recent fault injection and timing side channel attacks on cryptosystems may be interpreted in the context of optimisation-based search. The attacks on PP and PPP problems have shown the potential power of such interpretations. We make no claims to optimality for our results; extensive experimentation and profiling would allow more effective and efficient sets of cost functions to be determined.

Virtually all work using optimisation techniques attempts to solve the problem at hand in one go. We believe we should stop expecting optimisation to solve problems in this way, and view optimisation as a means of creating data ('failed' solutions) on which to do cryptanalysis! This suggests a new form of cryptanalysis – the profiling and interpretation of local optima obtained by optimisation-based searches. The authors are currently investigating the use of such techniques to block and public key algorithms. We recommend the area to researchers.

Acknowledgements

The authors would like to thank Susan Stepney for comments on a previous draft. The authors would also like to thank the anonymous Eurocrypt 2002 referees for their comments.

References

1. D. Boneh, R. A. DeMillo, and R. J. Lipton. On the importance of checking cryptographic protocols for faults (extended abstract). In Walter Fumy, editor, *Advances in Cryptology - EuroCrypt '97*, pages 37–51, Berlin, 1997. Springer-Verlag. Lecture Notes in Computer Science Volume 1233.
2. P Chardaire, J C Lutton, and A Sutter. Thermostatistical persistency: a powerful improving concept for simulated annealing. *European Journal of Operations Research*, 86:565–579, 1995.
3. A. Fiat and A. Shamir. How to Prove Yourself:Practical Solutions of Identification and Signature Problems. In Ed Dawson, Andrew Clark, and Colin Boyd, editors, *Advances in Cryptology – Crypto '86*, pages 186–194. Springer Verlag LNCS 263, july 1987.
4. D.E. Goldberg. *Genetic Algorithms in Search, Optimization and Machine Learning.* Addison-Wesley, 1989.

5. S. Goldwasser, S. Micali, and C. Rackoff. Knowledge Complexity of Identification Proof Schemes. In *17th ACM Symposium on the Theory of Computing STOC*, pages 291–304. SACM, 1985.
6. Giddy J.P. and Safavi-Naini R. Automated Cryptanalysis of Transposition Ciphers. *The Computer Journal*, XVII(4), 1994.
7. S. Kirkpatrick, Jr. C. D. Gelatt, and M. P. Vecchi. Optimization by simulated annealing. *Science*, 220(4598):671–680, May 1983.
8. Lars R. Knudsen and Willi Meier. Cryptanalysis of an Identification Scheme Based on the Permuted Perceptron Problem. In *Advances in Cryptology Eurocrypt '99*, pages 363–374. Springer Verlag LNCS 1592, 1999.
9. P. C. Kocher. Timing attacks on implementations of Diffie-Hellman, RSA, DSS, and other systems. In Neal Koblitz, editor, *Advances in Cryptology - Crypto '96*, pages 104–113, Berlin, 1996. Springer-Verlag. Lecture Notes in Computer Science Volume 1109.
10. Robert A J Mathews. The Use of Genetic Algorithms in Cryptanalysis. *Cryptologia*, XVII(2):187–201, April 1993.
11. David Pointcheval. A New Identification Scheme Based on the Perceptron Problems. In *Advances in Cryptology Eurocrypt '95*. Springer Verlag LNCS X, 1995.
12. A. Shamir. An Efficient Scheme Based On Permuted Kernels. In *Advances in Cryptology – Crypto '89*, pages 606–609. Springer Verlag LNCS 435, 1997.
13. Richard Spillman, Mark Janssen, Bob Nelson, and Martin Kepner. Use of A Genetic Algorithm in the Cryptanalysis of Simple Substitution Ciphers. *Cryptologia*, XVII(1):187–201, April 1993.
14. Jaques Stern. A New Identification Scheme Based On Syndrome Decoding. In *Advances in Cryptology – Crypto '93*, pages 13–21. Springer Verlag LNCS 773, 1997.
15. Jaques Stern. Designing Identification Schemes with Keys of Short Size. In *Crypto '93*, pages 164–173. Springer Verlag LNCS 839, 1997.
16. Forsyth W.S. and Safavi-Naini R. Automated Cryptanalysis of Substitution Ciphers. *Cryptologia*, XVII(4):407–418, 1993.

Speeding Up Point Multiplication on Hyperelliptic Curves with Efficiently-Computable Endomorphisms

Young-Ho Park[1]*, Sangtae Jeong[2], and Jongin Lim[3]

[1] Dept. of Information Security & System, Sejong Cyber Univ., Seoul, Korea,
youngho@cist.korea.ac.kr
[2] Dept. of Math., Seoul National Univ., Seoul, Korea,
stj@math.snu.ac.kr
[3] CIST, Korea Univ., Seoul, Korea,
jilim@korea.ac.kr

Abstract. As Koblitz curves were generalized to hyperelliptic Koblitz curves for faster point multiplication by Günter, *et al.* [10] we extend the recent work of Gallant, *et al.* [8] to hyperelliptic curves. So the extended method for speeding point multiplication applies to a much larger family of hyperelliptic curves over finite fields that have efficiently-computable endomorphisms. For this special family of curves, a speedup of up to 55 (59) % can be achieved over the best general methods for a 160-bit point multiplication in case of genus g =2 (3).

1 Introduction

The dominant cost operation in protocols based on the discrete logarithm problem on the Jacobians of hyperelliptic curves is point multiplication by an integer k, namely computing kD for a point D on the Jacobian. To speed up the main operation, a variety of techniques are now being in use by considering relevant objects involving curves and underlying base fields. Among other things, Koblitz [12] proposed the use of a certain family of elliptic curves, say Koblitz curves. These curves are ones defined over the binary field but considered over a suitably large extension field, with the advantage that point counting can be easily done with the help of the Frobenius endomorphism. Along this idea, Meier and Staffelbach [15], Müller [18], Smart [24], and Solinas [26,27] have thoroughly investigated elliptic curves defined over small finite fields. In addition, the idea of Koblitz curves was generalized to hyperelliptic curves of genus 2 by Günter, Lange and Stein [10]. We also refer the reader to [14] for a detailed investigation on hyperelliptic Koblitz curves of small genus defined over small base fields.

Recently, another improvement on faster point multiplication was carried out by Gallant, Lambert, and Vanstone [8] whose method is applicable to a family of elliptic curves having efficiently-computable endomorphisms. Their idea is to

* This work was supported by Korea Research Foundation Grant(KRF-01-003-D00012) and the second author acknowledges the support of the Brain Korea 21

L.R. Knudsen (Ed.): EUROCRYPT 2002, LNCS 2332, pp. 197–208, 2002.

decompose an integer k modulo n into two components whose bit-lengths are half that of k. A precise analysis of their method showed that a speedup of up to 50% could be achieved over the best general methods for a 160-bit point multiplication.

The purpose of this paper is to extend the method of Gallant,*et al.* [8] to the hyperelliptic setting. As is the case with elliptic curves, the extended method applies to a family of hyperellliptic curves having efficiently-computable endomorphisms since they also induce such endomorphisms on the Jacobians of the hyperellliptic curves. So what should be done here is to decompose an integer k modulo n into d components whose bit-lengths are $1/d$ that of n, where d is the degree of the characteristic polynomial of an efficiently-computable endomorphism on the Jacobian. Simultaneous multiple point multiplication then yields a significant speedup because of reduced bitlengths. A precise analysis shows that a speedup of up to 55 (59) % can be achieved over the best general methods for a 160-bit point multiplication when genus g =2 (3). The problem with this method is how efficiently a randomly chosen k can be decomposed into a sum of the required form. To resolve this problem we give two efficient algorithms for decomposing k. One method is a generalization of Gallant,*et al.* [8]. The other is an extension of the method developed in [20].

The rest of the paper is organized as follows. In Section 2 we shall briefly summarize some basics on the Jacobians of hyperelliptic curves. In Section 3, we list up a collection of hyperellipic curves with efficient endomorphisms and provide the characteristic polynomials of such endomorphisms. Section 4 contains how to use such endomorphisms for decomposing k and there we apply known simultaneous exponentiation methods to the hyperelliptic curves and compare them. In Section 5 we generalize two decomposing methods to hyperelliptic curves. For security considerations, in Section 6, we touch on all known attacks to the DLP on hyperelliptic curves. The final Section contains our conclusions to the present work.

2 Preliminaries

2.1 Jacobians of Hyperelliptic Curves

We begin by introducing basic facts on the Jacobians of hyperelliptic curves over finite fields. Let $\mathbb{F}_q$ be a finite field of q elements and let $\overline{\mathbb{F}}_q$ denote its algebraic closure. A hyperelliptic curve of genus g over $\mathbb{F}_q$ is given by the Weierstrass equation of the form

$$X : y^2 + h(x)y = f(x) \tag{1}$$

where $h \in \mathbb{F}_q[x]$ is a polynomial of degree at most g and $f(x) \in \mathbb{F}_q[x]$ is a monic polynomial of degree $2g+1$. Let $\mathbb{K}$ be an extension field of $\mathbb{F}_q$ in $\overline{\mathbb{F}}_q$. The set of $\mathbb{K}$-rational points on X consists of $\mathbb{K}$-solutions to the equation of X together with the point at infinity, denoted ∞.

In this Section we only mention (reduced) representations of elements on the Jacobian of a hyperelliptic curve X. We recommend the reader to consult an appendix in [13] for more details on the Jacobians. Indeed, the Jacobian of X

defined over $\mathbb{K}$, denoted $\mathbb{J}_X(\mathbb{K})$ is defined as the subgroup of $\mathbb{J}_X(\overline{\mathbb{F}})$ fixed by the Galois group $\mathrm{Aut}(\overline{\mathbb{F}}/\mathbb{K})$. It is well known by the Riemann-Roch theorem that every divisor D of degree 0 on X can be uniquely represented as an equivalence class in $\mathbb{J}_X(\mathbb{K})$ by a reduced divisor of the form $\sum m_i P_i - (\sum m_i)\mathcal{O}$ with $\sum m_i \leq g$. Thus, every element D of the Jacobian can be uniquely represented by a pair of polynomials $a, b \in \mathbb{K}[x]$ for which $deg(b) < deg(a) \leq g$, and $b(x)^2 + h(x)b(x) - f(x)$ is divisible by $a(x)$. Indeed, D is the equivalence class of the g.c.d. of the divisors of the functions $a(x)$ and $b(x) - y$. The element of $\mathbb{J}_X(\mathbb{K})$ will usually be abbreviated to $[a(x), b(x)] := \mathrm{div}(a, b)$.

As for addition in the Jacobian, it can be performed explicitly with Cantor's algorithm [4]. Here we do not go into details on the algorithms for composition and reduction but mention only the complexity of the generic operations in the Jacobian. Since operations in the Jacobian can be carried out using arithmetic in $\mathbb{K}[x]$, the generic addition needs $17g^2 + O(g)$ operations in $\mathbb{K}$ whereas doubling takes $16g^2 + O(g)$ field operations (see [28]). Another remark to complexity is that an inversion can be done for free, since the opposite of $D = [a(x), b(x)]$ is given by $-D = [(a(x), -h(x) - b(x)]$.

2.2 Counting Group Order of Jacobians

For cryptographic purposes, it is essentially necessary to know the group order of the Jacobian of a hyperelliptic curve in designing public schemes. Computing the group order of the Jacobian is believed to be a computationally hard task because it involves counting the number of rational points of a given hyperelliptic curve over an extension field of a base field of degree up to genus. However, it is rather easy to compute the group order of the Jacobians of hyperelliptic curves with extra properties such as complex multiplication. For example, Bulher and Koblitz [3] considered an especially simple family of hyperelliptic curves of genus g of the form $y^2 + y = x^{2g+1}$ defined over the field $\mathbb{F}_p$ of p elements such that $2g + 1$ is a prime < 9. They gave a procedure to determine the group order of the Jacobians of such curves by simply evaluating a Jacobi sum associated to a certain character. We mention here that these curves have efficiently-computable endomorphisms with which we can speed up point multiplication on the Jacobians(see Ex.5 in Section 3).

3 Hyperelliptic Curves with Efficient Endomorphisms

In this Section we first collect a family of hyperelliptic curves of genus g over $\mathbb{F}_q$ that have efficiently-computable endomorphisms ϕ. They also induce efficient endomorphisms on the Jacobians and then we compute their characteristic polynomial. Since every element of the Jacobian $\mathbb{J}_X(\mathbb{F}_q)$ can be uniquely represented by a reduced divisor with at most g points, we can explicitly give induced endomorphisms, denoted ϕ also, on reduced divisors to see that they can be efficiently computed. For simplicity, we assume for once and all that $\phi(\infty) = \infty$ for any morphism ϕ involved and that ζ_m is a primitive mth root of unity in the prime field $\mathbb{F}_p$ of p elements.

Example 1.([14]) Let X_1 be a hyperelliptic curve over $\mathbb{F}_q$ given by (1). The q-th power map, called the Frobenius, $\Phi : X_1 \rightarrow X_1$ defined by $(x, y) \rightarrow (x^q, y^q)$ then induces an endomorphism on the Jacobian. We note that the Frobenius can be computed with no further costly arithmetic over $\mathbb{F}_{q^n}$ because for a given divisor D, computing $\Phi(D)$ is just reduced to cyclic shifting provided that an extension field $\mathbb{F}_{q^n}$ is represented with respect to a normal basis. Indeed, it is computed by at most $2g$ cyclic shiftings. The characteristic polynomial of the Frobenius Φ is given by

$$P(t) = t^{2g} + a_1 t^{2g-1} + \cdots + a_g T^g + q a_{g-1} t^{g-1} + \cdots + q^{g-1} a_1 t + q^g,$$

where $a_0 = 1$, and $i a_i = S_i a_0 + S_{i-1} a_1 + \cdots + S_1 a_{i-1}$ for $S_i := N_i - (q^i + 1), 1 \leq i \leq g$ and $N_i = |X_1(\mathbb{F}_{q^i})|$.

Example 2. Let $p \equiv 1 \pmod 4$. Consider the hyperelliptic curve X_2 of genus g over the field $\mathbb{F}_p$ defined by

$$X_2 : y^2 = x^{2g+1} + a_{2g-1} x^{2g-1} + \cdots a_3 x^3 + a_1 x.$$

Then the morphism ϕ on X_2 defined by $P = (x, y) \mapsto \phi(P) := (-x, \zeta_4 y)$ induces an efficient endomorphism on the Jacobian. The characteristic polynomial of ϕ on the Jacobian is given by $P(t) = t^2 + 1$. The defining formulae for ϕ on the Jacobian are given by

$$\begin{aligned} \phi : [x^2 + a_1 x + a_0, b_1 x + b_0] &\mapsto [x^2 - a_1 x + a_0, -\zeta_4 b_1 x + \zeta_4 b_0] \\ [x + a_0, b_0] &\mapsto [x - a_0, \zeta_4 b_0] \\ 0 &\mapsto 0. \end{aligned}$$

We notice that ϕ can be easily computed using at most 2 field operations in $\mathbb{F}_p$, and the Jacobian has an automorphism of order 4, which follows from the composition of ϕ with the hyperelliptic involution.

Example 3. Let $p \equiv 1 \pmod 8$. Consider the hyperelliptic curve X_3 of genus 2 over the field $\mathbb{F}_p$ defined by

$$X_3 : y^2 = x^5 + ax.$$

Then the morphism ϕ on X_3 defined by $P = (x, y) \mapsto \phi(P) := (\zeta_8^2 x, \zeta_8 y)$ induces an efficient endomorphism. The characteristic polynomial of ϕ is given by $P(t) = t^4 + 1$.

The formulae for ϕ on the Jacobian are given by

$$\begin{aligned} \phi : [x^2 + a_1 x + a_0, b_1 x + b_0] &\mapsto [x^2 + \zeta_8^2 a_1 x + \zeta_8^4 a_0, \zeta_8^{-1} b_1 x + \zeta_8 b_0] \\ [x + a_0, b_0] &\mapsto [x + \zeta_8^2 a_0, \zeta_8 b_0] \\ 0 &\mapsto 0. \end{aligned}$$

It is easily seen that ϕ can be computed using at most 4 field operations in $\mathbb{F}_p$, and the Jacobian has an automorphism of order 8.

Example 4. Let $p \equiv 1 \pmod{12}$. Consider the hyperelliptic curve X_4 of genus 3 over the field $\mathbb{F}_p$ defined by

$$X_4 : y^2 = x^7 + ax.$$

Then the morphism ϕ on X_4 defined by $P = (x, y) \mapsto \phi(P) := (\zeta_{12}^2 x, \zeta_{12} y)$ induces an efficient endomorphism on the Jacobian as follows .

$$\phi : [x^3 + a_2x^2 + a_1x + a_0, b_2x^2 + b_1x + b_0] \mapsto$$
$$[x^3 + \zeta_{12}^2 a_2x^2 + \zeta_{12}^4 a_1x + \zeta_{12}^6 a_0, \zeta_{12}^{-3} b_2x^2 + \zeta_{12}^{-1} b_1x + \zeta_{12} b_0]$$
$$[x^2 + a_1x + a_0, b_1x + b_0] \mapsto [x^2 + \zeta_{12}^2 a_1x + \zeta_{12}^4 a_0, \zeta_{12}^{-1} b_1x + \zeta_{12} b_0]$$
$$[x + a_0, b_0] \mapsto [x + \zeta_{12}^2 a_0, \zeta_{12} b_0]$$
$$0 \mapsto 0.$$

It is easily seen that ϕ can be obtained using at most 6 field operations in $\mathbb{F}_p$, and the Jacobian has an automorphism of order 12. The characteristic polynomial of ϕ is given by $P(t) = t^4 - t^2 + 1$.

Example 5.([5],[3]) Let $m = 2g + 1$ be an odd prime and let $p \equiv 1 \pmod m$. Consider the hyperelliptic curve X_5 of genus g over the field $\mathbb{F}_p$ defined by

$$X_5 \colon y^2 = x^m + a.$$

The morphism ϕ defined by $P = (x, y) \mapsto \phi(P) := (\zeta_m x, y)$ induces an efficient endomorphism on the Jaconbian. It is easily seen that ϕ can be obtained using at most $2g - 1$ field operations in $\mathbb{F}_p$, and the Jacobian has an automorphism of order $2m$. The defining morphism for the action by ζ_m on the Jacobian is left to the reader. The characteristic polynomial of ϕ is given by $P(t) = t^{2g} + t^{2g-1} + \cdots + t + 1$.

4 Using an Efficient Endomorphism and Simultaneous Multi-exponentiation

4.1 Using an Efficient Endomorphism

Let X be a hyperelliptic curve over $\mathbb{F}_q$ having an efficiently-computable endomorphism ϕ on the Jacobian, $\mathbb{J}_X(\mathbb{F}_q)$. Let $D = [a(x), b(x)] \in \mathbb{J}_X(\mathbb{F}_q)$ be a reduced divisor of a large prime order n. The endomorphism ϕ acts as a multiplication map by λ on the subgroup $< D >$ of $\mathbb{J}_X(\mathbb{F}_q)$ where λ is a root of the characteristic polynomial $P(t)$ of ϕ modulo n. In what follows, let d denote the degree of the characteristic polynomial $P(t)$.

The problem we consider now is that of computing kD for k selected randomly from the range $[1, n-1]$. Suppose that one can write

$$k = k_0 + k_1\lambda + \cdots + k_{d-1}\lambda^{d-1} \pmod n, \tag{2}$$

where $k_i \approx n^{1/d}$. Then we compute

$$\begin{aligned} kD &= (k_0 + k_1\lambda + \cdots + k_{d-1}\lambda^{d-1})D \\ &= k_0D + k_1\lambda D + \cdots + k_{d-1}\lambda^{d-1}D \\ &= k_0D + k_1\phi(D) + \cdots + k_{d-1}\phi^{d-1}(D). \end{aligned} \tag{3}$$

Since $\phi(D)$ can be easily computed and the bitlengths of components are approximately $\frac{1}{d}$ that of k, various known methods for simultaneous multiple exponentiation can be applied to (3) to yield faster point multiplication. Thus we might expect to achieve a significant speedup because a great number of point doublings are eliminated at the expense of a few addition on the Jacobian.

4.2 Analysis on Simultaneous Multi-exponentiation

When simultaneous multi-exponentiation methods apply to hyperelliptic settings we here focus on determining the best method by comparing running times taken by these methods. In fact, seeking the optimal one involves various factors such as bitlengths of components, the number of decomposed components and absolute memory constrains.

There are two conventional methods for simultaneous multi-exponentiation: simultaneous 2^w-ary method and simultaneous sliding window method. Recently, Möller [17] presented a method, called wNAF-based interleaving method (for short, wNAF-IM), which is applicable to groups where inverting elements is easy (e.g. elliptic curves, hyperelliptic curves). It is analyzed there that his method usually wins over the conventional methods. One reason for this is that a speedup for simultaneous multi-exponentiation is affected by storage requirements, which are given by the formula concerning the expected number of generic operations in the precomputation stage. By the formula in Table 1 below, the expected number of generic operations by the wNAF-IM is linear in the number d of decomposed components but that by other methods is more or less exponential in d, so the wNAF-IM could be preferably chosen in practical implementations of point multiplication on the Jaconians of hyperelliptic curves.

We now give a precise analysis of speedup by comparing the expected number of doublings and additions taken by the three methods above. The comparison procedure we consider here consists of two stages, the precomputation and the evaluation. Since addition on the Jacobian takes $17g^2 + O(g)$ operations in $\mathbb{F}_p$ and doubling costs $16g^2 + O(g)$ operations, we may assume that one addition takes the same cost as one doubling because for security reasons, the genus g involved is relatively small, e.g. 2 or 3 (see Section 6 for security). In Table 1 we list the expected number of generic operations by three methods. Let b be the longest bitlength of components k_i and let w be the window size.

We compare the best algorithm to compute a single multiplication kD with that to compute a multi-exponentiation (3). In case of a single multiplication ($d = 1$), the NAF sliding window method [2] is known as one with best per-

Table 1. Expected number of generic operations by three methods for $\sum_{i=0}^{d-1} k_i D_i$ with multipliers up to b bits.

	Precomputation stage	Evaluation stage
Simultaneous 2^w-ary method	$2^{dw} - 1 - d$	$\lfloor \frac{b-1}{w} \rfloor w + b(1 - \frac{1}{2^{dw}})/w$
Simultaneous sliding window method	$2^{dw} - 2^{d(w-1)} - d$ $(w = 1)$ $2^{dw} - 2^{d(w-1)}$ $(w > 1)$	$b - 1 + b/(w + \frac{1}{2^d - 1})$
wNAF-based interleaving method	0 $(w = 1)$ $d2^{w-1}$ $(w > 1)$	$b + d\frac{b}{w+2}$

formance in general. The expected number of additions taken by this method with window size w is estimated at $b + \frac{b+1}{w+\nu(w)} + \frac{2^w-(-1)^w}{3} - 2$ where $\nu(w) = 4/3 - (-1)^w/(3 \cdot 2^{w-2})$.

In Table 2 below we give the minimum of the expected numbers of additions taken by all four methods(including a single point multiplication) for given k and d. Table 2 provides some indication of the relative benefits of simultaneous methods applied to our decomposition (3) to a single multiplication kD as in elliptic curves [8]. As shown in Table 2, the contribution to running times depends on the bitlength b of k and on the degree d of the characteristic polynomial $P(t)$ of ϕ. It also shows that the wNAF-IM turns out to be the best algorithm except for two cases where $d = 1$, denoted $(*)$ below. In those cases the NAF sliding window method is the best among the methods.

Table 2. Expected number of additions to compute $\sum_{i=0}^{d-1} k_i\phi^i(D)$ where k_i is b bits

$d = 1, b = 160$	$d = 2, b = 80$	$d = 4, b = 40$	$d = 6, b = 27$
193.7 $(*w = 4)$	120 $(w = 3)$	88 $(w = 2)$	79.5 $(w = 2)$
$d = 1, b = 256$	$d = 2, b = 128$	$d = 4, b = 64$	$d = 6, b = 43$
305.3 $(*w = 5)$	186.7 $(w = 4)$	131.2 $(w = 3)$	118.6 $(w = 3)$
$d = 1, b = 512$	$d = 2, b = 256$	$d = 4, b = 128$	$d = 6, b = 86$
601.1 $(w = 5)$	357.3 $(w = 4)$	245.3 $(w = 4)$	213.2 $(w = 3)$

Table 3. The ratio of the running times of the exended method to the conventional method.

d	ratio	Examples
2	0.62	Ex.2
4	0.45	Ex.1 (g=2), Ex.3 (g=2), Ex.5 (g=2)
6	0.41	Ex.1 (g=3), Ex.5 (g=3)

Table 3 contains the ratios of running times of the extended Gallant's method to the conventional method for a 160-bit single point multiplication. It also shows that the extended method improves multiplication reasonably compared to the conventional method. For example, when $d = 6$ and $b = 27$, the extended method improves a running time up to 59 % compared with the best general methods when $d = 1$ and $b = 160$.

5 Decomposition of an Integer k

We are now in a position to decompose an integer k into a sum of the form given by (2). To this end, we briefly describe a generalization of Gallant, *et al.*'s method to the hyperelliptic setting.

5.1 A General Method of Gallant *et al.*'s

We retain notation of Section 4.1. An extended method of Gallant *et al.*'s is composed of two steps. Consider the homomorphism

$$f : \prod_{i=0}^{d-1} \mathbb{Z} \to \mathbb{Z}_n, \qquad \prod_{i=0}^{d-1} a_i \mapsto \sum_{i=0}^{d-1} a_i \lambda^i \pmod{n}.$$

Firstly, we find d linearly independent short vectors $v_j \in \prod_{i=0}^{d-1} \mathbb{Z}$ such that $f(v_j) = 0$ for $0 \leq j \leq d-1$. As a stage of precomputations this process can be done by the LLL algorithm, independently of k.

Secondly, one needs to find a vector in $\mathbb{Z}v_0 + \cdots + \mathbb{Z}v_{d-1}$ that is close to $(k, 0, \cdots, 0)$ using linear algebra. Then $(k_0, \cdots, k_{d-1})$ is determined by the equation:

$$(k_0, \cdots, k_{d-1}) = (k, 0, \cdots, 0) - (\lfloor b_0 \rceil v_0 + \cdots + \lfloor b_{d-1} \rceil v_{d-1}),$$

where $(k, 0, 0, \cdots, 0) = b_0 v_0 + \cdots + b_{d-1} v_{d-1}$ is represented as an element in $\prod_{i=0}^{d-1} \mathbb{Q}$ and $\lfloor b \rceil$ denotes the nearest integer to b. Finally, we obtain a short vector $v = (k_0, \cdots, k_{d-1})$ such that $f(v) = f((k, 0, \cdots, 0)) - f((\lfloor b_0 \rceil v_0 + \cdots + \lfloor b_{d-1} \rceil v_{d-1})) = k$ and then we have (2) as desired.

The following Lemma shows that the vector v is indeed short.

Lemma 1. *The vector* $v = (k, 0, \cdots, 0) - (\lfloor b_0 \rceil v_0 + \cdots + \lfloor b_{d-1} \rceil v_{d-1})$ *constructed as above has norm at most* $\frac{d}{2} \max\{\|v_0\|, \cdots, \|v_{d-1}\|\}$.

Proof. The statement is a generalization of Lemma 1 (d=2) in [8], so the proof proceeds in a similar way.

5.2 Another Method Using a Division

We are now describing an alternate method for decomposing k using a division in the ring $\mathbb{Z}[\phi]$ generated by an efficiently-computable endomorphism ϕ.

Let us consider the map

$$g : \mathbb{Z}[\phi] \to \prod_{i=0}^{d-1} \mathbb{Z}, \qquad \sum_{i=0}^{d-1} a_i \phi^i \mapsto \prod_{i=0}^{d-1} a_i.$$

Then $f \circ g(\sum_{i=0}^{d-1} a_i \phi^i) = \sum_{i=0}^{d-1} a_i \lambda^i \pmod{n}$.

Firstly, we need to find $\alpha \in \mathbb{Z}[\phi]$ with short components such that $f \circ g(\alpha) = 0$. More precisely, we find a short vector $v \in \prod_{i=0}^{d-1} \mathbb{Z}$ such that $f(v) = 0$. (Note that in the Gallant's method one has to find d such short vectors which are linearly independent but here only one such vector.) Then we can obtain $\alpha = g^{-1}(v)$. Secondly, viewing an integer k as an element in $\mathbb{Z}[\phi]$ we divide k by α using Algorithm below and write

$$k = \beta\alpha + \rho$$

with $\beta, \rho \in \mathbb{Z}[\phi]$. Since $f \circ g(\alpha) = 0$ and $\alpha D = O$ for $D \in \mathbb{J}_X(\mathbb{F}_q)$, we compute

$$kD = (\beta\alpha + \rho)D = \beta\alpha D + \rho D = \rho D.$$

Writing $\rho = \sum_{i=0}^{d-1} k_i\phi^i \in \mathbb{Z}[\phi]$, the preceding equation alternately gives an desired decomposition of an integer k as in Eqn.(3). This decomposition makes use of the division process in the ring $\mathbb{Z}[\phi]$, so we now describe an efficient and practical algorithm to compute a remainder ρ of a given integer k divided by α. Let $\alpha = \sum_{i=0}^{d-1} a_i\phi^i \in \mathbb{Z}[\phi]$ with its minimal polynomial $g(t)$. Write $g(t) = t \cdot h(t) + N$ for some $h(t) \in \mathbb{Z}[t]$. It is then easy to see that $N = -\alpha h(\alpha)$ and $-h(\alpha) \in \mathbb{Z}[\phi]$. Put $\widehat{\alpha} = -h(\alpha) \in \mathbb{Z}[\phi]$.

Algorithm (**Divide** k **by** $\alpha = \sum_{i=0}^{d-1} a_i\phi^i$)

Input: $k \approx n$.
Output: $\rho = \sum_{i=0}^{d-1} k_i\phi^i$.

---

1) Precompute $\widehat{\alpha} = N/\alpha$ in $\mathbb{Z}[\phi]$ and put $\widehat{\alpha} = \sum_{i=0}^{d-1} b_i\phi^i$.
2) $x_i = k \cdot b_i$ (for $i = 0, , d-1$).
3) $y_i = \lfloor \frac{x_i}{N} \rceil$ (for $i = 0, , d-1$).
4) $\rho = k - \sum_{i=0}^{d-1} \sum_{j=0}^{d-1} a_i y_j \phi^{i+j}$.

Return: $\rho = \sum_{i=0}^{d-1} k_i\ \phi^i$.

Proof. Assume that $k = \beta\alpha + \rho$ for some $\beta, \rho \in \mathbb{Z}[\phi]$. Then we have $k/\alpha = \beta + \rho/\alpha$. Since $k\widehat{\alpha}/\alpha\widehat{\alpha} = k\widehat{\alpha}/N$, we have

$$k/\alpha = \sum_{i=0}^{d-1} (kb_i/N)\phi^i.$$

Putting $\beta = \sum_{i=0}^{d-1} \lfloor kb_i/N \rfloor \phi^i$ gives $\rho = k - \alpha\beta$. □

Giving explicit upper bounds for components of a remainder ρ depends on the characteristic polynomial of ϕ and so it is complicated to obtain good upper bounds in general. But, for a fixed ϕ one can give explicit upper bounds for components by analyzing the above algorithm further.

Now we compare two decomposition methods. For this we apply both methods to the hyperelliptic curves in Section 3. Our implementation results show that two decompositions of an integer $k \in [1, n]$ turn out to be identically same and the bitlengths of components are approximately $1/d$ that of n. More precisely, for each curve in Section 3 (Ex.2 - 5), we select 100 random primes n of size 160-bits and for each n we carried out decompositions of 10^5 random integers $k \in [1, n]$ by two methods and see that two decompositions coincide. But it is expected that two decompositions might not be the same, as in elliptic curves [20]. As for the bitlengths of components, in Table 4 we compute the maximum of ratios of A to B where A denotes the maximum of the absolute value of decomposed components and B denotes $n^{1/d}$. These maxima tell us that the bitlengths of components are approximately $1/d$ that of n because they are < 2, which implies that A and B are within one bit.

Table 4. Numerical experiments for decomposition

Examples	Characteristic polynomial of ϕ	Maxum of ratios of A to B
Ex. 2	$P(t) = t^2 + 1$	0.704
Ex. 3	$P(t) = t^4 + 1$	1.082
Ex. 4	$P(t) = t^4 - t^2 + 1$	1.247
Ex. 5 (g=2)	$P(t) = (t^5 - 1)/(t - 1)$	1.477
Ex. 5 (g=3)	$P(t) = (t^7 - 1)/(t - 1)$	1.682

6 Security Considerations

We described a method for speeding up point multiplication, which is applicable to hyperelliptic curves with efficiently-computable endomorphisms. Such endomorphisms could be also helpful to obtain a speedup for attacks to the discrete log problems on the Jacobians [9]. In this Section we shall touch on various attacks to public-key cryptosystems based on the DLP on the Jacobians of hyperelliptic curves. Most attacks are hyperelliptic variants extending those to elliptic curves and standard finite fields. Adleman, DeMarris and Hwang [1] came up with the first-published algorithm for computing DLP, which runs in subexponential time. This algorithm applies to hyperelliptic curves over finite fields whose genus is sufficiently large relative to the size of the underlying fields. Later, Enge [6] improved their algorithm and precisely evaluated the running time. Moreover Müller, Stein and Thiel [19] extends the results to real quadratic congruence function fields of large genus.

When selecting hyperelliptic curves $X/\mathbb{F}_q$ of small genus < 4, one has to avoid curves for which special attacks are known such as the Pohlig-Hellman and the Pollard rho method. For this reason, hyperelliptic curves are believed to be "cryptographically good" provided that the group order of the Jacobians is divisible by a large prime number $\approx$ 160-bit.

The hyperelliptic curves we have considered have a small number of automorphisms as in a family of hyperelliptic Koblitz curves. In applying the Pollard's ρ method, Duursma, Gaudry, and Morain [5] employed an equivalence relation on points of the Jacobian via automorphisms and could speed up the attack by a factor of $\sqrt{2l}$, where l is the order of an automorphism. Gaudry [9] also gave a variant of existing index-calculus methods like [1] to achieve a more speed-up of a factor of l^2. Indeed, this method is faster than the Pollard ρ for genus > 4, and its complexity $O(q^2)$ depends on the cardinality of the base field. So the public schemes based on our curves are still intractable to this attack since these curves are hyperelliptic ones of genus 2 or 3 defined over large prime fields.

We now mention other known attacks using special features on groups. Rück [21] extended an attack on anomalous curves to hyperelliptic curves. His method works for the groups whose order is divisible by a power of p, where p is the characteristic of the base field. On the other hand, anomalous curves are investigated by Semaev [23], Smart[25], and Satoh and Araki [22].

There is the Frey-Rück attack using Tate pairing [7]. It is an extension of an attack using the Weil pairing on elliptic curves. In fact, these attacks are applied to curves over $\mathbb{F}_q$ of which group order divides $q^k - 1$ for some $k \leq 20$.

The Weil descent attack on elliptic curves has a hyperelliptic variant. To avoid this attack one must choose curves defined over extension fields of prime degree for odd characteristic and of degree $\neq 2^l - 1$ for even characteristic. For a detailed analysis on the Weil descent we refer to [11], [16].

Finally we conclude that our hyperellipic curves of small genus < 4 are intractable to all known attacks even if efficient endomorphisms may result in speedy attacks by a factor, which depends upon the number of automorphisms groups on the Jacobians as shown by Gaudry [9].

7 Conclusion

Motivated by the work of [8], we presented an extended method for accelerating point multiplication on a family of hyperelliptic curves having efficiently-computable endomorphisms. One of advantages of this method is that it improves a running time by 55 (59) % compared with the best general ordinary methods for a 160-bit point multiplication when applied to such curves of g =2 (3). Another advantage is that there is a wide range of possibility of selecting hyperelliptic curves of genus $g \leq 3$ over large prime fields rather than elliptic curves. Also we presented two algorithms for decomposing a multiplier k so that the extended method can be applicable to such curves. Computer implementations of two algorithms showed that the bitlengths of decomposition components are roughly equal to $1/d$-bit of an integer k.

References

1. L. Adleman, J.DeMarrais, M-D. Hwang, *"A Subexponential algorithms for Discrete Logarithms over the Rational Subgroups of the Jacobians of Large Genus Hyperelliptic Curves over Finite fields"*, ANTS-I, LNCS **77** (Springer) 1994, 28-40.
2. I. Blake, G. Seroussi, N. Smart: 'Elliptic Curves in Cryptography', London Mathematical Society Lecture Note Series. 265, Cambridge University Press, 1999.
3. J. Buhler, N. Koblitz, *"Lattice Basis Reduction, Jacobi Sums and Hyperelliptic Cryptoststems"*, Bull. Austral. Math. Soc., **57**, 147-154.
4. D. Cantor, *"Computing in the Jacobian of A Hyperelliptic curve"*, Mathematics of Computation, **48**(1987), 95-101.
5. I. Duursma, P. Gaudry, F. Morain, *"Speeding up the discrete log computation on curves with automorphisms"*, Advances in Cryptology, Asiacrypt'99 LNCS **1716** (Springer), 1999, 103-121.
6. A. Enge, *"Computing discrete logarithms in high-genus hyperelliptic Jacobian in provably subexponential time"*, University of Waterloo Technical Report CORR99-04, 2000.

7. G. Frey, H.-G. Rück, *"A Reamrk concerning m-Divisibility and the Discrete logarithm Problems in the Divisor Class Group of Curves"*, Mathematics of Computation **62**, 1994, 865-874.
8. R. Gallant, R. Lambert, S. Vanstone, *"Faster Point Multiplication on Elliptic Curves with Efficient Endomorphisms"*, Advances in Cryptology-Crypto 2001, LNCS **2139** (Springer), 2001, 190-200.
9. P. Gaudry, *"An algorithm for solving the discrete log problems on hyperelliptic curves"*, Advances in Cryptology, Eurocrypt'2000, LNCS **1807** (Springer), 2000, 19-34.
10. C. Günter, T. Lange, A. Stein, *"Speeding up the Arithmetic on Koblitz Curves of Genus Two"*, SAC 2000, LNCS **2012** (Springer), 2001, 106-117.
11. P. Gaudry, F. Hess, N.P. Smart, *"Constructive and destructive facets of Weil Decent on elliptic curve"*, Preprint (2000)
12. N. Koblitz, *"CM-curves with good cryptographic properties,"* Advances in Cryptology-Crypto'91, 1992, 279-287.
13. N. Koblitz, 'Algebraic Aspects of Cryptography', Algorithms and Computations in Mathematics, **3**, Springer-Verlag, 1998.
14. T. Lange,'Efficient Arithmetic on Hyperelliptic Koblitz Curves', Ph.D. Thesis, University of Essen, 2001
15. W. Meier, O. Staffelbach, *"Efficient multiplication on certain nonsupersingular elliptic curves"*, Advances in Cryptology-Crypto'92, 1992, 333-344.
16. A. Menezes , M. Qu , *"Analysis of the Weil Descent Attack of Gaudry, Hess, and Smart"*, TCT-RSA 2001, LNCS **2020** (Springer) 2001, 308-318.
17. B. Möller, *"Algorithms for multi-exponentiation"*, SAC 2001, 179-194.
18. V. Müller, *"Fast multiplication in elliptic curves over small fields of characteristic two"*, Journal of Cryptology, **11**, 1998, 219-234.
19. V. Müller, A. Stein, C. Thiel, *"Computing Discrete Logarithms in Real Quadratic Congruence Function Fields of Large Genus"*, Mathematics of Computations **68**, 1999, 807-822.
20. Y.-H. Park, S. Jeong, C. Kim, J. Lim, *"An alternate decomposition of an integer for faster point multiplication on certain elliptic curves"*, to appear PKC2002.
21. R.G. Ruck, *"On the discrete logarithm in the divisor class group of curves"*, Mathematics of Computations, **68**, 1999, 805-806.
22. T. Satoh, K. Araki, *"Fermat quotients and the polynomial time discrete log algotithm for anamalous elliptic curves"*, Commentari Math. Univ. St. Pauli, **47**, 1998, 81-92.
23. I.A. Semaev, *"Evaluation of Discrete logathrims in a group of p-torsion points of an elliptic curves in characteristic p"*, Mathematics of Computations, **67**, 1998, 353-356.
24. N. Smart, *"Elliptic curve cryptosystems over small fields of odd characteristic"*, Journal of Cryptology, No 2 **12**, 1999, 141-145.
25. N. Smart, *"The Disrete Logarithm Problem on Elliptic Curves of Trace One"*, Journal of Cryptology, No 3 **12**, 1999, 193-196.
26. J. Solinas, *"An improved algorithm for arithmetic on a family of elliptic curves,"* Advances in Cryptology-Crypto '97, 1997, 357-371.
27. J. Solinas, *"Efficient arithmetic on Koblitz curves "*, Design , Codes and Cryptography, **19**, 2000, 195-249.
28. A. Stein, *"Sharp Upper Bounds for Arithmetics in Hyperelliptic Function Fields"*, Techn. Report. CORR 99-23, University of Waterloo (1999), 68 pages.

Fast Correlation Attacks: An Algorithmic Point of View

Philippe Chose, Antoine Joux, and Michel Mitton

DCSSI, 18 rue du Docteur Zamenhof, F-92131 Issy-les-Moulineaux cedex, France,
Philippe.Chose@ens.fr, Antoine.Joux@m4x.org, michelmitton@compuserve.com

Abstract. In this paper, we present some major algorithmic improvements to fast correlation attacks. In previous articles about fast correlations, algorithmics never was the main topic. Instead, the authors of these articles were usually addressing theoretical issues in order to get better attacks. This viewpoint has produced a long sequence of increasingly successful attacks against stream ciphers, which share a main common point: the need to find and evaluate parity-checks for the underlying linear feedback shift register. In the present work, we deliberately take a different point of view and we focus on the search for efficient algorithms for finding and evaluating parity-checks. We show that the simple algorithmic techniques that are usually used to perform these steps can be replaced by algorithms with better asymptotic complexity using more advanced algorithmic techniques. In practice, these new algorithms yield large improvements on the efficiency of fast correlation attacks.

Keywords. Stream ciphers, fast correlation attacks, match-and-sort, algorithmics, parity-checks, linear feedback shift registers, cryptanalysis.

1 Introduction

Stream ciphers are a special class of encryption algorithms. They generally encrypt plaintext bits one at a time, contrary to block ciphers that use blocks of plaintext bits. A *synchronous stream cipher* is a stream cipher where the ciphertext is produced by bitwise adding the plaintext bits to a stream of bits called the keystream, which is independent of the plaintext and only depends on the secret key and on the initialization vector. These synchronous stream ciphers are the main target of fast correlation attacks.

The goal in stream cipher design is to produce a pseudo-random keystream sequence, pseudo-random meaning indistinguishable from a truly random sequence by polynomially bounded attackers. A large number of stream ciphers use Linear Feedback Shift Registers (LFSR) as building blocks, the initial state of these LFSRs being related to the secret key and to the initialization vector. In nonlinear combination generators, the keystream bits are then produced by combining the outputs of these LFSRs through a nonlinear boolean function (see Fig. 1). Many variations exist where the LFSRs are multiplexed or irregularly clocked.

L.R. Knudsen (Ed.): EUROCRYPT 2002, LNCS 2332, pp. 209–221, 2002.

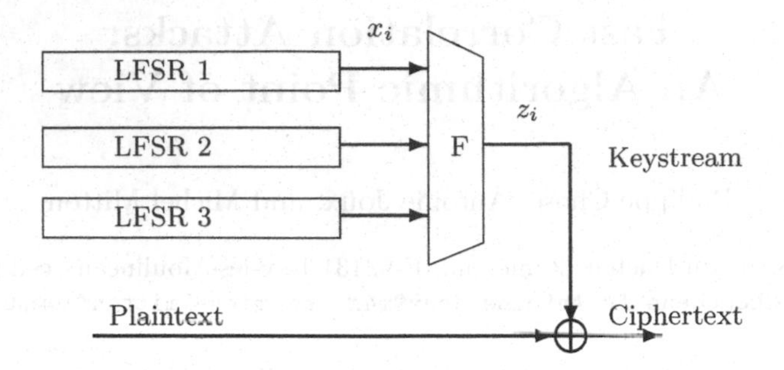

Fig. 1. Three LFSRs combined by a nonlinear boolean function

Among the different kinds of attacks against stream ciphers, correlation attacks are one of the most important [12, 13]. These cryptanalytic methods target nonlinear *combination* keystream generators. They require the existence of linear correlations between LFSR internal stages and the nonlinear function output, i.e. correlation between linear combinations of internal and output bits; these correlations need to be good enough for the attack to be successful. A very important fact about nonlinear functions is that linear correlations always exist [12]. Finding them can sometimes be the hardest part of the job, since the main method for finding them is by statistical experimentation. For the simplest ciphers, the correlations can be found by analyzing the nonlinear function with the well-known Walsh transform. Once a correlation is found, it can be written as a probability

$$p = \Pr(z_i = x_i^{j_1} \oplus x_i^{j_2} \oplus \ldots \oplus x_i^{j_M}) \neq 0.5$$

where z_i is the i-th keystream output bit and the $x_i^{j_1}, \ldots, x_i^{j_M}$ are the i-st output bits of some LFSRs $j_1, \ldots, j_M$ (see Fig. 1). The output can thus be considered as a noisy version of the corresponding linear combination of LFSR outputs. The quality of the correlation can be measured by the quantity $\varepsilon = |2p - 1|$. If ε is close to one, the correlation is a very good one and the cipher is not very strong. On the other hand, when ε is close to zero, the output is very noisy and correlation attacks will likely be inefficient. Since the LFSR output bits are produced by linear relations, we can always write this sum of output bits $x_i^{j_1} \oplus x_i^{j_2} \oplus \ldots \oplus x_i^{j_M}$ as the output of only one larger LFSR. Without loss of generality, the cipher can be presented as in Fig. 2, where this sum has been replaced by x_i the output of one LFSR, and the influence of the nonlinear function has been replaced by a BSC (binary symmetric channel), i.e. by a channel introducing noise with probability $1 - p$. Fast correlation attacks are an improvement of the basic correlation attacks. They essentially reduce the time complexity of the cryptanalysis by pre-computing data [3, 5, 7–9].

In this article, we present substantial algorithmic improvements of existing fast correlation attacks. The paper is organized as follows. In Sect. 2, we introduce the basics of fast correlation attacks and a sketch of our algorithmic

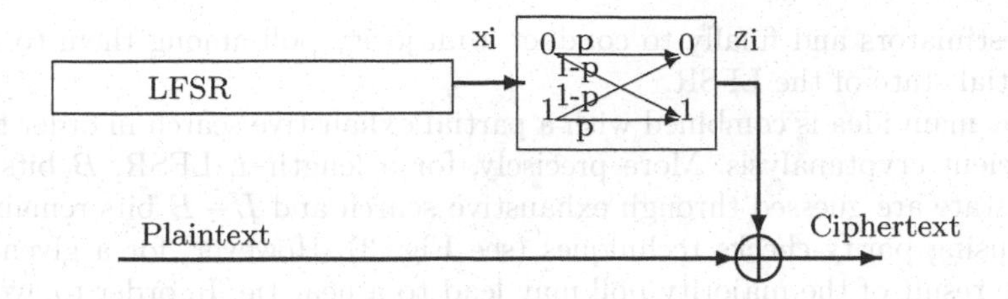

Fig. 2. Equivalent diagram where the output x_i of the LFSR is correlated with the keystream output z_i: $P(z_i = x_i) = p \neq 0.5$

improvements. A detailed description of the algorithm is given in Sect. 3, together with its complexity analysis and, finally, some comparisons with other algorithms are provided in Sect. 4.

2 Fast Correlation Attacks

Fast correlation attacks are usually studied in the binary symmetric channel model as shown on Fig. 2. In this model, we consider the output of the generator as a noisy version of the output of some of the linear registers. The cryptanalysis then becomes a problem of decoding: given a noisy output, find the exact output of the registers and, when needed, reconstruct the initial filling of the registers.

The common point between all fast correlation attacks algorithms is the use of the so-called parity-check equations, i.e. linear relations between register output bits x_i. Once found, these relations can be evaluated on the noisy outputs z_i of the register. Since they hold for the exact outputs x_i, the evaluation procedure on the noisy z_i leaks information and helps to reconstruct the exact output sequence of the LFSR.

Fast correlation algorithms are further divided into iterative algorithms and one-pass algorithms. In iterative algorithms, starting from the output sequence z_i, the parity-checks are used to modify the value of these output bits in order to converge towards the output x_i of the LFSR thus removing the noise introduced by the BSC. The reconstruction of the internal state is then possible [2, 7]. In one-pass algorithms, the parity-checks values enable us to directly compute the correct value of a small number of LFSR outputs x_i from the output bits z_i of the generator. This small number should be larger than the size of the LFSR in order to allow full reconstruction [3, 5, 8, 9].

2.1 Sketch of One-Pass Fast Correlation Attacks

The one-pass correlation attack presented here is a variation of the attacks found in [3] and [9]. The main idea is, for each LFSR's output bit to be predicted (henceforth called *target bits*), to construct a set of estimators (the parity-check equations) involving k output bits (including the target bit), then to evaluate these estimators and finally to conduct a majority poll among them to recover the initial state of the LFSR.

This main idea is combined with a partial exhaustive search in order to yield an efficient cryptanalysis. More precisely, for a length-L LFSR, B bits of the initial state are guessed through exhaustive search and $L - B$ bits remain to be found using parity-checks techniques (see Fig. 3). However, for a given target bit, the result of the majority poll may lead to a near tie. In order to avoid this problem, we target more than $L-B$ bits, namely D and hope that at least $L-B$ will be correctly recovered.

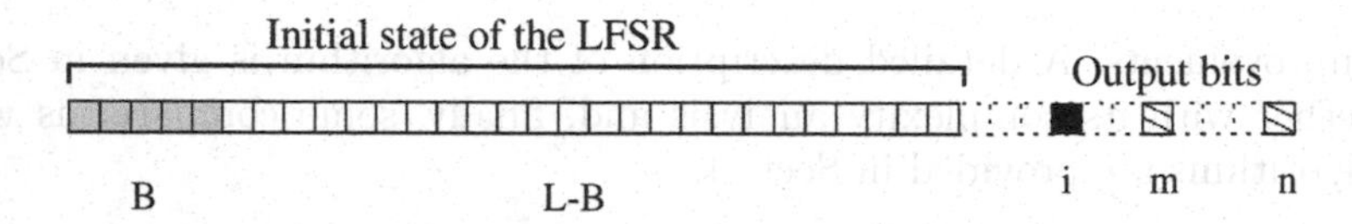

Fig. 3. Construction of parity-checks: In this example, the parity-check combines two bits of output (m and n) together with a linear combination of the B guessed bits in order to predict the target bit i

For each of these D target bits, we evaluate a large number Ω of estimators using the noisy z_i values and we count the number of parity-checks that are satisfied and unsatisfied, respectively N_s and $N_u = \Omega - N_s$. When the absolute value of the difference of these two numbers is smaller than some threshold θ, we forget this target bit. However when the difference is larger than the threshold, the majority poll is considered as successful. In that case, we predict $\hat{x}_i = z_i$ if $N_s > N_u$ and $\hat{x}_i = z_i \oplus 1$ otherwise (see Fig. 4). When the majority polls are successful and give the correct result for at least $L - B$ of the D target bits, we can recover the complete state of the LFSR using simple linear algebra.

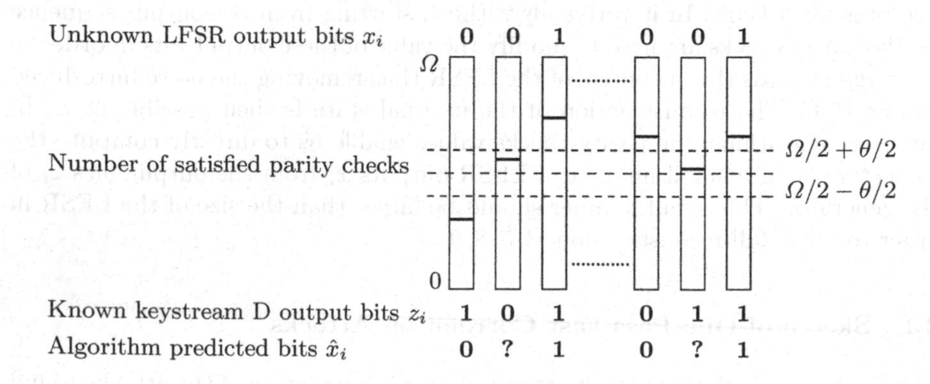

Fig. 4. Sketch of the decision procedure

2.2 New Algorithmic Ideas

When implementing the fast correlation attack from the previous section, several algorithmic issues arise. First we need to pre-compute the parity-checks for each target bit. Then we need to efficiently evaluate these parity-checks and recover the target bits. In previous papers, the latter step was performed using the straightforward approach, i.e. by evaluating parity-checks one by one for each possible guess of the first B bits, then by counting the number of positive and negative checks. For the preprocessing step, three algorithms were known, simple exhaustive search, square-root time-memory tradeoffs and Zech's logarithm technique [10]. The main contribution of this paper is to propose better algorithmic techniques for both tasks.

Pre-processing Stage The usual square-root algorithm for computing parity-checks on k bits (the target bit plus $k-1$ output bits) works as follows. For each target bit, compute and store in a table the formal expression of the sum of $\lfloor \frac{k-1}{2} \rfloor$ output bits and the target bit in term of the initial L-bit state. Then sort this table. Finally compute the formal sum of $\lceil \frac{k-1}{2} \rceil$ output bits and search for a partial collision (on the $L-B$ initial bits, excluding the B guessed bits). The time and memory complexity of this algorithm are respectively $\mathcal{O}(DN^{\lceil (k-1)/2 \rceil} \log N)$ and $\mathcal{O}(N^{\lfloor (k-1)/2 \rfloor})$ where N is the length of the considered output sequence and D is the number of target bits. For even values of k, a different tradeoff exists: it yields a respective time and memory complexity $\mathcal{O}(N^{k/2} \log N)$ and $\mathcal{O}(N^{k/2})$. This square-root algorithm is part of a family of algorithms which can be used to solve a large class of problems. In some cases, there exists an alternative algorithm with the same time complexity as the original and a much lower memory complexity. A few examples are:

- the knapsack problem and modular knapsack problem [11, 1],
- the match and sort stage of SEA elliptic curve point counting [6],
- the permuted kernel problem [4].

Our goal is to propose such an alternative for constructing parity-checks. According to known results, we might expect a time complexity of
$\mathcal{O}(\min(DN^{\lceil (k-1)/2 \rceil} \log N, N^{\lceil k/2 \rceil} \log N))$ and a memory complexity of
$\mathcal{O}(N^{\lfloor k/4 \rfloor})$. It turns out that such an alternative really exists for $k \geq 4$, the algorithmics being given in Sect. 3.

Decoding Stage When using the usual method for evaluating the parity-checks, i.e. by evaluating every parity-check for each target bit and every choice of the B guessed bits, the time complexity is $\mathcal{O}(D2^B \Omega)$ where Ω is the number of such parity-check equations. By grouping together every parity-check involving the same dependence pattern on a well chosen subset of the B guessed bits (of size around $\log_2 \Omega$), it is possible to evaluate these grouped parity-checks in a single pass through the use of a Walsh transform. This is much faster than restarting the computation for every choice of the B bits. The expected time complexity of this stage then becomes $\mathcal{O}(D2^B \log_2 \Omega)$.

3 Algorithmic Details and Complexity Analysis

We present in this section a detailed version of our algorithmic improvements. We consider here that the parameters N, L, D, B, θ and ε are all fixed. The optimal choice of these parameters is a standard calculation and is given in appendix A.

Let us recall the notations used in the following:

- N is the number of available output bits;
- L is the length of the LFSR;
- B is the number of guessed bits;
- D is the number of target bits;
- x_i is the i-th output bit of the LFSR;
- z_i is the i-th output bit of the generator;
- $p = Pr(x_i = z_i) = \frac{1}{2}(1+\varepsilon)$ is the probability of correct prediction;

3.1 Pre-processing Stage

During the pre-processing stage, we search for all parity-checks of weight k associated with one of the D target bits. Following [9] and [2], we construct the set Ω_i of parity-check equations associated with the target bit i. This set contains equations of the form:

$$x_i = x_{m_1} \oplus \ldots \oplus x_{m_{k-1}} \oplus \sum_{j=0}^{B-1} c_{\boldsymbol{m},i}^j x_j$$

where the m_j are arbitrary indices among all the output bits and the $c_{\boldsymbol{m},i}^j$ are binary coefficients characterizing the parity-check. $\boldsymbol{m}$ stands for $[m_1, \ldots, m_{k-1}]$. In these equations, we express x_i as a combination of $k-1$ output bits plus some combination of the B guessed bits. The expected number of such parity-check equations for a given i is:

$$\Omega \approx 2^{B-L} \binom{N}{k-1}$$

For $k \leq 4$, the basic square-root time-memory tradeoff gives us these parity-checks with a time and memory complexity respectively of $\mathcal{O}(DN^{\lceil (k-1)/2 \rceil} \log N)$ and $\mathcal{O}(N^{\lfloor (k-1)/2 \rfloor})$.

In the sequel, we first solve a slightly more general problem. We try to find equations of the form:

$$A(\boldsymbol{x}) = x_{m_1} \oplus \ldots \oplus x_{m_{k'}} \oplus \sum_{j=0}^{B-1} c_{\boldsymbol{m},i}^j x_j$$

where $A(\boldsymbol{x}) = \sum_{j=0}^{L-1} a_j x_j$, $\boldsymbol{x} = [x_0, \ldots, x_{L-1}]$ and the a_j are fixed constants. When k is even, $A(\boldsymbol{x})$ will not be used and will be set to 0. When k is odd, $A(\boldsymbol{x})$

will be set to the formal expresion of one of the target bits and the problem will be similar to the even case. For the time being, let k' be the weight of the parity-check equation.

The main idea of the match-and-sort algorithm alternative we are going to use here is to split the huge task of finding collisions among $N^{k'}$ combinations into smaller tasks: finding less restrictive collisions on smaller subsets, sort the results and then aggregate these intermediate results to solve the complete task.

Algorithm 1 Find parity-checks of weight k' for a given $A(\boldsymbol{x})$

Evenly split k' between l_1, l_2, l_3 and l_4 with $l_1 \geq l_2$ and $l_3 \geq l_4$
for all choice of l_2 bits $(j_1 \dots j_{l_2})$ **do**
  Formally compute $x_{j_1} \oplus \dots \oplus x_{j_{l_2}} = \sum_{k=0}^{L-1} u_k x_k$
  Store in $U[\boldsymbol{u}] = \{j_1, \dots, j_{l_2}\}$
end for
for all choice of l_4 bits $(m_1 \dots m_{l_4})$ **do**
  Formally compute $x_{m_1} \oplus \dots \oplus x_{m_{l_4}} = \sum_{k=0}^{L-1} v_k x_k$
  Store in $V[\boldsymbol{v}] = \{m_1, \dots, m_{l_4}\}$
end for
for all $s = 0 \dots 2^S - 1$ **do**
  for all choice of l_1 bits $(i_1 \dots i_{l_1})$ **do**
    Formally compute $A(\boldsymbol{x}) \oplus x_{i_1} \oplus \dots \oplus x_{i_{l_2}} = \sum_{k=0}^{L-1} c_k x_k$
    Search for $\boldsymbol{u}$ in U such that $\pi_S(\boldsymbol{u} \oplus \boldsymbol{c}) = s$
    Store in $C[\boldsymbol{u} \oplus \boldsymbol{c}] = \{i_1, \dots, i_{l_1}, j_1, \dots, j_{l_2}\}$
  end for
  for all choice of l_3 bits $(k_1 \dots k_{l_3})$ **do**
    Formally compute $x_{k_1} \oplus \dots \oplus x_{k_{l_3}} = \sum_{k=0}^{L-1} d_k x_k$
    Search for $\boldsymbol{v}$ in V such that $\pi_S(\boldsymbol{v} \oplus \boldsymbol{d}) = s$
    Let $\boldsymbol{t} = \boldsymbol{v} \oplus \boldsymbol{d}$
    Search for $\boldsymbol{c}$ in C such that $\pi_{L-B}(\boldsymbol{c} \oplus \boldsymbol{t}) = 0$
    Output $\{A(\boldsymbol{x}), i_{1 \dots l_1}, j_{1 \dots l_2}, k_{1 \dots l_3}, m_{1 \dots l_4}, \boldsymbol{c} \oplus \boldsymbol{t}\}$
  end for
end for

First evenly split k' between four integer l_1, l_2, l_3 and l_4 with $l_1 \geq l_2$ and $l_3 \geq l_4$, i.e. find l_1, l_2, l_3 and l_4 such that $l_1+l_2+l_3+l_4 = k'$ and for i from 1 to 4, $l_i = \lfloor \frac{k'}{4} \rfloor$ or $l_i = \lceil \frac{k'}{4} \rceil$. Compute the formal sums of l_2 output bits in terms of the initial L-bit state, $x_{j_1} \oplus \dots \oplus x_{j_{l_2}} = \sum_{k=0}^{L-1} u_k x_k$. Let us write $\boldsymbol{u} = \{u_0, \dots, u_{L-1}\}$. Store all these expressions in table U at entries $\boldsymbol{u}$. Do the same for a table V containing combinations of l_4 output bits. We will now try to match elements of table U with formal sums of l_1 output bits and elements of table V with formal sums of l_3 output bits (see Fig. 5). We only require the matching to be effective on a subset S of the $L - B$ bits. This S is chosen close to $\frac{k'}{4} \log_2 N$ in order to minimize the memory usage without increasing the time complexity. For each value s of the S bits, compute the formal sum $\boldsymbol{c}$ of $A(\boldsymbol{x})$ and of l_1 output bits in terms of initial bits and search for a partial collision in table U on S bits, i.e.

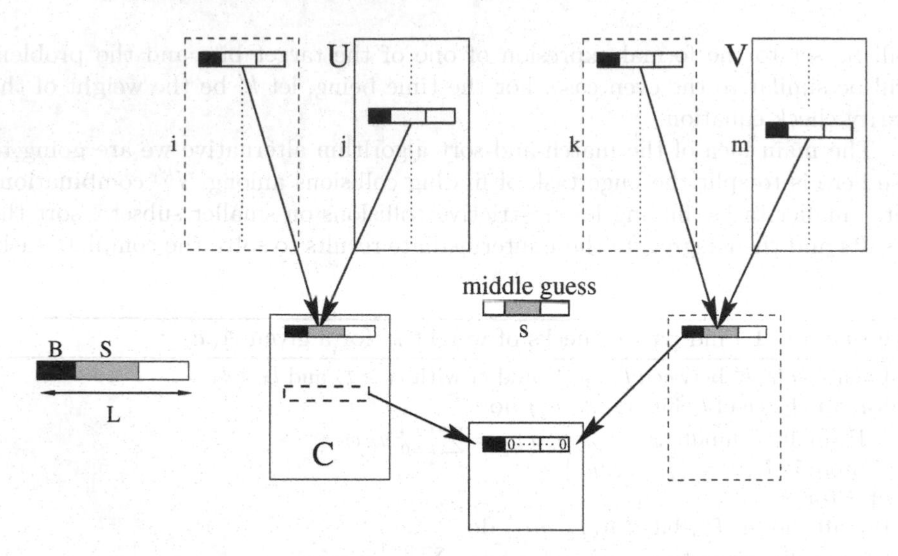

Fig. 5. Match-and-sort algorithm for finding parity-check equations

find a $\boldsymbol{u}$ in U such that $\pi_S(\boldsymbol{u} \oplus \boldsymbol{c}) = s$ where π_S is the projection on the subspace spanned by the S bits. Store these collisions on S bits in a table C. Repeat the same procedure (finding partial collisions) with l_3 and l_4 replacing l_1 and l_2. For each found collision, search for a new collision combining the just found collision with an entry of table C, this time not on the S bits but on the full set of $L - B$ bits. Every found collision is a valid parity-check since it only involves the B guessed bits and $l_1 + l_2 + l_3 + l_4 = k'$ output bits (see Algorithm 1).

The time complexity of this algorithm is $\mathcal{O}(N^{\max(l_1+l_2, l_3+l_4)} \log N)$ and the memory complexity is $\mathcal{O}(N^{\max(\min(l_1,l_2), \min(l_3,l_4))})$, which for an even division of k' are equal respectively to $\mathcal{O}(N^{\lceil k'/2 \rceil} \log N)$ and $\mathcal{O}(N^{\lfloor (k'+1)/4 \rfloor})$. In fact, using the above algorithm for the homogenous case when $A(\boldsymbol{x}) = 0$ is a simple matter.

Table 1. Time and memory complexities of the new algorithm compared to those of the square-root algorithm

	New algorithm		Square-root algorithm (tradeoff 1)		Square-root algorithm (tradeoff 2)	
k	*Time*	*Memory*	*Time*	*Memory*	*Time*	*Memory*
4	$N^2 \log N$	N	$DN^2 \log N$	N	$N^2 \log N$	N^2
5	$DN^2 \log N$	N	$DN^2 \log N$	N^2	$DN^2 \log N$	N^2
6	$N^3 \log N$	N	$DN^3 \log N$	N^2	$N^3 \log N$	N^3
7	$DN^3 \log N$	N	$DN^3 \log N$	N^3	$DN^3 \log N$	N^3
8	$N^4 \log N$	N^2	$DN^4 \log N$	N^3	$N^4 \log N$	N^4
9	$DN^4 \log N$	N^2	$DN^4 \log N$	N^4	$DN^4 \log N$	N^4

Depending on the parity of k, two alternatives are possible. When k is odd, we simply let $k' = k - 1$ and let $A(\boldsymbol{x})$ represent x_i, one of the target bits. Of course we need to run the algorithm D times. When k is even, we let $k' = k$ and $A(\boldsymbol{x}) = 0$. In that case, we get the parity-checks for all output bits of the LFSR instead of merely D. With these choices, the complexities are listed in Table 1.

3.2 Processing Stage

Decoding Part Let us write $B = B_1 + B_2$ where B_1 and B_2 are positive integers to be determined. These two integers define two sets of bits in the initial state of the LFSR. Let us guess the B_1 bits of the initial state of the LFSR and denote by $\boldsymbol{X}_1$ the value of this guess. We regroup together all parity-check equations that involve the same pattern of the B_2 initial bits; let us rewrite each parity-check equation as:

$$z_i = \underbrace{z_{m_1} \oplus \ldots \oplus z_{m_{k-1}} \oplus \sum_{j=0}^{B_1-1} c^j_{\boldsymbol{m},i} x_j}_{t^1_{\boldsymbol{m},i}} \oplus \underbrace{\sum_{j=B_1}^{B_1+B_2-1} c^j_{\boldsymbol{m},i} x_j}_{t^2_{\boldsymbol{m},i}}$$

We group them in sets $M_i(\boldsymbol{c}_2) = \{\boldsymbol{m} \mid \forall j < B_2,\ c^{B_1+j}_{\boldsymbol{m},i} = c^j_2\}$ where $\boldsymbol{c}_2$ is a length B_2 vector and we define the function f_i as follows:

$$f_i(\boldsymbol{c}_2) = \sum_{\boldsymbol{m} \in M_i(\boldsymbol{c}_2)} (-1)^{t^1_{\boldsymbol{m},i}}$$

The Walsh transform of f_i is:

$$F_i(\boldsymbol{X}_2) = \sum_{\boldsymbol{c}_2} f_i(\boldsymbol{c}_2)(-1)^{\boldsymbol{c}_2 \cdot \boldsymbol{X}_2}$$

When $\boldsymbol{X}_2 = [x_{B_1}, x_{B_1+1}, \ldots, x_{B-1}]$, we have $F_i(\boldsymbol{X}_2) = \sum_{\boldsymbol{m}} (-1)^{t^1_{\boldsymbol{m},i} \oplus t^2_{\boldsymbol{m},i}}$. So $F_i(\boldsymbol{X}_2)$ is the difference between the number of predicted 0 and the number of predicted 1 for the bit z_i, given the choice $\boldsymbol{X} = [\boldsymbol{X}_1, \boldsymbol{X}_2]$ for the B initial guessed bits. Thus a single Walsh transform can evaluate this difference for the 2^{B_2} choices of the B_2 bits.

The computation of $f_i(\boldsymbol{c}_2)$ for every $\boldsymbol{c}_2$ in $\mathbb{F}_2^{B_2}$ requires 2^{B_2} steps for the initialization and Ω steps for the evaluation of each parity-checks, whereas the Walsh transforms takes a time proportional to $2^{B_2} \log_2 2^{B_2} = 2^{B_2} B_2$. Since these calculations are done for every bit among the D considered ones and for each guess of the B_1 bits, the complexity of this part of the decoding is:

$$\begin{aligned} C_1 &= \mathcal{O}(2^{B_1} D(2^{B_2} + \Omega + 2^{B_2} B_2)) \\ &= \mathcal{O}(2^B D(\frac{\Omega}{2^{B_2}} + B_2)) \end{aligned}$$

Choosing $B_2 = \log_2 \Omega$, we get $C_1 = \mathcal{O}(2^B D \log_2 \Omega)$. This should be compared to the complexity using the straightforward approach: $C_1' = \mathcal{O}(2^B D \Omega)$.

Once $F_i(\boldsymbol{X}_2)$ is evaluated, predicting the corrected values $\hat{x}_i$ can be done with a simple procedure: for each i among the D considered bits, we have a function $F_i(\boldsymbol{X}_2)$. If the value of this function for a given value of $\boldsymbol{X}_2$ is far enough from zero, then the computed value of z_i that dominates among the $|\Omega_i|$ parity-check equations has a big probability to be the correct value of x_i (see Fig. 4). Let us call θ the threshold on the function $F_i(\boldsymbol{X}_2)$; we thus predict that, for a given $\boldsymbol{X}_2$:

$$x_i = \begin{cases} 0 \text{ if } F_i(\boldsymbol{X}_2) > \theta \\ 1 \text{ if } F_i(\boldsymbol{X}_2) < -\theta \end{cases}$$

Checking Part In order for our algorithm to be succesful, we need to have at least $L - B$ correctly predicted bits among the D considered bits. However, every predicted bit can sometimes be wrong. In order to increase the overall probability of success, let us introduce an extra step in the algorithm.

When the number of correctly predicted bits is less than $L-B$, the algorithm has failed. When this number is exactly equal to $L - B$, we can only hope that none of these bits is wrong. But when we have more than $L - B$ predicted bits, namely $L - B + \delta$, the probability that, among these bits, $L - B$ are correctly predicted greatly increase. So we add to the procedure an exhaustive search on all subset of size $L - B$ among the $L - B + \delta$ bits, in order to find at least one full correct prediction. Every candidate is then checked by iterating the LFSR and computing the correlation between the newly generated keystream x_i and the original one z_i.

If p_{err} is the probability that a wrong guess gives us at least $L - B + \delta$ predicted bits, then the checking part of the processing stage has a complexity of:

$$C_2 = \mathcal{O}((1 + p_{err}(2^B - 1)) \binom{L - B + \delta}{\delta} C_3)$$

since among the $2^B - 1$ wrong guesses, $p_{err}(2^B - 1)$ will be kept for checking and the 1 being there for the correct guess that should be kept. C_3 is the complexity of a single checking, i.e. $C_3 = \mathcal{O}(\frac{1}{\varepsilon^2})$.

The total complexity of the processing stage of the algorithm is then:

$$C = \mathcal{O}(2^B D \log_2 \Omega + (1 + p_{err}(2^B - 1)) \binom{L - B + \delta}{\delta} \frac{1}{\varepsilon^2})$$

4 Performance and Implementation

In this section, we present experimental and theoretical results of our algorithm applied to two LFSRs of lengths 40 and 89 bits. Optimal parameters were computed according to appendix A.

Table 2. Complexity of the cryptanalysis for a probability of success p_{succ} close to 1. The LFSR polynomial is $1 + x + x^3 + x^5 + x^9 + x^{11} + x^{12} + x^{17} + x^{19} + x^{21} + x^{25} + x^{27} + x^{29} + x^{32} + x^{33} + x^{38} + x^{40}$.

Algorithm	*Noise*	*Required Sample*	*Complexity*
FSE'2001 [9]	0.469	400000	$\sim 2^{42}$
FSE'2001 [9]	0.490	360000	$\sim 2^{55}$
Our algorithm	0.469	80000	$\sim 2^{31}$
Our algorithm	0.490	80000	$\sim 2^{40}$

4.1 40-Bit Test LFSR

The chosen LFSR is the standard register used in many articles. The attack on the LFSR with noise $1 - p = 0.469$ has been implemented in C on a Pentium III and provides results in a few days for the preprocessing stage and a few minutes for the decoding stage. After optimization of all the parameters ($D = 64$, $\delta = 3$, $B = 18$ and $k = 4$ for the first case, $D = 30$, $\delta = 3$, $B = 28$, $k = 4$ for the second one) the results are presented on Table 2. Results for $1 - p = 0.490$ are only theoretical. The gain on the complexity is at least equal to 2^{11}: it comes primarily from the Walsh transform. Moreover, the required length is five times smaller. This represents a major improvement on the time complexity of one-pass fast correlation attacks.

4.2 89-Bit LFSR Theoretical Result

For a 89-bit LFSR, only theoretical results are provided on Table 3. The expected time complexity is 2^8 times smaller than previous estimations. Moreover the required sample length has decreased in a large amount. The parameters for our algorithm are $D = 128$, $\delta = 4$, $B = 32$ and $k = 4$.

5 Conclusion

In this paper, we presented new algorithmic improvements to fast correlation attacks. These improvements yield better asymptotic complexity than previous techniques for finding and evaluating parity-checks, enabling us to cryptanalyze larger registers with smaller correlations. Experimental results clearly show the gain on efficiency that these new algorithmic techniques bring to fast correlation attacks.

Table 3. Complexity of the cryptanalysis for a probability of success p_{succ} close to 1 on a 89-bit LFSR

Algorithm	*Noise*	*Required Sample*	*Complexity*
FSE'2001 [9]	0.469	2^{38}	2^{52}
Our algorithm	0.469	2^{28}	2^{44}

References

1. D. Boneh, A. Joux, and P. Nguyen. Why textbook ElGamal and RSA encryption are insecure. In *Proceedings of ASIACRYPT'2000*, volume 1976 of *Lecture Notes in Computer Science*, pages 30–43. Springer, 2000.
2. A. Canteaut and M. Trabbia. Improved fast correlation attacks using parity-check equations of weight 4 and 5. In *Advances in Cryptology – EUROCRYPT'00*, volume 1807 of *Lecture Notes in Computer Science*, pages 573–588. Springer Verlag, 2000.
3. V. V. Chepyzhov, T. Johansson, and B. Smeets. A simple algorithm for fast correlation attacks on stream ciphers. In *Fast Software Encryption – FSE'00*, volume 1978 of *Lecture Notes in Computer Science.* Springer Verlag, 2000.
4. É. Jaulmes and A. Joux. Cryptanalysis of pkp: a new approach. In *Public Key Cryptography 2001*, volume 1992 of *Lecture Notes in Computer Science*, pages 165–172. Springer, 2001.
5. T. Johansson and F. Jönsson. Fast correlation attacks through reconstruction of linear polynomials. In *Advances in Cryptology – CRYPTO'00*, volume 1880 of *Lecture Notes in Computer Science*, pages 300–315. Springer Verlag, 2000.
6. A. Joux and R. Lercier. "Chinese & Match", an alternative to atkin's "Match and Sort" method used in the SEA algorithm. Accepted for publication in Math. Comp., 1999.
7. W. Meier and O. Staffelbach. Fast correlation attacks on certain stream ciphers. *Journal of Cryptology*, 1:159–176, 1989.
8. M. Mihaljević, M. P. C. Fossorier, and H. Imai. A low-complexity and high-performance algorithm for fast correlation attack. In *Fast Software Encryption – FSE'00*, pages 196–212. Springer Verlag, 2000.
9. M. Mihaljević, M. P. C. Fossorier, and H. Imai. Fast correlation attack algorithm with list decoding and an application. In *Fast Software Encryption – FSE'01*, pages 208–222. Springer Verlag, 2001. Pre-proceedings, final proceedings to appear in LNCS.
10. W. T. Penzhorn and G. J. Kuhn. Computation of low-weight parity checks for correlation attacks on stream ciphers. In *Cryptography and Coding – 5th IMA Conference*, volume 1025 of *Lecture Notes in Computer Science*, pages 74–83. Springer, 1995.
11. R. Schroeppel and A. Shamir. A $T = O(2^{n/2})$, $S = O(2^{n/4})$ algorithm for certain NP-complete problems. *SIAM J. Comput.*, 10(3):456–464, 1981.
12. T. Siegenthaler. Correlation-immunity of nonlinear combining functions for cryptographic applications. *IEEE Trans. on Information Theory*, IT-30:776–780, 1984.
13. T. Siegenthaler. Decrypting a class of stream ciphers using ciphertext only. *IEEE Trans. Comput.*, C-34:81–85, 1985.

A Optimal Parameters

In this appendix, we evaluate quantities needed in order to optimize parameters of the algorithm (B, D, θ and δ). We first look at the probability of successful decoding, i.e. the probability that the right guess gives us the right complete initial filling of the LFSR. Then we will evaluate the probability of false alarm, i.e. the probability that, having done a wrong guess, the algorithm outputs a full initial filling of the LFSR. This probability of false alarm enters in the complexity evaluation of our algorithm. In all our experimental results, the parameters were tuned to get a probability of success higher than 0.99.

A.1 Probability of Successful Decoding

Let us first suppose we have done the right guess for the B bits. Let us write $q = \frac{1}{2}(1+\varepsilon^{k-1})$ the probability for one parity-check equation to yield the correct prediction. Then the probability that at least $\Omega - t$ parity-check equations predict the correct result is:

$$P_1(t) = \sum_{j=\Omega-t}^{\Omega} (1-q)^{\Omega-j} q^j \binom{\Omega}{j}$$

Let t be the smallest integer such that $D\ P_1(t) \geq L - B + \delta$ (t is related to the former parameter θ by $\theta = \Omega - 2t$). Then we have, statistically, at least $L - B + \delta$ predicted bits in the selection of D bits, and we are able to reconstruct the initial state of the LFSR. The probability that at least $\Omega - t$ parity-check equations predict the wrong result is:

$$P_2(t) = \sum_{j=\Omega-t}^{\Omega} q^{\Omega-j}(1-q)^j \binom{\Omega}{j}$$

Let p_V be the probability that a bit is correctly predicted, knowing that we have at least $\Omega - t$ parity-check equations that predict the same value for this bit: $p_V = \frac{P_1(t)}{P_1(t)+P_2(t)}$. Then

$$p_{succ} = \sum_{j=0}^{\delta} \binom{L-B+\delta}{j} p_V^{L-B+\delta-j}(1-p_V)^j$$

is the probability that at most δ bits are wrong among the $L - B + \delta$ predicted bits, i.e. the probability of success of the first part of our algorithm.

A.2 Probability of False Alarm

The probability that a wrong guess gives at least $\Omega - t$ identical predictions among the Ω parity-check equations for a given bit i is

$$E(t) = \frac{1}{2^{\Omega-1}} \sum_{j=\Omega-t}^{\Omega} \binom{\Omega}{j}$$

since the probability that a parity-check equation is verified in this case is $\frac{1}{2}$. We can then deduce the probability that, with a wrong guess, more than $L - B + \delta$ bits are predicted, i.e. the probability of false alarm of the first part of our algorithm:

$$p_{err} = \sum_{j=L-B+\delta}^{D} \binom{D}{j} E(t)^j (1-E(t))^{D-j}$$

BDD-Based Cryptanalysis of Keystream Generators

Matthias Krause

Theoretische Informatik, Universität Mannheim, 68131 Mannheim, Germany,
krause@informatik.uni-mannheim.de

Abstract. Many of the keystream generators which are used in practice are LFSR-based in the sense that they produce the keystream according to a rule $y = C(L(x))$, where $L(x)$ denotes an internal linear bitstream, produced by a small number of parallel linear feedback shift registers (LFSRs), and C denotes some nonlinear compression function. We present an $n^{O(1)}2^{(1-\alpha)/(1+\alpha)n}$ time bounded attack, the FBDD-attack, against LFSR-based generators, which computes the secret initial state $x \in \{0,1\}^n$ from cn consecutive keystream bits, where α denotes the rate of information, which C reveals about the internal bitstream, and c denotes some small constant. The algorithm uses Free Binary Decision Diagrams (FBDDs), a data structure for minimizing and manipulating Boolean functions. The FBDD-attack yields better bounds on the effective key length for several keystream generators of practical use, so a $0.656n$ bound for the self-shrinking generator, a $0.6403n$ bound for the A5/1 generator, used in the GSM standard, a $0.6n$ bound for the E_0 encryption standard in the one level mode, and a $0.8823n$ bound for the two-level E_0 generator used in the Bluetooth wireless LAN system.

1 Introduction

A keystream generator is a finite automaton, E, determined by a set Q of inner states, a state transition function $\delta_E : Q \longrightarrow Q$, and an output function $a_E : Q \longrightarrow \{0,1\}$. The usual case is that $Q = \{0,1\}^n$ for some integer $n \geq 1$, n is called the keylength of E. Starting from an initial state $x^0 \in Q$, in each time unit i, E outputs a key bit $y_i = a_E(x^i)$ and changes the inner state according to $x^{i+1} = \delta_E(x^i)$. For each initial state $x \in \{0,1\}^n$ we denote by $y = E(x)$ the keystream $y = y_0, y_1, \ldots$ produced by E when starting on x.

Keystream generators are designed for fast online encryption of bitstreams which have to pass an insecure channel. A standard application is to ensure the over-the-air privacy of communicating via mobile cellular telephones. A plaintext bit stream $p_0, p_1, p_2, \ldots$ is encrypted into a ciphertext bitstream $e_0, e_1, e_2, \ldots$ via the rule $e_i = p_i \oplus y_i$, where $y = E(x)$. The legal receiver knows x and can decrypt the bitstream using the same rule. The only secret information is the initial state x, which is exchanged before starting the transmission using a suitable key-exchange protocol. It is usual to make the pessimistic assumption that an attacker knows not only the encrypted bitstream, but even some short piece of

L.R. Knudsen (Ed.): EUROCRYPT 2002, LNCS 2332, pp. 222–237, 2002.

the plaintext, and, therefore, can easily compute some piece of the keystream. Consequently, one necessary criterion for the security of a keystream generator is that there is no feasible way to compute the secret initial state x from $y = E(x)$. Observe that a trivial exhaustive search attack needs time $n^{O(1)}2^n$.

In this paper we suggest a new type of attack against keystream generators, which we will call FBDD-attack, and show that LFSR-based keystream generators are vulnerable against FBDD-attacks. We will call a generator to be LFSR-based if it consists of two components, a linear bitstream generator L which generates for each initial state $x \in \{0,1\}^n$ a linear bitstream $L(x)$ by one or more parallel LFSRs, and a compressor C which transforms the internal bitstream into the output keystream $y = C(L(x))$. Due to the ease of implementing LFSRs in hardware, and due to the nice pseudorandomness properties of bitstreams generated by maximal length LFSRs, many keystream generators occuring in practice are LFSR-based.

FBDD is the abbreviation for *free binary decision diagrams*, a data structure for representing and manipulating Boolean function, which were introduced by *Gergov* and *Meinel* in [9] and *Sieling* and *Wegener* in [16]. Due to specific algorithmic properties, FBDDs and in particular Ordered BDDs (OBDDs), a special kind of FBDDs, became a very usefull tool in the area of automatic hardware verification (see also the paper of *Bryant* [4] who initiated the study of graph-based data structures for Boolean function manipulation). The important properties of FBDDs are that they can be efficiently minimized, that they allow an efficient enumeration of all satisfying assignments, and that the minimal FBDD-size of a Boolean function is not much larger than the number of satisfying assignments. We show that these properties can also be successfully used for cryptanalysis. The problem of finding a secret key x fulfilling $y = E(x)$ for a given encryption algorithm E and a given ciphertext y can be reduced to finding the minimal FBDD P for the decision if x fulfils $y = E(x)$. If the length of y is close to the unicity distance of E then P is small, and x can be efficiently computed from P. The main result of this paper is that the special structure of LFSR-based keystream generators implies a nontrivial dynamic algorithm for computing this FBDD P.

In particular, the weakness of LFSR-based keystream generators is that the compressor C has to produce the keystream in an online manner. For getting a high bitrate, C should use only a small memory, and should consume only a few new internal bits for poducing the next output bit. These requirements imply that the decision if an internal bitstream z generates a prefix of a given keystream y via C can be computed by small FBDDs. This allows to compute dynamically a sequence of FBDDs P_m, $m \geq n$, which test a given initial state $x \in \{0,1\}^n$ whether $C(L_{\leq m}(x))$ is prefix of y, where $L_{\leq m}(x)$ denotes the first m bits of the internal linear bitstream generated via L on x. On average, the solution becomes unique for $m \geq \lceil \alpha^{-1} n \rceil$, where α denotes the rate of information which C reveals about the internal bitstream. The FBDDs P_m are small at the beginning and again small if m approaches $\lceil \alpha^{-1} n \rceil$, and we will show that all intermediate FBDDs have a size of at most $n^{O(1)}2^{(1-\alpha)/(1+\alpha)n}$. For all m the FBDD P_m has to read the first $\lceil \gamma m \rceil$ bits of the keystream, where γ denotes the best case compression ratio of C. Thus, our algorithm computes the secret initial state x from the first $\lceil \gamma \alpha^{-1} n \rceil$ bits of $y = E(x)$. Observe that $\gamma = \alpha$ if C consumes always the same number of internal bits for producing one output bit, and $\alpha < \gamma$ if not. It holds $\gamma \alpha^{-1} \leq 2.5$ in all our examples. Note that for gaining a high bit-rate, α and γ should be as large as possible. Our results say that a higher bit-rate has to be paid by a loss of security.

One advantage of the FBDD-attack is that it is a *short-keystream* attack, i.e., the number of keybits needed for computing the secret initial state $x \in \{0,1\}^n$ is at most cn for some small constant $c \geq 1$. We apply the FBDD-attack to some of the keystream generators which are most intensively discussed in the current literature, the A5/1 generator which is used in the GSM standard, the self-shrinking generator, and the E_0-encryption standard, which is included in the Bluetooth wireless LAN system. For all theses ciphers, the FBDD-attack has a better time behaviour than *all* short-keystream attacks known before. In some cases there have been obtained *long-keystream* attacks which have a better time behaviour. They use a time-memeory tradeoff technique suggested by *Golić* [8] and are based on the assumption that a long piece of keystream of length 2^{dn}, $d < 1$ some constant, is available. We give an overview on previous results and the relations to our results in section 7.

The paper is organized as follows. In section 2 we give some formal definitions concerning LFSR-based keystream generators and present the keystream generators which we want to cryptanalyze. In section 3 FBDDs are introduced. In section 4 we derive FBDD-relevant properties of LFSR-based keystream generators. In section 5 we define the relevant parameters of LFSR-based keystream generators and formulate the main result. Our cryptanalysis algorithm is presented in section 6 and is applied to particular generators in section 7. Due to space restrictions we cannot present all proofs and BDD-constructions in this extended abstract. The complete version of this paper can be received from [11].

At the first glance it may seem contradictory that we consider practical ciphers like the A5/1 generator with variable keylength n. But observe that definitions of LFSR-based keystream generators, even if they originally were designed for fixed keylength, can be generalized to variable keylength in a very natural way, simply by considering the internal LFSRs to have variable length. Considering variable key-length n allows to evaluate the security of ciphers in terms of how many polynomial time operations are necessary for breaking the cipher. This 'rough' way of security evaluation is sufficient in our context, since the aim of this paper is to present only the general algorithmic idea of the FBDD-attack, to give a rather rough estimation of the time behaviour, and to show an inherent weakness of LFSR-based generators. For practical implementations of FBDD attacks against real-life generators much more effort has to be invested for making the particular polynomial time operations as efficient as possible, see the discussion in section 8.

2 LFSR-Based Keystream Generators

Let us call a keystream generator to be LFSR-based if the generation rule $y = E(x)$ can be written as $y = C(L(x))$, where L denotes a linear bitstream generator consisting of one or more LFSRs, and $C : \{0,1\}^* \longrightarrow \{0,1\}^*$ a nonlinear compression function, which transforms the internal linear bitstream $L(x)$ into the nonlinear (compressed) output keystream $y = C(L(x))$.[1] Formally, an n-LFSR is a device which produces a bitstream

$$L(x) = L_0(x), L_1(x), \ldots, L_m(x), \ldots$$

on the basis of a public string $c = (c_1, \ldots, c_n) \in \{0,1\}^n$, the generator polynomial, and a secret initial state $x = (x_0, \ldots, x_{n-1}) \in \{0,1\}^n$, according to the relation $L_i(x) = x_i$ for $0 \leq i \leq n-1$ and

$$L_m(x) = c_1 L_{m-1}(x) \oplus c_2 L_{m-2}(x) \oplus \ldots \oplus c_n L_{m-n}(x) \qquad (1)$$

for $m \geq n$. Observe that for all $m \geq 1$, $L_m(x)$ is a GF(2)-linear Boolean function in $x_0, \ldots, x_{n-1}$ which can be easily determined via iteratively applying (1).

A linear bitstream generator L of keylength n is defined to be an algorithm which, for some $k \geq 1$, generates a linear bitstream $L(x)$

$$L(x) = L_0(x), L_1(x), L_2(x), \ldots$$

by k parallel LFSRs $L^0, \ldots, L^{k-1}$ of keylengths $n_0, \ldots, n_{k-1}$, where $n = n_0 + \ldots + n_{k-1}$. The initial states $x \in \{0,1\}^n$ for L are formed by the initial states $x^r \in \{0,1\}^{n_r}$, $r = 0, \ldots, k-1$, of $L^0, \ldots, L^{k-1}$. L produces in each time unit $j \geq 0$ the bit $L_j(x)$ according to the rule $L_j(x) = L^r_s(x^r)$, where $r = j \bmod k$ and $s = j \operatorname{div} k$. Observe that for all $j \geq 0$ $L_j(x)$ is a GF(2)-linear function in x.

The motivation for taking LFSRs as building blocks for keystream generators is that they can be easily implemented using n register cells connected by a feedback channel. Moreover, if the generator polynomial is primitive, they produce bit streams with nice pseudorandomness properties (maximal period, good auto correlation and local statistics). See, e.g., the monograph by *Golomb* [10] or the article by *Rueppel* [15] for more about the theory of shift register sequences. Clearly, LFSR-sequences alone do not provide any cryptographic security. Thus, the aim of the compression function $C : \{0,1\}^* \longrightarrow \{0,1\}^*$ is to destroy the low linear complexity of the internal linear bit stream while preserving its nice pseudorandomness properties. Many keystream generators occuring in practice are LFSR-based in the above sense. We investigate the following types of generators.

One classical construction (see, e.g., [10] and [15]) is to combine k parallele LFSRs $L^0, \ldots, L^{k-1}$ with an appropriate connection function $c : \{0,1\}^k \longrightarrow$

[1] C compresses the internal bit-stream in an online manner, i.e., $C(z')$ is prefix of $C(z)$ if z' is prefix of z, for all $z, z' \in \{0,1\}^*$. This justifies to write $y = C(L(x))$ despite of the fact that $L(x)$ is assumed to be infinitely long.

$\{0,1\}$. The keystream $y = y_0, y_1, y_2, \ldots$ is defined via the rule

$$y_j = c(L_j^0(x^0), \ldots, L_j^{k-1}(x^{k-1})),\ j \geq 0,$$

where x^r denotes the initial state for L^r, for $r = 0 \ldots, k-1$. Let us call generators of this type **Connect-k** constructions.

The Self-Shrinking Generator was introduced by *Meier* and *Staffelbach* in [14]. It consists of only one LFSR L. The compression is defined via the shrinking function *shrink* : $\{0,1\}^2 \longrightarrow \{0,1,\varepsilon\}$, defined as $shrink(ab) = b$, if $a = 1$, and $shrink(ab) = \varepsilon$, the empty word, otherwise. The shrinking-function is extended to bit-strings of even length r as

$$shrink(z_0 z_1 \ldots z_{r-1}) = y_0 y_1 \cdots y_{r/2-1},$$

where $y_i = shrink(z_{2i} z_{2i+1})$ for $i = 0, \ldots, r/2 - 1$. For each initial state x for L, the self-shrinking generator produces the keystream y according to $y = shrink(L(x))$.

The E_0 Generator is the keystream generator used in the Bluetooth wireless LAN system [3]. It is defined as $E_0(x) = C(L(x))$, where the linear bitstream generator L of E_0 consists of 4 LFSRs $L^0, \ldots, L^3$. The compression is organized by a finite automaton M with external input alphabet $\{0,1,2\}$, state space $Q = \{0,1,\ldots,15\}$ and output alphabet $\{0,1\}$, which is defined by an output function $a : Q \times \{0,1,2\} \longrightarrow \{0,1\}$ and a state transition function $\delta : Q \times \{0,1,2\} \longrightarrow Q$. The exact specification of M is published in [3] but does not matter for our purpose and is therefore omitted.

The compression $C(z) = y = y_0 y_1 \ldots y_{m-1}$ of an internal bit-stream

$$z = z_0^0 z_0^1 z_0^2 z_0^3 z_1^0 z_1^1 z_1^2 z_1^3 \ldots z_{m-1}^0 z_{m-1}^1 z_{m-1}^2 z_{m-1}^3$$

is defined as $y_j = a(q_j, s_j) \oplus t_j$, where $s_j = (z_j^0 + z_j^1 + z_j^2 + z_j^3)$ div 2 and $t_j = (z_j^0 + z_j^1 + z_j^2 + z_j^3)$ mod 2, for all $0 \leq j \leq m-1$. The actual inner state is updated in each cycle according to the rule $q_{j+1} = \delta(q_j, s_j)$, where q_0 denotes the initial state of M. In practice, the E_0 generator is used with key length 128, the four LFSRs have lengths 39, 33, 31, 25.

The E_0 Encryption Standard in the two-level mode (for short, E_0^2-Generator) of keylength n combines two E_0 devices of internal keylength $N \geq n$ in the following way. For $x \in \{0,1\}^n$ it holds $y = E_0^2(x) = E_0(z)$, where z denotes the prefix of length N of $E_0(u)$, and where

$$u = (x_0, \ldots, x_{n-1}, U_n(x) \ldots, U_{N-1}(x)).$$

U_i, $i = n \ldots N-1$, are public GF(2)-linear functions in $(x_0, \ldots, x_{n-1})$. In practice, the string u results from putting n secret bits together with $N - n$ known dummy bits into the LFSRs and running them a certain number of steps. The Bluetooth system uses $N = 128$, and n can be chosen as 8, 16, 32, or 64. The reason for choosing a larger internal key length N is to achieve a larger effective key length in n.

The A5/1 Generator is used in the GSM standard. The definition was discovered by *Briceno et. al.* [5] via reverse engineering. The A5/1 generator of key-length n consists of 3 LFSRs L^0, L^1, and L^2 of key-lengthes n_0, n_1, and n_2. In each time step i, the output key bit y_i is the XOR of the actual output bits of the 3 LFSRs. A clock control decides in each timestep which of the 3 LFSRs are shifted, and which not. The clock control takes for all $k \in \{0,1,2\}$ a control value c_k from the $N_k - th$ register cell of L^k, and computes the control value $m = maj_3(c_0, c_1, c_2)$. [2] LFSR L^k is shifted if $m = c_k$, for $k = 0,1,2$. The control positions N_k are fixed and fulfil $N_k \in \left\{\lceil \frac{n_k}{2} \rceil - 1, \lceil \frac{n_k}{2} \rceil\right\}$.

This keystream generation rule can be written down in a $y = C(L(x))$ fashion in the following way. Given an internal bitstream $z = (z_0^0, z_0^1, z_0^2, \ldots, z_m^0, z_m^1, z_m^2, \ldots)$ the keystream $y = C(z)$ is defined as follows. In each timestep, C holds 3 output positions $i[0], i[1], i[2]$ and 3 control positions $j[0], j[1], j[2]$. C outputs $x_{i[0]}^0 \oplus x_{i[1]}^1 \oplus x_{i[2]}^2$, computes the new control value $m = maj_3(x_{j[0]}^0, x_{j[1]}^1, x_{j[2]}^2)$, and updates the i- and j-values via $i[k] := i[k] + 1$ and $j[k] := j[k] + 1$, for those $k \in \{0,1,2\}$ for which $m = x_{j[k]}^k$. The output positions are initialized by 0. The control positions are initialized by N_0, N_1, N_2. Note that in the GSM standard the A5/1 generator is used with key length 64, the 3 LFSRs have lengthes 19, 22 and 23

3 Binary Decision Diagrams (BDDs)

For m a natural number let X_m denote the set of m Boolean variables $\{x_0, \ldots, x_{m-1}\}$. A BDD P over X_m is an acyclic directed graph with inner nodes of outdegree 2, a distinguished inner node of indegree 0, the source, and two sink nodes of outdegree 0, one 0-sink and one 1-sink. All inner nodes, i.e. nodes of outdegree > 0, are labelled with queries $x_i?$, $0 \leq i \leq m-1$, and are left by one edge labelled 0 (corresponding to the answer $x_i = 0$) and one edge labelled 1 (corresponding to the answer $x_i = 1$). Each assignment b to the X_m-variables defines a unique computational path in P, which will be called the b-path in P. The b-path starts at the source, answers always b_i on queries $x_i?$ and, thus, leads to a unique sink. The label of this sink is defined to be the output $P(b) \in \{0,1\}$ of P on input $b \in \{0,1\}^m$. We denote by $One(P) \subseteq \{0,1\}^m$ the set of inputs accepted by P, $One(P) = \{b \in \{0,1\}^m;\ P(b) = 1\}$. Each BDD P over X_m computes a unique function $f : \{0,1\}^m \longrightarrow \{0,1\}$, by $f(b) = 1 \iff b \in One(P)$. The size of P, $|P|$, is defined to be the number of inner nodes of P. Two BDDs are called equivalent if they compute the same function.

We call an BDD P to be a *free* binary decision diagram (FBDD) if along each computational path in P each variable occurs at most once. In [9] and [16] it was observed that FBDDs can be efficiently minimized with respect to all equivalent FBDDs which read the input variables in an equivalent order. The equivalence of orders of reading the input variables is expressed by using the notion of graph orderings.

[2] maj_3 is defined to output $c \in \{0,1\}$ iff at least 2 of its 3 arguments have value c.

Definition 1. *A graph ordering G of X_m is an FBDD over X_m with only one (unlabelled) sink, for which on each path from the root to the sink all m variables occur.*

Graph orderings are not designed for computing Boolean functions. Their aim is to define for each assignments $b = (b_0, \ldots, b_{m-1})$ to X_m a unique variable ordering $\pi_G(b) = (x_{i_1(b)}, \ldots, x_{i_m(b)})$, namely the ordering in which the variables are requested along the unique b-path in G.

Definition 2. *An FBDD is called G-driven, for short, G-FBDD, if the ordering in which the variables are requested along the b-path in P respects $\pi_G(b)$, for all assignments b. I.e., there do not exist assignments b, variables x_i and x_j such that x_i is requested above x_j at $\pi_G(b)$, but below x_j at the b-path in P.*

A special, extensively studied variant of FBDDs are Ordered Binary Decision Diagrams (OBDDs). An FBDD P is called OBDD with variable ordering π (for short π-OBDD) if all pathes in P respect π.

We need the following nice algorithmic properties of graph-driven FBDDs. Let $f, g : \{0,1\}^m \longrightarrow \{0,1\}$ be Boolean functions, let G be a graph ordering for X_m, and let P and Q be G-driven FBDDs for f and g, respectively.

Property 3.01 *There is an algorithm MIN which computes from P in time $O(|P|)$ the (uniquely defined) minimal G-driven FBDD* $\min(P)$ *for f.*

Property 3.02 *It holds that* $|\min(P)| \leq m|One(P)|$.

Property 3.03 *There is an algorithm $SYNTH$ which computes in time $O(|P||Q||G|)$ a G-driven FBDD $P \wedge Q$, $|P \wedge Q| \leq |P||Q||G|$, which computes $f \wedge g$.*

Property 3.04 *There is another algorithm SAT which enumerates all elements in $One(P)$ in time $O(|One(P)||P|)$.*

See, e.g., the book by *Wegener* [17] for a detailed description and analysis of the OBDD- and FBDD-algorithms. FBDDs together with the procedures MIN, $SYNTH$ and SAT will be the basic data structure used in our cryptanalysis. Instead giving explicite examples of FBDDs we refer to the following relation between algorithms and FBDDs, see, e.g. [12]. Let $F \subseteq \{0,1\}^*$ decision problem written as a sequence of Boolean functions $(F_m : \{0,1\}^m \longrightarrow \{0,1\})_{m \in \mathbb{N}}$.

Theorem 1. *Suppose that F can be decided by an algorithm A which reads each input bit at most once and uses for all $m \geq 1$ and all inputs of length m at most $s(m)$ memory cells. Then, for all $m \geq 1$, A provides an efficient construction of an FBDD P_m of size $O(m2^{s(m)})$ for F_m.* □

Scetch of a Proof. The inner nodes of P_m are organized in m disjoint levels. Each level j, $1 \leq j \leq m$, contains potentially one node v_r^j for each possible assignment $r \in \{0,1\}^{s(m)}$ to the memory cells used by A on inputs of length m. This implies the size bound claimed in the Theorem. P_m is constructed top down. The first inner node we insert to P_m is the root $v_{r_0}^1$ labelled by x_{i_1}, where r_0 denotes the initial assignment of the memory cells and i_1 the position of the first input bit read by A. If we have inserted an inner node $v = v_r^j$, labelled by x_i, to P_m than for all $b \in \{0,1\}$ the b-successor v' of v is constructed as follows. If A starting on configuration r after reading $x_i = b$ stops accepting (resp. rejecting) without reading another input bit then v' is defined to be the 1-sink (resp. the 0-sink). Otherwise let r' denote the first configuration when A is going to read a new input bit, say x_k. Then v' is defined to be $v_{r'}^{j+1}$ labelled by x_k. □

4 FBDD-Aspects of Key-Stream Generators

Let E be a LFSR-based keystream generator of key-length n with linear keystream generator L and compression function $C : \{0,1\}^* \longrightarrow \{0,1\}^*$. Let $x \in \{0,1\}^n$ denote an initial state for L.

Definition 3. *For all $m \geq 1$ let G_m^C denote the graph ordering, which assigns to each internal bitstream z the order in which C reads the first m bits of z.*

Observe that for the E_0 generator, the self-shrinking generator, as well as for Connect-k generators, the order in which the compressor reads the internal bits does not depend of the internal bitstream itself, i.e., G_m^C has size m and G_m^C-driven FBDDs are OBDDs. But in the case of the A5/1 generator, this order is governed by the clock control, and can be different for different inputs. The efficiency of our cryptanalysis algorithm is based on the following FBDD assumption on E.

FBDD Assumption. The graph ordering G_m^C has polynomial size in m. Moreover, for arbitrary keystreams y, the predicate if for given $z \in \{0,1\}^m$ $C(z)$ is prefix of y can be computed by G_m^C-FBDDs of polynomial size in m.

It is quite easy to see that the compression function of a Connect-k generators, defined by a function $c : \{0,1\}^k \longrightarrow \{0,1\}$, fulfils the FBDD-assumption. The compressor reads the internal bits in the canonical order $\pi = 0,1,2,3,\ldots$. Linear size π-OBDDs which decide whether $z \in \{0,1\}^m$ generates the first $\lfloor m/k \rfloor$ bits of a given keystream y via c can be constructed, via Theorem 1, according to the following algorithm.

1. For $j := 0$ to $\lfloor m/k \rfloor$
2.   if $c(z_j^0, \ldots, z_j^{k-1}) \neq y_j$ then stop(0)
3. stop(1)

Quadratic size π-OBDDs which decide for $z \in \{0,1\}^m$ whether $shrink(z)$ is prefix of a given keystream y can be constructed, via Theorem 1, according to the following algorithm

1. $k := 0, j := 0$
2. while $j < m-1$
3. if $z_j = 0$
4. then $j := j+2$
5. else
6. if $z_{j+1} = y_k$
7. then $j := j+2,\ k := k+1$
7. else stop(0)
8. stop(1)

The FBDD constructions for all the E_0-, the E_0^2-, and the A5/1 are sketched in section 7, resp. can be studied in the long version [11].

We still need to estimate the size of FBDDs which decide whether a given $z \in \{0,1\}^m$ is a linear bit-stream.

Lemma 1. *For all $m \geq n$, the decision whether $z \in \{0,1\}^m$ is generated via linear bitstream generator L of keylength n can be computed by a G_m^C-driven FBDD of size at most $|G_m^C| 2^{m-n}$.*

Proof: Let V_m denote the set of inner nodes of G_m^C. We construct a G_m^C-driven FBDD R_m with the set $W_m = V_m \times \{0,1\}^{m-n}$ of inner nodes.

For all initial states $x \in \{0,1\}^n$ and all internal positions $j, n \leq j \leq m-1$, write $L_j(x)$ as

$$L_j(x) = \bigoplus_{k=0}^{n-1} L_{k,j} x_k.$$

G_m^C ensures that x_k is always read before x_j if $L_{k,j} = 1$.

Let the root of R_m be the node $(v_0, \vec{0})$ where v_0 denotes the root of G_m^C. Let all nodes (v,b) have the same label as v does in G_m^C. The edges of R_m are defined according to the following rules. Let $v \in V_m$ and $b = (b_n, \ldots, b_{m-1}) \in \{0,1\}^{m-n}$ be arbitrarily fixed. For $c \in \{0,1\}$ let $v(c)$ be the c-successor of v in G_m^C. We have to distinguish two cases.

- v is labelled with some x_k, $0 \leq k \leq n-1$. Then, for all $c \in \{0,1\}$, the c-successor of (v,b) is $(v(c), b(c))$, where $b(c) = (b_0 \oplus L_{k,n} c, \ldots, b_{r-1} \oplus L_{k,m-1} c)$.
- v is labelled with some x_j, $n \leq j \leq m-1$. Then, for all $c \in \{0,1\}$, if $b_{j-n} \neq c$, the c-successor of (v,b) is the 0-sink. If $b_{j-n} = c$ and $v(c)$ is the *-sink, then let the c-successor of (v,b) be the 1-sink. Otherwise let the c-successor of (v,b) be $(v(c), b)$.

It can be easily checked that R_m (after removing non-reachable nodes) matches all requirements of the Lemma. □

5 The Main Result

We fix an LFSR-based keystream generator of key-length n with linear bit-stream generator L and a compression function C. We assume that for all

$m \geq 1$ the probability that $C(z)$ is prefix of y for a randomly chosen and uniformly distributed $z \in \{0,1\}^m$ is the same for all keystreams y. Observe that all generators occuring in this paper have this property. Let us denote this probability by $p_C(m)$.

The cost of our cryptanalysis algorithm depends on two parameters of C. The first is the information rate (per bit) which a keystream y reveales about the first m bits of the underlying internal bitstream. It can be computed as

$$\frac{1}{m}I(Z^{(m)},Y) = \frac{1}{m}\left(H(Z^{(m)}) - H(Z^{(m)}|Y)\right) =$$

$$= \frac{1}{m}\left(m - \log(p_C(m)2^m)\right) = -\frac{1}{m}\log(p_C(m)). \quad (2)$$

where $Z^{(m)}$ denotes a random $z \in \{0,1\}^m$ and Y a random keystream.

As the compression algorithm computes the keystream in an online manner, the time difference between two succeeding key bits should be small in the average, and not vary too much. This implies the following partition rule: Each internal bit-stream z can be divided into consecutive elementary blocks $z = z^0z^1\ldots z^{s-1}$, such that $C(z) = y_0y_1\ldots y_{s-1}$ with $y_j = C(z^j)$ for all $j = 0,\ldots,s-1$, and the average length of the elementary blocks is a small constant. This partition rule implies that $p_C(m)$ can be supposed to behave as $p_C(m) = 2^{-\alpha m}$, for a constant $\alpha \in (0,1]$. Due to (2), α coincides with the information rate of C.

The second parameter of C is the maximal number of output bits which C produces on internal bitstreams of length m. Due to the partition rule, this value can be supposed to behave as γm, for some constant $\gamma \in (0,1]$. We call γ to be the (best case) compression ratio of C.

Observe that if C always reads the same number k of internal bits for producing one output bit, then $\alpha = \gamma = \frac{1}{k}$. If this number is not a constant then α can be obtained by the formulae

$$2^{-\alpha m} = p_C(m) = \sum_{i=0}^{\lceil \gamma m \rceil} 2^{-i} Prob_z\left[|C(z)| = i\right], \quad (3)$$

where z denotes a random, uniformly distributed element from $\{0,1\}^m$. Observe that (3) yields $\gamma \geq \alpha$, i.e. $\gamma\alpha^{-1} \geq 1$.

For all $x \in \{0,1\}^n$ and $m \geq 1$ let $L_{\leq m}(x)$ denote the first m bits of $L(x)$. Note the following design criterion for well-designed keystream generators.

Pseudorandomness Assumption. For all keystreams y and all $m \leq \lceil \alpha^{-1} n \rceil$ it holds that

$$Prob_z\left[C(z) \text{ is prefix of y}\right] \approx Prob_x\left[C(L_{\leq m}(x)) \text{ is prefix of y}\right],$$

where z and x denote uniformly distributed random elements from $\{0,1\}^m$ and $\{0,1\}^n$, respectively.

Lemma 2. *If the keystream generator fulfils the above pseudorandomness assumption then for all keystreams y and $m \leq \alpha^{-1}n$ there are approximately $2^{n-\alpha m}$ initial states x for which $C(L_m(x))$ is prefix of y.* □

Observe that a severe violation of the pseudorandomness assumption implies the possibility of attacking the cipher via a correlation attack. Our main result can now be formulated as

Theorem 2. *Let E be an LFSR-based keystream generator of key-length n with linear bit-stream generator L, and compression function C of information rate α and (best case) compression ratio γ. Let C and L fulfil the BDD- and the pseudorandomness assumption. Then there is an $n^{O(1)}2^{(1-\alpha)/(1+\alpha)n}$-time bounded algorithm, which computes the secret initial state x from the first $\lceil \gamma\alpha^{-1}n \rceil$ consecutive bits of $y = C(L(x))$.*

As usual, we define the *effective key length* of a cipher of key length n to be the minimal number of polynomial time operations that are necessary to break the cipher. We obtain a bound of $\frac{1-\alpha}{1+\alpha}n$ for the effective key length of keystream generators which fulfil the above conditions.

6 The Algorithm

Let us fix n, E, L, C, α and γ as in Theorem 2. For all $m \geq 1$ let G_m denote the graph ordering defined by C on internal bitstreams of length m. Let y be an arbitrarily fixed keystream which was generated via E. For all $m \geq 1$ let Q_m denote a minimal G_m-FBDD which decides for $z \in \{0,1\}^m$ whether $C(z)$ is prefix of y. Observe that Q_m has to read the first $\lceil \gamma m \rceil$ bits of y. The FBDD-assumption yields that Q_m has polynomial size in m.

For $m \geq n$ let P_m denote the minimal G_m-driven FBDD which decides whether $z \in \{0,1\}^m$ is a linear bitstream generated via L and if $C(z)$ is prefix of y. Observe that by Property 3.03 and Lemma 1

Lemma 3. $|P_m| \leq |Q_m||G_m|^2 2^{m-n}$ *for all* $m \geq n$. □

The strategy of our algorithm is simple, it dynamically computes P_m for $m = n, \ldots, \lceil \alpha^{-1}n \rceil$. Lemma 2 implies that for $m = \lceil \alpha^{-1}n \rceil$ with high probability only one bit-stream z^* will be accepted by P_m. Due to property 3.04 this bit-stream can be efficiently computed. The first n components of z^* form the initial state that we are searching for.

For all $m \geq n$ let S_m denote a minimal G_m-FBDD which decides for $z = (z_0, \ldots, z_m)$ whether $z_m = L_m(z_0, \ldots, z_{n-1})$. From Lemma 1 we obtain that $|S_m| \leq 2|G_m|$. Now our algorithm can be formulated as

(1) $P := Q_n$
(2) For $m := n+1$ to $\lceil \alpha^{-1}n \rceil$
(3) $\quad\quad P := \min(P \wedge Q_m \wedge S_{m-1})$

For the correctness of the minimization in step (3) observe that the definition of G_m implies that G_m is $G_{m'}$-driven for all $m' \geq m$. It follows from the definitions that for all $m \geq n$ P coincides with P_m after iteration m.

The FBDD-operation $\min(P \wedge Q_m \wedge S_{m-1})$ takes time $p(m)|P_{m-1}|$ for some polynomial p. Consequently, the running time of the algorithm can be estimated by

$$n^{O(1)} \max\{|P_m|, \ m \geq n\}.$$

Observe that on the one hand, by Lemma 3, $|P_m| \leq p'(m)2^{m-n}$ for some polynomial p', while on the other hand, by Property 3.02 and Lemma 2, $|P_m| \leq m|One(P_m)|$, where

$$|One(P_m)| \approx 2^{n-\alpha m} = 2^{(1-\alpha)n-\alpha(m-n)}.$$

Consequently, $|P_m|$ does not exceed $n^{O(1)}2^{r(n)}$, where $r(n)$ is the solution of

$$2^{r(n)} = 2^{(1-\alpha)n-\alpha r(n)}$$

which yields $r(n) = \frac{1-\alpha}{1+\alpha}n$. We have proved Theorem 2. □

7 Applications

We apply Theorem 2 to the keystream generators introduced in section 2. We suppose that these generators fulfill the pseudorandomness assumption, otherwise the running time estimations of our cryptanalysis hold on average. It remains to determine the information rate and the compression ratio, and to prove that the FBDD-assumption is true. For the Connect-k construction it holds $\alpha = \gamma = \frac{1}{k}$. The FBDD-assumption has shown to be true in section 4.

Theorem 3. *For all $k \geq 2$ and all stream ciphers E of key-length n which are a Connect-k construction, our algorithm computes the secret initial state $x \in \{0,1\}^n$ from the first n bits of $y = E(x)$ in time $n^{O(1)}2^{\frac{k-1}{k+1}n}$.* □

This is, as far as we know, the best known general upper bound on the effective key-length of the Connect-k construction. Observe that the initial state of one LFSR can be computed efficiently if the initial states of the other LFSRs are known. This leads to a Divide-and-Conquer attack of time $n^{O(1)}2^{\frac{k-1}{k}n}$ which is slightly worth than our result.

For the E_0-encryption standard in the one-level mode we obtain $\alpha = \gamma = \frac{1}{4}$. The decision if a given internal keystream $z \in \{0,1\}^m$ yields a prefix of a given keystream y can be computed by π-OBDDs of linear size, see [11], i.e. E_0 fulfils the FBDD-assumption. We obtain

Theorem 4. *For the E_0-encryption standard with key-length n, our algorithm computes the secret initial state $x \in \{0,1\}^n$ from the first n bits of $y = E_0(x)$ in time $n^{O(1)}2^{0.6n}$.* □

Observe that $128 \cdot 0.6 \approx 77$. Note that the best known attacks against the E_0 generator of key length 128 were derived by *Fluhrer* and *Lucks* [7] and *Canniere* [6]. [7] contains a tradeoff result between time and length of available keystream. It varies from 2^{84} necessary encryptions if 132 bit are available to 2^{73} necessary encryptions if 2^{43} bits are available.

Let us now consider the E_0 generator in the two level mode with real key length n and internal key length $N \geq n$. Observe that E_0^2 needs $4 \cdot 4 = 16$ internal bits per key bit for producing the first $N/4$ key bits, while for later key bits only 4 internal bits per key bit are needed. Observe further that our algorithm reaches maximal FBDD-size in iteration $m^* := n + \frac{1-\alpha}{1+\alpha} n$. For $\alpha = 1/16$ this gives $m^* = 32/17n$. As $m^*/16 < N/4$ we obtain $\alpha = \gamma = 1/16$ as relevant parameters for our algorithm on E_0^2. The decision if a given internal keystream $z \in \{0,1\}^m$ yields a prefix of a given keystream y can be computed by π-OBDDs of size $O(m)$, where the constant hidden in O is quite large, see [11]. Taking into account that $\frac{1-\alpha}{1+\alpha} = \frac{15}{17} \approx 0.8824$ we get

Theorem 5. *For the E_0^2-encryption generator with key-length n, our algorithm computes the secret initial state $x \in \{0,1\}^n$ from the first n bits of $y = E_0^2(x)$ in time $n^{O(1)} 2^{0.8824n}$.* □

As far as we know this is the first nontrivial upper bound on the key length of the E_2^0 generator.

Concerning the self-shrinking generator observe that for all even m and all keystreams y, $shrink(z)$ is prefix of y for exactly $3^{m/2}$ strings z of length m. We obtain an information rate $\alpha = 1 - \log(3)/2 \approx 0.2075$ for the self-shrinking generator by evaluating the relation $2^{-\alpha m} 2^m = 3^{m/2}$. The (best case) compression ratio of the self-shrinking generator is obviously 0.5. That the self-shrinking generator fulfils the FBDD-condition was already shown in section 4. Taking into account that for $\alpha = 0.2075$ it holds $\frac{1-\alpha}{1+\alpha} \approx 0.6563$ and $0.5\alpha^{-1} \approx 2.41$ we get

Theorem 6. *For the self-shrinking generator of an n-LFSR L, our algorithm computes the secret initial state $x \in \{0,1\}^n$ from the first $\lceil 2.41n \rceil$ bits of $y = shrink(L(x))$ in time $n^{O(1)} 2^{0.6563n}$.* □

Observe that the best previously known short-keystream attacks against the self-shrinking generator were given by *Meier* and *Staffelbach* [14] ($2^{0.75n}$ polynomial time operations) and *Zenner et. al.* [19] ($2^{0.694n}$ polynomial time operations). *Mihaljević* [13] presented an attack which yields a tradeoff between time and length of available keystream. It gives $2^{0.5n}$ necessary polynomial time operations if $2^{0.5n}$ bits of keystream are available, and matches our bound of $2^{0.6563n}$ necessary polynomial time operation if $2^{0.3n}$ bits of keystream are available, which is a quite unrealistic assumption.

The difficulty in applying our algorithm to the A5/1 generator is that the compression algorithm reads most of the internal bits twice, one time for the clock control and a certain time later for producing an output key bit. Read-twice BDDs do not have any of the nice algorithmic properties 3.01 - 3.04, unless $P = NP$. For making the A5/1 generator accessable to our approach we have to modify the keystream generation rule. We define the internal bitstream to be mixed of 6 LFSR-sequences $L^0, \ldots, L^5$, instead of 3. The first 3 LFSR-sequences are generated by the 3 LFSRs of the A5/1 generator. They are used for producing the output bits. The sequences L^3, L^4, L^5 are used for computing the control values. They are shifted copies of the first 3 sequences, defined by the rules $L_j^{3+k} = L_{j+N_k}^k$, for $k = 0, 1, 2$. As theses rules are linear restrictions we get a linear bitstream generated by 6 LFSRs. For this modified version, the decision if a given internal keystream $z \in \{0,1\}^m$ yields a prefix of a given keystream y can be computed by G_m-FBDDs of size $O(m^3)$. G_m-denotes the graph ordering induced by the clock control. This is the only example where OBDDs do not suffice, we really need FBDDs which allow different variable orderings for different inputs.

The (best case) compression ratio of the modified version of A5/1 is $\gamma = \frac{1}{4}$, as either 4 or 6 new internal bits are used for producing the next output bit. It can be proved that the information rate α is the solution of

$$2^{1-4\alpha} = \frac{1}{4}\left(3 + 2^{2\alpha}\right),$$

which yields $\alpha \approx 0.2193$, see [11]. Taking into account that $\frac{1-\alpha}{1+\alpha} \approx 0.6403$ and $\gamma\alpha^{-1} \approx 1.14$ we obtain

Theorem 7. *For an A5/1 generator E of key length n, our algorithm computes the secret initial state x from the first $\lceil 1.14n \rceil$ bits of $y = E(x)$ in time $n^{O(1)}2^{0.6403n}$.* □

The best previously known short-keystream attack was given by *Golić* [8]. It is against a version of A5/1 generator with keylength 64, which slightly deviates from the specification discovered in [5]. A tight analysis of the time behaviour of *Golić's* attack, when applied to the real A5/1 generator, was given by *Zenner* in [18] and yields 2^{42} polynomial time operations. We get a marginal improvement, as $\lceil 64 \cdot 0.6403 \rceil = 41$. The best long-keystream attacks were given by *Biryukov*, *Shamir* and *Wagner* in [2], and *Biham* and *Dunkelman* in [1]. After a preprocessing of 2^{42} operations the first attack in [2] breaks the cipher within seconds on a modern PC if around 2^{20} bits of keystream are available. The second attack in [2] breaks the cipher within minutes after a preprocessing of 2^{48} operations and under the condition that around 2^{15} bits of keystream are available. The attack in [1] breaks the cipher within $2^{39.91}$ A5/1 clockings on the basis of $2^{20.8}$ available keystream bits.

8 Discussion

There are classical design criterions for keystream generators like a large period, a large linear complexity, correlation immunity and good local statistics. In this paper we suggest a new one: resistance against FBDD-attacks. We have seen that there are two strategies to achieve this resistance. The first is to highly compress the internal bitstream (as in the case of E_0^2). This implies a low bit-rate which is not desirable. The second strategy is to design the compression function C in such a way that the decision about the consistence of a given internal bitstream with a given output keystream requires exponential size FBDDs. It is an interesting challenge to look for such constructions. For demonstrating the universality of our approach we presented the FBDD-attack in a very general setting. The obvious disadvantage of this setting is that the algorithm needs a lot of space as all intermediate FBDDs have to be explicitely constructed. It is an interesting open question if the algorithmic idea of FBDD-minimization can be used in a more subtle way for getting, at least for some ciphers, an algorithm which is less space consuming. Another interesting direction of further research is to check whether the FBDD-attack could be successfully combined with other more sophisticated methods of cryptanalysis like the tradeoff attacks suggested in [8], [2] and [1]. Moreover, it would be interesting to clarify by experiments how much do the real sizes of the minimized intermediate FBDDs deviate from the pessimistic upper bounds proved in our analysis.

Acknowledgement

I would like to thank Stefan Lucks, Erik Zenner, Christoph Meinel, Ingo Wegener, Rüdiger Reischuk and some unknown referees for helpful discussions.

References

1. E. Biham, O. Dunkelman. Cryptanalysis of the A5/1 GSM Stream Cipher. Proc. of INDOCRYPT 2000, LNCS 1977, 43-51.
2. A. Biryukov, A. Shamir, D. Wagner. Real Time Cryptanalysis of A5/1 on a PC. Proc. of Fast Software Encryption 2000, LNCS 1978, 1-18.
3. Bluetooth SIG. Bluetooth Specification Version 1.0 B, http//:www.bluetooth.com/
4. R. E. Bryant. Graph-based algorithms for Boolean function manipulations. IEEE Trans. on Computers 35, 1986, 677-691.
5. M. Briceno, I. Goldberg, D. Wagner. A pedagogical implementation of A5/1. http//:www.scard.org, May 1999.
6. C. de Canniere. Analysis of the Bluetooth Stream Cipher. Master's Project COSIC, Leuven, 2001.
7. S. R. Fluhrer, S. Lucks. Analysis of the E_0 Encryption System. Technical Report, Universität Mannheim 2001.
8. J. D. Golić.
   Cryptanalysis of alleged A5/1 stream cipher. Proc. of EUROCRYPT'97, LNCS 1233, 239-255.
9. J. Gergov, Ch. Meinel. Efficient Boolean function manipulation with OBDDs can be generalized to FBDDs. IEEE Trans. on Computers 43, 1994, 1197-1209.
10. S. W. Golomb. Shift Register Sequences. Aegean Park Press, Laguna Hills, revised edition 1982.
11. M. Krause. BDD-based Cryptanalysis of Keystream Generators. Report 2001/092 in the Cryptology ePrint Archive (http://eprint.iacr.org/curr/).

12. Ch. Meinel. Modified Branching Programs and their Computational Power. LNCS 370, 1989.
13. M. J. Mihaljević. A faster Cryptanalysis of the Self-Shrinking Generator. Proc. of ACIPS'96, LNCS 1172, 182-189.
14. W. Meier, O. Staffelbach. The Self-Shrinking Generator. Proc. of EUROCRYPT'94, LNCS 950, 205-214.
15. R. A. Rueppel. Stream Ciphers. Contemporary Cryptology: The Science of Information Integrity. G.Simmons ed., IEEE Press New York, 1991.
16. D. Sieling, I. Wegener. Graph driven BDDs - a new data structure for Boolean functions. Theoretical Computer Science 141, 1995, 283-310.
17. I. Wegener. Branching Programs and Binary Decision Diagrams. SIAM Monographs on Discrete Mathematics and Applications. Philadelphia 2000.
18. E. Zenner. Kryptographische Protokolle im GSM Standard: Beschreibung und Kryptanalyse (in german). Master Thesis, University of Mannheim, 1999.
19. E. Zenner, M. Krause, S. Lucks. Improved Cryptanalysis of the Self-Shrinking Generator. Proc. of ACIPS'2001, LNCS 2119, 21-35.

Linear Cryptanalysis of Bluetooth Stream Cipher

Jovan Dj. Golić, Vittorio Bagini, and Guglielmo Morgari

Rome CryptoDesign Center, Gemplus, Via Pio Emanuelli 1, 00143 Rome, Italy,
{jovan.golic,vittorio.bagini,guglielmo.morgari}@gemplus.com

Abstract. A general linear iterative cryptanalysis method for solving binary systems of approximate linear equations which is also applicable to keystream generators producing short keystream sequences is proposed. A linear cryptanalysis method for reconstructing the secret key in a general type of initialization schemes is also developed. A large class of linear correlations in the Bluetooth combiner, unconditioned or conditioned on the output or on both the output and one input, are found and characterized. As a result, an attack on the Bluetooth stream cipher that can reconstruct the 128-bit secret key with complexity about 2^{70} from about 45 initializations is proposed. In the precomputation stage, a database of about 2^{80} 103-bit words has to be sorted out.

Key words: Linear cryptanalysis, linear correlations, iterative probabilistic decoding, reinitialization.

1 Introduction

Bluetooth™ is a standard for wireless short-range connectivity specified by the Bluetooth™ Special Interest Group in [1]. The specification defines a stream cipher algorithm E_0 to be used for point-to-point encryption within the Bluetooth network. The algorithm consists of a keystream generator, derived from the well-known summation generator, and an initialization scheme which is based on the keystream generator. The size of the secret key used for encryption is 128 bits, and the initialization vector (IV) consists of 74 bits, 26 of which are derived from a real-time clock, while the remaining 48 are the address bits depending on users. The internal state of the keystream generator is 132 bits long, and the keystream sequences produced are very short, that is, at most 2745 bits for each initialization vector. The description of the Bluetooth security protocol given in [1] is not quite clear and, according to some interpretations, a number of security weaknesses of the protocol are presented in [12].

The keystream generator is a binary combiner composed of four linear feedback shift registers (LFSR's) of total length 128 that are combined by a nonlinear function with 4 bits of memory which is a modified combining function of the summation generator. This modification turns out to be important as it reduces some correlation weaknesses of the summation generator identified in [17] and [10]. Some further interesting improvements to this end which require minor

L.R. Knudsen (Ed.): EUROCRYPT 2002, LNCS 2332, pp. 238–255, 2002.

modifications of the combining function are proposed in [14]. However, according to [1] and [14], the short keystream sequences should prevent the correlation attacks based on the correlation properties of the Bluetooth combiner.

Due to a large size of the internal state, the complexity of general time-memory or time-memory-data tradeoff attacks (e.g., see [9]) for realistic amounts of known keystream data seems to be higher than the complexities reported below. Besides, as such attacks aim at recovering an internal or the initial state of the keystream generator, they are not directly applicable to Bluetooth if the objective is to recover the secret key because of the initialization scheme used. The basic divide-and-conquer attack on the Bluetooth combiner directly follows from the similar attack [3] on the summation generator (also see [12]). In such an attack, 89 bits of the initial states of the three shortest LFSR's along with 4 initial memory bits are guessed. This allows to recover the output sequence of the longest LFSR from the keystream sequence. Altogether, about 132 keystream bits are needed to identify the correct guess. The same attack applies to the initialization scheme, so that the secret key can be reconstructed in about 2^{93} steps from just one IV, where the step complexity is the same as in the exhaustive search method. If one guesses 56 bits of the two shortest LFSR's and applies a sort of the branching method [9] for producing a system of linear equations, then the initial states of the other two LFSR's can be recovered in about 2^{84} steps, and some optimizations are possible [4]. The secret key can be obtained in a similar way.

The main objective of this paper is to identify a large class of linear correlations in the Bluetooth combiner which, in spite of the short keystream sequences, enable one to reconstruct not only the LFSR initial states, but also the secret key from a relatively small number of IV's. More precisely, we consider the unconditioned linear correlations, the linear correlations conditioned on the output, and the linear correlations conditioned on both the output and one guessed input. The resulting system of linear equations holding with probabilities different from one half can then be solved by a general linear iterative cryptanalysis method similar to iterative probabilistic decoding algorithms used in fast correlation attacks. The secret key can be recovered by a related linear cryptanalysis method from a number of IV's. The total complexity is about 2^{70} steps, with the step complexity comparable to one of the exhaustive search method, the required number of IV's is about 45, and the precomputation stage consists in sorting out a database of about 2^{80} 103-bit words.

Description of the Bluetooth stream cipher is provided in Section 2. The linear correlations are explained and characterized in Section 3, the general method for solving binary systems of approximate linear equations and its application to the Bluetooth keystream generator are presented in Section 4, and a linear cryptanalysis method for initialization schemes is proposed in Section 5. Optimal choices of parameters for concrete attacks are discussed in Section 6 and conclusions are given in Section 7. Analogous linear correlations computed for the modified Bluetooth combiner [14] are displayed in the Appendix.

2 Description of Bluetooth Stream Cipher

The description is based on [1], but only the details relevant for our linear cryptanalysis method will be presented. The main component of the Bluetooth stream cipher algorithm is the keystream generator (Bluetooth combiner) which is derived from the well-known summation generator with four input LFSR's. The LFSR lengths are 25, 31, 33, and 39 (128 in total) and all the feedback polynomials are primitive and have 5 nonzero terms each. All the LFSR's are regularly clocked and their binary outputs are combined by a nonlinear function with 4 bits of memory. Let $\mathbf{x}^i = (x_t^i)_{t=0}^{\infty}$ denote the output sequence of LFSR$_i$, $1 \leq i \leq 4$, where the LFSR's are indexed in order of increasing length. The internal memory of the combiner at time t consists of 4 memory bits $C_t = (c_t, c_{t-1})$, where 2 carry bits $c_t = (c_t^0, c_t^1)$ are defined in terms of 2 auxiliary carry bits $s_t = (s_t^0, s_t^1)$. Let $\mathbf{z} = (z_t)_{t=0}^{\infty}$ denote the output sequence of the combiner. Then the output sequence of the combiner is defined recursively by

$$z_t = x_t^1 \oplus x_t^2 \oplus x_t^3 \oplus x_t^4 \oplus c_t^0 \tag{1}$$

$$c_{t+1}^0 = s_{t+1}^0 \oplus c_t^0 \oplus c_{t-1}^0 \oplus c_{t-1}^1, \qquad c_{t+1}^1 = s_{t+1}^1 \oplus c_t^1 \oplus c_{t-1}^0 \tag{2}$$

$$(s_{t+1}^0, s_{t+1}^1) = \left\lfloor \frac{x_t^1 + x_t^2 + x_t^3 + x_t^4 + 2c_t^1 + c_t^0}{2} \right\rfloor \tag{3}$$

with integer summation in the last equation, where the initial 4 memory bits $(c_0^0, c_0^1, c_{-1}^0, c_{-1}^1)$ have to be specified. Note that in the summation generator the memory consists of only 2 bits of the carry s_t, i.e., $c_t = s_t$.

Due to frequent resynchronizations, the maximal keystream sequence length produced from a given initial state of the keystream generator is only 2745 bits. The initial state consists of 128 bits defining the initial LFSR states and 4 initial memory bits. They are produced by an initialization scheme from (at most) 128 secret key bits and the known 74-bit IV consisting of 48 address bits depending on users and of variable 26 bits derived from a real-time master clock. The secret key itself is derived from some secret and some known random information by another algorithm, which is irrelevant for our cryptanalysis.

The initialization scheme is the Bluetooth combiner initialized with some secret key bits and some IV bits, while the initial 4 memory bits are all set to 0. The remaining secret key bits and IV bits are added modulo 2, one at a time, to the feedback bits of individual LFSR's, for a number of times depending on the LFSR. The details are not important, except for the fact that the LFSR sequences in the initialization scheme linearly depend on the secret key and IV. The combiner is clocked 200 times and the last produced 128 output bits are permuted in a specified way to define the LFSR initial states, while the last 4 memory bits are used as the initial 4 memory bits for keystream generation.

3 Linear Correlations in Bluetooth Combiner

The basis of the linear cryptanalysis method to be developed are linear relations among the input bits to the Bluetooth combiner that hold with probabilities different from one half, in the probabilistic model in which the input sequences are modeled as purely random, i.e., as mutually independent sequences of independent and balanced (uniformly distributed) binary random variables. Such linear relations are called linear correlations since they are directly or indirectly dependent on the known output sequence. The first point to analyze is the asymptotic distribution of the 4 memory bits in this probabilistic model, if the initial 4 memory bits are either fixed or purely random. In the summation generator, due to the fact that the nonlinear function (3) is not balanced, it follows that the 2 carry bits are not balanced asymptotically, and this is the main source of a number of correlation weaknesses derived and exploited in [17] and [10]. However, in the case of Bluetooth, due to the introduced linear functions (2), C_{t+1} is a balanced function of C_t and $X_t = (x_t^1, x_t^2, x_t^3, x_t^4)$ and hence C_t is balanced for every t if it is balanced for $t = 0$. Moreover, this also holds asymptotically, when t increases, if the initial memory bits, C_0, are fixed, because the underlying Markov chain is ergodic, and the convergence to the stationary distribution is very fast.

Consider a block of m consecutive output bits, $Z_t^m = (z_t, z_{t-1}, \cdots, z_{t-m+1})$ as a function of the corresponding block of m consecutive inputs $X_t^m = (X_t, X_{t-1}, \cdots, X_{t-m+1})$ and the preceding memory bits C_{t-m+1}. Assume that X_t^m and C_{t-m+1} are balanced and mutually independent. Then, according to [7], if $m \geq 5$, then there must exist linear correlations between the output and input bits, but they may also exist if $m \leq 4$. As the correlations are time invariant, we introduce the notation $Z^m = F^m(X^m, C)$, where $Z^m = (z_j)_{j=0}^{m-1}$ and $X^m = (X_j)_{j=0}^{m-1}$. By virtue of the linear output function (1), it follows that $F^m(X^m, C)$ is a balanced function that is also balanced for any fixed C. The input block X^m of $4m$ bits can be rearranged into $X^m = (X_i^m)_{i=1}^4$, where $X_i^m = (x_j^i)_{j=0}^{m-1}$ is the i-th input block of m bits, corresponding to the output of LFSR$_i$. Then (1) implies that $F^m(X^m, C)$ is balanced for any fixed X_i^m and, also, for any fixed X_i^m and C combined.

Let f and g be two Boolean functions of an n-bit input vector which is assumed to be uniformly distributed. Then the correlation coefficient between f and g conditioned on a subset $\mathcal{X} \subseteq \{0,1\}^n$ is defined as

$$c(f, g \mid \mathcal{X}) = \Pr(f(X) = g(X) \mid X \in \mathcal{X}) - \Pr(f(X) \neq g(X) \mid X \in \mathcal{X})$$
$$= \frac{1}{|\mathcal{X}|} \sum_{X \in \mathcal{X}} (-1)^{f(X) \oplus g(X)} = \frac{1}{|\mathcal{X}|} \sum_{X \in \mathcal{X}} (-1)^{f(X)} (-1)^{g(X)}. \quad (4)$$

The correlation coefficients conditioned on $\mathcal{X}$ between f and all linear functions l are thus determined by the Walsh transform of a real-valued function defined as $(-1)^{f(X)}$ for $X \in \mathcal{X}$ and as 0 otherwise. They can be computed by the fast Walsh transform algorithm of complexity $O(n2^n)$.

All the correlations of interest to be described below correspond to (4) and were feasible to compute exhaustively for $m \leq 6$. All significant correlation coefficients were also tested by computer simulations on sufficiently long output sequences. It turns out that for $m \leq 3$ the correlation coefficients are equal to zero in all the cases. For $4 \leq m \leq 6$, it turns out that relatively large absolute values of the correlation coefficients along with the associated input linear functions can be characterized in terms of the underlying conditions. In addition, it also turns out that the Boolean functions specifying the signs of the correlation coefficients have relatively simple characterizations. The Boolean sign function is here defined as $\mathrm{sign}(c) = 0$ for $c > 0$ and $\mathrm{sign}(c) = 1$ for $c < 0$.

3.1 Unconditioned Linear Correlations

The first type of correlations to be considered are the correlations between linear functions of input bits and linear functions of output bits, as introduced in [7]. Namely, let $\mathbf{W} \cdot X^m = \bigoplus_{i=1}^{4} \bigoplus_{j=0}^{m-1} w_{ij} x_j^i$ and $\mathbf{v} \cdot Z^m = \bigoplus_{j=0}^{m-1} v_j z_j$ denote two such linear functions defined by a matrix $\mathbf{W}$ and a vector $\mathbf{v}$, respectively. We want to find all $\mathbf{W}$ and $\mathbf{v}$ such that the correlation coefficient $c(\mathbf{W} \cdot X^m, \mathbf{v} \cdot Z^m)$ is relatively large in absolute value. Define the (column) weights of $\mathbf{W}$ as $w_j = \sum_{i=1}^{4} w_{ij}$, $0 \leq j \leq m-1$. Then the main property, observed in [14], which follows from the symmetry of the combiner output and next-state functions with respect to 4 input variables, is that the correlation coefficient depends on $\mathbf{v}$ and only on the weights of $\mathbf{W}$, i.e., on the weight vector $\mathbf{w} = (w_j)_{j=0}^{m-1}$.

4-bit case There are 96 pairs of input/output linear functions that are mutually correlated, with nonzero correlation coefficients $\pm 1/16$. The output and input linear functions respectively have the weight patterns $(1, v_1, v_2, 1)$ and $(4, w_1, w_2, 4)$ such that $(w_1)_2 \neq v_1$, where $(w_1)_2 \stackrel{\text{def}}{=} w_1 \bmod 2$. Each of 2 output linear functions with $v_2 = 0$ is correlated to $16 = 8 \times 2$ input linear functions with $w_2 \in \{0, 4\}$. Each of 2 output linear functions with $v_2 = 1$ is correlated to $32 = 8 \times 4$ input linear functions with $w_2 = 3$. One of these 96 pairs was theoretically found out in [14]. For all the pairs,

$$\mathrm{sign}(c) = v_1 \oplus (\lfloor w_1/2 \rfloor)_2 \oplus \lfloor w_2/4 \rfloor. \tag{5}$$

5-bit case There are 8 nonzero correlation coefficients $\{\pm 25/256, \pm 5/256, \pm 1/64, \pm 1/256\}$. The largest absolute value is attained by the following 16 pairs of input/output linear functions: each of 2 output linear functions with pattern $(1, v_1, 1, 1, 1)$ is correlated to 8 input linear functions with weight pattern $(4, w_1, 4, 4, 4)$ such that $(w_1)_2 \neq v_1$. For such pairs, the sign function is given by the first two terms on the right-hand-side of (5).

6-bit case There are 12 nonzero correlation coefficients $\{\pm 25/256, \pm 25/1024, \pm 5/256, \pm 5/1024, \pm 1/256, \pm 1/1024\}$. The largest absolute value is attained by the following 16 pairs of input/output linear functions: each of 2 output linear functions with pattern $(1, v_1, 0, 0, 0, 1)$ is correlated to 8 input linear functions with weight pattern $(4, w_1, 0, 0, 0, 4)$ such that $(w_1)_2 = v_1$. For such pairs, the sign function is given by the second term on the right-hand-side of (5).

3.2 Linear Correlations Conditioned on Output

The second type of correlations to be considered are the correlations between linear functions of input bits and the all-zero function when conditioned on the output bits. Namely, for every output Z^m, we would like to find all $\mathbf{W}$ such that the conditioned correlation coefficient $c(\mathbf{W}\cdot X^m, 0 \mid F^m(X^m, C) = Z^m)$ is relatively large in absolute value. One can also prove that for any given Z^m, the conditioned correlation coefficient depends only on the weight vector $\mathbf{w}$. The conditioned correlation coefficients are generally larger than the unconditioned ones, because they fully exploit the information contained in the known output sequence. Recall that $Z^m = (z_0, z_1, \cdots, z_{m-1})$.

4-bit case For each output value Z^4, there are 96 input linear functions with nonzero correlation coefficients, equal to $\pm 1/16$, with weight pattern $(4, w_1, w_2, 4)$ where w_1 is arbitrary and $w_2 \in \{0, 3, 4\}$. For such functions,

$$\mathrm{sign}(c) = (1 \oplus z_1)(1 \oplus (w_1)_2) \oplus (\lfloor w_1/2 \rfloor)_2 \oplus \lfloor w_2/4 \rfloor \oplus z_0 \oplus z_2(w_2)_2 \oplus z_3. \quad (6)$$

5-bit case For each output value Z^5, there are 12 nonzero correlation coefficients with 6 different absolute values. The largest absolute value 29/256 is attained by 8 input linear functions with weight pattern $(4, w_1, 4, 4, 4)$ such that $(w_1)_2 = z_1$. The second largest absolute value 21/256 is attained by 8 input linear functions with weight pattern $(4, w_1, 4, 4, 4)$ such that $(w_1)_2 \neq z_1$. For all 16 functions,

$$\mathrm{sign}(c) = (1 \oplus z_1)(1 \oplus (w_1)_2) \oplus (\lfloor w_1/2 \rfloor)_2 \oplus z_0 \oplus z_2 \oplus z_3 \oplus z_4. \quad (7)$$

6-bit case For each output value Z^6, there are 100 nonzero correlation coefficients with 50 different absolute values. Except for the value 83/1024, corresponding to another type of input linear functions, the largest 7 absolute values are attained by exactly 16 input linear functions with weight pattern $(4, w_1, 0, 0, 0, 4)$ and depend on $(w_1)_2 \oplus z_1 \oplus z_4$ and (z_2, z_4) in a way shown in the following table. For such functions,

$$\mathrm{sign}(c) = (\lfloor w_1/2 \rfloor)_2 \oplus z_0 \oplus z_1(w_1)_2 \oplus z_5. \quad (8)$$

$(w_1)_2$	$\neg(z_1 \oplus z_4)$				$z_1 \oplus z_4$			
(z_2, z_4)	(0,0)	(1,1)	(1,0)	(0,1)	(0,0)	(1,0)	(1,1)	(0,1)
$\lvert c\rvert$	$\frac{139}{1024}$	$\frac{129}{1024}$	$\frac{119}{1024}$	$\frac{113}{1024}$	$\frac{79}{1024}$	$\frac{79}{1024}$	$\frac{73}{1024}$	$\frac{69}{1024}$

3.3 Linear Correlations Conditioned on Output and One Input

The third type of correlations to be considered are the correlations between linear functions of 3 inputs and the all-zero function when conditioned on the output and one assumed input. More precisely, with the notation $X^m_{2-4} = (X^m_i)^4_{i=2}$, let $\mathbf{W}{\cdot}X^m_{2-4} = \bigoplus_{i=2}^4 \bigoplus_{j=0}^{m-1} w_{ij} x^i_j$ denote such a linear function of 3 inputs. For every assumed input X^m_1 and every possible output Z^m, we would like to find all $\mathbf{W}$

such that the conditioned correlation coefficient $c(\mathbf{W}{\cdot}X_{2-4}^m, 0|X_1^m, F^m(X^m, C) = Z^m)$ is relatively large in absolute value. One can similarly prove that for any assumed X_1^m and any given Z^m, the conditioned correlation coefficient depends only on the weight vector $\mathbf{w} = (w_j)_{j=0}^{m-1}$, where now $w_j = \sum_{i=2}^4 w_{ij}$, $0 \leq j \leq m-1$. These correlation coefficients are generally larger than the ones conditioned only on the output, because of the information provided by one known input. Recall that $Z^m = (z_0, z_1, \cdots, z_{m-1})$ and $X_1^m = (x_0^1, x_1^1, \cdots, x_{m-1}^1)$.

<u>**4-bit case**</u> For each input value X_1^4, there are 4 nonzero correlation coefficients $\pm 1/4$ and $\pm 1/8$. The absolute value $1/4$ is attained by an average of 2 (out of 8) input linear functions with weight pattern $(3, w_1, 3, 3)$ such that $z_2 \neq x_2^1$ and $(w_1)_2 \neq z_1 \oplus x_1^1$. For such functions,

$$\mathrm{sign}(c) = 1 \oplus \lfloor w_1/2 \rfloor \oplus z_0 \oplus z_3 \oplus x_0 \oplus x_1 \oplus x_2 \oplus x_3 \oplus z_1 x_1. \tag{9}$$

The absolute value $1/8$ is attained for every output value Z^4 by 16 (out of 32) input linear functions with weight pattern $(3, w_1, w_2, 3)$ such that $(w_1)_2 \neq z_1 \oplus x_1^1$ and $(w_2)_2 = 0$. For such functions,

$$\mathrm{sign}(c) = \lfloor w_1/2 \rfloor \oplus z_0 \oplus z_2 \lfloor w_2/2 \rfloor \oplus z_3 \oplus x_0 \oplus x_1 \oplus x_2 \lfloor w_2/2 \rfloor \oplus x_3 \oplus z_1 x_1. \tag{10}$$

So, per each X_1^4 and Z^4, $|c| = 1/4$ is on average attained by 2 out of 8 input linear functions and $|c| = 1/8$ is attained by 16 out of 32 input linear functions.

<u>**5-bit case**</u> For each input value X_1^5, there are 12 nonzero correlation coefficients with 6 different absolute values. The largest 3 absolute values are attained for every output value Z^5 by 4 (out of 8) input linear functions with weight pattern $(3, w_1, 3, 3, 3)$ such that $(w_1)_2 \neq z_1 \oplus x_1^1$. The dependence of $|c|$ on (X_1^5, Z^5) along with the average number α of the corresponding input linear functions are shown in the following table. For the remaining 4 input linear functions such that $(w_1)_2 = z_1 \oplus x_1^1$, $|c| = 1/32$ for every (X_1^5, Z^5). For all 8 functions,

$$\begin{aligned} \mathrm{sign}(c) = {} & 1 \oplus \lfloor w_1/2 \rfloor \oplus (w_1)_2 \oplus z_0 \oplus z_1 \oplus z_1 (w_1)_2 \oplus z_2 \oplus z_3 \oplus z_4 \\ & \oplus x_0 \oplus x_1 (w_1)_2 \oplus x_2 \oplus x_3 \oplus x_4 \oplus z_1 x_1. \end{aligned} \tag{11}$$

$\lvert c\rvert$	α	$(z_2 \oplus x_2^1, z_3 \oplus x_3^1)$
9/32	1	(0,0)
3/16	2	(0,1) or (1,0)
1/8	1	(1,1)

<u>**6-bit case**</u> For each input value X_1^6, there are 50 nonzero correlation coefficients with 25 different absolute values. The most significant absolute values are attained by 4 (out of 8) input linear functions with weight pattern $(3, w_1, 0, 0, 0, 3)$ such that $(w_1)_2 = z_1 \oplus x_1^1$, but some large absolute values are also achieved by the remaining 4 input linear functions such that $(w_1)_2 \neq z_1 \oplus x_1^1$. For all 8 functions,

$$\begin{aligned}\text{sign}(c) &= (w_1)_2 \oplus \lfloor w_1/2 \rfloor \oplus z_0 \oplus z_1 \oplus z_1(w_1)_2 \oplus z_5 \oplus x_0 \oplus x_1(w_1)_2 \oplus x_2(w_1)_2 \\ &\oplus x_4(w_1)_2 \oplus x_5 \oplus z_1x_1 \oplus z_1x_2 \oplus z_1x_4 \oplus z_4x_2(w_1)_2 \oplus z_4x_4(w_1)_2 \oplus x_1x_2 \oplus x_1x_4 \\ &\oplus x_2x_4(w_1)_2 \oplus z_1z_4x_2 \oplus z_1z_4x_4 \oplus z_1x_2x_4 \oplus z_4x_1x_2 \oplus z_4x_1x_4 \oplus x_1x_2x_4. \quad (12)\end{aligned}$$

The dependence of $|c|$ on (X_1^6, Z^6) along with the average number α of the corresponding input linear functions are shown in the following table for $|c| > 1/8$ and an average number 4.125 of such functions. For an average number 3.875 of remaining input linear functions, $|c| < 1/8$. The displayed 7 values along with 21/128 and 19/128, corresponding to other input linear functions, are the largest possiblc.

$\lvert c\rvert$	α	CONDITIONS	FUNCTIONS
9/32	1	$z_2 = x_2^1, z_4 = x_4^1$	$(w_1)_2 = z_1 \oplus x_1^1$
13/64	1/2	$z_2 = x_2^1, z_4 \neq x_4^1 = 1$	$(w_1)_2 = z_1 \oplus x_1^1$
3/16	9/8	$z_2 \neq x_2^1, (z_4 = x_4^1 \text{ or } z_4 \neq x_4^1 = x_2^1 = 1, z_3 = x_3^1)$	$(w_1)_2 = z_1 \oplus x_1^1$
11/64	5/8	$z_4 \neq x_4^1, (z_2 = x_2^1, x_4^1 = 0 \text{ or } z_2 \neq x_2^1 = x_4^1 = 1, z_3 \neq x_3^1)$	$(w_1)_2 = z_1 \oplus x_1^1$
5/32	1/8	$z_2 \neq x_2^1, z_4 \neq x_4^1, x_2^1 = x_4^1 = 0, z_3 = x_3^1$	$(w_1)_2 = z_1 \oplus x_1^1$
9/64	5/8	$z_2 \neq x_2^1, z_4 \neq x_4^1, x_2^1 = x_4^1 = 0, z_3 \neq x_3^1$	$(w_1)_2 = z_1 \oplus x_1^1$
		$z_4 \neq x_4^1, z_2 = x_2^1, z_3 = x_3^1$	$(w_1)_2 \neq z_1 \oplus x_1^1$
17/128	1/8	$z_4 \neq x_4^1, z_2 \neq x_2^1 = x_4^1 = 1, z_3 = x_3^1$	$(w_1)_2 \neq z_1 \oplus x_1^1$

4 Linear Iterative Cryptanalysis of Keystream Generator

The objective of the linear cryptanalysis of the Bluetooth keystream generator is to reconstruct 128 bits of the LFSR initial states from a given segment of the keystream sequence of length at most 2745 bits, by using the linear correlations described in Section 3. Accordingly, the starting point of the cryptanalysis is a set of linear equations in the initial state bits which hold with probabilities different from one half. The aim is to find a solution to this systcm that is consistent with the given probabilities.

4.1 Solving Approximate Linear Systems

Let $\mathbf{x} = (x_j)_{j=1}^k$ be a vector of k binary variables and let $\mathbf{y} = (y_i)_{i=1}^n$ be a vector of n, $n \geq k$, binary variables that are defined as linear functions of $\mathbf{x}$, that is, $y_i = l_i(\mathbf{x})$, $1 \leq i \leq n$. In matrix notation, $\mathbf{y} = \mathbf{G}^{\mathrm{T}}\mathbf{x}$, where $\mathbf{G}$ is an $k \times n$ matrix whose columns correspond to linear functions l_i, and vectors are represented as one-column matrices. It is assumed that $\mathbf{G}$ has full rank k (linearly independent rows), which means that the linear transform defined by $\mathbf{G}^{\mathrm{T}}$ is injective, so that $\mathbf{y}$ uniquely determines $\mathbf{x}$.

It is further assumed that $\mathbf{y}$ is known only probabilistically, in terms of the marginal probabilities $\Pr(y_i = 0) = p_i$, $1 \leq i \leq n$. More precisely, a probabilistic model is assumed in which the variables y_i, $1 \leq i \leq n$, are mutually independent.

Since in this case they can take arbitrary values, define an event $\mathcal{L}$ that $\mathbf{y}$ belongs to the range of the linear transform determined by $\mathbf{G}^{\mathrm{T}}$, i.e., that y_i are linearly dependent according to $\mathbf{G}^{\mathrm{T}}$, i.e., that the linear system $\mathbf{y} = \mathbf{G}^{\mathrm{T}}\mathbf{x}$ has a (unique) solution in $\mathbf{x}$. Now, in this model the most likely solution to the linear system is the one that maximizes the conditioned block probability

$$\begin{aligned} \Pr(\mathbf{x}|\mathcal{L}) &= \Pr(\mathbf{y} = \mathbf{G}^{\mathrm{T}}\mathbf{x}|\mathcal{L}) \\ &= \frac{\Pr(\mathbf{y} = \mathbf{G}^{\mathrm{T}}\mathbf{x})}{\Pr(\mathcal{L})} = \frac{1}{\Pr(\mathcal{L})} \prod_{i=1}^{n} p_i^{1-l_i(\mathbf{x})}(1-p_i)^{l_i(\mathbf{x})}. \end{aligned} \tag{13}$$

It follows that $\Pr(\mathcal{L}) = \sum_{\mathbf{x}\in\{0,1\}^k} \prod_{i=1}^{n} p_i^{1-l_i(\mathbf{x})}(1-p_i)^{l_i(\mathbf{x})}$. Of course, 2^k steps are required to find the solution.

The problem is in fact directly related to a decoding problem for the binary linear (n,k) block code C with a generator matrix $\mathbf{G}$ on a time-varying memoryless binary symmetric channel (BSC) with error probabilities $1-p_i$, $1 \le i \le n$. Namely, in a probabilistic model in which the codewords are equiprobable, if the all-zero word is observed at the output of this BSC, then the posterior probability of an information word $\mathbf{x}$ is the same as the conditioned probability (13). However, our model is more appropriate as it directly deals with the problem considered and as such does not involve any communication channel (e.g., symmetry is not needed).

Another approach is to find a solution that maximizes each of the conditioned bit probabilities $\Pr(x_j|\mathcal{L})$, $1 \le j \le k$. This requires only k steps, but such probabilities have to be computed, and that requires 2^{n-k} steps if the well-known Hartmann-Rudolph algorithm [13] is applied. For linear codes, this algorithm minimizes the decoding error probability for individual symbols rather than blocks of symbols. It can also be used for computing the conditioned bit probabilities $\hat{p}_i = \Pr(y_i = 0|\mathcal{L})$, $1 \le i \le n$, which are important for iterative algorithms. In our problem, k is large and the probabilities p_i are rather close to one half, so that the decision error probabilities can be small only if $n-k$ is also large. Therefore, the Hartmann-Rudolph algorithm is computationally infeasible and numerical approximations are required.

Let $\mathbf{H}$ denote a parity-check matrix of the code C, i.e., a generator matrix of its dual code C^{d}. $\mathbf{H}$ is an $(n-k)\times n$ matrix of full rank $n-k$ such that $\mathbf{H}\mathbf{G}^{\mathrm{T}} = \mathbf{0}$. Recall that C^{d} is a binary linear $(n, n-k)$ code consisting of all the binary vectors $\mathbf{v} = (v_i)_{i=1}^{n}$ that are orthogonal to each codeword $\mathbf{y}$ from C ($\mathbf{v}\cdot\mathbf{y} = v_1y_1 \oplus\cdots\oplus v_ny_n = 0$). The dual codewords represent the linear relations among the codeword bits and are hence called the parity checks. Instead of taking into account all 2^{n-k} parity checks as in the Hartmann-Rudolph algorithm, one can only consider numerically more important parity checks having a relatively low weight, which is defined as the number of nonzero terms minus one.

Let V_i be a set of parity checks $\mathbf{v}$ involving the i-th codeword bit y_i, i.e., such that $v_i = 1$. Let $c_i = 2p_i - 1$ and $\hat{c}_i = 2\hat{p}_i - 1$ denote the corresponding unconditioned and conditioned correlation coefficients of y_i, respectively. Then, according to [11], we get an approximate expression

$$\hat{c}_i = /c_i + \sum_{\mathbf{v} \in V_i} \prod_{j=1:\ v_j=1, j \neq i}^{n} c_j / \tag{14}$$

where the clipping function $/(\cdot)/$ ensures that $|\hat{c}_i| \leq 1$. Interestingly, this expression can also be obtained as the limit form, when all c_i tend to zero, of the well-known expression (e.g., see [18])

$$\frac{1-\hat{c}_i}{1+\hat{c}_i} = \frac{1-c_i}{1+c_i} \prod_{\mathbf{v} \in V_i} \frac{1-\prod_{j=1:\ v_j=1, j \neq i}^{n} c_j}{1+\prod_{j=1:\ v_j=1, j \neq i}^{n} c_j} \tag{15}$$

which is used if the parity checks from each V_i are orthogonal, that is, if the i-th bit is the only bit that they share in common. Expression (14) appears to be more appropriate as the orthogonality is not required. In both expressions, the product term $\prod_{j=1:\ v_j=1, j \neq i}^{n} c_j$ represents the correlation coefficient of the binary sum of all the bits y_j other than y_i from the parity check $\mathbf{v}$ involving y_i. The absolute value of this correlation coefficient is a measure of information about y_i contained in the considered parity check $\mathbf{v}$. Accordingly, low-weight parity checks are more informative than the others.

The most effective way is to use (14) iteratively, in each iteration improving the conditioned correlation coefficients $\hat{c}_i$. The iterations are useful because (14) is only an approximate expression and because hard decisions based on conditioned bit probabilities generally do not result in codewords. Instead of directly recycling (14), by substituting $\hat{c}_i$ from the current iteration for c_i in the next iteration (e.g., see [16]), one can also use a more sophisticated and more effective belief propagation recycling [6] (e.g., see [15], [5], and [11]). According to both experimental and theoretical [11] arguments, the correlation coefficients will converge in a relatively small number of iterations to values ± 1 for most coordinates, and, in the case of success, the final hard decisions on individual bits will result in a binary word at a small Hamming distance from a codeword. A simple, information set decoding technique will then yield this codeword along with the corresponding information word $\mathbf{x}$, which is the desired solution to the approximate linear system under consideration.

It is important to point out the conditions for success for both the approaches described above. In accordance with the capacity argument, the decision error probability of the block-based approach using (13) can be made arbitrarily close to zero if

$$\sum_{i=1}^{n} c_i^2 \geq k. \tag{16}$$

This means that we can reliably distinguish a correct solution from the remaining $2^k - 1$ incorrect solutions if (16) is satisfied. On the other hand, provided that $|c_i| = c$, $1 \leq i \leq n$, it is theoretically argued in [11] that the average bit-decision error probability of the iterative approach using (14) will be close to zero if $\sum_w M_w c^{w-1} > 1$, where M_w is the average number per bit of the parity checks of weight w that are used in (14). In the limit, when c tends to zero, this condition coincides with the similar condition from [18] corresponding to (15) which is both theoretically derived and experimentally verified (see also [2]). Anyway, both conditions are also supported by numerous experimental results on fast correlation attacks on regularly clocked LFSR's (e.g., see [10]).

We will work with a stronger condition, resulting in higher complexity estimates,

$$\sum_w M_w \, c^w \; \geq \; 1 \tag{17}$$

where the exponent $w-1$ is conservatively replaced by w. The new condition can be given another interpretation, directly in terms of (14) or (15). Namely, if we assume that M_w is exactly the number of parity checks of weight w that are used for the i-th bit, then (14) reduces to

$$\hat{c}_i \; = \; /c_i + \sum_w (m_w^+ - m_w^-) c^w / \tag{18}$$

where m_w^+ and m_w^- denote the numbers of parity checks of weight w with the positive and negative sign of the product, respectively. Now, if $y_i = 0$ or $y_i = 1$, then the expected value of $m_w^+ - m_w^-$ is equal to $+M_w c^w$ or $-M_w c^w$, respectively. So, the contribution of the parity checks of weight w can be regarded (statistically) significant for the iterative process to converge to the most likely (correct) values of y_i for each $1 \leq i \leq n$ if $M_w c^w \geq 1$. Accordingly, by combining the contributions of parity checks of different weights we get the condition (17). This condition demonstrates the significant advantage of iterative over one-step algorithms which when applied to individual bits, in light of (16), will be successful if $\sum_w M_w c^{2w} \geq 1$.

4.2 Application to Bluetooth

The main approach to be pursued here is one in which the initial state of the shortest LFSR, LFSR_1, is guessed, so that linear correlations conditioned on both the output and one input can be utilized. This is needed in order to minimize the precomputation complexity to be given below. Let $n = \alpha n_0$ be the number of linear equations chosen out of those resulting from the 4-bit, 5-bit, and 6-bit linear correlations described in Section 3. Here, α is the average number of chosen equations per output bit and $n_0 = 2740$ is the maximum number of output bits that can be used if 6-bit linear correlations are exploited. For each output bit, the average number of α equations are chosen from a possibly larger set of $\beta = \gamma\alpha$ equations that is independent of the observed 6-bit output segment and of the known 6-bit input segment resulting from the guessed initial state of LFSR_1. By expressing each LFSR bit involved as a linear function of the initial state bits, each obtained linear equation becomes a linear equation in the unknown 103=128-25 initial state bits of all the LFSR's but the shortest.

The correlation coefficients associated with these linear equations depend on the known output segment, of maximal length 2745 bits, and on the guessed input segment of the same length. According to Section 3, the linear equations can be grouped in several types corresponding to 4-bit, 5-bit, and 6-bit linear correlations. For each such group, there are several absolute values of the correlation coefficients, each appearing with a given probability. Altogether, let the absolute value μ_j appear with probability ν_j, where $\nu_j = \alpha_j/\alpha$ and α_j is the average number of linear equations with the absolute value of the correlation coefficient equal to μ_j. For each output bit, the equations chosen and the absolute values and the signs of the associated correlation coefficients depend on the known output and guessed input. Two average values of the correlation coefficient magnitudes are important for measuring the success of the linear cryptanalysis to be applied. One, which is related to the iterative bit-based approach and (14) and (17), is the weighted geometric mean $\mu = \prod_j \mu_j^{\nu_j}$. Note that $\mu \leq \bar{\mu} = \sum_j \nu_j \mu_j$. The other, which is related to the block-based approach and (16), is the expected value of the squares $\bar{\mu^2} = \sum_j \nu_j \mu_j^2$.

For the iterative approach, it is necessary to find the linear dependencies among the obtained linear equations (codeword bits) that involve only a relatively small number of linear equations, that is, to determine the corresponding low-weight parity checks. The weights to be used should result in numbers of parity checks that should be sufficient for success according to the condition (17). The number of parity checks of a given weight, w, per codeword bit is a characteristic of the produced linear system (linear code), which depends on the observed output and one guessed input, and can be modeled by assuming that the system is randomly generated as the expected value

$$M_w = 2^{-103}\binom{n-1}{w} \approx 2^{-103}\frac{n^w}{w!} \tag{19}$$

where the approximation error is negligible if $w \ll n$.

Consequently, if we utilize all the parity checks of weight at most w, the success condition (17), with the geometric mean of the correlation coefficient magnitudes, becomes

$$\sum_{j=2}^{w} \frac{n^j}{j!}\,\mu^j \geq 2^{103} \tag{20}$$

(parity checks of weight 1 are impossible for the problem considered). As the term with the maximal weight, w, is dominant, we finally get the condition

$$w\,(\log_2 n_0 + \log_2 \alpha - \log_2 \frac{1}{\mu}) \;\geq\; 103 + \log_2 w! \tag{21}$$

which can be solved numerically to give the minimal required weight w. *This condition is conservative because we neglected the contribution of terms with weight lower than w and because M_w is expected to be larger than* (19) *due to the specific structure of the obtained linear equations for the Bluetooth keystream generator* *(i.e., each parity check gives rise to more parity checks through appropriate phase shifts).*

As the iterative algorithm has to be run for each of 2^{25} guesses about the LFSR_1 initial state, its complexity can be expressed as

$$C = 2^{25} \cdot n \cdot \frac{1}{\mu^w} = 2^{25+\log_2 n_0 + \log_2 \alpha + w \log_2 \frac{1}{\mu}} \tag{22}$$

where a computational step consists of all the computations per bit for a number of iterations, which on average is not greater than about 10. In each iteration, the computations are predominantly determined by the number of real multiplications needed to compute (14) for every bit and for parity checks of weight w. This number is given as $3n/\mu^w$, in view of a simple fact that only $3(w-1)$ real multiplications are needed to compute all $w+1$ products of w elements out of a set of $w+1$ elements. Accordingly, a computational step approximately consists of at most 30 real multiplications, where an 8-bit precision will suffice. As a real product of two 8-bit words can be performed by an average of 3 real additions, each requiring about $8 \cdot 3 = 24$ binary operations, the step consists of about 2160 binary operations. This is comparable with one step of the exhaustive search method which consists of about $128 \cdot 15 = 1920$ binary operations.

The next point to be explained is how to generate all the parity checks of weight at most w for a possibly larger set of γn linear equations. This can be done in precomputation time, by computing and sorting all the linear combinations of $\lceil (w+1)/2 \rceil$ linear equations, altogether about $(\gamma n)^{\lceil (w+1)/2 \rceil} / \lceil (w+1)/2 \rceil!$ of them. The matches obtained by sorting directly give all the linear combinations of at most $2\lceil (w+1)/2 \rceil$ linear equations that evaluate to zero identically (e.g., see [8]). More precisely, we have to sort out only

$$\begin{aligned} D &\approx \frac{\gamma^{\lceil (w+1)/2 \rceil}}{\lceil (w+1)/2 \rceil!} \left(2^{103/w} (w!)^{1/w} \frac{1}{\mu} \right)^{\lceil (w+1)/2 \rceil} \\ &\approx \frac{(w!)^{\lceil (w+1)/2 \rceil / w}}{\lceil (w+1)/2 \rceil!} 2^{103 \lceil (w+1)/2 \rceil / w + \lceil (w+1)/2 \rceil (\log_2 \gamma + \log_2 \frac{1}{\mu})} \end{aligned} \tag{23}$$

randomly chosen linear combinations, represented as 103-bit words. The total obtained number of matches per bit, i.e., the total number of parity checks per each equation is then γ^w/μ^w. They are all stored as the final result of precomputation.

Now, given an output segment and each guessed input, we have to filter $1/\mu^w$ parity checks out of a set of γ^w/μ^w collected parity checks, for each of n linear equations. If $\gamma > 1$, then γ^w/μ^w parity checks can be sorted out, with respect to n_0 bit positions and β indexes of linear equations per each bit position, so that the filtering takes only about $1/\mu^w$ steps. The complexity of filtering is then given by (22), but the corresponding step complexity is negligible in comparison with one of the iterative algorithm.

After the iterative algorithm has converged to probabilities close to 0 or 1, if the guess about the LFSR_1 initial state was correct, then the 103 bits of the remaining LFSR initial states, along with the initial 4 memory bits, can be reconstructed by information set decoding (e.g., by looking for error-free sets of 103 linearly independent equations), with complexity much smaller than (22).

5 Linear Cryptanalysis of Initialization Scheme

The objective of the linear cryptanalysis of the Bluetooth initialization scheme is to reconstruct 128 bits of the secret key from a given number of 128-bit (or 132-bit) outputs of the Bluetooth initialization scheme obtained from the same secret key and different IV's. Such outputs can be obtained by the linear iterative cryptanalysis method described in Section 4. As the initialization scheme is essentially the same as the keystream generator, for each IV we will again use the linear correlations described in Section 3 to produce another approximate system of linear equations. Other approaches, possibly requiring a smaller number of IV's, may also exist (e.g., see [4]).

The main point facilitating the linear cryptanalysis is that the secret key and IV are linearly combined together to form the initial state of the Bluetooth keystream generator used for initialization. Therefore, each equation linear in LFSR bits can be expressed as the binary sum of an equation linear in secret key bits and an equation linear in IV bits which itself can be evaluated as IV is known. If the same linear equation in LFSR bits is used with different, say q, IV's, one thus effectively obtains q independent observations of the same linear function, say y, of secret key bits. If the correlation coefficient associated with the i-th equation is c_i and if s_i is the value of the linear function of the corresponding IV_i, then the correlation coefficient associated with the i-th observation is $(-1)^{s_i}c_i$, $1 \le i \le q$. In view of (15), the combined correlation coefficient, $\hat{c}$, of y is then determined by

$$\frac{1-\hat{c}}{1+\hat{c}} = \prod_{i=1}^{q} \left(\frac{1-c_i}{1+c_i}\right)^{(-1)^{s_i}} \tag{24}$$

or, approximately, for small c_i, by

$$\hat{c} = /\sum_{i=1}^{q}(-1)^{s_i}c_i/. \tag{25}$$

Assume that $|c_i| = c$, $1 \le i \le q$. Then, if $y = 0$ or $y = 1$, the expected value of $\hat{c}$ is equal to $+qc^2$ or $-qc^2$, respectively. So, the combined correlation coefficient will be close to ± 1 if $q \ge 1/c^2$. In general, in view of (16), it will be close to ± 1 if

$$\sum_{i=1}^{q} c_i^2 \ge 1. \tag{26}$$

This condition determines the minimal q required for reconstructing the correct value of y with a small probability of decision error.

To minimize the required number of IV's, we will again use linear correlations conditioned on the output and one input, which has to be guessed. Assuming that the sizes of the secret subkeys controlling indivudual LFSR's are the same as their respective lenghts, we have to guess 25 secret key bits controlling LFSR_1. Let an average number of α out of a set of $\beta = \gamma\alpha$ linear equations be chosen for each of $n_0 = 123$ available output bits, provided that 6-bit linear correlations are exploited. For q IV's, each of the resulting βn_0 linear functions of the remaining 103 secret key bits is then treated in the way explained above. Note that each of the functions will on average appear q/γ instead of q times. Then the condition (26) reduces to

$$q \geq \gamma \frac{1}{\bar{\mu^2}} \tag{27}$$

where $\bar{\mu^2} = \sum_j \nu_j \mu_j^2$ is the mean square value of the used correlation coefficients μ_j appearing with probabilities ν_j. The linear correlations to be used should be chosen so as to minimize q. The resulting βn_0 linear equations in 103 secret key bits which hold with probabilities close to 1 can then be solved by information set decoding if the guess about the 25 secret key bits is correct. As the complexity of reconstructing the secret key from given q outputs of the initialization scheme is much smaller than the complexity of reconstructing these outputs, the total complexity is determined by the latter, and is q times larger than (22).

6 Optimal Complexities

There are many possible choices of the linear correlations described in Section 3 to be used in the linear iterative cryptanalysis of the keystream generator in order to reconstruct the LFSR initial states. The objective is to minimize the computation complexity C given by (22) and the precomputation complexity D given by (23). However, this is not possible to achieve simultaneously, as there is a tradeoff between the two criteria.

In general, C is minimal if one uses the linear correlations conditioned on the output, described in Section 3.2, but D is then relatively large. In this case, we do not have to guess one input and the complexity analysis is the same as in Section 4.2 except that the number of LFSR initial state bits to be reconstructed is now 128 instead of 103. For example, if we choose to use the largest 4 6-bit correlation coefficients and the corresponding 8 input linear functions, from the condition (21) get $w = 15$ and hence $C \approx 2^{60}$ and $D \approx 2^{98.5}$. By guessing one input we generally decrease D and increase C. Two illustrative examples are explained in more detail below.

First, choose the 2 largest 5-bit and the 3 largest 6-bit conditioned correlation coefficients to work with. In this case, we get $\alpha = (1+2)+(1+1/2+9/8) = 5.625$, $\beta = 8+8 = 16$, $\gamma = 16/5.625 \approx 2.8444$, and

$$\mu = \left(\left(\frac{9}{32}\right)^2 \left(\frac{13}{64}\right)^{1/2} \left(\frac{3}{16}\right)^{25/8}\right)^{1/5.625} \approx 0.2181. \tag{28}$$

Then (21) yields $w = 11$ and hence we get $C \approx 2^{63.07}$ and $D \approx 2^{82.68}$.

Second, choose the largest 2, 3, and 7 conditioned correlation coefficients from 4-bit, 5-bit, and 6-bit linear correlations to work with, respectively. In this case, we get $\alpha = (2+16)+(1+2+1)+(1+1/2+9/8+5/8+1/8+5/8+1/8) = 26.125$, $\beta = (8+32)+8+8 = 56$, $\gamma = 56/26.125 \approx 2.1435$, and

$$\begin{aligned}\mu &= \left(\left(\frac{9}{32}\right)^2\left(\frac{1}{4}\right)^2\left(\frac{13}{64}\right)^{1/2}\left(\frac{3}{16}\right)^{25/8}\left(\frac{11}{64}\right)^{5/8}\left(\frac{5}{32}\right)^{1/8}\left(\frac{9}{64}\right)^{5/8}\left(\frac{17}{128}\right)^{1/8}\left(\frac{1}{8}\right)^{17}\right)^{1/26.125}\\ &\approx 0.1504. \end{aligned} \tag{29}$$

Then (21) yields $w = 9$ and hence we get $C \approx 2^{65.73}$ and $D \approx 2^{79.74}$.

There are also different possible choices of the linear correlations to be used in the linear cryptanalysis of the initialization scheme in order to reconstruct the secret key. The objective is to minimize the required number q of IV's given by (27). To this end, the 25 secret key bits controlling the shortest LFSR are guessed. It is slightly better to work with 6-bit than 5-bit linear correlations conditioned on the output and one input. If we use the 7 largest 6-bit conditioned correlation coefficients, we get $\gamma = 8/4.125 \approx 1.9394$ and

$$\begin{aligned}\overline{\mu^2} &= \frac{1}{4.125}\left(\left(\frac{9}{32}\right)^2+\frac{1}{2}\left(\frac{13}{64}\right)^2+\frac{9}{8}\left(\frac{3}{16}\right)^2+\frac{5}{8}\left(\frac{11}{64}\right)^2+\frac{1}{8}\left(\frac{5}{32}\right)^2+\frac{5}{8}\left(\frac{9}{64}\right)^2+\frac{1}{8}\left(\frac{17}{128}\right)^2\right)\\ &\approx 0.04251. \end{aligned} \tag{30}$$

Then (27) yields $q \approx 45.2 \approx 2^{5.51}$, so that the total complexity of the secret key reconstruction increases to $C \approx 2^{68.58}$ and $C \approx 2^{71.24}$ for the two cases described above, respectively.

7 Conclusions

The developed linear cryptanalysis method shows that correlation attacks may also be applicable to stream ciphers producing very short keystream sequences which are reinitialized frequently by using a cryptographically strong initialization scheme. The complexity analysis concentrates on mathematical rather than practical implementation arguments. The obtained attack complexities for the Bluetooth stream cipher are overestimated as they are based on a conservative assumption about the underlying parity-check weight distribution. It is in principle possible that the complexity can be further decreased by exploiting m-bit linear correlations for $m > 6$ if they are feasible to compute. It may also be possible that the actual precomputation complexity is lower than predicted.

Consequently, at least from the theoretical standpoint, there is a need to redesign the Bluetooth stream cipher, maybe by using the improvement suggested in [14]. This modified Bluetooth stream cipher appears to be more resistant to the linear cryptanalysis, but might not be the optimal choice (see the Appendix).

Appendix

A Linear Correlations in Modified Bluetooth Combiner

We exhaustively computed all the m-bit linear correlations for $m \leq 6$ in the modified Bluetooth combiner proposed in [14]. The distributions of the largest correlation coefficients are determined and displayed here.

The only modification relates to the linear update functions for the 4 memory bits. Namely, instead of (2), we now have

$$c_{t+1}^0 = s_{t+1}^0 \oplus c_t^0 \oplus c_{t-1}^0, \qquad c_{t+1}^1 = s_{t+1}^1 \oplus c_t^1 \oplus c_{t-1}^1. \tag{31}$$

The stationary distribution of the 4 memory bits remains to be uniform.

First of all, there are no nonzero correlation coefficients for $m \leq 4$. The largest absolute values of the correlation coefficients, $|c|$, and the (average) numbers, α, of linear functions attaining them are shown in the following tables. In the last table we also show the total numbers, β, of linear functions out of which the desired α are chosen. In this respect, note that for each m, each smaller set is contained in the next larger. For $m = 5$, there are no other nonzero correlation coefficients. For $m = 6$, there are also 46080 and 57600 pairs of input/output linear functions with $|c| = 1/512$ and $|c| = 1/1024$, respectively, whereas $\alpha = 65024$ for $1/1024 \leq |c| \leq 13/1024$, conditioned on output, and $\alpha = 1260$ for $1/64 \leq |c| \leq 3/64$, conditioned on output and one input, and there are no other nonzero correlation coefficients. A general conclusion is that the absolute values of the correlation coefficients are smaller than in the Bluetooth combiner, but their numbers are considerably larger.

Unconditioned Linear Correlations

m	5			6			
$\|c\|$	$\frac{5}{128}$	$\frac{1}{64}$	$\frac{1}{128}$	$\frac{25}{1024}$	$\frac{5}{512}$	$\frac{5}{1024}$	$\frac{1}{256}$
α	256	1536	3840	256	3072	7680	9216

Linear Correlations Conditioned on Output

m	5			6			
$\|c\|$	$\frac{7}{128}$	$\frac{3}{128}$	$\frac{1}{128}$	$\frac{33}{1024}$	$\frac{25}{1024}$	$\frac{21}{1024}$	$\frac{15}{1024}$
α	128	768	3200	64	128	256	64

Linear Correlations Conditioned on Output and One Input

m	5			6		
$\|c\|$	$\frac{3}{16}$	$\frac{1}{8}$	$\frac{1}{16}$	$\frac{9}{64}$	$\frac{3}{32}$	$\frac{1}{16}$
α	8	16	120	4	16	16
β	64	128	512	64	128	128

References

1. Bluetooth™, *Bluetooth Specification*, Version 1.1, Feb. 2001.
2. V. Chepyzhov and B. Smeets, "On a fast correlation attack on stream ciphers," Advances in Cryptology – EUROCRYPT '91, *Lecture Notes in Computer Science*, vol. 547, pp. 176–185, 1991.
3. E. Dawson and A. Clark, "Divide and conquer attacks on certain classes of stream ciphers," *Cryptologia*, vol. 18, pp. 25–40, 1994.
4. S. Fluhrer and S. Lucks, "Analysis of the E_0 encryption system," Selected Areas in Cryptography – SAC 2001, *Lecture Notes in Computer Science*, vol. 2259, pp. 38–48, 2001.
5. M. P. C. Fossorier, M. J. Mihaljević, and H. Imai, "Reduced complexity iterative decoding of low-density parity check codes based on belief propagation," *IEEE Trans. Commun.*, vol. 47, pp. 673–680, May 1999.
6. R. G. Gallager, "Low-density parity-check codes," *IRE Trans. Inform. Theory*, vol. 8, pp. 21–28, Jan. 1962.
7. J. Dj. Golić, "Correlation properties of a general binary combiner with memory," *Journal of Cryptology*, vol. 9, pp. 111–126, 1996.
8. J. Dj. Golić, "Computation of low-weight parity-check polynomials," *Electronics Letters*, vol. 32, pp. 1981–1982, Oct. 1996.
9. J. Dj. Golić, "Cryptanalysis of alleged A5 stream cipher," Advances in Cryptology – EUROCRYPT '97, *Lecture Notes in Computer Science*, vol. 1233, pp. 239–255, 1997.
10. J. Dj. Golić, M. Salmasizadeh, and E. Dawson, "Fast correlation attacks on the summation generator," *Journal of Cryptology*, vol. 13, pp. 245–262, 2000.
11. J. Dj. Golić, "Iterative optimum symbol-by-symbol decoding and fast correlation attacks," *IEEE Trans. Inform. Theory*, vol. 47, pp. 3040–3049, 2001.
12. M. Jakobsson and S. Wetzel, "Security weaknesses in Bluetooth," Topics in Cryptology – CT-RSA 2001, *Lecture Notes in Computer Science*, vol. 2020, pp. 176–191, 2001.
13. C. R. P. Hartmann and L. D. Rudolph, "An optimum symbol-by-symbol decoding rule for linear codes," *IEEE Trans. Inform. Theory*, vol. 22, pp. 514–517, Sept. 1976.
14. M. Hermelin and K. Nyberg, "Correlation properties of the Bluetooth combiner," Information Security and Cryptology – ICISC '99, *Lecture Notes in Computer Science*, vol. 1787, pp. 17–29, 1999.
15. D. J. C. MacKay, "Good error-correcting codes based on very sparse matrices," *IEEE Trans. Inform. Theory*, vol. 45, pp. 399–431, Mar. 1999.
16. W. Meier and O. Staffelbach, "Fast correlation attacks on certain stream ciphers," *Journal of Cryptology*, vol. 1, pp. 159–176, 1989.
17. W. Meier and O. Staffelbach, "Correlation properties of combiners with memory in stream ciphers," *Journal of Cryptology*, vol. 5, pp. 67–86, 1992.
18. M. J. Mihaljević and J. Dj. Golić, "A method for convergence analysis of iterative probabilistic decoding," *IEEE Trans. Inform. Theory*, vol. 46, pp. 2206–2211, Sept. 2000.

Generic Lower Bounds for Root Extraction and Signature Schemes in General Groups

Ivan Damgård and Maciej Koprowski*

BRICS**, Aarhus University

Abstract. We study the problem of root extraction in finite Abelian groups, where the group order is unknown. This is a natural generalization of the problem of decrypting RSA ciphertexts. We study the complexity of this problem for generic algorithms, that is, algorithms that work for any group and do not use any special properties of the group at hand. We prove an exponential lower bound on the generic complexity of root extraction, even if the algorithm can choose the "public exponent" itself. In other words, both the standard and the strong RSA assumption are provably true w.r.t. generic algorithms. The results hold for arbitrary groups, so security w.r.t. generic attacks follows for any cryptographic construction based on root extracting. As an example of this, we revisit Cramer-Shoup signature scheme [10]. We modify the scheme such that it becomes a generic algorithm. This allows us to implement it in RSA groups without the original restriction that the modulus must be a product of safe primes. It can also be implemented in class groups. In all cases, security follows from a well defined complexity assumption (the strong root assumption), without relying on random oracles, and the assumption is shown to be true w.r.t. generic attacks.

1 Introduction

The well known RSA assumption says, informally speaking, that given a large RSA modulus N, exponent e and $x \in Z_N^*$, it is hard to find y, such that $y^e = x \bmod N$. The strong RSA assumption says that given only n, x it is hard to find any root of x, i.e., to find $y, e > 1$ such that $y^e = x \bmod N$. Clearly the second problem is potentially easier to solve, so the assumption that it is hard is potentially stronger.

Both these assumptions generalize in a natural way to any finite Abelian group: suppose we are given a description of some large finite Abelian group G allowing us to represent elements in G and compute inversion and multiplication in G efficiently. For RSA, the modulus N plays the role of the description. Another example is class groups, where the discriminant Δ serves as the description. Then we can define the *root assumption* which says that given $x \in G$ and

* Part of the work done while visiting IBM Zurich Research Laboratory.

** Basic Research in Computer Science,
Centre of the Danish National Research Foundation.

L.R. Knudsen (Ed.): EUROCRYPT 2002, LNCS 2332, pp. 256–271, 2002.

number $e > 1$, it is hard to find y such that $y^e = x$. The *strong root assumption* says that given x, it is hard to find $y, e > 1$ such that $y^e = x$.

Clearly, it is essential for these assumptions to hold, that the order of G is hard to compute from the description of G. It must also be difficult to compute discrete logarithms in G. Otherwise, root extraction would be easy: we would be able to find a multiple M of the order of the element x, and then computing $x^{e^{-1} \bmod M}$ would produce the desired root (if $\gcd(e, |G|) = 1$).

In this paper, we study the complexity of root finding for *generic* algorithms, that is, algorithms that work in any group because they use no special properties of the group G and only use the basic group operations for computing in G. The generic complexity of discrete logarithms was showed to be exponential by Nechaev [18] and Shoup[24] who introduced a broader model for generic algorithms. In both models, however, the algorithm knows the order of the group, which means we cannot use them for studying the (strong) root assumption as we defined it here[1]. Generic algorithms based on those models were studied and discussed in [16,17,22,12,23,21]. A similar concept of algebraic algorithms was introduced in [3].

We propose a new model, in which the group and its order are hidden from the algorithm (as in RSA and class groups setting). More precisely, the group is chosen at random according to a known probability distribution D, and the algorithm only knows that the group order is in a certain interval. For instance, if D is chosen to be the output distribution produced by an RSA key generation algorithm, then this models generic attacks against RSA. But the same model also incorporates generic attacks against schemes based on class groups (we elaborate later on this).

We show that if D is a so called *hard* distribution, then both the root and strong root assumptions are true for generic algorithms, namely we show exponential lower bounds for both standard and strong root extraction. Roughly speaking, D is hard if, when you choose the group order n according to D, then the uncertainty about the largest prime factor in n is large (later in the paper, we define what this precisely means). For instance, if the distribution is such that the largest prime factor in n is a uniformly chosen $k + 1$-bit prime, then, ignoring lower order contributions, extracting roots with non-negligible probability requires time $\Omega(2^{k/4})$ for a generic algorithm. More precise bounds are given later in the paper. Our proof technique resembles that of Shoup [24], however, we need some new ideas in order to take advantage of the uncertainty about group order (which is known and fixed in Shoup's case).

Thus the distribution of the order is the only important factor. Knowledge of the group structure, such as the number of cyclic components, does not help the algorithm. Standard RSA key generation produces distributions that are

[1] We note that it *is* possible to study the generic complexity of root finding in Shoups model, but in this setting the problem can only be hard if the "public exponent" is not relatively prime to the order of the group. This was done by Maurer and Wolf [16]. They show that no efficient generic algorithm can compute p'th roots efficiently if p^2 divides the order of the group G and p is a large prime.

indeed hard in our sense, so this means that both the standard and the strong RSA assumption are provably true w.r.t. generic algorithms. The results hold for arbitrary groups, so security w.r.t. generic attacks follows for any cryptographic construction based solely on root extracting being hard.

As an example of this, we revisit the Cramer-Shoup signature scheme [10]. We modify it such that it becomes a generic algorithm. This allows us to implement it in RSA groups without the original restriction that the modulus must be a product of safe primes. It can also be implemented in class groups. Here, however, we need a "flat tree" signature structure to make the scheme be efficient.

In all cases, security follows from a well defined complexity assumption (the strong root assumption), without relying on random oracles (in contrast to Biehl et al. [1]). We stress that the security proof is general, i.e., if our complexity assumption holds, then the scheme is secure against *all polynomial-time attacks*, not just generic ones. In addition, however, our lower bounds imply that the assumption is true w.r.t. generic attacks.

Before discussing the meaning and significance of our results, we note that one must always be careful in estimating the value of a generic lower bound. This has been demonstrated recently where a server aided RSA protocol proposed in [15] was proved generically secure in [17], but was later broken in [19]. This is unrelated to our results because neither the generic proof nor the break applied to root extraction, but to the problem of breaking a particular protocol. We believe this demonstrates that a generic lower bound for a very specific and not so well-studied problem is of highly questionable value: the problem may turn out to be very easy after all, or to be so specialised that there are no variants of it for which generic solutions are the best. From this point of view, it seems reasonable to consider generic lower bounds for root extraction.

Nevertheless, we have of course NOT proved that RSA is hard to break, or that root finding in class groups is hard: for both problems, there are non-generic algorithms known that solve the problem in sub-exponential time. It must also be mentioned that while there are groups known for which only generic algorithms seem to be able to find discrete logs (namely elliptic curve groups), similar examples are not known for root extraction. Despite this, we believe that the results are useful and interesting for other reasons:

- They shed some light on the question whether the strong RSA assumption really is stronger than the standard RSA assumption: since we show exponential generic lower bounds for both problems, it follows that this question must be studied separately for each type of group: only an algorithm directed specifically against the group at hand can separate the two problems.
- They give extra credibility to the general belief that the best RSA key generation is obtained by choosing the prime factors large enough and as randomly as possible, the point being that this implies that $\phi(n)$ contains at least one large and random prime factor, and so the distribution of the group order is hard in our sense. Note that a generic algorithm, namely a straightforward variant of Pollards $p-1$ method, will be able to find roots if all primes factors of $\phi(n)$ are too small. When $\phi(n)$ contains large prime factors this attack is

trivially prevented, but our results say that in fact all generic attacks will fail.
- The fact that no examples are known of groups where only generic root finding works, does not mean that such examples do not exist. We hope that our results can motivate research aimed at finding such examples.
- The generic point of view allows us to look at constructions such as the Cramer-Shoup signature scheme in an abstract way, and "move" them to other groups. The generic security of the scheme will be automatically inherited, on the other hand the real, non-generic complexity of root finding may be completely different in different groups. Therefore, having a construction work in arbitrary groups makes it more robust against non-generic attacks.

1.1 Open Problems

It can be argued that the RSA example can be more naturally considered as a ring than as a group. One may ask if root extraction is also hard for generic algorithms allowed to exploit the full ring structure, i.e., it may do additions as well as multiplications. Boneh and Lipton [2] have considered this problem for fields, in a model called black-box fields (with known cardinality of the field), and have shown that in this model, one cannot hide the identity of field elements from the algorithm using random encodings: the algorithm will be able figure out in subexponential time which element is hidden behind an encoding, and so exponential lower bounds in this generic model cannot be shown. The black-box field model can easily be changed to a black-box ring model. But if we do not give the algorithm the order of the multiplicative group of units in the ring, it is unclear whether the results of Boneh and Lipton [2] extend to black-box rings, and on the other hand also unclear if our lower bound still holds.

2 Lower Bound on Generic Root Extraction Algorithms

2.1 Model

In our model, we have public parameters B, C and D. Here, B, C are natural numbers and D is a probability distribution on Abelian groups with order in the interval $]B, B+C]$.

Let G be a finite abelian group of order n chosen according to D and let S denote the set of bit strings of cardinality at least $B+C$. We define *encoding function* as an injective map σ from G into S.

A generic algorithm A on S is a probabilistic algorithm which gets as input an *encoding list* $(\sigma(x_1), \ldots .\sigma(x_k))$, where $x_i \in G$ and σ is an encoding function of G on S. Note that unlike in Shoups' model [24] this algorithm does not receive the order n of the group as a part of its input.

The algorithm can query a *group oracle* $\mathcal{O}$, specifying two indices i and j from the encoding list, and a sign bit. The oracle returns $\sigma(x_i \pm x_j)$ according to the given sign bit, and this bit string is appended to the encoding list.

The algorithm can also ask the group oracle $\mathcal{O}$ for a random element. Then the oracle chooses a random element $r \in_R G$ and returns $\sigma(r)$, which is appended to the encoding list. After its execution the algorithm returns a bit string denoted by $A(\sigma, x_1, \ldots, x_k)$.

Note that this model forces the algorithm to be generic by hiding the group elements behind encodings. In the following we will choose a random encoding function each time the algorithm is executed, and we will show that root extraction is hard for most encoding functions.

2.2 Lower Bound

The distribution D induces in a natural way a distribution D_p over the primes, namely we choose a group of order n according to D and look at the largest prime factor in n. We let $\alpha(D)$ be the maximal probability occurring in D_p. One can think of $\alpha(D)$ as a measure for how hard it is to guess the largest prime factor of n, since clearly the best strategy is to guess at some prime that has a maximal probability of being chosen. We also need to measure how large p can be expected to be. So for an integer M, let $\beta(D, M)$ be the probability that $p \leq M$. As we shall see, a good distribution D (for which root extraction is hard) has small $\alpha(D)$ and small $\beta(D, M)$, even for large values of M[2]. Since, as we shall see, the only important property of D is how the order of the group is distributed, we will sometimes in the following identify D with the distribution of group orders it induces.

First we state a number of observations.

Lemma 1. *Let n be a number chosen randomly from the interval $]B, B + C]$ according to the probability distribution D and let p be its biggest prime divisor. Let a be any fixed integer satisfying $|a| \leq 2^m$. Then the probability that $p|a$ is at most*

$$m\alpha(D).$$

Proof. There are at most m prime numbers dividing a, each of these can be the largest prime factor in n with probability at most $\alpha(D)$.

The following lemma was introduced by Shoup [24]:

Lemma 2. *Let p be prime and let $t \geq 1$. Let $F(X_1, \ldots, X_k) \in \mathbb{Z}/p^t[X_1, \ldots, X_k]$ be a nonzero polynomial of total degree d. Then for random $x_1, \ldots, x_k \in \mathbb{Z}/p^t$, the probability that $F(x_1, \ldots, x_k) = 0$ is at most d/p.*

Lemma 3. *Let $0 < M < B$ and $F(X_1, \ldots, X_k) = a_1X_1 + \cdots + a_kX_k \in \mathbb{Z}[X_1, \ldots, X_k]$ be a nonzero polynomial, whose coefficients satisfy $|a_i| \leq 2^m$. Let the integer n be chosen randomly from the range $]B, B+C]$ according to the probability distribution D. If p is the biggest prime divisor of n and $t \geq 1$, then*

[2] The two measures $\alpha(D), \beta(D, M)$ are actually related, we elaborate below

for random $x_1, \ldots, x_k \in \mathbb{Z}/p^t$, *the probability that* $F(x_1, \ldots, x_k) = 0$ *in* Z/p^t *is at most*

$$m\alpha(D) + \beta(D, M) + \frac{1}{M},$$

for any M.

Proof. Wlog we can assume that $a_1 \neq 0$. Let E be the event that $F(x_1, \ldots, x_k) = 0$, and A the event that $p > M$ and $p \nmid a_1$. We have $P(E) = P(E, \neg A) + P(E, A) \leq P(\neg A) + P(E|A)$. By lemma 1, and by definition of $\beta(D, M)$ we have that

$$P(\neg A) \leq m\alpha(D) + \beta(D, M).$$

On the other hand, assuming that $p \nmid a_1$ (and $p > M$), we can consider $F(X_1, \ldots, X_k)$ as a nonzero linear polynomial in the ring $\mathbb{Z}/p^t[X_1, \ldots, X_k]$. Let $x_1, \ldots, x_k$ be chosen at random in $\mathbb{Z}/p^t$. Then by lemma 2 we have

$$P(E|A) \leq \frac{1}{p} < \frac{1}{M}.$$

Now we are able to prove the main theorem.

Theorem 1. *Let a probability distribution* D *on the range* $]B, B + C]$ *be given. Let* $S \subset \{0,1\}^*$ *be a set of cardinality at least* $B + C$ *and* A *a generic algorithm on* S *that makes at most* m *oracle queries. Let* M *be any number such that* $2m \leq M$. *Suppose* A *is an algorithm for solving the strong root problem, that is,* A *gets an element from* S *as input and outputs a number* e *and an element from* S. *Assume* $\log e \leq v$.

Now choose a group G *according to* D *and let* n *be its order. Further choose* $x \in G$ *and an encoding function* σ *at random. Then the probability that* $A(\sigma, x) = e, \sigma(y)$ *where* $e > 1$ *and* $ey = x$ *is at most*

$$(m^3 + m^2 + mv + 2v + m)\alpha(D) + (m^2 + m + 2)\beta(D, M) + (m^2 + m + 3)/M + \frac{m^2}{B}.$$

Proof. First observe that A may output a new encoding from S that it was not given by the oracle, or it may output one of the encodings it was given by the oracle. Without loss of generality, we may assume that A always outputs a new encoding $s \in S$, if we define that A has success if this new encoding represents an e'th root of x, or if *any* of the encodings it was given represents such a root.

We define a new group oracle $\mathcal{O}'$ that works like $\mathcal{O}$, except that when asked for a random element, it chooses at random among elements for which the algorithm does *not* already know encodings. The executions of A that can be produced using $\mathcal{O}'$ are a subset of those one can get using $\mathcal{O}$. For each query, the probability that $\mathcal{O}$ chooses a "known" element is at most (number of encodings known)/(group order). Since at most one new encoding is produced by every oracle quiry, this probability is at most m/n. Hence the overall probability that $\mathcal{O}$ generates something $\mathcal{O}'$ could not is at most $m^2/n \leq m^2/B$. It follows that the success probability of A using $\mathcal{O}$ is at most the sum of the success probability of A using $\mathcal{O}'$ and m^2/B.

To estimate the success probability of A using $\mathcal{O}'$, we will simulate the oracle $\mathcal{O}'$ as it interacts with the algorithm. However, we will not choose the group G or x until after the algorithm halts. While simulating, we keep track of what values the algorithm has computed, as follows: Let $X, Y_1, \dots, Y_m$ be indeterminants. At any step, we say that the algorithm has computed a list $F_1, \dots, F_k$ of linear integer polynomials in variables $X, Y_1, \dots, Y_m$. The algorithm knows also a list $\sigma_1, \dots, \sigma_k$ of values in S, that the algorithm "thinks" encodes corresponding elements in G. When we start, $k = 1$; $F_1 = X$ and σ_1 is chosen at random and given to A. Whenever the algorithm queries two indices i and j and a bit sign, we compute $F_{k+1} = F_i \pm F_j \in \mathbb{Z}[X, Y_1, \dots, Y_m]$ according to the bit sign. If $F_{k+1} = F_l$ for some ℓ with $1 \leq l \leq k$, we define $\sigma_{k+1} = \sigma_l$; otherwise we choose σ_{k+1} at random from S distinct from $\sigma_1, \dots, \sigma_k$.

When the algorithm asks for a random element from the group G, we choose at random σ_{k+1} from $S \setminus \{\sigma_1, \dots, \sigma_k\}$. We define $F_{k+1} = Y_k$.

When the algorithm terminates, it outputs an exponent e, such that $\log e \leq v$ and $e > 1$ (and also outputs some new element s in S).

We then choose at random a group G according to the probability distribution D and hence also order n from the interval $]B, B + C]$.

Let p be the biggest prime dividing n. We know that for some integer $r > 0$

$$G \cong \mathbb{Z}_{p^r} \times H.$$

We choose at random $x, y_1, \dots, y_m \in_R G$, which is equivalent to random independent choices of $x', y_1', \dots, y_m' \in_R \mathbb{Z}_{p^r}$ and $x'', y_1'', \dots, y_m'' \in_R H$.

We say that *the algorithm wins* if either

- $F_i(x', y_1', \dots, y_m') \equiv F_j(x', y_1', \dots, y_m') \bmod p^r$ and $F_i \neq F_j$ for some $1 \leq i \neq j \leq m+1$, or:
- The new encoding s output by A is an e'th root of x, or:
- $eF_i(x', y_1', \dots, y_m') \equiv x' \bmod p^r$ for some $1 \leq i \leq m+1$.

To estimate the probability that the algorithm wins, we simply estimate the probability that each of the above three cases occur.

For this first case, recall that $F_i(X, Y_1, \dots, Y_m) = a_{i,0}X + a_{i,1}Y_1 + \dots + a_{i,m}Y_m$, where $F_1(X, Y_1, \dots, Y_m) = X$. Since at each oracle query we could only either introduce a new polynomial Y_k, add or subtract polynomials, then $|a_{i,j}| \leq 2^m$ for $i = 1, \dots, m+1, j = 0, \dots, m$. Lemma 3 implies that $F_i(x', y_1', \dots, y_m') \equiv F_j(x', y_1', \dots, y_m') \bmod p^r$ for given $i \neq j$, $F_i \neq F_j$ with probability at most $m\alpha(D) + \beta(D, M) + \frac{1}{M}$. Since we have at most m^2 pairs $i \neq j$, the probability that $F_i(x', y_1', \dots, y_m') \equiv F_j(x', y_1', \dots, y_m') \bmod p^r$ for some $i \neq j$, $F_i \neq F_j$ is at most

$$m^2 \left(m\alpha(D) + \beta(D, M) + \frac{1}{M} \right).$$

For the second case, first observe that the probability that $p|e$ or that $p < M$ is at most $v\alpha(D) + \beta(D, M)$, by lemma 1. On the other hand, assuming that $p \nmid e$ and $p \geq M$, a random group element different from the at most m elements the algorithm has been given encodings of is an e'th root of x with probability at most $\frac{1}{p-m} \leq \frac{1}{M-m} \leq \frac{2}{M}$. We conclude: the new encoding output by A represents an e'th root of x with probability at most

$$v\alpha(D) + \beta(D, M) + 2/M.$$

Finally let us consider the event $eF_i(x', y'_1, \ldots, y'_m) \equiv x' \bmod p^r$ for given $1 \leq i \leq m+1$. Since $F_i(X, Y_1, \ldots, Y_m)$ is an integer polynomial and $e > 1$, then $eF_i(X, Y_1, \ldots, Y_m) \neq X$ in $\mathbb{Z}[X, Y_1, \ldots, Y_m]$. We know that $\log e \leq v$ and coefficients $a_{i,j}$ of the polynomial F_i satisfy the condition $|a_{i,j}| \leq 2^m$. Therefore by lemma 3 the probability that $eF_i(x', y'_1, \ldots, y'_m) \equiv x' \bmod p^r$ is at most $(v+m)\,\alpha(D) + \beta(D, M) + \frac{1}{M}$. Hence the event, that for some $1 \leq i \leq m+1$ the equivalence $eF_i(x', y'_1, \ldots, y'_m) \equiv x' \bmod p^r$ holds, happens with the probability

$$(m+1)\left((v+m)\,\alpha(D) + \beta(D, M) + \frac{1}{M}\right).$$

Summarizing, the probability that A wins when we simulate $\mathcal{O}'$ is at most

$$(m^3 + m^2 + mv + 2v + m)\alpha(D) + (m^2 + m + 2)\beta(D, M) + (m^2 + m + 3)/M.$$

Given this bound on the probability that A wins the simulation game, we need to argue the connection to actual executions of A (using $\mathcal{O}'$). To this end, note that since the only difference between simulation and execution is the time at which the group, group elements and encodings are chosen, the event that

$$F_i(x', y'_1, \ldots, y'_m) \equiv F_j(x', y'_1, \ldots, y'_m) \bmod p^r \text{ and } F_i \neq F_j$$

for some $1 \leq i \neq j \leq m+1$ is well defined in both scenarios. Let BAD_{sim}, respectively BAD_{exec} be the events that this happens in simulation, respectively in an execution. We let a *history* of execution or simulation be the choice of group and group elements followed by the sequence of oracle calls and responses, followed by A's final output. We then make the following

Claim: Every history in which BAD_{sim} does not happen in simulation of $\mathcal{O}'$ occur with the same probability as history in which BAD_{exec} does not happen in execution of $\mathcal{O}'$. In particular, $P(BAD_{sim}) = P(BAD_{exec})$.

Recall that we defined that A wins the simulation game if BAD_{sim} occurs, or A finds a root. Therefore the claim clearly implies that our bound on the probability that A wins the simulation game is also a bound on A's success probability in a real execution.

The proof of the claim is available in the full version of the paper [11]. □

We now consider the case, where we have a family of values of the parameters B, C, D, i.e., they are functions of a security parameter k. As usual we say that a function of k is negligible if it is at most $1/p(k)$ for any polynomial $p()$ and all large enough k.

Definition 1. *Let* $\{D(k)|\ k = 1, 2, \ldots\}$ *be a family of probability distributions, where* $D(k)$ *ranges over Abelian groups of order in the interval* $]B(k), B(k) + C(k)]$. *The family is said to be* hard *if* $\alpha(D(k))$ *and* $\frac{1}{B(k)}$ *are negligible in* k.

In the following, we will sometimes write just D in place of $D(k)$ when the dependency on k is clear. We can observe:

Fact 1 *If $\{D(k)|\ k = 1, 2, ...\}$ is a hard family, then there exists $M(k)$, such that $\beta(D(k), M(k))$ and $\frac{1}{M(k)}$ are negligible in k.*

Proof. Let $M(k) = \frac{1}{\sqrt{\alpha(D(k))}}$. Clearly $\frac{1}{M(k)} = \sqrt{\alpha(D(k))}$ is negligible.

Now let n be the order of a group chosen according to $D(k)$. What is the probability that the biggest prime factor of n is smaller than $M(k)$? We have at most $M(k)$ prime factors smaller than $M(k)$. Each can be chosen with probability at most $\alpha(D(k))$. Therefore

$$\beta(D(k), M(k)) \leq M(k)\alpha(D(k)) = \frac{1}{\sqrt{\alpha(D(k))}}\alpha(D(k)) = \sqrt{\alpha(D(k))},$$

which is negligible.

Corollary 1. *Let the family $\{D(k)|\ k = 1, 2, ...\}$ be hard. We choose randomly an abelian group G of order n according to the distribution $D(k)$. Let $x \in G$ be chosen at random. Consider any probabilistic polynomial time generic algorithm A. The probability that A given x outputs a natural number $e > 1$ and an encoding of an element $y \in G$ such that $y^e = x$ is negligible.*

Proof. Let M be such that $\alpha(D)$, $\beta(D, M)$ and $\frac{1}{M}$ are negligible in k. Let m be the number of group queries made by the algorithm A. Then m and $\log e$ are polynomial in k, so clearly $2m < M$ for all large enough values of the security parameter. It follows from Theorem 1 that the probability that A succeeds is negligible.

One may note that if $\alpha(D(k))$ and $1/B(k)$ are exponentially small in k (as it is in the concrete examples we know), say $\alpha(D(k)) \leq 2^{-ck}$ for a constant $c > 0$ and $B(k) \geq 2^{dk}$ for a constant $d > 0$. Suppose the running time of a generic algorithm A is $O\left(2^{c'k}\right)$ where $c' < \min(c/4, d/2)$. Then it follows from the concrete expression in Theorem 1 and the proof of Fact 1 that A has negligible success probability. Thus for such examples, we get an exponential generic lower bound. For instance, if $D(k)$ is such that the largest prime factor in the order is a uniformly chosen $k+1$-bit prime, then certainly $B(k) \geq 2^k$, and also by the prime number theorem $\alpha(k) \approx 2^{-k+\log k} \ln 2$. Now, if we ignore lower order contributions we get that root extraction requires time $\Omega(2^{k/4})$. We present some proven or conjectured examples of hard families of distributions below in next subsections.

Applying the results to the standard root problem. What we have said so far concerns only the strong root assumption. For the standard root assumption, the scenario is slightly different: A group G and a public exponent $e > 1$ are chosen according to some joint distribution. Then e and an encoding of a random element x is given as input to the (generic) algorithm, which tries to find an e'th root of x. Clearly, if we let D be the distribution of G given e, Theorem 1 can be applied to this situation as well, as it lower bounds the complexity of finding *any* root of x.

Now consider a family of distributions as above $\{D(k) | k = 1, 2, ...\}$, where each $D(k)$ now outputs a group G and an exponent e, let $D(k, e)$ be the distribution of G given e, and $]B(k, e), B(k, e) + C(k, e)]$ be the interval in which $|G|$ lies, given e. The family is said to be *hard* if $Max_e(\alpha(D(k, e)))$ and $Max_e(\frac{1}{B(k,e)})$ are negligible in k, where the maximum is over all e that can be output by $D(k)$. One can then show in a straightforward way a result corresponding to Corollary 1 for the standard root problem.

2.3 RSA

For the RSA case, it is clear that any reasonable key generation algorithm would produce hard distributions because it is already standard to demand from good RSA key generation that the order of the group produced contains at least one large and random prime factor.

Furthermore, the fact that a public exponent e with $(e, |G|) = 1$ is given does not constrain seriously the distribution of the group, i.e., it remains hard. Consider, for instance, the distribution of random RSA moduli versus random RSA moduli where 3 is a valid public exponent. Since essentially half of all primes p satisfy that 3 does not divide $p - 1$, the uncertainty about the largest prime factor in the group order remains large.

2.4 The Uniform Distribution

Assume our distribution D is such that the group order n is chosen from the interval $]B, B + C]$ according to the uniform distribution, and such that both B and C are at least 2^k, where k is the security parameter. We prove that this uniform distribution is *hard*.

Let $u = \sqrt{k}$. Wlog we can assume that u is a natural number. Let $M = 2^u$. Clearly $\frac{1}{M}$ is negligible in k.

We prove that in this case $\alpha(D)$ and $\beta(D, M)$ are also negligible.

2.5 Class Groups

Buchmann and Williams [4] proposed a first crypto system based on class groups of imaginary quadratic orders (IQC). We describe briefly class groups and their application in cryptography.

Let Δ be a negative integer such that $\Delta \equiv 0, 1 \bmod 4$. The *discriminant* Δ is called *fundamental* if $\frac{\Delta}{4}$ or Δ is square free for $\Delta \equiv 0 \bmod 4$ or $\Delta \equiv 1 \bmod 4$, respectively. Given Δ it is possible to construct an abelian finite group called *class group* and denoted by $Cl(\Delta)$. The order of the class group is denoted by

$h(\Delta)$ and called *class number*. We denote the cardinality of the odd part of a class group $Cl(\Delta)$ by $h_{\text{odd}}(\Delta)$.

There is no known efficient algorithm to compute class number $h(\Delta)$ if the discriminant Δ is fundamental. See [14]) for the discussion of the problem of computing class numbers.

Hamdy and Möller [13] analyze the smoothness probabilities of class numbers based on heuristics of Cohen and Lenstra [7] and Buell [5]. Hamdy and Möller provide examples of discriminants with nice smoothness properties.. One of them is $\Delta = -8pq$ where p and q are primes such that $p \equiv 1 \bmod 8$, $p+q \equiv 8 \bmod 16$ and the Legendre symbol $\left(\frac{p}{q}\right)$ is equal -1.

We assume that a discriminant Δ is chosen randomly from some interval (under the restriction in above example) and that results in a distribution D of class numbers. In provided example we have that $\beta(D, M)$ and $\frac{1}{M}$ are negligible ($\frac{1}{B}$ is trivially negligible).

We believe it is reasonable to conjecture that the distribution induced by choosing discriminant as described above is indeed hard. For more detailed discussion of this problem see the full version of the paper [11].

3 Generic Cryptographic Protocols

Based on our model of generic adversary, we look at the notion of generic cryptographic protocols. Such protocols do not exploit any specific properties of group element representation. They are based on a simple model of finite abelian group with unknown order.

One advantage of this approach is an easy application of generic protocols in any family of groups of difficult to compute order. As a first example we note the family of RSA groups (without necessarily restricting the RSA modulus to a product of safe primes). Another example is class groups.

This generic approach has been used earlier: Shoup [24] proved the security of Schnorr's identification scheme [20] in the generic model with public order of the group. Biehl et al. [1] introduced some generic cryptographic protocols and applied them in class groups.

In this section we propose a generic signature scheme which can be seen as an abstraction of Cramer-Shoup signature scheme [10]. We will show security of this protocol against arbitrary attacks, given certain intractability assumption (strong root assumption), which we have proven in our generic model. This immediately implies provable security with respect to generic attacks.

3.1 Protocol

For this signature scheme, we assume that we are given D, a probability distribution over Abelian groups with order in the interval $]B, B+C]$. We also assume for this version of the scheme that one can efficiently choose a group according to D, such that the order of the group is known (the assumption about known order is removed later). Moreover, we assume a collision-resistant hash function H whose output is a natural number smaller than $\frac{B}{2}$.

Key Generation. We choose G of order n according to the probability distribution D. A description of the group G is announced, but its order remains hidden (for RSA, for instance, this amounts to publishing the modulus).

We choose randomly elements $h \in_R G$ and $x \in_R < h >$(let $< h >$ denote the cyclic subgroup generated by h). Let e' be a prime satisfying $B + C < e' < 2(B + C)$.

The public key is (h, x, e').

The private key is n.

Signing Protocol

1. When a message M (it can be a bit string of arbitrary size) is requested to be signed, a random prime $e \neq e'$ is chosen such that $B+C < e < 2(B+C)$. A random element $y' \in_R G$ is chosen.
2. We compute
$$x' = \frac{(y')^{e'}}{h^{H(M)}}$$
and
$$y = \left(xh^{H(x')}\right)^{\frac{1}{e}}.$$
Note that the signer can invert e, since he knows the order n of the group G.
3. The signature is defined to be (e, y, y').

Signature Verification. To verify that (e, y, y') is a signature on a message M, we perform the following steps.

1. We check that e is an integer from interval $]B + C, 2(B + C)[$ and different from e'.
2. We compute $x' = (y')^{e'} h^{-H(M)}$.
3. We verify that $x = y^e h^{-H(x')}$.

3.2 Security Proof

First we state our intractability assumption:

Conjecture 1. We choose randomly an abelian group G of order n, where $n \in]B, B + C]$, according to the probability distribution D. Let $x \in G$ be chosen at random. Consider any probabilistic polynomial time algorithm A. The probability that A, given the description of G and x, outputs $e > 1$ and $y \in G$ such that $y^e = x$ is negligible.

In our generic model, where A knows only D and encodings of group elements, and assuming that D is from a *hard* family, it follows from the results we proved above that this assumption is true.

The following fact is easy to proof:

Fact 2 *For any $E > 0$, given natural numbers $e_1, \ldots, e_t$ smaller than E and an element g in a finite abelian multiplicative group G, we are able to compute $g^{\prod_{i \neq j} e_i}$ for all $j = 1, \ldots, t$ using $O(t \log t \log E)$ multiplications in G.*

We prove that our signature scheme is secure against chosen message attacks. We stress that this security proof holds for any type of polynomial-time attack (not just generic ones), as long as the conjecture holds.

Theorem 2. *The above signature scheme is secure assuming the conjecture 1 and that H is collision-resistant.*

The proof is somehow standard and it is available in the full version of the article [11].

Corollary 2. *The above signature scheme is secure against chosen message attack performed by any generic algorithm if we assume that D is a hard probability distribution.*

3.3 Concrete Instances of the Generic Scheme

Known order. If we are able to choose a random group of order known for us and hidden for others, then we have a very efficient signature scheme, secure assuming our conjectures.

The original Cramer-Shoup signature scheme [10] is a special case of this, where the RSA keys are restricted to safe prime products. But our generic scheme can also be instantiated with general RSA groups without restricting the keys to safe primes.

Unknown order. If we do not know how to generate a group of known for us order, we cannot use the signing algorithm we specified above. Instead, we can use the method based on Fact 2 and given in the simulation from the proof in order to sign messages. This requires $O(kt \log t)$ group operations, where t is the number of the messages to be signed and k the security parameter. Moreover we need to keep t exponents e in our memory.

Our signature scheme in such form can be applied, for instance, in class groups where we do not know how to generate efficiently a discriminant together with the corresponding class number. However, as t becomes large, this scheme becomes rather inefficient. In the following section we describe a solution to this.

4 Tree Structured Signature Scheme in Groups with Unknown Order

Here, we describe a generic signature scheme, which can be applied, e.g, in class groups, is more efficient than the previous scheme and still provably secure without assuming the random oracle model.

We recall the idea of authentication "flat tree" [8,9,6] and put it on top of our signature scheme.

4.1 Description of the Scheme

Let l and d be new integer parameters. We will construct a tree of degree l and depth d.

As in the original protocol we choose an abelian group G of a random order n according to the distribution D.

We generate a list of l primes $e_1, \ldots, e_l$ such that $B + C < e_i < 2(B + C)$ for each $1 \leq i \leq l$.

Let us consider the root of the authentication tree. As in the simulation in the security proof of the original protocol we choose random $z_0, \omega_0 \in G$ and a prime e' satisfying $B + C < e' < 2(B + C)$ and $e' \neq e_i$ for all $1 \leq i \leq l$. We compute $h_0 = z_0^{\prod_{1 \leq i \leq l} e_i}$ and $x_0 = \omega_0^{\prod_{1 \leq i \leq l} e_i}$.

The values h_0, x_0 and e' are publicly known.

For each child i of the root we choose random elements $z_{1,i}, \omega_{1,i} \in G$ and compute $h_{1,i} = z_{1,i}^{\prod_{1 \leq j \leq l} e_j}$ and $x_{1,i} = \omega_{1,i}^{\prod_{1 \leq j \leq l} e_j}$. Now as in the simulation we can sign the pair $h_{1,i}, x_{1,i}$ using h_0, x_0 and the exponent e_i. Let $\sigma_{1,i}$ be that signature.

If we repeat that authentication procedure for each node using its signed values h and x together with exponent e', we can construct a tree of degree l and depth d. We will have l^d leaves. Instead of constructing new values h and x for each leaf we will sign messages using leaves. Hence we will be able to sign l^d messages (one message for each leaf).

It is important to note that we don't need to remember the whole tree in order to sign messages. If we use leaves in order "from the left", the path from the root to the leaf is sufficient to construct a signature of a new message (for details see [8,9,6]). Let the nodes on the path be (h_0, x_0, h_1, x_1 with signature $\sigma_1, \ldots, h_{d-1}, x_{d-1}$ with signature σ_{d-1}). Using the values h_{d-1}, x_{d-1}, the public exponent e' and appropriate exponent e_i we sign a message m. That gives us the signature σ_d. Finally the signature of message m is the list $\sigma = (h_1, x_1, \sigma_1, \ldots, h_{d-1}, x_{d-1}, \sigma_{d-1}, \sigma_d)$.

To verify the signature σ we verify partial signatures $\sigma_1, \ldots, \sigma_d$ consecutively using the original scheme with appropriate values of x, h and e_i.

4.2 Security Proof

Theorem 3. *Let l^d be polynomial in the security parameter k. Assume that the signature scheme from previous scheme was unforgeable under adaptively chosen message attack. Then our tree authentication scheme is unforgeable under adaptive chosen message attack.*

4.3 Efficiency Analysis

Each signature contains d partial signatures. Hence the size of the signature is $O(dk)$.

The signer has to keep the list of l primes, which corresponds to $O(kl)$ bits and a path of size $O(dk)$ bits.

For signing a new signature the signer has to move to a new path. It requires computing at most d partial signatures, on average less than $1+2/\ell$. Computing a new partial signature in a straightforward way requires $O(lk)$ group operations (generally computing roots takes this time).

It follows from Fact 2 that we could have done some precomputation for each node requiring $O\left(kl \log \ell\right)$ multiplications in G. That would give us on average $O\left(k \log \ell\right)$ group operations for each partial signature. Then it would be sufficient to use $O(k)$ group multiplications for each partial signature. However this solution requires keeping $O(kl)$ bits in memory for each node. If we store the results of precomputation for each node on the path, it will occupy $O(kld)$ bits in the memory of signer.

The verifier just checks d partial signatures (see previous section for details).

4.4 Realistic Parameters

Let $l = 1000$ and $d = 3$. Then we are able to sign one billion messages, which should be sufficient in real life applications. Signing messages will be still fast and the signer's system will require a reasonable amount of data to be stored.

References

1. Ingrid Biehl, Johannes Buchmann, Safuat Hamdy, and Andreas Meyer. A signature scheme based on the intractability of computing roots. Technical Report 1/00, Darmstadt University of Technology, 2000.
2. D. Boneh and R. J. Lipton. Algorithms for black-box fields and their application to cryptography. *Lecture Notes in Computer Science*, 1109:283–297, 1996.
3. D. Boneh and R. Venkatesan. Breaking RSA may not be equivalent to factoring. *Lecture Notes in Computer Science*, 1403:59–71, 1998.
4. Johannes Buchmann and H. C. Williams. A key-exchange system based on imaginary quadratic fields. *Journal of Cryptology: the journal of the International Association for Cryptologic Research*, 1(2):107–118, 1988.
5. Duncan A. Buell. The expectation of success using a Monte Carlo factoring method—some statistics on quadratic class numbers. *Mathematics of Computation*, 43(167):313–327, July 1984.
6. Marc Bütikofer. An abstraction of the Cramer-Damgård signature scheme based on tribes of q-one-way-group-homomorphisms. ETH Zürich, 1999.
7. H. Cohen and Jr. H.W. Lenstra. Heuristics on class groups of number fields. In *Number Theory, Noordvijkerhout 1983*, volume 1068 of *Lecture Notes in Math.*, pages 33–62, 1984.
8. R. Cramer and I. Damgaard. Secure signature schemes based on interactive protocols. *Lecture Notes in Computer Science*, 963:297–310, 1995.
9. R. Cramer and I. Damgaard. New generation of secure and practical RSA-Based signatures. *Lecture Notes in Computer Science*, 1109:173–185, 1996.
10. Ronald Cramer and Victor Shoup. Signature schemes based on the strong RSA assumption. In *ACM Conference on Computer and Communications Security*, pages 46–51, 1999.

11. Ivan Damgård and Maciej Koprowski. Generic lower bounds for root extraction and signature schemes in general groups (extended version). Cryptology ePrint Archive, Report 2002/013, 2002. `http://eprint.iacr.org/`.
12. Marc Fischlin. A note on security proofs in the generic model. In T. Okamoto, editor, *Advances in Cryptology – ASIACRYPT ' 2000*, volume 1976 of *Lecture Notes in Computer Science*, pages 458–469, Kyoto, Japan, 2000. International Association for Cryptologic Research, Springer-Verlag, Berlin Germany.
13. Safuat Hamdy and Bodo Möller. Security of cryptosystems based on class groups of imaginary quadratic orders. In T. Okamoto, editor, *Advances in Cryptology - ASIACRYPT 2000*, pages 234–247. Springer-Verlag, 2000.
14. M. Jacobson. *Subexponential class group computation in quadratic orders.* PhD thesis, Technische Universitat Darmstadt, Darmstadt, Germany, 1999.
15. Tsutomu Matsumoto, Koki Kato, and Hideki Imai. Speeding up secret computations with insecure auxiliary devices. In S. Goldwasser, editor, *Advances in Cryptology—CRYPTO '88*, volume 403 of *Lecture Notes in Computer Science*, pages 497–506. Springer-Verlag, 1990, 21–25 August 1988.
16. U. Maurer and S. Wolf. Lower bounds on generic algorithms in groups. *Lecture Notes in Computer Science*, 1403:72–84, 1998.
17. J. Merkle and R. Werchner. On the security of server-aided RSA protocols. *Lecture Notes in Computer Science*, 1431:99–116, 1998.
18. V. I. Nechaev. Complexity of a determinate algorithm for the discrete logarithm. *Mathematical Notes*, 55(2):165–172, 1994. Translated from Matematicheskie Zametki, 55(2):91–101, 1994.
19. P. Q. Nguyen and I.E. Shparlinski. On the insecurity of a server-aided RSA protocol. In C. Boyd, editor, *Advances in Cryptology—Asiacrypt'2001*, volume 2248 of *Lecture Notes in Computer Science*, pages 21–35. Springer-Verlag, 2001.
20. C.-P. Schnorr. Efficient signature generation by smart cards. *Journal of Cryptology: the journal of the International Association for Cryptologic Research*, 4(3):161–174, 1991.
21. Claus Peter Schnorr. Security of DL-encryption and signatures against generic attacks - a survey. In K.Alster, H.C.Williams, and J.Urbanowicz, editors, *Proceedings of Public-Key Cryptography and Computational Number Theory Conference, Warsaw, September, 2000.* Walter De Gruyter, 2002.
22. Claus Peter Schnorr and Markus Jakobsson. Security of discrete log cryptosystems in the random oracle + generic model. In *Conference on The Mathematics of Public-Key Cryptography*, The Fields Institute, Toronto, Canada, 1999.
23. Claus Peter Schnorr and Markus Jakobsson. Security of signed ElGamal encryption. In T. Okamoto, editor, *Advances in Cryptology – ASIACRYPT ' 2000*, volume 1976 of *Lecture Notes in Computer Science*, pages 73–89, Kyoto, Japan, 2000. International Association for Cryptologic Research, Springer-Verlag, Berlin Germany.
24. V. Shoup. Lower bounds for discrete logarithms and related problems. In *Advances in Cryptology: Eurocrypt '97*, pages 256–266, 1997.

Optimal Security Proofs for PSS and Other Signature Schemes

Jean-Sébastien Coron

Gemplus Card International,
34 rue Guynemer, Issy-les-Moulineaux, F-92447, France,
coron@ens.fr

Abstract. The Probabilistic Signature Scheme (PSS) designed by Bellare and Rogaway is a signature scheme provably secure against chosen message attacks in the random oracle model, whose security can be tightly related to the security of RSA. We derive a new security proof for PSS in which a much shorter random salt is used to achieve the same security level, namely we show that $\log_2 q_{sig}$ bits suffice, where q_{sig} is the number of signature queries made by the attacker. When PSS is used with message recovery, a better bandwidth is obtained because longer messages can now be recovered. In this paper, we also introduce a new technique for proving that the security proof of a signature scheme is optimal. In particular, we show that the size of the random salt that we have obtained for PSS is optimal: if less than $\log_2 q_{sig}$ bits are used, then PSS is still provably secure but it cannot have a tight security proof. Our technique applies to other signature schemes such as the Full Domain Hash scheme and Gennaro-Halevi-Rabin's scheme, whose security proofs are shown to be optimal.

Key-words: Probabilistic Signature Scheme, Provable Security.

1 Introduction

Since the invention of public-key cryptography in the seminal Diffie-Hellman paper [9], significant research endeavors were devoted to the design of practical and provably secure schemes. A proof of security is usually a computational reduction from solving a well established problem to breaking the cryptosystem. Well established problems of cryptographic relevance include factoring large integers, computing discrete logarithms in prime order groups, or extracting roots modulo a composite integer.

For digital signature schemes, the strongest security notion was defined by Goldwasser, Micali and Rivest in [13], as *existential unforgeability under an adaptive chosen message attack*. This notion captures the property that an attacker cannot produce a valid signature, even after obtaining the signature of (polynomially many) messages of his choice.

Goldwasser, Micali and Rivest proposed in [13] a signature scheme based on signature trees that provably meets this definition. The efficiency of the scheme

L.R. Knudsen (Ed.): EUROCRYPT 2002, LNCS 2332, pp. 272–287, 2002.

was later improved by Dwork and Naor [10], and Cramer and Damgård [7]. A significant drawback of those signature schemes is that the signature of a message depends on previously signed messages: the signer must thus store information relative to the signatures he generates as time goes by. Gennaro, Halevi and Rabin presented in [12] a new hash-and-sign scheme provably secure against adaptive chosen message attacks which is both state-free and efficient. Its security is based on the strong-RSA assumption. Cramer and Shoup presented in [8] a signature scheme provably secure against adaptive chosen message attacks, which is also state-free, efficient, and based on the strong-RSA assumption.

The random oracle model, introduced by Bellare and Rogaway in [1], is a theoretical framework allowing to prove the security of hash-and-sign signature schemes. In this model, the hash function is seen as an oracle that outputs a random value for each new query. Bellare and Rogaway defined in [2] the Full Domain Hash (FDH) signature scheme, which is provably secure in the random oracle model assuming that inverting RSA is hard. [2] also introduced the Probabilistic Signature Scheme (PSS), which offers better security guarantees than FDH. Similarly, Pointcheval and Stern [19] proved the security of discrete-log based signature schemes in the random oracle model (see also [16] for a concrete treatment). However, security proofs in the random oracle are not real proofs, since the random oracle is replaced by a well defined hash function in practice; actually, Canetti, Goldreich and Halevi [4] showed that a security proof in the random oracle model does not necessarily imply that a scheme is secure in the real world.

For practical applications of provably secure schemes, the tightness of the security reduction must be taken into account. A security reduction is tight when breaking the signature scheme leads to solving the well established problem with probability close to one. In this case, the signature scheme is almost as secure as the well established problem. On the contrary, if the above probability is too small, the guarantee on the signature scheme will be weak; in which case larger security parameters must be used, thereby decreasing the efficiency of the scheme.

The security reduction of [2] for Full Domain Hash bounds the probability ε of breaking FDH in time t by $(q_{hash} + q_{sig}) \cdot \varepsilon'$ where ε' is the probability of inverting RSA in time t' close to t and where q_{hash} and q_{sig} are the number of hash queries and signature queries performed by the forger. This was later improved in [5] to $\varepsilon \simeq q_{sig} \cdot \varepsilon'$, which is a significant improvement since in practice q_{sig} happens to be much smaller than q_{hash}. However, FDH's security reduction is still not tight, and FDH is still not as secure as inverting RSA.

On the contrary, PSS is almost as secure as inverting RSA ($\varepsilon \simeq \varepsilon'$). Additionally, for PSS to have a tight security proof in [2], the random salt used to generate the signature must be of length at least $k_0 \simeq 2 \cdot \log_2 q_{hash} + \log_2 1/\varepsilon'$, where q_{hash} is the number of hash queries requested by the attacker and ε' the probability of inverting RSA within a given time bound. Taking $q_{hash} = 2^{60}$ and $\varepsilon' = 2^{-60}$ as in [2], we obtain a random salt of size $k_0 = 180$ bits. In this paper, we show that PSS has actually a tight security proof for a random salt as short as $\log_2 q_{sig}$ bits, where q_{sig} is the number of signature queries made by the attacker. For example, for an application in which at most one billion signatures will be generated, $k_0 = 30$ bits of random salt are actually sufficient to guarantee the same level of security as RSA, and taking a longer salt *does not* increase the security level. When PSS is used with message recovery, we obtain a better bandwidth because a larger message can now be recovered when verifying the signature.

Moreover, we show that this size is optimal: if less than $\log_2 q_{sig}$ bits of random salt are used, PSS is still provably secure, but PSS cannot have exactly the same security level as RSA. First, using a new technique, we derive an upper bound for the security of FDH, which shows that the security proof in [5] with $\varepsilon \simeq q_{sig} \cdot \varepsilon'$ is optimal. In other words, it is not possible to further improve the security proof of FDH in order to obtain a security level equivalent to RSA. This answers the open question raised by Bellare and Rogaway in [2], about the existence of a better security proof for FDH: as opposed to PSS, FDH *cannot* be proven as secure as inverting RSA. The technique also applies to other signature schemes such as Gennaro-Halevi-Rabin's scheme [12] and Paillier's signature scheme [17]. To our knowledge, this is the first result concerning optimal security proofs. Then, using the upper bound for the security of FDH, we show that our size k_0 for the random salt in PSS is optimal: if less than $\log_2 q_{sig}$ bits are used, no security proof for PSS can be tight.

2 Definitions

In this section we briefly present some notations and definitions used throughout the paper. We start by recalling the definition of a signature scheme.

Definition 1 (signature scheme). *A signature scheme* (`Gen`, `Sign`, `Verify`) *is defined as follows:*

- *The key generation algorithm* `Gen` *is a probabilistic algorithm which given* 1^k*, outputs a pair of matching public and private keys,* (pk, sk).
- *The signing algorithm* `Sign` *takes the message* M *to be signed, the public key* pk *and the private key* sk*, and returns a signature* $x = \mathtt{Sign}_{pk,sk}(M)$. *The signing algorithm may be probabilistic.*
- *The verification algorithm* `Verify` *takes a message* M*, a candidate signature* x' *and* pk. *It returns a bit* $\mathtt{Verify}_{pk}(M, x')$*, equal to one if the signature is accepted, and zero otherwise. We require that if* $x \leftarrow \mathtt{Sign}_{pk,sk}(M)$*, then* $\mathtt{Verify}_{pk}(M, x) = 1$.

In the previously introduced *existential unforgeability under an adaptive chosen message attack* scenario, the forger can dynamically obtain signatures of messages of his choice and attempts to output a valid forgery. A *valid forgery* is a message/signature pair (M, x) such that $\mathtt{Verify}_{pk}(M, x) = 1$ whereas the signature of M was never requested by the forger.

A significant line of research for proving the security of signature schemes is the previously introduced random oracle model, where resistance against adaptive chosen message attacks is defined as follows [1]:

Definition 2. *A forger $\mathcal{F}$ is said to $(t, q_{hash}, q_{sig}, \varepsilon)$-break the signature scheme* (Gen, Sign, Verify) *if after at most $q_{hash}(k)$ queries to the hash oracle, $q_{sig}(k)$ signatures queries and $t(k)$ processing time, it outputs a valid forgery with probability at least $\varepsilon(k)$ for all $k \in \mathbb{N}$.*

and quite naturally:

Definition 3. *A signature scheme* (Gen, Sign, Verify) *is $(t, q_{sig}, q_{hash}, \varepsilon)$-secure if there is no forger who $(t, q_{hash}, q_{sig}, \varepsilon)$-breaks the scheme.*

The RSA cryptosystem, invented by Rivest, Shamir and Adleman [20], is the most widely used cryptosystem today:

Definition 4 (The RSA cryptosystem). *The RSA cryptosystem is a family of trapdoor permutations, specified by:*

- *The RSA generator $\mathcal{RSA}$, which on input 1^k, randomly selects two distinct $k/2$-bit primes p and q and computes the modulus $N = p \cdot q$. It randomly picks an encryption exponent $e \in \mathbb{Z}^*_{\phi(N)}$ and computes the corresponding decryption exponent d such that $e \cdot d = 1 \mod \phi(N)$. The generator returns (N, e, d).*
- *The encryption function $f : \mathbb{Z}^*_N \to \mathbb{Z}^*_N$ defined by $f(x) = x^e \mod N$.*
- *The decryption function $f^{-1} : \mathbb{Z}^*_N \to \mathbb{Z}^*_N$ defined by $f^{-1}(y) = y^d \mod N$.*

FDH was the first practical and provably secure signature scheme based on RSA. It is defined as follows: the key generation algorithm, on input 1^k, runs $\mathcal{RSA}(1^k)$ to obtain (N, e, d). It outputs (pk, sk), where the public key pk is (N, e) and the private key sk is (N, d). The signing and verifying algorithms use a hash function $H : \{0,1\}^* \to \mathbb{Z}^*_N$ which maps bit strings of arbitrary length to the set of invertible integers modulo N.

$\texttt{SignFDH}_{N,d}(M)$	$\texttt{VerifyFDH}_{N,e}(M, x)$
$y \leftarrow H(M)$	$y \leftarrow x^e \mod N$
return $y^d \mod N$	if $y = H(M)$ then return 1 else return 0.

FDH is provably secure in the random oracle model, assuming that inverting RSA is hard. An *inverting algorithm* $\mathcal{I}$ for RSA gets as input (N, e, y) and tries to find $y^d \mod N$. Its success probability is the probability to output $y^d \mod N$ when (N, e, d) are obtained by running $\mathcal{RSA}(1^k)$ and y is set to $x^e \mod N$ for some x chosen at random in $\mathbb{Z}^*_N$.

Definition 5. *An inverting algorithm $\mathcal{I}$ is said to (t, ε)-break RSA if after at most $t(k)$ processing time its success probability is at least $\varepsilon(k)$ for all $k \in \mathbb{N}$.*

Definition 6. *RSA is said to be (t, ε)-secure if there is no inverter that (t, ε)-breaks RSA.*

The following theorem [5] proves the security of FDH in the random oracle model.

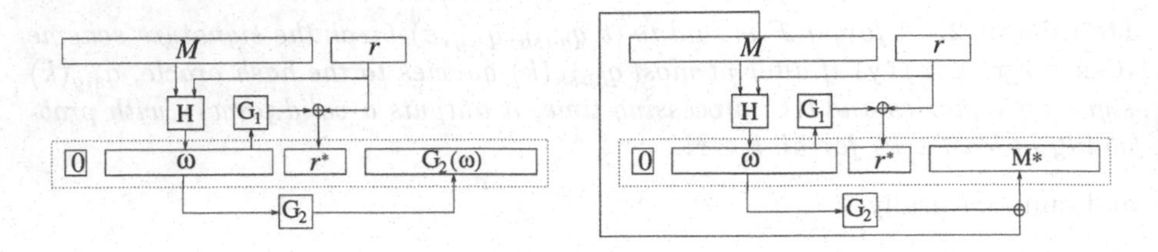

Fig. 1. PSS (left) and PSS-R (right)

Theorem 1. *Assuming that RSA is* (t_I, ε_I)*-secure, FDH is* $(t_F,\ q_{hash},\ q_{sig},\ \varepsilon_F)$*-secure, with:*

$$t_I = t_F + (q_{hash} + q_{sig} + 1) \cdot \mathcal{O}(k^3) \tag{1}$$

$$\varepsilon_I = \frac{\varepsilon_F}{q_{sig}} \cdot \left(1 - \frac{1}{q_{sig}+1}\right)^{q_{sig}+1} \tag{2}$$

The technique described in [5] can be used to obtain an improved security proof for Gennaro-Halevi-Rabin's signature scheme [12] in the random oracle model and for Paillier's signature scheme [17]. From a forger which outputs a forgery with probability ε_F, the reduction succeeds in solving the hard problem with probability roughly ε_F/q_{sig}, in approximately the same time bound.

The security reduction of FDH is not tight: the probability ε_F of breaking FDH is smaller than roughly $q_{sig} \cdot \varepsilon_I$ where ε_I is the probability of inverting RSA, whereas the security reduction of PSS is tight: the probability of breaking PSS is almost the same as the probability of inverting RSA ($\varepsilon_F \simeq \varepsilon_I$).

3 New Security Proof for PSS

Several standards include PSS [2], among these are IEEE P1363a [14], a revision of ISO/IEC 9796-2, and the upcoming PKCS#1 v2.1 [18]. The signature scheme PSS is parameterized by the integers k, k_0 and k_1. The key generation is identical to FDH. The signing and verifying algorithms use two hash functions $H : \{0,1\}^* \to \{0,1\}^{k_1}$ and $G : \{0,1\}^{k_1} \to \{0,1\}^{k-k_1-1}$. Let G_1 be the function which on input $\omega \in \{0,1\}^{k_1}$ returns the first k_0 bits of $G(\omega)$, whereas G_2 is the function returning the remaining $k - k_0 - k_1 - 1$ bits of $G(\omega)$. A random *salt* r of k_0 bits is concatenated to the message M before hashing it. The scheme is illustrated in figure 1. In this section we obtain a better security proof for PSS, in which a shorter random salt is used to generate the signature.

$\texttt{SignPSS}(M)$:	$\texttt{VerifyPSS}(M, x)$:
$r \overset{R}{\leftarrow} \{0,1\}^{k_0}$	$y \leftarrow x^e \mod N$
$\omega \leftarrow H(M\|r)$	Break up y as $b\|\omega\|r^*\|\gamma$
$r^* \leftarrow G_1(\omega) \oplus r$	Let $r \leftarrow r^* \oplus G_1(\omega)$
$y \leftarrow 0\|\omega\|r^*\|G_2(\omega)$	if $H(M\|r) = \omega$ and $G_2(\omega) = \gamma$ and $b = 1$
return $y^d \mod N$	then return 1 else return 0

The following theorem [2] proves the security of PSS in the random oracle model:

Theorem 2. *Assuming that RSA is (t', ε')-secure, the scheme* PSS$[k_0, k_1]$ *is* $(t, q_{sig}, q_{hash}, \varepsilon)$*-secure, where :*

$$t = t' - (q_{hash} + q_{sig} + 1) \cdot k_0 \cdot \mathcal{O}(k^3) \tag{3}$$

$$\varepsilon = \varepsilon' + 3 \cdot (q_{sig} + q_{hash})^2 \cdot \left(2^{-k_0} + 2^{-k_1}\right) \tag{4}$$

Theorem 2 shows that for PSS to be as secure as RSA (*i.e.* $\varepsilon' \simeq \varepsilon$), it must be the case that $(q_{sig} + q_{hash})^2 \cdot (2^{-k_0} + 2^{-k_1}) < \varepsilon'$, which gives $k_0 \geq k_{min}$ and $k_1 \geq k_{min}$, where:

$$k_{min} = 2 \cdot \log_2(q_{hash} + q_{sig}) + \log_2 \frac{1}{\varepsilon'} \tag{5}$$

Taking $q_{hash} = 2^{60}$, $q_{sig} = 2^{30}$ and $\varepsilon' = 2^{-60}$ as in [2], we obtain that k_0 and k_1 must be greater than $k_{min} = 180$ bits.

The following theorem shows that PSS can be proven as secure as RSA for a much shorter random salt, namely $k_0 = \log_2 q_{sig}$ bits, which for $q_{sig} = 2^{30}$ gives $k_0 = 30$ bits. The minimum value for k_1 remains unchanged.

Theorem 3. *Assuming that RSA is (t', ε')-secure, the scheme* PSS$[k_0, k_1]$ *is* $(t, q_{sig}, q_{hash}, \varepsilon)$*-secure, where :*

$$t = t' - (q_{hash} + q_{sig}) \cdot k_1 \cdot \mathcal{O}(k^3) \tag{6}$$

$$\varepsilon = \varepsilon' \cdot \left(1 + 6 \cdot q_{sig} \cdot 2^{-k_0}\right) + 2 \cdot (q_{hash} + q_{sig})^2 \cdot 2^{-k_1} \tag{7}$$

In appendix A, we give a security proof for a variant of PSS, for which the proof is simpler. The proof of theorem 3 is very similar and can be found in the full version of the paper [6]. The difference with the security proof of [2] is the following: in [2], a new random salt r is randomly generated for each signature query, and if r has appeared before, the inverter stops and has failed. Since at most $q_{hash} + q_{sig}$ random salts can appear during the reduction, the inverter stops after a given signature query with probability less than $(q_{hash} + q_{sig}) \cdot 2^{-k_0}$. There are at most q_{sig} signature queries, so this gives an error probability of:

$$q_{sig} \cdot (q_{hash} + q_{sig}) \cdot 2^{k_0}$$

which accounts for the term $(q_{hash} + q_{sig})^2 \cdot 2^{-k_0}$ in equation (4). On the contrary, in our new security proof, we generate for each new message M_i a list of q_{sig} random salts. Those random salts are then used to answer the signature queries for M_i, so there is no error probability when answering the signature queries.

3.1 Discussion

Theorem 3 shows that PSS is actually provably secure for any size k_0 of the random salt. In figure 2 we plot $\log_2 \varepsilon'/\varepsilon$ as a function of the size k_0 of the salt,

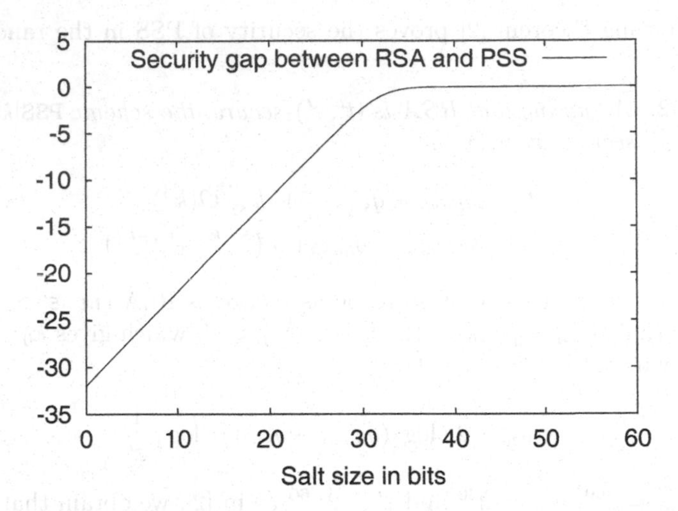

Fig. 2. Security gap between PSS and RSA: $\log_2 \varepsilon'/\varepsilon$ as a function of the salt size k_0 for $q_{sig} = 2^{30}$ signature queries.

which depicts the relative security of PSS compared to RSA, for $q_{sig} = 2^{30}$ and $k_1 > k_{min}$. For $k_0 = 0$, we reach the security level of FDH, where approximately $\log_2 q_{sig}$ bits of security are lost compared to RSA. For k_0 comprised between zero and $\log_2 q_{sig}$, we gain one bit of security when k_0 increases by one bit. And for k_0 greater than $\log_2 q_{sig}$, the security level of PSS is almost the same as inverting RSA. This shows that PSS has a tight security proof as soon as the salt size reaches $\log_2 q_{sig}$, and using larger salts does not further improve security. For the signer, q_{sig} represents the maximal number of signatures which can be generated for a given public-key. For example, for an application in which at most one billion signatures will be generated, $k_0 = 30$ bits of random salt are actually sufficient to guarantee the same level of security as RSA, and taking a larger salt does not increase the security level.

PSS-R is a variant of PSS which provides message recovery; the scheme is illustrated in figure 1. The goal is to save on the bandwidth: instead of transmitting the message separately, the message is recovered when verifying the signature. The security proof for PSS-R is almost identical to the security proof of PSS, and PSS-R achieves the same security level as PSS. Consequently, using the same parameters as for PSS with a 1024-bits RSA modulus, 813 bits of message can now be recovered when verifying the signature (instead of 663 bits with the previous security proof).

4 Optimal Security Proof for FDH

In section 2 we have seen that the security proof of theorem 1 for FDH is still not tight: the probability ε_F of breaking FDH is smaller than roughly $q_{sig} \cdot \varepsilon_I$ where ε_I is the probability of inverting RSA. In this section we show that the security proof of theorem 1 for FDH is optimal, *i.e.* there is no better reduction from inverting RSA to breaking FDH, and one cannot avoid loosing the q_{sig} factor in the probability bound. We use a similar approach as Boneh and Venkatesan in [3] for disproving the equivalence between inverting low-exponent RSA and factoring. They show that any efficient algebraic reduction from factoring to inverting low-exponent RSA can be converted into an efficient factoring algorithm. Such reduction is an algorithm $\mathcal{A}$ which factors N using an e-th root oracle for N. They show how to convert $\mathcal{A}$ into an algorithm $\mathcal{B}$ that factors integers without using the e-th root oracle. Thus, unless factoring is easy, inverting low-exponent RSA cannot be equivalent to factoring under algebraic reductions.

Similarly, we show that any better reduction from inverting RSA to breaking FDH can be converted into an efficient RSA inverting algorithm. Such reduction is an algorithm $\mathcal{R}$ which uses a forger as an oracle in order to invert RSA. We show how to convert $\mathcal{R}$ into an algorithm $\mathcal{I}$ which inverts RSA without using the oracle forger. Consequently, if inverting RSA is hard, there is no such better reduction for FDH, and the reduction of theorem 1 must be optimal.

Our technique is the following. Recall that resistance against adaptive chosen message attacks is considered, so the forger is allowed to make signature queries for messages of its choice, which must be answered by the reduction $\mathcal{R}$. Eventually the forger outputs a forgery, and the reduction must invert RSA. Therefore we first ask the reduction to sign a message M and receive its signature s, then we rewind the reduction to the state in which it was before the signature query, and we send s as a forgery for M. This is a true forgery for the reduction, because after the rewind there was no signature query for M, so eventually the reduction inverts RSA. Consequently, we have constructed from $\mathcal{R}$ an algorithm $\mathcal{I}$ which inverts RSA without using any forger. Actually, this technique allows to simulate a forger with respect to $\mathcal{R}$, without being able to break FDH. However, the simulation is not perfect, because it outputs a forgery only for messages which can be signed by the reduction, whereas a real forger outputs the forgery of a message that the reduction may or may not be able to sign.

We quantify the efficiency of a reduction by giving the probability that the reduction inverts RSA using a forger that $(t_F, q_{hash}, q_{sig}, \varepsilon_F)$-breaks the signature scheme, within an additional running time of t_R:

Definition 7. *We say that a reduction algorithm* $\mathcal{R}$ $(t_R, q_{hash}, q_{sig}, \varepsilon_F, \varepsilon_R)$*-reduces inverting RSA to breaking FDH if upon input* (N, e, y) *and after running any forger that* $(t_F, q_{hash}, q_{sig}, \varepsilon_F)$*-breaks FDH, the reduction outputs* $y^d \bmod N$ *with probability greater than* ε_R*, within an additional running time of* t_R.

In the above definition, t_R is the running time of the reduction algorithm only and does not include the running time of the forger. Eventually, the time needed to invert RSA is $t_F + t_R$, where t_F is the running time of the forger. For example, the reduction of theorem 1 for FDH $(t_R, q_{hash}, q_{sig}, \varepsilon_F, \varepsilon_R)$-reduces inverting RSA to breaking FDH with $t_R(k) = (q_{hash} + q_{sig}) \cdot \mathcal{O}(k^3)$ and $\varepsilon_R = \varepsilon_F/(4 \cdot q_{sig})$.

The following theorem, whose proof is given in appendix B, shows that from any such reduction $\mathcal{R}$ we can invert RSA with probability greater than roughly $\varepsilon_R - \varepsilon_F/q_{sig}$, in roughly the same time bound.

Theorem 4. *Let $\mathcal{R}$ be a reduction that $(t_R, q_{hash}, q_{sig}, \varepsilon_R, \varepsilon_F)$-reduces inverting RSA to breaking FDH. $\mathcal{R}$ runs the forger only once. From $\mathcal{R}$ we can construct an algorithm that (t_I, ε_I)-inverts RSA, with:*

$$t_I = 2 \cdot t_R \tag{8}$$

$$\varepsilon_I = \varepsilon_R - \varepsilon_F \cdot \frac{\exp(-1)}{q_{sig}} \cdot \left(1 - \frac{q_{sig}}{q_{hash}}\right)^{-1} \tag{9}$$

Theorem 4 shows that from any reduction $\mathcal{R}$ that inverts RSA with probability ε_R when interacting with a forger that outputs a forgery with probability ε_F, we can invert RSA with probability roughly $\varepsilon_R - \varepsilon_F/q_{sig}$, in roughly the same time bound, without using a forger. For simplicity, we omit here the factors $\exp(-1)$ and $(1 - q_{sig}/q_{hash})$ in equation (9). Moreover we consider a forger that makes q_{sig} signature queries, and with probability $\varepsilon_F = 1$ outputs a forgery[1].

Theorem 4 implies that from a polynomial time reduction $\mathcal{R}$ that succeeds with probability ε_R when interacting with this forger, we obtain a polynomial time RSA inverter $\mathcal{I}$ that succeeds with probability $\varepsilon_I = \varepsilon_R - 1/q_{sig}$, without using the forger. If inverting RSA is hard, the success probability ε_I of the polynomial time inverter must be negligible. Consequently, the success probability ε_R of the reduction must be less than $1/q_{sig} + \texttt{negl}$. This shows that from a forger that outputs a forgery with probability one, a polynomial time reduction cannot succeed with probability greater than $1/q_{sig} + \texttt{negl}$. On the contrary, a tight security reduction would invert RSA with probability close to one. Here we cannot avoid the q_{sig} factor in the security proof: the security level of FDH cannot be proven equivalent to RSA, and the security proof of theorem 1 for FDH is optimal.

5 Extension to Any Signature Scheme with Unique Signature

We have introduced a new technique that enables to simulate a forger with respect to a reduction. It consists in making a signature query for a message M, rewinding the reduction, then sending the signature of M as a forgery. Actually, this technique stretches beyond FDH and can be generalized and applied to any signature scheme in which each message has a unique signature. Moreover, the technique can be generalized to reductions running a forger more than once. The following theorem shows that for a hash-and-sign signature scheme with unique signature, a reduction allowed to run or rewind a forger at most r times cannot succeed with probability greater than roughly $r \cdot \varepsilon_F/q_{sig}$. The definitions and the proof of the theorem are given in the full version of the paper [6].

[1] Such forger can be constructed by first factoring the modulus N, then computing a forgery using the factorisation of N.

Theorem 5. *Let $\mathcal{R}$ be a reduction that $(t_R, q_{hash}, q_{sig}, \varepsilon_F, \varepsilon_R)$-reduces solving a problem Π to breaking a hash-and-sign signature scheme with unique signature. $\mathcal{R}$ is allowed to run or rewind a forger at most r times. From $\mathcal{R}$ we can construct an algorithm that (t_A, ε_A)-solves Π, with:*

$$t_A = (r+1) \cdot t_R \tag{10}$$

$$\varepsilon_A = \varepsilon_R - \varepsilon_F \cdot \frac{\exp(-1) \cdot r}{q_{sig}} \cdot \left(1 - \frac{q_{sig}}{q_{hash}}\right)^{-1} \tag{11}$$

6 Security Proofs for Signature Schemes in the Standard Model

The same technique can be applied to security reductions in the standard model, and we obtain the same upper bound in $1/q_{sig}$ for signature schemes with unique signature. The definitions and the proof of the following theorem are given in the full version of the paper [6].

Theorem 6. *Let $\mathcal{R}$ be a reduction that $(t_R, q_{sig}, \varepsilon_F, \varepsilon_R)$-reduces solving Π to breaking a signature scheme with unique signature. $\mathcal{R}$ can run or rewind the forger at most r times. Assume that the size of the message space is at least 2^ℓ. From $\mathcal{R}$ we can construct an algorithm that (t_A, ε_A)-solves Π, with:*

$$t_A = (r+1) \cdot t_R \tag{12}$$

$$\varepsilon_A = \varepsilon_R - \varepsilon_F \cdot \frac{\exp(-1) \cdot r}{q_{sig}} \cdot \left(1 - \frac{q_{sig}}{2^\ell}\right)^{-1} \tag{13}$$

In [6] we give an example of a signature scheme with unique signature, provably secure in the standard model, and reaching the the above bound in $1/q_{sig}$.

7 Optimal Security Proof for PSS

In section 3 we have seen that $k_0 = \log_2 q_{sig}$ bits of random salt are sufficient for PSS to have a security level equivalent to RSA, and taking a larger salt does not further improve the security. In this section, we show that that this length is optimal: if a shorter random salt is used, the security level of PSS cannot be proven equivalent to RSA. Our technique described in section 4 does not apply directly because PSS is not a signature scheme with unique signature. We extend our technique to PSS using the following method.

We consider PSS in which the random salt is fixed to 0^{k_0}, and we denote this signature scheme $\mathsf{PSS0}[k_0, k_1]$. Consequently, $\mathsf{PSS0}[k_0, k_1]$ is a signature scheme with unique signature. First, we show how to convert a forger for $\mathsf{PSS0}[k_0, k_1]$ into a forger for $\mathsf{PSS}[k_0, k_1]$. A reduction $\mathcal{R}$ from inverting RSA to breaking $\mathsf{PSS}[k_0, k_1]$ uses a forger for $\mathsf{PSS}[k_0, k_1]$ in order to invert RSA. Consequently, from a forger for $\mathsf{PSS0}[k_0, k_1]$, we can invert RSA using the reduction $\mathcal{R}$. This means that from $\mathcal{R}$ we can construct a reduction $\mathcal{R}_0$ from inverting RSA to breaking $\mathsf{PSS0}[k_0, k_1]$.

Since $\mathtt{PSS0}[k_0, k_1]$ is a signature scheme with unique signature, theorem 5 gives an upper bound for the success probability of $\mathcal{R}_0$, from which we derive an upper bound for the success probability of $\mathcal{R}$.

Theorem 7. *Let $\mathcal{R}$ a reduction that $(t, q_{hash}, q_{sig}, \varepsilon_F, \varepsilon_R)$-reduces inverting RSA to breaking $\mathtt{PSS}[k_0, k_1]$, with $q_{hash} \geq 2 \cdot q_{sig}$. The reduction can run or rewind the forger at most r times. From $\mathcal{R}$ we can construct an inverting algorithm for RSA that (t_I, ε_I)-inverts RSA, with:*

$$t_I = (r+1) \cdot (t_R + q_{sig} \cdot \mathcal{O}(k)) \tag{14}$$

$$\varepsilon_I = \varepsilon_R - r \cdot \varepsilon_F \cdot \frac{2^{k_0+2}}{q_{sig}} \tag{15}$$

Proof. The proof is given in the full version of the paper [6].

Let consider as in section 4 a forger for $\mathtt{PSS}[k_0, k_1]$ that makes q_{sig} signature queries and outputs a forgery with probability $\varepsilon_F = 1/2$. Then, from a polynomial time reduction $\mathcal{R}$ that succeeds with probability ε_R when running once this forger, we obtain a polynomial time inverter that succeeds with probability $\varepsilon_I = \varepsilon_R - 2^{k_0+1}/q_{sig}$, without using the forger. If inverting RSA is hard, the success probability ε_I of the polynomial time inverter must be negligible, and therefore the success probability ε_R of the reduction must be less than $2^{k_0+1}/q_{sig} + \mathtt{negl}$. Consequently, in order to have a tight security reduction ($\varepsilon_R \simeq \varepsilon_R$), we must have $k_0 \simeq \log_2 q_{sig}$. The reduction of theorem 3 is consequently optimal.

8 Conclusion

We have described a new technique for analyzing the security proofs of signature schemes. The technique is both general and very simple and allows to derive upper bounds for security reductions using a forger as a black box, both in the random oracle model and in the standard model, for signature schemes with unique signature. We have also obtained a new criterion for a security reduction to be optimal, which may be of independent interest: we say that a security reduction is optimal if from a better reduction one can solve a difficult problem, such as inverting RSA. Our technique enables to show that the Full Domain Hash scheme, Gennaro-Halevi-Rabin's scheme and Paillier's signature scheme have an optimal security reduction in that sense. In other words, we have a matching lower and upper bound for the security reduction of those signature schemes: one cannot do better than losing a factor of q_{sig} in the security reduction.

Moreover, we have described a better security proof for PSS, in which a much shorter random salt is sufficient to achieve the same security level. This is of practical interest, since when PSS is used with message recovery, a better bandwidth is obtained because larger messages can be embedded inside the signature. Eventually, we have shown that this security proof for PSS is optimal: if a smaller random salt is used, PSS remains provably secure, but it cannot have the same level of security as RSA.

Acknowledgements

I would like to thank Burt Kaliski, Jacques Stern and David Pointcheval for helpful discussions and the anonymous referees for their comments.

References

1. M. Bellare and P. Rogaway, *Random oracles are practical: a paradigm for designing efficient protocols.* Proceedings of the First Annual Conference on Computer and Commmunications Security, ACM, 1993.
2. M. Bellare and P. Rogaway, *The exact security of digital signatures – How to sign with RSA and Rabin.* Proceedings of Eurocrypt'96, LNCS vol. 1070, Springer-Verlag, 1996, pp. 399-416.
3. D. Boneh and R. Venkatesan, *Breaking RSA may not be equivalent to factoring.* Proceedings of Eurocrypt' 98, LNCS vol. 1403, Springer-Verlag, 1998, pp. 59–71.
4. R. Canetti, O. Goldreich and S. Halevi, *The random oracle methodology, revisited*, STOC' 98, ACM, 1998.
5. J.S. Coron, *On the exact security of Full Domain Hash*, Proceedings of Crypto 2000, LNCS vol. 1880, Springer-Verlag, 2000, pp. 229-235.
6. J.S. Coron, *Security proofs for PSS and other signature schemes*, Cryptology ePrint Archive, Report 2001/062, 2001. `http://eprint.iacr.org`
7. R. Cramer and I. Damgård, *New generation of secure and practical RSA-based signatures*, Proceedings of Crypto'96, LNCS vol. 1109, Springer-Verlag, 1996, pp. 173-185.
8. R. Cramer and V. Shoup, *Signature schemes based on the Strong RSA Assumption*, May 9, 2000, revision of the extended abstract in Proc. 6th ACM Conf. on Computer and Communications Security, 1999; To appear, ACM Transactions on Information and System Security (ACM TISSEC). Available at `http://www.shoup.net/`
9. W. Diffie and M. Hellman, *New directions in cryptography*, IEEE Transactions on Information Theory, IT-22, 6, pp. 644-654, 1976.
10. C. Dwork and M. Naor, *An efficient existentially unforgeable signature scheme and its applications*, In J. of Cryptology, 11 (3), Summer 1998, pp. 187-208.
11. FIPS 186, *Digital signature standard*, Federal Information Processing Standards Publication 186, U.S. Department of Commerce/NIST, 1994.
12. R. Gennaro, S. Halevi and T. Rabin, *Secure hash-and-sign signatures without the random oracle*, proceedings of Eurocrypt '99, LNCS vol. 1592, Springer-Verlag, 1999, pp. 123-139.
13. S. Goldwasser, S. Micali and R. Rivest, *A digital signature scheme secure against adaptive chosen-message attacks*, SIAM Journal of computing, 17(2), pp. 281-308, April 1988.
14. IEEE P1363a, *Standard Specifications For Public Key Cryptography: Additional Techniques*, available at `http://www.manta.ieee.org/groups/1363`
15. A. Lenstra and H. Lenstra (eds.), *The development of the number field sieve*, Lecture Notes in Mathematics, vol 1554, Springer-Verlag, 1993.
16. K. Ohta and T. Okamoto, *On concrete security treatment of signatures derived from identification.* Prooceedings of Crypto '98, Lecture Notes in Computer Science vol. 1462, Springer-Verlag, 1998, pp. 354-369.

17. P. Paillier, *Public-key cryptosystems based on composite degree residuosity classes.* Proceedings of Eurocrypt'99, Lecture Notes is Computer Science vol. 1592, Springer-Verlag, 1999, pp. 223-238.
18. PKCS #1 v2.1, *RSA Cryptography Standard (draft)*, available at `http://www.rsa security.com /rsalabs/pkcs`.
19. D. Pointcheval and J. Stern, *Security proofs for signature schemes.* Proceedings of Eurocrypt'96, LNCS vol. 1070, Springer-Verlag, pp. 387-398.
20. R. Rivest, A. Shamir and L. Adleman, *A method for obtaining digital signatures and public key cryptosystems*, CACM 21, 1978.

A Security Proof of a Variant of PSS

We describe a variant of PSS that we call PFDH, for Probabilistic Full Domain Hash, for which the security proof is simpler. The scheme is similar to Full Domain Hash except that a random salt of k_0 bits is concatenated to the message M before hashing it. The difference with PSS is that the random salt is not recovered when verifying the signature; instead the random salt is transmitted separately. As FDH, the scheme uses a hash function $H : \{0,1\}^* \to \mathbb{Z}_N^*$.

$\texttt{SignPFDH}(M)$:	$\texttt{VerifyPFDH}(M, s, r)$:
$r \stackrel{R}{\leftarrow} \{0,1\}^{k_0}$	$y \leftarrow s^e \mod N$
$y \leftarrow H(M\|r)$	if $y = H(M\|r)$ then return 1
return $(y^d \mod N, r)$	else return 0

The following theorem proves the security of PFDH in the random oracle model, assuming that inverting RSA is hard. It shows that PFDH has a tight security proof for a random salt of length $k_0 = \log_2 q_{sig}$ bits.

Theorem 8. *Suppose that RSA is (t', ε')-secure. Then the signature scheme $\texttt{PFDH}[k_0]$ is $(t, q_{hash}, q_{sig}, \varepsilon)$-secure, where:*

$$t = t' - (q_{hash} + q_{sig}) \cdot \mathcal{O}(k^3) - q_{hash} \cdot q_{sig} \cdot \mathcal{O}(k) \tag{16}$$

$$\varepsilon = \varepsilon' \cdot \left(1 + 6 \cdot q_{sig} \cdot 2^{-k_0}\right) \tag{17}$$

Proof. Let $\mathcal{F}$ be a forger that $(t, q_{sig}, q_{hash}, \varepsilon)$-breaks PFDH. We construct an inverter I that (t', ε')-breaks RSA. The inverter receives as input (N, e, η) and must output $\eta^d \mod N$. We assume that the forger never repeats a hash query. However, the forger may repeat a signature query, in order to obtain the signature of M with distinct integers r. The inverter $\mathcal{I}$ maintains a counter i, initially set to zero.

When a message M appears for the first time in a hash query or a signature query, the inverter increments the counter i and sets $M_i \leftarrow M$. Then, the inverter generates a list L_i of q_{sig} random integers in $\{0,1\}^{k_0}$.

When the forger makes a hash query for $M_i\|r$, we distinguish two cases. If r belongs to the list L_i, the inverter generates a random $x \in \mathbb{Z}_N^*$ and returns $H(M_i\|r) = x^e \mod N$. Otherwise, the inverter generates a random $x \in \mathbb{Z}_N^*$ and returns $\eta \cdot x^e \mod N$. Consequently, for each message M_i, the list L_i contains the integers $r \in \{0,1\}^{k_0}$ such that the inverter knows the signature x corresponding to $M_i\|r$.

When the forger makes a signature query for M_i, the inverter takes the next random r in the list L_i. Since the list contains initially q_{sig} integers and there are at most q_{sig} signature queries, this is always possible. If there was already a hash query for $M_i\|r$, we have $H(M_i\|r) = x^e \mod N$ and the inverter returns the signature x. Otherwise the inverter generates a random $x \in \mathbb{Z}_N^*$, sets $H(M_i\|r) = x^e \mod N$ and returns the signature x.

When the forger outputs a forgery (M, s, r), we assume that it has already made a hash query for M, so $M = M_i$ for a given i. Otherwise, the inverter goes ahead and makes the hash query for $M\|r$. Then if r does not belong to the list L_i, we have $H(M_i\|r) = \eta \cdot x^e \mod N$. From $s = H(M_i\|r)^d = \eta^d \cdot x \mod N$, we obtain $\eta^d = s/x \mod N$ and the inverter succeeds in outputting $\eta^d \mod N$.

Since the forger has not made any signature query for the message M_i in the forgery (M_i, s, r), the forger has no information about the q_{sig} random integers in the list L_i. Therefore, the probability that r does not belong to L_i is $(1-2^{-k_0})^{q_{sig}}$. If the size k_0 of the random salt is greater than $\log_2 q_{sig}$, we obtain if $q_{sig} \geq 2$:

$$\left(1 - 2^{-k_0}\right)^{q_{sig}} \geq \left(1 - \frac{1}{q_{sig}}\right)^{q_{sig}} \geq \frac{1}{4}$$

Since the forger outputs a forgery with probability ε, the success probability ε' of the inverter is then at least $\varepsilon/4$, which shows that for $k_0 \geq \log_2 q_{sig}$ the probability of breaking PFDH is almost the same as the probability of inverting RSA.

For the general case, *i.e.* if we do not assume $k_0 \geq \log_2 q_{sig}$, we generate fewer than q_{sig} random integers in the list L_i, so that the salt r in the forgery (M_i, s, r) belongs to L_i with lower probability. More precisely, starting from an empty list L_i, the inverter generates with probability β a random $r \leftarrow \{0,1\}^{k_0}$, adds it to L_i, and starts again until the list L_i contains q_{sig} elements. Otherwise (so with probability $1 - \beta$) the inverter stops adding integers to the list. The number a_i of integers in L_i is then a random variable following a geometric law of parameter β:

$$\Pr[a_i = j] = \begin{cases} (1-\beta) \cdot \beta^j & \text{if } j < q_{sig} \\ \beta^{q_{sig}} & \text{if } j = q_{sig} \end{cases} \tag{18}$$

The inverter answers a signature query for M_i if the corresponding list L_i contains one more integer, which happens with probability β (otherwise the inverter must abort). Consequently, the inverter answers all the signature queries with probability greater than $\beta^{q_{sig}}$. Note that if $\beta = 1$, the setting boils down to the previous case: all the lists L_i contain exactly q_{sig} integers, and the inverter answers all the signature queries with probability one.

The probability that r in the forgery (M_i, s, r) does not belong to the list L_i is then $(1 - 2^{-k_0})^j$, when the length a_i of L_i is equal to j. The probability that

r does not belong to L_i is then:

$$f(\beta) = \sum_{j=0}^{q_{sig}} \Pr[a_i = j] \cdot \left(1 - 2^{-k_0}\right)^j \tag{19}$$

Since the forger outputs a forgery with probability ε, the success probability of the inverter is at least $\varepsilon \cdot \beta^{q_{sig}} \cdot f(\beta)$. We select a value of β which maximizes this success probability; in [6], we show that for any (q_{sig}, k_0), there exists β_0 such that:

$$\beta_0^{q_{sig}} \cdot f(\beta_0) \geq \frac{1}{1 + 6 \cdot q_{sig} \cdot 2^{-k_0}} \tag{20}$$

which gives (17). The running time of $\mathcal{I}$ is the running time of $\mathcal{F}$ plus the time necessary to compute the integers $x^e \mod N$ and to generate the lists L_i, which gives (16).

B Proof of Theorem 4

From $\mathcal{R}$ we build an algorithm $\mathcal{I}$ that inverts RSA, without using a forger for FDH. We receive as input (N, e, y) and our goal is to output $y^d \mod N$ using $\mathcal{R}$. We select q_{hash} distinct messages $M_1, \ldots, M_{q_{hash}}$ and start running $\mathcal{R}$ with (N, e, y).

First we ask $\mathcal{R}$ to hash the q_{hash} messages $M_1, \ldots, M_{q_{hash}}$, and obtain the hash values $h_1, \ldots, h_{q_{hash}}$. We select a random integer $\beta \in [1, q_{hash}]$ and a random sequence α of q_{sig} integers in $[1, q_{hash}] \setminus \{\beta\}$, which we denote $\alpha = (\alpha_1, \ldots, \alpha_{q_{sig}})$. We select a random integer $i \in [1, q_{sig}]$ and define the sequence of i integers $\alpha' = (\alpha_1, \ldots, \alpha_{i-1}, \beta)$. Then we make the i signature queries corresponding to α' to $\mathcal{R}$ and receive from $\mathcal{R}$ the corresponding signatures, the last one being the signature s_β of M_β. For example, if $\alpha' = (3, 2)$, this corresponds to making a signature query for M_3 first, and then for M_2.

Then we rewind $\mathcal{R}$ to the state it was after the hash queries, and this time, we make the q_{sig} signature queries corresponding to α. If $\mathcal{R}$ has answered all the signature queries, then with probability ε_F, we send (M_β, s_β) as a forgery to $\mathcal{R}$. This is a true forgery for $\mathcal{R}$ because after the rewind of $\mathcal{R}$, there was no signature query for M_β. Eventually $\mathcal{R}$ inverts RSA and outputs $y^d \mod N$.

We denote by $\mathcal{Q}$ the set of sequences of signature queries which are correctly answered by $\mathcal{R}$ after the hash queries, in time less than t_R. If a sequence of signature queries is correctly answered by $\mathcal{R}$, then the same sequence without the last signature query is also correctly answered, so for any $(\alpha_1, \ldots, \alpha_j) \in \mathcal{Q}$, we have $(\alpha_1, \ldots, \alpha_{j-1}) \in \mathcal{Q}$. Let us denote by **ans** the event $\alpha \in \mathcal{Q}$, which corresponds to $\mathcal{R}$ answering all the signature queries after the rewind, and by **ans'** the event $\alpha' \in \mathcal{Q}$, which corresponds to $\mathcal{R}$ answering all the signature queries before the rewind.

Let us consider a forger that makes the same hash queries, the same signature queries corresponding to α, and outputs a forgery for M_β with probability ε_F. By definition, when interacting with such a forger, $\mathcal{R}$ would output $y^d \mod N$

with probability at least ε_R. After the rewind, $\mathcal{R}$ sees exactly the same transcript as when interacting with this forger, except if event ans is true and ans' is false: in this case, the forger outputs a forgery with probability ε_F, whereas our simulation does not output a forgery. Consequently, when interacting with our simulation of a forger, $\mathcal{R}$ outputs $y^d \mod N$ with probability at least:

$$\varepsilon_R - \varepsilon_F \cdot \Pr[\texttt{ans} \wedge \neg\texttt{ans'}] \tag{21}$$

The proof of the following lemma is given in the full version of the paper [6].

Lemma 1. *Let $\mathcal{Q}$ be a set of sequences of at most n integers in $[1,k]$, such that for any sequence $(\alpha_1,\ldots,\alpha_j) \in \mathcal{Q}$, we have $(\alpha_1,\ldots,\alpha_{j-1}) \in \mathcal{Q}$. Then the following holds:*

$$\Pr_{\substack{i \leftarrow [1,n] \\ (\alpha_1,\ldots,\alpha_n,\beta) \leftarrow [1,k]^{n+1}}} [(\alpha_1,\ldots,\alpha_n) \in \mathcal{Q} \wedge (\alpha_1,\ldots,\alpha_{i-1},\beta) \notin \mathcal{Q}] \leq \frac{\exp(-1)}{n}$$

Using lemma 1 with $n = q_{sig}$ and $k = q_{hash}$, we obtain:

$$\Pr[\texttt{ans} \wedge \neg\texttt{ans'}] \leq \frac{\exp(-1)}{q_{sig}} \left(1 - \frac{q_{sig}}{q_{hash}}\right)^{-1} \tag{22}$$

The term $(1 - q_{sig}/q_{hash})$ in equation (22) is due to the fact that we select $\alpha_1,\ldots,\alpha_{q_{sig}}$ in $[1,q_{hash}] \setminus \{\beta\}$ whereas in lemma 1 the integers are selected in $[1,q_{hash}]$. From equations (21) and (22) we obtain that $\mathcal{I}$ succeeds with probability greater than ε_I given by (9). Because of the rewind, the running time of $\mathcal{I}$ is at most twice the running time of $\mathcal{R}$, which gives (8) and terminates the proof.

Cryptanalysis of SFLASH

Henri Gilbert and Marine Minier

France Télécom R&D, 38-40, rue du Général Leclerc,
92794 Issy les Moulineaux Cedex 9 – France,
henri.gilbert@francetelecom.com

Abstract. SFLASH [Spec] is a fast asymmetric signature scheme intended for low cost smart cards without cryptoprocessor. It belongs to the family of multivariate asymmetric schemes. It was submitted to the call for cryptographic primitives organised by the European project NESSIE, and successfully passed the first phase of the NESSIE selection process in September 2001. In this paper, we present a cryptanalysis of SFLASH which allows an adversary provided with an SFLASH public key to derive a valid signature of any message. The complexity of the attack is equivalent to less than 2^{38} computations of the public function used for signature verification. The attack does not appear to be applicable to the FLASH companion algorithm of SFLASH and to the modified (more conservative) version of SFLASH proposed in October 2001 to the NESSIE project by the authors of SFLASH in replacement of [Spec].

Keywords: asymmetric signature, cryptanalysis, multivariate polynomials, SFLASH.

1 Introduction

SFLASH [Spec] is a an asymmetric signature scheme which was submitted to the call for cryptographic primitives organized by the European project NESSIE, together with a more conservative companion algorithm named FLASH. SFLASH was selected in September 2001 for phase II of the NESSIE project (whereas FLASH was not, probably because its longer key size makes it less attractive than SFLASH, assuming equivalent security levels [Nes01a]). No weakness of the SFLASH and FLASH algorithms was reported in the NESSIE security evaluation [Nes01b].

SFLASH and FLASH both belong to the family of multivariate asymmetric schemes [Pa00, Cou01], and do both represent particular instances of C^{*--}, a variant of the C^* scheme [MI88] in which a sufficient number r of public equations of the C^* trapdoor permutation are withdrawn in order to withstand Patarin's cryptanalysis of C^* [Pa95]. Both schemes are based on the difficulty of solving large systems of quadratic multivariate polynomials over a finite field K. Their trapdoor essentially consists in hiding a monomial transformation over an extension L of K, using two affine transformations s and t of the K-vector space K^n.

L.R. Knudsen (Ed.): EUROCRYPT 2002, LNCS 2332, pp. 288–298, 2002.

One of the distinctive properties of SFLASH and FLASH is that unlike most standard public key signature schemes (e.g. RSA, DSA, ECDSA, etc.), they are sufficiently fast to be well suited for implementation on low cost smart cards without cryptographic coprocessor. SFLASH and FLASH produce rather short signatures (259 bits in the case of SFLASH). The moderate public key size of SFLASH (2.2 Kbytes, versus 18 Kbytes for FLASH) represents an additional advantage for such applications.

In this paper, we present an attack of SFLASH which takes advantage of some special features introduced in SFLASH in order to save a substantial factor in the public key size as compared with more general instances of C^{*--} such as FLASH. This attack allows an adversary provided with an SFLASH public key to derive a valid signature, for that public key, of any message M. The complexity of the attack is well under the security target of 2^{80}: it is equivalent to less than 2^{38} computations of the SFLASH public function used for signature verification. Although the attack was not fully implemented, the essential parts were confirmed by computer experiments.

Our attack does not appear to be applicable to FLASH and to the modified (more conservative, at the expense of a larger public key size) version of SFLASH proposed in October 2001 to the NESSIE project by the authors of SFLASH, in replacement of [Spec].

This paper is organised as follows. Section 2 describes SFLASH and its connection to C^*. Section 3 gives an overview of the attack. Section 4 details the two most essential steps of the attack.

2 Outline of C^*, C^{*--}, and SFLASH

In this Section, we briefly outline those features of C^* and its cryptanalysis which are relevant for the attack presented here, and then provide a short description of SFLASH.

2.1 C^*

C^* is a trapdoor permutation based on hidden monomial field equations proposed by Matsumoto and Imai in 1988 [MI88]. An efficient attack of C^* was found by Patarin [Pa95]. It is sufficient for the sequel to only consider the basic version of C^*, which can be summarised as follows:

- K denotes a finite field of characteristic 2: $K = F_{2^m} = F_q$, where $q = 2^m$.
- L denotes an extension of K of degree n: $L = F_{q^n}$. The representation of L associated with a $P(X)$ irreducible polynomial of degree n of $K[X]$ is used in the various computations. Thus any element a of L can be represented as the $\sum_{i=0}^{n-1} a_i X^i$ element of $K[X]/P(X)$. We will denote in the sequel by φ the one to one mapping from K^n to L associated with this representation: $\forall c = (c_0, c_1, .., c_{n-1}) \in K^n$, $\varphi(c) = \sum_{i=0}^{n-1} c_i X^i \bmod P(X)$.

- The public key of C^* consists of a set of n quadratic functions from K^n to K which together define a G function from K^n to K^n

$$\begin{array}{rcl} G: & K^n & \rightarrow & K^n \\ & x = (x_0, .., x_{n-1}) & \mapsto & y = (y_0, .., y_{n-1}) \end{array}$$

where

$$y_i = \sum_{0 \le j < k \le n-1} \rho_{ijk} x_j x_k + \sum_{0 \le j \le n-1} \sigma_{ij} x_j + \tau_i$$

(in other words, the public key is made up of the ρ_{ijk}, σ_{ij} and τ_i coefficients in K of the n public equations).
- The private key consists of two secret affine one to one functions of K^n: s and t (each determined by $n(n+1)$ K coefficients). The knowledge of s and t provides a secret representation of G as:

$$G = t \circ \varphi^{-1} \circ F \circ \varphi \circ s$$

where

$$\begin{array}{rcl} F: L & \rightarrow & L \\ & a & \mapsto b = a^{q^\theta + 1} \end{array}$$

(θ being a public or private integer such that $q^\theta + 1$ be co-prime with $q^n - 1$). Note that since F is the pointwise product of the two L automorphisms $a \mapsto a$ and $a \mapsto a^{q^\theta}$, $\varphi^{-1} \circ F \circ \varphi$ (and thus G) is quadratic. Moreover, F is one to one, and its inverse F^{-1} is the monomial function $a \mapsto a^h$, where h is the inverse of $q^\theta + 1$ modulo $q^n - 1$.

The knowledge of the private key allows to compute the inverse of the G function. Thus G is a trapdoor permutation which was initially conjectured to be one way, and proposed as a public key encryption or signature function.

2.2 Attack of C^*

The main attack of C^* described in [Pa95] is based upon the following observation: the $b = a^{q^\theta + 1}$ equation of the F function implies $a^{q^{2\theta}} \cdot b = a \cdot b^{q^\theta}$ as can be seen is multiplying the former equation by $a^{q^{2\theta}}$. But the latter equation has the property that both the left and the right terms are "bilinear" in a and b. As a consequence, there exists "bilinear" equations of the form

$$\sum_{0 \le j \le n-1, 0 \le k \le n-1} \gamma_{jk} x_j y_k + \sum_{0 \le j \le n-1} \delta_j x_j + \sum_{0 \le j \le n-1} \epsilon_j y_j + \eta = 0$$

relating the $(x_0, .., x_{n-1})$ and $(y_0, .., y_{n-1})$ K^n input and output vectors of the G public function (i.e. equations of total degree 2 without any $x_j x_k$ or $y_j y_k$ term). It is shown in [Pa95] that the linear equations in γ_{jk}, δ_j, ϵ_j and η provided by a sufficient number of G input-output pairs allow to recover these unknown coefficients, and that once this has been done, the obtained vector space of solutions can be used to compute the inverse by G of any K^n element y at the expense of solving a small K-linear system. The complexity of the attack is about $m^2 n^4 \log n$.

J. Patarin, L. Goubin and N. Courtois investigated in [PGC98] the simple variant of C^* obtained by removing r of the public equations, say the r last ones. Thus the public key now consists of a G function from K^n to K^{n-r} given by $n-r$ quadratic equations over K. They came to the conclusion that the obtained variant of C^* (denoted by C^{*-}) can still be attacked if r is sufficiently small. However, attacks investigated in [PGC98] are not applicable when q^r is larger than say 2^{64}. The C^{*--} name was introduced to refer to C^{*-} variants for which q and r satisfy this condition. Unlike C^*, C^{*--} can only be used for signature purposes, not for encryption purposes.

2.3 Description of SFLASH

SFLASH is a special instance of C^{*--}, in which a particular choice of the s and t functions (and of the polynomials associated with the representation of K and L) enables to considerably shorten the public key size.

More precisely:

- K is chosen equal to $GF(2^7)$, i.e. $m = 7$ and $q = 2^7$. We denote by K' the $GF(2) = \{0,1\}$ subfield of K. K elements are represented as 7-tuples of K' elements, using the representation of $GF(2^7)$ associated with the X^7+X+1 irreducible polynomial of $K'[X]$.
- L is chosen equal to $K[X]/P(X)$, where $P(X)$ is publicly known and equal to the $X^{37}+X^{12}+X^{10}+X^2+1$ irreducible polynomial of $K[X]$. (Note that all coefficients of $P(X)$ belong to K'). Thus n is equal to 37 and L elements can be represented as 37-tuples of K elements.
- The F monomial function of L involved in the secret representation of G is taken equal to $a \mapsto a^{128^{11}+1}$; in other words, θ is public and equal to 11.
- The number r of withdrawn equations is equal to 11. Thus $q^r = 2^{77} > 2^{64}$ and the C^{*--} condition is satisfied.
- The two secret affine functions s and t of $K^n = K^{37}$ are taken from a small subset of the bijective affine functions from K^n to K^n, namely those which can be represented by an $n \times n$ matrix and a $n \times 1$ column vector which $n \times (n+1)$ coefficients do all belong to the K' subfield.

It is easy to see that as a consequence of the special choice of the s and t functions (and of the K and L representations), all the coefficients of the $n - r = 26$ public quadratic equations of the public function G belong to the $K' = GF(2)$ subfield. This results in a gain by a factor of approximately $m = 7$ in the length of the SFLASH public key.

In addition to the above mentioned s and t affine mappings, an SFLASH private key also contains a 80-bit secret key Δ, which acts as a pseudo-random generation seed in the signature generation process.

In order to sign a message M, the owner of a (s, t, Δ) private key performs the following operations.

- The $M1 = SHA-1(M)$ and $M2 = SHA-1(M1)$ 160-bit strings, the 182-bit string $V = M1_{0\to159}||M2_{0\to21}$ and the 77-bit string $W = SHA-1(V||\Delta)_{0\to76}$ are computed.
- V is divided into $n-r=26$ strings $y_0,...,y_{25}$ of length 7 bits each, representing 26 elements of K, and W is divided into $r=11$ strings $y_{26},..,y_{36}$ of length 7 bits each representing 11 elements of K. Let us denote the $(y_0,..,y_{25})$ 26-tuple by y, and the $(y_0,..,y_{25},y_{26},\cdots y_{36})$ 37-tuple by y^*.
- The secret function $s^{-1}\circ\varphi^{-1}\circ F^{-1}\circ\varphi\circ t^{-1}$ is applied to y^*. The obtained 37-tuple x of K elements represents the signature of M. In order to check that the x signature of an M message is valid, a verifier just needs to compute $G(x)$, using the 26 public quadratic equations of G, and to make sure that the obtained value is equal to y.

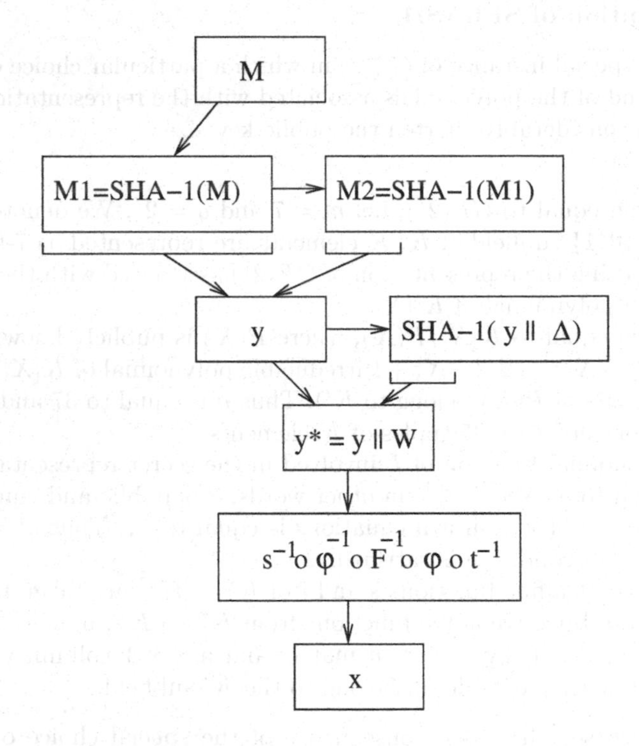

Fig. 1. SFLASH signature scheme

3 Overview of Our Attack

The following simple observation represents the starting point for our attack. Let us consider the $G^* = t\circ\varphi^{-1}\circ F\circ\varphi\circ s$ untruncated SFLASH transformation of K^n from which G is derived (G^* is given by the $n-r$ quadratic equations of G and r additional quadratic equations). Since the s, t and $\varphi^{-1}\circ F\circ\varphi$ mappings are constructed as to leave the $K'^n = GF(2)^{37}$ subset of K^n invariant (this is the price to pay for having very compact public key equations), G^* and its secret inverse $s^{-1} \circ \varphi^{-1} \circ F^{-1} \circ \varphi \circ t^{-1}$ used in the signature computations also leave K'^n invariant. In other words, the restriction of G^* (resp G) to K'^n induces a g^* (resp g) mapping of K'^n to K'^n (resp K'^n to K'^{n-r}) and since G^* is one to one, g^* is also one to one. [1] It is also worth noticing that G^* and g^*, though they are defined over distinct vector spaces (K^n and K'^n), are described by exactly the same set of n quadratic equations which coefficients belong by construction to K'.

Moreover, due to the fact that $K'^n = GF(2)^{37}$ is a small set, it is computationally easy to "invert" the public function g, i.e. given any 26-tuple y of K' elements, to determine the $class(y)$ set of all the $2^r = 2^{11}$ x values in K'^{37} such that $g(x) = y$. Our attack makes an extensive use of this property.

The purpose of our attack is to find r additional quadratic equations of the form

$$z_i(x) = \sum_{0 \le j < k \le n-1} \alpha_{ijk} x_j x_k + \sum_{0 \le j \le n-1} \beta_{ij} x_j$$

(where the α_{ijk} and β_{ij} coefficients are K' elements) which, together with the $n - r$ G quadratic equations

$$y_i(x) = \sum_{0 \le j < k \le n-1} \rho_{ijk} x_j x_k + \sum_{0 \le j \le n-1} \sigma_{ij} x_j + \tau_i$$

represent a full C^* instance consistent with G. More formally, we want to find r additional quadratic equations such that there exists a t' one to one affine mapping of K^n with coefficients in K' such that

$$\forall x = (x_0, x_1, .., x_{n-1}) \in K^n,$$
$$(y_0(x), .., y_{n-r-1}(x), z_0(x), .., z_{r-1}(x)) = t' \circ \varphi^{-1} \circ F \circ \varphi \circ s(x) \quad (1)$$

Once any such set of n equations over K^n satisfying (1) have been determined, then the C^* attack of [Pa95] can be applied to compute the preimage of any K^n element in few operations, so that a valid signature of any message M can then be computed by the adversary, using the following procedure:

---

[1] The following even stronger property of g deserves being mentioned: the public function g represents a "restricted SFLASH" induced over K'^n by the initial SFLASH , with distinct parameters ($q' = 2$ whereas $q = 2^7$, $\theta' = 3$ whereas $\theta = 11$, $n' = n = 37$, $r' = r = 11$). The $q'^{r'} > 2^{64}$ condition of C^{*--} is not satisfied by this restricted SFLASH, since q'^r is only equal to 2^{11}. This mere property is sufficient to make the security of SFLASH suspicious, as first pointed out by Nicolas Courtois, Louis Goubin and Jacques Patarin in a discussion we had with them at an early stage of this work. However, we did not manage to apply the attacks of C^{*-} described in [PGC98] to the g function, so we are unsure that this property is sufficient to draw firm conclusions concerning the security of SFLASH. Therefore we mounted a different attack dedicated to SFLASH, which takes advantage of the small value of $q'^n = 2^{37}$, as explained in the rest of this paper.

- $V = SHA-1(M)_{0\to159}||SHA-1(M1)_{0\to21}$, is computed and is divided into $n-r=26$ 7-bit strings $y_0,...,y_{25}$, and $r=11$ arbitrary additional 7-bit values $z_0,...,z_{10}$ are selected ;
- The preimage of $(y_0,...,y_{25},z_0,...,z_{10})$, which is computed using the C^* attack of [Pa95], is a valid signature of M.

It is easy to see (one simply to consider t' and t) that the

$$z_i(x) = \sum_{0\leq j<k\leq n-1} \alpha_{ijk}x_jx_k + \sum_{0\leq j\leq n-1} \beta_{ij}x_j$$

quadratic equations satisfying requirement (1) are those linear combinations of the $n-r$ public quadratic equations $y_i(x)$ (without their τ_i constants) and the r additional hidden quadratic equations (again without their τ_i constants) such that in addition the n quadratic functions $y_0(x),\cdots,y_{n-r-1}(x)$, $z_0(x),\cdots,z_{r-1}(x)$ be linearly independent.

So, each of the $r=11$ additional quadratic functions z_i we are trying to determine, belongs to the same 37-dimensional K'-vector space E of quadratic functions, generated by the 37 public and hidden (constant less) quadratic equations. Our attack from now on consists in determining this partly unknown vector space E. (Once E has been found, any $z_0(x),..,z_{r-1}(x)$ functions of E such the n quadratic equations $y_0(x),..,y_{n-r-1}(x),z_0(x),..,z_{r-1}(x)$ be linearly independent can be used to mount the rest of the attack, using the C^* cryptanalysis of [Pa95].) There are two main steps in the determination of E:

The first step consists of an initial (partial) characterization of the coefficients of the $z_i(x)$ equations by expressing the fact that g^* is one to one. This first phase allows to reduce the set of $z_i(x)$ candidates from the K'-vector space of all quadratic functions constant less with K' coefficients, which dimension is $n(n-1)/2+n=703$, to a smaller K'-vector space E' of dimension $4*37=148$.

The second step consists of an enhanced characterization of the $z_i(x)$ coefficients. We are using the knowledge of E' to express additional conditions reflecting the a priori knowledge by the adversary of the degree in the y_0 to y_{n-1} variables of the quadratic functions of E'. Our computer experiments indicated that these additional conditions allow to fully determine the E set.

4 Detail of the Two Main Steps of the Attack

As said before, we attempt to characterize the 703 $GF(2)$-coefficients of any quadratic functions of E

$$z(x) = \sum_{0\leq j\leq k\leq n-1} \alpha_{jk}x_jx_k + \sum_{0\leq j\leq n-1} \beta_jx_j$$

(representing any of the $z_0(x)$ to $z_{10}(x)$ functions we are try to determine in order to extend the G set of 26 public equations to a complete set G^* of 37 equations representing a C^* instance.)

4.1 First Step of the Attack: Derivation of E'

For that purpose, we are expressing the fact that since g^* is one to one, each $class(y)$ of 2^{11} x preimages by g of any arbitrary element $y = (y_0, .., y_{26})$ of K'^{26} necessarily contains exactly $2^{r-1} = 2^{10}$ x values such that $z(x) = 0$ and $2^{r-1} = 2^{10}$ x values such that $z(x) = 1$, so that

$$\sum_{x \in class(y)} z(x) \equiv 0 \bmod 2$$

So any arbitrary y value provides one $GF(2)$-linear equation in the 703 coefficients of the quadratic function $z(x)$.

In order to compute the coefficients of the equation associated with y, one first needs to determine $class(y)$. This can be done with a total of less than 2^{37} computations and a limited amount of memory if we first select once for all the N arbitrary $y = (y_0, \cdots, y_{25})$ values for which we want to determine $class(y)$ and if we then perform an exhaustive computation of the g public function for all 2^{37} possible x input values, and store the $N2^r$ x preimages of the N selected y values. Once $class(y)$ has been determined, the $GF(2)$-coefficients of the corresponding equation are easy to compute, and equal to $\sum_{x \in class(y)} x_j x_k$ for each α_{jk} coefficient, and to $\sum_{x \in class(y)} x_j$ for each β_j coefficient.

We collect a little bit more than 703 such equations (say $N = 1000$ for instance) thus obtaining a $N \times 703$ matrix representing a system of N $GF(2)$-linear equations which right terms are equal to zero, and compute the kernel of this matrix using gaussian elimination.

Instead of the initially anticipated 37-dimensional $GF(2)$ vector space E spanned by the 26 public equations and the 11 hidden public equations without their constant terms, we found a much larger $GF(2)$-vector space E' of solutions, of dimension $37 * 4 = 148$. Unsurprisingly, E' is a superset of E.

4.2 Explanation of the Above Phenomenon

The reason why E' contains parasitic solutions distinct from the quadratic functions of the E set appears to be the following: $z(x) = \sum_{x \in class(y)} z(x)$ can be regarded as a $z(y^*)$ function of the actual (partly hidden) $y^* = (y_0, ..., y_{36}) = g^*(x)$ value. For any fixed $y = (y_0, \cdots, y_{25})$ value, let us denote by $V_{11}(y)$ the $(y_0, \cdots, y_{25}) \times GF(2)^{11}$ affine subset of $GF(2)^{37}$. We can write

$$\sum_{x \in class(y)} z(x) = \sum_{y^* \in V_{11}} z(s^{-1} \circ \varphi^{-1} \circ F^{-1} \circ \varphi \circ t^{-1}(y^*)) =_{def} \sum_{y^* \in V_{11}} z(y^*)$$

In other words, $\sum_{x \in class(y)} z(x)$ can be expressed as an 11th order derivative of the $z(y^*)$ function of y^* induced by $z(x)$. Therefore, the equations of the previous Section are satisfied if $z(y^*)$ can be expressed as a boolean function of total degree at most 10 of the components of y^*.

Now, let us consider any $g_i(x) = \varphi^{-1} \circ f_i \circ \varphi \circ s(x)$ quadratic function of K'^{37} associated with any f_i monomial function $a \mapsto a^{2^i+1}$ of $L' = GF(2^{37})$. Let us use the q' and θ' notation of the footnote of Section 3 to refer to the parameters of the restricted SFLASH g. Since $b = t^{-1}(y^*)$ is equal to $g_3(x)$, x is equal to $b^{h'}$, where h' is the inverse of $q'^{\theta'} + 1 = 2^3 + 1$ modulo $2^{37} - 1$. Therefore, if $z(x)$ is equal to any linear combination of the outputs of $g_i(x)$, $z(x)$ can be expressed as a linear combination of the 37 $GF(2)$-components of $b^{h'.(2^i+1)}$. Thus the degree of $z(x)$ as seen as a $z(y^*)$ function of y^* is then bounded above by the Hamming weight of $h'.(2^i + 1) \bmod 2^{37} - 1$.

We computed $h'_i = h'.(2^i + 1) \bmod 2^{37} - 1$ for the i values between 0 and 36, and found exactly 4 h'_i values of weight at most 10 [2], of weights 1, 4, 7 and 10 respectively, namely $i = 3$, 9, 15 and 21. Thus the output bits of $g_3(x)$, $g_9(x)$ $g_{15}(x)$ and $g_{21}(x)$ are quadratic in x and can all be expressed as functions of degree at most 10 of y^* , so that any $z(x)$ linear combination of these 148 output bits satisfies the equations of the previous Section.

So in summary E' is the 148-dimensional vector space spanned by the 37 components of each of the g_3, g_9 g_{15} and g_{21} functions of $GF(2)^{37}$.

4.3 Second Step of the Attack: Derivation of E

We select an arbitrary $B = (\zeta_0(x), .., \zeta_{147}(x))$ basis of E' provided by the gaussian elimination of step 1, and now attempt to characterize the 148 $GF(2)$-coordinates γ_i in this basis of any $z(x) = \sum_{0 \le i \le 147} \gamma_i \zeta_i(x)$ element of E, in order to eliminate the $E' \backslash E$ set of "parasitic solutions".

As said in the previous Section, each $\zeta_i(x)$ quadratic function and their $z(x)$ linear combination can be seen as a $\zeta_i(y^*)$ and a $z(y^*)$ boolean function of the y^* 37-tuple. We can notice that due to the structure of E', the total degree in $y_0, .., y_{36}$ of each of the $\zeta_i(y^*)$ functions is very likely to be equal to 10. If $z(x) \in E$ the total degree in $y_0, .., y_{36}$ of $z(y^*)$ is by definition equal to 1. Therefore, if $z(x)$ belongs to E, then any 12th degree derivative of each of the $z(y^*) \cdot \zeta_i(y^*)$ functions (which degree is at most $10 + 1 = 11$) is equal to zero. On the other hand, if $z(x)$ belongs to $E' \backslash E$, the degree of at least one of the $z(y^*) \cdot \zeta_i(y^*)$ functions (in practice of one of the $z(y^*) \cdot \zeta_0(y^*)$ and $z(y^*) \cdot \zeta_1(y^*)$ functions) can be expected to be at least $4 + 10 = 14$, due to the structure of E', so that the 12th degree derivative of $z(y^*) \cdot \zeta_0(y^*)$ or $z(y^*) \cdot \zeta_1(y^*)$ (or both) can then be expected to differ from the null function. This provides one non trivial linear equation in the γ_i unknown coefficients of $z(x)$.

For any fixed $y = (y_0, \cdots, y_{24})$ value, let us denote by $V_{12}(y)$ the affine subset of $GF(2)^{37}$ defined by $(y_0, \cdots, y_{24}) \times GF(2)^{12}$. For any arbitrary $(y_0, \cdots, y_{24})$ value we have

$$\sum_{y^* \in V_{12}(y)} z(y^*) \cdot \zeta_0(y^*) = 0 \text{ and } \sum_{y^* \in V_{12}(y)} z(y^*) \cdot \zeta_1(y^*) = 0.$$

For each y value, each of these two equations provides the cryptanalyst with a $GF(2)$-linear equation in the 148 unknown $GF(2)$ coefficients γ_i, as can be

[2] up to circular rotations of h'_i. Indeed, if $i1$ and $i2$ are such that $h'_{i1} = 2^\delta h'_{i2} \bmod 2^{37} - 1$, then g_{i1} and g_{i2} are equal up to a linear monomial transformation, and span the same set of quadratic functions.

seen in rewriting the first equation (associated with ζ_0) as

$$\sum_{0\leq i\leq 147}\gamma_i\left(\sum_{x\in class(y_0,..,y_{24},0)\cup class(y_0,..,y_{24},1)}\zeta_i(x)\cdot\zeta_0(x)\right)\equiv 0 \bmod 2$$

We collect a little bit more than 148 such equations (say $N' = 200$, some of which being associated with $z(x)\cdot\zeta_0(x)$ and the other being associated with $z(x)\cdot\zeta_1(x)$), thus obtaining a $N'\times 148$ matrix representing a system of N' 148-bit vectors corresponding to $GF(2)$-linear equations which right terms are equal to zero. We compute the kernel of this matrix using gaussian elimination. It was confirmed by computer experiments that we obtain a kernel of dimension only 37, equal to the E subspace of E'. This completes step 2 of our attack.

Once E has been recovered with the above method, a complete G^* set of 37 K'^n-quadratic functions with K' coefficient can be obtained (one just needs to complete the 26 public equations of G as to obtain a basis of E), and the C^* attack of [Pa95] can be applied to compute the inverse by G^* of any K^{37} element, so that a valid signature of any message M can be produced by the adversary.

4.4 Complexity of the Attack

The most complex calculation required by the attack is the exhaustive computation of the 2^{37} values of the public function g, which is needed to obtain the (at most) $N+2N'$ sets of 2^{11} preimages required for the computations of step 1 and step 2.

The computations of step 1 are essentially the derivation of the $N=1000$ linear equations in 703 variables and the gaussian elimination of the resulting $N\times 703$ system in step 1. So, the complexity of step 1 is bounded above by $N.703.2^{11}+\frac{N^3}{3}\leq 2^{31}$. In the same way, the complexity of the derivation of the $N'=148$ linear equations in 703 variables and the gaussian elimination of the resulting $N'\times 148$ system in step 2 are bounded above by 2^{27}. Both complexities are far lower than 2^{37} computations of the SFLASH public function. Moreover the complexity of the attack of C^* presented in [Pa95] is here about 2^{27} computations. In summary, the overall complexity of the attack is bounded above by 2^{38}.

5 Conclusion

The attack presented in this paper uses extensively the fact that the SFLASH public function over K^{37} induces a restricted scale function over the much smaller vector space $GF(2)^{37}$.

Our attack does not seem applicable to more conservative instances of C^{*--} such as FLASH, because a more sophisticated method than the one used in our attack would then have to be found to determine complete sets of 2^r preimages of some C^{*--} outputs.

Acknowledgements

We thank Mehdi-Laurent Akkar, Nicolas Courtois, Louis Goubin, and Jacques Patarin for interesting discussions and helpful remarks, and for having provided us with the reference implementation of SFLASH we used for computer experiments.

References

[MI88] T. Matsumoto and H. Imai, "Public Quadratic Polynomial-tuples for efficient signature-verification and message encryption". In *Advances in Cryptology - Eurocrypt'88*, pp. 419-453, LNCS 330, Springer Verlag, May 1988.

[Nes01a] NESSIE Phase I: Selection of Primitives, september 2001, available at `http://www.cryptonessie.org/`.

[Nes01b] Security Evaluation of NESSIE First Phase, september 2001, available at `http://www.cryptonessie.org/`.

[Pa95] J. Patarin, "Cryptanalysis of the Matsumoto and Imai Public Key Scheme of Eurocrypt'88". In *Advances in Cryptology - Crypto'95*, pp. 248-261, LNCS 963, Springer Verlag, August 1995.

[Pa00] J. Patarin, "La Cryptographie Multivariable", mémoire d'habilitation à diriger des recherches, Université Paris VII, France, 2000.

[PGC98] N. Courtois, L. Goubin and J. Patarin, "C^{*-+} and HM: Variations around two Schemes of T. Matsumoto and H. Imai". In *Advances in Cryptology - Asiacrypt 98*, pp. 35-49, LNCS 1514, Springer-Verlag, October 1998.

[Cou01] N. Courtois, "La Sécurité des Primitives Cryptographiques Basées sur des Problèmes Algébriques Multivariables: MQ, IP, MinRank, HFE", PhD. dissertation, Université Paris VI, France, September 2001, available at `http://www.minrank.org/phd.pdf`.

[Spec] Specifications of SFLASH, NESSIE documentation, available at `https://www.cosic.esat.kuleuven.ac.be/nessie/workshop/`.

Cryptanalysis of the Revised NTRU Signature Scheme

Craig Gentry[1] and Mike Szydlo[2]

[1] DoCoMo USA Labs, San Jose, CA, USA,
cgentry@docomolabs-usa.com
[2] RSA Laboratories, Bedford, MA, USA,
mszydlo@rsasecurity.com

Abstract. In this paper, we describe a three-stage attack against Revised NSS, an NTRU-based signature scheme proposed at the Eurocrypt 2001 conference as an enhancement of the (broken) proceedings version of the scheme. The first stage, which typically uses a transcript of only 4 signatures, effectively cuts the key length in half while completely avoiding the intended hard lattice problem. After an empirically fast second stage, the third stage of the attack combines lattice-based and congruence-based methods in a novel way to recover the private key in polynomial time. This cryptanalysis shows that a passive adversary observing only a few valid signatures can recover the signer's entire private key. We also briefly address the security of NTRUSign, another NTRU-based signature scheme that was recently proposed at the rump session of Asiacrypt 2001. As we explain, some of our attacks on Revised NSS may be extended to NTRUSign, but a *much* longer transcript is necessary. We also indicate how the security of NTRUSign is based on the hardness of several problems, not solely on the hardness of the usual NTRU lattice problem.

Keywords: NSS, NTRU, NTRUSign, Signature Scheme, Lattice Reduction, Cryptanalysis, Orthogonal Lattice, Cyclotomic Integer, Galois Congruence.

1 Introduction

The Revised NTRU Signature Scheme (R-NSS) and "NTRUSign" are the two most recent of several signature schemes related to the NTRU encryption scheme (now called NTRUEncrypt). NTRUEncrypt and the related signature schemes are not based on traditional hard problems such as factoring or computing discrete logarithms, like much of today's cryptography. Instead, NTRUEncrypt was originally conceived as a cryptosystem based on polynomial arithmetic. Based on an early attack found by Coppersmith and Shamir [7], however, the underlying hard problem was soon reformulated as a lattice problem. See [22] for an update on how lattices have recently been used both as a cryptanalytic tool and as a potential basis for cryptography.

L.R. Knudsen (Ed.): EUROCRYPT 2002, LNCS 2332, pp. 299–320, 2002.

There are two reasons for seeking alternative hard problems on which cryptography may be based. First, it is prudent to hedge against the risk of potential breakthroughs in factoring and computing discrete logarithms. A second and more significant reason is efficiency. NTRU-based algorithms, for example, are touted to run hundreds of times faster while providing the same security as competing algorithms. The drawback in using alternative hard problems is that they may not be as well understood. Although lattice theory has been studied for over 100 years,[1] the *algorithmic* nature of hard lattice problems such the "shortest vector problem" (SVP) was not really studied *intensively* until Lenstra, Lenstra and Lovász discovered a polynomial-time lattice basis reduction algorithm in 1982. Moreover, NTRU-based schemes use specific types of lattices based on an underlying polynomial ring, and these lattices generate specific types of lattice problems that may be easier to solve than general lattice problems. Since these specific lattice problems have been studied intensively only since NTRUEncrypt's introduction in 1996, we can expect plenty of new results. This paper is a case in point: we use a new polynomial-time algorithm to find the shortest vector in certain lattices that arise in R-NSS, allowing us to break the scheme.

1.1 History of NTRU-Based Signature Schemes

Since the invention of NTRUEncrypt in 1996, several related identification and signature schemes have been proposed. These include an identification scheme invented by Kaliski, et.al. in 1997 [12], a "preliminary" version of NSS presented at the rump session of Crypto 2000, and the scheme described in the proceedings of Eurocrypt 2001 [17]. (See also [16] and [4].) All of these have been broken. (See [21] and [9].) In their Eurocrypt presentation, the authors of NSS sketched a revised scheme. They described these revisions in more detail in a technical note entitled "Enhanced Encoding and Verification Methods for the NTRU Signature Scheme" [14], which was revised several months later [15]. They finally committed to a scheme, which we will call "R-NSS," by publishing it in the preliminary cryptographic standard document EESS [5]. They also published analysis and research showing how the new scheme defeated previous attacks. Although R-NSS does indeed appear to be a significantly stronger scheme than previous versions, this paper describes how it can be broken.

Since the initial submission of this paper, NTRU has proposed a new NTRU-based signature scheme called NTRUSign. Although our primary focus is R-NSS, we also provide security analysis of NTRUSign, as requested by the Program Committee.

1.2 Our Cryptanalysis

In our cryptanalysis of R-NSS, we use for concreteness the parameters suggested in the technical note [15] and standards document [5]. We show how a passive

[1] This includes early work by Hermite and Minkowski, the latter calling the topic "Geometrie der Zahlen" (Geometry of Numbers) in 1910.

adversary who observes only a few valid signatures can recover the signer's entire private key. Although some might consider R-NSS to be even more *ad hoc* than previous NTRU-based signature schemes, our attacks against it are more fundamental than previous attacks, in that they target the basic tenets of the scheme rather than its peculiarities.

The rest of this paper is organized as follows: In Section 2, we provide background mathematics, and then in Section 3, we describe R-NSS. In Section 4, we survey the previous attacks on NTRU-based signature schemes that are relevant to our cryptanalysis of R-NSS. In Section 5, we detail the first stage of our attack: the lifting procedure. Next, in Section 6, we describe how to obtain the polynomial $f * \overline{f}$, which we use in the final stage of the attack. In Section 7, we introduce novel techniques for circulant lattices which enable a surprising algorithm to obtain the private key f in polynomial time. We give a summary of our R-NSS cryptanalysis in Section 8. Finally, in Section 9, we describe NTRUSign, consider attacks against it and describe alternative hard problems that underlie its security.

2 Background Mathematics

As with NTRUEncrypt and previous NTRU-based signature schemes, the key underlying structure of R-NSS is the polynomial ring

$$R = \mathbb{Z}[X]/(X^N - 1) \quad (1)$$

where N is a prime integer (e.g., 251) in practice. In some steps, R-NSS uses the quotient ring $R_q = \mathbb{Z}_q[X]/(X^N - 1)$, where the coefficients are reduced modulo q, and normally taken in the range $(-q/2, q/2]$, where q is typically a power of 2 (e.g., 128). Multiplication in R and in R_q is also called convolution. We define $\|f\|$ to be the Euclidean norm of a polynomial in the basis $\{1, x, \ldots\}$ and the convolution matrix of f to be the matrix whose rows correspond to $\{f, xf, x^2 f \ldots\}$.

At times, we will refer to the reversal $\overline{a}$ of a polynomial a, defined by $\overline{a}_k = a_{N-k}$ (with $\overline{a}_0 = a_0$). The mapping $a \mapsto \overline{a}$ is an automorphism of R, since applying the map twice yields the original polynomial. We use the term "palindromes" in referring to polynomials that are fixed under the reversal mapping on R – i.e., polynomials a such that $a = \overline{a}$. For any $a \in R$, it is easy to see that the product $a * \overline{a}$ is a palindrome. This fact, as well as the reversal mapping, may be described in elementary terms, but also in terms of an automorphism of the underlying cyclotomic field $\mathbb{Q}(\zeta_N)$. We refer the reader to the full version of the paper for further details on the Galois theory of R and $\mathbb{Q}(\zeta_N)$.

2.1 Lattices

The analysis of R-NSS will make frequent use of *lattices*. Formally, a lattice is a discrete subgroup of a vector space, but concretely, a lattice may be presented as the integral span of some set $B = \{b_0, \ldots, b_{m-1}\}$ of linearly independent vectors in $\mathbb{R}^N$ - that is,

$$L = \{v | v = \sum a_i \, b_i \mid a_i \in \mathbb{Z}\} \; . \tag{2}$$

We call m the dimension of the lattice, and B a basis of L. Bases will often be presented as a matrix in which the rows are the basis vectors $\{b_i\}$. Each lattice has an infinite number of bases, related by $B' = UB$ where U is a unimodular matrix, but some bases are more useful than others. The goal of *lattice reduction* is to find useful bases, typically ones with reasonably short, reasonably orthogonal basis vectors. The most celebrated lattice reduction algorithm is LLL [19], which has found many uses in cryptology. The contemporary survey [22] provides an overview of lattice techniques and [2] provides detailed descriptions of LLL variants.

The most famous lattice problem is the shortest vector problem (SVP): given a basis of a lattice L, find the shortest nonzero vector in L. Although LLL and its variants manage to find *somewhat short* vectors in lattices, they do not necessarily find the *shortest* vector. In fact, SVP is an NP-hard problem (under randomized reductions) [1]. In previous cryptanalysis of NTRU and NSS, LLL's inability to recover the shortest (or a very, very short) vector was a significant shortcoming. In some of our attacks, however, we will construct lattices in which even vectors that are only somewhat short reveal information about the signer's private key, and then we will use LLL and its variants as black box algorithms to find these vectors. We will explain other aspects of lattice theory as they become relevant.

2.2 Ideals

Since R-NSS operates with polynomials in the ring R, we will need to consider multiplication in R, as well as ideals in this ring. Recall that an *ideal* is an additive subgroup of a commutative ring which is also closed under multiplication by any element in R, and a *principal ideal* is an ideal of R consisting of all R-multiples of a single element. We write (a) to denote the principal ideal consisting of all R-multiples of a, and say that the ideal is generated by a. We remark that not all ideals are principal, and furthermore, a generator of a principal ideal is not unique since there are infinitely many units $u \in R$, and for each u, both M_f and $M_{(f*u)}$ define the same ideal. We naturally extend these notions to lattices by defining a *lattice ideal* to be a lattice which is also closed under "multiplication" by polynomials in R. Each lattice $L(M_a)$ is a principal lattice ideal, and we will exploit this extra structure in our novel attacks on R-NSS.

3 Description of R-NSS

The signature scheme R-NSS is a triplet (*keygen, sign, verify*) of algorithms operating on polynomials in $R = \mathbb{Z}[X]/(X^N - 1)$ and $R_q = \mathbb{Z}_q[X]/(X^N - 1)$, where N is prime, and $q < N$, (e.g. $N = 251$, $q = 128$). Other parameters in

R-NSS include the modulus p, which is relatively prime to q and is typically chosen to be 3, as well as the integers d_u, d_f, d_g, d_m and d_z, whose suggested values are respectively 88, 52, 36, 80, and 58. The latter parameters are used to define several families of *trinary* polynomials as follows: $\mathcal{L}(d_1, d_2)$ denotes the set of polynomials in R_q, with d_1 coefficients 1, d_2 coefficients -1 and all other coefficients 0.

Key generation: Two polynomials f and g are randomly generated according to the equations

$$f = u + pf_1$$

$$g = u + pg_1$$

where $u \in \mathcal{L}(d_u, d_u + 1)$, $f_1 \in \mathcal{L}(d_f, d_f)$ and $g_1 \in \mathcal{L}(d_g, d_g)$. The signer keeps these two polynomials secret, with f serving as the signer's private key. The public key h is computed as $f^{-1} * g$ in R_q, and it is therefore necessary that f be invertible in R_q (i.e., $f * f^{-1} = 1$ for some $f^{-1} \in R_q$). This is true with very high probability (see [24]); in any case the preceding step may be repeated by choosing a different polynomial f_1.

As in previous versions of NSS, the coefficients of f and g are small – i.e., they lie in a narrow range ($[-4, 4]$ assuming $p = 3$) of $\mathbb{Z}_q$. However, R-NSS introduces a new secret polynomial – namely, u – into the private key generation process. In the previous version of NSS, f (mod p) and g (mod p) were public, allowing a statistical attack on a transcript of signatures. In the full version of the paper we briefly describe this transcript attack, and explain how using u defeats it.

Signature generation: To sign a message, one transforms the message to be signed into a message representative according to a hash function-based procedure such as that described in [4]. We do not base any attack on this encoding, which can be made as safe as for any signature scheme. This message representative m is a polynomial in $\mathcal{L}(d_m, d_m)$. The signer then computes the following temporary variables:

$$y = u^{-1} * m \bmod p$$

$$z \in \mathcal{L}(d_z, d_z)$$

$$w = y + pz ,$$

where u^{-1} is computed in $R_p = \mathbb{Z}_p[X]/(X^N - 1)$. Notice that in R, $f * w \equiv m$ (mod p). This is not necessarily the case in R_q, however, since reduction modulo q causes "deviations" in the modulo p congruence, given that p and q are relatively prime. During the rest of the signing process, the signer will try to keep the number of these deviations to a minimum. The signer computes more temporary variables:

$$s = f * w \bmod q$$

$$t = g * w \bmod q$$

$$Dev_s = (s - m) \bmod p$$

$$Dev_t = (t - m) \bmod p \ .$$

Dev_s and Dev_t represent the deviations that the signer would like to correct, but, unfortunately for the signer, correcting a coefficient in s may cause additional deviations in t. Therefore, the signer limits his corrections to coefficient positions j such that $(Dev_s)_j = (Dev_t)_j$. He initializes a polynomial e to 0 and sets e_j to $-(Dev_s)_j$ when $(Dev_s)_j = (Dev_t)_j$. He then lets $e' = u^{-1} * e \pmod p$, adds e' to w, and recomputes $s = f * w$ in R_q. The pair (m, s) is the signer's signature of m.

Signature verification: To avoid the forgery attacks presented in [9], verification has become a rather complicated process involving up to 20 distinct steps, detailed in [6] and [15], which fall into three broad categories: the *Quartile Distribution* tests, the *Mod 3 Distribution* tests, and the *L2 Norm* tests. In essence, the verifier checks, respectively, that

1. the coefficients of s and $t = s * h \pmod q$ have a roughly normal distribution;
2. the coefficients of s and t deviating from m are few and have a certain distribution; and
3. the L2 norms of $s' = p^{-1}(s - m) \pmod q$, $t' = p^{-1}(t - m) \pmod q$ and $(s' \| t')$ (concatenated) are below certain thresholds.

Since we do not focus on forgery attacks in this paper, we do not describe the verification process in full detail here. On average, the signer has to run the signing process two or three times to produce a valid signature.

4 Previous Attacks on NSS

In this section, we review some relevant known attacks against NTRUEncrypt and previous NTRU-based signature schemes. This review will help us explain our cryptanalysis of R-NSS, which will occasionally leverage pieces of the attacks mentioned here.

4.1 Coppersmith-Shamir Attack

As with NTRUEncrypt, the security of R-NSS is claimed to be based on a hard lattice problem. Coppersmith and Shamir [7] were the first to present a lattice-based attack against NTRUEncrypt, an attack which is also relevant to R-NSS. Let L_{CS} be the lattice generated by the rows of the following matrix:

$$B_{CS} = \begin{bmatrix} I_{(N)} & M_h \\ 0 & qI_{(N)} \end{bmatrix} , \tag{3}$$

where $I_{(N)}$ is the N-dimensional identity matrix. This lattice clearly contains the vector $(f \| g)$, since $f * h = g \pmod q$. Moreover, for technical reasons [7], it is highly probable that $(f \| g)$ is the shortest nonzero vector in this lattice (up to rotation, sign, and excluding trivial vectors such as $(1^N, 1^N)$, the vector of all 1's.) Therefore, recovering the private key is simply a matter of recovering the shortest vector in L_{CS}, which we can presumably do using a lattice reduction algorithm. This attack is very effective when N is small (e.g., 107).

The problem with this approach (and lattice attacks, in general) is that no known lattice reduction algorithm is both very fast and very effective. More specifically, the LLL algorithm is "polynomial-time" – i.e., it terminates in time polynomial in the dimension m of the lattice – but, for high-dimensional lattices, such as those used in NTRU-based schemes, it almost certainly will *not* find the shortest vector. Rather, LLL only guarantees finding a vector that is no more than $2^{(m-1)/2}$ times as long as the shortest vector. Even though, in practice, LLL performs significantly better than this worst case bound, its performance is not sufficient for this lattice; we need the shortest vector or a *very* small multiple thereof. Other lattice reduction algorithms can find shorter vectors, but they naturally have greater time-complexity. The bottom line, based on current knowledge and on extensive empirical tests run by NTRU [23], seems to be that the time necessary to find $(f\|g)$ in L_{CS} grows at least exponentially in the dimension of the lattice ($2N$). This apparently hard lattice problem of recovering $(f\|g)$ from the $2N$-dimensional lattice is claimed to underlie the security of both NTRUEncrypt and R-NSS.

Remark 1. As mentioned previously, the fact that SVP is an NP-hard problem for general lattices does not necessarily mean that finding short vectors in L_{CS} is hard. May [20] exploited the specifics of NTRUEncrypt's private key structure to construct lower-dimensional "zero-run" lattices and "dimension-reducing" lattices from which an attacker could quickly recover an NTRUEncrypt-107 private key. Gentry [8] used a ring homomorphism from R to $\mathbb{Z}[X]/(X^{N/d}-1)$ to "fold" L_{CS} into a more manageable lattice of dimension $2N/d$ for N having a nontrivial divisor d.

4.2 GCD Lattice Attack

Since reducing L_{CS} does indeed appear to be infeasible, the natural inclination of the cryptanalyst is to look for smaller lattices that contain the private key. The authors of R-NSS mention one such lattice in [16]. They observe that if an attacker is able to recover the values of several $f*w$'s in R – "unreduced" modulo q – then f will likely be the shortest vector in the N-dimensional lattice formed by the rows of the several M_{f*w}'s.

Recall from section 2.2 that, for any polynomial a, there is an equivalence between the ideal (a) of a-multiples and the lattice generated by M_a. Similarly, the lattice spanned by the rows of M_{f*w_1} and M_{f*w_2} corresponds to the ideal $I=(f*w_1, f*w_2)$. Every polynomial in I is a multiple of f. Moreover, if (w_1) and (w_2) are relatively prime – i.e., there exist $a,b\in R$ with $a*w_1+b*w_2=1$ – then $f\in I$, and we may say that $\mathrm{GCD}(f*w_1, f*w_2)=(f)$. This "lattice attack" is, in fact, the standard ideal-GCD algorithm, and is among the lattice ideal operations discussed in [2].

Given how the w_i are produced in R-NSS, (f) often is indeed the GCD of two unreduced signatures, and it is even more likely to be the GCD of several unreduced signatures. Therefore, given a few unreduced signatures, we can construct an N-dimensional lattice whose shortest vector is likely f. The authors of R-NSS note that, although reducing this N-dimensional lattice is still not a trivial problem for R-NSS parameters, it is much easier than reducing the $2N$-dimensional L_{CS}, given the exponential relationship between dimension and running time. R-NSS uses "masking" techniques to prevent recovery of $f * w$'s in R to avoid this lattice attack. (In section 5 we show that recovering $f * w$'s is nonetheless quite easy.)

4.3 Averaging Attack

The so-called "averaging attack" was first considered by Kaliski during collaboration with Hoffstein in the context of an early precursor to NSS (see patent [12]). In this work, it was observed that in the ring R, the value $f * \overline{f}$ could be obtained by an averaging attack. This attack was fatal, since in the scheme [12], f itself is a palindrome (unlike in R-NSS), thus $f * \overline{f} = f^2$. There is an efficient algorithm for taking the square root in R (see, e.g., [13]), so $f * \overline{f}$ revealed f.

In [16], the authors of R-NSS do mention this averaging attack, but also remark that knowledge of $f * \overline{f}$ does not appear to be useful. See [16], [9], [21] for a discussion of this and other ways in which the attacker may average a transcript of signatures in such a way as to get information about the private key.

Here is a description of how the averaging attack works. Suppose we can obtain a set of unreduced $f * w$'s in R. Now, consider the average

$$A_r = (1/r) \sum_{i=1}^{r} (f * \overline{f}) * (w_i * \overline{w_i}) \;..$$

For each i, $(w_i * \overline{w_i})_0 = \|w_i\|^2$, which is a large positive quantity. However, for $k \neq 0$, $(w_i * \overline{w_i})_k$ has a random distribution of positive and negative quantities that averaging essentially cancels out. Thus, as r increases, A_r essentially converges to a scalar multiple of $f * \overline{f}$. This convergence is quite fast: after a few thousand signatures a close estimate of $f * \overline{f}$ can be computed, meaning that we obtain an estimate z such that for most coefficients, $|z_i - (f * \overline{f})_i| \leq 2$. Even if reduced signatures are used, with some corrections, there is still a convergence to $f * \overline{f}$, albeit about 10 times as slow. Clearly, the more signatures, the better the estimate. In section 6, we explain how this averaging attack may be combined with a lattice attack to recover $f * \overline{f}$ quickly and completely.

5 Lifting the Signatures

In this section, we present our first (and arguably the most important) attack with which we obtain R-multiples of the private key f. More specifically, we assume that, as passive adversaries, we are given a transcript of legitimate signatures $\{(m_1, s_1), \ldots, (m_r, s_r)\}$. Using this transcript and the signer's public key, we directly compute the following elements in R_q:

$$\{f * w_1 \bmod q, \ldots, f * w_r \bmod q\} \text{ and } \{g * w_1 \bmod q, \ldots, g * w_r \bmod q\},$$

where the w_i are computed according to the signing process above. We will then lift these signatures to the ring R, obtaining a list of multiples of f and g:

$$\{f * w_1, \ldots, f * w_r\} \text{ and } \{g * w_1, \ldots, g * w_r\},$$

which are "unreduced" modulo q. This is a devastating attack against R-NSS, because undoing the q modular reduction of the signatures allows us to use the N-dimensional GCD lattice attack, described in section 4.2, rather than the $2N$-dimensional Coppersmith-Shamir attack, reducing the key recovery time by a factor exponential in N. It also permits the other, more efficient attacks discussed later in this paper.

5.1 The Principle

From the signing procedure in the ring R_q, we get the following equations for each i:

$$f * w_i \equiv s_i \bmod q \text{ and } g * w_i \equiv t_i \bmod q, \tag{4}$$

$$f * w_i \equiv g * w_i \equiv m_i + e_i \bmod p. \tag{5}$$

If we knew e_i, we would be able to compute $f * w_i$ and $g * w_i$ modulo pq via the Chinese Remainder Theorem. We could then recover $f * w_i$ and $g * w_i$ over R without too much difficulty, since almost all of the coefficients of $f * w_i$ and $g * w_i$ will lie in the interval $(-pq/2, pq/2]$.[2]

The use of e_i in the signing process makes lifting the signatures much less straightforward. Since e_i will have about 20 nonzero coefficients for the suggested parameters of R-NSS, we cannot simply guess e_i from among $\binom{251}{20}2^{20}$ possibilities. Later, we will mention how specific properties of the signing process make some possibilities much more probable than others, but even with these refinements the guessing approach remains infeasible.

Instead, we will use an iterative approach that, at each step, attempts to improve upon previous approximations. Let S_i denote our approximation of $f * w_i$, and T_i our approximation of $g * w_i$. We initialize S_i and T_i to be the polynomials in R_{pq} that satisfy

$$S_i \equiv s_i \bmod q \text{ and } T_i \equiv t_i \bmod q, \tag{6}$$

$$S_i \equiv T_i \equiv m_i \bmod p. \tag{7}$$

[2] See [7] for an analysis of the coefficient distributions of convolution products. The key points here are that L2 norms $\|f\|$, $\|g\|$ and $\|w_i\|$ are small, $\|f * w_i\| \approx \|f\|\|w_i\|$ and the coefficients of $f * w_i$ have a roughly normal distribution.

Our approximations will have two different types of errors. First, the kth coefficients of S_i and T_i will be wrong if the kth coefficient of e_i is nonzero. Second, a coefficient of S_i (resp. T_i) may be correct modulo pq but incorrect in R if the corresponding coefficient of $f*w_i$ (resp. $g*w_i$) lies outside the interval $(-pq/2, pq/2]$. On average, about 25 (out of 251) coefficients of our initial approximations will be incorrect.

In refining our approximations, we begin with the following observation: For any (i, j),

$$(f * w_i) * (g * w_j) - (f * w_j) * (g * w_i) = 0 . \tag{8}$$

Based on this observation, we would expect $S_i * T_j - S_j * T_i$ to tend towards 0 as our approximations improve. In fact, this is the case. We can use the norms $n_{ij} = \|S_i * T_j - S_j * T_i\|$ to decide what adjustments we should make, and to know when our approximations are finally correct. The rest of the lifting procedure is simply an elaboration of this basic idea, and one can imagine a variety of different methods through which an attacker could use these norms to create an effective lifting procedure. In our particular implementation against R-NSS, we used the norms n_{ij}, together with some R-NSS-specific heuristics, to create a very fast lifting procedure that worked almost all the time with a transcript of only four signatures.

5.2 Our Implementation of the Lifting Procedure

For each approximation pair (S_i, T_i), we computed a "norm product" with the other approximation pairs according to the formula $P_i = \prod_{j \neq i} n_{ij}$. Preferring approximation pairs with higher norm products, we then picked a random pair (S_i, T_i) to be corrected. For each coefficient position, we temporarily added or subtracted certain multiples of q to S_i and T_i (since the approximations are already correct modulo q), and recomputed P_i. With a little bookkeeping, this step can be made extremely fast. We preserved the adjustment that reduced P_i by the greatest amount. Finally, we terminated this process when the norm product of some approximation pair reached zero, at which point we would have two correct approximation pairs.

We note that not every adjustment made during the lifting procedure is actually a correction. Often, this procedure will make a previously correct coefficient incorrect, and then switch it back later on. In other words, it behaves somewhat like a "random walk". This fact, together with the heuristic nature of the overall algorithm, admittedly makes the lifting procedure difficult to analyze. For the specified parameters of R-NSS, however, it works quickly and reliably. For a transcript of four signatures, it is able to lift two signatures 90% of the time in an average of about 25 seconds (on a desktop computer). In the remaining 10%, the number of errors never converges to zero. For three signatures, it still works 70% of the time, typically finishing in about 15 seconds.

6 Obtaining $f * \overline{f}$

An important ingredient of the final algorithm is the product of f with its reversal, $\overline{f}$. In order to recover $f * \overline{f}$ in the context of R-NSS, we used a combination of the averaging attack mentioned in section 4.3 and a lattice attack on $f * \overline{f}$ noticed by the authors and Jonsson, Nguyen and Stern.

The lattice attack is a derivative of the Coppersmith-Shamir attack described in section 4.2. Since sending a polynomial to its reversal is an automorphism, $(f * \overline{f}) * (h * \overline{h}) \equiv (g * \overline{g}) \pmod{q}$. This means that the vector $(f * \overline{f} \| g * \overline{g})$ is contained in the lattice L_{norm} generated by

$$B_{norm} = \begin{bmatrix} I_{(N)} & M_{h*\overline{h}} \\ 0 & qI_{(N)} \end{bmatrix} . \tag{9}$$

This lattice has dimension $2N$, but it has an $(N+1)$-dimensional sublattice of palindromes, which contains $(f * \overline{f} \| g * \overline{g})$. Conceivably, recovering $(f * \overline{f} \| g * \overline{g})$ from this sublattice could give us useful information about f and g. However, this attack fails for typical NTRU or NSS parameters, since $(f * \overline{f} \| g * \overline{g})$ is normally not the shortest vector.[3]

For R-NSS, we combine ideas from the above attack with the GCD attack in 4.2 and the averaging technique in 4.3. First, we use our unreduced signatures to form the ideal $(f * \overline{f})$ from a few unreduced signature products $s * \overline{s} = f * \overline{f} * w_i * \overline{w_i}$, exactly as described in Section 4.2. Then, we take the subring of $(f * \overline{f})$ consisting of palindromes, which forms a lattice of dimension $(N+1)/2$. In fact, this lattice is generated by $f * \overline{f}$, and $(N-1)/2$ vectors $(X^k + X^{N-k}) * f * \overline{f}$. For the same reason as above, $f * \overline{f}$ might not be the shortest vector in the lattice. However, we may use the averaging attack to obtain a good estimate t of $f * \overline{f}$, modify the lattice to include t, and then use lattice reduction to obtain the (shortest) vector $t - f * \overline{f}$.[4] In practice, this attack is amazingly effective for two reasons: the lattice problem is only $(N+1)/2$ dimensional, and $\|t - f * \overline{f}\|$ will be much less than $\|f * \overline{f}\|$ for even a *very* poor estimate of t. We found that we needed only 10 signatures to obtain a sufficiently accurate estimate t of $f * \overline{f}$ (even though only a handful of coefficients in t were actually correct). With these 10 signatures, we consistently recovered $f * \overline{f}$ in less than 10 seconds.

7 Orthogonal Congruence Attack

In this section, we describe a polynomial-time algorithm for recovering the private key f from $f * \overline{f}$ and one other multiple of f, such as $f * w$, when w is

[3] By the "Gaussian heuristic," the expected length of the shortest vector in a lattice of determinant d and dimension n is $d^{1/n}\sqrt{n/(2\pi e)}$. For L_{norm}, this length is $\sqrt{qN/(\pi e)}$. On the other hand, since $\|f\| > \sqrt{N}$ in NSS and $\|f * \overline{f}\| \geq \|f\|^2$ (the latter inequality following from $(f * \overline{f})_0 = \|f\|^2$), the norm of $f * \overline{f}$ is greater than N and hence greater than the Gaussian heuristic, since N is typically chosen to be greater than q.

[4] One could also use Babai's algorithm to solve the closest vector problem (CVP).

relatively prime to $\overline{f}$. In other words, this algorithm requires $f * \overline{f}$ and a basis B_f of the ideal (f).[5] This algorithm is quite surprising, and uses novel ideas combining orthogonal lattices with number theoretic congruence arising from the cyclotomic field $\mathbb{Q}(\zeta_N)$.

The complete algorithm is rather complex, but here is a brief (and not entirely accurate) sketch: We begin by choosing a large prime number $P \equiv 1 \pmod{N}$. (For now, we defer discussing how large P must be.) Then, using $f * \overline{f}$ and our basis for (f), we use a series of lattice reductions to obtain $f^{P-1} * a$ for some polynomial a, and a guarantee that $\|a\| < P/2$. Using the congruence $f^{P-1} \equiv 1 \pmod{P}$, we will be able to compute $a \pmod{P}$ and hence a exactly, from which we will be able to compute f^{P-1} exactly. We will then use this power of f to recover f.

We describe this algorithm and the theory behind it in more detail below. Our first task will be to find a tool that ensures that when LLL gives us $f^{P-1} * a$, there is a definite bound on $\|a\|$.

7.1 Orthogonal Lattices

Certain lattices possess a basis of N equal length, mutually perpendicular basis vectors. We denote such lattices *orthogonal lattices*. Two lattices are called *homothetic* if up to a constant stretching factor, λ, there is a distance preserving map from one lattice to the other. That is, all orthogonal lattices are homothetic to the trivial lattice $\mathbb{Z}^N$. Similarly, we define f to be an *orthogonal polynomial* if the circulant matrix M_f is the orthogonal basis of an orthogonal lattice. We are interested in orthogonal lattices because they possess a multiplicative norm property.

We note that for randomly chosen polynomials a and f, the norm is quasi-multiplicative, $\|f * a\| \approx \|f\| \cdot \|a\|$. However, if one of the polynomial factors, say f, is *orthogonal*, then equality will hold

$$\|f * a\| = \|f\| \cdot \|a\| . \tag{10}$$

For general polynomials f, applying LLL to (f) is guaranteed to find a multiple of f, say $f * a$, such that the norm $\|f * a\|$ is less than a specific factor times the norm of the shortest vector in the lattice. In the case where f is orthogonal, we can additionally bound $\|a\|$ by this factor, since $\|f * a\| = \|f\| \cdot \|a\|$. In the case of LLL, this means that we can be certain that $\|a\| < 2^{(N-1)/2}$.

7.2 Using $f * \overline{f}$ to Construct an Implicit Orthogonal Lattice

What do we do when f is not an orthogonal polynomial, but our objective is to find $f * a$ with small $\|a\|$ (given only $f * \overline{f}$ and lattice basis B_f of (f))? Of course, we may apply LLL to B_f and just hope that the output vector $f * a$ has

[5] Yet another characterization of the algorithm is that it recovers f from B_f and the relative norm of f over the index 2 subfield of $\mathbb{Q}(\zeta_N)$.

short a, but this may not work even if f is only slightly nonorthogonal[6]. This section describes how we can accomplish this task by using knowledge of $f * \overline{f}$ to *implicitly* define an orthogonal lattice.

Since B_f and M_f are both bases of (f), they are related by $B_f = U \cdot M_f$ for some unimodular matrix U. Notice that each row of B_f is of the form $f * u_i$ where u_i is the ith row of U. This means that the objective of finding $f * a$ with bounded $\|a\|$ is equivalent to bounding the norms of the rows of U. So, in some sense, we would like to apply lattice reduction to U. How can we reduce the rows of U when we only know $B_f = U \cdot M_f$, and not U itself?

Supposing that f is a not a zero divisor in R, we can also divide by $f * \overline{f}$. Allowing denominators in our notation, we let $D = M_{(1/(f*\overline{f}))}$ and compute

$$B_f \cdot D \cdot B_f^T = U \cdot U^T , \tag{11}$$

which is the Gram matrix of our unknown unimodular matrix U. Although we do not know U explicitly, $U \cdot U^T$ has all the information that LLL needs to know about U in order to reduce it – namely, the mutual dot products $u_i \cdot u_j$ of each pair of row vectors. We can therefore apply a Gram matrix version of LLL to $U \cdot U^T$, which outputs the unimodular transformation matrix V, and the Gram matrix of the reduced lattice: $(V \cdot U) \cdot (V \cdot U)^T$. By the LLL bound, the norms of the rows of (the unknown) basis $V \cdot U$ will be bounded by $2^{(N-1)/2}$. Now, we can compute a new basis of (f) – namely, $(V \cdot U) \cdot M_f = V \cdot B_f$ – and be certain that each row of this basis equals $f * a_i$ for $\|a_i\| < 2^{(N-1)/2}$. Effectively, we have reduced the orthogonal lattice defined by U, without even knowing an explicit basis for it.

7.3 Galois Congruence

In addition to the orthogonal lattices technique, we use some interesting congruences on the ring R. The first congruence states that for any prime P such that $P \equiv 1 \pmod{N}$,

$$f^P = f \pmod{P} . \tag{12}$$

This implies that for any f which is not a zero divisor[7] in $R_P = \mathbb{Z}_P[X]/(X^N - 1)$ that

$$f^{P-1} = 1 \pmod{P} . \tag{13}$$

We may generalize these equations to arbitrary primes P by using a *Galois Conjugation* function (written as a superscript) $\sigma(r) : R \to R$, defined by

$$f^{\sigma(r)}(x) = f(x^r). \tag{14}$$

For any r not divisible by N, $\sigma(r)$ defines an automorphism on R. There are $N-1$ such automorphisms, since two values of r which differ by a factor of N define

6 The notion of nonorthogonality is made precise with concept called *orthogonality defect.*

7 See the full version of the paper for a discussion of the zero divisor issue in R_P.

the same automorphism. We call $\sigma(r)$ the rth Galois conjugation mapping, and have the congruence

$$f^P = f^{\sigma(P)} \pmod{P} . \tag{15}$$

For elementary proofs of equations 13 and 15 and their relationship with the the Galois theory of $Q(\zeta)$, we refer the reader to the the full version of the paper.

As mentioned above, our motivation for considering such congruences is that given a multiple of f^{P-1}, say $f^{P-1} * a$, we may use the congruence $f^{P-1} = 1 \pmod{P}$ to conclude that

$$f^{P-1} * a = a \pmod{P} . \tag{16}$$

Now, if a is so small that all of its coefficients lie in the interval $(-P/2, P/2]$, then the representatives for $f^{P-1} * a \pmod{p}$ in this interval reveal a exactly. Assuming that a is not a zero divisor in R, dividing the product by a then yields the exact value of f^{P-1}. With this observation, we are in a position to use orthogonal lattice theory with lattice reduction to obtain small multiples of powers of f and thus exact powers of f.

However, there is a technical difficulty arising from the fact that LLL only guarantees that $\|a\| < 2^{(N-1)/2}$. To ensure that a has coefficients in $(-P/2, P/2]$, P has to be quite large – about the same order of magnitude as $2^{(N-1)/2}$. This means that f^{P-1} has bit-length exponential in N, which makes it impossible for us to even store the value of this polynomial. Initially, this might appear to be a fatal problem. It also indicates that the "brief sketch" given at the beginning of Section 7 could not have been entirely accurate.

Fortunately, we will not need to work with f^{P-1} directly; it will be sufficient for our purposes to be able to compute f^{P-1} modulo some primes. As discussed further in section 7.4, we can make such computations using a process similar to repeated squaring. However, to recover f, we will need some power of f that we can work with directly (unreduced). To solve this problem, we may choose a second prime $P' \equiv 1 \pmod{N}$ with $\mathrm{GCD}(P-1, P'-1) = 2N$ for which we can compute $f^{P'-1}$ modulo some primes. Then, using the Euclidean Algorithm, we may compute f^{2N} modulo these primes, and ultimately f^{2N} exactly via the Chinese Remainder Theorem. We may work with f^{2N} directly, since its bit-length is merely polynomial in N. A more elegant solution (in the full version) computes f^{2N} directly from f^{P-1} if $\mathrm{GCD}(P-1, 2N-1) = 1$. In Section 7.5, we describe how to recover f from f^{2N}.

7.4 Ideal Power Algorithms

In practice, it might be the case that smaller P would be sufficient, allowing us to work with the ideal (f^{P-1}) directly, but some tricks are needed to handle the very large primes P that the LLL bound imposes on us. Suppose we had the chains of polynomials $\{v_0^2 * \overline{v_1}, \ldots, v_{r-1}^2 * \overline{v_r}\}$ and $\{v_0 * \overline{v_0}, \ldots, v_{r-1} * \overline{v_{r-1}}\}$. Then, we could make the following series of computations:

$$v_0^4 * \overline{v_2} = (v_0^2 * \overline{v_1})^2 * (v_1 * \overline{v_1})^{-2} * (v_1^2 * \overline{v_2}) \pmod{P} , \tag{17}$$

$$v_0^8 * \overline{v_3} = (v_0^4 * \overline{v_2})^2 * (v_2 * \overline{v_2})^{-2} * (v_2^2 * \overline{v_3}) \pmod{P} , \tag{18}$$

and so on, ending with the equation:

$$v_0^{2^r} * \overline{v_r} = (v_0^{2^{r-1}} * \overline{v_{r-1}})^2 * (v_{r-1} * \overline{v_{r-1}})^{-2} * (v_{r-1}^2 * \overline{v_r}) \pmod{P}. \quad (19)$$

In other words, we could compute $v_0^{2^r} * \overline{v_r} \pmod{P}$ efficiently even though the exponent 2^r may be quite large. If we could use this approach to get $v_0^{P-1} * \overline{v_r} \pmod{P}$ where $\|\overline{v_r}\| < P/2$, then we could recover $\overline{v_r}$ exactly, and then we could use the same chains of polynomials to compute v_0^{P-1} modulo other primes.

The main tool that we use is the multiplication of ideals (see [2]), and we adapt this technique to use the orthogonal lattice theory above. The algorithm described in the paragraph above is essentially a repeated squaring algorithm, so we first review how to multiply ideals, and in particular, obtain the ideal (f^2) from the ideal (f).

Remark 2. Ideal multiplication in R: If $A = (f)$ and $B = (g)$ are principal ideals generated as $\mathbb{Z}$-modules by $(f * a_1, \ldots, f * a_n)$ and $(g * b_1, \ldots, g * b_n)$, then the ideal product AB is generated as a $\mathbb{Z}$-module by the n^2 elements of the set $\{a_i * b_j * f * g\}$, which defines $(f * g)$.

Note that we describe the ideals as modules – i.e., as the $\mathbb{Z}$-span of a set of polynomials (rather than by giving generators over R). This is because our algorithms represent ideals as lattices – i.e., as lists of polynomials.

We use ideal multiplication as follows: Given the ideal (f) in terms of the basis $B_f = U \cdot M_f$, we can generate (f^2) from the rows $b_i * b_j = u_i * u_j * f^2$. Since we know $f^2 * \overline{f}^2$, we can use the orthogonal lattice method described in section 7.2 to obtain a reduced basis $B_{f^2} = U' \dot{M}_{f^2}$ where the rows of U' have norm less than $2^{(N-1)/2}$.[8] Next, we pick a row of B_{f^2} and name it $f^2 * \overline{v_1}$. We can directly compute $v_1 * \overline{v_1}$ and a basis for v_1: $U' \cdot M_{v_1}$. Now, we can compute a basis for (v_1^2) to begin another iteration of this process.

Thus, we have the chains of polynomials previously introduced, and with a modification to take into account the binary representation of $P - 1$, we may obtain $f^{P-1} * \overline{v_r} \pmod{P}$. The polynomial complexity of the algorithm follows from the norm bound on the v_i's and the polynomial running time of LLL. By selecting a second prime P', as described in Section 7.3, we may obtain f^{2N} in polynomial time. In Appendix A, we write this algorithm in terms of pseudocode in an effort to clarify its details as well as its polynomial-time complexity.

7.5 Computing f from f^{2N}

Our final task is to compute the private key f from f^{2N}. We use the following theorem.

[8] In the worst case, this involves reduction of an N-dimensional lattice defined by N^2 generators. This reduction would only have polynomial time-complexity, but normally we will be able to do much better, since we will usually be easy to find far fewer (on the order of N) polynomials that span the ideal.

Theorem 1. *Galois Polynomial*
Given f^{2N} and $b \in \mathbb{Z}_N^*$, we may compute $z = f^{\sigma(-b)} f^b$ in time polynomial in b, N and bit-length of f.

Proof. Let $P_2 > 2\|f^{\sigma(-b)} f^b\|$ be a prime number such that $P_2 = 2c_2N - b$ for some integer c_2. Then, $(f^{2N})^{c_2} \equiv f^{P_2} * f^b \equiv f^{\sigma(-b)} f^b \pmod{P_2}$. Since $P_2 > 2\|f^{\sigma(-b)} f^b\|$, we recover $z = f^{\sigma(-b)} f^b$ exactly.

Now, in terms of recovering f, we first note that f^{2N} uniquely defines f only up to sign and rotation – i.e., up to multiplication by $\pm X^k$, the $2N$th roots of unity in R. The basic idea of our approach is that, given f^{2N}, fixing $f(\zeta)$ for one (complex) Nth root of unity ζ completely determines $f(\zeta^d)$ for all exponents d. Then, we may use the $N-1$ values of $f(\zeta^d)$, together with $f(1)$ (which we will know up to sign), to solve for f using Gaussian elimination. If we set $-b$ to be a primitive root modulo N, the polynomial z given in Theorem 1 will help us iteratively derive the $f(\zeta^d)$ from $f(\zeta)$ as follows:

$$f(\zeta^{(-b)^{i+1}}) = z(\zeta^{(-b)^i}) / f^b(\zeta^{(-b)^i}) . \tag{20}$$

Repeated exponentiation will not result in a loss of precision, since the value may be corrected at each stage. Since $-b$ is a primitive root modulo N, these evaluations give us $N-1$ linearly independent equations in the coefficients of f, which together with $f(1)$, allow us to recover the private key f completely, up to sign and rotation.

8 Summary and Generalizations of R-NSS Cryptanalysis

For the reader's convenience, we briefly review the main points of the attack. The first two stages of the attack are fast in practice, but they are both heuristic and have no proven time bounds. The lifting procedure lifts a transcript of signatures from R_q to R, obtaining unreduced f-multiples in R. The second stage uses an averaging attack to approximate $f * \overline{f}$, and then solves the closest vector problem (CVP) to recover $f * \overline{f}$ exactly. The algorithm of the final stage, which we have not fully implemented, uses output from the previous two stages to recover the private key in polynomial time. By combining lattice-based methods and number-theoretic congruences, the algorithm of the final stage can be used to:

1. Recover f from $f * \overline{f}$ and a basis B_f of (f);
2. Recover f from only B_f when f is an orthogonal polynomial; and
3. Recover $f/\overline{f}$ from B_f whether f is an orthogonal polynomial or not.

We anticipate that this algorithm could be generalized to recover f given a basis of (f) and the relative norm of f over an index 2 subfield where the degree-2 extension is complex conjugation. In Section 9, we discuss another possible generalization of this algorithm that may be an interesting area of research.

9 NTRUSign

NTRUSign was proposed at the rump session of Asiacrypt 2001 as a replacement of R-NSS [11], and, as requested by the Program Committee, we provide some preliminary security analysis. The scheme is more natural than previous NTRU-based signature schemes, particularly in terms of its sole verification criterion: the signer (or forger) must solve an "approximate CVP problem" in the NTRU lattice – i.e., produce a lattice point that is sufficiently close to a message digest point, à la Goldreich, Goldwasser and Halevi [10]. Similar to the GGH cryptosystem, the signer has private knowledge of a "good" basis of the NTRU lattice having short basis vectors, and publishes the usual "bad" NTRU lattice basis:

$$B_{priv} = \begin{bmatrix} M_f & M_g \\ M_F & M_G \end{bmatrix} \qquad B_{pub} = \begin{bmatrix} I_{(N)} & M_h \\ 0 & qI_{(N)} \end{bmatrix} .$$

Unlike GGH, these bases may be succinctly represented as polynomials: (f,g,F,G) for the private basis, and h for the public basis. (Recall that M_f denotes the circulant matrix corresponding to the polynomial f; see Section 2.) In terms of key generation, the signer first generates short polynomials f and g and computes the public key as $h = f^{-1} * g \pmod{q}$, as in NTRUEncrypt or R-NSS. Due to lack of space, we refer the reader to [11] for details on how the signer generates his second pair of short polynomials (F, G), but we note the following properties: 1) $f * G - g * F = q$ and 2) $\|F\|$ and $\|G\|$ are 2 to 3 times greater than $\|f\|$ and $\|g\|$.

To sign, the message is hashed to create a random message digest vector (m_1, m_2) with $m_1, m_2 \in R_q$. The signer then computes:

$$G * m_1 - F * m_2 = A + q * C , \tag{22a}$$

$$-g * m_1 + f * m_2 = a + q * c , \tag{22b}$$

where A and a have coefficients in $(-q/2, q/2]$, and sends his signature:

$$s \equiv f * C + F * c \pmod{q} . \tag{23}$$

The verifier computes $t \equiv s * h \pmod{q}$ and checks that (s,t) is "close enough" to (m_1, m_2) – specifically,

$$\|s - m_1\|^2 + \|t - m_2\|^2 \leq \text{Normbound} . \tag{24}$$

We can see why verification works when we write, say, s in terms of Equations 22a and 22b:

$$s \equiv m_1 - (A * f + a * F)/q \pmod{q} , \tag{25}$$

where $\|A * f + a * F\|$ will be reasonably short since f and F are short.

In the absence of a transcript, the forgery problem is provably as difficult as the approximate CVP problem in the NTRU lattice. However, it is clear that

NTRUSign signatures leak some information about the private key. The mapping involution $\mathbb{Z}[X]/(q, X^N - 1)$ sending $m \mapsto s$ is not a permutation, and the associated identification protocol is not zero knowledge (even statistically or computationally). Below, we describe concrete transcript attacks using ideas from our cryptanalysis of R-NSS. We have had fruitful discussions with NTRU regarding these attacks, and they have begun running preliminary tests to determine their efficacy. Based on these tests, some of these attacks appear to require a very long transcript that may make them infeasible in practice. This is a subject of further research. In any case, these attacks show that NTRUSign cannot have any formal security property, since it is not secure against passive adversaries.

9.1 Second Order Attack

Using Equation 25, we may obtain polynomials of the form $(A * f + a * F)$ (similarly, $(A * g + a * G)$). Consider the following average:

$$Avg_{ff}(r) = (1/r) \sum_{i=1}^{r} (a_i * F + A_i * f) * \overline{(a_i * F + A_i * f)} \tag{26}$$

$$= (1/r) \sum_{i=1}^{r} (a_i * \overline{a_i}) * (F * \overline{F}) + (A_i * \overline{A_i}) * (f * \overline{f}) + \text{other terms} . \tag{27}$$

The "other terms" will converge to 0, since A and a are uniformly distributed at random modulo q and, though dependent, have small statistical correlation. (See Remark 3 below.) The explicit portion of the average will converge essentially to a scalar multiple of $f * \overline{f} + F * \overline{F}$, for the same reasons as discussed in Section 4.3. Thus,

$$\lim Avg_{ff}(r) = \gamma(f * \overline{f} + F * \overline{F}) . \tag{28}$$

Because the signatures in a transcript are random variables, the limit converges as $1/\sqrt{r}$ where r is the length of a transcript. We may use this averaging to obtain a sufficiently close approximation of $f * \overline{f} + F * \overline{F}$ to obtain the exact value by solving the CVP in the $(N + 1)$-dimensional lattice given in Equation 9 using lattice reduction. Thus, we recover a polynomial that is quadratic in the private key, and we can obtain $f * \overline{g} + F * \overline{G}$ and $g * \overline{g} + G * \overline{G}$ in a similar fashion.

Remark 3. One may artificially construct situations where $(1/r) \sum_{i=1}^{r} a_i * \overline{A_i}$ does not converge to 0. For example, if we let $f = F$, then $A_i * f + a_i * F \equiv 0 \pmod q$ basically implies $A_i = -a_i$ and hence $a_i * \overline{A_i} = -a_i * \overline{a_i}$, the average of which does not converge to 0. Conceivably, NTRUSign could be modified so as to constrain $f^{-1} * F \pmod q$, but this would likely allow alternative attacks.

It is worth noting that these second order polynomials give us the Gram matrix $B_{priv}^T \cdot B_{priv}$:

$$B_{priv}^T \cdot B_{priv} = \begin{bmatrix} M_{\overline{f}} & M_{\overline{F}} \\ M_{\overline{g}} & M_{\overline{G}} \end{bmatrix} \begin{bmatrix} M_f & M_g \\ M_F & M_G \end{bmatrix} = \begin{bmatrix} M_{f*\overline{f}+F*\overline{F}} & M_{g*\overline{f}+G*\overline{F}} \\ M_{f*\overline{g}+F*\overline{G}} & M_{g*\overline{g}+G*\overline{G}} \end{bmatrix} .$$

This Gram matrix gives us the "shape" of the parallelepiped defined by B_{priv}^T, but the "orientation" of this parallelepiped is unclear. An interesting (and open) question is: Can an attacker recover B_{priv} from $B_{priv}^T \cdot B_{priv}$ and $B_{pub} = U \cdot B_{priv}$, where U is a unimodular matrix? We answered a similar question in the affirmative in Section 7; we showed that an attacker, in polynomial time, can recover M_f from $M_f^T \cdot M_f$ and $U \cdot M_f$, where U is a unimodular matrix. We have not found a way to extend the orthogonal congruence attack to solve the NTRUSign Gram matrix problem, however, where the bi-circulant (rather than purely circulant) nature of the matrices in question (such as B_{priv}) destroys the commutativity that our orthogonal congruence attack appears to require, but this does not imply that the NTRUSign Gram matrix problem is necessarily hard. We note that it would more than suffice to find an algorithm that factors $U \cdot U^T$ for unimodular U. (This factorization is unique up to a signed permutation matrix.) Further research in this area would be interesting, if only because it is relevant to NTRUSign's security.

9.2 Second Order Subtranscript Attack

It is clear that the second order polynomials recovered above contain information not contained in the public key, but using this information to create an effective attack is not so straightforward. Our approach has been to use the second order polynomials to recover, say, $f * \overline{f}$, so that we may then apply the orthogonal congruence attack to recover f. One way to get $f * \overline{f}$ is to use the following subtranscript attack.

First, we notice that since $\|F\| > \|f\|$, the norm $\|A * f + a * F\|$ is dictated more by $\|a\|$ than by $\|A\|$. More relevantly, for our purposes, an $A * f + a * F$ that is longer than normal will usually have $\|a\| > \|A\|$. This suggests a subtranscript attack, including in Equation 26 only those polynomials $A_i * f + a_i * F$ for which $\|A_i * f + a_i * F\|$ is greater than some bound.[9] Then, we have:

$$\text{Subtranscript: } \lim Avg_{ff}(r) = \gamma_1(f * \overline{f}) + \gamma_2(F * \overline{F}) , \qquad (30)$$

for $\gamma_1 < \gamma_2$. Since this linear combination of $f * \overline{f}$ and $F * \overline{F}$ is distinct from that in Equation 28, we may compute $f * \overline{f}$ and $F * \overline{F}$. The convergence of this subtranscript averaging will be affected by the proportional size of the subtranscript, but more importantly, by the fact that γ_1 may be only a few percentage points greater than γ_2 (in our preliminary experiments using the longest 50% of the $A_i * f + a_i * F$'s). Further experiments are necessary to determine the effectiveness of this attack. Another consideration is that in [11], a possible modification of NTRUSign was proposed in which one chooses the transpose of B_{priv} to be the private key. The basis vectors are then (f, F) and (g, G) with $\|f\| \approx \|g\|$ and $\|F\| \approx \|G\|$. Choosing the private basis in this way appears to defeat this subtranscript attack.

[9] This selection criterion might be refined, as by also considering the norm of $A * \overline{F} - a * \overline{f}$, which may be computed using the above second order polynomials.

9.3 Fourth Order Attack

An alternative way to get $f*\overline{f}$ is to use the following fourth order attack. Viewing the average in Equation 26, one may consider the corresponding variance and conclude, under the assumption of the statistical independence of a and A, that:

$$\lim_{r\to\infty}(1/r)\sum_{i=1}^{r}(s*s_{rev})^2-(1/r)\beta(\lim_{r\to\infty}\sum_{i=1}^{r}(s*s_{rev}))^2=\gamma f*\overline{f}*F*\overline{F}\,. \quad (31)$$

The adjustment value of β depends on the scheme parameters n and q, and so the above value may not be exactly the variance. The factor γ also depends on the scheme parameters, and is a constant that slows convergence by a factor $\frac{1}{\gamma^2}$. This limit does converge more slowly than the second order averaging, but, as above, we may use a close approximation in conjunction with the lattice of Equation 9 to obtain the exact value. Preliminary tests show that it may not be practical to obtain an error lower than the Gaussian estimate with a reasonable number of signatures. Assuming we do obtain the value $f*\overline{f}*F*\overline{F}$, we may use it in combination with $(f*\overline{f}+F*\overline{F})^2$ to obtain $(f*\overline{f}-F*\overline{F})^2$, and then $f*\overline{f}-F*\overline{F}$ (using, perhaps, the algorithm given in Section 7.5). Then, $f*\overline{f}+F*\overline{F}$ and $f*\overline{f}-F*\overline{F}$ give us $f*\overline{f}$ and $F*\overline{F}$.

9.4 Preliminary Conclusion

The approach of these initial attacks was to reduce the breaking problem to the orthogonal congruence attack using the results of various averagings. We showed how this could be done, but the practical feasibility of these attacks has yet to be determined. In our experiments, we have found that the second order attack is feasible; for example, by averaging 20000 signatures, an attacker may obtain an approximation whose squared error is about 88, about 1/20 of the squared Gaussian heuristic of the $(N+1)$-dimensional CVP lattice that would be used to correct this approximation. Although we have not tested this CVP lattice, we believe its reduction would be feasible. Alternatively, more signatures could first be used to obtain a better approximation. We have also shown how the security of NTRUSign rests on the hardness of several new hard problems. These attacks will continue to be analyzed by the authors, NTRU corporation, and the cryptographic community.

Acknowledgments

The authors would like to thank Burt Kaliski, Alice Silverberg and Yiqun Lisa Yin for helpful discussions, Jakob Jonsson, Phong Nguyen, and Jacques Stern for discussions and collaboration on the precursor of this article, and Jeffrey Hoffstein, Nick Howgrave-Graham, Jill Pipher, Joseph Silverman and William Whyte, who have given us valuable feedback on our cryptanalysis, particularly on our preliminary cryptanalysis of NTRUSign.

References

1. M. Ajtai, *The shortest vector problem in L_2 is NP-hard for randomized reductions*, in Proc. 30th ACM Symposium on Theory of Computing, 1998, 10–19.
2. H. Cohen, A Course in Computational Algebraic Number Theory, Graduate Texts in Mathematics, 138. Springer, 1993.
3. H. Cohen, *Advanced Topics in Computational Number Theory*, Graduate Texts in Mathematics 138 ,1993.
4. Consortium for Efficient Embedded Security. Efficient Embedded Security Standard (EESS) # 1: Draft 1.0. Previously on `http://www.ceesstandards.org`.
5. Consortium for Efficient Embedded Security. Efficient Embedded Security Standard (EESS) # 1: Draft 2.0. Previously on `http://www.ceesstandards.org`.
6. Consortium for Efficient Embedded Security. Efficient Embedded Security Standard (EESS) # 1: Draft 3.0. Available from `http://www.ceesstandards.org`.
7. D. Coppersmith and A. Shamir, *Lattice Attacks on NTRU*, in Proc. of Eurocrypt '97, LNCS 1233, pages 52–61. Springer-Verlag, 1997.
8. C. Gentry, *Key Recovery and Message Attacks on NTRU-Composite*, in Proc. of Eurocrypt '01, LNCS 2045, pages 182–194. Springer-Verlag, 2001.
9. C. Gentry, J. Jonsson, J. Stern, M. Szydlo, *Cryptanalysis of the NTRU signature scheme*, in Proc. of Asiacrypt '01, LNCS 2248, pages 1–20. Springer-Verlag, 2001.
10. O. Goldreich, S. Goldwasser, S. Halevi, *Public-key Cryptography from Lattice Reduction Problems*, in Proc. of Crypto '97, LNCS 1294, pages 112–131. Springer-Verlag, 1997.
11. J. Hoffstein, N. Howgrave-Graham, J. Pipher, J.H. Silverman, W. Whyte, *NTRUSign: Digital Signatures Using the NTRU Lattice*, December, 2001. Available from `http://www.ntru.com`.
12. J. Hoffstein, B.S. Kaliski, D. Lieman, M.J.B. Robshaw, Y.L. Yin, *Secure user identification based on constrained polynomials*, US Patent 6,076,163, June 13, 2000.
13. J. Hoffstein, D. Lieman, J.H. Silverman, *Polynomial Rings and Efficient Public Key Authentication*, in Proc. International Workshop on Cryptographic Techniques and E-Commerce (CrypTEC '99), Hong Kong, (M. Blum and C.H. Lee, eds.), City University of Hong Kong Press.
14. J. Hoffstein, J. Pipher, J.H. Silverman, *Enhanced Encoding and Verification Methods for the NTRU Signature Scheme*, NTRU Technical Note #017, May 2001. Available from `http://www.ntru.com`.
15. J. Hoffstein, J. Pipher, J.H. Silverman. *Enhanced encoding and verification methods for the NTRU signature scheme (ver. 2)*, May 30, 2001. Available from `http://www.ntru.com`.
16. J. Hoffstein, J. Pipher, J.H. Silverman, *NSS: The NTRU Signature Scheme*, preprint, November 2000. Available from `http://www.ntru.com`.
17. J. Hoffstein, J. Pipher, J.H. Silverman, *NSS: The NTRU Signature Scheme*, in Proc. of Eurocrypt '01, LNCS 2045, pages 211–228. Springer-Verlag, 2001.
18. J. Hoffstein, J. Pipher, J.H. Silverman, *NSS: The NTRU Signature Scheme: Theory and Practice*, preprint, 2001. Available from `http://www.ntru.com`.
19. A.K. Lenstra, H.W. Lenstra Jr., L. Lovász, *Factoring Polynomials with Rational Coefficients*, Mathematische Ann. 261 (1982), 513–534.
20. A. May, *Cryptanalysis of NTRU-107*, preprint, 1999. Available from `http://www.informatik.uni- frankfurt.de/~alex/crypto.html`.
21. I. Mironov, *A Note on Cryptanalysis of the Preliminary Version of the NTRU Signature Scheme*, IACR preprint server, `http://eprint.iacr.org/2001/005`.

22. P. Nguyen and J. Stern, *Lattice Reduction in Cryptology: An Update*, in Proc. of Algorithm Number Theory (ANTS IV), LNCS 1838, pages 85–112. Springer-Verlag, 2000.
23. J.H. Silverman, *Estimated Breaking Times for NTRU Lattices*, NTRU Technical Note #012, March 1999. Available from http://www.ntru.com.
24. J.H. Silverman, *Invertibility in Truncated Polynomial Rings.*, NTRU Technical Note #009, October 1998. Available from http://www.ntru.com.
25. L. Washington, *Introduction to Cyclotomic Fields*, Graduate Texts in Mathematics 83, 1982.

A Orthogonal Congruence Attack: The Key Algorithm

Here, we present in pseudocode the key algorithm of the orthogonal congruence attack, through which an attacker computes chains of polynomials that allow him to derive modular information about f^{P-1}. This algorithm, as well as other algorithms from the orthogonal congruence attack, are discussed in more detail in the full version of the paper. Note that this algorithm will compute $f^{2^r} * \overline{v_r}$ (mod P) instead of $f^{P-1} * \overline{v_r}$ (mod P) for $r = \lfloor \log_2 P \rfloor$; more general powers require only minor adjustments. Let $v_0 = f$.

Input: $v_0 * \overline{v_0}$; basis B_0 of (v_0); prime $P \equiv 1 \pmod N$, $r = \lfloor \log_2 P \rfloor = (N+1)/2$.
Output: $\{v_0^2 * \overline{v_1}, \ldots, v_{r-1}^2 * \overline{v_r}\}$ and $\{v_0 * \overline{v_0}, \ldots, v_{r-1} * \overline{v_{r-1}}\}$ with $\|v_i\| < 2^{(N-1)/2}$; $v_0^{2^r} * \overline{v_r} \pmod P$.

1. Set $r' := 0$.
2. While $r' < r$ do
   (a) Use $B_{r'}$ to construct a set G of vectors generating $(v_{r'}^2)$. Let $G = H \cdot M_{v_{r'}^2}$.
   (b) Compute the Gram matrix $H \cdot H^T = G \cdot M^{-2}_{v_{r'} * \overline{v_{r'}}} \cdot G^T$.
   (c) Reduce the Gram matrix $H \cdot H^T$ to get $(A \cdot H) \cdot (A \cdot H)^T$ for known A.
   (d) Compute the basis $(A \cdot H) \cdot M_{v_{r'}^2} = A \cdot G$ of $(v_{r'}^2)$. ($A \cdot H$ has short rows.)
   (e) Pick, say, the jth row of $A \cdot G$, call it $v_{r'}^2 * \overline{v_{r'+1}}$. Note: $\overline{v_{r'+1}}$ is the jth row of $A \cdot H$.
   (f) Output $v_{r'}^2 * \overline{v_{r'+1}}$.
   (g) If $r' + 1 \neq r$
   – Compute $v_{r'+1} * \overline{v_{r'+1}} = (v_{r'}^2 * \overline{v_{r'+1}}) * \overline{(v_{r'}^2 * \overline{v_{r'+1}})} * (v_{r'} * \overline{v_{r'}})^{-2}$.
   – Output $v_{r'+1} * \overline{v_{r'+1}}$.
   – Compute $(A \cdot H) \cdot M_{v_{r'+1}} = A \cdot G \cdot M^{-2}_{v_{r'} * \overline{v_{r'}}} \cdot M_{\overline{v_{r'}}^2 * v_{r'+1}}$. This is our basis $B_{r'+1}$ of $(v_{r'+1})$.
   (h) Increment r'.
3. Set $r' := 1$; Set $y := v_0^2 * \overline{v_1} \pmod P$.
4. While $r' < r$ do
   (a) Compute $y := y^2 \pmod P$.
   (b) Compute $y := y * (v_{r'} * \overline{v_{r'}})^{-2} * (v_{r'}^2 * \overline{v_{r'+1}}) \pmod P$. At this point, $y = v_0^{2^{r'+1}} * \overline{v_{r'+1}} \pmod P$.
   (c) Increment r'.
5. Output y.

Dynamic Group Diffie-Hellman Key Exchange under Standard Assumptions

Emmanuel Bresson[1], Olivier Chevassut[2,3,*], and David Pointcheval[1]

[1] École normale supérieure, 75230 Paris Cedex 05, France,
http://www.di.ens.fr/~{bresson,pointche},
{Emmanuel.Bresson,David.Pointcheval}@ens.fr.
[2] Lawrence Berkeley National Laboratory, Berkeley, CA 94720, USA,
http://www.itg.lbl.gov/~chevassu, OChevassut@lbl.gov.
[3] Université Catholique de Louvain, 31348 Louvain-la-Neuve, Belgium

Abstract. Authenticated Diffie-Hellman key exchange allows two principals communicating over a public network, and each holding public/private keys, to agree on a shared secret value. In this paper we study the natural extension of this cryptographic problem to a group of principals. We begin from existing formal security models and refine them to incorporate major missing details (e.g., strong-corruption and concurrent sessions). Within this model we define the execution of a protocol for authenticated dynamic group Diffie-Hellman and show that it is provably secure under the decisional Diffie-Hellman assumption. Our security result holds in the standard model and thus provides better security guarantees than previously published results in the random oracle model.

1 Introduction

Authenticated Diffie-Hellman key exchange allows two principals A and B communicating over a public network and each holding a pair of matching public/private keys to agree on a shared secret value. Protocols designed to deal with this problem ensure A (B resp.) that no other principals aside from B (A resp.) can learn any information about this value; the so-called authenticated key exchange with "implicit" authentication (AKE). These protocols additionally often ensure A and B that their respective partner has actually computed the shared secret value (i.e. authenticated key exchange with explicit key confirmation). A natural extension to this protocol problem would be to consider a scenario wherein a pool of principals agree on a shared secret value. We refer to this extension as authenticated group Diffie-Hellman key exchange.

Consider scientific collaborations and conferencing applications [5,11], such as data sharing or electronic notebooks. Applications of this type usually involve

* The second author was supported by the Director, Office of Science, Office of Advanced Scientific Computing Research, Mathematical Information and Computing Sciences Division, of the U.S. Department of Energy under Contract No. DE-AC03-76SF00098. This document is report LBNL-49087.

L.R. Knudsen (Ed.): EUROCRYPT 2002, LNCS 2332, pp. 321–336, 2002.

users aggregated into small groups and often utilize multiple groups running in parallel. The users share responsibility for parts of tasks and need to coordinate their efforts in an environment prone to attacks. To reach this aim, the principals need to agree on a secret value to implement secure multicast channels. Key exchange schemes suited for this kind of application clearly needs to allow concurrent executions between parties.

We study the problem of authenticated group Diffie-Hellman key exchange when the group membership is dynamic – principals join and leave the group at any time – and the adversary may generate cascading changes in the membership for subsets of principals of his choice. After the initialization phase, and throughout the lifetime of the multicast group, the principals need to be able to engage in a conversation after each change in the membership at the end of which the session key is updated to be sk'. The secret value sk' should be only known to the principals in the multicast group during the period when sk' is the session key.

(2-party) Diffie-Hellman key exchange protocols also usually achieve the property of forward-secrecy [15,16] which entails that corruption of a principal's long-term key does not threaten the security of previously established session keys. Assuming the ability to *erase* a secret, some of these protocols achieve forward-secrecy even if the corruption also releases the principal's internal state (i.e. strong-corruption [24]). In practice secret erasure is, for example, implemented by hardware devices which use physical security and tamper detection to not reveal any information [12,22,21,28]. Protocols for group Diffie-Hellman key exchange need to achieve forward-secrecy even when facing strong-corruption.

Contributions. This paper is the third tier in the formal treatment of the group Diffie-Hellman key exchange using public/private key pairs. The first tier was provided for a scenario wherein the group membership is static [7] and the second, by extension of the latter for a scenario wherein the group membership is dynamic [8]. We start from the latter formal model and refine it to add important attributes. In the present paper, we model instances of players via oracles available to the adversary through queries. The queries are available to use at any time to allow model attacks involving multiple instances of players activated concurrently and simultaneously by the adversary. In order to model two modes of corruption, we consider the presence of two cryptographic devices which are made available to the adversary through queries. Hardware devices are useful to overcome software limitations however there has thus far been little formal security analysis [12,23].

The types of crypto-devices and our notion of forward-secrecy leads us to modifications of existing protocols to obtain a protocol, we refer to it as $\mathsf{AKE1}^+$, secure against strong corruptions. Due to the very limited computational power of a smart card chip, smart card is used as an authentication token while a secure coprocessor is used to carry out the key exchange operations. We show that within our model the protocol $\mathsf{AKE1}^+$ is secure assuming the decisional Diffie-Hellman problem and the existence of a pseudo-random function family. Our security theorem does not need a random oracle assumption [4] and thus holds in the standard model. A proof in the standard model provides better security guarantees than one in an idealized model of computation [8,7]. Furthermore we exhibit a security reduction with a much tighter bound than [8], namely we suppress the exponential factor in the size of the group. Therefore the security result is meaningful even for large groups. However the protocols are not practical for groups larger than 100 members.

The remainder of this paper is organized as follows. We first review the related work and then introduce the building blocks which we use throughout the paper. In Section 3, we present our formal model and specify through an abstract interface the standard functionalities a protocol for authenticated group Diffie-Hellman key exchange needs to implement. In Section 4, we describe the protocol $\mathsf{AKE1}^+$ by splitting it down into functions. This helps us to implement the abstract interface. Finally, in Section 5 we show that the protocol $\mathsf{AKE1}^+$ is provably secure in the standard model.

Related Work. Several papers [1,10,19,14,27] have extended the 2-party Diffie-Hellman key exchange [13] to the multi-party setting however a formal analysis has only been proposed recently. In [8,7], we defined a formal model for the authenticated (dynamic) group Diffie-Hellman key exchange and proved secure protocols within this model. We use in both papers an ideal hash function [4], without dealing with dynamic group changes in [7], or concurrent executions of the protocol in [8].

However security can sometimes be compromised even when using a proven secure protocol: the protocol is incorrectly implemented or the model is insufficient. Cryptographic protocols assume, and do not usually explicitly state, that secrets are *definitively and reliably erased* (only the most recent secrets are kept) [12,18]. Only recently formal models have been refined to incorporate the cryptographic action of erasing a secret, and thus protocols achieving forward-secrecy in the strong-corruption sense have been proposed [3,24].

Protocols for group Diffie-Hellman key exchange [7] achieve the property of forward-secrecy in the strong-corruption sense assuming that "ephemeral" private keys are erased upon completion of a protocol run. However protocols for dynamic group Diffie-Hellman key exchange [8] do not, since they reuse the "ephemeral" keys to update the session key. Fortunately, these "ephemeral" keys can be embedded in some hardware cryptographic devices which are at least as good as erasing a secret [22,21,28].

2 Basic Building Blocks

We first introduce the pseudo-random function family and the intractability assumptions.

Message Authentication Code. A *Message Authentication Code* MAC= (MAC.SGN,MAC.VF) consists of the following two algorithms (where the key space is uniformly distributed) [2]:

- The *authentication algorithm* MAC.SGN which, on a message m and a key K as input, outputs a tag μ. We write $\mu \leftarrow$ MAC.SGN(K, m). The pair (m, μ) is called an authenticated message.
- The *verification algorithm* MAC.VF which, on an authenticated message (m, μ) and a key K as input, checks whether μ is a valid tag on m with respect to K. We write `True/False` $\leftarrow$ MAC.VF(K, m, μ).

A (t, q, L, ϵ)-MAC-forger is a probabilistic Turing machine $\mathcal{F}$ running in time t that requests a MAC.SGN-oracle up to q messages each of length at most L, and outputs an authenticated message (m', μ'), without having queried the MAC.SGN-oracle on message m', with probability at least ϵ. We denote this success probability as $\mathsf{Succ}^{\mathsf{cma}}_{\mathsf{mac}}(t, q, L)$, where CMA stands for (adaptive) Chosen-Message Attack. The MAC scheme is (t, q, L, ϵ)-**CMA-secure** if there is no (t, q, L, ϵ)-MAC-forger.

Group Decisional Diffie-Hellman Assumption (G-DDH)**.** Let $\mathbb{G} =< g >$ be a cyclic group of prime order q and n an integer. Let I_n be $\{1, \ldots, n\}$, $\mathcal{P}(I_n)$ be the set of all subsets of I_n and Γ be a subset of $\mathcal{P}(I_n)$ such that $I_n \notin \Gamma$.

We define the *Group Diffie-Hellman distribution* relative to Γ as:

$$\mathsf{G\text{-}DH}_\Gamma = \left\{ \left(J, g^{\prod_{j \in J} x_j} \right)_{J \in \Gamma} \mid x_1, \ldots, x_n \in_R \mathbb{Z}_q \right\}.$$

Given Γ, a (T, ϵ)-G-DDH$_\Gamma$-distinguisher for $\mathbb{G}$ is a probabilistic Turing machine Δ running in time T that given an element X from either $\mathsf{G\text{-}DH}^{\$}_\Gamma$, where the tuple of $\mathsf{G\text{-}DH}_\Gamma$ is appended a random element g^r, or $\mathsf{G\text{-}DH}^\star_\Gamma$, where the tuple is appended $g^{x_1 \ldots x_n}$, outputs 0 or 1 such that:

$$\left| \Pr\left[\Delta(X) = 1 \mid X \in \mathsf{G\text{-}DH}^{\$}_\Gamma\right] - \Pr\left[\Delta(X) = 1 \mid X \in \mathsf{G\text{-}DH}^\star_\Gamma\right] \right| \geq \epsilon.$$

We denote this difference of probabilities by $\mathsf{Adv}^{\mathsf{gddh}_\Gamma}_{\mathbb{G}}(\Delta)$. The G-DDH$_\Gamma$ problem is (T, ϵ)-**intractable** if there is no (T, ϵ)-G-DDH$_\Gamma$-distinguisher for $\mathbb{G}$.

If $\Gamma = \mathcal{P}(I) \backslash \{I_n\}$, we say that G-DH$_\Gamma$ is the **Full** *Generalized Diffie-Hellman distribution* [6,20,26]. Note that if $n = 2$, we get the classical DDH problem, for which we use the straightforward notation $\mathsf{Adv}^{\mathsf{ddh}}_{\mathbb{G}}(\cdot)$.

Lemma 1. *The* DDH *assumption implies the* G-DDH *assumption.*

Proof. Steiner, Tsudik and Waidner proved it in [26]. □

Multi Decisional Diffie-Hellman Assumption (M-DDH). We introduce a new decisional assumption, based on the Diffie-Hellman assumption. Let us define the *Multi Diffie-Hellman* M-DH and the *Random Multi Diffie-Hellman* $\mathsf{M\text{-}DH}^\$$ distributions of size n as:

$$\mathsf{M\text{-}DH}_n = \left\{ \left(\{g^{x_i}\}_{1 \le i \le n}, \{g^{x_i x_j}\}_{1 \le i < j \le n} \right) \mid x_1, \ldots, x_n \in_R \mathbb{Z}_q \right\}$$

$$\mathsf{M\text{-}DH}_n^\$ = \left\{ \left(\{g^{x_i}\}_{1 \le i \le n}, \{g^{r_{j,k}}\}_{1 \le j < k \le n} \right) \mid x_i, r_{j,k} \in_R \mathbb{Z}_q, \forall i, 1 \le j < k \le n \right\}.$$

A (T, ϵ)-$\mathsf{M\text{-}DDH}_n$-distinguisher for $\mathbb{G}$ is a probabilistic Turing machine Δ running in time T that given an element X of either $\mathsf{M\text{-}DH}_n$ or $\mathsf{M\text{-}DH}_n^\$$ outputs 0 or 1 such that:

$$\left| \Pr[\Delta(X) = 1 \mid X \in \mathsf{M\text{-}DH}_n] - \Pr[\Delta(X) = 1 \mid X \in \mathsf{M\text{-}DH}_n^\$] \right| \ge \epsilon.$$

We denote this difference of probabilities by $\mathsf{Adv}_{\mathbb{G}}^{\mathsf{mddh}_n}(\Delta)$. The $\mathsf{M\text{-}DDH}_n$ problem is (T, ϵ)-**intractable** if there is no (T, ϵ)-$\mathsf{M\text{-}DDH}_n$-distinguisher for $\mathbb{G}$.

Lemma 2. *For any group $\mathbb{G}$ and any integer n, the $\mathsf{M\text{-}DDH}_n$ problem can be reduced to the* DDH *problem and we have: $\mathsf{Adv}_{\mathbb{G}}^{\mathsf{mddh}_n}(T) \le n^2 \mathsf{Adv}_{\mathbb{G}}^{\mathsf{ddh}}(T)$.*

3 Model

In this section, we model instances of players via oracles available to the adversary through queries. These oracle queries provide the adversary a capability to initialize a multicast group via Setup-queries, add players to the multicast group via Join-queries, and remove players from the multicast group via Remove-queries. By making these queries available to the adversary at any time we provide him an ability to generate concurrent membership changes. We also take into account hardware devices and model their interaction with the adversary via queries.

Players. We fix a nonempty set $\mathcal{U}$ of N players that can participate in a group Diffie-Hellman key exchange protocol P. A player $U_i \in \mathcal{U}$ can have many *instances* called oracles involved in distinct concurrent executions of P. We denote instance t of player U_i as Π_i^t with $t \in \mathbb{N}$. Also, when we mean a not fixed member of $\mathcal{U}$ we use U without any index and denote an instance of U as Π_U^t with $t \in \mathbb{N}$.

For each concurrent execution of P, we consider a nonempty subset $\mathcal{I}$ of $\mathcal{U}$ called the *multicast group*. And in $\mathcal{I}$, the group controller $\mathsf{GC}(\mathcal{I})$ initiates the addition of players to the multicast group or the removal of players from the multicast group. The group controller is trusted to do only this.

In a multicast group $\mathcal{I}$ of size n, we denote by $\mathcal{I}_i$, for $i = 1, \ldots, n$, the index of the player related to the i-th instance involved in this group. This i-th instance is furthermore denoted by $\Pi(\mathcal{I}, i)$. Therefore, for any index $i \in \{1, \ldots, n\}$, $\Pi(\mathcal{I}, i) = \Pi_{\mathcal{I}_i}^t \in \mathcal{I}$ for some t.

Each player U holds a long-lived key LL_U which is a pair of matching public/private keys. LL_U is specific to U not to one of its instances.

Abstract Interface. We define the basic structure of a group Diffie-Hellman protocol. A group Diffie-Hellman scheme GDH consists of four algorithms:

- The *key generation algorithm* GDH.KeyGen(1^ℓ) is a probabilistic algorithm which on input of a security parameter 1^ℓ, provides each player in $\mathcal{U}$ with a long-lived key LL_U. The structure of LL_U depends on the particular scheme.

The three other algorithms are interactive multi-party protocols between players in $\mathcal{U}$, which provide each principal in the new multicast group with a new session key SK.

- The *setup algorithm* GDH.Setup($\mathcal{J}$), on input of a set of instances of players $\mathcal{J}$, creates a new multicast group $\mathcal{I}$, and sets it to $\mathcal{J}$.
- The *remove algorithm* GDH.Remove($\mathcal{I}, \mathcal{J}$) creates a new multicast group $\mathcal{I}$ and sets it to $\mathcal{I} \backslash \mathcal{J}$.
- The *join algorithm* GDH.Join($\mathcal{I}, \mathcal{J}$) creates a new multicast group $\mathcal{I}$, and sets it to $\mathcal{I} \cup \mathcal{J}$.

An execution of P consists of running the GDH.KeyGen algorithm once, and then many concurrent executions of the three other algorithms. We will also use the term *operation* to mean one of the algorithms: GDH.Setup, GDH.Remove or GDH.Join.

Security Model. The security definitions for P take place in the following game. In this game $\mathbf{Game}^{\mathsf{ake}}(\mathcal{A}, P)$, the adversary $\mathcal{A}$ plays against the players in order to defeat the security of P. The game is initialized by providing coin tosses to GDH.KeyGen($\cdot$), $\mathcal{A}$, any oracle Π_U^t; and GDH.KeyGen(1^ℓ) is run to set up players' LL-key. A bit b is as well flipped to be later used in the Test-query (see below). The adversary $\mathcal{A}$ is then given access to the oracles and interacts with them via the queries described below. We now explain the capabilities that each kind of query captures:

Instance Oracle Queries. We define the oracle queries as the interactions between $\mathcal{A}$ and the oracles only. These queries model the attacks an adversary could mount through the network.

- Send(Π_U^t, m): This query models $\mathcal{A}$ sending messages to instance oracles. $\mathcal{A}$ gets back from his query the response which Π_U^t would have generated in processing message m according to P.
- Setup($\mathcal{J}$), Remove($\mathcal{I}, \mathcal{J}$), Join($\mathcal{I}, \mathcal{J}$): These queries model adversary $\mathcal{A}$ initiating one of the operations GDH.Setup, GDH.Remove or GDH.Join. Adversary $\mathcal{A}$ gets back the flow initiating the execution of the corresponding operation.
- Reveal(Π_U^t): This query models the attacks resulting in the loss of session key computed by oracle Π_U^t; it is only available to $\mathcal{A}$ if oracle Π_U^t has computed its session key $\mathrm{SK}_{\Pi_U^t}$. $\mathcal{A}$ gets back $\mathrm{SK}_{\Pi_U^t}$ which is otherwise hidden. When considering the *strong-corruption model* (see Section 5), this query also reveals the flows that have been exchanged between the oracle and the secure coprocessor.

– $\mathsf{Test}(\Pi_U^t)$: This query models the semantic security of the session key $\mathrm{SK}_{\Pi_U^t}$. It is asked only once in the game, and is only available if oracle Π_U^t is Fresh (see below). If $b = 0$, a random ℓ-bit string is returned; if $b = 1$, the session key is returned. We use this query to define $\mathcal{A}$'s advantage.

Secure Coprocessor Queries. The adversary $\mathcal{A}$ interacts with the secure coprocessors by making the following two queries.

– $\mathsf{Send}_c(\Pi_U^t, m)$: This query models $\mathcal{A}$ *directly* sending and receiving messages to the secure coprocessor. $\mathcal{A}$ gets back from his query the response which the secure coprocessor would have generated in processing message m. The adversary could directly interact with the secure coprocessor in a variety of ways: for instance, the adversary may have broken into a computer without being detected (e.g., bogus softwares, trojan horses and viruses).
– $\mathsf{Corrupt}_c(\Pi_U^t)$: This query models $\mathcal{A}$ having access to the private memory of the device. $\mathcal{A}$ gets back the internal data stored on the secure coprocessor. This query can be seen as an attack wherein $\mathcal{A}$ gets physical access to a secure coprocessor and bypasses the tamper detection mechanism [29]. This query is only available to the adversary when considering the *strong-corruption model* (see Section 5). The $\mathsf{Corrupt}_c$-query also reveals the flows the secure coprocessor and the smart card have exchanged.

Smart Card Queries. The adversary $\mathcal{A}$ interacts with the smart cards by making the two following queries.

– $\mathsf{Send}_s(U, m)$: This query models $\mathcal{A}$ sending messages to the smart card and receiving messages from the smart card.
– $\mathsf{Corrupt}_s(U)$: This query models the attacks in which the adversary gets access to the smart card and gets back the player's LL-key. This query models attacks like differential power analysis or other attacks by which the adversary bypasses the tamper detection mechanisms of the smart card [29].

When $\mathcal{A}$ terminates, it outputs a bit b'. We say that $\mathcal{A}$ *wins* the AKE game (see in Section 5) if $b = b'$. Since $\mathcal{A}$ can trivially win with probability 1/2, we define $\mathcal{A}$'s advantage by $\mathsf{Adv}_P^{\mathsf{ake}}(\mathcal{A}) = 2 \times \Pr[b = b'] - 1$.

4 An Authenticated Group Diffie-Hellman Scheme

In this section, we describe the protocol $\mathsf{AKE1}^+$ by splitting it into functions that help us to implement the GDH abstract interface. These functions specify how certain cryptographic transformations have to be performed and abstract out the details of the devices (software or hardware) that will carry out the transformations. In the following we identify the multicast group to the set of indices of players (instances of players) in it. We use a security parameter ℓ and, to make the description easier see a player U_i not involved in the multicast group as if his private exponent x_i were equal to 1.

4.1 Overview

The protocol $\mathsf{AKE1}^+$ consists of the $\mathsf{Setup1}^+$, $\mathsf{Remove1}^+$ and $\mathsf{Join1}^+$ algorithms. As illustrated in Figures 1, 2 and 3, in $\mathsf{AKE1}^+$ the players are arranged in a ring and the instance with the highest-index in the multicast group $\mathcal{I}$ is the group controller $\mathsf{GC}(\mathcal{I})$: $\mathsf{GC}(\mathcal{I}) = \Pi(\mathcal{I}, n) = \Pi^t_{\mathcal{I}_n}$ for some t. This is also a protocol wherein each instance saves the set of values it receives in the down-flow of $\mathsf{Setup1}^+$, $\mathsf{Remove1}^+$ and $\mathsf{Join1}^+$[1].

The session-key space **SK** associated with the protocol $\mathsf{AKE1}^+$ is $\{0,1\}^\ell$ equipped with a uniform distribution. The arithmetic is in a group $\mathbb{G} =< g >$ of prime order q in which the DDH assumption holds. The key generation algorithm $\text{GDH.KeyGen}(1^\ell)$ outputs ElGamal-like LL-keys $LL_i = (s_i, g^{s_i})$.

4.2 Authentication Functions

The authentication mechanism supports the following functions:

- $\text{Auth\_Key\_Derive}(i, j)$. This function derives a secret value K_{ij} between U_i and U_j. In our protocol, $K_{ij} = F_1(g^{s_i s_j})$, where the map F_1 is specified in Section 4.4. (K_{ij} is never exposed.)
- $\text{Auth\_Sig}(i, j, m)$. This function invokes $\text{MAC.Sgn}(K_{ij}, m)$ to obtain tag μ, which is returned.
- $\text{Auth\_Ver}(i, j, m, \mu)$. This function invokes $\text{MAC.Vf}(K_{ij}, m, \mu)$ to check if (m, μ) is correct w.r.t. key K_{ij}. The boolean answer is returned.

The two latter functions should of course be called after initializing K_{ij} via $\text{Auth\_Key\_Derive}(\cdot)$.

4.3 Key-Exchange Functions

The key-exchange mechanism supports the following functions:

- $\text{GDH\_Picks}(i)$. This function generates a new private exponent $x_i \stackrel{R}{\leftarrow} \mathbb{Z}_q^\star$. Recall that x_i is never exposed.
- $\text{GDH\_Picks}^\star(i)$. This function invokes $\text{GDH\_Picks}(i)$ to generate x_i but do not delete the previous private exponent x'_i. x'_i is only deleted when explicitly asked for by the instance.
- $\text{GDH\_Up}(i, j, k, \mathsf{Fl}, \mu)$. First, if $j > 0$, the authenticity of tag μ on message Fl is checked with $\text{Auth\_Ver}(j, i, \mathsf{Fl}, \mu)$. Second, Fl is decoded as a set of intermediate values $(\mathcal{I}, Y, Z)$ where $\mathcal{I}$ is the multicast group and

$$Y = \bigcup_{m \neq i} \{Z^{1/x_m}\} \text{ with } Z = g^{x_t}.$$

[1] In the subsequent removal of players from the multicast group any oracle Π could be selected as the group controller GC and so will need these values to execute $\mathsf{Remove1}^+$.

The values in Y are raised to the power of x_i and then concatenated with Z to obtain these intermediate values

$$Y' = \bigcup \{Z'^{1/x_m}\}, \text{ where } Z' = Z^{x_i} = g^{x_t}.$$

Third, $\mathsf{Fl}' = (\mathcal{I}, Y', Z')$ is authenticated, by invoking $\text{AUTH\_SIG}(i, k, \mathsf{Fl}')$ to obtain tag μ'. The flow (Fl', μ') is returned.

- $\text{GDH\_DOWN}(i, j, \mathsf{Fl}, \mu)$. First, the authenticity of (Fl, μ) is checked, by invoking $\text{AUTH\_VER}(j, i, \mathsf{Fl}, \mu)$. Then the flow Fl' is computed as in GDH_UP, from $\mathsf{Fl} = (\mathcal{I}, Y, Z)$ but without the last element Z' (i.e. $\mathsf{Fl}' = (\mathcal{I}, Y')$). Finally, the flow Fl' is appended tags $\mu_1, \ldots, \mu_n$ by invoking $\text{AUTH\_SIG}(i, k, \mathsf{Fl}')$, where k ranges in $\mathcal{I}$. The tuple $(\mathsf{Fl}', \mu_1, \ldots, \mu_n)$ is returned.
- $\text{GDH\_UP\_AGAIN}(i, k, \mathsf{Fl} = (\mathcal{I}, Y'))$. From Y' and the previous random x'_i, one can recover the associated Z'. In this tuple (Y', Z'), one replaces the occurrences of the old random x'_i by the new one x_i (by raising some elements to the power x_i/x'_i) to obtain Fl'. The latter is authenticated by computing via $\text{AUTH\_SIG}(i, k, \mathsf{Fl}')$ the tag μ. The flow (Fl', μ') is returned. From now the old random x'_i is no longer needed and, thus, can be erased.
- $\text{GDH\_DOWN\_AGAIN}(i, \mathsf{Fl} = (\mathcal{I}, Y'))$. In Y', one replaces the occurrences of the old random x'_i by the new one x_i, to obtain Fl'. This flow is appended tags $\mu_1, \ldots, \mu_n$ by invoking $\text{AUTH\_SIG}(i, k, \mathsf{Fl}')$, where k ranges in $\mathcal{I}$. The tuple $(\mathsf{Fl}', \mu_1, \ldots, \mu_n)$ is returned. From now the old random x'_i is no longer needed and, thus, can be erased.
- $\text{GDH\_KEY}(i, j, \mathsf{Fl}, \mu)$ produces the session key sk. First, the authenticity of (Fl, μ) is checked with $\text{AUTH\_VER}(j, i, \mathsf{Fl}, \mu)$. Second, the value $\alpha = g^{\prod_{j \in \mathcal{I}} x_j}$ is computed from the private exponent x_i, and the corresponding value in Fl. Third, sk is defined to be $F_2(\mathcal{I} \| \mathsf{Fl} \| \alpha)$, where the map $F_2(\cdot)$ is defined below.

4.4 Key Derivation Functions

The key derivation functions F_1 and F_2 are implemented via the so-called "entropy-smoothing" property. We use the left-over-hash lemma to obtain (almost) uniformly distributed values over $\{0,1\}^\ell$.

Lemma 3 (Left-Over-Hash Lemma [17]). *Let $\mathcal{D}_s : \{0,1\}^s$ be a probabilistic space with entropy at least σ. Let e be an integer and $\ell = \sigma - 2e$. Let $h : \{0,1\}^k \times \{0,1\}^s \to \{0,1\}^\ell$ be a universal hash function. Let $r \in_{\mathcal{U}} \{0,1\}^k$, $x \in_{\mathcal{D}_s} \{0,1\}^s$ and $y \in_{\mathcal{U}} \{0,1\}^\ell$. Then the statistical distance δ is:*

$$\delta(h_r(x) \| r, y \| r) \leq 2^{-(e+1)}.$$

Any universal hash function can be used in the above lemma, provided that y is uniformly distributed over $\{0,1\}^\ell$. However, in the security analysis, we need an additional property from h. This property states that the distribution $\{h_r(\alpha)\}_\alpha$ is *computationally undistinguishable* from the uniform one, for any r. Indeed, we need there is no "bad" parameter r, since such a parameter may be chosen by the adversary.

The map $F_1(\cdot)$ is implemented as follows through *public certified random strings*. In a Public-Key Infrastructure (PKI), each player U_i is given $N-1$ random strings $\{r_{ij}\}_{j\neq i}$ each of length k when registering his identity with a Certification Authority (CA). Recall that $N = |\mathcal{U}|$. The random string $r_{ij} = r_{ji}$ is used by U_i and U_j to derivate from input value x a symmetric-key $K_{ij} = F_1(x) = h_{r_{ij}}(x)$.

The map $F_2(\cdot)$ is implemented as follows. First, GDH_DOWN($\cdot$) is enhanced in such a way that it also generates a random value $r_\alpha \in \{0,1\}^k$, which is included in the subsequent broadcast. Then, player U_i derives from input value x a session key $sk = F_2(x) = h_{r_\alpha}(x)$.

One may note that in both cases, the random values are used only once, which gives almost uniformly and independently distributed values, according to the lemma 3.

4.5 Scheme

We correctly deal with concurrent sessions running in an adversary-controlled network by creating a new instance for each player in a multicast group. We in effect create an instance of a player via the algorithm Setup1$^+$ and then create new instances of this player through the algorithms Join1$^+$ and Remove1$^+$.

Setup1$^+$($\mathcal{I}$): This algorithm consists of two stages, up-flow and down-flow (see Figure 1). On the up-flow oracle $\Pi(\mathcal{I}, i)$ invokes GDH_PICKS($\mathcal{I}_i$) to generate its private exponent $x_{\mathcal{I}_i}$ and then invokes GDH_UP($\mathcal{I}_i, \mathcal{I}_{i-1}, \mathcal{I}_{i+1}, \mathsf{Fl}_{i-1}, \mu_{i-1,i}$) to obtain both flow Fl_i and tag $\mu_{i,i+1}$ (by convention, $\mathcal{I}_0 = 0$, $\mathsf{Fl}_0 = \mathcal{I}\|g$ and $\mu_{0,i} = \emptyset$). Then, $\Pi(\mathcal{I}, i)$ forwards $(\mathsf{Fl}_i, \mu_{i,i+1})$ to the next oracle in the ring. The down-flow takes place when GC($\mathcal{I}$) receives the last up-flow. Upon receiving this flow, GC($\mathcal{I}$) invokes GDH_PICKS($\mathcal{I}_n$) and GDH_DOWN($\mathcal{I}_n, \mathcal{I}_{n-1}, \mathsf{Fl}_{n-1}, \mu_{n-1,n}$) to compute both Fl_n and the tags $\mu_1, \ldots, \mu_n$. GC($\mathcal{I}$) broadcasts $(\mathsf{Fl}_n, \mu_1, \ldots, \mu_n)$. Finally, each oracle $\Pi(\mathcal{I}, i)$ invokes GDH_KEY($\mathcal{I}_i, \mathcal{I}_n, \mathsf{Fl}_n, \mu_i$) and gets back the session key $\mathrm{SK}_{\Pi(\mathcal{I},i)}$.

Remove1$^+$($\mathcal{I}, \mathcal{J}$): This algorithm consists of a down-flow only (see Figure 2). The group controller GC($\mathcal{I}$) of the new set $\mathcal{I} = \mathcal{I} \backslash \mathcal{J}$ invokes GDH_PICKS*($\mathcal{I}_n$) to get a *new* private exponent and then GDH_DOWN_AGAIN($\mathcal{I}_n, \mathsf{Fl}'$) where Fl' is the saved previous broadcast. GC($\mathcal{I}$) obtains a new set of intermediate values from which it deletes the elements related to the removed players (in the set $\mathcal{J}$) and updates the multicast group. This produces the new broadcast flow Fl_n. Upon receiving the down-flow, $\Pi(\mathcal{I}, i)$ invokes GDH_KEY($\mathcal{I}_i, \mathcal{I}_n, \mathsf{Fl}_n, \mu_i$) and gets back the session key $\mathrm{SK}_{\Pi(\mathcal{I},i)}$. Here, is the reason why an oracle must store its private exponent and only erase its internal data when it leaves the group.

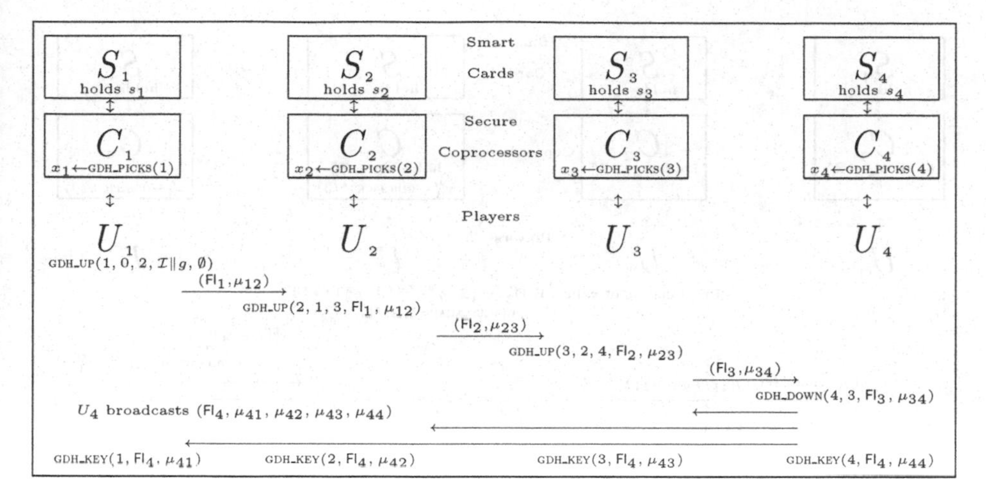

Fig. 1. Algorithm Setup1$^+$. A practical example with 4 players $\mathcal{I} = \{U_1, U_2, U_3, U_4\}$.

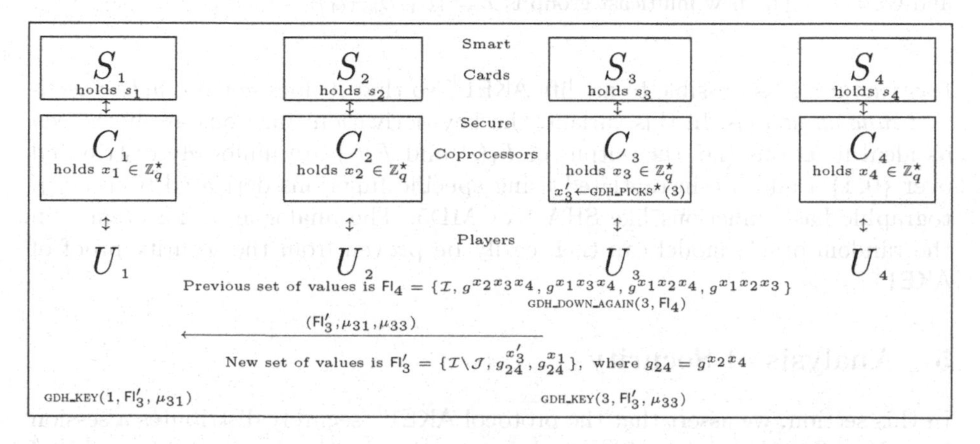

Fig. 2. Algorithm Remove1$^+$. A practical example with 4 players: $\mathcal{I} = \{U_1, U_2, U_3, U_4\}$ and $\mathcal{J} = \{U_2, U_4\}$. The new multicast group is $\mathcal{I} = \{U_1, U_3\}$ and $\mathsf{GC} = U_3$.

Join1$^+$($\mathcal{I}, \mathcal{J}$): This algorithm consists of two stages, up-flow and down-flow (see Figure 3). On the up-flow the group controller $\mathsf{GC}(\mathcal{I})$ invokes GDH_PICKS*($\mathcal{I}_n$), and then GDH_UP_AGAIN($\mathcal{I}_n, j, \mathsf{Fl}'$) where Fl', j are respectively the saved previous broadcast and the index of the first joining player. One updates $\mathcal{I}$, and forwards the result to the first joining player. From that point in the execution, the protocol works as the algorithm Setup1$^+$, where the group controller is the highest index player in $\mathcal{J}$.

4.6 Practical Considerations

When implementors choose a protocol, they take into account its security but also its ease of integration. For a minimal disruption to a current security in-

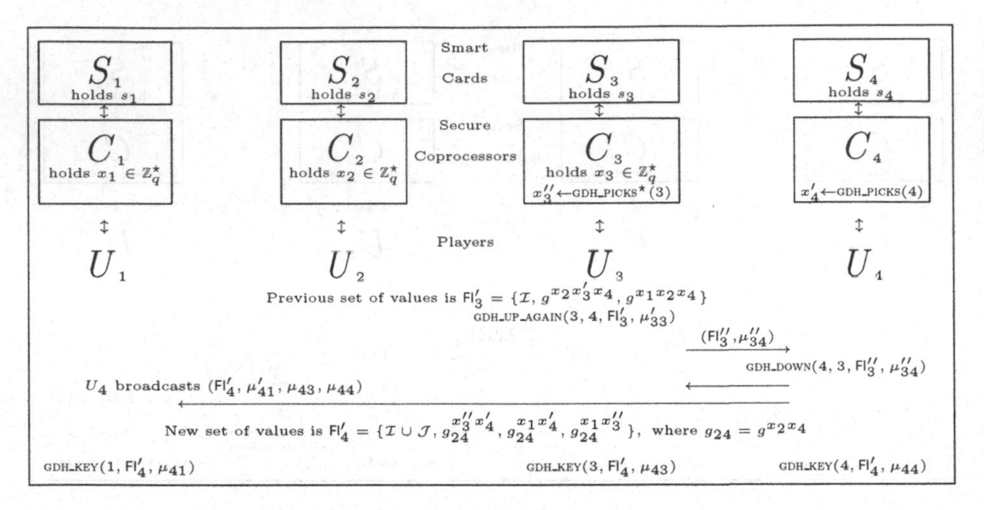

Fig. 3. Algorithm $\mathsf{Join1}^+$. A practical example with 4 players: $\mathcal{I} = \{U_1, U_3\}$, $\mathcal{J} = \{U_4\}$ and $\mathsf{GC} = U_3$. The new multicast group is $\mathcal{I} = \{U_1, U_3, U_4\}$.

frastructure, it is possible to modify $\mathsf{AKE1}^+$ so that it does not use *public certified random strings*. In this variant, the key derivation functions are both seen as ideal functions (i.e. the output of $F_1(\cdot)$ and $F_2(\cdot)$ are uniformly distributed over $\{0,1\}^\ell$) and are instantiated using specific functions derivated from cryptographic hash functions like SHA-1 or MD5. The analogue of Theorem 1 in the random oracle model can then easily be proven from the security proof of $\mathsf{AKE1}^+$.

5 Analysis of Security

In this section, we assert that the protocol $\mathsf{AKE1}^+$ securely distributes a session key. We refine the notion of forward-secrecy to take into account two modes of corruption and use it to define two notions of security. We exhibit a security reduction for $\mathsf{AKE1}^+$ that holds in the standard model.

5.1 Security Notions

Forward-Secrecy. The notion of forward-secrecy entails that the corruption of a (static) LL-key used for authentication does not compromise the semantic security of previously established session keys. However while a corruption may have exposed the static key of a player it may have also exposed the player's internal data. That is either the LL-key or the ephemeral key (private exponent) used for session key establishment is exposed, or both. This in turn leads us to define two modes of corruption: the weak-corruption model and the strong-corruption model.

In the weak-corruption model, a corruption only reveals the LL-key of player U. That is, the adversary has the ability to make $\mathsf{Corrupt}_s$ queries. We then talk about *weak-forward secrecy* and refer to it as wfs. In the strong-corruption model, a corruption will reveal the LL-key of U and additionally all internal data that his instances did not explicitly erase. That is, the adversary has the ability to make $\mathsf{Corrupt}_s$ and $\mathsf{Corrupt}_c$ queries. We then talk about *strong-forward secrecy* and refer to it as fs.

Freshness. As it turns out from the definition of forward-secrecy two flavors of freshness show up. An oracle Π_U^t is wfs-Fresh, in the current execution, (or holds a wfs-Fresh SK) if the following conditions hold. First, no $\mathsf{Corrupt}_s$ query has been made by the adversary since the beginning of the game. Second, in the execution of the current operation, U has accepted and neither U nor his partners has been asked for a Reveal-query.

An oracle Π_U^t is fs-Fresh, in the current execution, (or holds a fs-Fresh SK) if the following conditions hold. First, neither a $\mathsf{Corrupt}_s$-query nor a $\mathsf{Corrupt}_c$-query has been made by the adversary since the beginning of the game. Second, in the execution of the current operation, U has accepted and neither U nor his partners have been asked for a Reveal-query.

AKE Security. In an execution of P, we say an adversary $\mathcal{A}$ *wins* if she asks a single Test-query to a Fresh player U and correctly guesses the bit b used in the game $\mathbf{Game}^{\mathsf{ake}}(\mathcal{A}, P)$. We denote the AKE advantage as $\mathsf{Adv}_P^{\mathsf{ake}}(\mathcal{A})$. Protocol P is an **$\mathcal{A}$-secure AKE** if $\mathsf{Adv}_P^{\mathsf{ake}}(\mathcal{A})$ is negligible.

By notation $\mathsf{Adv}(t, \ldots)$, we mean the maximum values of $\mathsf{Adv}(\mathcal{A})$, over all adversaries $\mathcal{A}$ that expend at most the specified amount of resources (namely time t).

5.2 Security Theorem

A theorem asserting the security of some protocol measures how much computation and interactions helps the adversary. One sees that $\mathsf{AKE1}^+$ is a secure AKE protocol provided that the adversary does not solve the group decisional Diffie-Hellman problem G-DDH, does not solve the multi-decisional Diffie-Hellman problem M-DDH, or forges a Message Authentication Code MAC. These terms can be made negligible by appropriate choice of parameters for the group $\mathbb{G}$. The other terms can also be made "negligible" by an appropriate instantiation of the key derivation functions.

Theorem 1. *Let $\mathcal{A}$ be an adversary against protocol P, running in time T, allowed to make at most Q queries, to any instance oracle. Let n be the number of players involved in the operations which lead to the group on which $\mathcal{A}$ makes the* Test*-query. Then we have:*

$$\mathsf{Adv}_P^{\mathsf{ake}}(\mathcal{A}, q_{se}) \leq 2nQ \cdot \mathsf{Adv}_{\mathbb{G}}^{\mathsf{gddh}_{\Gamma_n}}(T') + 2\mathsf{Adv}_{\mathbb{G}}^{\mathsf{mddh}_n}(T)$$
$$+ n(n-1) \cdot \mathsf{Succ}_{\mathsf{mac}}^{\mathsf{cma}}(T) + n(n-1) \cdot \delta_1 + 2nQ \cdot \delta_2$$

where δ_i denotes the distance between the output of $F_i(\cdot)$ and the uniform distribution over $\{0,1\}^\ell$, $T' \leq T + QnT_{exp}(k)$, where $T_{exp}(k)$ is the time of computation required for an exponentiation modulo a k-bit number, and Γ_n corresponds to the elements adversary $\mathcal{A}$ can possibly view:

$$\Gamma_n = \bigcup_{2\leq j\leq n-2} \{\{i \,|\, 1 \leq i \leq j, i \neq l\} \,|\, 1 \leq l \leq j\}$$
$$\bigcup \{\{i \,|\, 1 \leq i \leq n, i \neq k, l\} \,|\, 1 \leq k, l \leq n\}.$$

Proof. The formal proof of the theorem is omitted due to lack of space and can be found in the full version of this paper [9]. We do, however, provide a sketch of the proof here.

Let the notation $\mathbf{G}_0$ refer to $\mathbf{Game}^{\mathsf{ake}}(\mathcal{A}, P)$. Let b and b' be defined as in Section 3 and S_0 be the event that $b = b'$. We incrementally define a sequence of games starting at $\mathbf{G}_0$ and ending up at $\mathbf{G}_5$. We define in the execution of $\mathbf{G}_{i-1}$ and $\mathbf{G}_i$ a certain "bad" event E_i and show that as long as E_i does not occur the two games are identical [25]. The difficulty is in choosing the "bad" event. We then show that the advantage of $\mathcal{A}$ in breaking the **AKE** security of P can be bounded by the probability that the "bad" events happen. We now define the games $\mathbf{G}_1, \mathbf{G}_2, \mathbf{G}_3, \mathbf{G}_4, \mathbf{G}_5$. Let S_i be the event $b = b'$ in game $\mathbf{G}_i$.

Game $\mathbf{G}_1$ is the same as game $\mathbf{G}_0$ except we abort if a MAC forgery occurs before any Corrupt-query. We define the MAC forgery event by Forge. We then show: $|\Pr[\mathsf{S}_0] - \Pr[\mathsf{S}_1]| \leq \Pr[\mathsf{Forge}]$.

Lemma 4. *Let δ_1 be the distance between the output of the map F_1 and the uniform distribution. Then, we have (proof appears in full version of the paper [9]):*

$$\Pr[\mathsf{Forge}] \leq \mathsf{Adv}_{\mathbb{G}}^{\mathsf{mddh}_n}(T) + \frac{n(n-1)}{2}\mathsf{Succ}_{\mathsf{mac}}^{\mathsf{cma}}(T) + \frac{n(n-1)}{2}\delta_1.$$

Game $\mathbf{G}_2$ is the same as game $\mathbf{G}_1$ except that we add the following rule: we choose at random an index i_0 in $[1, n]$ and an integer c_0 in $[1, Q]$. If the Test-query does not occur at the c_0-th operation, or if the very last broadcast flow before the Test-query is not operated by player i_0, the simulator outputs "`Fail`" and sets b' randomly. Let E_2 be the event that these guesses are not correct. We show: $\Pr[\mathsf{S}_2] = \Pr[\mathsf{E}_2]/2 + \Pr[\mathsf{S}_1](1 - \Pr[\mathsf{E}_2])$, where $\Pr[\mathsf{E}_2] = 1 - 1/nQ$.

Game $\mathbf{G}_3$ is the same as game $\mathbf{G}_2$ except that we modify the way the queries made by $\mathcal{A}$ are answered; the simulator's input is $\mathcal{D}$, a $\mathsf{G\text{-}DH}^\star_{\Gamma_n}$ element, with $g^{x_1 \cdots x_n}$. During the attack, based on the two values i_0 and c_0, the simulator injects terms from the instance such that the Test-ed key is derived from the G-DH-secret value relative to that instance. The simulator appears in the full version of the paper: briefly, the simulator is responsible for embedding (by random self-reducibility) in the protocol the elements of the instance $\mathcal{D}$ so that the Test-ed key is derived from $g^{x_1 \cdots x_n}$. We then show that: $\Pr[\mathsf{S}_2] = \Pr[\mathsf{S}_3]$.

Game $\mathbf{G_4}$ is the same as game $\mathbf{G_3}$ except that the simulator is given as input an element $\mathcal{D}$ from $\mathsf{G\text{-}DH}^{\$}_{\Gamma_n}$, with g^r. And in case $b = 1$, the value random value g^r is used to answer the Test-query. The, the difference between $\mathbf{G_3}$ and $\mathbf{G_4}$ is upper-bounded by the computational distance between the two distributions $\mathsf{G\text{-}DH}^{\star}_{\Gamma_n}$ and $\mathsf{G\text{-}DH}^{\$}_{\Gamma_n}$: $|\Pr[\mathsf{S}_3] - \Pr[\mathsf{S}_4]| \leq \mathsf{Adv}^{\mathsf{gddh}_{\Gamma_n}}_{\mathbb{G}}(T')$, where T' takes into account the running time of the adversary, and the random self-reducibility operations, and thus $T' \leq T + QnT_{exp}(k)$.

Game $\mathbf{G_5}$ is the same as $\mathbf{G_4}$, except that the Test-query is answered with a completely random value, independent of b. It is then straightforward that $\Pr[\mathsf{S}_5] = 1/2$. Let δ_2 be the distance between the output of $F_2(\cdot)$ and the uniform distribution, we have: $|\Pr[\mathsf{S}_5] - \Pr[\mathsf{S}_4]| \leq \delta_2$.

The theorem then follows from putting altogether the above equations. □

6 Conclusion

This paper represents the third tier in the treatment of the group Diffie-Hellman key exchange using public/private keys. The first tier was provided for a scenario wherein the group membership is static [7] and the second, by extension of the latter to support membership changes [8]. This paper adds important attributes (strong-corruption, concurrent executions of the protocol, tighter reduction, standard model) to the group Diffie-Hellman key exchange.

Acknowledgements

The authors thank Deborah Agarwal and Jean-Jacques Quisquater for many insightful discussions and comments on an early draft of this paper. The authors also thank the anonymous referees for their useful comments.

References

1. G. Ateniese, M. Steiner, and G. Tsudik. Authenticated group key agreement and friends. In *ACM CCS '98*, pp. 17–26. 1998.
2. M. Bellare, R. Canetti, and H. Krawczyk. Keying hash functions for message authentication. In *Proc. of Crypto '96*, LNCS 1109, pp. 1–15. Springer, 1996.
3. M. Bellare, D. Pointcheval, and P. Rogaway. Authenticated key exchange secure against dictionary attacks. In *Proc. of Eurocrypt '00*, LNCS 1807, pp. 139–155. Springer, 2000.
4. M. Bellare and P. Rogaway. Random oracles are practical: a paradigm for designing efficient protocols. In *ACM CCS '93*, pp. 62–73. 1993.
5. K. P. Birman. A review experience with reliable multicast. *Software – Practice and Experience*, 29(9):741–774, 1999.
6. D. Boneh. The decision Diffie-Hellman problem. In *Proc. of ANTS III*, LNCS 1423, pp. 48–63. Springer, 1998.

7. E. Bresson, O. Chevassut, D. Pointcheval, and J.-J. Quisquater. Provably authenticated group Diffie-Hellman key exchange. In *ACM CCS '01*, pp. 255–264. 2001.
8. E. Bresson, O. Chevassut, and D. Pointcheval. Provably authenticated group Diffie-Hellman key exchange – the dynamic case. In *Proc. of Asiacrypt '01*, LNCS 2248, pp. 290–309. Springer, 2001.
9. E. Bresson, O. Chevassut, and D. Pointcheval. Dynamic group Diffie-Hellman key exchange under standard assumptions. In *Proc. of Eurocrypt '02*, LNCS. Springer, 2002. Full version of this paper available at `http://www.di.ens.fr/~pointche`.
10. M. Burmester and Y. G. Desmedt. A secure and efficient conference key distribution system. In *Proc. of Eurocrypt '94*, LNCS 950, pp. 275–286. Springer, 1995.
11. G. V. Chockler, I. Keidar, and R. Vitenberg. Group communication specifications: A comprehensive study. *ACM Computing Surveys*, 33(4):1–43, 2001.
12. G. Di Crescenzo, N. Ferguson, R. Impagliazzo, and M. Jakobsson. How to forget a secret. In *Proc. of STACS '99*, LNCS 1563, pp. 500–509. Springer, 1999.
13. W. Diffie and M. E. Hellman. New directions in cryptography. *IEEE Transactions on Information Theory*, IT-22(6):644–654, 1976.
14. W. Diffie, D. Steer, L. Strawczynski, and M. Wiener. A secure audio teleconference system. In *Proc. of Crypto '88*, LNCS 403, pp. 520–528. Springer, 1988.
15. W. Diffie, P. van Oorschot, and W. Wiener. Authentication and authenticated key exchange. In *Designs, Codes and Cryptography*, vol. 2(2), pp. 107–125, 1992.
16. C. G. Gunter. An identity-based key exchange protocol. In *Proc. of Eurocrypt '89*, LNCS 434, pp. 29–37. Springer, 1989.
17. J. Håstad, R. Impagliazzo, L. Levin, and M. Luby. A pseudorandom generator from any one-way function. *SIAM Journal of Computing*, 28(4):1364–1396, 1999.
18. M. Joye and J.-J. Quisquater. On the importance of securing your bins: The garbage-man-in-the-middle attack. In *ACM CCS'97*, pp. 135–141. 1997.
19. M. Just and S. Vaudenay. Authenticated multi-party key agreement. In *Proc. of Asiacrypt '96*, LNCS 1163, pp. 36–49. Springer, 1996.
20. M. Naor and O. Reingold. Number-theoretic constructions of efficient pseudo-random functions. In *FOCS '97*, pp. 458–467. IEEE, 1997.
21. NIST. *FIPS 140-1: Security Requirements for Cryptographic Modules*. U. S. National Institute of Standards and Technology, 1994.
22. E. R. Palmer, S. W. Smith, and S. Weingart. Using a high-performance, programmable secure coprocessor. In *Financial Crypto '98*, LNCS 1465, pp. 73–89. Springer, 1998.
23. A. Rubin and V. Shoup. Session-key disribution using smart cards. In *Proc. of Eurocrypt '96*, LNCS 1070, pp. 321–331. Springer, 1996.
24. V. Shoup. On formal models for secure key exchange. Technical Report RZ 3120, IBM Zürich Research Lab, 1999.
25. V. Shoup. OAEP reconsidered. In J. Kilian, editor, *Proc. of Cryto' 01*, volume 2139 of *LNCS*, pages 239–259. Springer-Verlag, 2001.
26. M. Steiner, G. Tsudik, and M. Waidner. Diffie-Hellman key distribution extended to group communication. In *ACM CCS '96*, pp. 31–37. 1996.
27. W. G. Tzeng. A practical and secure fault-tolerant conference-key agreement protocol. In *Proc. of PKC '00*, LNCS 1751, pp. 1–13. Springer, 2000.
28. K. Vedder and F. Weikmann. Smart cards requirements, properties, and applications. In *State of the Art in Applied Cryptography*, LNCS 1528. Springer, 1997.
29. S. H. Weingart. Physical security devices for computer subsystems: A survey of attacks and defenses. In *Proc. of CHES '00*, LNCS 1965, pp. 302–317. Springer, 2000.

Universally Composable Notions of Key Exchange and Secure Channels*
(Extended Abstract)

Ran Canetti[1] and Hugo Krawczyk[2],**

[1] IBM T.J. Watson Research Center,
canetti@watson.ibm.com.
[2] EE Department, Technion,
hugo@ee.technion.ac.il

Abstract. Recently, Canetti and Krawczyk (Eurocrypt'2001) formulated a notion of security for key-exchange (KE) protocols, called SK-security, and showed that this notion suffices for constructing secure channels. However, their model and proofs do not suffice for proving more general composability properties of SK-secure KE protocols.
We show that while the notion of SK-security is strictly weaker than a fully-idealized notion of key exchange security, it is sufficiently robust for providing secure composition with arbitrary protocols. In particular, SK-security guarantees the security of the key for *any application* that desires to set-up secret keys between pairs of parties. We also provide new definitions of secure-channels protocols with similarly strong composability properties, and show that SK-security suffices for obtaining these definitions.
To obtain these results we use the recently proposed framework of "universally composable (UC) security." We also use a new tool, called "non-information oracles," which will probably find applications beyond the present case. These tools allow us to bridge between seemingly limited indistinguishability-based definitions such as SK-security and more powerful, simulation-based definitions, such as UC security, where general composition theorems can be proven. Furthermore, based on such composition theorems we reduce the analysis of a full-fledged multi-session key-exchange protocol to the (simpler) analysis of individual, stand-alone, key-exchange sessions.

Keywords: Key Exchange, Cryptographic Protocols, Proofs of Security, Composition of protocols.

1 Introduction

Authenticated Key-Exchange protocols (KE, for short) are protocols by which two parties that communicate over an adversarially controlled network can generate a common secret key. These protocols are essential for enabling the use of

* Full version appears in [CK02].
** Supported by Irwin and Bethea Green & Detroit Chapter Career Development Chair.

L.R. Knudsen (Ed.): EUROCRYPT 2002, LNCS 2332, pp. 337–351, 2002.

shared-key cryptography to protect transmitted data. As such they are a central piece for building secure communications, and are perhaps the most commonly used cryptographic protocols. (Popular examples include SSH, SSL, IPSec. Many others exist.)

Capturing the security requirements from a key exchange protocol has proven to be non-trivial. On the one hand, a definition should be strong enough to guarantee the desired functionality within the protocol settings under consideration. On the other hand, it should not be overly strong and should not impose unnecessary restrictions on key-exchange protocols. Moreover, it should be simple and easy to work with as much as possible.

Numerous works studying the cryptographic security for key-exchange protocols have been carried out in the past, and some quite different definitional approaches were proposed. A very partial list includes [B+91, DOW92, BR93, BJM97, BCK98, S99, CK01,] See [MOV96, Chapter 12] for more background information. Most recently, Canetti and Krawczyk [CK01], building on several prior works (most notably, [BR93,BCK98]) have proposed a definition that has several attractive properties. It is simple and permissive, and yet it was shown to suffice for the quintessential application of key-exchange, namely providing keys to symmetric encryption and authentication algorithms in order to obtain secure communication channels. In other words, the [CK01] notion of security, called SK-security, guarantees that *composing* a key exchange protocol with symmetric encryption and authentication suffices for the specific purpose of providing secure channels.

This specific composability property of SK-security is indeed an important one. However, we would like to be able to guarantee more general composability properties of key-exchange protocols. Specifically, we would like to be able to guarantee that a key exchange protocol remains secure for *any* application protocol that may wish to set-up secret keys between pairs of parties, and even when the key-exchange protocols runs concurrently with an arbitrary set of other protocols. In addition, we would like to have definitions of the task of providing secure channels with similar composability properties.

In order to provide such strong composability properties one needs a general framework for representing and arguing about arbitrary protocols. We use the recently proposed framework of [C01]. This framework allows formulating definitions of security of practically any cryptographic task. Furthermore, it is shown that protocols proven secure in this framework maintain their security under a very general composition operation, called universal composition. Following [C01], we refer to notions of security formulated in this framework as universally composable (UC).

Our main result is a universally composable notion of security for key exchange protocols that is *equivalent* to SK-security. This allows us to combine the relative simplicity and permissiveness of SK-security with the strong composability properties of the UC framework. We also provide a UC definition of secure channels and demonstrate that our notion of UC key exchange suffices for realizing UC secure channels.

An additional advantage of the new definitions is that they treat key-exchange and secure-channel protocols as protocols for a *single session* between two parties (i.e., a single exchange of a key, or a single pairwise communication session). In contrast, previous works treated such protocol as multi-party protocols where a single instance of the protocol handles multiple pairwise sessions in a multi-party network. On top of being conceptually simpler, the single-session treatment simplifies the analysis of protocols.

Obtaining these definitions and especially proving equivalence with SK-security proves to be non-trivial, and required some new definitional techniques that may well be useful elsewhere. Let us elaborate.

Bridging two approaches to defining security. The definition of SK-security follows a definitional approach which is often called "security by indistinguishability". This approach proceeds roughly as follows. In order to define the security of some cryptographic task, formulate two *games*, G_0 and G_1, in which an adversary interacts with the protocol under consideration. The protocol is considered secure if no feasible adversary can distinguish between the case where it interacts in game G_1 and the case where it interacts in game G_2. This definitional approach was first used to define semantic security of encryption schemes [GM84] and was used in many definitions since. It was first applied to the context of key-exchange protocols in [BR93].

In contrast, definitions in the UC framework follow a different definitional approach, which is often referred to as "security by emulation of an ideal process". This approach proceeds roughly as follows. In order to define the security of some cryptographic task, formulate an "ideal process" that captures the desired functionality of the task. Typically, this ideal process involves adding an idealized "trusted party" to the system and having the trusted party compute the desired output for the parties. A protocol is considered secure if for any adversary that attacks the protocol there exists another adversary (a "simulator") that causes essentially the same effect on the system by interacting with the ideal process for the task. This implies that the real protocol is essentially as secure as the ideal process. This approach is used to formulate an alternative (and equivalent) notion of semantic security of encryption schemes (see e.g., [G01]). In addition, it is used for capturing security of tasks such as zero-knowledge [GMRa89] and general cryptographic protocols (e.g., [GL90,MR91,B91,C00,PSW00,DM00,C01]).

Typical advantages of definitions that take the first approach are relative simplicity and permissiveness. On the other hand, definitions that take the second approach usually appear to capture the security requirements in a more "convincing" way. More importantly, the second approach (and in particular the UC framework) enables demonstrating the general "secure composability" properties discussed above.

One case where definitions that follow the two approaches were shown to be equivalent is the case of semantically secure encryption against chosen plaintext attacks [GM84,G01]. However, in most other cases the two approaches seem to result in distinct notions of security, where the emulation approach typically results in a strictly more restrictive definition. One example is the case of Zero-

Knowledge versus Witness-Indistinguishable protocols (see, e.g., [FS90,G01]). Another example is key exchange protocols: There are protocols that are SK-secure but do not satisfy the emulation-based definitions of [BCK98,S99]. Interestingly, these include natural protocols that do not exhibit any exploitable security weaknesses. A quintessential example is the original two-round Diffie-Hellman protocol in the case of ideally authenticated communication.

Indeed, an initial attempt at formalizing a UC definition of secure key-exchange protocols results in a definition that is even more restrictive than [BCK98,S99], and thus strictly more restrictive than SK-security. We thus *relax* the UC notion so that it becomes *equivalent* to SK-security. To do that, we modify the "ideal process" to provide the simulator with additional information that the simulator can use to complete its simulation. This information, which we call a "non-information oracle", has the property that it is "computationally independent" from the exchanged key and therefore does not over-weaken the notion of security. In fact, we show that the resultant, relaxed definition is *equivalent* to SK-security. See the overview section (Section 2.4) for a sketch of our relaxation technique via non-information oracles.

On defining and realizing secure channels. Another contribution of this work is a universally composable definition of secure channels. This provides a notion of secure channels with strong composability guarantees. As in the case of key-exchange protocols, we first formulate the intuitively-immediate version of UC secure channels. We demonstrate that SK-secure key-exchange can be combined with any message authentication function and a certain class of encryption mechanisms in order to realize secure channels. This provides further assurance in the adequacy of the notion of SK-security.

However, it turns out that some natural encryption mechanisms result in protocols that do *not* realize this notion of UC secure channels. Here we encounter a similar problem as in the case of UC-secure key exchange: Some of these "insecure" protocols do not seem to have any exploitable security weakness. We remark that here the problem stems from a different source, namely the use of (symmetric) encryption in a setting where the adversary adaptively corrupts parties. In particular, the problem persists even if we use strong UC key exchange protocols. We formulate a relaxed version of UC secure channels (called weak UC secure-channels) based on a variant of non-information oracles. We demonstrate that SK-secure key-exchange, combined with any encryption scheme (that is semantically secure against chosen plaintext attacks) and any message authentication function, results in a weak UC secure-channels protocol.

Organization. Section 2 provides an overview of the definition of SK-security, the UC framework, and our results. For lack of space, we do not include a detailed technical exposition of our results in this extended abstract. For full details, the reader is referred to [CK02].

2 Overview

Sections 2.1 and 2.2 provide some background on the definitions of [CK01] and the UC framework, respectively. Later sections present our contributions. In Section 2.3 we describe the methodology for reducing the analysis of a multi-session protocol to the analysis of a simplified single-session protocol. Section 2.4 overviews our results regarding universally composable key exchange protocols. Finally, Section 2.5 overviews our results regarding the definition and construction of UC secure channels. Throughout, the presentation remains high-level and ignores many important details, such as the syntax of key-exchange protocols, session-id's, and many others. Full details are presented in [CK02].

2.1 On the Definitions of [CK01]

We very briefly sketch the [CK01] definition of SK-security and its applicability to secure channels (refer to that paper for a precise description). Following [BCK98], two models of computation are first defined: the unauthenticated-links model (UM) and the authenticated-links model (AM). In both models the communication is public, asynchronous, and without guaranteed delivery of messages. In both models the protocols are message-driven (i.e., a party is activated with an incoming message, performs some internal computation, possibly generates outgoing messages, and then waits for the next activation). Furthermore, in both models the adversary may adaptively corrupt parties, or individual key-exchange sessions within parties, and obtain their internal states. In the UM the adversary can arbitrarily modify messages before delivering them. In the AM the adversary can deliver only messages that were sent by parties, and must deliver them unmodified.

Key-exchange protocols are treated as multiparty protocols where multiple pairwise exchanges are handled within a single instance of the protocol. That is, each instance of a key exchange protocol (running within some party) consists of multiple KE-sessions, where each KE-session is an invocation of some subroutine which handles a single exchange of a key with a given peer and a given session ID. To define SK-security, the following game between an adversary and parties running the protocol is formulated (following [BR93]). The adversary is allowed to invoke multiple KE-sessions within parties to exchange keys with each other. It can then deliver messages and corrupt parties (and expose KE-sessions within parties) as in the UM (resp., AM). When the adversary chooses to, it announces a specific KE-session to be the "test session". Once the session completes, a random bit b is chosen and the adversary receives a "test value." If $b = 0$ then the test value is the key generated in this session. If $b = 1$ then the test value is a random value. The adversary can then continue the usual interaction except that it is not allowed to expose the test session or the "matching" session held by the test session's partner. At the end of its run the adversary outputs a guess (a single bit). The adversary wins if it manages to guess the value of b.

A key exchange protocol is secure in the UM (resp., AM) if no adversary can cause two partners of an exchange to output different values of the session key, and in addition no adversary can win in the above game with probability non-negligibly more than one half.

A secure channels protocol is defined to be a protocol which is a secure network authenticator and also a secure network encryptor. The definition of secure network authenticators follows that of [BCK98]: A protocol α is a secure network authenticator if any protocol π in the AM, when composed with protocol α (i.e., when each sending of a message by π is replaced with an activation of protocol α) results in a composed protocol that has essentially the same functionality as π — but in the UM.

The definition of network encryptors follows the style of definition by indistinguishability. That is, first a game between an adversary and parties running the protocol is formulated. The game captures the requirement that the adversary should be unable to distinguish between encryptions of two adversarially chosen test-messages in some session, even after the adversary sees encryptions of messages of its choice and decryptions of ciphertexts of its choice (except for decryptions that result in the test-message itself). A network encryptor is deemed secure if no adversary can win the game with non-negligible probability.

Consider the following generic protocol for realizing secure channels, given a key-exchange protocol, an encryption scheme and a message authentication function: In order to set-up a secure channel, the two partners first run a key-exchange protocol and obtain a key. Then, the sender encrypts each message and then sends the ciphertext together with a tag computed by applying the message authentication function to the ciphertext. Encryption and authentication are done using different portions of the obtained key (or via keys derived from the exchanged session key). Verification and decryption are done analogously. It is shown that if the key exchange protocol is secure, the encryption scheme is semantically secure against chosen plaintext attacks, and the message authentication function is secure, then this protocol is a secure secure-channels protocol. (A counter-based mechanism is added to avoid replay of messages.)

2.2 On Universally Composable Definitions

Providing meaningful security guarantees under composition with arbitrary protocols requires using an appropriate framework for representing and arguing about such protocols. Our treatment is based in a recently proposed such general framework [C01]. This framework builds on known definitions for function evaluation and general tasks [GL90,MR91,B91,C00,DM00,PSW00], and allows defining the security properties of practically any cryptographic task. Most importantly, in this framework security of protocols is maintained under a general composition operation with an unbounded number of copies of arbitrary protocols running concurrently in the system. This composition operation is called universal composition. Similarly, definitions of security in this framework are called universally composable (UC). We briefly summarize the relevant properties of this framework. See more details in [C01,CK02].

As in other general definitions, the security requirements of a given task (i.e., the functionality expected from a protocol that carries out the task) are captured via a set of instructions for a "trusted party" that obtains the inputs of the participants and provides them with the desired outputs (in one or more iterations). Informally, a protocol securely carries out a given task if running the protocol amounts to "emulating" an ideal process where the parties hand their inputs to a trusted party with the appropriate functionality and obtain their outputs from it, without any other interaction. We call the algorithm run by the trusted party an ideal functionality.

In order to allow proving the composition theorem, the notion of emulation in this framework is considerably stronger than previous ones. Traditionally, the model of computation includes the parties running the protocol and an adversary, $\mathcal{A}$. "Emulating an ideal process" means that for any adversary $\mathcal{A}$ there should exist an "ideal process adversary" (or, simulator) $\mathcal{S}$ that results in similar distribution on the outputs for the parties. Here an additional adversarial entity, called the environment $\mathcal{Z}$, is introduced. The environment generates the inputs to all parties, reads all outputs, and in addition interacts with the adversary in an arbitrary way throughout the computation. (Arbitrary interaction between $\mathcal{Z}$ and $\mathcal{A}$ is essential for proving the universal composition theorem.) A protocol is said to securely realize a given ideal functionality $\mathcal{F}$ if for any adversary $\mathcal{A}$ there exists an "ideal-process adversary" $\mathcal{S}$, such that *no environment* $\mathcal{Z}$ can tell whether it is interacting with $\mathcal{A}$ and parties running the protocol, or with $\mathcal{S}$ and parties that interact with $\mathcal{F}$ in the ideal process. (In a sense, here $\mathcal{Z}$ serves as an "interactive distinguisher" between a run of the protocol and the ideal process with access to $\mathcal{F}$. See [c01] for more motivating discussion on the role of the environment.)

The following universal composition theorem is proven in [c01]. Consider a protocol π that operates in a *hybrid* model of computation where parties can communicate as usual, and in addition have ideal access to an unbounded number of copies of some ideal functionality $\mathcal{F}$. (This model is called the $\mathcal{F}$-hybrid model.) Let ρ be a protocol that securely realizes $\mathcal{F}$ as sketched above, and let π^ρ be the "composed protocol". That is, π^ρ is identical to π with the exception that each interaction with some copy of $\mathcal{F}$ is replaced with a call to (or an invocation of) an appropriate instance of ρ. Similarly, ρ-outputs are treated as values provided by the appropriate copy of $\mathcal{F}$. Then, π and π^ρ have essentially the same input/output behavior. In particular, if π securely realizes some ideal functionality $\mathcal{G}$ given ideal access to $\mathcal{F}$ then π^ρ securely realizes $\mathcal{G}$ from scratch.

We also make use of an additional composition operation, called universal composition with joint state (JUC), proposed in [cr02]. This operation is similar to universal composition, with the important difference that multiple instances of the subroutine protocol, ρ, may have some amount of joint state. (In contrast, if universal composition is used then each instance of ρ has its own separate local state.) This becomes useful in the case of key-exchange protocols where multiple protocol instances (sessions) have access to the same long-term authentication module (realized, for instance, via a signature scheme that uses the same signature key for authenticating multiple sessions of the key-exchange protocol run under a party; in this case the signature key represents a joint state).

Extensions to the UC model. As a preliminary step for our study, we cast the unauthenticated-links model (UM) and the authenticated-links model (AM) of [CK01] in the UC framework. This is done by casting these models as "hybrid models" with access to the appropriate ideal functionality. In both cases the ideal functionality is $\mathcal{F}_{auth}$ from [C01], which allows an ideally authenticated transmission of a single message. In the AM the parties have access to an unlimited number of copies of $\mathcal{F}_{auth}$. In the UM each party can send only a single message to each other party using $\mathcal{F}_{auth}$. We also extend the UC framework to accommodate the session-state corruption operation of [CK01] that allows the adversary to obtain the internal data of individual sessions within parties.

2.3 Single-Session vs. Multi-session Protocols

In contrast to previous works, we treat key exchange and secure channel protocols as protocols where each instance handles a single pairwise session (i.e., a single exchange of a key or a single pairwise communication session). This results in greater conceptual and analytical simplicity. However, it requires taking care of the following two issues.

Multi-session extensions. In order to be able to compare the definitions here to the definitions of [CK01], we define the multi-session extension of a (single session) key exchange protocol π to be the protocol $\hat{\pi}$ that handles multiple exchanges of a key, where each exchange consists of running an instance of the original protocol π. The multi-session extension of a (single session) secure session protocol is defined analogously. This way, we are able to state and prove results of the sort "A single-session protocol π is secure according to some UC definition if and only if the multi-session extension $\hat{\pi}$ is secure according to [CK01]".

The long-term authentication module. In typical key exchange protocols multiple pairwise sessions use the same instance of a long-term authentication mechanism. (For instance, this mechanism may be a long-term shared key between parties, or a public-key infrastructure based either on digital signatures or asymmetric encryption.) Thus, pairwise key-exchange and secure channels sessions are not completely disjoint from each other. Still, the "main bulk" of the state of each such pairwise session is disjoint from all other sessions and can be treated separately. In order to do that, we proceed as follows.

First, we restrict attention to (single session) key exchange protocols that have an explicitly specified long-term authentication module. This module represents the part of the key-exchange protocol that handles the long-lived information used to bind each generated key to an identity of a party in the network. Typically, this part consists of a standard cryptographic primitive with a well-defined interface. Next, we analyze key exchange protocols under the assumption that the functionality of the long-term authentication module is ideally provided. (That is, we work in a hybrid model with access to the appropriate ideal functionality.) This in particular means that in a setting where multiple instances of a key-exchange protocols are being used, each such instance uses its own separate copy of the idealized long-term authentications module. We then use universal composition with joint state (see above) to replace all copies of the idealized long-term authentications module *single instance* of a protocol that realizes this module.

For concreteness, we further specify the functionality of the long-term authentication module, when based on digital signatures, and describe the use of universal composition with joint state for this case. (Here we basically use the modeling and results of [CR02].) Similar treatment can be done for other types of long-term authentication. We stress, however, that the results in this work are general and apply regardless of the specific long-term authentication module in use.

2.4 UC Key-Exchange

In order to establish the relationship between the notion of SK-security and UC notions, we first rephrase the definition of SK-security in the UC framework. This is done as follows. We formulate a specific environment machine, $\mathcal{Z}_{test}$, which carries out the game of the definition of SK-security. That is, $\mathcal{Z}_{test}$ expects to interact with a protocol $\hat{\pi}$ which is the multi-session extension of some key-exchange protocol π. Whenever the adversary $\mathcal{A}$ asks $\mathcal{Z}_{test}$ to invoke a session within some party to exchange a key with another party, $\mathcal{Z}_{test}$ does so. When the adversary asks to obtain the session key generated in some session, $\mathcal{Z}_{test}$ reveals the key to $\mathcal{A}$. When $\mathcal{A}$ announces a test session, $\mathcal{Z}_{test}$ flips a coin b. If $b = 0$ then $\mathcal{Z}_{test}$ hands $\mathcal{A}$ the real session key of that session. If $b = 1$ then $\mathcal{A}$ gets a random value. $\mathcal{Z}_{test}$ outputs 1 if $\mathcal{A}$ managed to guess b. (If in any session the partners output different values for the key then $\mathcal{Z}_{test}$ lets $\mathcal{A}$ determine the output.) A (single session) protocol π is said to be SK-secure if no adversary $\mathcal{A}$ can skew the output of $\mathcal{Z}_{test}$ non-negligibly away from fifty-fifty, when interacting with $\hat{\pi}$, the multi-session extension of π. In all, $\mathcal{Z}_{test}$ is designed so that this formulation of SK-security remains only a rewording of the formulation in [CK01]. We later refer to this notion of security as security by indistinguishability.

We then turn to defining UC-secure key exchange. This is done by formulating an ideal functionality that captures the security requirements from a single exchange of a key between a pair of parties. We first formulate a functionality, $\mathcal{F}_{ke}$, that simply waits to receive requests from two uncorrupted parties to exchange a key with each other, and then privately sends a randomly chosen value to both parties, and halts. (If one of the partners to an exchange is corrupted then it gets to determine the value of the key.) We first show that known protocols satisfy this notion:

Theorem 1. *The ISO 9798-3 Diffie-Hellman key exchange protocol authenticated via digital signatures (see* [CK01]*) securely realizes functionality $\mathcal{F}_{ke}$ in the* UM*, under the Decisional Diffie-Hellman assumption, and assuming security of the signature scheme in use.*

Next we show that any protocol that securely realizes $\mathcal{F}_{ke}$ is secure by indistinguishability:

Theorem 2. *Any protocol that securely realizes $\mathcal{F}_{ke}$ is secure by indistinguishability. This holds both in the* UM *and in the* AM*, and both with and without forward secrecy.*

The converse, however, is not true. Specifically, we show that, surprisingly, the "classic" two-move Diffie-Hellman protocol does not securely realize $\mathcal{F}_{ke}$ in the AM, whereas this protocol is secure by indistinguishability in the AM. (Other examples are also given.) Moreover, the proof of "insecurity" of this protocol does not point out to any exploitable security weakness of this protocol. Rather, it seems to point to a technical "loophole" in the UC definition. Specifically, the problem arises when a party gets corrupted, and the real-life adversary expects to see an internal state of the party. This information needs to match other information, such as the value of the session key and the messages sent by the party in the past. Mimicking such an activity in the ideal process is problematic, since the simulator needs to "commit" to messages sent by the party before knowing the value of the key, which is randomly chosen (by $\mathcal{F}_{ke}$) only later.

With the above discussion in mind, we relax the ideal key exchange functionality as follows. We define a special type of probabilistic interactive Turing machine, called a non-information oracle. Essentially, a non-information oracle has the property that its local output is computationally independent from its communication with the outside world. Now, when the functionality is asked to hand a key to a pair of uncorrupted parties, it invokes the non-information oracle $\mathcal{N}$, and lets $\mathcal{N}$ interact with the simulator. The key provided to the parties is set to be the local output of $\mathcal{N}$. When the adversary corrupts one of the partners to the session, it is given the internal state of $\mathcal{N}$.

On the one hand, we are guaranteed that the additional information provided to the simulator (i.e., to the adversary in the ideal process) does not compromise the security of the session key as long as both partners of the session remain uncorrupted. (This follows from the fact that $\mathcal{N}$ is a non-information oracle.) On the other hand, when the simulator corrupts a partner, it obtains some additional information (the internal state of $\mathcal{N}$). With an adequate choice of $\mathcal{N}$, this information allows the simulator to complete its task (which is to mimic the behavior of the real-life adversary, $\mathcal{A}$).

We call the relaxed ideal key-exchange functionality, parameterized by a non-information oracle $\mathcal{N}$, $\mathcal{F}_{wke}^{\mathcal{N}}$. A protocol π is called weakly UC secure if there exists a non-information oracle $\mathcal{N}$ such that π securely realizes $\mathcal{F}_{wke}^{\mathcal{N}}$.

Let us exemplify the use of non-information oracles by sketching a proof for the security of the classic two-move Diffie-Hellman protocol. (Let 2DH *denote this protocol.) Assume that a prime p and a generator g of a large subgroup of Z_p^* of prime order are given. Recall that the protocol instructs the initiator I to choose $x \stackrel{R}{\leftarrow} Z_q$ and send $\alpha = g^x$ to the responder R, who chooses $y \stackrel{R}{\leftarrow} Z_q$ and sends $\beta = g^y$ to I. Both parties locally output g^{xy} (and erase x and y if forward secrecy is desired). Simulating this interaction with access to $\mathcal{F}_{ke}$ (which only chooses a random session key and gives it to the parties) is not possible. Let us informally reason why this is the case. The simulator has to first come up with values α' and β' as the messages sent by the parties. Next, when, say, I gets* *corrupted before receiving R's message, the simulator learns the random value k that $\mathcal{F}_{ke}$ chose to be the session key, and has to come up with a value x' such that $g^{x'} = \alpha$ and $\beta'^{x'} = k$. However, since α', β' were chosen independently of k, such a value x' exists only with negligible probability $1/q$.*

To solve this problem we define the following non-information oracle, $\mathcal{N}$. Upon receiving p, q, g as defined above, $\mathcal{N}$ chooses $x, y \stackrel{R}{\leftarrow} Z_q$, sends out a message containing $\alpha = g^x, \beta = g^y$, locally outputs $k = g^{xy}$ and halts. It is easy to see that, under the Decisional Diffie-Hellman assumption, the local output of $\mathcal{N}$ is computationally indistinguishable from random even given the communication of $\mathcal{N}$ with the outside world. Now, having access to $\mathcal{F}^{\mathcal{N}}_{wke}$, we can simulate protocol 2DH using the following strategy. Recall that, in order to provide I and R with a session key in the ideal process, functionality $\mathcal{F}^{\mathcal{N}}_{wke}$ first runs $\mathcal{N}$, lets the simulator obtain the messages sent by $\mathcal{N}$, and sets the session key to be the local output of $\mathcal{N}$. In our case, this means that the simulator obtains $\alpha = g^x, \beta = g^y$, while the session key is set to $k = g^{xy}$. Therefore, all the simulator has to do is say that the messages sent by I and R are α and β, respectively. Now, if either I or R is corrupted, the simulator receives from $\mathcal{F}^{\mathcal{N}}_{wke}$ the internal state of $\mathcal{N}$, which contains x, y.

We show that a key-exchange protocol is secure by indistinguishability *if and only if* it is weakly UC secure:

Theorem 3. *A key-exchange protocol π is secure by indistinguishability if and only if there exists a non-information oracle $\mathcal{N}$ such that π securely realizes $\mathcal{F}^{\mathcal{N}}_{wke}$. This holds both in the* UM *and in the* AM*, and both with and without forward secrecy.*

Theorem 3 provides a *characterization* of the composability properties of security by indistinguishability: Using a key exchange protocol that is secure by indistinguishability is essentially the same as using an ideal key-exchange functionality that provides the adversary with some randomized information that is computationally independent from the exchanged key.

2.5 UC Secure Channels

The main application of key-exchange protocols is for providing secure channels. Indeed, [CK01] provide a definition of secure-channels protocols, and demonstrate that SK-secure key exchange suffices for realizing their definition of secure channels. (See more details in Section 2.1.)

However, the secure-channels notion of [CK01] does not provide any secure composability guarantees. For example, there is no guarantee that a secure-channels protocol remains secure when used within general "application protocols" that assume "idealized secure channels" between pairs of parties.

We formulate universally composable notions of secure channels. Such notions carry strong composability guarantees with any application protocol and with any number of other protocols that may be running concurrently in the system. In addition, in contrast with [CK01], here we treat a secure channel protocol as a single session protocol (i.e., a protocol that handles only a single communication session between two parties). The extension to the multi-session case is obtained via the general composition theorems.

Formulating a UC notion of secure channels requires formulating an ideal functionality that captures the security requirements from a secure channels protocol. We first formulate an ideal functionality, $\mathcal{F}_{sc}$, that captures these requirements in a straightforward way: Upon receiving a request by two peers to establish a secure channel between them, functionality $\mathcal{F}_{sc}$ lets the adversary know that the channel was established. From now on, whenever one of the peers asks $\mathcal{F}_{sc}$ to deliver a message m to the other peer, $\mathcal{F}_{sc}$ privately sends m to the other peer, and lets the adversary know that a message of $|m|$ bits was sent on the channel. As soon as one of the parties requests to terminate the channel, $\mathcal{F}_{sc}$ no longer transmits information on the channel. A protocol that securely realizes the functionality $\mathcal{F}_{sc}$ is called a UC secure channels protocol.

We wish to show that any weak UC key-exchange protocol suffices to build UC secure channels. More specifically, recall the generic protocol for realizing secure channels, given a key-exchange protocol, an encryption scheme and a message authentication function: In order to set-up a secure channel, the two partners first run a key-exchange protocol and obtain a key. Then, the sender encrypts each message and then sends the ciphertext together with a tag computed by applying the message authentication function to the ciphertext. Encryption and authentication are done using different portions of the obtained key. Verification and decryption are done analogously. We want to show that if the key exchange protocol is weak UC secure, the encryption scheme is semantically secure against chosen plaintext attacks, and the message authentication function is secure against chosen message attacks, then this protocol constitutes a UC secure channels protocol (i.e., it securely realizes $\mathcal{F}_{sc}$). We prove this result for a special case where the encryption function is of a certain form. That is:

Theorem 4. *Let* MAC *be a secure Message Authentication Code function, and let* π *be a weakly UC secure key exchange protocol. Then there exist symmetric encryption schemes* ENC *such that the above sketched protocol, based on* π, MAC, *and* ENC, *securely realizes* $\mathcal{F}_{sc}$ *in the* UM.

Unfortunately, this statement is not true for any semantically secure symmetric encryption scheme. There exist natural encryption protocols that are semantically secure and where the resulting protocol does *not* securely realize $\mathcal{F}_{sc}$, regardless of which message authentication function and which key-exchange protocol are in use. (In fact, most practical encryption protocols are such. This holds even if the key-exchange protocol is a strong UC secure one, i.e. if it securely realizes $\mathcal{F}_{ke}$.) As in the case of key-exchange protocols, some of the protocols that fail to realize $\mathcal{F}_{sc}$ do not seem to have any exploitable security weakness. Rather, the failure to realize $\mathcal{F}_{sc}$ stems from a technical "loophole" in the definition. As there, the problem arises when the real-life adversary adaptively corrupts a party or a session and wishes to see the plaintexts that correspond to the previously transmitted ciphertexts. As there, mimicking such behavior in the ideal process is problematic since the simulator (i.e., the ideal process adversary) has to "commit" to the ciphertext before knowing either the plaintext or the decryption key.

We thus proceed to formulate a relaxed version of the secure channels functionality. Also here we let the relaxed functionality use a non-information oracle in order to provide the simulator with "randomized information on the plaintexts" at the time when these are secretly transmitted to its recipient. More specifically, if the two partners of the secure channel are uncorrupted, then the relaxed functionality, $\mathcal{F}_{wsc}^{\mathcal{N}}$, invokes the non-information oracle $\mathcal{N}$. Whenever one party wishes to send a message m on the channel, $\mathcal{F}_{wsc}^{\mathcal{N}}$ secretly transmits m to the other party, and in addition feeds m to $\mathcal{N}$. The output of $\mathcal{N}$ is then forwarded to the adversary. When the channel or one of its peers is corrupted, $\mathcal{F}_{wsc}^{\mathcal{N}}$ reveals the internal state of $\mathcal{N}$ to the adversary.

Here the security requirement from a non-information oracle is slightly different from the case of key-exchange. Specifically, we require that the messages generated by a non-information oracle $\mathcal{N}$ be "computationally independent" from the messages received by $\mathcal{N}$. That is, we require that an interaction with $\mathcal{N}$ will be indistinguishable from a "modified interaction" where each message sent to $\mathcal{N}$ is replaced with an all-zero string of the same length before it is handed to $\mathcal{N}$.

The rationale of using non-information oracles in $\mathcal{F}_{wsc}$ is the same as in the case of $\mathcal{F}_{wke}$: the fact that $\mathcal{N}$ is a non-information oracle guarantees that the information gathered by the simulator is essentially independent from the secretly transmitted messages. However, when a party gets corrupted, the simulator received additional information which, for an appropriately chosen non-information oracle, is helpful in completing the simulation.

We say that a protocol π is weak UC secure channels if there exists a non-information oracle $\mathcal{N}$ as defined here such that π securely realizes $\mathcal{F}_{wsc}^{\mathcal{N}}$. We show that the above generic protocol is a weak UC secure channels protocol, as long as the key-exchange protocols is weak UC secure, the encryption scheme is semantically secure against chosen message attacks, and the message authentication function is secure:

Theorem 5. *Let* MAC *be a secure Message Authentication Code function, let* ENC *be a semantically secure symmetric encryption scheme, and let π be a weakly UC secure key exchange protocol. Then there exists a non-information oracle $\mathcal{N}$ for encryption such that the above-sketched protocol, based on π,* MAC, *and* ENC, *securely realizes $\mathcal{F}_{sc}^{\mathcal{N}}$ in the* UM.

Finally, as further assurance in the adequacy of this weaker notion of secure channels, we note that any weak UC secure channels protocol is also secure according to [CK01]. (Recall however that the definition of [CK01] only addresses secure channel protocols where each request to transmit a message from one party to another over the channel results in a single actual protocol message. Consequently, the implication holds only for such protocols.)

References

[B91] D. Beaver, "Secure Multi-party Protocols and Zero-Knowledge Proof Systems Tolerating a Faulty Minority", J. Cryptology (1991) 4: 75–122.

[BCK98] M. Bellare, R. Canetti and H. Krawczyk, "A modular approach to the design and analysis of authentication and key-exchange protocols", *30th STOC*, 1998.

[BR93] M. Bellare and P. Rogaway, "Entity authentication and key distribution", *Advances in Cryptology, – CRYPTO'93*, Lecture Notes in Computer Science Vol. 773, D. Stinson ed, Springer-Verlag, 1994, pp. 232–249.

[BR95] M. Bellare and P. Rogaway, "Provably secure session key distribution – the three party case," *Annual Symposium on the Theory of Computing (STOC)*, 1995.

[B+91] R. Bird, I. Gopal, A. Herzberg, P. Janson, S. Kutten, R. Molva and M. Yung, "Systematic design of two-party authentication protocols," *IEEE Journal on Selected Areas in Communications* (special issue on Secure Communications), 11(5):679–693, June 1993. (Preliminary version: Crypto'91.)

[BJM97] S. Blake-Wilson, D. Johnson and A. Menezes, "Key exchange protocols and their security analysis," *Proceedings of the sixth IMA International Conference on Cryptography and Coding*, 1997.

[C00] R. Canetti, "Security and Composition of Multiparty Cryptographic Protocols", *Journal of Cryptology,* Winter 2000. On-line version at http://philby.ucsd.edu/cryptolib/1998/98-18.html.

[C01] R. Canetti, "Universally Composable Security: A New paradigm for Cryptographic Protocols", *42nd FOCS,* 2001. Full version available at http://eprint.iacr.org/2000/067.

[CK01] R. Canetti and H. Krawczyk, "Analysis of Key-Exchange Protocols and Their Use for Building Secure Channels", *Eurocrypt 01,* 2001. Full version at http://eprint.iacr.org/2001.

[CK02] R. Canetti and H. Krawczyk, "Universally Composable Notions of Key Exchange and Secure Channels", IACR's Eprint archive, http://eprint.iacr.org/2002.

[CR02] R. Canetti and T. Rabin, "Universal Composition with Join State", available on the Eprint archive, eprint.iacr.org/2002, 2002.

[DH76] W. Diffie and M. Hellman, "New directions in cryptography," *IEEE Trans. Info. Theory* IT-22, November 1976, pp. 644–654.

[DOW92] W. Diffie, P. van Oorschot and M. Wiener, "Authentication and authenticated key exchanges", *Designs, Codes and Cryptography*, 2, 1992, pp. 107–125.

[DM00] Y. Dodis and S. Micali, "Secure Computation", *CRYPTO '00,* 2000.

[FS90] U. Feige and A. Shamir. Witness Indistinguishability and Witness Hiding Protocols. In *22nd STOC*, pages 416–426, 1990.

[G01] O. Goldreich, *"Foundations of Cryptography"*, Cambridge University Press, 2001. Prelim. version available at `http://philby.ucsd.edu/cryptolib.html`

[GL90] S. Goldwasser, and L. Levin, "Fair Computation of General Functions in Presence of Immoral Majority", *CRYPTO '90, LNCS 537,* Springer-Verlag, 1990.

[GM84] S. Goldwasser and S. Micali, Probabilistic encryption, *JCSS,* Vol. 28, No 2, April 1984, pp. 270–299.

[GMRa89] S. Goldwasser, S. Micali and C. Rackoff, "The Knowledge Complexity of Interactive Proof Systems", *SIAM Journal on Comput.*, Vol. 18, No. 1, 1989, pp. 186–208.

[GMRi88] S. Goldwasser, S. Micali, and R.L. Rivest. A Digital Signature Scheme Secure Against Adaptive Chosen-Message Attacks. *SIAM J. Comput.*, April 1988, pages 281–308.

[MOV96] A. Menezes, P. Van Oorschot and S. Vanstone, “Handbook of Applied Cryptography,” CRC Press, 1996.

[MR91] S. Micali and P. Rogaway, “Secure Computation”, unpublished manuscript, 1992. Preliminary version in *CRYPTO 91.*

[PSW00] B. Pfitzmann, M. Schunter and M. Waidner, “Provably Secure Certified Mail”,IBM Research Report RZ 3207 (#93253), IBM Research, Zurich, August 2000.

[S99] V. Shoup, “On Formal Models for Secure Key Exchange” Theory of Cryptography Library, 1999. Available at: http://philby.ucsd.edu/cryptolib/1999/99-12.html.

On Deniability in Quantum Key Exchange

Donald Beaver

Syntechnica, LLC

Abstract. We show that claims of "perfect security" for keys produced by quantum key exchange (QKE) are limited to "privacy" and "integrity." Unlike a one-time pad, QKE does not necessarily enable Sender and Receiver to pretend later to have established a different key. This result is puzzling in light of Mayers' "No-Go" theorem showing the impossibility of quantum bit commitment. But even though a simple and intuitive application of Mayers' protocol transformation appears sufficient to provide deniability (else QBC would be possible), we show several reasons why such conclusions are ill-founded. Mayers' transformation arguments, while sound for QBC, are insufficient to establish deniability in QKE.

Having shed light on several unadvertised pitfalls, we then provide a candidate deniable QKE protocol. This itself indicates further shortfalls in current proof techniques, including reductions that preserve privacy but fail to preserve deniability. In sum, purchasing undeniability with an off-the-shelf QKE protocol is significantly more expensive and dangerous than the mere optic fiber for which "perfect security" is advertised.

1 Introduction

Privacy and integrity are the cornerstones of security. But a third, more subtle property is often overlooked: deniability, or the ability to pretend, after sending a message, that a different message was sent (perhaps by pretending a different key was used). The ability to deny a message is important in settings such as voting (to inhibit selling or coercion) and free, private speech.

A one-time pad is private, supports integrity, and easily provides deniability: after actually sending $c = m_0 \oplus k$, one can pretend that the cleartext was m_1 by pretending the key was $k' = k \oplus m_0 \oplus m_1$. But OTP's cannot be generated from scratch, and the length of an OTP must be at least as long as the total cleartext. Otherwise, the key equivocation – Shannon's pioneering measure of information-theoretic security – will not be sufficient, and information will be leaked, ultimately limiting the range of alternate, fake keys.

Private and public key cryptography address these issues by ensuring that finding the key or cleartext is computationally difficult, under widely-accepted complexity assumptions. But it is amply clear that they provide no key equivocation whatsoever. Even though it is difficult to find m from $m^e \bmod n$, or $g^{ab} \bmod p$ from g^a and g^b, it is obvious that there is a unique solution in each case. (Pretending that $m' \neq m$ was sent is impossible, since $(m')^e \not\equiv m^e$.) Moreover, the

L.R. Knudsen (Ed.): EUROCRYPT 2002, LNCS 2332, pp. 352–367, 2002.

mere fact that the keys are used to encrypt long messages makes it immediately obvious that, from an information-theoretic viewpoint, equivocation is limited.

With limited equivocation, and with the obviously unique mathematical solutions behind m^e or g^{ab}, it is unsurprising that RSA and Diffie-Hellmann key exchange are *undeniable.* There is no false message or key that can possibly match the public record, even though finding the real message or key might be difficult.

1.1 Perfect, Unconditional Security

Along comes quantum key exchange (QKE), offering to establish a key with "perfect, unconditional security." How is this possible in a world where each party has unlimited computing power? Unlike the classical information-theoretic world, determining the precise state of a particle prepared in an unknown fashion is generally impossible. Thus, one can obtain asymmetries in knowledge that are not achievable in classical settings.

Wiesner pioneered the cryptographic application of this principle in a proposal to authenticate money, and Bennett and Brassard showed the first key exchange protocol based on it [Wi83,BB84]. The canonical example is conjugate coding: a polarized photon represents a bit in one of two ways: either using a $+$ basis where $0°$ indicates 0 and $1°$ indicates 1, or using a $\times$ basis where $45°$ indicates 0 and $135°$ indicates 1. Knowing the right basis, it is easy to discover the bit. Without knowing the basis, any attempted observation is likely to extract possibly-incorrect information at the cost of irreversibly leaving the photon in a changed state. (This is not an *engineering* issue but a fundamental corollary of physics.)

Thus it is possible to detect attempted eavesdropping, unlike the classical world, where complete information about a given transmission can be obtained by direct inspection (in ideal principle). Moreover, if the level of eavesdropping is low, then the extracted information is low, ultimately enabling successful key exchange (and quantum money). Best of all, this can be done "from scratch," assuming that a Sender and Receiver also have a classical, authenticated public line.

A series of papers show that QKE is "perfectly, unconditionally secure." Want to avoid Shannon's key equivocation bounds? Just generate more key. Since there are no computational issues, this looks like a convenient way (engineering aside) to create a OTP of unlimited length.

The current work offers a strong note of caution: advertised "perfect, unconditional security" is not the same as equivalence to a OTP. In particular, while privacy and integrity have been provably established for QKE, deniability is not covered, nor is it implied.

1.2 How to Bind the Message?

Imagine that Eve measures only one photon. With probability 1/2, she chooses the correct basis, obtains complete information on that photon, and transmits an unchanged photon to R. (Or, she uses an incorrect basis, thereby "disturbing" the photon, but R's later measurement coincidentally restores it.) No secrecy is compromised; no theorems are violated.

As for *deniability*, however, things are different. If S and R later try to open up their accounting records to show that a different key was established, then they must change *something* in their actual records. There is some nonzero chance that they decide to pretend a different bit was used for the one photon Eve measured. With significantly nonzero probability, the false record they provide will not match Eve's observation.

1.3 Quantum Subtleties

The issue is clouded by related results in quantum cryptography. Although it was thought that the asymmetries in knowledge might also enable bit commitment,[1] Mayers showed that quantum bit commitment (QBC) is in fact impossible. One subtle insight is that the programming command "(step n) Party P measures particle X privately" is not enforceable against an adversary. Intuitively speaking, a cheater can postpone certain required measurements (or more general "collapses" of quantum systems), thereby keeping her options open.

At face value, this serves only to add optimism to the QKE setting, where one might now happily conclude that QKE is fully deniable, since otherwise it would enable QBC. Although we will extract inspiration from Mayers' insightful work, we also show that such off-the-shelf conclusions are logically unfounded.

It is easy to invoke Mayers' no-commit theorem blindly to (1) dismiss the significance of a positive result ("it follows easily") or (2) dismiss correctness of negative results ("it contradicts no-commitment"). Ordinarily, one would imagine that a demonstration that BB84 is binding would be sufficient *prima facie* to show that the no-commit theorem does not apply.

But nothing quantum is *prima facie*. Instead, without deeper investigation, many have been tempted to challenge the "counterexample" (namely, the assertion of undeniability). Our deeper investigation displays why Mayers' result is true against QBC but insufficient against QKE. In fact, it is somewhat surprising that deniability might in fact be achievable through LOCC (local operations and classical communication), since the no-commitment proof demands potentially nonlocal operations on S and R.

1.4 Contributions

Our work is directed at (1) making the properties of QKE fully apparent for existing and newly-designed protocols, and (2) analyzing the extent to which existing techniques (such as Mayers' methods for QBC) apply to QKE, and extending them where needed. In sum:

---

[1] In bit commitment, a bit to be committed must be kept secret from the "receiving" party, at least until much later, when it is to be "decommitted/unveiled.".

- The merely "perfectly, unconditionally secure" key established through quantum key exchange is not completely equivalent to a one-time-pad.
- Perfect privacy and integrity do not imply deniability.
- Quantum protocols are not necessarily deniable, even if "perfectly secure."
- The BB84 quantum protocol is binding.
- Mayers' "no-commitment" theorem is not sufficient to imply deniability.
- Deniability can be achieved through extensions that require a quantum computer.

Our positive result balances the need for purification with the need to use the public, authenticated, classical channel of QKE. It represents the first *deniable* quantum key exchange protocol.

2 Background, Notation, Definitions

We employ standard terminology from quantum computing and cryptography. Let $|0\rangle$ and $|1\rangle$ denote an orthonormal basis for a two-dimensional complex Hilbert space $\mathcal{H}_2$. Using the Dirac bra-ket notation, $\langle\phi| \equiv_{def} |\phi\rangle^\dagger$. The Pauli matrices are

$$X = \tfrac{1}{2}\begin{pmatrix} 0 & 1 \\ 1 & 0 \end{pmatrix},\ Y = \tfrac{1}{2}\begin{pmatrix} 0 & -i \\ i & 0 \end{pmatrix},\ Z = \tfrac{1}{2}\begin{pmatrix} 1 & 0 \\ 0 & -1 \end{pmatrix}$$

Conjugate coding uses two bases: $B_+ = \{|0\rangle, |1\rangle\}$ and $B_\times = \{(1/\sqrt{2})(|0\rangle + |1\rangle), (1/\sqrt{2})(|0\rangle - |1\rangle)\}$. (N.b.: subscript "+" is interchangeable with "0", and subscript "×" with "1.") The Bell basis describes entanglement:

$$\beta_{00} = \tfrac{1}{\sqrt{2}}(|00\rangle + |11\rangle)\ ,\ \ \beta_{10} = \tfrac{1}{\sqrt{2}}(|00\rangle - |11\rangle)$$
$$\beta_{01} = \tfrac{1}{\sqrt{2}}(|01\rangle + |10\rangle)\ ,\ \ \beta_{11} = \tfrac{1}{\sqrt{2}}(|01\rangle - |10\rangle)$$

A state can be represented as a *density matrix* over $\mathcal{H}_n$ for some n. A density matrix ρ is a weighted sum of projectors, with $\mathrm{Tr}(\rho) = 1$. A density matrix can represent a mixed state or, in the case that $\rho = |\phi\rangle\langle\phi|$, a pure state ϕ. We typically consider "binary" Hilbert spaces, expressible as $(\mathcal{H}_2)^{\otimes n}$.

Let A be Hermitian. Traditionally, a measurement of A is seen as "collapsing" the state ρ to an eigenvector of A. The expected value of A will be $\langle A\rangle = \mathrm{Tr}(\rho A)$. We allow parties to perform generalized measurements through the standard toolkit: (1) appending an unentangled ancillary subsystem; (2) applying a unitary transformation; (3) making an orthogonal measurement; (4) tracing out a local part of the system.[2]

Let C_1 and C_2 be $[n, k_1]$ and $[n, k_2]$ binary codes, respectively, with

$$\{0\} \subset C_2 \subset C_1 \subset \mathrm{GF}(2)^n.$$

[2] This is tantamount to discarding part of the system. In this paper, we accept this approach at face value.

The $\mathrm{CSS}_{x,z}$ quantum encoding [CSS96] maps $v \in C_1$ to the following codeword:

$$v \mapsto \frac{1}{\sqrt{|C_2|}} \sum_{w \in C_2} (-1)^{z \cdot w} |x + v + w\rangle.$$

In the context of BB84, phase errors turn out to be irrelevant and the straightforward protocol purification of BB84 (see *infra*) is more like $\mathrm{CSS}_{x,z}$ with z omitted or averaged out.

2.1 Protocol Execution

We use a circuit model for protocol execution. A global state Φ, described over a basis B_+^k for some k, is advanced through applying each party P (*i.e.* each circuit) to a collection of registers, where a register is a local subset of the "wires" of the overall circuit. We use superscripts to indicate location of given registers; thus, *e.g.*, $\Phi = \sum_{v_1..v_5} \alpha_{v_1..v_5} |v_1 v_2 v_3\rangle^A \otimes |v_4 v_5\rangle^B$ describes a state with Alice holding the first three registers/wires and Bob holding the rest. The tensor product sign $\otimes$ is omitted when clear from context. We also use the shorthand $\mathcal{H}_A \otimes \mathcal{H}_B$ to express the state space.

A transition of the system thus consists of applying a local unitary transformation $U_P \otimes \mathbf{1}$ to the registers held by party P, along with a possible orthogonal measurement. More precisely, U_P applies to the Hilbert subspace $\mathcal{H}_2^m$ at indices $k_1, \ldots, k_m$ that are labelled as under P's control. Subsequent communication is modelled by reassignments of those labels.

Initialization. We say that a protocol is *properly initialized* if each party starts with quantum registers in unentangled states $|0\rangle$. In particular, there is no entanglement with other parties, nor with the auxiliary-parties that comprise the "environment."

Communication. As noted, communication is generally just a reassignment of the register labels. When an eavesdropper is present, the register is assigned to the eavesdropper first before being reassigned to the destination party. Generally, the eavesdropper can forward any transformation or substitution she pleases.

This suffices to model QBC, but in QKE there is an additional channel: the reliable public classical channel. One way to regard this channel is as a separate party who first measures the input, then broadcasts the result to the source, destination, and eavesdropper (who is not allowed to alter the result). This irrevocable measurement induces a mixture among several outcomes of the overall protocol.

The collection of such auxiliary parties consists the "environment" and is described by an environment space $\mathcal{H}_{env}$. Thus a two-party protocol is executed over $\mathcal{H}_A \otimes \mathcal{H}_B \otimes \mathcal{H}_{env}$.

Views and Outputs. The distribution produced by running a quantum protocol is obtained by tracing out each party. The *view* of party P includes her state in $\mathcal{H}_P$ along with the classical strings in any environmental/classical channels she used or saw. In particular, direct *erasure* of classical information is not allowed. For some purposes, we may more generally regard the *view* as the collection of registers along with the measurement history of a party, rather than tracing out the final state.

Parameters. We generally use κ to denote a security parameter, k to denote a key (generated or exchanged), and we sometimes overload k when describing error correction: $[n,k]$ connotes an error correcting code mapping k logical bits to n representation bits.

2.2 Deniability

There are several variants on the meaning of *deniability* [CDNO97,Be96] and *binding*, depending on the parties attempting to equivocate and what their success rates may be.

Let m_1 and m_2 be arbitrary messages. Run $S(m_1)$, R and E, obtaining global state $\rho(m_1)$ whose registers are $|\phi_S\rangle$, $|\phi_R\rangle$, $|\phi_E\rangle$, $|\phi_{env}\rangle$. Let D_S and D_R be local computations (not necessarily unitary operations). Let $\rho(m_1,m_2) = \sum |D_S(m_1,m_2,\phi_S)\rangle|D_R(m_1,m_2,\phi_R)\rangle|\phi_E\rangle|\phi_{env}\rangle$, representing an attempt at denial: pretending that m_2 was really sent.

Let J be a judge, who has inputs for registers ϕ_S, ϕ_R, ϕ_E, and ϕ_{env}. J's final state is described in registers d and J', where d is a single-bit, "decision" register. Flip a coin c in the environment to determine whether denial will be attempted. If $c=0$, run J on $\rho(m_1)$; if $c=1$, run J on $\rho(m_1,m_2)$ (namely apply D_R and D_S before submitting to the judge). The final result is of the form

$$\rho(m_1,m_2,c) = \sum |cd\rangle|\phi_{J'}\rangle|\phi_{env}\rangle.$$

Tracing out J' and the environment gives a mixture over $|cd\rangle$'s.

A judge is *safe* if $|01\rangle$ has zero probability, namely the judge makes no false accusations.

Definition 1. *Let (S,R) be a quantum key exchange protocol with denial programs (D_S,D_R). For eavesdropper E and judge J, let $P_{J,E}(m_1,m_2,\kappa)$ be the probability J gives $|11\rangle$, on security parameter κ. We say the protocol is* deniable *if, for any E, any safe J, and for any m_1,m_2: $P_{J,E}(m_1,m_2,\kappa) = \kappa^{-\omega(1)}$.*

Simplifying, let $P(\kappa)$ be the maximal probability of $|11\rangle$ over all messages of size $O(\kappa)$. A protocol family indexed by integers C is *perilously deniable* if $P(\kappa,C) = O(\kappa^{-C})$, namely S and R can reduce their vulnerability to a small (but non-negligible) polynomial fraction. A protocol is *weakly binding* if $P(\kappa,c) = \Omega(\kappa^{-c})$ for some $c>0$. (Thus a protocol family can be perilously deniable while each given value of C produces a weakly binding protocol.) A protocol is *binding* if $P(\kappa) = 1-\kappa^{-\omega(1)}$.

(Variants on this approach include sender-only and receiver-only deniability, single-bit messages, unsafe judges, and many others. Note that when an exchanged key is used like a one-time pad, "key" can often be interchanged with "message" to simplify the discussion.)

BB84

1. S selects $2(4+\delta)n$ random bits $\{b[i,0], b[i,1]\}$.
2. S encodes $\{b[i,0]\}$ as qubits $\{p[i]\}$, each in basis $+$ or $\times$ depending on $\{b[i,1]\}$.
3. S sends $\{p[i]\}$.
4. S chooses random $v_k \in C_1$.
5. E forwards $\{q[i]\}$ to R (possibly unchanged).
6. R measures each $\{q[i]\}$ in random basis $\{c[i]\}$.
7. S announces $\{b[i,1]\}$ on the classical channel.
8. S and R discard indices wherever $b[i,0] \neq c[i]$. S selects and announces a random remaining subset of $2n$ bits, along with a random n-subset π of check indices. (Abort if impossible.)
9. S and R reveal $p[i]$ and $q[i]$ classically for $i \in \pi$ and abort if any (resp. more than t) disagree.
10. S announces $x \oplus v_k$, where x is the n-bit remaining string in $\{b[i,0]\}$.
11. R computes $y \oplus x \oplus v_k$, where y is the n-bit remaining string in $\{q[i]\}$, and applies C_1 to correct it to v_k (presumably).
12. S and R calculate k from the coset $v_k + C_2$.

Fig. 1. BB84 protocol for n-bit key k, in modern conventions, with C_1 used for reconciliation and cosets of C_2 used for privacy amplification.

3 Quantum Key Exchange

The relevant steps of the BB84 protocol, with eavesdropper, are sketched in Fig. 1. While arbitrary hashing and privacy amplification techniques can be used with the basic BB84 approach, we have illustrated the typical approach of employing binary codes. More particularly, we follow the conventions of [ShPr01] so that we can connect to related work.

It can be shown that an eavesdropper gains information only with $O(2^{-k})$ probability, or gains at most $O(2^{-k})$ as measured by entropy, where $k = k_1 - k_2$.

3.1 Variants of BB84

In the original [BB84] protocol, R measured the photons before knowing the proper bases; the mistaken bases were discarded. Since photons were difficult to store without measuring, this made the theoretical protocol more feasible. As discussed in [BBM92], however, if S and R simply try to establish EPR pairs, the result is similar to having R postpone her measurements until S announces the bases on the classical channel. Since S will wait until he has confirmed that R has the particles, namely that Eve no longer has a chance to change what she has forwarded, this is "okay." [BBM92] suggest that such a protocol has security equivalent to the original [BB84].

Further degrees of purification are possible. Note that the specific resulting protocol depends on the particular reconciliation and amplification routines.

– (BB84) No entanglement or purification.

- (BB84-EPR/Ekert) Use entangled qubits, then measure; no other purification.
- (PQECC) Use entangled qubits; measure check subset; measure key; leave other registers in superposition; (more details later).
- (BB84-Key) Purify completely according to [Ma96]; measure key.
- (BB84-Pur) Purify completely according to [Ma96].

Apart from PQECC, these variants turn out to be either binding, vacuous (non-transmitting), or or unimplementable within the communication model.

The common instantiations of universal hashing and privacy amplification [BBCM95,Ma93,BBR88] correspond to error correcting codes over GF(2). Although any number of reconciliation protocols are available, let us expand and simplify a natural path. S and R will statistically sample half the particles to put a cap on how much interfering Eve did. As long as it is sub-threshold (say, less than 1%), they continue. This means that (with exponentially-high probability) Eve has only measured a small fraction (say $<< 1/100$) of the remaining indices. (More precisely, she has applied a measurement/alteration with "small" quantum entropy; it may affect any number of particles and in superposition.)

For single-bit messages, there are two canonical superpositions that S will have sent, depending on whether $k = 0$ or $k = 1$:

$$\phi_k = \frac{1}{\sqrt{|C_2|}} \sum_{z \in P(k)} |z\rangle$$

where $P(k)$ denotes all strings in the coset $v_k + C_2$ (with v_k being a fixed member corresponding to k).

For example, let BB-EPR-PAR(k) be the variant of BB84 in which $P(m)$ contains all k-bit strings whose parity is m. To alter a given word requires reversing at least one bit.

4 Eavesdropping and Binding BB84

We now consider what happens when an eavesdropper listens in on BB84. Taking a cue from standards bodies, we let "must" indicate an apparent intuitive requirement (not necessarily necessary), and let *MUST* indicate a requirement that we do assert as factual.

The intuitive observation is that S and R "must" change a bit somewhere in order to pretend to have sent the opposite cleartext. While this classical reasoning is not justifiable in the quantum setting, it approaches the same final conclusions.

Even though (S, R) face an unknown adversary, they *MUST* select a strategy (D_S, D_R) in advance. Clearly, the actual computation can depend dynamically on the results of the transmission. But there can be no argument: the encryption/denial programs (S, R, D_S, D_R) *MUST* be written down by a designer.

Let eavesdropper $\mathrm{M}(a, k)$ trade a qubit at position a:

$$|z\rangle^S |00\rangle^E |0^k\rangle^R \rightarrow |z\rangle^S |z(a)0\rangle^E |z(a{:}0)\rangle^R$$

where $z(a)$ indicates the ath bit in z, and $z(a{:}0)$ indicates replacing bit a in z by 0. Like measuring a photon, this action disturbs the stream of bits, although it does not give M information about the overall "key." (Recall that the notation was simplified by adjusting it after S announces the bases. When M acts, she does not know the correct basis, and simply uses +. We now consider only those paths in which S announced + at index a.)

Consider BB-EPR-PAR (parity-based) with odd $k \geq 3$. Let $\Phi(E, m)$ be the state obtained after sending bit m. There are a variety of equivocation transforms U available to patch $\Phi(\emptyset, 0)$ to $\Phi(\emptyset, 1)$ against a passive eavesdropper. Some simple ones are (*cf.* one-time pad) to negate all qubits; a given qubit; a random qubit. Luckily, these are all local transforms, too. We take U_{neg} which negates all qubits.

We now show this leads to catastrophe, as do other similar choices. Let $P(m; a, b) = \{z \in P(m) \mid z(a) = b\}$, and let $\rho(E, m) = \mathrm{Tr}_{\mathcal{H}_{E,env}} \Phi(E, m)$.

$$\rho(\mathrm{M}(a,k), 0) = \sum_{P(0;a,0)} |z\rangle^S |z\rangle^R \langle Z|^R \langle Z|^S + \sum_{P(0;a,1)} |z\rangle^S |z(a{:}0)\rangle^R \langle Z(a{:}0)|^R \langle Z|^S$$

$$\rho(\mathrm{M}(a,k), 1) = \sum_{P(1;a,0)} |z\rangle^S |z\rangle^R \langle Z|^R \langle Z|^S + \sum_{P(1;a,1)} |z\rangle^S |z(a{:}0)\rangle^R \langle Z(a{:}0)|^R \langle Z|^S$$

$$U_{neg}\rho(\mathrm{M}(a,k), 0)U_{neg}^{\dagger} = \sum_{P(1;a,0)} |\bar{z}\rangle^S |\bar{z}\rangle^R \langle \bar{Z}|^R \langle \bar{Z}|^S + \sum_{P(1;a,1)} |\bar{z}\rangle^S |\bar{z}(a{:}1)\rangle^R \langle \bar{Z}(a{:}1)|^R \langle \bar{Z}|^S$$

$$= \sum_{P(1;a,0)} |z\rangle^S |z(a{:}1)\rangle^R \langle Z(a{:}1)|^R \langle Z|^S + \sum_{P(1;a,1)} |z\rangle^S |z\rangle^R \langle Z|^R \langle Z|^S$$

There is clearly 0 fidelity between $\rho(\mathrm{M}(a,k), 1)$ and $U_{neg}\rho(\mathrm{M}(a,k), 0)$. A judge simply calculates $d = (\oplus_i z^S(i)) \oplus (z^R(k))$. Even if the protocol designer opts to use U_{neg} only some of the time, then against a passive adversary or against $\mathrm{M}(a,k)$, equivocation is detected with probability at least 1/2 (again, in at least 1/2 of all paths). By inspection, the judge is both safe and binding.

If the protocol designer uses "$U_{neg,p}$", in which qubit at location p is negated, then $\mathrm{M}(p,k)$ (versus passive) presents a problem. If "$U_{neg,r}$" is used, in which a randomly selected qubit is negated, then $\mathrm{M}(1,k)$ leads to $(1-1/k)$ fidelity. This is better for S and R, but still binding.

Observations. We avoided constructing a single eavesdropper who randomly chooses to be passive or to invade. Other objections aside, this could pose the risk of enabling an equivocation strategy (insofar as E's program were to become public), as mentioned earlier (§6.1). This is a subtle but important issue that differentiates classical cryptography from quantum.

It is straightforward to imagine extensions to BB84 that apparently provide *perilous deniability*. For instance, by lengthening the "ciphertext" to contain κ^{2C} bits, arranging equivocation could be as simple as changing a "small" $(1/\kappa^C)$

fraction of bits. For each fixed C, there remains a non-negligible chance to get caught (hence perilous and still weakly binding). But even this weak goal is much harder to prove in the quantum setting than initial intuition suggests, since Eve can spread her eavesdropping among many bits, not just focus on a particular small subset. Space does not permit a full analysis here.

5 Mayers' Theorem

Mayers takes great care to state the model precisely, and we follow his descriptions [Ma96].

5.1 Defining Bit Commitment

In a bit commitment protocol, Alice encodes her input bit b into a state $|\psi_b\rangle$ of $\mathcal{H}_A \otimes \mathcal{H}_B \otimes \mathcal{H}_{env}$, using an initial protocol, *commit*(b). A second protocol, *unveil*($|\psi_b\rangle$), is used to give Bob b or a "refusal" string $\perp$.

Alice has the power to choose $p(b \mid \text{not } \perp)$ by behaving honestly. The goal of the *commit* protocol is to prevent her from subsequently changing $p(b \mid \text{not } \perp)$ even if she was dishonest. Let *unveil'* denote running *unveil* with possibly dishonest A. A state $|\psi\rangle$ is *perfectly committing* if every attack *unveil'* either returns $\perp$ with probability 1 or returns b with probability $p'(b \mid \text{not } \perp) = p(b \mid \text{not } \perp)$.

Bob may attempt to gain b prematurely. Let η_B be Bob's classical information from $\mathcal{H}_{env \cap B}$, as the state is collapsed into $|\psi_{b,\eta}\rangle$ of $\mathcal{H}_A \otimes \mathcal{H}_B \otimes \mathcal{H}_{env}$. The reduced density matrix of Bob given η is:

$$\rho_B(|\psi_{b,\eta}\rangle) = \mathrm{Tr}_{\mathcal{H}_{env \cap A}}(|\psi_{b,\eta}\rangle\langle\psi_{b,\eta}|).$$

If b is fixed by η, we let $F(\eta) = 0$. Otherwise, we let $F(\eta)$ be the *fidelity* between $\rho_B(|\psi_{0,\eta}\rangle)$ and $\rho_B(|\psi_{1,\eta}\rangle)$. (The fidelity $F(\alpha, \beta)$ is the supremum of $|\ \psi_\alpha^\dagger \psi_\beta\ |$ over all purifications ψ_α, ψ_β of α, β. Note that $F(\alpha, \beta) = 1$ only when $\alpha = \beta$.) A state is *perfectly concealing* if η is independent of b and the expected value of $F'(\eta)$ is 1, where F' refers to executions with a possibly cheating Bob.

A commitment protocol is *perfectly secure* if it is both perfectly committing and perfectly concealing. One can replace "perfect" by a tolerance of ϵ (or $\epsilon(k)$ for some security parameter k) in the preceding definitions.

5.2 Previous Work: No Bit Commitment

Mayers' Theorem states that quantum bit commitment is impossible:

Theorem 1. [Mayers' Theorem, or No-Commitment] *No properly initialized quantum bit commitment protocol is unconditionally secure.*

The proof is based on what we call an *equivocation strategy* that allows A to change the bit even after protocol *commit.* We sketch some of the ideas here.

Consider a properly initialized quantum bit commitment protocol. Let dishonest A' refrain from making measurements. Mayers shows that there is a purification $|\psi_{01}\rangle$ of $\rho_B(|\psi'_{0,\eta}\rangle)$ such that $\langle\psi_{01} \mid \psi'_{1,\eta}\rangle \geq F'(\eta)$. Moreover, there is a unitary transformation $U = U_A \otimes \mathbf{1}$ mapping $|\psi'_{0,\eta}\rangle \to |\psi_{01}\rangle$.

Because an ϵ-concealing protocol must have fidelity $F(\eta) \geq 1-\epsilon$, this implies an equivocation strategy for A'. If she wishes to set $b = 0$, she makes the measurements required of A and continues with *unveil.* To set $b = 1$, she applies U, performs the measurements required of A in *commit*, and continues with *unveil.* The result is that Bob accepts this unveiling with probability approaching 1.

We refer to this strategy as *Mayers Equivocation.* Unfortunately, there is an intuitive but incorrect way to paraphrase the theorem, which goes something like this:

Claim 2 [NoGo Folk "Theorem"] *In any quantum protocol, whenever* $F(\rho_B(|\psi_{0,\eta}\rangle), \rho_B(|\psi_{1,\eta}\rangle)) \approx 1$, *then Alice can equivocate successfully with probability* ≈ 1. *(to be disproved)*

6 Limitations on Applying No-Commitment

The No-Commitment theorem is obtained through two methods: (1) abstaining from private measurement, followed by (2) applying a unitary transformation to change $|\psi_0\rangle$ to $|\psi_1\rangle$.

There are several aspects of the model and the result that make it insufficient to apply automatically to QKE. These include the quantifiers, colocation, and the impact of generic abstinence from measurement.

6.1 Quantifier Problems

Mayers' result essentially says, $(\forall B)(\exists U_A(B))$ such that a cheating committer A can employ $U_A(B)$ to equivocate. The strategy can depend on B's program, which is acceptable because the honest programs for A and B must be declared. Naturally, the cheating programs need not be disclosed, but to disprove security, it suffices to show that no honest program is protected.

Likewise, it is hard to imagine an encryption protocol in which S and R are allowed to know what E's program is. Their attempts to communicate, and to deny, must be successful even without being given details of E's program. (There may be some dynamic deductions to make about E based on her behavior, but this is vastly different than knowing her full program.) A quantification, "$(\forall E)(\exists S, R)$ such that S and R successfully communicate," is insufficient.

Yet the natural generalization of Mayers' result to QKE is precisely backwards. S and R are not given E's program and then allowed to equivocate.

The "folk no-commit theorem" fails to hold: an arbitrary E does indeed gain no information, and indeed there mathematically exists a Mayers equivocating transform on the joint (S, R) state, but S and R have no way to determine what it is, since E is arbitrary and inaccessible. Further problems occur; see below.

6.2 Colocation

To equivocate a one-time pad, one merely needs to reverse a bit. This can be done individually and locally by S and R, without communication (apart from knowing they must equivocate).

The direct application of No-Commitment treats (S, R) as the committer, who, for any trustee/eavesdropper, has an equivocation transformation U. (Forget about the order of quantifiers.) There is no mathematical guarantee that U can be factored into local transforms U_S and U_R. Therefore, the (nevertheless correct) proof given in [Ma96] does not provide sufficient grounds to apply to QKE. There may indeed be local strategies for S and R for particular QKE protocols, but they are not implied directly by [Ma96].

If S and R need to communicate or be co-located in order to equivocate, the deniability property is weakened. A vote-coercing Mafioso simply interrogates them separately. While deniability that requires colocation may be better than nothing, it is not always sufficient.

6.3 Abstinence Makes the Bit Grow Weaker

An extremely critical (and clever) aspect of Mayers' approach is the demand that parties refrain from making internal measurements. This gives them the flexibility to equivocate later.

In bit commitment, abstaining seems to have no obvious impact. There is no particular reason why A should do any measurements. She already knows her bit and doesn't stand to discover anything much.

But [BB84] specifically requires measurements to be peformed, both to check how invasive E is and to discern what the message is. In the full protocol purification, BB84-Pur, R cannot receive the cleartext privately. Depending on interpretation, either no measurement is made at all, or the only actionable information is whatever was transmitted over the clear classic channel.

In the BB84-Key variant, close inspection reveals that the previously classical channel is now used to send check-information as qubits to R. But this presupposes the end result: a secure quantum channel. Thus BB84-Key is not implementable within the rules of the model.

Even the limited purification to BB84-EPR/Ekert does not buy anything. An attack similar to the one against BB84 in §4 will work against entanglement purification of a tainted EPR source.

7 PQECC: Deniable QKE

We now propose a protocol to achieve deniability, although it requires a quantum computer. Normally, quantum error-correcting codes are useful to protect against decoherence. Typically, a syndrome is measured and then used to restore the original state. We applied QECC in an unusual twist: we avoid measuring the relative syndrome. Instead, a postulated quantum computer applies the QECC

PQECC

1. S constructs duplicate registers (K, B, Π, Y, V, W) with respective lengths k, $2n$, $2n \log 2n$, n, n, n, in superposition $\sum |K', B', \Pi', Y', V', W'\rangle |K, B, \Pi, Y, V, W\rangle$, (except $W' = W = |0^n\rangle$). All computations are quantum (except explicit measurements later), including error correction.
2. S quantum-computes the Y-selected ECC representation of K and places it in W, leaving $K = |0^n\rangle$.
3. S applies interleaving Π against the 2n qubits in (V, W), leaving $\Pi = |0^{2n \log 2n}\rangle$.
4. Let T be the joint $2n$-qubit register (V, W). S applies Hadamards based on B to corresponding qubits in T.
5. S sends T to E.
6. E attacks T via operator U_E.
7. R receives the manipulated T register and reports receipt.
8. S measures $(b, \pi, v, y) \leftarrow (B, \Pi, V', Y')$ and announces them.
9. R applies Hadamards to T according to b, then applies π^{-1} to T. Consider T again as registers (V, W).
10. R measures register V and aborts if there are any (or more than a threshold t) mismatches with v.
11. R sets new register E to $|0^n\rangle$ and applies error correction (using y) to (W, E).
12. S measures key $k_S \leftarrow K'$ and R measures key $k_R \leftarrow W$. Encryption is as a OTP: $ciph \leftarrow k_S \oplus m$, $m_{rcv} \leftarrow ciph \oplus k_R$.

Fig. 2. PQECC protocol for n-bit key k, rewritten to be compatible with modern conventions. The ECC steps are quantum-level calculations for the random hashing and error correction in BB84 *et seq.*

without measuring the relative syndrome at all. The goal is to purify the BB84 protocol as much and as carefully as possible. Only the most necessary measurements are actually performed. (Note that the key/message is emphatically not the only register that is measured.)

This *purified* QECC, or PQECC protocol, is the basis for establishing deniability. Although it was investigated several years ago [Be96d], it is closely related by convergent evolution to the "modified Lo-Chau" and "QKD-CSS" protocols [ShPr01], discussed below. We have tried to present it in a form that illustrates its connections to those protocols.

First, we comment on the registers and computations. The B and Π registers correspond to random axis choices and random interleaving of key bits with check bits. (Π can certainly be a permutation, as suggested by [ShPr01]. The random hashing and amplification in BB84 is purified to a random ECC selected by Y (corresponding to the (x, z) choice in [ShPr01]). The decoding procedure decouples the signal (W) from the noise (E).

7.1 Related Protocols

Two related protocols, called QKD-CSS and "modified Lo-Chau," appeared subsequent to the first consideration of PQECC as the maximal effective protocol purification of BB84 [Be96d]. If one imagines a hierarchy of protocol classes depending on purification and/or specific error-correction codes, PQECC is more or less subsumed by "modified Lo-Chau" while it more or less subsumes QKD-CSS.

The motivation for PQECC was to obtain deniability, while the motivation for Lo-Chau and QKD-CSS was to find a proof of privacy for any kind of QKE, and especially QKE without quantum computation. Hindsight shows that this independent convergence to similar protocols is natural given the drive to use protocol purification as (1) a tool for proving privacy and (2) a tool for achieving deniability.

7.2 Privacy and Deniability

Recently, simplified proofs of privacy have appeared for BB84 [LC99, GoPr00, ShPr01]. There are two important ingredients that concern us.

It is useful to try to establish the secret key as a sequence of k ebits shared by S and R, namely $\Phi = \beta_{00}^{\otimes k}$, starting with a noisy pair of n-bit registers in state ρ. Let ρ' describe the state of the k-bit key registers after the full protocol. Lo and Chau's approach employs a result shown by Gottesman and Preskill:

$$\langle \beta_{00}^{\otimes k} | \rho' | \beta_{00}^{\otimes k} \rangle \geq \mathrm{Tr}(\Pi \rho),$$

where Π projects onto Bell states differing from $\beta_{00}^{\otimes n}$ by at most t bit-flip errors (applications of Pauli X) and at most t phase-flip errors (applications of Pauli Z). Indirectly argued, the sampling test gives an accurate bound on $\mathrm{Tr}(\Pi \rho)$, ultimately implying that the fidelity between ρ' and $\beta_{00}^{\otimes k}$ is exponentially close to 1. This demonstrates privacy for the Lo-Chau protocol, although a quantum computer is necessary.

In a second stage, Shor and Preskill apply this to a protocol (QKD-CSS) that uses CSS codes. By then instructing S and R to perform measurements sooner rather than later, the QKD-CSS protocol "reduces to" BB84, and the requisite quantum computer can be avoided without introducing any information leak to Eve.

For deniability, the first ingredient is essential. It ultimately allows us to conclude that Eve is entangled negligibly with K' (and instead overwhelmingly with register E) in protocol PQECC. This means that the simple OTP denial strategy is overwhelmingly effective: pretend $k' = k \oplus m_0 \oplus m_1$. Without sufficient space for proof, we merely assert:

Claim. PQECC is a deniable quantum cryptosystem.

This analysis should also extend to other cryptosystems such as the "modified Lo-Chau" protocol.

8 Conclusions

The BB84 protocol is weakly binding on S and R. Despite the reaction of some, Mayers' no-commit theorem [Ma96] does not suffice to turn BB84 into a deniable QKE. There are formal improprieties with quantifiers, insufficient support for equivocation without co-location, conflicts between measurement abstinence and correct or allowable transmission, and counterintuitive adversarial binding arguments. (None of this impugns the correct work on bit commitment.)

This paper seeks to make practitioners aware of the incomplete analysis of QKE, despite claims of "perfect, unconditional security." An off-the-shelf optic-fiber-based QKE might enable a private electronic vote but it will support coercion and vote-selling.

Using protocol purification in a refined manner can provide a deniable QKE, in the form of the PQECC protocol, but a quantum computer is required. But there remains a great deal of turbidity in current quantum "security reductions."

Acknowledgements

I thank Claude Crépeau and Jeroen van de Graaf for helpful conversations (long ago!), and Joern Müller-Quade for more recent ones. Several referees made extensive remarks (over many years) and I thank them for their efforts. All mistakes remain mine.

References

[ADH97] L. Adleman, J. Demarrais, M. Huang. "Quantum Computability." SIAM J. Comput., **26**:5, 1997, 1524–1540.

[Ba95] A. Barenco. "A Universal Two-Bit Gate for Quantum Computation." Proc. Royal Society of London, **449**, 1995, 679–683.

[BDM+95] A. Barenco, C. Bennett, R. Cleve, D. DiVincenzo, N. Margolus, P. Shor, T. Sleator, J. Smolin, H. Weinfurter. "Elementary Gates for Quantum Computation." Phys. Rev. Letters A, **52**, 1995, 3457–3467.

[Be96] D. Beaver. "Plausible Deniability." Proc. of PragoCrypt 1996, J. Prybl, Ed., CTU Publishing House, Prague, 1996, 272–288.

[Be96d] D. Beaver. Unpublished manuscript, 1996.

[Be99] D. Beaver. "Imperfections in Perfectly Secure Key Exchange." IEEE Information Theory and Networking Workshop, Metsovo, 1999.

[Be92] C. Bennett. "Quantum Cryptography Using Any Two Orthogonal States." Phys. Rev. Letters, **67**:21, 1992, 2121–2124.

[BBBSS92] C.H. Bennett, F. Bessette, G. Brassard, L. Salvail, J. Smolin. "Experimental Quantum Cryptography." *Journal of Cryptography*, **5**:1, 1992, 3–28.

[BB84] C. Bennett, G. Brassard. "Quantum Cryptography: Public-Key Distribution and Coin-Tossing." Proceedings of IEEE CSSP, Bangalore, India, 1984, 175–179.

[BBCM95] C. Bennett, G. Brassard, C. Crépeau, U. Maurer. "Generalized Privacy Amplification." IEEE Trans. Information Theory, **41**:6, 1995.

[BBM92] C. Bennett, G. Brassard, D. Mermin. "Quantum Cryptography Without Bell's Theorem." Phys. Rev. Letters, **68**:5, 1992, 557–559. Also see Manuscript, March 6, 1995.

[BBR88] C. Bennett, G. Brassard, J.M. Robert. "Privacy Amplification by Public Discussion." SIAM J. Computing, **16**:2, 1988, 210–229.

[BCMS97] G. Brassard, C. Crépeau, D. Mayers, L. Salvail. "A Brief Review on the Impossibility of Quantum Bit Commitment." Los Alamos Preprint Archive quant-ph/9712023, 1997.

[BCMS98] G. Brassard, C. Crépeau, D. Mayers, L. Salvail. "Defeating Classical Bit Commitments with a Quantum Computer." Los Alamos Preprint Archive quant-ph/9806031, 1998.

[BCJL93] G. Brassard, C. Crépeau, R. Josza, D. Langlois. "A Quantum Bit Commitment Scheme Provably Unbreakable by Both Parties." *Proc. of* 34th *FOCS*, IEEE, 1993, 362–371.

[BS93] G. Brassard, L. Salvail. "Secret-Key Reconciliation by Public Discussion." *Advances in Cryptology – EuroCrypt '93*, Springer Verlag LNCS **765**, 1993, 410–423.

[CSS96] A. R. Calderbank, P. Shor, "Good Quantum Error Correcting Codes Exist." Phys. Rev. A **54**, 1996, 1098–1105. A. M. Steane, "Multiple Particle Interference and Error Correction." Proc. R. Soc. London A **452**, 1996, 2551–2577.

[CDNO97] R. Canetti, C. Dwork, M. Naor, R. Ostrovsky. "Deniable Encryption." *Advances in Cryptology – Crypto '97*, Springer-Verlag LNCS **1294**, 1997, 90–104.

[De89] D. Deutsch. "Quantum Computational Networks." Proc. Royal Society of London, **425**, 1989, 73–90.

[Di95] D. DiVincenzo. "Two-Bit Gates are Universal for Quantum Computation." Phys. Rev. A, **50**, 1995, 1015-1022.

[Ek91] A. Ekert. "Quantum Cryptography Based on Bell's Theorem." Phys. Rev. Letters, **67**:6, 1991, 661-663.

[ERTP92] "Practical Quantum Cryptography Based on Two-Photon Interferometry." Phys. Rev. A, **48**:1, 1993, R5–R8.

[Fe86] R. Feynman. "Quantum Mechanical Computers." Found. Phys. **16**, 1986, 507-531.

[GoPr00] D. Gottesman, J. Preskill. "Secure Quantum Key Distribution using Squeezed States." Los Alamos Preprint Archive quant-ph/0008046, 2000.

[HJW93] L. Hughson, R. Josza, W. Wooters. "A Complete Classification of Quantum Ensembles Having a Given Density Matrix." *Phys Letters A*, **183**, 1993, 14–18.

[LC96] H.K. Lo, H.F. Chau. "Is Quantum Bit Commitment Really Possible?" Los Alamos Preprint Archive quant-ph/9603004, 1996.

[LC97] H.K. Lo, H.F. Chau. "Why Quantum Bit Commitment and Ideal Quantum Coin Tossing are Impossible." Los Alamos Preprint Archive quant-ph/9711065, 1997.

[LC99] H.-K. Lo, H. F. Chau, "Unconditional Security of Quantum Key Distribution over Arbitrarily Long Distances." Science **283**, 1999, 2050–2056.

[Ma93] U. Maurer. "Secret Key Agreement by Public Discussion from Common Information." IEEE Trans. Information Theory, **39**:3, 1993, 733–742.

[Ma96t] D. Mayers. "The Trouble with Quantum Bit Commitment." Los Alamos Preprint Archive quant-ph/9603015, 1996.

[Ma96] D. Mayers. "Unconditionally Secure Quantum Bit Commitment is Impossible." *PhysComp '96*, Boston, November 1996.

[Ma97] D. Mayers. "Unconditionally Secure Quantum Bit Commitment is Impossible." *Phys. Rev. Letters*, **78**, 1997, 3414–3417.

[ShPr01] P. Shor, J. Preskill. "Simple Proof of Security of the BB84 Quantum Key Distribution Protocol." Los Alamos Preprint Archive quant-ph/0003004, 2000.

[Wi83] S. Wiesner. "Conjugate Coding." SIGACT News, **15**:1, 1983, 78–88; orig. manuscript circa 1970.

A Practice-Oriented Treatment of Pseudorandom Number Generators

Anand Desai[1], Alejandro Hevia[2], and Yiqun Lisa Yin[1]

[1] NTT Multimedia Communications Laboratories, Palo Alto, California 94306, USA,
{desai,yiqun}@nttmcl.com

[2] University of California, San Diego, La Jolla, California 92093, USA,
ahevia@cs.ucsd.edu

Abstract. We study Pseudorandom Number Generators (PRNGs) as used in practice. We first give a general security framework for PRNGs, incorporating the attacks that users are typically concerned about. We then analyze the most popular ones, including the ANSI X9.17 PRNG and the FIPS 186 PRNG. Our results also suggest ways in which these PRNGs can be made more efficient and more secure.

1 Introduction

Random numbers or bits are essential for virtually every cryptographic application. For example, seeds for key generation in both secret-key and public-key algorithms, session keys used for encryption and authentication, salts to be hashed with passwords, and challenges used in identification protocols are all assumed to be "random" by system designers. However, since generating enough randomness is expensive, most applications rely on a cryptographic mechanism, known as a Pseudorandom Number Generator (PRNG), to stretch a short string of random bits to a longer string of "random-looking" bits.

Background and Motivation. Cryptographic theory provides us with a number of constructions for PRNGs, such as the Blum-Blum-Shub generator [14] and the Blum-Micali generator [15], that are provably-secure under reasonable number-theoretic assumptions. However, for practical reasons, the PRNGs, that are in prevalent use today, are typically based on efficient cryptographic primitives such as block ciphers and hash functions. The two most widely-used PRNGs are the ANSI X9.17 PRNG [3] and the FIPS 186 PRNG [22]. The ANSI X9.17 PRNG is a part of a popular banking standard and was suggested (nearly the same time as DES) as a mechanism to generate DES keys and nonces. The FIPS 186 PRNG was standardized for generating randomness in DSA. These two constructions remain the only standardized PRNGs, despite there being many subsequent security-related standards. As a result, both of them are now being used as general-purpose PRNGs in various systems and applications.

There has been some analysis of the ANSI X9.17 PRNG and the FIPS 186 PRNG [29,26], but it has been mostly ad hoc and based on heuristic arguments. There have been several theory-oriented treatments of PRNGs [34,15]

L.R. Knudsen (Ed.): EUROCRYPT 2002, LNCS 2332, pp. 368–383, 2002.

but these do not fully capture the way PRNGs are used in practice. Thus, despite their being around for some time now and their wide-spread usage, these PRNGs have yet to be validated in the tradition of provable security. Several cryptographic toolkits, both in commercial products and in free libraries, include general-purpose PRNG implementations that are different from the ANSI X9.17 and FIPS 186 PRNGs. None of these other PRNGs have been analyzed, in any rigorous sense, either.

Our Contributions. Our main goal is to study PRNGs as used in practice. We focus on the ANSI X9.17 and FIPS 186 PRNGs, not only because they are the most widely-used ones, but also because they represent the two typical design approaches for practical PRNGs – one is based on a keyed block cipher while the other is based on a keyless hash function. We also suggest ways of enhancing these PRNGs and look at some other practical ones.

A PRNG can be modeled as an iterative algorithm [1,12]. Each iteration takes a single input called *state* (initial seed or intermediate state) and produces a random output of some fixed length and the next state. All the states are assumed to be hidden at all times. Although such a model seems sufficient for theoretical PRNGs, it does not capture all the nuances of a PRNG as used in practice. Indeed, a PRNG used in practice are often more complicated: (1) There are usually "auxilliary" inputs, such as time-stamps or counters, that the user (or even the attacker) may be able to control. (2) Some state information may be leaked out over time or modified by an attacker. (3) There is a wide range of cryptographic primitives on which they can be based, some that are secret-key based and others that are keyless.

We model a PRNG as an iterative algorithm, that in each iteration take *three* inputs: a key, a current state, and an auxiliary input. It generates two outputs: a PRNG output and a new state. "Pseudorandomness" roughly means that the output sequence should be "indistinguishable from a truly random sequence" to an attacker. To formalize the types of attacks on the above PRNG model, we look at the model from an attacker's point of view – the inputs can be hidden, known or chosen, while the outputs can be hidden or known. This leads to several different attacks, ranging from the "strongest" one where the attacker is allowed to choose all three inputs and see both the outputs to the "weakest" one where it does not see the key or the states and does not get to control the auxiliary input. Each attack coupled with the notion of "pseudorandomness" gives rise to a different definition of security. It is easy to see that there is not a strict hierarchy in the strengths of the attacks. Most of the attacks, that practical PRNGs resist, lie somewhere between the strongest and the weakest. We will be interested in the *strongest* attacks that a PRNG can be secure against since this gives us a complete picture of its security in our framework.

Our analysis of the ANSI X9.17 and the FIPS 186 PRNGs characterize the conditions on the inputs, states, and the underlying primitive that would ensure the security of the construction. They are also instructive as to what would or would not constitute secure-usage of these PRNGs. For the ANSI X9.17 we show that the construction is secure as long as the key is kept secret from the attacker and it is not allowed to control both the input and the state (though, neither needs to be kept secret from it). For the FIPS 186 PRNG we show that the construction is secure if the attacker cannot control the input. Unlike with the ANSI X9.17 PRNG, the key in this case is known to all, including the attacker. The states, however, must be kept secret. Besides validating the security of the PRNGs, our results have several useful practical implications. First, they suggest guidelines on how the PRNGs should be properly used to achieve their security objectives. Second, they suggest how to obtain some efficiency gains. For example, our results on the ANSI X9.17 PRNG imply that the construction can be made twice as efficient, by outputting intermediate states along with the output, without comprising security. Third, they suggest where the use of "good" randomness is critical and where it is not. For example, our results on the ANSI X9.17 PRNG, suggest that it is better to put all the randomness in the key rather than in the initial state. Finally, the analysis provides some insight on what security properties the underlying primitives should possess. This allows implementers to choose the appropriate primitives and to examine whether any newly-discovered weakness on the primitive can have a potential impact on the security of the PRNG.

A CLOSER LOOK. One important aspect of our framework is to allow for an auxiliary input to model things like time-stamps and counters. Auxiliary inputs are a common feature in practical PRNGs since they are a means of injecting something "random" into the PRNG at regular intervals and to prevent repeated seeds from causing repeated outputs. It is tempting to think that if a PRNG is secure without any auxiliary input then it must also be secure with it present. This, however, is not the case, particularly if this input can be controlled by the attacker. Indeed, in practice the auxiliary input may be supplied by a timer that an attacker may have some control over. Another aspect in which our model differs from most others is in making a distinction between the key and the state. When such a distinction is not made (and the key is simply understood to be a part of the state) then we miss out on being able to exactly characterize the security of the PRNG. The key and what we call as state play very different roles in a PRNG and moreover, they may have significantly different impacts on the overall security. Understanding this allows the user to make better use of the short seed it possesses.

A critical part of our analysis is to formalize reasonable security assumptions on the underlying primitives for the manner they are used in the ANSI X9.17 and the FIPS 186 PRNGs. There are no assumptions stated on the underlying primitive in the original ANSI X9.17 standard [3] or in the updated version (ANSI X9.31) [4]. In the FIPS 186 standard [22], it stated that the underlying primitive should be a "one-way" function. However, it is easy to see the one-way property itself is not enough to ensure the security of the construction. Going to an extreme, one may model the underlying primitive as a Random Oracle. The proof of security, in this case, would indeed be straightforward, but there are some results [16] that cast doubt on whether this would be a reasonable assumption. Our assumption for both the constructions is that the underlying primitives are (finite) pseudorandom functions (PRFs). This is quite a natural assumption for the block-cipher-based ANSI X9.17 PRNG, following the work of Bellare, Kilian and Rogaway [10]. It is less so for the hash-function-based FIPS 186 PRNG, since there are no "secret keys" as such associated with the underlying function in this PRNG. However, we propose a different "view" of the PRNG under which it can be seen as based on a secret-keyed hash function. Under this view it seems reasonable to model the underlying function as a PRF. We note here that similar assumptions have been made before [9,7,6,2] and that none of the known attacks on hash functions [18,21,32] suggest any weakness of their being used in this manner.

Starting with the strongest attacks in our framework, we look at the ability of the ANSI X9.17 and the FIPS 186 PRNGs to withstand them. For every attack stronger than the ones we eventually proved these PRNGS secure against, we identify the attack. Our approach helps us identify the criticality of each feature in these PRNGs. The superfluous ones can be eliminated leading to an improvement in the overall efficiency. It turns out that the designers of both these PRNGs anticipated many of the attacks in our framework and factored these into their designs – but not necessarily in the most efficient manner. We also suggest ways of making these PRNGs secure against attacks that, in their present form, they are insecure against. We remark that although the PRNGs we consider seem to incorporate many of the design features of better-known constructions, most notably of the CBC and Counter Mode constructs, their security analyses are quite different. This is partly due to the goal of the attacker being different and since the attacks we must consider are unlike any of those considered before. We show that if the primitives underlying the PRNGs are secure as PRFs then the PRNGs must be secure as well. Our analysis is *exact* rather than asymptotic.

Related Work. There have been numerous works on the theory of PRNGs, such as, Blum, Blum and Shub [14], Blum and Micali [15] and Hastad, Impagliazzo, Levin and Luby [27], to name just a few. However there have been comparatively few analyses of PRNGs of the type we study here, namely those based on efficient cryptographic primitives, such as block ciphers and hash functions.

Aeillo, Rajagopalan and Venkatesan [2] give a provably-secure design based on block ciphers. Bellare and Yee [12] and Abdalla and Bellare [1] also studied such PRNGs from the point of bringing forward-security to bear on them. They give generic methods of converting any PRNG into one that is forward-secure. We build on their security model to get a more general one.

While the above works did look at PRNGs from a provable-security viewpoint, there has not been anything similar done for the two most prevalent ones, namely the ANSI X9.17 and the FIPS 186 PRNGs. Kelsey, Schneier, Wagner and Hall [29] have, however, analysed these PRNGs, from an attackers viewpoint. That is, they give several different types of attacks and investigate the ability of these PRNGs to withstand them. While such analyses do not give the same level of guarantees as those we get, they are useful in identifying weaknesses. Indeed, many of the attacks on the PRNGs that become obvious in our framework were identified there as well. The only other work that we are aware of, which looks at one of these PRNGs, is the recent work of Bleichenbacher [13]. A weakness in the FIPS 186 PRNG, as used to generate DSA parameters, is identified. It does not apply to the general-purpose FIPS 186 PRNG we consider. The standard has anyway subsequently been revised [23] to function like the general-purpose one we consider.

There have been analyses of constructions bearing some similarity to the PRNGs we consider, namely the CBC MAC [10] and CBC and Counter Mode Encryption [8]. As mentioned earlier, the security of the PRNGs we consider do not follow from these analyses but the general provable-security techniques used therein are, however, applicable.

2 Security Framework and Definitions

PRNG MODEL. A PRNG $\mathcal{GE} = (\mathcal{K}, \mathcal{G})$ consists of two algorithms. The *seed generation* algorithm $\mathcal{K}$ takes as input a security parameter $k \in \mathsf{N}$ and returns a *key* K and an *initial state* s_0. For $i \geq 1$, the *generation* algorithm $\mathcal{G}$ takes as input the key K, the *current state* s_{i-1} and an *auxiliary input* t_i and returns a *PRNG output* y_i and the *next state* s_i. We refer to the length of the PRNG output in each iteration $n = |y_i|$ as the block length of the PRNG. Note that syntax requires the PRNG to have the "online" property. That is, it should be possible to generate y_i, s_i before t_{i+1} is known. We will sometimes refer to the auxiliary input simply as the "input" and the PRNG-output simply as the "output", when it is unambiguous from context.

We assume that there is a "good" source of randomness that accumulates in an *entropy pool* and that $\mathcal{K}$ has access to it. The process by which randomness is collected into the pool can be considered as orthogonal to the above-defined pseudorandom number generation process, and will not be considered further in our analysis. A couple of practical PRNGs [31,28] do specify both processes as part of the PRNG definition. It is quite easy, however, to separate the two and analyze them independently.

ATTACK MODELS. Consider the normative operation of a PRNG as described above. At the start, a seed consisting of a key K and an initial state s_0 is generated from the entropy pool. After seed generation, the operation of the PRNG can be viewed as an application-controlled process – outputs are generated only as many as are required, using the *current* state and an auxiliary input. The auxiliary input may be a time-stamp, a counter or samples from a low entropy source. When the application stops the PRNG, the last state is usually computed and saved for use as the initial-state in the next invocation of the PRNG. The application may also choose to "re-seed" the PRNG at any time, if enough entropy has accumulated in the pool.

From an attacker's point of view, each of the three inputs to $\mathcal{G}$ can be hidden, known or chosen, while each of the two outputs from $\mathcal{G}$ can be hidden or known. Given the nature of a PRNG, it is not necessary to consider all the cases. We allow the *key* to be hidden or known, but not chosen. We view a key as defining the PRNG and hence a "chosen-key" attack would not help in "breaking" the PRNG being attacked as such. The *state* needs to be maintained between invocations of the PRNG. There is thus an increased probability that an attacker may learn of it. Moreover it may be able to "change" the value of the stored state. For these reasons we model attacks that allow the state to be hidden, known, or even chosen. The *auxiliary inputs* may not always be known a priori, even to a legitimate user. However, it is reasonable to assume that they are quite "predictable" by an attacker due to their low entropy. (Otherwise the generation of pseudorandom outputs would be trivial). Also, it is possible that the auxiliary inputs may be supplied by a timer or source over which an attacker may exercise some control. So we model the auxiliary inputs to be known or chosen. The *PRNG output* could conceivably be hidden, particularly if the PRNG is being used as a key-derivation function. However, since our focus is on general-purpose PRNGs, we will assume that the outputs become known. We further note that the choice of the primitive to realize the PRNG often rules out additional cases. For example, for PRNGs based on unkeyed hash functions the "key" is understood to be known. For PRNGs based on block ciphers the key is typically required to be hidden, though it is also be possible that it is known. Given the above, one can see why it is advantageous to consider the state and key separately in a practical setting. In most PRNGs the role played by the key and what we call the state are quite different. The key typically has a much longer lifetime and may be repeatedly used for different invocations of the PRNG. The part that we call the state has a more transient nature, since it is usually updated during every iteration of the generation algorithm.

SECURITY DEFINITIONS. We adopt a standard notation with respect to probabilistic algorithms and sets. If $A(\cdot,\cdot,\ldots)$ is any probabilistic algorithm then $a \leftarrow A(x_1, x_2, \ldots)$ denotes the experiment of running A on inputs $x_1, x_2, \ldots$ and letting a be the outcome, the probability being over the coins of A. Similarly, if A is a set then $a \stackrel{R}{\leftarrow} A$ denotes the experiment of selecting a point uniformly from A and assigning a this value.

"Pseudorandomness" roughly means that the outputs should be "indistinguishable from random" to an attacker. Each attack in our framework coupled with the notion of "pseudorandomness" gives rise to a different security definition. Here we give more precise definitions for three of the attacks. We denote these attacks as CIA, for Chosen-Input Attack, CSA, for Chosen-State Attack and KKA, for Known-Key Attack. Under CIA, the key is hidden, the states are known, but not chosen, and the auxiliary input may be chosen by the attacker. CSA is similar, except that the auxiliary inputs are not allowed to be chosen while the states may now be chosen. KKA is different in that it allows the key to be known. However, under KKA, the states are hidden and the auxiliary inputs are not allowed to be chosen.

We imagine a distinguisher D running in two stages, an iterated find stage and a guess stage. In each iteration of the find stage D may get to choose some of the inputs (depending on the attack being modeled) that will be used to generate the PRNG output. In cases modeling a "chosen" input D's choice "overwrites" the existing value of that input. Depending on the bit b, D either gets the true PRNG output from the previous iteration or a random string. The find stage is iterated i times until $i = m$ or until D sets a flag p_i, indicating that it is ready to move on to the next stage. In the guess stage D receives some state information c_i gathered in the find stage and outputs its guess for b.

Definition 1. [PRG-CIA, PRG-CSA, PRG-KKA] *Let $\mathcal{GE} = (\mathcal{K}, \mathcal{G})$ be a PRNG with block-length n. (For simplicity, assume that the security parameter $k = n$.) Let $b \in \{0,1\}$. Let D be a distinguisher that runs in two stages. For* $\text{atk} \in \{\text{cia}, \text{csa}, \text{kka}\}$, *we consider the following experiment:*

> *Experiment* $\mathbf{Exp}^{\text{prg-atk-}b}_{\mathcal{GE},m,D}(k)$
> $(K, s_0) \stackrel{R}{\leftarrow} \mathcal{K}(k)$ // *A key and an initial state are generated*
> $c_0 \leftarrow \{t_1, t_2, \ldots, t_m\}$ // *D is initialized with auxiliary input values*
> *for* $i = 1$ *to* m : $y_i^0 \stackrel{R}{\leftarrow} \{0,1\}^n$ // *A random mn-bit string is chosen*
> $i \leftarrow 0$; $y_0^0 \leftarrow \epsilon$; $y_0^1 \leftarrow \epsilon$
> *repeat*
> $i \leftarrow i + 1$
> *if* $\text{atk} = \text{cia}$: $(p_i, t_i, c_i) \leftarrow D(\mathsf{find}, y_{i-1}^b, s_{i-1}, c_{i-1})$
> *if* $\text{atk} = \text{csa}$: $(p_i, s_{i-1}, c_i) \leftarrow D(\mathsf{find}, y_{i-1}^b, s_{i-1}, c_{i-1})$
> *if* $\text{atk} = \text{kka}$: $(p_i, c_i) \leftarrow D(\mathsf{find}, K, y_{i-1}^b, c_{i-1})$
> $(y_i^1, s_i) \leftarrow \mathcal{G}_K(s_{i-1}, t_i)$ // *PRNG output and next state are generated*
> *until* $(p_i = \mathsf{guess})$ *or* $(i = m)$
> $d \leftarrow D(\mathsf{guess}, c_i)$
> *return* d

We define the advantage of the distinguisher via

$$\mathbf{Adv}^{\text{prg-atk}}_{\mathcal{GE},m,D} = \Pr[\mathbf{Exp}^{\text{prg-atk-1}}_{\mathcal{GE},m,D} = 1] - \Pr[\mathbf{Exp}^{\text{prg-atk-0}}_{\mathcal{GE},m,D} = 1] .$$

We define the advantage function of $\mathcal{GE}$ as follows. For any integer t,

$$\mathbf{Adv}^{\text{prg-atk}}_{\mathcal{GE},m}(t) = \max_D \{\mathbf{Adv}^{\text{prg-atk}}_{\mathcal{GE},m,D}\}$$

where the maximum is over all D with "time complexity" t. ∎

The "time complexity" is the worst case total execution time of the experiment, plus the size of the code of the distinguisher, in some fixed RAM model of computation.

Note that the notions of security captured here are very strong but not the strongest in our framework. The reason for explicitly defining these and not the other possible ones was to give only as many definitions as were needed to give the complete picture about the security of PRNGs we will consider in this work. However, using the above as examples, it should be easy to come up with definitions modeling any attack in our framework.

It is instructive to see where these notions stand in our framework. When the key is hidden, the strongest notion would be one against a chosen-state, chosen-input and known-next-state attack. The next strongest notions, when the key is hidden, are PRG-CIA and PRG-CSA. When the key is known, the states must be hidden (otherwise there is no secret). This leaves us with only two notions: one allowing the input to be chosen and the other only allowing it to be known. PRG-KKA is the latter notion (which is the weaker of the two). Note that these three notions are mutually incomparable (ie. they are neither stronger nor weaker than one another).

3 Analysis of the ANSI X9.17 PRNG

SPECIFICATIONS. The ANSI X9.17 PRNG $\mathcal{GE}^F_{\text{ANSI}} = (\mathcal{K}_{\text{ANSI}}, \mathcal{G}_{\text{ANSI}})$, for a block cipher F, is described as in Figure 1. The key K, output by the key-generation algorithm, is used to key the block cipher, thereby specifying a function F_K that maps n bits to n bits. ATTACKS. The PRNG is insecure under any attack

$\mathcal{K}_{\text{ANSI}}(n)$	$\mathcal{G}_{\text{ANSI}}(s_{i-1}, t_i)$
$K \stackrel{R}{\leftarrow} \{0,1\}^n$	$y_i \leftarrow F_K(s_{i-1} \oplus F_K(t_i))$
$s_0 \stackrel{R}{\leftarrow} \{0,1\}^n$	$s_i \leftarrow F_K(y_i \oplus F_K(t_i))$
return (K, s_0)	return (y_i, s_i)

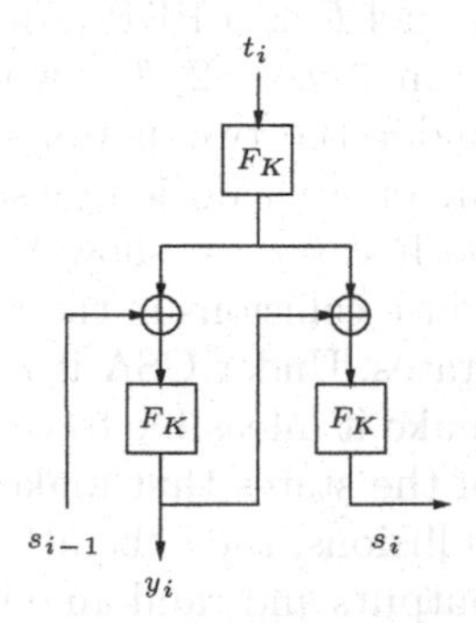

Fig. 1. Specifications of the ANSI X9.17 PRNG. The picture depicts the function $\mathcal{G}_{\text{ANSI}}(\cdot,\cdot)$.

in which the key is known. An attacker, knowing the key and a single output, can completely determine subsequent outputs and states. The PRNG is also not secure under any attack where both the state and the inputs may be chosen. It is easy to see that an attacker can cause outputs to repeat by simply repeating the same combination of the state and input. Given this, we focus on the setting where the key is hidden and where either the states or the inputs are not under the control of the attacker.

FINITE PRF FAMILIES. We model block ciphers as finite pseudorandom function families [10], a concrete security version of the original notion of pseudorandom functions [25]. A finite function family F is pseudorandom if, for a random K, the input-output behavior of F_K is indistinguishable from that of a random function of the same domain and range. Following [10], we associate an advantage function $\mathbf{Adv}^{\text{prf}}_F(t, q)$ with F which denotes the maximum advantage of any distinguisher, with time complexity t and query complexity q. The (standard) formalization of this can be found in the full version of this paper [19].

SECURITY RESULTS. The following theorem implies that, if F is a PRF, then the ANSI X9.17 PRNG based on it is secure in the PRG-CSA sense and also in the PRG-CIA sense. Note that the two notions are mutually incomparable and thus proving the PRNG secure under one does not imply security under the other.

Theorem 1. *Let $\mathcal{GE}$ be the ANSI X9.17 PRNG based on a function family $F = \{F_K\}_{K \in \mathrm{Keys}(F)}$ where F_K maps n bits to n bits. Then*

$$\mathbf{Adv}^{\text{prg-csa}}_{\mathcal{GE},m}(t) \leq 2 \cdot \mathbf{Adv}^{\text{prf}}_{F}(t, 3m) + \frac{m(2m-1)}{2^n}$$

$$\mathbf{Adv}^{\text{prg-cia}}_{\mathcal{GE},m}(t) \leq 2 \cdot \mathbf{Adv}^{\text{prf}}_{F}(t, 3m) + \frac{(2m-1)^2}{2^n} \quad \blacksquare$$

A proof of the theorem is given in the full version of this paper [19]. If we model F as a PRP (a more suitable model for a block cipher) then an additional term $(3m)^2 \cdot 2^{-n-1}$ would appear in the bounds above. The proofs of security under the two notions share much in common. We first analyze the PRNG, under each notion, assuming that the underlying function is truly random. In both cases we show that, with overwhelming probability, an attacker cannot cause collisions in the inputs to the functions computing the outputs or the next states. Under CSA it is the inputs, which essentially function as a counter, that make it infeasible to cause these collisions. Under CIA, it is the unpredictability of the states that makes it infeasible. It is easy to see that, if there are no such collisions, then the attacker has no advantage in distinguishing between PRNG outputs and random ones. The next part of the analysis is a reduction argument that shows that, if the random function in the above analysis were replaced by a PRF, then the PRNG must remain secure.

REMARKS. Our results suggest how the PRNG could be used more efficiently without compromising security. Notice that in the analysis, under either of the notions, we assume that the attacker learns of the states. Moreover, the states themselves are indistinguishable from random to the attacker. (In fact, there is complete symmetry in the way the outputs and the states are computed.) This means that the states could be used as "outputs", effectively doubling the throughput of the PRNG, without any loss in security. Our analysis also offers some insights into how the PRNG can be extended to counter the attacks that it does not in its present form. We discuss this further in Section 5.

4 Analysis of the FIPS 186 PRNG

SPECIFICATIONS. The FIPS 186 PRNG $\mathcal{GE}^{H}_{\text{FIPS}} = (\mathcal{K}_{\text{FIPS}}, \mathcal{G}_{\text{FIPS}})$, for a function H, is described as in Figure 2. Note that in this PRNG, unlike in the ANSI X9.17 one, the key K is *known*, and the values for K are *fixed* in the original

$\mathcal{K}_{\mathrm{FIPS}}(n)$	$\mathcal{G}_{\mathrm{FIPS}}(s_{i-1}, t_i)$
$K \leftarrow \{0,1\}^n$	$y_i \leftarrow H_K((s_{i-1} + t_i) \bmod 2^n)$
$s_0 \stackrel{R}{\leftarrow} \{0,1\}^n$	$s_i \leftarrow (s_{i-1} + y_i + 1) \bmod 2^n$
return (K, s_0)	return (y_i, s_i)

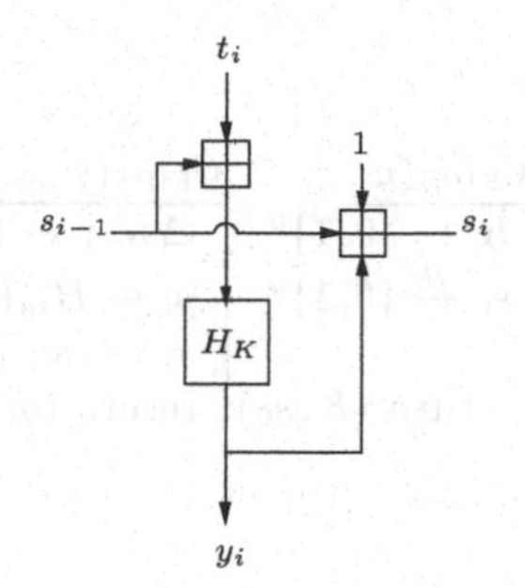

Fig. 2. Specifications of the FIPS PRNG. The picture depicts the function $\mathcal{G}_{\mathrm{FIPS}}(\cdot,\cdot)$.

specifications [22].[1] To fully specify this PRNG, we must describe the underlying primitive H_K. However, for the sake of simplicity of exposition, we defer this to the end of this section, viewing H_K, for now, simply as a function mapping n bits to n bits.

ATTACKS. Since the key K is known, the only secret in the PRNG is the initial state s_0. Clearly, if s_0 is known to the attacker then the outputs become completely predictable. In general, such a PRNG cannot be secure in any attack where the state is known. It also turns out that this PRNG is not secure under any attack where the inputs are chosen since otherwise it becomes possible for an attacker to cause outputs to repeat. Given this, we focus on the setting where the states are hidden and the inputs are not under the control of an attacker.

AN ALTERNATIVE VIEW. We first need to formalize a security assumption on the underlying primitive H_K that is reasonable for the manner in which it is realized and used in the specifications. Towards this, we will consider an alternative view of FIPS. We start with some notation. Let

$$\hat{H}_{s_0}(x) \stackrel{\text{def}}{=} H_K((s_0 + x) \bmod 2^n).$$

For simplicity of notation we do not have an explicit reference to K in the definiton of $\hat{H}_{s_0}$. Nevertheless, one should keep in mind that K is a part of the definition of $\hat{H}_{s_0}$. Now, we can rewrite the specifications of the FIPS 186 PRNG as in Figure 3. Note that the alternative specifications are simply another way of viewing the original FIPS 186 PRNG. The two specifications interpreted literally will have different-looking implementations but they will, nevertheless, have exactly the same input-output characteristics. We view it in our "secret-key" function-based form for sake of analysis.

[1] In the original FIPS specifications [22], the underlying primitive is denoted by $G(K, x)$, where K is a constant and x is the input. It specifies two 160-bit values for the constant K, with each constant instantiating a different PRNG process for different use in DSA. Although the specifications do not interpret K as a key, the construction fits naturally into our PRNG model if we view K as a known key. The FIPS 186 construction can be generalized to have an arbitrary 160-bit value (even secret) for K.

$\mathcal{K}_{\mathrm{FIPS}}(n)$	$\hat{\mathcal{G}}_{\mathrm{FIPS}}(s_{i-1}, t_i)$
$K \leftarrow \{0,1\}^n$	$\Delta s_{i-1} \leftarrow (s_{i-1} - s_0) \bmod 2^n$
$s_0 \stackrel{R}{\leftarrow} \{0,1\}^n$	$y_i \leftarrow \hat{H}_{s_0}((\Delta s_{i-1} + t_i) \bmod 2^n)$
	$s_i \leftarrow (s_{i-1} + y_i + 1) \bmod 2^n$
return (K, s_0)	return (y_i, s_i)

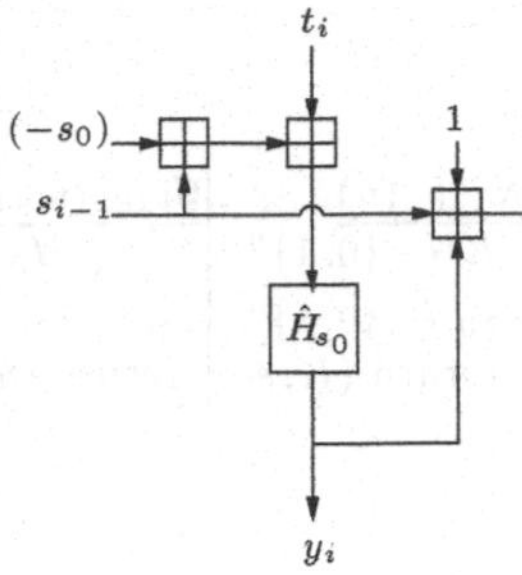

t_i
$(-s_0)$
1
s_{i-1}
s_i
$\hat{H}_{s_0}$
y_i

Fig. 3. Alternative specifications of the FIPS 186 PRNG. The picture depicts the function $\hat{\mathcal{G}}_{\mathrm{FIPS}}(\cdot,\cdot)$, which is an equivalent view of $\mathcal{G}_{\mathrm{FIPS}}(\cdot,\cdot)$.

SECURITY RESULTS. The following theorem implies that, if $\hat{H}$ is a PRF family, then the FIPS 186 PRNG based on it is secure in the PRG-KKA sense.

Theorem 2. *Let $\mathcal{GE}$ be the FIPS 186 PRNG based on a function family $\hat{H} = \{\hat{H}_{s_0}\}_{s_0 \in \{0,1\}^n}$. Then*

$$\mathbf{Adv}^{\text{prg-kka}}_{\mathcal{GE},m}(t) \leq 2 \cdot \mathbf{Adv}^{\text{prf}}_{\hat{H}}(t,m) + \frac{m(m-1)}{2^{n-1}} \quad \blacksquare$$

A proof of the theorem is given in the full version of this paper [19]. The approach in the analysis is much like the one for the ANSI X9.17 PRNG, where we first analyze the PRNG assuming the underlying function to be random and then give a reduction argument to show that if the PRNG is secure with the random function then it must also be secure with a PRF. The interesting part for this particular analysis is in the reduction argument. Recall that for this argument, we must construct a distinguisher A for the PRF that simulates the PRNG experiment by calling its oracle f on the input $(\Delta s_{i-1} + t_i) \bmod 2^n$. It is easy to see that $\Delta s_i = (\Delta s_{i-1} + y_i + 1) \bmod 2^n$ can be computed without knowledge of s_0. This makes it possible for A to make the necessary queries and the argument goes through.

THE FUNCTION H_K AND THE FUNCTION FAMILY $\hat{H}$. It remains to give some details about the function H_K and to comment on the security of the function family $\hat{H}$. The FIPS 186 specifications provides two constructions for the function H_K, one based on SHA1 and the other based on DES.

The SHA1-based construction is fairly simple – $H_K(x)$ is defined to be the compression function of SHA1 where x is the input to the function and K is the fixed IV in SHA1. Our assumption is that one can view $\hat{H}_{s_0}(x) = H_K((s_0 + x) \bmod 2^n)$ for a known K and a secret random s_0 of length n, as a PRF. As mentioned earlier, this assumption is not entirely new. Bellare et al [9,7,6] make an assumption that, in our context, amounts to viewing $H_K(s_0, x)$ as a PRF. It appears as though all existing attacks on hash functions need to either fix certain bits of the input to the compression function or exploit the underlying iterative structure of the functions. Since neither of these seem possible in our setting, we believe that it is reasonable to view the compression function of SHA1, as used in the FIPS 186 PRNG, as a PRF.

The DES-based construction is more involved – DES encryption is performed five times with a different secret key and input each time. The secret keys and the inputs are both derived from K and x. It is significantly more complicated to gauge whether a DES-based $\hat{H}$ function family can be considered to be a PRF family. Anyway, it appears that the DES-based construction is rarely, if at all, used in practice.

5 Forward Security for PRNGs

The notion of forward security has been applied to a range of cryptographic problems. It is a particularly desirable goal for PRNGs. Indeed, the idea itself seems to have originated in the PRNG context. Informally, a PRNG is said to be forward secure if the compromise of the current state *and* key does not compromise the security of any previously generated PRNG-output. Two different formulations of the notion of forward security in PRNGs have appeared in the literature. We will look at each in turn.

The first, that we will refer to as *weak* forward security, assumes that the PRNG-outputs, prior to the compromise, are *hidden* from the attacker. The goal of the attacker is to derive information about the earlier outputs. This notion seems to have been considered by Kelsey et al [29]. Although the term "forward security" was not explicitly used, the "backtracking attack" introduced therein is precisely the type of attack implicit in the notion of weak forward security.

A more comprehensive study of the notion of forward security in the context of PRNGs was initiated in the works of Bellare and Yee [12] and Abdalla and Bellare [1]. They considered a stronger formulation of the notion, that we will refer to as *strong* forward security. In this notion it is assumed that the outputs prior to the compromise are *known* to the attacker. The goal of the attacker is to distinguish the earlier output sequence from a truly random one.

Bellare and Yee [12] suggest ways of making a strong forward-secure PRNG out of any generic PRF-based PRNG. The main idea there is to keep part of the PRNG output secret and use it to get a new state and key. The idea applied quite literally to the PRNGs we consider result in unnecessarily inefficient constructions. Moreover, the fact that their analysis did not consider auxilliary inputs or make a distinction between the key and the state, makes it somewhat difficult to see if their method would be secure in our setting. Nevertheless, we believe that one could use their basic idea involving re-keying to get forward security for both the PRNGs we consider.

We now look at some alternate approaches to this problem, specifically for the ANSI X9.17 and the FIPS 186 PRNGs. We want to avoid, as far as possible, necessitating major changes to the constructions (such as using a different underlying primitive), so that our "fixes" can be applied to existing implementations of the PRNGs. For efficiency reasons we look for solutions that do not involve re-keying the underlying functions.

THE CASE OF THE ANSI X9.17 PRNG. It is easy to see that this PRNG is not forward secure, under even the weaker form, since revealing the (secret) key would make the PRNG process completely reversible. As a first step towards bringing forward security to bear upon this PRNG, we must transform the underlying function F_K into a non-invertible function (even when the key K is known). While such a change may be sufficient for weak forward security, a more subtle attack implies that the modified version will still be insecure under the stronger notion. The attack is to check, using the last output, time-stamp and key, if the state-update function does in fact give the current state.

The above attack suggests that a way to get strong forward security is to require that the state-update function use a value which remains *secret* even when the key and the current state are compromised (in addition to using a non-invertible function). One possibility is to change the state-update function from $s_i \leftarrow F_K(y_i \oplus F_K(t_i))$ to $s_i \leftarrow F_K(y_i \oplus s_{i-1})$. It remains to discuss how we can get a non-invertible function. What we need specifically is a function F', such that given $F'_K(x)$ and K, it is infeasible to get information about x. One can build such a function using ideas from converting PRPs to PRFs [11]. However these require rekeying. One possible candidate that avoids this and may suffice for our purposes is the function F' defined as $F'_K(x) = F_K(x) \oplus x$, where F is a block cipher.

THE CASE OF THE FIPS 186 PRNG. Observe that the "feedback" structure (ie. using output y_i as an input to the state-update function) in this PRNG does not really play a role in preventing against any of the attacks considered in our framework. However, it is precisely this feature that makes this PRNG weak forward-secure. The PRNG, however, is completely insecure under the stronger notion. Given the current state and the outputs, every previous state, going back to the initial state can be determined. Note that the standard attacks on this PRNG anyway make it necessary for the output-computation function to be "perfectly" one-way. Thus if we want to make it strong forward secure, then we only need to modify the state-update function. One possibility is to use a hash function in the state update. (The function H_K used in the PRNG may be used as this hash function.)

6 Other Practical PRNGs

We have, so far, concentrated on just the ANSI X9.17 and the FIPS 186 PRNGs. Other than being the two most popular PRNGs, they are also representative of the two main types – secret-keyed block-cipher-based ones and keyless hash-function-based ones. Most other practical PRNGs are similar to one of these and it is relatively straightforward to analyze them using the same framework and approach. We discuss some of these below.

PRNGS IN STANDARDS. We have reviewed the PRNGs in most security-related standards of ANSI, FIPS, IETF, IEEE and ISO. Not surprisingly, the ANSI X9.17 and FIPS 186 PRNGs have been chosen by subsequent ANSI and FIPS standards. In addition, both PRNGs have been re-validated and cross-validated by the two standards bodies (e.g., ANSI X9.31 [4] and FIPS 186-2 (change notice 1) [23]). Most other standards do not specify any concrete PRNG construction but merely provide guidelines on how random numbers should be generated. The ANSI X9.17 and FIPS 186 constructions are, as far as we know, the only standardized PRNGs.

The only relevant PRNG-like construction that has appeared in standards is the PRF(k, *seed*) function in the Transport Layer Security Protocol (TLS) [20] (TLS is part of the IETF effort to standardize SSL [24].) Despite its name, PRF is used for key derivation rather than generating random numbers needed in the protocol. The latter was assumed to be generated by some "secure PRNG." Nevertheless, PRF is of interest to us since it does represent a standalone pseudorandom number generating process. At a high level, PRF is a secret-key based construction using HMAC [6] as the underlying primitive. The initial state is assumed a "random" but public value, and there is no user-supplied input. HMAC is used to both generate the output and update the state. Our preliminary analysis suggests that PRF is secure under known-state attacks, but not secure under chosen-state attacks. We remark that this apparent weakness does not have any immediate impact on the security of TLS itself.

PRNGs in Commercial Products and Free Libraries. There are several cryptographic toolkits, both in commercial products and in free libraries, which are likely to be used when implementing cryptographic solutions. Among commercial products, NAI's PGP [31] and RSA's BSAFE [5] are perhaps the two most widely-used ones. Among popular free libraries, we considered CryptoLib [30], OpenSSL [35], Crypto++ [17], RSAREF [33] and Yarrow [28], all of which include general-purpose PRNG implementations.

Recently NAI released the source code for PGPsdk [31] for peer review. This version of PGP specifies the default PRNG to be ANSI X9.17 with the block cipher CAST5. It also implements a quite complicated process for generating the initial seed and time-stamps required by the ANSI X9.17 PRNG. There seems to be some uncertainty as to whether the ANSI X9.17 PRNG can produce pseudorandom output if the seed and time-stamps do not have enough entropy. Our analysis, however, shows that the PRNG is secure as long as the key is secret and random and the attacker does not have control over both the state and time-stamps. Thus, it seems that the complication involved in generating pseudorandom inputs in PGP may be unnecessary.

The PRNG included in the BSAFE toolkit [5] is a hash-function-based construction with no secret key and with user-inputs processed as they become available. The construction bears some resemblence to the forward-secure variant of the FIPS PRNG we suggest in Section 5, without the output feedback. Our preliminary analysis shows that the BSAFE PRNG has similar security properties as the FIPS PRNG. That is, it is secure under known-input attacks but not secure under chosen-input attacks. The BSAFE PRNG also has the additional feature of allowing multiple output-computation between two state updates.

The Crypto++ library includes a PRNG based on a (less efficient) variant of the ANSI X9.17 construction. The PRNG implemented in OpenSSL bears some similarities with the FIPS 186 construction but is a lot more complicated and is less efficient. The PRNGs in CryptoLib, RSAREF and Yarrow, on the other hand, have a quite simple Counter-Mode structure. CryptoLib's PRNG implements a scheme that combines the encryption and hash of a counter. The RSAREF PRNG uses a hash function to compute digests of the counter and inputs. Poor handling of inputs, however, have raised concerns about its security [29]. The Yarrow PRNG seems to be a carefully-designed secret-key-based construction using Counter-Mode encryption. It provides some level of forward security by performing frequent (hash-based) re-seeding and re-keying operations.

Acknowledgements

We thank Mihir Bellare and the members of the Program Committee for their helpful comments.

References

1. M. Abdalla and M. Bellare, "Increasing the Lifetime of a Key: A Comparative Analysis of the Security of Re-keying Techniques," ASIACRYPT 2000.
2. W. Aiello, S. Rajagopalan, and R. Venkatsan, "High-Speed Pseudorandom Number Generation with Small Memory," FSE 1999.
3. ANSI X9.17 (Revised), "American National Standard for Financial Institution Key Management (Wholesale)," America Bankers Association, 1985.
4. ANSI X9.31, "American National Standard for Financial Institution Key Management (Wholesale)," America Bankers Association, 2001.
5. R. Baldwin, "Preliminary Analysis of the BSAFE 3.x Pseudorandom Number Generators," *RSA Laboratories*' Bulletin No. 8, 1998.
6. M. Bellare, R. Canetti and H.Krawczyk, "Keying Hash Functions for Message Authentication," CRYPTO 1996.
7. M. Bellare, R. Canetti and H. Krawczyk, "Pseudorandom Functions Revisited: The Cascade Construction and its Concrete Security," FOCS 1996.
8. M. Bellare, A. Desai, E. Jokipii and P. Rogaway, "A Concrete Security Treatment of Symmetric Encryption," FOCS 1997.
9. M. Bellare, R. Guérin, and P. Rogaway, "XOR MACs: New Methods for Message Authentication using Finite Pseudorandom Functions," CRYPTO 1995.
10. M. Bellare, J. Kilian and P. Rogaway, "The Security of the Cipher Block Chaining Message Authentication Code," CRYPTO 1994.
11. M. Bellare, T. Krovetz, and P. Rogaway, "Luby-Rackoff Backwards: Increasing Security by making Block Ciphers Non-Invertible," EUROCRYPT 1998.
12. M. Bellare and B. Yee, "Forward Security in Private-Key Cryptography," *Cryptology ePrint Archive*, Report 2001/035.
13. D. Bleichenbacher, Lucent Technologies Press Release, `http://www.lucent.com/press/0201/010205.bla.html`.
14. L. Blum, M. Blum, and M. Shub, "A Simple Unpredictable Pseudorandom Number Generator." *SIAM J. Computing*, 15(2), 1986.

15. M. Blum and S. Micali, "How to Generate Cryptographically Strong Sequences of Pseudorandom Bits." *SIAM J. Computing*, 13, 1984.
16. R. Canetti, O. Goldreich, and S. Halevi, "The Random Oracle Model, Revisited." STOC 1998.
17. W. Dai, Crypto++ Library. `http://www.eskimo.com/~weidai/cryptlib.html`
18. B. den Boer and A. Bosselaers, "Collisions for the Compression Function of MD5," CRYPTO 1993.
19. A. Desai, A. Hevia and Y.L. Yin, "A Practice-Oriented Treatment of Pseudorandom Number Generators," `http://www.cs.ucsd.edu/users/adesai`.
20. T. Dierks, and C. Allen, "The TLS Protocol Version 1.0," RFC 2246, *Internet Request for Comments*, 1999.
21. H. Dobbertin, "The Status of MD5 After a Recent Attack," RSA Labs' CryptoBytes, Vol.2 No.2, 1996.
22. FIPS PUB 186-2, "Digital Signature Standard," National Institute of Standards and Technologies, 1994.
23. FIPS PUB 186-2 (Change Notice 1), "Digital Signature Standard," National Institute of Standards and Technologies, 2001.
24. A. Frier, P. Karlton, and P. Kocher, "The SSL 3.0 Protocol," Netscape Communications Corp., Nov 18, 1996.
25. O. Goldreich, S. Goldwasser, and S. Micali, "How to Construct Random Functions," *Journal of the ACM*, Vol. 33, N0. 4, 1986.
26. P. Gutmann, "Software Generation of Practically Strong Random Numbers," USENIX Security Symposium 1998.
27. J. Hastad, R. Impagliazzo, L. A. Levin, and M. Luby, "Pseudorandom Generation from One-Way Functions," STOC 1989.
28. J. Kelsey, B. Schneier, and N. Ferguson, "Yarrow-160: Notes on the Design and Analysis of the Yarrow Cryptographic Pseudorandom Number Generator," SAC 1999.
29. J. Kelsey, B. Schneier, D. Wagner, and C. Hall, "Cryptanalytic Attacks on Pseudorandom Number Generators," FSE 1998.
30. J.B. Lacy, D.P. Mitchell, and V.M. Schell, "Cryptolib: Cryptography in Software," USENIX Security Symposium, 1993.
31. Network Associates, Inc., "PGPsdk 2.1.1 Source Code for Peer Review".
32. P. van Oorschot and M. Wiener "Parallel Collision Search with Applications to Hash Functions and Discrete Logarithms," ACM CCS 1994.
33. RSA Laboratories, "RSAREF Cryptographic Library," 1994.
34. A.C. Yao, "Theory and Applications of Trapdoor Functions," FOCS 1982.
35. E.A. Young and T.J. Hudson, "OpenSSL Library v.0.9.6c," 2001.

A Block-Cipher Mode of Operation for Parallelizable Message Authentication

John Black[1] and Phillip Rogaway[2]

[1] Dept. of Computer Science, University of Nevada, Reno NV 89557, USA,
jrb@cs.unr.edu, www.cs.unr.edu/~jrb

[2] Dept. of Computer Science, University of California, Davis, CA 95616, USA,
rogaway@cs.ucdavis.edu, www.cs.ucdavis.edu/~rogaway

Abstract. We define and analyze a simple and fully parallelizable block-cipher mode of operation for message authentication. Parallelizability does not come at the expense of serial efficiency: in a conventional, serial environment, the algorithm's speed is within a few percent of the (inherently sequential) CBC MAC. The new mode, PMAC, is deterministic, resembles a standard mode of operation (and not a Carter-Wegman MAC), works for strings of any bit length, employs a single block-cipher key, and uses just $\max\{1, \lceil |M|/n \rceil\}$ block-cipher calls to MAC a string $M \in \{0,1\}^*$ using an n-bit block cipher. We prove PMAC secure, quantifying an adversary's forgery probability in terms of the quality of the block cipher as a pseudorandom permutation.

1 Introduction

Background. Many popular message authentication codes (MACs), like the CBC MAC [17] and HMAC [1], are inherently sequential: one cannot process the i-th message block until all previous message blocks have been processed. This serial bottleneck becomes increasingly an issue as commodity processors offer up more and more parallelism, and as increases in network speeds outpace increases in the speed of cryptographic hardware. By now there would seem to be a significant interest in having a parallelizable MAC which performs well in both hardware and software, built from a block cipher like AES.

There are several approaches to the design of such an MAC. One is to generically construct a more parallelizable MAC from an arbitrary one. For example, one could begin with breaking the message $M[1] \cdots M[2m]$ into $M' = M[1]M[3] \cdots M[2m-1]$ and $M'' = M[2]M[4] \cdots M[2m]$ then separately MAC each half. But such an approach requires one to anticipate the maximal amount of parallelism one aims to extract. In the current work we are instead interested in *fully parallelizable* MACs: the amount of parallelism that can be extracted is effectively unbounded.

One idea for making a fully parallelizable MAC is to use the Carter-Wegman paradigm [13, 23], as in [12, 16, 19], making sure to select a universal hash-function family that is fully parallelizable. In fact, most universal hash functions that have been suggested *are* fully parallelizable. This approach is elegant

L.R. Knudsen (Ed.): EUROCRYPT 2002, LNCS 2332, pp. 384–397, 2002.

and can lead to a nice MAC, but constructions for fast universal hash-functions have proven to be quite complex to specify or to implement well [7, 9], and may be biased either towards hardware or towards software. Twenty years after the paradigm was introduced, we still do not know of a single Carter-Wegman MAC that actually gets used. So the current work goes back to giving a conventional mode, but one designed, this time around, for serial *and* parallel efficiency.

THE XOR MAC. Bellare, Guérin and Rogaway introduced a parallelizable MAC in their XOR MACs [3]. The message M is divided into pieces $M[1] \cdots M[\ell]$ of length less than the blocksize; for concreteness, think of each $M[i]$ as having 64 bits when the blocksize is $n = 128$ bits. Each piece $M[i]$ is preceded by $[i]$, the number i encoded as a 64-bit number, and to each $[i] \,\|\, M[i]$ one applies the block cipher E, keyed by the MAC key K. One more block is enciphered, it having a first bit of 1 and then a counter or random value in the remaining $n-1$ bits. The MAC is that counter or random value together with the XOR of all $\ell + 1$ ciphertext blocks. Thus the MAC uses $\ell + 1 \approx 2m + 1$ block-cipher invocations to authenticate a message of m blocks of n bits; one has paid for parallelizability at a cost of about a factor of two in serial speed. Further disadvantages include the need for randomness or state (conventional MACs are deterministic) and in the increased length of the MAC (because of the counter or random value that has to be included).

PMAC. Unlike the XOR MAC, our new algorithm, PMAC, doesn't waste any block-cipher invocations because of block-indices (nor for a counter or random values). Also, in the spirit of [11], we optimally deal with short final blocks; we correctly MAC messages of arbitrary and varying bit lengths. The result is that PMAC makes do with just $\lceil |M|/n \rceil$ block-cipher calls to MAC a non-empty message M using an n-bit block cipher. PMAC is deterministic, freeing the user from having to provide a counter or random value, and making the MAC shorter. Overhead beyond the block-cipher calls has been aggressively optimized, so a serial implementation of PMAC runs just a few percent slower than the CBC MAC.

Besides the efficiency measures already mentioned, PMAC uses very little key-setup: one block-cipher call. (A few shifts and conditional xors are also used.) The PMAC key is a single key for the underlying blocks cipher; in particular, we forgo the need for key-separation techniques. Avoiding multiple block-cipher keys saves time because many block ciphers have significant key-setup costs.

Being so stingy with keys and block-cipher invocations takes significant care; note that even the traditional CBC MAC uses between one and four additional block-cipher calls, as well as additional key material, once it has been enriched to take care of messages of arbitrary lengths [6, 11, 17, 21]. Of course avoiding this overhead doesn't matter much on long messages, but it is significant on short ones. And in many environments, short messages are common.

We prove PMAC secure, in the sense of reduction-based cryptography. Specifically, we prove that PMAC approximates a random function (and is therefore a good MAC) as long as the underlying block cipher approximates a random permutation. The actual results are quantitative; the security analysis is in the concrete-security paradigm.

PMAC was proposed in response to NIST's call for contributions for a first modes-of-operation workshop. Earlier versions of this writeup were submitted to NIST and posted to their website (Oct 2000, Apr 2001).

ADDITIONAL RELATED WORK. Building on [3], Gligor and Donescu describe a MAC they call the XECB MAC [14]. That MAC is not deterministic, it uses more block-cipher invocations, and it was not designed for messages of arbitrary bit length. But, like PMAC, it goes beyond the XOR MAC by combining a message index and a message block in a way other than encoding the two. In particular, [14] combines i and $M[i]$ by adding to $M[i]$, modulo 2^n, a secret multiple i. We combine i and $M[i]$ by different means, to reduce overhead and obtain a better bound.

PMAC was also influenced by the variant of the XOR MAC due to Bernstein [8]. His algorithm is deterministic, and the way that the XOR MAC was made deterministic in [8] is similar to the way that PMAC has been made deterministic. Finally, there is also some similarity in appearance between PMAC and Jutla's IAPM encryption mode [18].

2 Mathematical Preliminaries

NOTATION. If $i \geq 1$ is an integer then $\mathsf{ntz}(i)$ is the number of trailing 0-bits in the binary representation of i. So, for example, $\mathsf{ntz}(7) = 0$ and $\mathsf{ntz}(8) = 3$. If $A \in \{0,1\}^*$ is a string then $|A|$ denotes its length in bits while $\|A\|_n = \max\{1, \lceil |A|/n \rceil\}$ denotes its length in n-bit blocks (where the empty string counts as one block). If $A = a_{n-1} \cdots a_1 a_0 \in \{0,1\}^n$ is a string (each $a_i \in \{0,1\}$) then $\mathsf{str2num}(A)$ is the number $\sum_{i=0}^{n-1} 2^i a_i$. If $A, B \in \{0,1\}^*$ are equal-length strings than $A \oplus B$ is their bitwise xor. If $A \in \{0,1\}^*$ and $|A| < n$ then $\mathsf{pad}_n(A)$ is the string $A\,10^{n-|A|-1}$. If $A \in \{0,1\}^n$ then $\mathsf{pad}_n(A) = A$. With n understood we write $\mathsf{pad}(A)$ for $\mathsf{pad}_n(A)$. If $A = a_{n-1}a_{n-2} \cdots a_1 a_0 \in \{0,1\}^n$ then $A \ll 1 = a_{n-2}a_{n-3} \cdots a_1 a_0 0$ is the n-bit string which is the left shift of A by 1 bit while $A \gg 1 = 0 a_{n-1} a_{n-2} \ldots a_2 a_1$ is the n-bit string which is the right shift of A by one bit. In pseudocode we write "Partition M into $M[1] \cdots M[m]$" as shorthand for "Let $m = \|M\|_n$ and let $M[1], \ldots, M[m]$ be strings such that $M[1] \cdots M[m] = M$ and $|M[i]| = n$ for $1 \leq i < m$."

THE FIELD WITH 2^n POINTS. The field with 2^n points is denoted $\mathrm{GF}(2^n)$. We interchangeably think of a point a in $\mathrm{GF}(2^n)$ in any of the following ways: (1) as an abstract point in the field; (2) as an n-bit string $a_{n-1} \ldots a_1 a_0 \in \{0,1\}^n$; (3) as a formal polynomial $a(\mathtt{x}) = a_{n-1}\mathtt{x}^{n-1} + \cdots + a_1\mathtt{x} + a_0$ with binary coefficients; (4) as a nonnegative integer between 0 and $2^n - 1$, where $a \in \{0,1\}^n$ corresponds to $\mathsf{str2num}(a)$. We write $a(\mathtt{x})$ instead of a if we wish to emphasize that we are thinking of a as a polynomial. To add two points in $\mathrm{GF}(2^n)$, take their bitwise xor. We denote this operation by $a \oplus b$. To multiply two points, fix some irreducible polynomial $p(\mathtt{x})$ having binary coefficients and degree n. To be concrete, choose the lexicographically first polynomial among the irreducible degree n polynomials having a minimum number of coefficients. To multiply points $a, b \in \mathrm{GF}(2^n)$, which we denote $a \cdot b$, regard a and b as polynomials $a(\mathrm{x}) = a_{n-1}\mathrm{x}^{n-1} + \cdots + a_1\mathrm{x} + a_0$ and $b(\mathrm{x}) = b_{n-1}\mathrm{x}^{n-1} + \cdots + b_1\mathrm{x} + b_0$, form their product $c(\mathrm{x})$ where one adds and multiplies coefficients in GF(2), and take the remainder when dividing $c(\mathrm{x})$ by $p(\mathrm{x})$. Note that it is particularly easy to multiply a point $a \in \{0,1\}^n$ by x. We illustrate the method for $n = 128$, where $p(\mathrm{x}) = \mathrm{x}^{128} + \mathrm{x}^7 + \mathrm{x}^2 + \mathrm{x} + 1$. Then multiplying $a = a_{n-1} \cdots a_1 a_0$ by x yields

$$a \cdot \mathrm{x} = \begin{cases} a{\ll}1 & \text{if firstbit}(a) = 0 \\ (a{\ll}1) \oplus 0^{120}10000111 & \text{if firstbit}(a) = 1 \end{cases} \tag{1}$$

It is similarly easy to divide a by x (meaning to multiply a by the multiplicative inverse of x). To illustrate, assume that $n = 128$. Then

$$a \cdot \mathrm{x}^{-1} = \begin{cases} a{\gg}1 & \text{if lastbit}(a) = 0 \\ (a{\gg}1) \oplus 10^{120}1000011 & \text{if lastbit}(a) = 1 \end{cases} \tag{2}$$

If $L \in \{0,1\}^n$ and $i \geq -1$, we write $L(i)$ to mean $L \cdot \mathrm{x}^i$. To compute $L(-1), L(0), \ldots, L(\mu)$, where μ is small, set $L(0) = L$ and then, for $i \in [1..\mu]$, use Equation (1) to compute $L(i) = L(i-1) \cdot \mathrm{x}$ from $L(i-1)$; and use Equation (2) to compute $L(-1)$ from L.

We point out that *huge* $= \mathrm{x}^{-1}$ will be an enormous number (when viewed as a number); in particular, *huge* starts with a 1 bit, so *huge* $> 2^{n-1}$. In the security proof this fact is relevant, so there we use *huge* as a synonym for x^{-1} when this seems to add to clarity.

GRAY CODES. For any $\ell \geq 1$, a Gray code is an ordering $\gamma^\ell = \gamma_0^\ell\ \gamma_1^\ell\ \ldots\ \gamma_{2^\ell-1}^\ell$ of $\{0,1\}^\ell$ such that successive points differ (in the Hamming sense) by just one bit. For n a fixed number, PMAC makes use of the "canonical" Gray code $\gamma = \gamma^n$ constructed by $\gamma^1 = 0\ 1$ while, for $\ell > 0$,

$$\gamma^{\ell+1} = 0\gamma_0^\ell\ 0\gamma_1^\ell\ \cdots\ 0\gamma_{2^\ell-2}^\ell\ 0\gamma_{2^\ell-1}^\ell\ 1\gamma_{2^\ell-1}^\ell\ 1\gamma_{2^\ell-2}^\ell\ \cdots\ 1\gamma_1^\ell\ 1\gamma_0^\ell.$$

It is easy to see that γ is a Gray code. What is more, for $1 \leq i \leq 2^n - 1$, $\gamma_i = \gamma_{i-1} \oplus (0^{n-1}1{\ll}\mathsf{ntz}(i))$. This makes it easy to compute successive points. Note that $\gamma_1, \gamma_2, \ldots, \gamma_{2^n-1}$ are distinct, different from 0, and $\gamma_i \leq 2i$.

Let $L \in \{0,1\}^n$ and consider the problem of successively forming the strings $\gamma_1 \cdot L, \gamma_2 \cdot L, \gamma_3 \cdot L, \ldots, \gamma_m \cdot L$. Of course $\gamma_1 \cdot L = 1 \cdot L = L$. Now, for $i \geq 2$, assume one has already produced $\gamma_{i-1} \cdot L$. Since $\gamma_i = \gamma_{i-1} \oplus (0^{n-1}1{\ll}\mathsf{ntz}(i))$ we know that $\gamma_i \cdot L = (\gamma_{i-1} \oplus (0^{n-1}1{\ll}\mathsf{ntz}(i))) \cdot L = (\gamma_{i-1} \cdot L) \oplus (0^{n-1}1{\ll}\mathsf{ntz}(i)) \cdot L = (\gamma_{i-1} \cdot L) \oplus (L \cdot \mathrm{x}^{\mathsf{ntz}(i)}) = (\gamma_{i-1} \cdot L) \oplus L(\mathsf{ntz}(i))$. That is, the ith word in the sequence $\gamma_1 \cdot L, \gamma_2 \cdot L, \gamma_3 \cdot L, \ldots$ is obtained by xoring the previous word with $L(\mathsf{ntz}(i))$.

3 Definition of PMAC

PMAC depends on two parameters: a block cipher and a tag length. The block cipher is a function $E\colon \mathcal{K} \times \{0,1\}^n \to \{0,1\}^n$, for some number n, where each

```
Algorithm PMAC_K (M)
1.   L ← E_K(0^n)
2.   if |M| > n2^n then return 0^τ
3.   Partition M into M[1] ··· M[m]
4.   for i ← 1 to m − 1 do
5.       X[i] ← M[i] ⊕ γ_i · L
6.       Y[i] ← E_K(X[i])
7.   Σ ← Y[1] ⊕ Y[2] ⊕ ··· ⊕ Y[m − 1] ⊕ pad(M[m])
8.   if |M[m]| = n then X[m] = Σ ⊕ L · x^{-1}
9.                 else X[m] ← Σ
10.  Tag = E_K(X[m]) [first τ bits]
11.  return Tag
```

Fig. 1. Definition of PMAC. The message to MAC is M and the key is K. The algorithm depends on a block cipher E: $\mathcal{K} \times \{0,1\}^n \to \{0,1\}^n$ and a number $\tau \in [1..n]$. Constants $\gamma_1, \gamma_2, \ldots$, the meaning of the multiplication operator, and the meaning of pad() are all defined in the text.

$E(K, \cdot) = E_K(\cdot)$ is a permutation on $\{0,1\}^n$. Here $\mathcal{K}$ is the set of possible keys and n is the block length. The tag length is an integer $\tau \in [1..n]$. By trivial means the adversary will be able to forge a valid ciphertext with probability $2^{-\tau}$. With E: $\mathcal{K} \times \{0,1\}^n \to \{0,1\}^n$ and $\tau \in [1..n]$, we let PMAC$[E, \tau]$ denote PMAC using block cipher E and tag length τ. We simplify to PMAC-E when τ is irrelevant. PMAC$[E, \tau]$ is a function taking a key $K \in \mathcal{K}$ and a message $M \in \{0,1\}^*$ and returning a string in $\{0,1\}^\tau$. The function is defined in Figure 1 and illustrated in Figure 2. We comment that line 2 of Figure 1 is simply to ensure that PMAC is well-defined even for the highly unrealistic case that $|M| > n2^n$ (by which time our security result becomes vacuous anyway). Alternatively, one may consider PMAC's message space to be strings of length at most $n2^n$ rather than strings of arbitrary length.

4 Comments

As we shall soon prove, PMAC is more than a good MAC: it is good as a pseudorandom function (PRF) having variable-input-length and fixed-output-length. As long as the underlying block cipher E is secure, no reasonable adversary will be able to distinguish PMAC$_K(\cdot)$, for a random and hidden key K, from a random function ρ from $\{0,1\}^*$ to $\{0,1\}^\tau$. It is a well-known observation, dating to the introduction of PRFs [15], that a good PRF is necessarily a good MAC.

Conceptually, the key is (K, L). But instead of regarding this as the key (and possibly defining K and L from an underlying key), the value L is defined from K and then K is still used as a key. Normally such "lazy key-derivation" would get one into trouble, in proofs if nothing else. For PMAC we prove that this form of lazy key-derivation works fine.

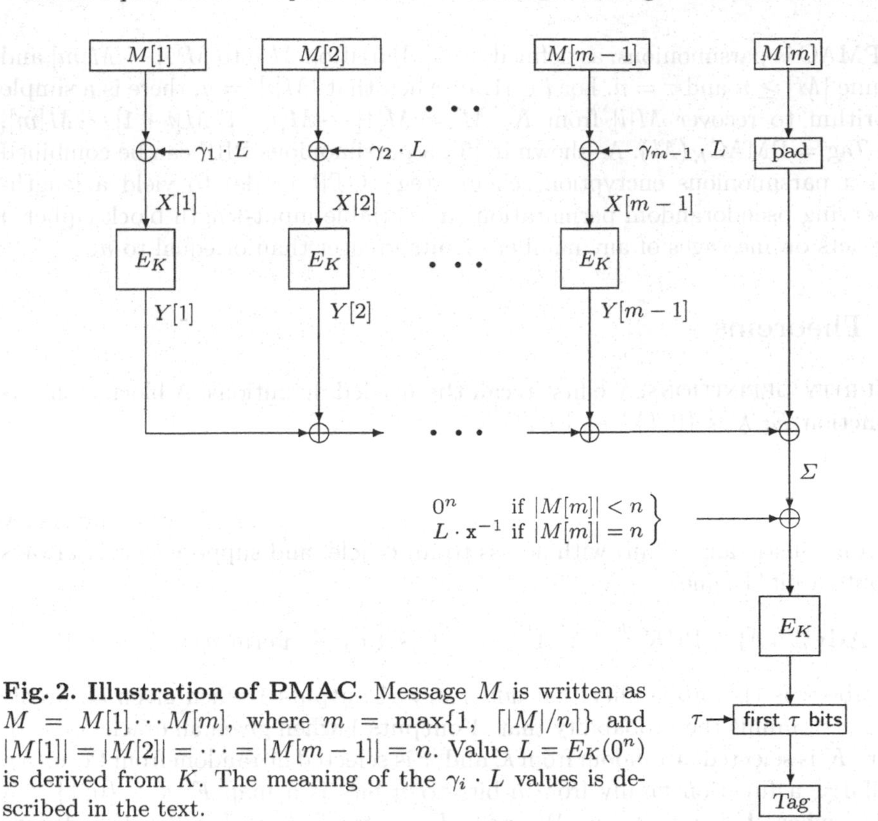

Fig. 2. Illustration of PMAC. Message M is written as $M = M[1] \cdots M[m]$, where $m = \max\{1, \lceil |M|/n \rceil\}$ and $|M[1]| = |M[2]| = \cdots = |M[m-1]| = n$. Value $L = E_K(0^n)$ is derived from K. The meaning of the $\gamma_i \cdot L$ values is described in the text.

Any string $M \in \{0,1\}^*$ can be MACed, and messages which are not a multiple of the block length are handled without the need for obligatory padding, which would increase the number of block-cipher calls.

MAC generation is "on line," meaning that one does not need to know the length of the message M in advance. Instead, the message can be MACed as one goes along, continuing until there is an indication that the message is now complete. The work of Petrank and Rackoff brought out the importance of this property [21].

In contrast to a scheme based on mod p arithmetic (for a prime p) or mod 2^n arithmetic, there is almost no endian-favoritism implicit in the definition of PMAC. (The exception is that the left shift used for forming $L(i+1)$ from $L(i)$ is more convenient under a big-endian convention, as is the right shift used for forming $L(-1) = L \cdot \mathrm{x}^{-1}$ from L.)

If $\tau = n$ (or one retains a constant amount of extra information) PMAC is incremental in the sense of [15] with respect to operations $\mathtt{append}(M, x) = M \| x$, $\mathtt{truncate}(M, \Delta) = M$ [first $|M| - \Delta$ bits], for $|M| \geq \Delta$, and $\mathtt{replace}(M, i, x) = M$ [first $i-1$ bits] $\| x \| M$[last $|M| - i - |x| + 1$ bits], where $|M| \geq i + |x| - 1$). For each operation it is easy to see how to update the MAC of M in time proportional to $|x|$, Δ, or $|x|$, respectively.

PMAC is parsimonious, as defined in [5]. Partition M into $M[1]\cdots M[m]$ and assume $|M|\geq n$ and $\tau=n$. For $i\in[1..m]$ such that $|M[i]|=n$, there is a simple algorithm to recover $M[i]$ from K, $M'=M[1]\cdots M[i-1]\,M[i+1]\cdots M[m]$, and $Tag=\mathrm{PMAC}_K(M)$. As shown in [5], a parsimonious PRF can be combined with a parsimonious encryption scheme (eg., CTR mode) to yield a length-preserving pseudorandom permutation (a “variable-input-length block cipher”) that acts on messages of any number of bits greater than or equal to n.

5 Theorems

SECURITY DEFINITIONS. We first recall the needed definitions. A block cipher is a function $E\colon \mathcal{K}\times\{0,1\}^n\to\{0,1\}^n$ where $\mathcal{K}$ is a finite set and each $E_K(\cdot)=E(K,\cdot)$ is a permutation on $\{0,1\}^n$. Let $\mathrm{Perm}(n)$ denote the set of all permutations on $\{0,1\}^n$. This set can be regarded as a block cipher by imagining that each permutation is named by a unique element of $\mathcal{K}$. Let A be an adversary (a probabilistic algorithm) with access to an oracle, and suppose that A always outputs a bit. Define

$$\mathbf{Adv}_E^{\mathrm{prp}}(A)=\Pr[K\stackrel{R}{\leftarrow}\mathcal{K}:\ A^{E_K(\cdot)}=1]-\Pr[\pi\stackrel{R}{\leftarrow}\mathrm{Perm}(n):\ A^{\pi(\cdot)}=1]$$

The above is the probability that adversary A outputs 1 when given an oracle for $E_K(\cdot)$, minus the probability that A outputs 1 when given an oracle for $\pi(\cdot)$, where K is selected at random from $\mathcal{K}$ and π is selected at random from $\mathrm{Perm}(n)$. Similarly, a function family from n-bits to n-bits is a map $F\colon \mathcal{K}\times\{0,1\}^n\to\{0,1\}^n$ where $\mathcal{K}$ is a finite set. We write $F_K(\cdot)$ for $F(K,\cdot)$. Let $\mathrm{Rand}(n)$ denote the set of all functions from $\{0,1\}^n$ to $\{0,1\}^n$. This set can be regarded as a function family as above. Define

$$\mathbf{Adv}_F^{\mathrm{prf}}(A)=\Pr[K\stackrel{R}{\leftarrow}\mathcal{K}:\ A^{F_K(\cdot)}=1]-\Pr[\rho\stackrel{R}{\leftarrow}\mathrm{Rand}(n):\ A^{\rho(\cdot)}=1]$$

Finally, a function family from $\{0,1\}^*$ to $\{0,1\}^\tau$ is a map $f\colon \mathcal{K}\times\{0,1\}^*\to\{0,1\}^\tau$ where $\mathcal{K}$ is a set with an associated distribution. We write $f_K(\cdot)$ for $f(K,\cdot)$. Let $\mathrm{Rand}(*,\tau)$ denote the set of all functions from $\{0,1\}^*$ to $\{0,1\}^\tau$. This set is given a probability measure by asserting that a random element ρ of $\mathrm{Rand}(*,\tau)$ associates to each string $x\in\{0,1\}^*$ a random string $\rho(x)\in\{0,1\}^\tau$. Define

$$\mathbf{Adv}_f^{\mathrm{prf}}(A)=\Pr[K\stackrel{R}{\leftarrow}\mathcal{K}:\ A^{f_K(\cdot)}=1]-\Pr[g\stackrel{R}{\leftarrow}\mathrm{Rand}(*,\tau):\ A^{g(\cdot)}=1]$$

MAIN RESULT. We now give an information-theoretic bound on the security of our construction.

Theorem 1. **[Security of PMAC]** *Fix $n,\tau\geq 1$. Let A be an adversary with an oracle. Suppose that A asks its oracle q queries, these having aggregate length of σ blocks. Let $\bar\sigma=\sigma+1$. Then*

$$\mathbf{Adv}^{\mathrm{prf}}_{\mathrm{PMAC}[\mathrm{Perm}(n),\tau]}(A)\leq\frac{\bar\sigma^2}{2^{n-1}}$$

```
10   L ←R {0,1}^n
11   for i ← 1 to m − 1 do { X[i] ← M[i] ⊕ γ_i · L;  Y[i] ←R {0,1}^n }
12   Σ ← Y[1] ⊕ ··· ⊕ Y[m − 1] ⊕ pad(M[m])
13   if |M[m]| = n then X[m] ← Σ ⊕ huge · L else X[m] ← Σ
14   𝒳 ← {X[1], ..., X[m]}
15   if there is a repetition in {0^n} ∪ 𝒳 then Mcoll ← true

16   Equal ← {i ∈ [1..min{m, m̄} − 1] : M[i] = M̄[i]}
17   Unequal ← [1..m̄] \ Equal
18   for i ← 1 to m̄ − 1 do
19       if i ∈ Equal then { X̄[i] ← X[i];  Ȳ[i] ← Y[i] }
20       if i ∈ Unequal then { X̄[i] ← M̄[i] ⊕ γ_i · L;  Ȳ[i] ←R {0,1}^n }
21   Σ̄ ← Ȳ[1] ⊕ ··· ⊕ Ȳ[m̄ − 1] ⊕ pad(M̄[m̄])
22   if |M̄[m̄]| = n then X̄[m̄] ← Σ̄ ⊕ huge · L else X̄[m̄] ← Σ̄
23   𝒳̄ ← {X̄[i] : i ∈ Unequal}
24   if 𝒳 ∩ 𝒳̄ ≠ ∅ then MMcoll ← true
```

Fig. 3. Defining the collision probabilities. Functions $\mathrm{Mcoll}_n(\cdot)$ and $\mathrm{MMcoll}_n(\cdot,\cdot)$ are defined using this game. In lines 14 and 23, $\mathcal{X}$ and $\bar{\mathcal{X}}$ are understood to be multisets. The union in line 15 is a multiset union. Recall that *huge* is a synonym for $\mathtt{x}^{-1}$.

In the theorem statement and from now on, the aggregate length of messages $M_1, \ldots, M_q$ asked by A is the number $\sigma = \sum_{r=1}^{q} \|M_r\|_n$.

From the theorem above it is standard to pass to a complexity-theoretic analog. Fix parameters $n, \tau \geq 1$ and block cipher $E: \mathcal{K} \times \{0,1\}^n \to \{0,1\}^n$. Let A be an adversary with an oracle and suppose that A asks queries having aggregate length of σ blocks. Let $\bar{\sigma} = \sigma + 1$. Then there is an adversary B for attacking E that achieves advantage $\mathbf{Adv}_E^{\mathrm{prp}}(B) \geq \mathbf{Adv}_{\mathrm{PMAC}[E,\tau]}^{\mathrm{prf}}(A) - \bar{\sigma}^2/2^{n-1}$. Adversary B asks at most $\bar{\sigma}$ oracle queries and has a running time equal to A's running time plus the time to compute E on $\bar{\sigma}$ points, plus additional time of $cn\bar{\sigma}$ for a constant c that depends only on details of the model of computation.

It is a standard result that being secure in the sense of a PRF implies an inability to forge with good probability. See [4, 15].

STRUCTURE OF THE PROOF. The proof combines two lemmas. The first, the *structure lemma*, measures the pseudorandomness of PMAC using two functions: the M-collision probability, denoted $\mathrm{Mcoll}_n(\cdot)$, and the MM-collision probability, denoted $\mathrm{MMcoll}_n(\cdot,\cdot)$. The second lemma, the *collision-bounding lemma*, upper-bounds $\mathrm{Mcoll}_n(m)$ and $\mathrm{MMcoll}_n(m, \bar{m})$.

We begin by defining $\mathrm{Mcoll}_n(\cdot)$ and $\mathrm{MMcoll}_n(\cdot,\cdot)$. Fix n and choose M and $\bar{M}$, partitioning them into $M[1] \cdots M[m]$ and $\bar{M}[1] \cdots \bar{M}[\bar{m}]$. Consider the experiment of Figure 3. When M is a string, let $\mathrm{Mcoll}_n(M)$ denote the probability that *Mcoll* gets set to `true` in line 15 when the program of Figure 3 is run on M. When m is a number, $\mathrm{Mcoll}_n(m)$ is the maximum value of $\mathrm{Mcoll}_n(M)$ over all strings M such that $\|M\|_n = m$. Similarly, when M and $\bar{M}$ are strings, let $\mathrm{MMcoll}_n(M, \bar{M})$ denote the probability that *MMcoll* gets set to `true` when the program of Figure 3 is run on strings $M, \bar{M}$. When m and $\bar{m}$ are numbers, let $\mathrm{MMcoll}_n(m, \bar{m})$ denote the maximum value of $\mathrm{MMcoll}_n(M, \bar{M})$ over all strings $M, \bar{M}$ such that $\|M\|_n = m$ and $\|\bar{M}\|_n = \bar{m}$. We can now state the structure lemma.

Lemma 1. [Structure lemma] *Fix $n, \tau \geq 1$. Let A be an adversary that asks q queries, these having an aggregate length of σ blocks. Then*

$$\mathbf{Adv}^{\mathrm{prf}}_{\mathrm{PMAC}[\mathrm{Perm}(n),\tau]}(A) \leq$$

$$\max_{\substack{m_1,\ldots,m_q \\ \sigma=\sum m_i \\ m_i \geq 1}} \left\{ \sum_{1\leq r\leq q} \mathrm{Mcoll}_n(m_r) + \sum_{1\leq r<s\leq q} \mathrm{MMcoll}_n(m_r, m_s) \right\} + \frac{(\sigma+1)^2}{2^{n+1}}$$

The proof of this lemma is found in [10].

EXPLANATION. Informally, $\mathrm{Mcoll}_n(m)$ measures the probability of running into trouble when the adversary asks a single question M having block length m. Trouble means a collision among the values $X[0], X[1], \ldots, X[m]$, where $X[0] = 0^n$ and each $X[i]$ is the block-cipher input associated to message block i. Informally, $\mathrm{MMcoll}_n(m, \bar{m})$ measures the probability of running into trouble across two messages, M and $\bar{M}$, having lengths m and $\bar{m}$. This time trouble means a "non-trivial" collision. That is, consider the $m+\bar{m}+1$ points at which the block cipher is applied in processing M and $\bar{M}$. There are m points $X[1], \ldots, X[m]$, another $\bar{m}$ points $\bar{X}[1], \ldots, \bar{X}[\bar{m}]$, and then there is the point 0^n (the block cipher was applied at this point to define L). Some pairs of these $m + \bar{m} + 1$ points could coincide for a "trivial" reason: namely, we know that $X[i] = \bar{X}[i]$ if $i < m$ and $i < \bar{m}$ and $M[i] = \bar{M}[i]$. We say that there is a nontrivial collision between M and $\bar{M}$ if some *other* $X[i]$ and $\bar{X}[j]$ happened to coincide. Note that M-collisions include collisions with 0^n, while MM-collisions do not. Also, MM-collisions do not include collisions within a single message (or collisions with 0^n) because both of these possibilities are taken care of by way of M-collisions.

The structure lemma provides a simple recipe for measuring the maximum advantage of any adversary that attacks the pseudorandomness of PMAC: bound the collision probabilities $\mathrm{Mcoll}_n(\cdot)$ and $\mathrm{MMcoll}_n(\cdot,\cdot)$ and then use the formula. The lemma simplifies the analysis of PMAC in two ways. First, it allows one to excise adaptivity as a concern. Dealing with adaptivity is a major complicating factor in proofs of this type. Second, it allows one to concentrate on what happens to single messages and to a fixed pair of messages. It is easier to think about what happens with one or two messages than what is happening with all q of them.

BOUNDING THE COLLISION PROBABILITIES. The following lemma indicates that the two types of collisions we have defined rarely occur. The proof shall be given shortly.

Lemma 2. **[Collision-bounding lemma]** *Let* $\mathrm{Mcoll}_n(\cdot)$ *and* $\mathrm{MMcoll}_n(\cdot,\cdot)$ *denote the M-collision probability and the MM-collision probability. Then*

$$\mathrm{Mcoll}_n(m) \le \binom{m+1}{2} \cdot \frac{1}{2^n} \quad \text{and} \quad \mathrm{MMcoll}_n(m,\bar{m}) \le \frac{m\bar{m}}{2^n}$$

Concluding the theorem. Pseudorandomness of PMAC, Theorem 1, follows by combining Lemmas 1 and 2. Namely,

$$\begin{aligned}
&\mathbf{Adv}^{\mathrm{prf}}_{\mathrm{PMAC[Perm}(n),\tau]} \\
&\le \max_{\substack{m_1,\dots,m_q\\ \sigma=\sum m_i\\ m_i\ge 1}} \left\{ \sum_{1\le r\le q} \mathrm{Mcoll}_n(m_r) + \sum_{1\le r<s\le q} \mathrm{MMcoll}_n(m_r,m_s) \right\} + \frac{(\sigma+1)^2}{2^{n+1}} \\
&\le \max_{\substack{m_1,\dots,m_q\\ \sigma=\sum m_i\\ m_i\ge 1}} \left\{ \sum_{1\le r\le q} \mathrm{Mcoll}_n(m_r) \right\} \\
&\qquad + \max_{\substack{m_1,\dots,m_q\\ \sigma=\sum m_i\\ m_i\ge 0}} \left\{ \sum_{1\le r<s\le q} \mathrm{MMcoll}_n(m_r,m_s) \right\} + \frac{(\sigma+1)^2}{2^{n+1}} \\
&\le \max_{\substack{m_1,\dots,m_q\\ \sigma=\sum m_i\\ m_i\ge 0}} \left\{ \sum_{1\le r\le q} \binom{m_r+1}{2} \cdot \frac{1}{2^n} \right\} \\
&\qquad + \max_{\substack{m_1,\dots,m_q\\ \sigma=\sum m_i\\ m_i\ge 0}} \left\{ \sum_{1\le r<s\le q} \frac{m_r m_s}{2^n} \right\} + \frac{(\sigma+1)^2}{2^{n+1}} \\
&\le \frac{(\sigma+1)^2}{2^n} + \frac{(\sigma^2/2)}{2^n} + \frac{(\sigma+1)^2}{2^{n+1}} \qquad (3) \\
&\le \frac{2\,(\sigma+1)^2}{2^n}
\end{aligned}$$

where (3) follows because the first sum is maximized with a single message of length σ, while the second sum is maximized by q messages of length σ/q. (These claims can be justified using the method of Lagrange multipliers.) This completes the proof of Theorem 1.

Proof of Lemma 2. We now bound $\mathrm{Mcoll}_n(m)$ and $\mathrm{MMcoll}_n(m,\bar{m})$. To begin, let $\mathit{Unequal}' = \mathit{Unequal} \setminus \{\bar{X}[\bar{m}]\}$ (multiset difference: remove one copy of $\bar{X}[\bar{m}]$) and define

$$D_1 = \{0^n\} \qquad D_2 = \{X[1],\dots,X[m-1]\} \qquad D_3 = \{X[m]\}$$
$$D_4 = \{\bar{X}[j] : j \in \mathit{Unequal}'\} \qquad D_5 = \{\bar{X}[\bar{m}]\}$$

We first show that for any two points $X[i]$ and $X[j]$ in the multiset $D_1 \cup D_2 \cup D_3$, where $i < j$ and $X[0] = 0^n$, the probability that these two points collide is at most 2^{-n}. The inequality follows because there are $m+1$ points in $D_1 \cup D_2 \cup D_3$. Afterwards, we show that for any point $X[i]$ in $D_2 \cup D_3$ and any point in $\bar{X}[j]$ in $D_4 \cup D_5$, the probability that they collide is at most 2^{-n}. The inequality $\mathrm{MMcoll}_n(m, \bar{m}) \leq \frac{m\bar{m}}{2^n}$ follows because $|D_2 \cup D_3| \cdot |D_4 \cup D_5| \leq m\bar{m}$.

To show $\mathrm{Mcoll}_n(m) \leq \binom{m+1}{2} \cdot \frac{1}{2^n}$ consider the following four cases:

Case (D_1, D_2): $\Pr[0^n = X[i]] = \Pr[M[i] \oplus \gamma_i \cdot L = 0^n] = \Pr[L = \gamma_i^{-1} \cdot M[i]] = 2^{-n}$. We have used that γ_i is nonzero and we are working in a field. (We will continue to use this without mention.)

Case (D_1, D_3): If $|M[m]| < n$ and $m \geq 2$ then Σ is a random n-bit string and so $\Pr[0^n = X[m]] = \Pr[0^n = \Sigma] = \Pr[0^n = Y[1] \oplus \cdots Y[m-1] \oplus \mathsf{pad}(M[m])] = 2^{-n}$. If $|M[m]| = n$ and $m \geq 2$ then Σ is a random n-bit string that is independent of L and so $\Pr[0^n = X[m]] = \Pr[0^n = \Sigma \oplus \mathit{huge} \cdot L] = 2^{-n}$. If $|M[m]| < n$ and $m = 1$ then $\Pr[0^n = X[1]] = \Pr[0^n = \mathsf{pad}(M[m])] = 0$. If $|M[m]| = n$ and $m = 1$ then $\Pr[0^n = X[1]] = \Pr[0^n = \mathsf{pad}(M[m]) \oplus \mathit{huge} \cdot L] = 2^{-n}$.

Case (D_2, D_2): For $i, j \in [1..m-1]$, $i < j$, $\Pr[X[i] = X[j]] = \Pr[M[i] \oplus \gamma_i \cdot L = M[j] \oplus \gamma_j \cdot L] = \Pr[M[i] \oplus M[j] = (\gamma_i \oplus \gamma_j) \cdot L] = 2^{-n}$ because $\gamma_i \neq \gamma_j$ for $i \neq j$. (Here one assumes that $j < 2^n$ because the lemma gives a non-result anyway if j were larger.)

Case (D_2, D_3): Assume that $m \geq 2$, for otherwise there is nothing to show. Suppose first that $|M[m]| < n$. Then $\Pr[X[i] = X[m]] = \Pr[M[i] \oplus \gamma_i \cdot L = \Sigma]$. The value Σ is uniformly random and independent of L, so this probability is 2^{-n}. Suppose next that $|M[m]| = n$. Then $\Pr[X[i] = X[m]] = \Pr[M[i] \oplus \gamma_i \cdot L = \Sigma \oplus \mathit{huge} \cdot L] = \Pr[M[i] \oplus \Sigma = (\gamma_i \oplus \mathit{huge}) \cdot L]$. This value is 2^{-n} since $\gamma_i \neq \mathit{huge}$. Here we are assuming that $i < 2^{n-1}$, which is without loss of generality since a larger value of i, and therefore m, would give a non-result in the theorem statement.

Moving on, to show that $\mathrm{MMcoll}_n(m, \bar{m}) \leq \frac{m\bar{m}}{2^n}$ we verify the following four cases:

Case (D_2, D_4): Let $i \in [1..m-1]$ and $j \in \mathit{Unequal}'$ and consider $\Pr[X[i] = \bar{X}[j]] = \Pr[M[i] \oplus \gamma_i \cdot L = \bar{M}[j] \oplus \gamma_j \cdot L] = \Pr[M[i] \oplus \bar{M}[j] = (\gamma_i \oplus \gamma_j) \cdot L$. If $i \neq j$ then $\gamma_i \neq \gamma_j$ and this probability is 2^{-n}. If $i = j$ then the probability is 0 since, necessarily, $M[i] \neq \bar{M}[j]$.

Case (D_2, D_5): Suppose that $|\bar{M}[\bar{m}]| < n$. Then $\Pr[X[i] = \bar{X}[\bar{m}]] = \Pr[M[i] \oplus \gamma_i \cdot L = \bar{\Sigma}] = 2^{-n}$ because $\bar{\Sigma}$ is independent of L. Suppose that $|\bar{M}[\bar{m}]| = n$. Then $\Pr[X[i] = \bar{X}[\bar{m}]] = \Pr[M[i] \oplus \gamma_i \cdot L = \bar{\Sigma} \oplus \mathit{huge} \cdot L] = \Pr[M[i] \oplus \bar{\Sigma} = (\gamma_i \oplus \mathit{huge}) \cdot L] = 2^{-n}$ because $\bar{\Sigma}$ is independent of L and $\gamma_i \neq \mathit{huge}$.

Case (D_3, D_4): Suppose that $|M[m]| < n$. Then $\Pr[X[m] = \bar{X}[j]] = \Pr[\Sigma = \bar{M}[j] \oplus \gamma_j \cdot L] = 2^{-n}$ because Σ is independent of L. Suppose that $|M[m]| = n$. Then $\Pr[X[m] = \bar{X}[j]] = \Pr[\Sigma \oplus \mathit{huge} \cdot L = \bar{M}[j] \oplus \gamma_j \cdot L] = \Pr[\Sigma \oplus \bar{M}[j] = (\gamma_j \oplus \mathit{huge}) \cdot L] = 2^{-n}$ because $\gamma_j \neq \mathit{huge}$.

Algorithm	16 B	128 B	2 KB
PMAC-AES128	22.1	18.7	18.4
CBCMAC-AES128	18.9	17.4	17.1

Fig. 4. Performance results. Numbers are in cycles per byte (cpb) on a Pentium 3, for three message lengths, the code written in assembly.

CASE (D_3, D_5): Suppose that $|M[m]| < n$ and $|\bar{M}[\bar{m}]| < n$. If $m > \bar{m}$ then $\Pr[X[m] = \bar{X}[\bar{m}]] = \Pr[\Sigma = \bar{\Sigma}] = 2^{-n}$ because of the contribution of $Y[m-1]$ in Σ – a random variable that is not used in the definition of $\bar{\Sigma}$. If $m < \bar{m}$ then $\Pr[X[m] = \bar{X}[\bar{m}]] = \Pr[\Sigma = \bar{\Sigma}] = 2^{-n}$ because of the contribution of $\bar{Y}[\bar{m}-1]$ in $\bar{\Sigma}$ – a random variable that is not used in the definition of Σ. If $m = \bar{m}$ and there is an $i < m$ such that $M[i] \neq \bar{M}[i]$ then $\Pr[X[m] = \bar{X}[\bar{m}]] = \Pr[\Sigma = \bar{\Sigma}] = 2^{-n}$ because of the contribution of $\bar{Y}[i]$ in $\bar{\Sigma}$ – a random variable that is not used in the definition of Σ. If $m = \bar{m}$ and for every $i < m$ we have that $M[i] = \bar{M}[i]$, then, necessarily, $M[m] \neq \bar{M}[m]$. In this case $\Pr[\Sigma = \bar{\Sigma}] = 0$, as the two checksums differ by the nonzero value $\mathsf{pad}(M[m]) \oplus \mathsf{pad}(\bar{M}[m])$.

Suppose that $|M[m]| = n$ and $|\bar{M}[\bar{m}]| = n$. Then $X[m]$ and $\bar{X}[m]$ are offset by the same amount, $huge \cdot L$, so this offset is irrelevant in computing $\Pr[X[m] = \bar{X}[\bar{m}]$. Proceed as above.

Suppose that $|M[m]| < n$ and $|\bar{M}[\bar{m}]| = n$. Then $\Pr[X[m] = \bar{X}[m]] = \Pr[\Sigma = \bar{\Sigma} \oplus huge \cdot L] = 2^{-n}$ since Σ and $\bar{\Sigma}$ are independent of L. Similarly, if $|M[m]| = n$ and $|\bar{M}[\bar{m}]| < n$, then $\Pr[X[m] = \bar{X}[m]] = 2^{-n}$. This completes the proof.

6 Performance

A colleague, Ted Krovetz, implemented PMAC-AES128 and compared its performance in an entirely sequential setting to that of CBCMAC-AES128. The latter refers to the "basic" CBC MAC; nothing is done to take care of length-variability or the possibility of strings which are not a multiple of the block length. The code was written in modestly-optimized assembly under Windows 2000 sp1 and Visual C++ 6.0 sp4. All data fit into L1 cache. Disregarding the one-block message in Figure 4 we see that, in a serial environment, PMAC-AES128 was about 8% slower than CBCMAC-AES128. A more aggressively optimized implementation of CBCMAC-AES128, due to Helger Lipmaa, achieves 15.5 cpb for 1 KByte message lengths [20]. Adding the same 8%, we expect that this code could be modified to compute PMAC at a rate of about 16.7 cpb. In general, differences in implementation quality would seem to be a more significant in determining implementation speed than the algorithmic difference between PMAC and the CBC MAC.

Though some or all of the $L(i)$-values are likely to be pre-computed, calculating all of these values "on the fly" is not expensive. Starting with 0^n we form successive offsets by xoring the previous offset with L, $2 \cdot L$, L, $4 \cdot L$, L, $2 \cdot L$, L, $8 \cdot L$, and so forth, making the expected number of $a \cdot$ x-operations to compute an offset at most $\sum_{i=1}^{\infty} i/2^{i+1} = 1$. For $n = 128$, each $a \cdot \mathrm{x}$ instruction requires a 128-bit shift and a conditional 32-bit xor.

Acknowledgments

We would like to thank several people for helpful comments on drafts of this manuscript: Michael Amling, Mihir Bellare, Johan Håstad, Ted Krovetz, David McGrew, David Wagner, and the Eurocrypt program committee. Special thanks to Ted, who wrote reference code, prepared test vectors, and collected performance data. Virgil Gligor described [14] to Rogaway at Crypto '00 and that event, along with the announcement of a NIST modes-of-operation workshop in October 2000, inspired the current work.

This paper was written while Rogaway was on leave from UC Davis, visiting the Department of Computer Science, Faculty of Science, Chiang Mai University.

References

1. M. Bellare, R. Canetti, and H. Krawczyk. Keying hash functions for message authentication. *Advances in Cryptology – CRYPTO '96.* Lecture Notes in Computer Science, vol. 1109, Springer-Verlag, pp. 1–15, 1996. Available at URL www-cse.ucsd.edu/users/mihir
2. M. Bellare, S. Goldwasser, and O. Goldreich. Incremental cryptography and applications to virus protection. *Proceedings of the 27th Annual ACM Symposium on the Theory of Computing* (STOC '95). ACM Press, pp. 45–56, 1995. Available at URL www.cs.ucdavis.edu/~rogaway
3. M. Bellare, R. Guérin and P. Rogaway. "XOR MACs: New methods for message authentication using finite pseudorandom functions." *Advances in Cryptology – CRYPTO '95.* Lecture Notes in Computer Science, vol. 963, Springer-Verlag, pp. 15–28, 1995. Available at URL www.cs.ucdavis.edu/~rogaway
4. M. Bellare, J. Kilian, and P. Rogaway. The security of the cipher block chaining message authentication code. *Journal of Computer and System Sciences,* vol. 61, no. 3, Dec 2000. (Full version of paper from *Advances in Cryptology – CRYPTO '94.* Lecture Notes in Computer Science, vol. 839, pp. 340–358, 1994.) Available at URL www.cs.ucdavis.edu/~rogaway
5. M. Bellare and P. Rogaway. Encode-then-encipher encryption: How to exploit nonces or redundancy in plaintexts for efficient encryption. *Advances in Cryptology – ASIACRYPT '00.* Lecture Notes in Computer Science, vol. 1976, Springer-Verlag, 2000. Available at URL www.cs.ucdavis.edu/~rogaway
6. A. Berendschot, B. den Boer, J.P. Boly, A. Bosselaers, J. Brandt, D. Chaum, I. Damgård, M. Dichtl, W. Fumy, M. van der Ham, C.J.A. Jansen, P. Landrock, B. Preneel, G. Roelofsen, P. de Rooij, and J. Vandewalle. Integrity primitives for secure information systems, Final report of RACE integrity primitives evaluation (RIPE-RACE 1040). Lecture Notes in Computer Science, vol. 1007, Springer-Verlag, 1995.
7. D. Bernstein. Floating-point arithmetic and message authentication. Unpublished manuscript. Available at URL http://cr.yp.to/papers.html#hash127
8. D. Bernstein. How to stretch random functions: the security of protected counter sums. *Journal of Cryptology,* vol. 12, no. 3, pp. 185–192 (1999). Available at URL cr.yp.to/djb.html

9. J. BLACK, S. HALEVI, H. KRAWCZYK, T. KROVETZ, and P. ROGAWAY. UMAC: Fast and secure message authentication. *Advances in Cryptology – CRYPTO '99*. Lecture Notes in Computer Science, Springer-Verlag, 1999. Available at URL www.cs.ucdavis.edu/~rogaway
10. J. BLACK and P. ROGAWAY. A block-cipher mode of operation for parallelizable message authentication. Full version of this paper. Available at URL www.cs.ucdavis.edu/~rogaway
11. J. BLACK and P. ROGAWAY. CBC MACs for arbitrary-length messages: The three-key constructions. Full version of paper from *Advances in Cryptology – CRYPTO '00*. Lecture Notes in Computer Science, vol. 1880, pp. 197–215, 2000. Available at URL www.cs.ucdavis.edu/~rogaway
12. G. BRASSARD. On computationally secure authentication tags requiring short secret shared keys. *Advances in Cryptology – CRYPTO '82*. Plenum Press, pp. 79–86, 1983.
13. L. CARTER and M. WEGMAN. Universal hash functions. *J. of Computer and System Sciences*. vol. 18, pp. 143–154, 1979.
14. V. GLIGOR and P. DONESCU. Fast encryption and authentication: XCBC encryption and XECB authentication modes. *Fast Software Encryption*, Lecture Notes in Computer Science, Springer-Verlag, April 2001. Available at URL www.eng.umd.edu/~gligor
15. O. GOLDREICH, S. GOLDWASSER, and S. MICALI. How to construct random functions. *Journal of the ACM*, vol. 33, no. 4, pp. 210–217, 1986.
16. S. HALEVI and H. KRAWCZYK. MMH: Software message authentication in the Gbit/second rates. *Fast Software Encryption* (FSE 4), Lecture Notes in Computer Science, vol. 1267, Springer-Verlag, pp. 172–189, 1997. Available at URL www.research.ibm.com/people/s/shaih
17. ISO/IEC 9797. Information technology – Security techniques – Data integrity mechanism using a cryptographic check function employing a block cipher algorithm. International Organization for Standards (ISO), Geneva, Switzerland, 1994 (second edition).
18. C. JUTLA. Encryption modes with almost free message integrity. *Advances in Cryptology – EUROCRYPT 2001*. Lecture Notes in Computer Science, vol. 2045, B. Pfitzmann, ed., Springer-Verlag, 2001.
19. H. KRAWCZYK. LFSR-based hashing and authentication. *Advances in Cryptology – CRYPTO '94*. Lecture Notes in Computer Science, vol. 839, Springer-Verlag, pp 129–139, 1994.
20. H. LIPMAA. Personal communication, July 2001. Further information available at www.tcs.hut.fi/~helger
21. E. PETRANK and C. RACKOFF. CBC MAC for real-time data sources. *Journal of Cryptology*, vol. 13, no. 3, pp. 315–338, Nov 2000. Available at URL www.cs.technion.ac.il/~erez/publications.html. Earlier version as 1997/010 in the Cryptology ePrint archive, eprint.iacr.org
22. B. PRENEEL. Cryptographic primitives for information authentication – State of the art. *State of the Art in Applied Cryptography*, COSIC '97, LNCS 1528, B. Preneel and V. Rijmen, eds., Springer-Verlag, pp. 49–104, 1998.
23. M. WEGMAN and L. CARTER. New hash functions and their use in authentication and set equality. *J. of Comp. and System Sciences*. vol. 22, pp. 265–279, 1981.

Rethinking PKI: What's Trust Got to Do with It?

Stephen Kent

Chief Scientist – Information Security, BBN Technologies, USA

Much of the literature related to public key infrastructure (PKI) uses terms such as "trust" extensively and assumes that certification authorities (CAs) are trusted third parties (TTPs). It is certainly true that the best known CAs today are commercial TTPs, and such CAs have played an important role in making the general public aware of PKIs. But, not all PKIs need adopt this sort of CA model, in which relying parties are required to make value judgments about the trustworthiness of the organizations that operate CAs. PKIs are not intrinsically valuable. They are infrastructures that, if successful, facilitate authentication and authorization services based on the use of public key cryptography. Thus it is appropriate to ask questions about these services:

- In what context are these services being employed?
- What forms of identifiers are meaningful for the context?
- Does the context relate to existing physical world, or does it exist only in cyberspace?
- Are the services offered to anyone, or are they intended for identifiable user populations?
- Are their existing organizational entities that are authoritative for the authentication or authorization information contained in the certificates issued by the CAs?

In many of the situations in which PKIs are being used today, or which have been proposed, the contexts have been transplanted from the physical world to cyberspace. Often, the users (subjects) of these PKIs already have been assigned identifiers in the physical world, identifiers managed by organizational entities that are considered authoritative for assigning these identifiers to these users. In such contexts it would seem natural for these organizational entities to act as CAs, identifying users in cyberspace in the same fashion as they have identified them in the physical world.

An implicit form of trust relationship exists here, between relying parties and these organizations, but this trust is often based on long established business or social relationships, contracts, or by statute. Trust in this context is not created out of thin air in cyberspace. It is a very different form of trust from what is usually described in papers on "trust management."

This presentation argues that, in the best circumstances, CAs should not have to be trusted explicitly. Rather, CAs of the sort noted above merit an implied trust due to their position as authoritative entities responsible for name spaces, authorization information, etc. The presentation describes why this style of PKI has numerous advantages relative to the models commonly described

L.R. Knudsen (Ed.): EUROCRYPT 2002, LNCS 2332, pp. 398–399, 2002.

in the literature, and promoted by commercial TTPs. It examines examples of authoritative CAs in many aspects of everyday life, and analyzes the nature of the relationships that give rise to these CAs. It describes how standard X.509 constructs can be used to facilitate cross- certification among organizations in the context this model, how several forms of cyberspace identities could be certified consistent with this model, and how organizations can issue certificates easily to users with whom they have existing, client relationships.

A variety of photographs (shot by the speaker) will punctuate the presentation.

Efficient Generic Forward-Secure Signatures with an Unbounded Number of Time Periods

Tal Malkin[1], Daniele Micciancio[2,⋆], and Sara Miner[2,⋆⋆]

[1] AT&T Labs Research, 180 Park Avenue, Florham Park, NJ 07932, USA,
tal@research.att.com

[2] University of California, San Diego,
Department of Computer Science and Engineering, Mail Code 0114,
9500 Gilman Drive, La Jolla, CA 92093-0114, USA,
{daniele, sminer}@cs.ucsd.edu

Abstract. We construct the first efficient forward-secure digital signature scheme where the total number of time periods for which the public key is used does not have to be fixed in advance. The number of time periods for which our scheme can be used is bounded only by an exponential function of the security parameter (given this much time, any scheme can be broken by exhaustive search), and its performance depends (minimally) only on the time elapsed so far. Our scheme achieves excellent performance overall, is very competitive with previous schemes with respect to all parameters, and outperforms each of the previous schemes in at least one parameter. Moreover, the scheme can be based on any underlying digital signature scheme, and does not rely on specific assumptions. Its forward security is proven in the standard model, without using a random oracle. As an intermediate step in designing our scheme, we propose and study two general composition operations that can be used to combine *any* existing signature schemes (whether standard or forward-secure) into new forward-secure signature schemes.

1 Introduction

The standard notion of digital signature security is extremely vulnerable to leakage of the secret key, which, over the lifetime of the scheme, may be quite a realistic threat. Indeed, if the secret key is compromised, any message can be forged. All future signatures are invalidated as a result of such a compromise, and, furthermore, no *previously issued* signatures can be trusted. Once a leakage has been identified, some key revocation mechanism may be invoked, but this does not solve the problem of forgeability for past signatures. Asking the signer to reissue all previous signatures is very inefficient, and, moreover, requires trusting the signer. For example, it is very easy for a dishonest signer to leak

⋆ Supported in part by NSF Career Award CCR-0093029

⋆⋆ Supported in part by a Graduate Diversity Fellowship from the San Diego Supercomputer Center, and Mihir Bellare's 1996 Packard Foundation Fellowship in Science and Engineering

L.R. Knudsen (Ed.): EUROCRYPT 2002, LNCS 2332, pp. 400–417, 2002.

his secret key in order to repudiate a previously signed document. Furthermore, changing the schemes keys very frequently is also not a practical solution, since frequently registering new public keys and maintaining them in a place that is both publicly accessible and trusted is a difficult task.

To mitigate the consequences of possible key leaks, the notion of forward security for digital signatures was initially proposed by Anderson [1] and formalized by Bellare and Miner [3]. The basic idea is to extend a standard digital signature algorithm with a *key update* algorithm, so that the *secret key* can be changed frequently, while the public key stays the same. The resulting scheme is forward-secure if the knowledge of the secret key at some point in time does not help forge signatures relative to some previous time period. Thus, if the secret key is compromised during time period t, then the key can be revoked without invalidating signatures issued during earlier time periods.

Since the introduction of the concept of forward security, several such schemes have been suggested. These schemes exhibit varying performance both in terms of space and time. Typically, these schemes achieve efficiency in some parameters at the price of making other parameters significantly worse than in standard signature schemes. For example, some schemes have faster signature generation and verification algorithms, but slower key generation and update procedures, while other schemes have the opposite behavior. In other cases, time improvements are obtained at the cost of larger signatures or larger secret and public keys. Moreover, in essentially all previous schemes, although the number of time periods can be arbitrarily large, its value must be set in advance, and passed as a parameter during the key generation process. The performance of the algorithms then depends on the security parameter as well as the a priori maximum number of time periods T. So, setting T to an unnecessarily large number results in a considerable efficiency loss.

Clearly, there is a trade-off between the efficiency parameters, and which scheme is most practical depends on the requirements of the specific application. Consider, for example, a scenario where keys are updated once a day. Then, a slower key update algorithm might be acceptable, if this helps make signature verification faster or signatures themselves shorter. On the other hand, consider an electronic checkbook (e-check) application where the time period corresponds to the check serial number, rather than physical time. If your electronic wallet is stolen (with the current e-check secret key in it), you want to revoke all compromised checks without invalidating the checks that were legitimately issued. This can be achieved if the secret key is updated after signing every check. In this case, fast key update is as important as fast signature generation, as the two algorithms are always used together.

1.1 Our Results

In this paper, we design a new tree-like digital signature scheme which we call the MMM scheme. Our construction uses the idea of Merkle trees [10], which were suggested for use in the forward security context by Krawczyk [9]. Our scheme is a *generic* construction, namely it can be based on any underlying signature scheme, and does not rely on specific computational assumptions like discrete log or factoring. Moreover, the security of MMM can be proved in the standard complexity model, without resorting to the random oracle methodology. This scheme is very efficient in all parameters (simultaneously), and outperforms all previous constructions in at least some parameters (the price is never more than a constant increase in other parameters, and in most cases there is no price at all). In fact, our scheme is asymptotically essentially optimal, while in practice it is competitive even with standard signature schemes. For example, signing requires a single signature in the underlying scheme, and verifying requires little more than two verifications of the underlying scheme. (More performance details are given later.) Furthermore, the MMM scheme is the first efficient forward-secure scheme which does not require its maximal number of time periods to be fixed in advance. Instead, the user can keep calling the update procedure indefinitely. The only (theoretical) barrier to the number of time periods is exponential in the security parameter, and therefore cannot be feasibly reached. Unlike all previous schemes, the efficiency parameters of the MMM scheme do not depend at all on the *maximal* number of time periods, but rather on the number of time periods elapsed *so far* (and this dependence is minimal). To summarize, the significance of our scheme stems from the following properties:

- **Practical:** MMM has excellent performance for practical parameters. This is further improved by the fact that the maximal number of time periods is not a relevant parameter. Any standard signature scheme can be used as a building block, providing trade-offs between the parameters, thus allowing greater flexibility in accommodating the requirements of particular applications, and meeting practical constraints.
- **Strong security guarantee:** The scheme is secure under the minimal necessary assumption, namely that (standard) digital signature schemes exist (which is equivalent to the existence of one-way functions [15]). This security is proved without relying on the random oracle model.

As an intermediate step in developing the MMM scheme, we propose and study two general composition operations that can be used to combine *any* existing signature schemes (whether standard or forward-secure) into new forward-secure schemes. We prove their security in the standard complexity model, based on the security of the underlying schemes, and without requiring any additional assumption.[1] To obtain our main result, we then extend these composition operations to generate the MMM construction. Separately, in Section 5, we present several additional constructions (some new and some already known) derived via these composition operations, which illustrate some of the efficiency tradeoffs possible.

[1] We also use pseudo-random generators [4,16] and universal one way (a.k.a., target collision resistant) hash functions [13], but their existence is known to be equivalent to the existence of digital signature schemes.

1.2 Related Work

Recently, several forward-secure signature schemes have been proposed in the literature. Some of the first solutions described in [1] had the undesirable property that keys or signature size were linear in the (maximal) number of time periods T. (For example, one can generate T different secret/public key pairs using a conventional signature scheme, and delete a secret key from memory each time the update algorithm is invoked.) Subsequent efforts found schemes where the size parameters (key and signature size) were independent of, or at most logarithmic in, the number of time periods, possibly at the price of slower signing, verifying or update algorithms. (E.g., the solutions proposed by Bellare and Miner [3] and Abdalla and Reyzin [2] had signing and verification time linear in T.) Here, "size independent of T" means that the size depends only on the value of the security parameters. However, it should be noted that the security parameters must be superlogarithmic in T to avoid exhaustive search attacks, so having no explicit dependency on $\log T$ does not necessarily give shorter keys. For a fair comparison of different schemes, the actual dependency of the space and time complexity of all algorithms on the security parameters must be taken into account.

In this section, we briefly highlight these dependencies for all previously proposed schemes. We use two different security parameters, as follows:

- l: a security parameter such that exhaustive search over l-bit strings is infeasible. This is the security parameter of conventional (symmetric) cryptographic operations, e.g., the seed length of pseudo-random generators, or the output length of cryptographic hash functions.
- k: a security parameter such that k-bit numbers are hard to factor, or such that more generally common number theoretic problems (such as inverting the RSA function) become infeasible.

It is important to distinguish between the two values because private key cryptography is typically much more efficient than public key. Factoring can be solved in sub-exponential time ($\exp(k^{1/3} \log^{2/3} k)$), so asymptotically one needs $k \approx O(l^3)$ to be much bigger than l. In practice, $k = 1000$ and $l = 100$ are acceptable values. In the analysis of all constructions we assume that modular multiplications of k-bit numbers are performed in k^2 time[2] and hash functions and pseudo-random generators run in l^2 time. For $k = 1000$ and $l = 100$ this means that a block cipher application is 100 times faster than a modular arithmetic operation.

In the tables below, we omit small constant factors and some of the lower order terms from the running time, e.g., if an algorithm requires one modular exponentiation plus one hashing, we simplify the running time expression $O(k^3) + O(l^2)$, and write only the most significant term $O(k^3)$. The only assumptions used in these simplifications are that $\log T = o(l)$ and $l = o(k)$.

[2] Although faster multiplication algorithms are known, we feel that our assumptions accurately represent the relationship between the speeds of private and public key applications in practice.

The previous schemes we analyze below can be divided into two categories: those which are generic constructions built using any standard signature scheme as a black box, and those based on specific number theoretic assumptions, e.g., hardness of factoring large integers.

GENERIC CONSTRUCTIONS. Generic constructions are those proposed by Anderson [1], the tree scheme of Bellare and Miner [3], and the Krawczyk variant of the Anderson scheme with reduced secret storage [9]. Their performance is summarized in Figure 1. As usual, T stands for the total number of time periods in the scheme. When a standard signature scheme is required, we assume key generation, signing and verifying can all be performed in $O(lk^2)$ time, and public keys, secret keys and signatures are $O(k)$ bits long. For example, this is the performance of the signature schemes of Guillou and Quisquater [7], and of Micali [11]. Note that we have broken up the storage required of the signer into "Secret key size" and "Non-secret storage", to better illustrate the differences between these schemes. This was indeed the main improvement of Krawczyk scheme, which otherwise is essentially the same as the one originally proposed by Anderson. However, it should be remarked that even the non-secret storage must still be secure in the sense that it must be stored in (publicly readable) tamper-proof memory. Changing or altering the non-secret storage would disrupt the signing algorithm, making signature generation impossible. For simplicity, in the rest of this paper we will drop the distinction between secret and non-secret storage.

	Anderson	Binary Tree (BM)	Krawczyk
Key gen time	lk^2T	$lk^2 \log T$	lk^2T
Signing time	lk^2	lk^2	lk^2
Verification time	lk^2	$lk^2 \log T$	lk^2
Key update time (*)	$O(1)$	lk^2	lk^2
Secret key size	kT	$k \log T$	k
Non-secret storage	kT	$k \log T$	kT
Public key size	k	k	k
Signature size	k	$k \log T$	k

Fig. 1. Comparing parameters of generic constructions [1,3,9]. (*)The running time of the Binary Tree update algorithm is amortized over all update operations. If no amortization is used, the worst running time can be bigger by a factor of $\log T$.

As shown in Figure 1, the Binary Tree scheme has much faster key generation, and smaller secret/non-secret key storage than Anderson and Krawczyk schemes, with the linear dependency on T replaced by a logarithmic function. In exchange, the price paid is a slower verification procedure and longer signatures, which both increase by a factor $\log T$. For a detailed description of the schemes the reader is referred to the original papers [1,3,9].

	BM	AR	IR	IR'
Key gen time	lk^2T	lk^2T	$k^5 + (k + l^3)lT$	$k^5 + T\log^4 T + k^2 l + klT$
Signing time	$(T + l)k^2$	lk^2T	$k^2 l$	$k^2 l$
Verification time	$(T + l)k^2$	lk^2T	$k^2 l$	$k^2 l$
Key update time	lk^2	lk^2	$(k^2 + l^3)lT$	$(k^2 l + l^2 + \log^4 T)\log T$
Secret key size	lk	k	k	$(1 + \log T)k$
Public key size	lk	k	k	k
Signature size	k	k	k	k

Fig. 2. Comparing parameters of schemes built on specific security assumptions [3,2,8]. IR' denotes the Itkis-Reyzin variant using the two optimizations from [8].

The above constructions are generic, meaning they may be instantiated with any signature scheme. Their proofs of security rely only on the security of the underlying signature scheme (in particular, none of these security proofs rely on random oracles).[3]

CONSTRUCTIONS BASED ON SPECIFIC SECURITY ASSUMPTIONS. Number theoretic schemes were proposed by Bellare and Miner [3], Abdalla and Reyzin [2], and very recently (and independently of our work), by Itkis and Reyzin [8]. All these are proven to be secure *in the random oracle model*, assuming that factoring is hard (for [3,2]) or that taking any root modulo a composite is hard (for [8]). The performance of these schemes is summarized in Figure 2.

Except for the key generation and key update time, the Itkis-Reyzin (IR) scheme is essentially optimal among the number theoretic schemes. However, key update is very slow (linear in T), making it impractical for some applications (e.g., the e-check application described in the introduction). Itkis and Reyzin [8] also suggest a variant of their scheme where the update time is only proportional to $\log T$, but the secret key size increases by a factor $\log T$, thus matching generic constructions like the binary tree scheme. (In fact, generic constructions have potentially smaller keys because they can be based on any signature scheme with possibly shorter keys than factoring based ones.) Interestingly, the seemingly "optimal" (i.e., independent of T) verification time of the IR scheme, is not necessarily better than other schemes in which this time is proportional to $\log T$. For example, by properly instantiating the binary tree scheme with a basic signature algorithm with fast verification procedure (e.g., Rabin's signature scheme [14] which has verification time k^2), one can obtain a forward-secure signature scheme where verification takes only $O(k^2 \log T)$, beating the $O(lk^2)$ running time of the IR scheme because $\log T = o(l)$. (In the full version of this paper [12], we show a different tree construction, in which even faster verification times are possible.)

[3] Of course, when instantiated with a specific base signature scheme which requires other assumptions, such as a random oracle, the resulting forward-secure scheme also requires the same assumptions.

Comparison to Our Results. All previous known schemes (with the exception of the very inefficient "long signature scheme" described in [3]) require the total (maximal) number of time periods T to be fixed in advance and passed as a parameter to the key generation algorithm. The value of T then contributes to the overall performance of these schemes (at least proportionally to $\log T$, but in most cases linearly for some parameters). We propose a new scheme, MMM, whose performance avoids any dependence on T, and depends only on the number of time periods elapsed so far (even this dependence is minimal). In fact, T need never be fixed. In addition, MMM also combines the best of both types of previous constructions in terms of efficiency and security. Indeed, we construct it based on generic assumptions, yet we achieve competitive performance and stronger security guarantees, even when compared to the best previous schemes which used specific number theoretic assumptions.

In the left half of Figure 3, we demonstrate the efficiency of the scheme when instantiated with the Guillou-Quisquater signature scheme.[4] In the right half of that same figure, we select specific parameter values, and show an actual efficiency comparison between our scheme and the best previous schemes in each category, namely the two Itkis-Reyzin schemes, and the Bellare-Miner binary tree scheme. Relative to IR, note that we achieve the same or better improvements over the binary tree scheme, without paying the price that IR pays in other parameters. Also note that while we instantiated $t = 1000$ in MMM (the same value as the instantiated *maximal* T in the other schemes), most of the time t will be much smaller, implying better performance. More details about the performance of our scheme can be found in Section 4.

2 Definitions

As formalized in [3], a *key-evolving* signature scheme $\mathcal{S}$ consists of four algorithms: a key generation algorithm KeyGen, a secret key evolution algorithm Update, a signature generation algorithm Sign, and a signature verification algorithm Verify. It differs from a standard digital signature scheme in that the secret key is subject to an update (or evolution) algorithm, and the time for which the scheme is in use is divided into *time periods.* The public and secret keys are denoted *pk* and *sk* respectively, and the secret key *sk* changes via each invocation of the key update algorithm. We write $sk^{(j)}$ when we want to emphasize that a particular value of the secret key is relative to the jth time period. It is important to notice that the public key for the scheme remains constant

[4] In order to compare a generic construction with a construction using specific security assumptions, we need to select a specific base scheme for the generic construction. Here, GQ is a good selection for comparison, as it is efficient, and it is the one underlying the Itkis-Reyzing constructions as well. Because the GQ scheme is proven secure in the random oracle model, the proof of forward security of the resulting scheme also requires random oracles. This particular example does not change the fact that our security proof does not need them itself.

	MMM
Key gen time	k^2l^2
Signing time	k^2l
Verification time	$k^2l + l^2 \log l$
Key update time (*)	$k^2l + (k + l^2) \log t$
Secret key size	$k + l \log l$
Public key size	l
Signature size	$k + l \log l$

	MMM	IR	IR'	B.T.
Key gen time	10^{10}	10^{15}	10^{15}	10^9
Signing time	10^8	10^8	10^8	10^8
Verification time	10^8	10^8	10^8	10^9
Key update time	10^8	10^{11}	10^9	10^8
Secret key size	10^3	10^3	10^4	10^4
Public key size	10^2	10^3	10^3	10^3
Signature size	10^3	10^3	10^3	10^4

Fig. 3. Analyzing the MMM scheme instantiated with GQ. We first present an asymptotic analysis. Then, assuming parameter values $k = 1000$, $l = 100$, and $T = 1000$, we compare estimated times and sizes for MMM with the IR schemes and the Bellare-Miner binary tree scheme. In the case of MMM, we provide worst-case values, by setting $t = 1000$. (*) The running time of the update algorithm is amortized. The worst case update may take tlk^2. (Later, we provide another version of our scheme which achieves uniformly fast update time, at the price of additional $l \log^2 t + k \log l$ in the secret key size.)

throughout, and each signature is verified using the same public key pk, regardless of which $sk^{(j)}$ was used to generate it.

For a key-evolving signature scheme to be *forward-secure*, it must be computationally infeasible for an adversary, even after learning the secret key of the scheme for a particular time period, to forge a signature on a new message for a time period earlier than the secret key was leaked. Bellare and Miner formalized the adversary model for forward-secure signature schemes, and the experiment used to define the *insecurity function* for a scheme. The reader is referred to [3] for details.

Typically, the total number of time periods T for a particular instantiation of a forward-secure signature scheme must be fixed in advance, and passed as a parameter to the key generation algorithm KeyGen. Moreover, T is usually included in the public key pk, secret keys $sk^{(j)}$, and signatures σ, as it is required by the update, signing and verification algorithms. The secret keys $sk^{(j)}$ and signatures σ also include the current or issuing time period t. In this paper, we use the convention that both T and t are passed as *external* parameters to Update, Sign and Verify, instead of being included in the key and signature strings. This is to avoid unnecessary duplications when signature schemes are combined using our composition operations. However, most update, sign and verify algorithms need t (and perhaps even T) to work properly, and therefore these values should be thought as integral parts of keys and signatures. Below, we summarize the general input and output parameters of each algorithm of a forward-secure digital signature scheme. We note that for the MMM scheme in particular, an input T is not needed for any of its algorithms, and thus can be omitted below.

- The key generation algorithm KeyGen takes as input the total number of time periods T, and outputs a pair $(pk, sk^{(0)})$ of public and secret keys.
- The key update algorithm Update takes as input T, the current time period t, and the secret key $sk^{(t)}$ for time period t, and changes $sk^{(t)}$ into $sk^{(t+1)}$. If

$t + 1 = T$ then Update completely erases the secret key *sk*, and returns the empty string.
- The signing algorithm Sign takes as input T, the current time period t, the corresponding secret key $sk^{(t)}$, and a message M, and outputs a signature (σ, t) on M.
- The verification algorithm Verify takes as input T, *pk*, M, σ, and t, and accepts if and only if σ is a valid signature tag for message M and secret key $sk^{(t)}$. Note that here t represents the time period during which σ was issued, which is not necessarily the current one.

Additionally, all algorithms implicitly take as input one or more security parameters.

3 Composition Methods

In this section, we describe two general composition methods that can be used to combine both standard and forward-secure signature schemes into forward-secure schemes with more time periods. These are methods which have been used implicitly in previous constructions (e.g., in [10,3], see Sect. 5), and we formalize them here. Although these compositions are interesting in their own right, we present them here as a means to lead into our main construction, which can be found in Section 4. To illustrate the flexibility afforded by these composition operations, however, we present additional constructions in Section 5.

In order to unify the presentation, we regard standard signature schemes as forward-secure signature schemes with one time period, namely $T = 1$. Let Σ_0 and Σ_1 be two forward-secure signature schemes with T_0 and T_1 time periods respectively. We consider two methods to combine Σ_0 and Σ_1 into a new forward-secure signature scheme Σ with a larger number of time periods T. The first composition method results in a new scheme $\Sigma = \Sigma_0 \oplus \Sigma_1$ with $T = T_0 + T_1$ time periods. The second composition method results in a new scheme $\Sigma = \Sigma_0 \otimes \Sigma_1$ with $T = T_0 \cdot T_1$ time periods. We call these operations the "sum" and the "product" compositions, respectively. In the following two subsections, we give descriptions of the operations, and give theorems analyzing their security and performance.

Note that the sum and product procedures make use of specialized versions of the key generation algorithms KeyGen_i $(i = 0, 1)$ that produce only the *secret* or the *public* key of the original schemes. These specialized versions are called $\mathrm{SKeyGen}_i$ and $\mathrm{PKeyGen}_i$ respectively. As we shall see, the diversification of KeyGen to PKeyGen and SKeyGen plays a critical role for the performance of forward-secure signature schemes obtained by the iterated application of the basic composition operations.

3.1 The "Sum" Composition

Given any two schemes Σ_0 and Σ_1 as described above, we define a new scheme $\Sigma = \Sigma_0 \oplus \Sigma_1$ (called the sum of Σ_0 and Σ_1) with $T = T_0 + T_1$ time periods.

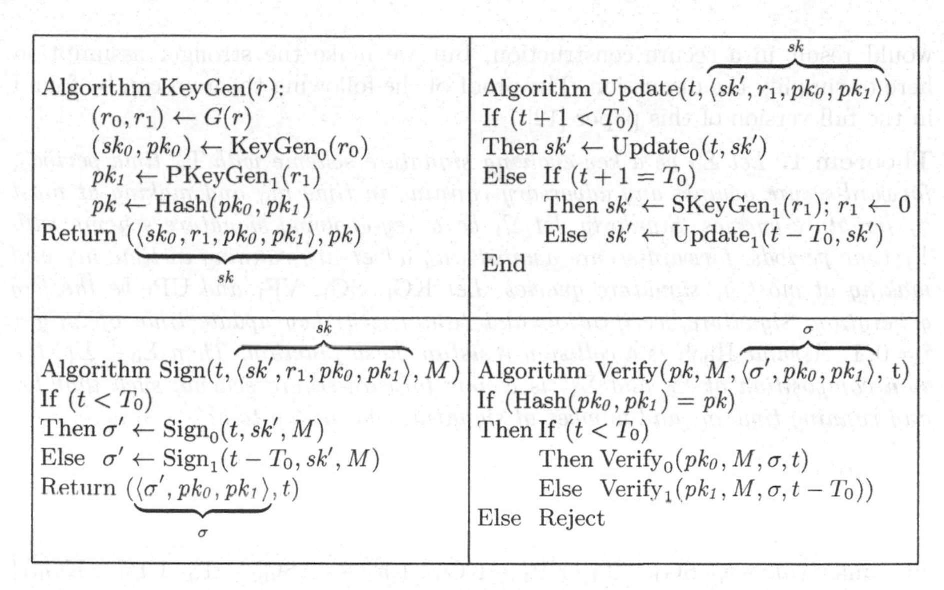

Fig. 4. The algorithms defining the sum composition.

The public key for the sum scheme pk is the (collision-resistant) hash of the public keys pk_0, pk_1 of the two constituent schemes. The composition works by first expanding a random seed r into a pair of seeds (r_0, r_1) using a length-doubling pseudorandom generator, and generating keys for both Σ_0 and Σ_1 using pseudorandom seeds r_0 and r_1 respectively. Then, the secret key for the Σ_1 scheme is deleted, while its public key and the randomness r_1 are saved. (Deleting the second secret key, and later recomputing it using the seed r_1, is essential to keep the size of the secret key small when the composition operation is iterated many times.) Signatures are generated using secret keys from the Σ_0 scheme during the first T_0 time periods. Then the Σ_1 key generation process is run again with the random string r_1, to produce the same secret key obtained earlier. (Notice that the public key will also be the same.) From this point forward, the Σ_0 secret key is deleted and only the Σ_1 keys are used. Signatures during any time period include the public keys for both schemes, so that the tag generated can be checked against the relevant public key, and so the authenticity of the signature can be checked against the hash value of both public keys, which was published. The details of the scheme are given in Figure 4. We denote the composition of the sum algorithm with itself iteratively to achieve T time periods, by $S^{\oplus}_{\log T}$ (since it looks like a tree with $\log T$ levels). This scheme will later be used (with varying T's) to construct the MMM scheme (and it is further discussed in Section 5).

Security Analysis. Here we state a claim about the security of the sum composition. We assume only that the underlying signature schemes are forward-secure, and that the hash function Hash used in the construction is collision-resistant. In reality, the weaker assumption of a target-collision-resistant hash function would result in a secure construction, but we make the stronger assumption here to simplify the discussion. The proof of the following theorem can be found in the full version of this paper [12].

Theorem 1. *Let Σ_0 be a key-evolving signature scheme with* T_0 *time periods, forward-secure against any adversary running in time m_0 and making at most q_0 signature queries. Similarly, let Σ_1 be a key-evolving signature scheme with* T_1 *time periods, forward-secure against any adversary running in time m_1 and making at most q_1 signature queries. Let* $\mathrm{KG_i}$, $\mathrm{SG_i}$, $\mathrm{VF_i}$ *and* $\mathrm{UP_i}$ *be the key generation, signature, verification and (amortized) key update time of Σ_i for $i = 0, 1$. Assume* Hash *is a collision-resistant hash function. Then $\Sigma_0 \oplus \Sigma_1$, the sum composition of Σ_0 and Σ_1, is a new forward-secure scheme such that for any running time m, and number of signature queries up to q:*

$$\mathbf{InSec}^{\mathrm{fs}}(\Sigma_0 \oplus \Sigma_1, m, q) \leq \mathbf{InSec}^{\mathrm{fs}}(\Sigma_0, m_0, q_0) + \mathbf{InSec}^{\mathrm{fs}}(\Sigma_1, m_1, q_1)$$

where both of the following hold:

$$m = \max\{(m_0 - q \cdot \mathrm{SG}_1 - \mathrm{T}_1 \cdot \mathrm{UP}_1 - \mathrm{KG}_1), (m_1 - q \cdot \mathrm{SG}_0 - \mathrm{T}_0 \cdot \mathrm{UP}_0 - \mathrm{KG}_0)\}$$
$$q \leq \max\{q_0, q_1\}$$

The theorem above demonstrates that the sum composition is security-preserving. That is, the sum of two schemes will have roughly the same security as the individual schemes' security, relative to the resulting number of time periods.

Performance Analysis. We now give an analysis of key and signature sizes, as well as running times for the sum composition. The relations below can be verified by inspection of Figure 4.

Theorem 2. *Let* $\mathrm{SK_i}$, $\mathrm{PK_i}$ *and* $\mathrm{SIG_i}$ *be the secret key, public key and signature sizes for the forward-secure signature scheme Σ_i for $i = 0, 1$. Then, the key and signature sizes of the sum scheme $\Sigma_0 \oplus \Sigma_1$ are:*

$$\begin{aligned} \mathrm{PK} &= l \\ \mathrm{SK} &= \max(\mathrm{SK}_0, \mathrm{SK}_1) + \mathrm{PK}_0 + \mathrm{PK}_1 + l \\ \mathrm{SIG} &= \max(\mathrm{SIG}_0, \mathrm{SIG}_1) + \mathrm{PK}_0 + \mathrm{PK}_1 \end{aligned}$$

Theorem 3. *Let* $\mathrm{KG_i}$, $\mathrm{SG_i}$, $\mathrm{VF_i}$ *and* $\mathrm{UP_i}$ *be the key generation, signature, verification and (amortized) key update time of the forward-secure signature scheme Σ_i for $i = 0, 1$. Then the running times of the sum scheme $\Sigma_0 \oplus \Sigma_1$ are:* [5]

$$\begin{aligned} \mathrm{KG} &= \mathrm{KG}_0 + \mathrm{KG}_1 + l^2 \\ \mathrm{SG} &= \max\{\mathrm{SG}_0, \mathrm{SG}_1\} \\ \mathrm{VF} &= \max\{\mathrm{VF}_0, \mathrm{VF}_1\} + l^2 \\ \mathrm{UP} &= (\mathrm{SKGtime}_1 + (T_0 - 1)\mathrm{UP}_0 + (T_1 - 1)\mathrm{UP}_1)/(T_0 + T_1 - 1) \end{aligned}$$

[5] As explained in Section 1.2, we assume that a call to a hash function or PRG takes time $O(l^2)$. Also, the update time here is the result of amortized analysis; it represents the average case.

Note that the update algorithm is usually very fast, but its worst case can be quite slow, as it includes key generation for Σ_1. This can be amortized, and, at a small price, made uniformly small. In this case, the update algorithm will be more complex, as partial computation of a future update will be performed at each stage. We discuss this below.

AMORTIZED UPDATE ALGORITHM. While the amortized update time in the sum construction is quite good, the worst case update time can be very bad (linear in T). This bad case happens when the left child is finished (at time T_0), and a new secret key for the right child needs to be generated. To make the update procedure uniformly fast, we can distribute the computation of this key over the time periods of the left child, so that by the time the left child is finished, the key for the right child is ready. Distributing the work (time) is straight-forward, but space becomes an issue, since the intermediate results of the computation need to be kept. So, the space necessary to generate the secret key for the right child will be added to size of the secret key of the scheme. When composing the sum algorithm with itself to achieve T time periods, the space required to generate the initial secret key for the right child is logarithmic. This means that, after iterating, the total secret key size will be the size of the original secret key (say k), plus $O(l \log^2 T)$. The other parameters of $S^{\oplus}_{\log T}$ (see full version [12]) do not change, and the update time becomes efficient even in the worst case.

3.2 The "Product" Composition

Beginning with any Σ_0 and Σ_1 as described in the beginning of Section 3, we now define a new scheme $\Sigma = \Sigma_0 \otimes \Sigma_1$ (called the product of Σ_0 and Σ_1) with $T = T_0 \cdot T_1$ time periods. The idea is to combine signatures of Σ_0 and Σ_1 in a chaining construction. In the process, we will generate several instances of the Σ_1 scheme, one for every time period of Σ_0, so as to achieve forward security. Specifically, for each time period in the Σ_0 scheme, which we will call an *epoch*, an instance of the Σ_1 scheme is generated, and the public key of the Σ_1 scheme is signed using the Σ_0 scheme. The public key of the entire scheme is simply the public key of the Σ_0 scheme, and a signature on a particular message M includes the signing time period, the public key of an instantiation of the Σ_1 scheme, the Σ_0 scheme signature on the Σ_1 public key, and the Σ_1 signature on the message. The scheme is precisely defined in Figure 5.

Security Analysis. Next, we give a security claim about the product composition, assuming only that the underlying signature schemes themselves are forward-secure. Its proof can be found in the full version of this paper [12].

Theorem 4. *Let Σ_0 be a forward-secure signature scheme with* T_0 *time periods, and let Σ_1 be a forward-secure signature scheme with* T_1 *time periods. Let* $\mathrm{KG_i}$, $\mathrm{SG_i}$, $\mathrm{VF_i}$ *and* $\mathrm{UP_i}$ *be the key generation, signature, verification and (amortized) key update time of Σ_i for $i = 0, 1$. Then $\Sigma_0 \otimes \Sigma_1$, the product composition of Σ_0 and Σ_1, is a new forward-secure scheme such that for any running time m, and*

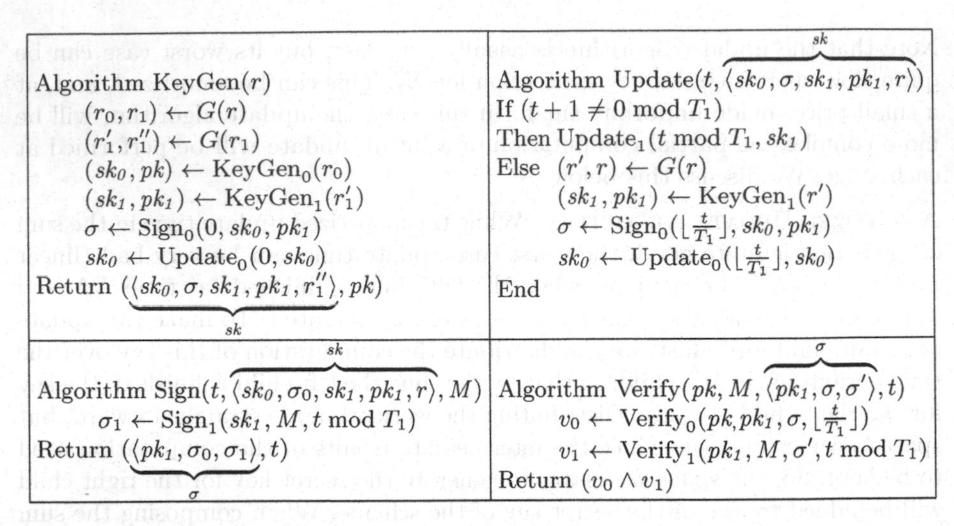

Algorithm KeyGen(r) $(r_0, r_1) \leftarrow G(r)$ $(r'_1, r''_1) \leftarrow G(r_1)$ $(sk_0, pk) \leftarrow \text{KeyGen}_0(r_0)$ $(sk_1, pk_1) \leftarrow \text{KeyGen}_1(r'_1)$ $\sigma \leftarrow \text{Sign}_0(0, sk_0, pk_1)$ $sk_0 \leftarrow \text{Update}_0(0, sk_0)$ Return $(\underbrace{\langle sk_0, \sigma, sk_1, pk_1, r''_1\rangle}_{sk}, pk)$	Algorithm Update$(t, \overbrace{\langle sk_0, \sigma, sk_1, pk_1, r\rangle}^{sk})$ If $(t+1 \neq 0 \bmod T_1)$ Then $\text{Update}_1(t \bmod T_1, sk_1)$ Else $(r', r) \leftarrow G(r)$ $(sk_1, pk_1) \leftarrow \text{KeyGen}_1(r')$ $\sigma \leftarrow \text{Sign}_0(\lfloor \frac{t}{T_1} \rfloor, sk_0, pk_1)$ $sk_0 \leftarrow \text{Update}_0(\lfloor \frac{t}{T_1} \rfloor, sk_0)$ End
Algorithm Sign$(t, \overbrace{\langle sk_0, \sigma_0, sk_1, pk_1, r\rangle}^{sk}, M)$ $\sigma_1 \leftarrow \text{Sign}_1(sk_1, M, t \bmod T_1)$ Return $(\underbrace{\langle pk_1, \sigma_0, \sigma_1\rangle}_{\sigma}, t)$	Algorithm Verify$(pk, M, \overbrace{\langle pk_1, \sigma, \sigma'\rangle}^{\sigma}, t)$ $v_0 \leftarrow \text{Verify}_0(pk, pk_1, \sigma, \lfloor \frac{t}{T_1} \rfloor)$ $v_1 \leftarrow \text{Verify}_1(pk_1, M, \sigma', t \bmod T_1)$ Return $(v_0 \wedge v_1)$

Fig. 5. The algorithms defining the product composition.

number of signature queries up to q:

$$\mathbf{InSec}^{\text{fs}}(\Sigma_0 \otimes \Sigma_1, m, q) \leq \mathbf{InSec}^{\text{fs}}(\Sigma_0, m_0, q_0) + \mathrm{T}_0 \cdot \mathbf{InSec}^{\text{fs}}(\Sigma_1, m_1, q_1)$$

where both of the following hold:

$$\begin{aligned} m \leq \max \{ & (m_0 - q \cdot \mathrm{SG}_1 - \mathrm{T}_0 \cdot (\mathrm{KG}_1 + \mathrm{T}_1 \cdot \mathrm{UP}_1)), \\ & (m_1 - q \cdot \mathrm{SG}_1 - \mathrm{T}_0 \cdot (\mathrm{KG}_1 + \mathrm{SG}_0 + \mathrm{UP}_0 + \mathrm{T}_1 \cdot \mathrm{UP}_1))\} \\ q \leq \max \{ & \mathrm{T}_0, q_1\} \end{aligned}$$

The above theorem demonstrates that a scheme generated using the product composition will have roughly the same insecurity as the underlying two schemes, relative to the resulting number of time periods. That is, the product construction is security-preserving.

Performance Analysis. For the product construction, we now give an analysis of key and signature sizes, as well as running times. The relations can be verified by inspection of Figure 5.

Theorem 5. *Let* $\mathrm{SK_i}$, $\mathrm{PK_i}$ *and* $\mathrm{SIG_i}$ *be the secret key, public key and signature sizes for the forward-secure signature scheme* Σ_i *for* $i = 0, 1$. *Then, the key and signature sizes of the product scheme* $\Sigma_0 \otimes \Sigma_1$ *are:*

$$\begin{aligned} \mathrm{PK} &= \mathrm{PK}_0 \\ \mathrm{SK} &= \mathrm{SK}_0 + \mathrm{SK}_1 + \mathrm{PK}_1 + \mathrm{SIG}_0 + l \\ \mathrm{SIG} &= \mathrm{SIG}_0 + \mathrm{SIG}_1 + \mathrm{PK}_1 \end{aligned}$$

Theorem 6. *Let* $\mathrm{KG_i}$, $\mathrm{SG_i}$, $\mathrm{VF_i}$ *and* $\mathrm{UP_i}$ *be the key generation, signature, verification and (amortized) key update time of the forward-secure scheme* Σ_i *for* $i = 0, 1$. *Then the running times of the product scheme* $\Sigma_0 \otimes \Sigma_1$ *are:* [6]

$$\begin{aligned}
\mathrm{KG} &= \mathrm{KG_0} + \mathrm{KG_1} + \mathrm{SG_0} + \mathrm{UP_0} + 2l^2 \\
\mathrm{SG} &= \mathrm{SG_1} \\
\mathrm{VF} &= \mathrm{VF_0} + \mathrm{VF_1} \\
\mathrm{UP} &= (T_0(T_1 - 1)\mathrm{UP_1} + (T_0 - 1)(l^2 + \mathrm{KG_1} + \mathrm{SG_0} + \mathrm{UP_0}))/(T_0 T_1 - 1) \\
&\leq \mathrm{UP_1} + (l^2 + \mathrm{KG_1} + \mathrm{SG_0} + \mathrm{UP_0} - \mathrm{UP_1})/T_1
\end{aligned}$$

4 The MMM Scheme

Here we present a new scheme obtained by the iterated application of the sum composition operation together with an asymmetric variant of the product construction. This scheme makes use of Merkle tree-type certification chains [10], combined with ideas from the binary tree scheme of Bellare and Miner [3]. The main feature of the MMM scheme is that the number of time periods is essentially unbounded: the number of available time periods is limited only by the security offered by security parameter l, i.e., we cannot use more than 2^l time periods because otherwise the scheme can be broken. Moreover, the scheme exhibits excellent performance, with almost instantaneous key generation, and update, signing and verification speed that depend on the *current* time period t, instead of being functions of the *maximum* time period T. This way, the performance of the signing, verifying and update operations in the initial time periods is the same as if we had chosen a small bound on T in one of the previously known schemes. Performance degrades only slightly if a large number of time periods is used, and still remains competitive with all previously known schemes.

The idea is the following. We start from a regular, non-forward-secure, digital signature scheme S and build a forward-secure signature scheme $L = S^{\oplus}_{\log l}$ with l time periods, iterating the sum composition $\log l$ times. We then take the product of L with another scheme of the form $S^{\oplus}_i$, but with a twist (see later). Remember, in the product construction we build a tree where the top part is given by an instance of the L scheme, and at every leaf of L we attach an instance of the $S^{\oplus}_i$ scheme. The twist here is that we use a different i for every leaf (see Figure 6), $S^{\oplus}_0$ for the first epoch, $S^{\oplus}_1$ for the second, and so on up to $S^{\oplus}_{l-1}$ for the last epoch. We see that the total number of available time periods is $T = \sum_{i=0}^{l-1} 2^i = 2^l - 1$; that is, practically unbounded, because 2^l must be much bigger than T anyway to avoid exhaustive search attacks.

Security Analysis. Because of space limitations, a formal analysis of security is omitted in this extended abstract. The security of the MMM scheme follows from

[6] As explained in Section 1.2, we assume that a call to a hash function or PRG takes time $O(l^2)$. Also, the update time here is the result of amortized analysis; it represents the average case.

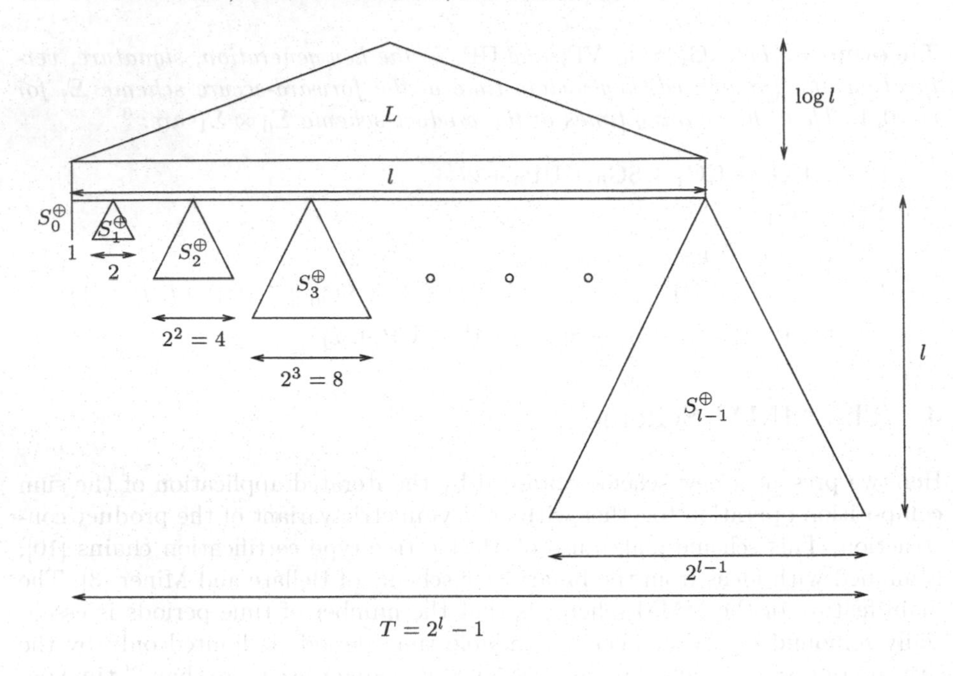

Fig. 6. The MMM construction, which uses an asymmetric product composition.

the security of the underlying composition operations, which we have shown to be security-preserving. However, because in the MMM scheme, the number of time periods is not fixed in advance, the insecurity will grow linearly not with T, the total number of periods in the scheme, but rather with t, the number of periods in the scheme *thus far*, or equivalently the number of update requests made by the adversary. (Since the adversary is constrained to be polynomial time, this number is always bounded by a polynomial, and the new scheme is guaranteed to satisfy asymptotic security.)

Performance Analysis. We now analyze the performance of the MMM scheme.[7]

KEY SIZES. The public key is just a hash value (l bits), while the secret key has roughly the same size as a digital signature: $O(k + (\log l + \log t)l)$ bits.

SIGNATURE SIZE. During time period t, each signature consists of $\log l$ hash values, a public key and a digital signature for the top level tree, and $\log t$ hash values, a public key and a digital signature for the bottom level tree. So, the total size of the signature is $4k + (\log l + \log t)l$ bits. For typical values of the security parameters, this exceeds the length k of a regular digital signature only by a constant factor.

[7] A summary of its performance when instantiated with the GQ scheme was given in Section 1.2.

SIGNATURE GENERATION TIME. Signing only requires computing a signature using the secret key at the leaf node corresponding to the current time period. So, it is as efficient as in standard signature schemes.

VERIFICATION TIME. Verification consists of 2 regular signature verifications, and $\log l + \log t$ hash function evaluations. If the Guillou-Quisquater [7] signature scheme is used, this is $4k^2 l + (\log l + \log t) l^2 = O(k^2 l)$, i.e., just twice as much as the regular signature scheme.

UPDATE TIME. Update consists of updating within the subtree S_i, or, between subtrees, it consists of generating the initial keys for the next subtree S_{i+1}, and updating the key of L. If we proceed in a straight forward manner, amortized analysis of the update time after t periods yields $O(\log t(k+l^2)+k^2 l)$. The worst case, however, is proportional to tlk^2, since in the beginning of a new epoch we need to generate keys for a new subtree S_{i+1} where $i = \log t$. As was discussed earlier, this update can be amortized across many time periods to achieve uniformly fast update, at the cost of larger secret keys. A similar amortization is described in detail by Merkle [10], so we give only a brief description here. The idea is to distribute the computation of this secret/public key pair across the 2^i time periods of the S_i scheme. It is easy to see that the dominant part of the S_{i+1} key generation is the generation of the 2^{i+1} keys associated to the leaves of the tree. If at each update operation of the S_i subtree we generate two leaves for the S_{i+1} scheme, then by the time the ith epoch is over, the key for the new epoch will be ready. The price paid for uniformly fast update is adding the space required to generate the initial keys for the next subtree to the size of the secret key. This results in an addition of $O(l \log^2 t + k \log l)$ to the secret key size.

This brief analysis shows that the MMM scheme is competitive with all other existing schemes with respect to all parameters and has the added advantage of a practically unbounded number of time periods. Moreover, MMM outperforms each of the previously proposed schemes in at least one parameter. For example, when compared to the recent scheme of [8], we see that our scheme has much faster key generation and update procedures, while increasing the other parameters only by a small constant factor. Even when [8] is implemented using their "pebbling" technique, our update procedure is superior because it requires $\log t$ hash function computations as opposed to $\log t$ modular exponentiations, and secret key is also much shorter, being only $O(k + l \log t)$ instead of $O(k \log t)$. It should be noted that since our scheme is generic, we can get different performance trade-offs by changing the underlying signature scheme. (Similar trade-offs are possible also for the binary tree scheme, or the scheme of Krawczyk [9], but not for the number theoretic constructions of [3,2,8].) For example, if the Rabin signature scheme is used, the verification time drops to $O(k^2)$, outperforming all non-generic schemes. The price in this example is making signature generation slightly slower, going from $O(k^2 l)$ to $O(k^3)$. Similarly, signature size can be reduced by choosing an underlying signature scheme with short signatures, possibly at the expense of signing or verification time.

5 Alternate Constructions

We used the sum and product composition operations discussed in Section 3 to generate the MMM construction above. Here, we briefly mention several other constructions possible using these operations, in order to highlight that different performance tradeoffs can be achieved by selecting the proper base schemes and the proper sequence of composition operations. (In fact, we can even achieve previously known schemes using our composition operations, as mentioned below.) In the full version of this paper [12], we analyze the following constructions in more detail:

- BM $\otimes$ BM (where BM denotes the main scheme of Bellare and Miner)
- the iterated product $\mathbf{S}^{\otimes}_{\log \log \mathbf{T}}$, exactly the binary tree scheme of Bellare and Miner
- the iterated sum $\mathbf{S}^{\oplus}_{\log \mathbf{T}}$
- $\mathbf{IR}^{\oplus}_{\mathbf{10}}$ (where IR denotes the standard construction of Itkis and Reyzin)

Acknowledgments

We thank Michel Abdalla and Leo Reyzin for helpful discussions regarding previous works, and comments on earlier drafts. We are also grateful to the anonymous reviewers for their detailed comments.

References

1. R. Anderson. *Two remarks on public-key cryptology.* Manuscript, Sep. 2000. Relevant material presented by the author in an invited lecture at the Fourth ACM Conference on Computer and Communications Security (Apr. 1997).
2. M. Abdalla and L. Reyzin. *A new forward-secure digital signature scheme.* In Advances in Cryptology - Asiacrypt 2000, LNCS **1976** (Dec. 2000), pp. 116-129.
3. M. Bellare and S. Miner. *A forward-secure digital signature scheme.* In Advances in Cryptology - CRYPTO '99, LNCS **1666** (Aug. 1999), pp. 431-448.
4. M. Blum and S. Micali. *How to generate cryptographically strong sequences of pseudorandom bits.* SIAM Journal of Computing **13(4)**(Nov. 1984), pp. 850-864.
5. O. Goldreich, S. Goldwasser, and S. Micali. *How to construct random functions.* Journal of the ACM, **33(4)**(Oct. 1986), pp. 281–308. Preliminary version in the Proceedings of the IEEE Symposium on the Foundations of Computer Science, 1984, pp 464–479.
6. S. Goldwasser, S. Micali and R. Rivest. *A digital signature scheme secure against adaptive chosen-message attacks.* SIAM Journal of Computing **17(2)**(Apr. 1988), pp. 281–308.
7. L.C. Guillou and J.J. Quisquater. *A "paradoxical" identity-based signature scheme resulting from zero-knowledge.* In Advances in Cryptology - CRYPTO '88, LNCS **403** (Aug. 1988), pp. 216–231.
8. G. Itkis and L. Reyzin. *Forward-secure signatures with optimal signing and verifying.* In Advances in Cryptology - CRYPTO '01, LNCS **2139** (Aug. 2001), pp. 332–354.

9. H. Krawczyk. *Simple forward-secure signatures from any signature scheme.* In Seventh ACM Conference on Computer and Communications Security (Nov. 2000), pp. 108–115.
10. R. C. Merkle. *A certified digital signature.* In Advances in Cryptology - CRYPTO '89, (Aug. 1989), pp. 218–238.
11. S. Micali. *A secure and efficient digital signature algorithm.* Technical Report MIT/LCS/TM-501, Massachusetts Institute of Technology, March 1994.
12. T. Malkin, D. Micciancio and S. Miner. *Efficient generic forward-secure signatures with and unbounded number of time periods.* Full version of this paper, available at http://eprint.iacr.org/2001/034/.
13. M. Naor and M. Yung. *Universal one-way hash functions and their cryptographic applications.* In Proceedings of the ACM Symposium on Theory of Computing, 1989, pp. 33–43.
14. M. Rabin. *Digital signatures and public key functions as intractable as factorization.* MIT Laboratory for Computer Science Report TR-212, January 1979.
15. J. Rompel. *One-way functions are necessary and sufficient for secure signatures.* In Proceedings of the ACM Symposium on Theory of Computing, 1990, pp. 387–394.
16. A. Yao. *Theory and applications of trapdoor functions.* In Proceedings of the IEEE Symposium on the Foundations of Computer Science, 1982, pp. 80–91.

From Identification to Signatures via the Fiat-Shamir Transform: Minimizing Assumptions for Security and Forward-Security

Michel Abdalla[1], Jee Hea An[2], Mihir Bellare[3], and Chanathip Namprempre[3]

[1] Magis Networks, Inc., 12651 High Bluff Drive, San Diego, CA 92130, USA,
mabdalla@cs.ucsd.edu, www.michelabdalla.net.

[2] SoftMax, Inc., 10760 Thornmint Road, San Diego, CA 92128, USA,
jeehea@cs.ucsd.edu

[3] Dept. of Computer Science & Engineering, University of California San Diego,
9500 Gilman Drive, La Jolla, California 92093, USA,
{mihir,meaw}@cs.ucsd.edu, www-cse.ucsd.edu/users/{mihir,cnamprem}

Abstract. The Fiat-Shamir paradigm for transforming identification schemes into signature schemes has been popular since its introduction because it yields efficient signature schemes, and has been receiving renewed interest of late as the main tool in deriving forward-secure signature schemes. We find minimal (meaning necessary and sufficient) conditions on the identification scheme to ensure security of the signature scheme in the random oracle model, in both the usual and the forward-secure cases. Specifically we show that the signature scheme is secure (resp. forward-secure) against chosen-message attacks in the random oracle model *if and only if* the underlying identification scheme is secure (resp. forward-secure) against impersonation under *passive* (i.e.. eavesdropping only) attacks, and has its commitments drawn at random from a large space. An extension is proven incorporating a random seed into the Fiat-Shamir transform so that the commitment space assumption may be removed.

1 Introduction

The Fiat-Shamir method of transforming identification schemes into signature schemes [11] is popular because it yields efficient signature schemes, and has been receiving renewed interest of late as the main tool in deriving forward-secure signature schemes. We find minimal (meaning necessary and sufficient) conditions on the identification scheme to ensure security of the signature scheme in the random oracle model. The conditions are simple and natural. Below we begin with some background and discussion of known results, and then move to our results, considering first the usual and then the forward-secure case.

CANONICAL ID SCHEMES. The Fiat-Shamir (FS) transform applies to identification (ID) schemes having a three-move format that we call *canonical.* The prover, holding a secret key *sk*, sends a message CMT called a *commitment* to

L.R. Knudsen (Ed.): EUROCRYPT 2002, LNCS 2332, pp. 418–433, 2002.

the verifier. The verifier returns a *challenge* CH consisting of a random string of some length. The prover provides a *response* RSP. Finally, the verifier applies a verification algorithm V to the prover's public key pk and the conversation CMT||CH||RSP to obtain a *decision* bit, and accepts iff Dec = 1. The length of the challenge is $c(k)$ where k is the security parameter and c is a function associated to the scheme. A large number of canonical ID schemes are known (e.g., [11,14,6,17,24,7,12,20,19,26,21]) and are candidates for conversion to signature schemes via the FS transform.

THE FS TRANSFORM. The signer has the public and secret keys pk, sk of the prover of the ID scheme. To sign a message M it computes CMT just as the prover would, hashes CMT$\|M$ using a public hash function $H\colon \{0,1\}^* \to \{0,1\}^{c(k)}$ to obtain a "challenge" CH $= H(\text{CMT}\|M)$, computes a response RSP just as the prover would, and sets the signature of M to CMT||RSP. To verify that CMT||RSP is a signature of M, one first computes CH $= H(\text{CMT}\|M)$ and then checks that the verifier of the identification scheme would accept, namely $V(pk, \text{CMT}\|\text{CH}\|\text{RSP}) = 1$. Fiat and Shamir's suggestion that one model H as a random oracle [11] is adopted by previous security analyses, both in the standard setting [23,18] and in the forward-secure setting [4,2,15], and also by this paper.

TARGET SECURITY GOAL FOR SIGNATURES. Focusing first on the standard setting (meaning where forward-security is not a goal), the target is to prove that the signature scheme is unforgeable under chosen-message attack [13] in the random oracle model [5]. This requires that it be computationally infeasible for an adversary to produce a valid signature of a new message even after being allowed a chosen-message attack on the signer and provided oracle access to the random hash function.

NON-TRIVIALITY. Previous works [23,18] have assumed that the ID scheme has the property that the space from which the prover draws its commitments is large, meaning super-polynomial. We refer to a scheme with this property as non-trivial. (A more general definition, in terms of min-entropy, is Definition 3.) We point out in Section 6 that non-triviality of the ID scheme is *necessary* for the security of the signature scheme derived via the FS transform, and thus all discussions related to the FS transform below will assume it. (We will see however that this assumption can be removed by considering a randomized generalization of the FS transform.)

1.1 Main Result

In this work we find simple and natural assumptions on the ID scheme that are both sufficient and necessary for the security of the signature scheme, and are related to the security of the underlying ID scheme for the purpose for which it was presumably designed, namely identification.

STATEMENT. We prove the following: The signature scheme resulting from applying the FS transform to a non-trivial ID scheme is secure against chosen-message attack in the random oracle model *if and only if* the underlying identification scheme is secure against impersonation under passive attack. A precise statement is Theorem 1. Let us recall the notion of security used here, following [10], and then compare this to previous work.

Security of identification schemes. As with any primitive, a notion of security considers adversary goals (what it has to do to win) and adversary capability (what attacks it is allowed). Naturally, for an ID scheme, the adversary goal is impersonation: it wins if it can interact with the verifier in the role of a prover and convince the latter to accept. There are two natural attacks to consider: passive and active. Passive attacks correspond to eavesdropping, meaning the adversary is in possession of transcripts of conversations between the real prover and the verifier. Active attacks mean that it gets to play the role of a verifier, interacting with the real prover in an effort to extract information. Security against impersonation under active attack is the attribute usually desired of an ID scheme to be used in practice for the purpose of identification. It is however the weaker attribute of security against impersonation under passive attack that we show is tightly coupled to the security of the derived signature scheme.

1.2 Comparison with Previous Work

Past security analyses identify assumptions on a non-trivial ID scheme that suffice to prove that corresponding the FS-transform based signature scheme is secure, as follows. The pioneering work of Pointcheval and Stern [23] assumes that the identification scheme is honest verifier zero-knowledge and also, in their Forking Lemma, assume a property that implies that it is a "proof of knowledge" [10,3], namely that there is an algorithm that can produce two transcripts which start with the same commitment (Cmt, Ch, Rsp), (Cmt, Ch′, Rsp′) such that, if both are accepted by the verifier V, the underlying secret key can be determined. (This property is called *collision intractability* in [9].) We refer to an ID scheme meeting these conditions as **PS**-secure.

Ohta and Okamoto [18] assume that the identification scheme is honest-verifier (perfect) zero-knowledge and that it is computationally infeasible for a cheating prover to convince the verifier to accept. We refer to such an ID scheme as **OO**-secure.

Relations. Figure 1 puts our result in context with previous works. It considers the three assumptions made on non-trivial identification schemes for the purpose of proving security of the corresponding FS-transform based signature scheme: **PS**-security [23]; **OO**-security [18]; and the assumption of security against impersonation under passive attacks. As the picture indicates, all three suffice to prove security of the signature scheme in the random oracle model. However, the assumption we make is not only necessary but also sufficient, while the others are provably not necessary. Furthermore, our assumption is weaker than the other assumptions, shown to imply them but not be implied by them. Let us discuss this further.

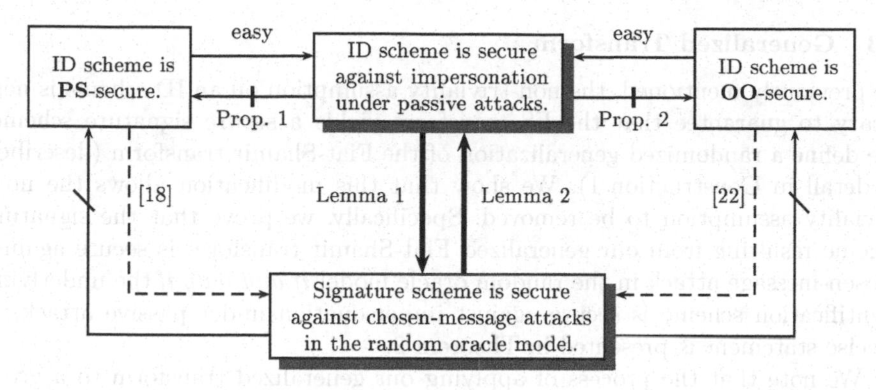

Fig. 1. We depict relations among assumptions on non-trivial ID schemes that have been used to prove security of the corresponding signature scheme. An arrow denotes an implication while a barred arrow denotes a separation. The dotted arrows are existing relations, annotated with citations to the papers establishing them. The full arrows are either relations established in this paper, or are easy.

It is well known that **PS** or **OO** security imply security against impersonation under passive attacks. The converse, however, is not true: in Section 4, we present examples that show that a non-trivial ID scheme could be secure against impersonation under passive attack yet be neither **PS** nor **OO** secure. Thus, our assumption on the ID scheme is weaker than previous ones. On the other hand, the fact that this assumption is necessary says that it is minimal. A consequence is that there exist (non-trivial) ID schemes that are neither **PS**-secure nor **OO**-secure, yet the corresponding signature scheme is secure, showing that the previous assumptions are not necessary conditions for the security of the signature scheme.

In practice, these gaps may not be particularly limiting, because practical ID schemes for the most part are **PS**-secure or **OO**-secure. However our result can simplify future or even existing constructions of identification based signature schemes, and clarifies the theoretical picture.

Assumptions related to the problem. Fiat and Shamir [11] suggested that their transform be applied to an ID scheme. However, previous security analyses have made assumptions that are in fact not inherent to the notion of identification itself. By this we mean assumptions such as honest verifier zero-knowledge or that underlying the forking lemma. These types of properties are convenient tools in the analysis of ID schemes, but not the end goals of identification. In particular, as we show in Section 4, there exist ID schemes, secure even against active attack, that are not honest verifier zero-knowledge and fail to meet the conditions of the forking lemma. In contrast, our necessary and sufficient condition, namely security against impersonation under passive attacks, is a natural end goal of identification. Our results thus support the original intuition that seems to have guided [11], namely that the security of the signature scheme stems from the security of the identification scheme relative to the job for which the latter was intended.

1.3 Generalized Transform

As previously mentioned, the non-triviality assumption on an ID scheme is necessary to guarantee that the FS transform yields a secure signature scheme. We define a randomized generalization of the Fiat-Shamir transform (described in detail in Construction 1). We show that this modification allows the non-triviality assumption to be removed. Specifically, we prove that the signature scheme resulting from our generalized Fiat-Shamir transform is secure against chosen-message attack in the random oracle model *if and only if* the underlying identification scheme is secure against impersonation under passive attack. A precise statement is presented in Theorem 2.

We note that the process of applying our generalized transform to a given ID scheme can be alternatively viewed as first modifying the ID scheme by enhancing its commitment space and then applying the FS transform.

1.4 Results for Forward Security

An important paradigm in the construction of forward-secure signature schemes, beginning with [4] and continuing with [2,15], has been to first design a forward-secure identification scheme and then obtain a forward-secure signature scheme via the FS transform. The analyses in these works are however ad hoc.

We prove an analogue of our main result that says that the signature scheme resulting from applying FS transform to a non-trivial ID scheme is forward-secure against chosen-message attacks in the random oracle model *if and only if* the underlying identification scheme is forward-secure against impersonation under passive attack. An extension based on the generalized FS transform, analogous to that mentioned above, also holds. This brings the characterization described above to forward-secure signature schemes, and helps to unify previous results [4,2,15]. Our result can simplify future or even existing constructions of identification based forward-secure signature schemes, saving repetition in the analytical work. (One should note however that non-modular analyses may have the benefit of yielding better concrete security than is obtained by our general result [2,15].)

1.5 Discussion and Remarks

Other Transforms. There are other methods of transforming ID schemes into signature schemes. A variant of the FS transform suggested by Micali and Reyzin [16] applies only to a subclass of canonical ID schemes. A transform suggested by Cramer and Damgård [9] has the advantage of not requiring random oracles in the analysis, but is relatively inefficient. Overall the FS transform has remained the most attractive, due to its wide applicability, the efficiency of the resulting signature scheme, and its robustness in the face of extra goals such as forward security, and thus is our focus.

The proofs. This abstract outlines proof ideas where space permits. Full proofs can be found in [1]. We note that our proofs appear to be simpler than previous ones even though our results are stronger. We believe that this is true because our assumptions, although weaker, have extracted more of the properties of the ID scheme that are truly relevant to the security of the signature scheme, thereby leaving less to be proven.

2 Definitions

NOTATION. If $A(\cdot,\cdot,\ldots)$ is a randomized algorithm, then $y \leftarrow A(x_1, x_2, \ldots; R)$ means y is assigned the unique output of the algorithm on inputs $x_1, x_2, \ldots$ and coins R, while $y \leftarrow A(x_1, x_2, \ldots)$ is shorthand for first picking R at random and then setting $y \leftarrow A(x_1, x_2, \ldots; R)$. We let $\mathsf{Coins}_A(k)$ denote the space from which R is drawn – it is a set of binary strings of some appropriate length – where k is the underlying security parameter. If S is a set then $s \stackrel{R}{\leftarrow} S$ indicates that s is chosen uniformly at random from S. If $x_1, x_2, \ldots$ are strings then $x_1 \| x_2 \| \cdots$ denotes an encoding under which the constituent strings are uniquely recoverable. It is assumed any string x can be uniquely parsed as an encoding of some sequence of strings. The empty string is denoted ε.

CANONICAL IDENTIFICATION SCHEMES. We use the term *canonical* to describe a three-move protocol in which the verifier's move consists of picking and sending a random string of some length, and the verifier's final decision is a deterministic function of the conversation and the public key (cf. Figure 2). The specification of a *canonical identification scheme* will take the form $\mathcal{ID} = (K, P, V, c)$ where K is the *key generation* algorithm, taking input a security parameter $k \in \mathsf{N}$ and returning a public and secret key pair (pk, sk); P is the *prover* algorithm taking input sk and the current conversation prefix to return the next message to send to the verifier; c is a function of k indicating the length of the verifier's challenge; V is a deterministic algorithm taking pk and a complete conversation transcript to return a boolean decision Dec on whether or not to accept. We associate to $\mathcal{ID}$ and each (pk, sk) a randomized *transcript generation oracle* which takes no inputs and returns a random transcript of an "honest" execution, namely:

Function $\mathsf{Tr}^{\mathcal{ID}}_{pk,sk,k}$
  $R_P \stackrel{R}{\leftarrow} \mathsf{Coins}_P(k)$
  $\textsc{Cmt} \leftarrow P(sk; R_P)$; $\textsc{Ch} \stackrel{R}{\leftarrow} \{0,1\}^{c(k)}$; $\textsc{Rsp} \leftarrow P(sk, \textsc{Cmt}\|\textsc{Ch}; R_P)$;
  Return $\textsc{Cmt}\|\textsc{Ch}\|\textsc{Rsp}$

The scheme must obey a standard completeness requirement, namely that for every k, we have $\Pr[V(pk, \textsc{Cmt}\|\textsc{Ch}\|\textsc{Rsp}) = 1] = 1$, the probability being over $(pk, sk) \leftarrow K(k)$ and $\textsc{Cmt}\|\textsc{Ch}\|\textsc{Rsp} \leftarrow \mathsf{Tr}^{\mathcal{ID}}_{pk,sk,k}$.

Security against impersonation under passive attacks considers an adversary – here called an impersonator – whose goal is to impersonate the prover without the knowledge of the secret key. In practice, such an adversary generally has access not only to the public key but also to conversations between the real prover and an honest verifier, possibly via eavesdropping over the network. We model this setting by viewing an impersonator as a probabilistic algorithm I and

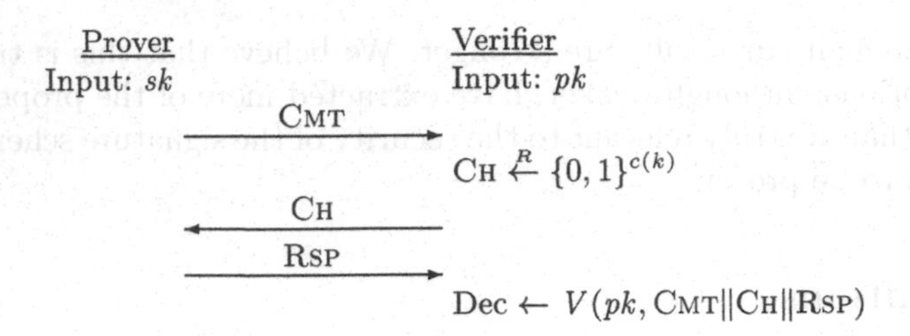

Fig. 2. A canonical identification protocol.

giving to it the public key and the transcript-generation oracle defined above. This oracle gives I the ability to obtain some number of transcripts of honest executions of the protocol. After reviewing the transcripts, the impersonator must then participate in the three-move protocol with an honest verifier and try to get the verifier to accept.

Definition 1 (Security of an identification scheme under passive attacks). Let $\mathcal{ID} = (K, P, V, c)$ be a canonical identification scheme, and let I be an impersonator, st be its state, and k be the security parameter. Define the *advantage* of I as

$$\mathbf{Adv}^{\text{imp-pa}}_{\mathcal{ID},I}(k) = \Pr[\,\mathbf{Exp}^{\text{imp-pa}}_{\mathcal{ID},I}(k) = 1\,]\,,$$

where the experiment in question is

$\mathbf{Exp}^{\text{imp-pa}}_{\mathcal{ID},I}(k)$
$(pk, sk) \leftarrow K(k)$; $st\|\text{CMT} \leftarrow I^{\mathsf{Tr}^{\mathcal{ID}}_{pk,sk,k}}(pk)$; $\text{CH} \stackrel{R}{\leftarrow} \{0,1\}^{c(k)}$
$\text{RSP} \leftarrow I(st, \text{CH})$; $\text{Dec} \leftarrow V(pk, \text{CMT}\|\text{CH}\|\text{RSP})$; Return Dec

We say that $\mathcal{ID}$ is *polynomially-secure against impersonation under passive attacks* if $\mathbf{Adv}^{\text{imp-pa}}_{\mathcal{ID},I}(\cdot)$ is negligible for every probabilistic poly(k)-time impersonator I. ∎

SIGNATURE SCHEMES. We recall the standard definition of security of a digital signature scheme under chosen-message attacks (cf. [13]) adapted to the random oracle model as per [5].

The specification of a *digital signature scheme* will take the form $\mathcal{DS} = (K, S, Vf, c)$ where: K is the *key generation* algorithm, taking input a security parameter $k \in \mathsf{N}$ and returning a public and secret key pair (pk, sk); S is the *signing* algorithm taking input sk and a message $M \in \{0,1\}^*$ to be signed and returning a signature; Vf is the *verification* algorithm taking input pk, a message M and a candidate signature σ for M and returning a boolean decision. The signing and verifying algorithms have oracle access to a function $H\colon \{0,1\}^* \to \{0,1\}^{c(k)}$ (which in the random oracle model will be a random function) so that c in the scheme description is a function of k whose value is the output-length of the hash function being used. The signing algorithm may be randomized, drawing coins from a space $\mathsf{Coins}_S(k)$, but the verification algorithm is deterministic. It is required that valid signatures are always accepted.

The adversary F – called a forger in this setting – gets the usual signing oracle plus direct access to the random oracle and wins if it outputs a valid signature of a new message. Below, we let $[\{0,1\}^* \to \{0,1\}^c]$ denote the set of all maps from $\{0,1\}^*$ to $\{0,1\}^c$. The notation $H \stackrel{R}{\leftarrow} [\{0,1\}^* \to \{0,1\}^c]$ is used to mean that we select a hash function H at random from this set. The discussion following the definition clarifies how this random selection from an infinite space is implemented.

Definition 2 (Security of a digital signature scheme). Let $\mathcal{DS}=(K,S,\mathcal{V},c)$ be a digital signature scheme, let F be a forger and k the security parameter. Define the experiment

$\mathbf{Exp}^{\text{frg-cma}}_{\mathcal{DS},F}(k)$

$H \stackrel{R}{\leftarrow} [\{0,1\}^* \to \{0,1\}^c]$

$(pk, sk) \leftarrow K(k)$; $(M,\sigma) \leftarrow F^{S^H_{sk}(\cdot),H(\cdot)}(pk)$; $\text{Dec} \leftarrow \mathcal{V}f^H(pk, M, \sigma)$

`If` M was previously queried to $S^H_{sk}(\cdot)$ `then return` 0 `else return` Dec

Define the *advantage* of F as

$$\mathbf{Adv}^{\text{frg-cma}}_{\mathcal{DS},F}(k) \;=\; \Pr[\,\mathbf{Exp}^{\text{frg-cma}}_{\mathcal{DS},F}(k) = 1\,]\;.$$

$\mathcal{DS}$ is *polynomially-secure against chosen-message attacks* if $\mathbf{Adv}^{\text{frg-cma}}_{\mathcal{DS},F}(\cdot)$ is negligible for every probabilistic poly(k)-time forger F. ∎

A special convention is needed with regard to how one can measure the time taken by the first step of $\mathbf{Exp}^{\text{frg-cma}}_{\mathcal{DS},F}(k)$ where one picks at random a function H from an infinite space. This selection of the hash function is not viewed as being performed all at once. Rather, the hash function is built dynamically using a table. In particular, for each hash-oracle query M, we check if the entry $H(M)$ exists. If so, we return it. Otherwise, we pick a random element y from $\{0,1\}^c$, make a table entry $H(M) = y$, and return y.

CONCRETE SECURITY ISSUES. In addition to our main results which speak in the usual language of polynomial security, we make concrete security statements so as to better gauge the practical impact of our reductions. Below, we discuss the parameters and conventions used.

When we refer to the running time of an adversary such as an impersonator or forger, we mean the time-complexity of the *entire* associated experiment, including the time taken to pick keys, compute replies to oracle queries, implement a random hash function as described above, and even compute the final outcome of the experiment.

For identification, the parameters of interest are the running time of the adversary and the number of queries q it makes to its transcript oracle. For signatures, the parameters of interest are the forger's running time, the number of sign-oracle queries, denoted q_s, and the number of hash-oracle queries, denoted q_h. All of these are functions of the security parameter k.

All query parameters are bounded by the running time, so if the adversary is polynomial time, all the other parameters are poly(k)-bounded. Thus, they can be ignored in the polynomial-time setting.

3 Equivalence Results

To save space (and avoid repetition), we present straightaway our randomized generalization of the Fiat-Shamir transform. The standard Fiat-Shamir transformation is the special case of the construction below in which the seed length is $s(k) = 0$.

Construction 1 (Generalized Fiat-Shamir Transform). Let $\mathcal{ID} = (K, P, V, c)$ be a canonical identification scheme and let $s\colon \mathsf{N} \rightarrow \mathsf{N}$ be a function which we call the *seed length*. We associate to these a digital signature scheme $\mathcal{DS} = (K, S, \mathit{Vf}, c)$. It has the same key generation algorithm as the identification scheme, and the output length of the hash function equals the challenge length of the identification scheme. The signing and verifying algorithms are defined as follows:

Algorithm $S^H(sk, M)$	Algorithm $\mathit{Vf}^H(pk, M, \sigma)$
$R \stackrel{R}{\leftarrow} \{0,1\}^{s(k)}$; $R_P \stackrel{R}{\leftarrow} \mathsf{Coins}_P(k)$	Parse σ as $R\|\mathrm{CMT}\|\mathrm{RSP}$
$\mathrm{CMT} \leftarrow P(sk; R_P)$	$\mathrm{CH} \leftarrow H(R\|\mathrm{CMT}\|M)$
$\mathrm{CH} \leftarrow H(R\|\mathrm{CMT}\|M)$	$\mathrm{Dec} \leftarrow V(pk, \mathrm{CMT}\|\mathrm{CH}\|\mathrm{RSP})$
$\mathrm{RSP} \leftarrow P(sk, \mathrm{CMT}\|\mathrm{CH}; R_P)$	Return Dec
Return $R\|\mathrm{CMT}\|\mathrm{RSP}$	

Note that the signing algorithm is randomized, using a random tape whose length is $s(k)$ plus the length of the random tape of the prover. Furthermore, the chosen random seed is included as part of the signature, to make verification possible. ∎

We use the concept of min-entropy [8] to measure how likely it is for a commitment generated by the prover of an identification scheme to collide with a fixed value. This is used to provide a more precise definition of what in the Introduction was referred to as a non-trivial ID scheme.

Definition 3 (Min-Entropy of Commitments). Let $\mathcal{ID} = (K, P, V, c)$ be a canonical identification scheme. Let $k \in \mathsf{N}$, and let (pk, sk) be a key pair generated by K on input k. Let $\mathcal{C}(sk) = \{P(sk; R_P) : R_P \in \mathsf{Coins}_P(k)\}$ be the set of commitments associated to sk. We define the maximum probability that a commitment takes on a particular value via

$$\alpha(sk) = \max_{\mathrm{CMT} \in \mathcal{C}(sk)} \left\{ \Pr\left[P(sk; R_P) = \mathrm{CMT} \ : \ R_P \stackrel{R}{\leftarrow} \mathsf{Coins}_P(k) \right] \right\}$$

Then, the *min-entropy* function associated to $\mathcal{ID}$ is defined as follows:

$$\beta(k) = \min_{sk} \left\{ \log_2 \frac{1}{\alpha(sk)} \right\},$$

where minimum is over all (pk, sk) generated by K on input k. We say that $\mathcal{ID}$ is *non-trivial* if $\beta(\cdot) = \omega(\log(\cdot))$ is super-logarithmic. ∎

We remark that for practical identification schemes, the commitment is drawn uniformly from some set. If the size of this set is $\gamma(\cdot)$ then the min-entropy of the scheme is $\log_2(\gamma(\cdot))$. Non-triviality means that this set has super-polynomial size.

The following theorem considers Construction 1 above in the special case where $s(k) = 0$. This case is exactly the Fiat-Shamir transform.

Theorem 1 (Equivalence Under Standard Fiat-Shamir Transform). *Let $\mathcal{ID} = (K, P, V, c)$ be a non-trivial, canonical identification scheme, and let $\mathcal{DS} = (K, S, Vf, c)$ be the associated signature scheme as per Construction 1 with $s(k) = 0$. Then $\mathcal{DS}$ is polynomially-secure against chosen-message attacks in the random oracle model if and only if $\mathcal{ID}$ is polynomially-secure against impersonation under passive attacks.* ∎

The non-triviality assumption above can be removed if one applies the generalized FS transform with a seed length that is not zero but which, when added to the min-entropy, results in a super-logarithmic function.

Theorem 2 (Equivalence Under Generalized Fiat-Shamir Transform). *Let $\mathcal{ID} = (K, P, V, c)$ be a canonical identification scheme, let $s(\cdot)$ be a seed length, and let $\mathcal{DS} = (K, S, Vf, c)$ be the associated signature scheme as per Construction 1. Let $\beta(\cdot)$ be the min-entropy function associated to $\mathcal{ID}$. Assume $s(\cdot) + \beta(\cdot) = \omega(\log(\cdot))$. Then $\mathcal{DS}$ is polynomially-secure against chosen-message attacks in the random oracle model if and only if $\mathcal{ID}$ is polynomially-secure against impersonation under passive attacks.* ∎

Theorem 1 is the special case of Theorem 2 in which $s(\cdot) = 0$ and $\beta(\cdot)$ is super-logarithmic. Accordingly, it suffices to prove Theorem 2. The proof of Theorem 2 follows easily from the two lemmas below. The first lemma relates the exact security of the signature scheme to that of the underlying identification scheme.

Lemma 1 (ID ⇒ SIG). *Let $\mathcal{ID} = (K, P, V, c)$ be a canonical identification scheme, let $s(\cdot)$ be a seed length, and let $\mathcal{DS} = (K, S, Vf, c)$ be the associated signature scheme as per Construction 1. Let $\beta(\cdot)$ be the min-entropy function associated to $\mathcal{ID}$. Let F be an adversary attacking $\mathcal{DS}$ in the random oracle model, having time-complexity $t(\cdot)$, making $q_s(\cdot)$ sign-oracle queries and $q_h(\cdot)$ hash-oracle queries. Then there exists an impersonator I attacking $\mathcal{ID}$ such that*

$$\mathbf{Adv}^{\text{frg-cma}}_{\mathcal{DS},F}(k) \leq \left(1+q_h(k)\right) \cdot \mathbf{Adv}^{\text{imp-pa}}_{\mathcal{ID},I}(k) + \frac{[1+q_h(k)+q_s(k)] \cdot q_s(k)}{2^{s(k)+\beta(k)}} . \quad (1)$$

Furthermore, I has time-complexity $t(\cdot)$ and makes at most $q_s(\cdot)$ queries to its transcript oracle. ∎

The full proof of Lemma 1 is presented in the full version of the paper [1], but we give a brief sketch of it here. We use a standard approach, namely assuming that a forger F can break the signature scheme, we construct an impersonator I that has access to a transcript generation oracle. The goal of I is to convince an honest verifier that it is a prover without knowing the secret key. I achieves its goal by running the forger F as a subroutine, answering its hash and sign oracle queries. When F outputs a forgery, I can make use of it in its interaction with the verifier. In order to do so, I guesses the "forgery point," at which F makes a hash query (of the form $R \| \textsc{Cmt} \| M$) that contains the message M on which F will attempt to forge, and uses CMT as its commitment to the verifier. The verifier then replies with a challenge, and I uses this value in its response to F's hash query at the forgery point. I simulates the response to F's other hash and sign queries using the transcript generation oracle and randomness. When F finally outputs a forgery, I uses it to respond to the verifier's challenge. If I guessed F's forgery point correctly and if F's forgery was successful, then the impersonator succeeds. Note that "enough" randomness or min-entropy is needed to successfully simulate the responses to the forger's hash and sign queries.

Going in the opposite direction, the following lemma relates the security of the identification scheme to that of the signature scheme derived from it. In fact, it says that if the signature scheme is secure then so is the identification scheme (regardless of the min-entropy of the ID scheme).

Lemma 2 (ID ⇐ SIG). *Let $\mathcal{ID} = (K, P, V, c)$ be a canonical identification scheme, let $s(\cdot)$ be a seed length, and let $\mathcal{DS} = (K, S, Vf, c)$ be the associated signature scheme as per Construction 1. Let I be an adversary attacking $\mathcal{ID}$, having time-complexity $t(\cdot)$ and making $q(\cdot)$ queries to its transcript oracle. Then, in the random oracle model, there exists a forger F attacking $\mathcal{DS}$ such that*

$$\mathbf{Adv}^{\text{imp-pa}}_{\mathcal{ID},I}(k) \leq \mathbf{Adv}^{\text{frg-cma}}_{\mathcal{DS},F}(k) . \tag{2}$$

Furthermore, F has time-complexity $t(\cdot)$, makes at most $q(\cdot)$ queries to its sign-oracle and at most $q(\cdot)$ queries to its hash-oracle. ▮

The proof of the lemma above uses a standard reduction technique and is straightforward. We assume that an impersonator mounting a passive attack can break the identification scheme, and build a forger who runs it as a subroutine. Transcript queries are answered by the forger using its signature oracle, and a successful impersonation attempt translates easily into a successful forgery. The proof details can be found in the full version of the paper [1].

4 Separations among Security Assumptions

In this section, we justify the claimed separations among the security conditions in Figure 1. Specifically, we give an example of an ID scheme that is secure against impersonation under passive attack but is not honest-verifier zero-knowledge, and also an example of an ID scheme that is secure against impersonation under passive attack and is not a proof of knowledge. (In this section, proof of knowledge means proof of knowledge of the secret key. More precisely, it refers to some underlying witness-relation $R(pk, sk)$ depending on the protocol.) Since the **PS** and **OO** assumptions include either an assumption of honest verifier zero-knowledge or an assumption of proof of knowledge, this implies that there exists an identification scheme secure against impersonation under passive attack that is not **PS** secure, and there exists an identification scheme secure against impersonation under passive attack that is not **OO** secure, justifying two of the claimed separations in Figure 1, and showing that our assumption on the ID scheme is strictly weaker than previous ones used to prove security of the signature scheme.

Furthermore, this also justifies two more separations claimed in Figure 1, namely that the signature scheme could be secure even if the ID scheme is not **PS** secure or **OO** secure. This follows simply by logic, because if we assume that security of the signature scheme implies, say, **PS**-security of the ID scheme, the existing arrows say that security against impersonation under passive attack implies **PS**-security, which we know from the above to not be true. The analogous argument applies in the case of **OO**.

We now proceed to the examples. Shoup notes that the 2^m-th root identification (a special case of the identification scheme of Ong and Schnorr [20]) is provably not a proof of knowledge if factoring is hard [25]. However, he shows that this scheme is secure against impersonation under active (and hence certainly under passive) attacks if factoring is hard. This yields the following:

Proposition 1. *If factoring is hard, then there exists a non-trivial canonical identification scheme that is secure against impersonation under passive attacks but is not a proof of knowledge.* ∎

Similarly, we show that there exists an identification scheme that is secure against impersonation under passive attacks yet is not honest verifier zero-knowledge. We take the following approach in constructing such an identification scheme. We begin with a canonical identification secure against impersonation under passive attacks and modify it so that it remains secure against impersonation under passive attacks but is not zero-knowledge. A detailed construction is presented in the full version of the paper [1]. The example we construct, though contrived, makes the point that zero-knowledge is not strictly necessary in a secure identification scheme. The following proposition states this more precisely.

Proposition 2. *If factoring is hard, then there exists a non-trivial canonical identification scheme that is secure against impersonation under passive attacks but is not honest-verifier zero-knowledge.* ∎

5 Extension to Forward Security

We prove an extension of Theorem 2 to the case where the security requirement is forward security.

CANONICAL FORWARD-SECURE IDENTIFICATION SCHEMES. We consider key-evolving identification schemes. The operation of the scheme is divided into time periods, where a different secret is used in each time period. The public key remains the same in every time period. A canonical key-evolving identification scheme is a three-move protocol in which the verifier's only move is to pick and send a random challenge to the prover. Unlike canonical identification schemes with fixed keys, the verifier's final decision, though still deterministic, is not only a function of the conversation with the prover and the public key, but also a function of the the index of the current time period. We say that a canonical key-evolving identification scheme is *forward-secure* if it is infeasible for a passive adversary, even with access to the current secret key, to impersonate the prover with respect to an honest verifier in any of the prior time periods.

As pointed out by Bellare and Miner [4], forward-secure identification schemes are artificial constructs since, due to the online nature of identification protocols, the kind of attack we withstand in this case cannot exist in reality. Nevertheless, the schemes are still very useful in the design of efficient forward-secure signature schemes. Please refer to the full version of the paper [1] for a formal definition of a key-evolving identification scheme and what it means for it to be forward-secure.

FORWARD-SECURE SIGNATURE SCHEMES. A forward-secure signature scheme is in essence a key-evolving signature scheme in which the secret key is updated periodically. As in standard signature schemes, the public key remains the same throughout the lifetime of the scheme. In each time period, a different secret key is used to sign messages. The verification algorithm checks not only the validity of a signature, but also the particular time period in which it was generated. At the end of each time period, an update algorithm is run to compute the new secret key from the current one, which is then erased. Informally, we say that a key-evolving signature scheme is *forward-secure* under chosen-message attack if it is infeasible for an adversary, even with access to the secret key for the current period and to previously signed messages of its choice, cannot forge signatures for a past time period. For a formal definition of a key-evolving signature scheme and what it means for it to be forward-secure, see the full version of the paper [1].

THE EQUIVALENCE. Our transformation of key-evolving ID schemes into key-evolving signature schemes follows the same paradigm of Construction 1, in which the challenge becomes the output of a hash function H. The main difference with respect to that construction is that the secret key is no longer fixed but varies according to the time period. As a result, the current time index j is also given as input to the signing algorithm and attached to the signature to allow for correct verification. The current time index j is also added to the input of the hash function, which now becomes $j\|R\|\text{CMT}\|M$. The update algorithm of the key-evolving signature scheme is exactly the same as that of the identification scheme on which it is based. The following theorem, where min-entropy is defined in a manner similar to that for canonical identification schemes, states precisely the equivalence with regard to forward security of the key-evolving ID scheme and the associated key-evolving signature scheme.

Theorem 3 (Forward security equivalence theorem). *Let $\mathcal{FID} = (K, P, Vid, c, T)$ be a canonical key-evolving identification scheme, let $s(\cdot)$ be a seed length, and let $\mathcal{FSDS} = (K, S, VSig, c, T)$ be the associated key-evolving signature scheme as per the new construction described above. Let $\beta(\cdot)$ be the min-entropy function associated to $\mathcal{FID}$ and assume $s(\cdot) + \beta(\cdot) = \omega(\log(\cdot))$. Then $\mathcal{FSDS}$ is polynomially-forward-secure against chosen-message attack in the random oracle model if and only if $\mathcal{FID}$ is polynomially-forward-secure against impersonation under passive attacks.* ∎

The full paper [1] states and proves a pair of lemmas, one for each direction of the "if and only if". These indicate the concrete security of the underlying reductions. The theorem follows.

As in the case of standard signature and ID schemes, if we consider key-evolving ID schemes in which the commitment is chosen from a large space (i.e., $\beta(\cdot) = \omega(\log(\cdot))$), then the key-evolving signature scheme resulting from the Fiat-Shamir transform (i.e., $s(k) = 0$) is forward-secure against chosen-message attack in the random oracle model *if and only if* the underlying identification scheme is forward-secure against impersonation under passive attacks.

6 The Non-triviality Condition

We show that applying the FS transform to a trivial identification scheme can result in an insecure signature scheme, which supports our claim in the Introduction that non-triviality of the ID scheme is necessary for security of the signature scheme obtained via the FS transform. This is implied by the following, whose proof is presented in the full version of the paper [1].

Proposition 3. *If factoring Williams integers is hard, then there exists a trivial, canonical identification scheme that is secure against impersonation under passive attacks, but the signature scheme resulting from applying the standard Fiat-Shamir transform is insecure.* ∎

This example also shows why the generalized FS transform that we have introduced is useful. Since the ID scheme is secure against impersonation under passive attacks, the generalized transform does yield a secure signature scheme, even though the triviality of the ID scheme prevented the FS transform from doing so.

Acknowledgments

Work done while the first two authors were at the University of California at San Diego. Third author supported in part by NSF Grant CCR-0098123, a 1996 Packard Foundation Fellowship in Science and Engineering, and an IBM Faculty Partnership Development Award. First and last authors supported in part by third author's grants.

References

1. M. Abdalla, J. H. An, M. Bellare, and C. Namprempre. From identification to signatures via the Fiat-Shamir transform: Minimizing assumptions for security and forward-security. Full version of this paper, available via `http:// www-cse. ucsd. edu/ users/ mihir`.
2. M. Abdalla and L. Reyzin. A new forward-secure digital signature scheme. In T. Okamoto, editor, *Advances in Cryptology – ASIACRYPT 2000*, volume 1976 of *Lecture Notes in Computer Science*, pages 116–129, Berlin, Germany, Dec. 2000. Springer-Verlag.

3. M. Bellare and O. Goldreich. On defining proofs of knowledge. In E. Brickell, editor, *Advances in Cryptology – CRYPTO ' 92*, volume 740 of *Lecture Notes in Computer Science*, pages 390–420, Berlin, Germany, Aug. 1992. Springer-Verlag.
4. M. Bellare and S. Miner. A forward-secure digital signature scheme. In M. Wiener, editor, *Advances in Cryptology – CRYPTO ' 99*, volume 1666 of *Lecture Notes in Computer Science*, pages 431–448, Berlin, Germany, Aug. 1999. Springer-Verlag.
5. M. Bellare and P. Rogaway. Random oracles are practical: A paradigm for designing efficient protocols. In V. Ashby, editor, *1st ACM Conference on Computer and Communications Security*. ACM Press, Nov. 1993.
6. T. Beth. Efficient zero-knowledge identification scheme for smart cards. In C. Guenther, editor, *Advances in Cryptology – EUROCRYPT ' 1988*, volume 330 of *Lecture Notes in Computer Science*, pages 77–86, Berlin, Germany, May 1988. Springer-Verlag.
7. E. Brickell and K. McCurley. An interactive identification scheme based on discrete logarithms and factoring. In I. Damgård, editor, *Advances in Cryptology – EUROCRYPT ' 90*, volume 473 of *Lecture Notes in Computer Science*, pages 63–71, Berlin, Germany, May 1991. Springer-Verlag.
8. B. Chor and O. Goldreich. Unbiased bits from sources of weak randomness and probabilistic communication complexity. In *26th Annual Symposium on Foundations of Computer Science*, pages 429–442, Los Angeles, CA, USA, Oct. 1985. IEEE Computer Society Press.
9. R. Cramer and I. Damgård. Secure signature schemes based on interactive protocols. In D. Coppersmith, editor, *Advances in Cryptology – CRYPTO '95*, volume 963 of *Lecture Notes in Computer Science*, pages 297–310, Berlin, Germany, 1995. Springer-Verlag.
10. U. Feige, A. Fiat, and A. Shamir. Zero knowledge proofs of identity. *Journal of Cryptology*, 1(2):77–94, 1988.
11. A. Fiat and A. Shamir. How to prove yourself: Practical solutions to identification and signature problems. In A. Odlyzko, editor, *Advances in Cryptology – CRYPTO ' 86*, volume 263 of *Lecture Notes in Computer Science*, pages 186–194, Berlin, Germany, Aug. 1986. Springer-Verlag.
12. M. Girault. An identity-based identification scheme based on discrete logarithms modulo a composite number. In I. Damgård, editor, *Advances in Cryptology – EUROCRYPT ' 90*, volume 473 of *Lecture Notes in Computer Science*, pages 481–486, Berlin, Germany, May 1991. Springer-Verlag.
13. S. Goldwasser, S. Micali, and R. Rivest. A digital signature scheme secure against adaptive chosen-message attacks. *SIAM Journal of Computing*, 17(2):281–308, Apr. 1988.
14. L. Guillou and J. J. Quisquater. A "paradoxical" identity-based signature scheme resulting from zero-knowledge. In S. Goldwasser, editor, *Advances in Cryptology – CRYPTO ' 88*, volume 403 of *Lecture Notes in Computer Science*, pages 216–231, Berlin, Germany, 21–25 Aug. 1988. Springer-Verlag.
15. G. Itkis and L. Reyzin. Forward-secure signatures with optimal signing and verifying. In J. Kilian, editor, *Advances in Cryptology – CRYPTO ' 01*, volume 2139 of *Lecture Notes in Computer Science*, pages 332–354, Berlin, Germany, Aug. 2001. Springer-Verlag.
16. S. Micali and L. Reyzin. Improving the exact security of digital signature schemes. *Journal of Cryptology*, 15(1):1–18, 2002.
17. S. Micali and A. Shamir. An improvement of the Fiat-Shamir identification and signature scheme. In S. Goldwasser, editor, *Advances in Cryptology – CRYPTO*

'88, volume 403 of *Lecture Notes in Computer Science*, pages 244–248, Berlin, Germany, Aug. 1990. Springer-Verlag.

18. K. Ohta and T. Okamoto. On concrete security treatment of signatures derived from identification. In H. Krawczyk, editor, *Advances in Cryptology – CRYPTO '98*, volume 1462 of *Lecture Notes in Computer Science*, pages 354–370, Berlin, Germany, Aug. 1998. Springer-Verlag.
19. T. Okamoto. Provably secure and practical identification schemes and corresponding signature schemes. In E. Brickell, editor, *Advances in Cryptology – CRYPTO '92*, volume 740 of *Lecture Notes in Computer Science*, pages 31–53, Berlin, Germany, Aug. 1993. Springer-Verlag.
20. H. Ong and C. Schnorr. Fast signature generation with a Fiat Shamir–like scheme. In I. Damgård, editor, *Advances in Cryptology – EUROCRYPT ' 90*, volume 473 of *Lecture Notes in Computer Science*, pages 432–440, Berlin, Germany, May 1990. Springer-Verlag.
21. D. Pointcheval. A new identification scheme based on the perceptrons problem. In J. Quisquater and L. Guillou, editors, *Advances in Cryptology – EUROCRYPT ' 95*, volume 921 of *Lecture Notes in Computer Science*, pages 319–328, Berlin, Germany, May 1995. Springer-Verlag.
22. D. Pointcheval and J. Stern. Security proofs for signature schemes. In U. Maurer, editor, *Advances in Cryptology – EUROCRYPT ' 96*, volume 1070 of *Lecture Notes in Computer Science*, pages 387–398, Berlin, Germany, May 1996. Springer-Verlag.
23. D. Pointcheval and J. Stern. Security arguments for digital signatures and blind signatures. *Journal of Cryptology*, 13(3):361–396, 2000.
24. C. Schnorr. Efficient signature generation by smart cards. *Journal of Cryptology*, 4(3):161–174, 1991.
25. V. Shoup. On the security of a practical identification scheme. In U. Maurer, editor, *Advances in Cryptology – EUROCRYPT ' 96*, volume 1070 of *Lecture Notes in Computer Science*, pages 344–353, Berlin, Germany, May 1996. Springer-Verlag.
26. J. Stern. A new identification scheme based on syndrome decoding. In D. Stinson, editor, *Advances in Cryptology – CRYPTO '93*, volume 773 of *Lecture Notes in Computer Science*, pages 13–21, Berlin, Germany, 1994. Springer-Verlag.

Security Notions for Unconditionally Secure Signature Schemes

Junji Shikata[1], Goichiro Hanaoka[1], Yuliang Zheng[2], and Hideki Imai[1]

[1] Institute of Industrial Science, University of Tokyo,
4-6-1 Komaba, Meguro-ku, Tokyo 153-8505, Japan,
{shikata,hanaoka}@imailab.iis.u-tokyo.ac.jp,imai@iis.u-tokyo.ac.jp

[2] Department of Software and Information Systems, UNC Charlotte,
9201 University City Blvd. Charlotte, NC 28223, USA,
yzheng@uncc.edu

Abstract. This paper focuses on notions for the security of digital signature schemes whose resistance against forgery is not dependent on unproven computational assumptions. We establish successfully a sound and strong notion for such signature schemes. We arrive at the sound notion by examining carefully the more established security notions for digital signatures based on public-key cryptography, and taking into account desirable requirements of signature schemes in the unconditional security setting. We also reveal an interesting relation among relevant security notions which have appeared in the unconditionally setting, and significantly, prove that our new security notion is the strongest among all those for unconditionally secure authentication and signature schemes known to date. Furthermore, we show that our security notion encompasses that for public-key signature schemes, namely, existential unforgeability under adaptive chosen-message attack. Finally we propose a construction method for signature schemes that are provably secure in our strong security notion.

1 Introduction

In this paper, we address security notions for signature schemes that do not depend on any computational assumption.

Since the discovery of public-key cryptography [10], significant advances have been reported on digital signature schemes [21][11]. Although it is shown in [10] that a trapdoor function allows to create digital signature schemes in the public-key setting, a number of technical problems arise if digital signatures are implemented using a general trapdoor function as suggested in [10]. Thus it is important to have a formal notion of what a secure digital signature scheme is, and to construct a digital signature scheme which can be proven to be secure in the formal notion. The current standard security notion was established by Goldwasser, Micali and Rivest [14]. In the same paper the authors also demonstrated the first digital signature scheme that was proven to be secure against a very general attack, called adaptive chosen message attack. Since then, many

L.R. Knudsen (Ed.): EUROCRYPT 2002, LNCS 2332, pp. 434–449, 2002.

provable secure digital signature schemes have been proposed by researchers [2][23][7][12][1].

These schemes and the infrastructure within which they operate have a limitation in that their underlying security relies on the presumed computational difficulty of certain number-theoretic problems such as the integer factoring problem and the (elliptic curve) discrete logarithm problem. Thus should future progress in computers as well as discoveries of revolutionary algorithms make it computationally feasible to solve larger size number-theoretic problems, such a presumption would not be able to assure the security of current digital signatures. This situation is disturbing considering that there are many cases where documents, such as court and government records, long-term leases and contracts, are required by law to be kept intact for a long period of time, say over 50 years.

In attempting to solve this problem, researchers have introduced unconditionally secure digital signature schemes and authentication codes which do not rely on any unproven assumption such as the discrete logarithm problem. Like many other areas in security, there is clearly a need to identify a kind of benchmarks that one can employ to analyze and compare various signature schemes in the unconditional security setting. A major contribution of this research is to establish a strong security notion for all digital signature schemes including unconditionally secure ones. Additionally, we will show a concrete construction of unconditionally secure digital signature schemes which satisfies the requirements of the strong security notion.

Let us briefly survey existing unconditionally secure schemes. The first unconditionally secure signature was proposed by Chaum and Roijakkers [5]. There have been many attempts to enhance conventional unconditionally secure authentication codes [13][27] with extra security-properties that are required by signature schemes. Major extensions of conventional authentication codes include the so-called A^2-codes [28][29][19][20][18], A^3-codes [3][8][30][17][18][31] and multi-receiver authentication codes (with dynamic senders) [9][24][25][26][18]. Recently, the first unconditionally secure signature scheme that admits provably secure transfer of signatures has been proposed in [15]. These schemes, however, have all been proven to be secure against some specific attacks. This raises a number of interesting questions: what are other possible attacks? More importantly, are these signature schemes secure against other yet to be identified attacks?

As mentioned earlier, the focus of this research is to establish a strong security notion for signature schemes whose security does not depend on any computational assumption. It is discussed by taking into account the security notions for public-key signature schemes and additional requirements for signature schemes in the unconditional security setting. Furthermore we examine relations among all the security notions which have been proposed in the context of unconditionally secure signature schemes. It turns out that our security notion is the strongest among all the security notions for unconditionally secure authentication and signature schemes known so far, and it encompasses the security notion for public-key signature schemes, namely existential unforgeability under adaptive chosen-message attack. Finally we propose a construction method for signature schemes that are secure in our strong security notion.

2 Approaches to the Notion of Unconditional Security

2.1 Discussion

In this section, we consider how *unconditionally secure signature schemes* should be defined. By *unconditionally secure* one generally means that security must not depend on any computational assumption. To address the question, there are two issues to be discussed. The first is how to establish a proper model for signature schemes, and the second is to define, in a formal way, unconditional security notion in that model.

When introducing a model for unconditionally secure signature schemes, care should be taken so that properties of public-key signature schemes are captured. In addition, the model should be as simple as possible.

We start with the following typical model for signature schemes.

Definition 1 A *signature scheme* $\Pi = (Gen, Sig, Ver)$ consists of a key generation algorithm, Gen, a signing algorithm, Sig, and a verification algorithm, Ver.

1. **Key Generation:** The key generation algorithm outputs a signing-key x for a signer and a verification-key y for a verifier, respectively.
2. **Signature Generation:** For a message m, the signer creates a signature $a := Sig(x, m)$ using his signing key x. The pair (m, a) is a resultant signed message.
3. **Verification:** The verifier checks whether (m, a) is created by the signer using his verification key. More precisely, the verifier accepts it as having originated from the signer if $Ver(y, m, a) = true$, and rejects it if $Ver(y, m, a) = false$.

Definition 2 Let x be a signing-key of a signer. A signed message (m, a) is said to be *valid* if $a = Sig(x, m)$. Likewise, a signature a of a message m is said to be *valid* if $a = Sig(x, m)$. Otherwise, (m, a) is said to be *invalid*.

To simplify our discussions, we consider a model of signature schemes in which there are a single signer S and multiple verifiers $V_1, V_2, \ldots$. We wish a signature scheme to fulfill the following requirement.

Requirement 1

1. *Verifiability*: Any verifier can non-interactively check whether a signed message received from a signer is valid with his own verification-algorithm. In other words, he can check the validity of a received signed message without

communicating with others after receiving the signed message. More precisely, for any verifier V with his verification-key y, (m, a) is regarded as a valid signed message if and only if $Ver(y, m, a) = true$. In other words, if (m, a) is valid, $Ver(y, m, a) = true$; and if (m, a) is invalid, $Ver(y, m, a) = false$.

2. *Resolution for Dispute by a Third Party*: If a dispute occurs among users, a third party (called an arbiter) can resolve the dispute in a reasonable way: The third party has his own verification-key, and he resolves a dispute among users following the resolution-rule below.
   - *Resolution-Rule*: Let T be the third party and y_T be his verification-key. If a signer S denies the fact that he has created a signed message (m, a) held by a verifier V, then V should be able to present (m, a) to T. T rules in favor of V if $Ver(y_T, m, a,) = true$ and in favor of S otherwise.

   Here, we assume that the third party honestly follows the resolution-rule and honestly outputs its result when a dispute occurs. However, we assume that the third party is not always fully trusted. Namely, we assume that the third party might forge a signature.
3. *Security (unforgeability)*: It is infeasible for any adversary to forge a signature. Here, we assume that not only a verifier may be dishonest but also the signer and a third party may be dishonest. Each of them may become an adversary who may wish to forge a signature.

The level of security we require will be discussed in greater details in Section 2.2.

Requirement 1 can be relaxed in such a way that a *small error probability* is allowed.

Requirement 2 *Verifiability* and *Resolution for Disputes by a Third Party* in Requirement 1 can be relaxed as follows:

1. *Verifiability*: For any verifier V with his verification-key y, if (m, a) is valid, the verifier always accepts it (i.e. $Ver(y, m, a) = true$); and if (m, a) is invalid, the probability that the verifier erroneously accepts it is at most ϵ_1, where ϵ_1 is a very small quantity.
2. *Resolution for Disputes by a Third Party*: If a dispute between a signer and a verifier occurs, the resolution-rule in Requirement 1 is applied. However, we admit the following: If (m, a) is valid, T always accepts it (i.e. $Ver(y_T, m, a,) = true$); and if (m, a) is invalid, the probability that T erroneously accepts it is at most ϵ_2, where ϵ_2 is a very small quantity.

In a digital signature scheme based on public-key cryptography, a verification-key for a verifier can be public and shared among all verifiers. The following theorem indicates that such a signature scheme cannot be secure against an adversary with unlimited computing power.

Theorem 1 *Consider a signature scheme which satisfies Requirement 1. If it is infeasible for an adversary with unlimited computing power to succeed in forging a signature, then the verification-key for each verifier must be kept secret from all* *other verifiers. Similarly, consider a signature scheme which satisfies Requirement 2 with $\epsilon_i \neq 0$ $(i = 1, 2)$. If it is infeasible for an adversary with unlimited computing power to succeed in forging a signature, then the verification-key for each verifier must be kept secret not only from all other verifiers but also from a signer.*

A proof for the above theorem will be provided in the full version of this paper.

A consequence of Theorem 1 is that with a signature scheme that allows an adversary to have unlimited computing power, its key generation algorithm must generate verification-keys for all verifiers, and more importantly, distribute the verification-keys to verifiers separately in a secure way. For this reason we have to assume that the number of verifiers is limited. This is in contrast with a public-key signature scheme in which a single public verification-key is adequate and there is no limit placed on the number of verifiers.

To further simplify our discussions, we introduce into our model a trusted authority, denoted by TA. The roles of TA are to generate a signing-key and verification-keys by using a key generation algorithm, and to distribute the signing-key to the signer and verification-keys to each verifier, in a secure way.

2.2 Unforgeability

We now discuss security notions in our signature model. Let $\mathcal{U}:=\{S,V_1,V_2,\ldots,V_n\}$ be a set of users, where S is a signer and V_i $(1 \leq i \leq n)$ are verifiers.

We note that the signer has information-theoretic advantage over other verifiers since the signing-key is secret information known only to the signer. We also note that each verifier has information-theoretic advantage over other users, since his verification-key is secret information known only to the verifier. From these facts it follows that we should take into account not only the secrecy of the signer's signing-key but also the secrecy of each verifier's verification-key. This is different from public-key signature schemes in which we need not to consider information-theoretic advantages of a verifier.

On the secrecy of the signer's signing-key, the following security notion can be considered, in conjunction with security notions for public-key signature schemes [14]:

Definition 3 (Forgery and Attacks against a Signer)[14]: Consider an *adversary* who can be either a dishonest verifier or an outsider in our model.

- Types of Forgery:
  1. *Total Break*: An adversary is able either to extract the signing key, or to find an efficient signing algorithm that is functionally equivalent to the signing algorithm equipped with the genuine signing key.
  2. *Selective Forgery*: An adversary is able to create a valid signature for a particular message or a class of messages chosen a priori.
  3. *Existential Forgery*: An adversary is able to forge a valid signed message that signer has not created, but the adversary has little or no control over which message will be the target.

- Types of Attacks:
  1. *Key-Only Attack*: If a dishonest receiver is an adversary, the only key information he knows is the information on his verification-key. If an outsider is an adversary, he knows no secret key information, other than publicly available information on the scheme.
  2. *Message Attacks*: An adversary is able to examine signatures corresponding either to known or chosen messages. Message attacks can be further subdivided into three classes:
     (a) *Known-Message Attack*: An adversary has valid signatures for a set of messages which are known to the adversary but not chosen by him.
     (b) *Chosen-Message Attack*: An adversary obtains valid signatures from a chosen list of messages before attempting to forge another signed message.
     (c) *Adaptive Chosen-Message Attack*: An adversary is allowed to use the signer as an oracle; the adversary may request signatures of messages which may depend on the signer's signing key and previously obtained signed messages. That is, at any time the adversary can query the signer with messages chosen at his will, except for the target message.

The strongest signature scheme is one that is secure against existential forgery under adaptive chosen message attack.

Next we consider the secrecy of a verifier's verification-key.

Definition 4 (Forgery and Attacks against a Verifier): Let V be a verifier. In the following, an *adversary* means a dishonest signer, a dishonest verifier, or an outsider in our model.

- Types of Forgery:
  1. *Total Acceptance Forgery for* V: An adversary is able either to compute the verification-key information of the verifier V, or find an efficient verification algorithm that is functionally equivalent to the verification algorithm equipped with the genuine verification-key.
  2. *Selective Acceptance Forgery for* V: An adversary is able to make a signature, which will be accepted by V, for a particular message or a class of messages chosen a priori.
  3. *Existential Acceptance Forgery for* V: An adversary is able to make a signed message that has not been created by the signer but will be accepted by V. The adversary has little or no control over which signed message will be targeted.
- Types of Attacks:
  1. *Key-Only Attack*: The only key information which an adversary knows is the adversary's secret key. In a case that the adversary is a signer in our model, the only key information available to him is that of his signing key. Otherwise if the adversary is a verifier, the only key information known to him is that of his verification-key.

2. *Signature Attacks for V*: An adversary is able to examine verification results of V corresponding either to known or chosen signatures. Signature attacks can be further subdivided into three classes:
   (a) *Known-Signature Attack for V*: An adversary has some signed messages and he knows whether these will be accepted by the verifier V or not. However, these are not chosen by him.
   (b) *Chosen-Signature Attack for V*: An adversary obtains some signed messages whose verification results (i.e. the results whether these are accepted or not by V) are known to him. These are chosen before attempting to forge a signed message.
   (c) *Adaptive Chosen-Signature Attack for V*: An adversary is allowed to use the verifier V as an oracle; the adversary may request for an answer as to whether a signed message will be accepted by V. The signed message may be dependent on V's verification-key and verification-results obtained previously from V. That is, at any time the adversary can query the verifier with any signed messages, except for the target.

Finally, some clarifications on the types of forgery and attacks on verifiers follow.

Definition 5 (Forgery Range among Verifiers)

1. *Forgery for All Verifiers*: An adversary can forge a signature for all verifiers.
2. *Forgery for Selective Verifiers*: An adversary can forge a signature for a particular verifier selected by the adversary.
3. *Forgery for Existential Verifiers*: An adversary can forge a signature for a verifier, but the adversary has little or no control over which verifier will be the victim.

The above discussions suggest that a strong security notion be considered along the following line: Under adaptive chosen-message and adaptive chosen-signature attacks, it is infeasible for an adversary to succeed in not only existential forgery but also existential acceptance forgery against any verifier. The following theorem whose proof is straightforward is helpful, as it shows that it will be sufficient to consider only existential acceptance forgery, rather than both existential forgery and existential acceptance forgery.

Theorem 2 *Let Π be a signature scheme. If Π is existentially acceptance unforgeable for any verifier under adaptive chosen-message and adaptive chosen-signature attacks, then it is also existentially unforgeable under adaptive chosen-message and adaptive chosen-signature attacks.*

Based on Theorem 2, we can define a strong security notion as follows:

Definition 6 (Strong Security) Let Π be a signature scheme. Then Π is called *secure* if it is existential acceptance unforgeable for any verifier under adaptive chosen-message and adaptive chosen-signature attacks.

2.3 Some Remarks on Security Notions

In this subsection we consider some conditions that should be met when discussing security notions for signature schemes in unconditional security setting.

- **The security parameter:** In signature schemes with computational security in public-key cryptography, the security parameter is introduced to govern the overall security of a scheme, the length and number of messages, and the running time of algorithms. Similarly, a security parameter k for unconditional secure signature schemes can be defined. This parameter determines the overall security, the key-length of signing-keys and that of verification-keys, the length of messages and that of signatures, and the running time of algorithms.
- **The number of colluders:** There may exist dishonest users, and some dishonest users might collude in order to succeed in forgery. In this paper we adopt the idea of threshold schemes. Namely, we assume that there exists at most ω colluders among the users $\mathcal{U} = \{S, V_1, V_2, \ldots, V_n\}$. In discussing signature schemes with unconditional security, at least from a theoretical viewpoint, introducing the pre-defined number of colluders does not pose a problem in practice when compared with digital signature schemes with computational security, because even in the latter case at most polynomially many colluders are implicitly assumed when discussing security.
- **The numbers of signing and verifying operations:** In order to describe security notions in a more formal way, we should introduce a number up to which an adversary can have access to the signing oracle, and a number up to which the adversary can have access to the verification oracle. We introduce a number up to which a signer is allowed to generate signatures, denoted by ψ, and a number up to which each verifier is allowed to check received signatures, denoted by ψ'. This implies that an adversary can obtain at most ψ valid signed message from the signer, and at most $\psi' - 1$ verification results on signed messages from the target verifier. This should be contrasted to public key signature schemes in which an adversary is allowed to obtain at most $poly(k)$, where k is a security parameter, valid signed messages, and an unlimited number of verification results using a publicly known verification-key.

3 Security Notions and Their Relations

3.1 The Model

As mentioned in the previous section, we consider the following simplified model of signature schemes:

Definition 7 A *signature scheme Π* consists of ($\mathcal{U}$,TA,$\mathcal{M}$,$\mathcal{X}$,$\mathcal{Y}$,$\mathcal{A}$,Gen,Sig,Ver):

1. **Notation:**
   - $\mathcal{U} = \{S, V_1, V_2, \ldots, V_n\}$ is a finite set of users, where S is a signer and $V_i (1 \leq i \leq n)$ are verifiers,

- TA is a trusted authority,
- $\mathcal{M} = \{\mathcal{M}_k\}_{k \in \mathbf{N}}$ is a sequence of finite sets of possible messages, where $\mathcal{M}_k \subset \{0,1\}^{l_M(k)}$, and $l_M(k)$ is a polynomial of k. Hereafter, k means a security parameter.
- $\mathcal{X} = \{\mathcal{X}_k\}_{k \in \mathbf{N}}$ is a sequence of finite sets of possible signing-keys. Here, $\mathcal{X}_k \subset \{0,1\}^{l_X(k)}$, and $l_X(k)$ is a polynomial of k,
- $\mathcal{Y} = \{\mathcal{Y}_k\}_{k \in \mathbf{N}}$ is a sequence of finite sets of possible verification-keys. Here, $\mathcal{Y}_k \subset \{0,1\}^{l_Y(k)}$, and $l_Y(k)$ is a polynomial of k,
- $\mathcal{A} = \{\mathcal{A}_k\}_{k \in \mathbf{N}}$ is a sequence of finite sets of possible signatures. Here, $\mathcal{A}_k \subset \{0,1\}^{l_A(k)}$, and $l_A(k)$ is a polynomial of k,
- Gen is a *key generation algorithm* which on input a security parameter 1^k, outputs a signing-key and verification-keys,
- $Sig : \mathcal{X} \times \mathcal{M} \longrightarrow \mathcal{A}$ is a *signing algorithm*,
- $Ver : \mathcal{Y} \times \mathcal{M} \times \mathcal{A} \longrightarrow \{true, false\}$ is a *verification algorithm.*

2. **Key Generation and Distribution by TA:** The TA generates a signing-key x for the signer S, and a verification-key y_{V_i} for each verifier V_i using Gen. Here, Gen is a probabilistic algorithm which produces, on input 1^k, where k is a security parameter, keys $(x, y_{V_1}, y_{V_2}, \dots, y_{V_n})$ of matching signing and verifying keys, where $x \in \mathcal{X}_k$ and $y_{V_i} \in \mathcal{Y}_k$ for $1 \le i \le n$. TA then transmits the signing-key x to the signer S and the verification-key y_{V_i} to the verifier V_i in a secure way. After delivering these keys, TA may erase the keys $(x, y_{V_1}, y_{V_2}, \dots, y_{V_n})$ from his memory. The signer keeps secret his signing-key, and each verifier keeps secret his verification-key.
3. **Signature Generation:** For a message $m \in \mathcal{M}_k$, the signer S generates a signature $a = Sig(x, m) \in \mathcal{A}_k$ by using the signing-key x in conjunction with Sig. The pair (m, a) is regarded as a signed message. Here, we assume that Sig is deterministic, but in general it might be randomized. If it is deterministic, for a message m and a signing-key x, the signature $a = Sig(x, m)$ is uniquely determined, while in the case of a randomized algorithm, each time a different signature can be produced for the same message.
4. **Signature Verification:** On receiving (m, a) from the signer S, a verifier V_j checks whether a is valid by using his verification-key $y_{V_j} \in \mathcal{Y}_k$. More precisely, V_j accepts (m, a) as a valid signed message if and only if $Ver(y_{V_j}, m, a) = true$. Here, we assume that Ver is deterministic.

In addition, in the above model a trusted party (or an arbiter) is selected among verifiers. When a dispute occurs, the trusted party can resolve the dispute with his verification-key by following the resolution-rule described in Requirement 1.

Let ψ be a number up to which the signer is allowed to generate signatures, and ψ' be a number up to which each verifier is allowed to check received signatures, respectively, and let ω be the number of possible colluders among users. Let $\mathcal{W} := \{W \subset \mathcal{U} |\ |W| \le \omega\}$. Each element of $\mathcal{W}$ represents a group of possibly collusive users. For a set $\mathcal{T}$ and a non-negative integer t, let $\wp_t^{\mathcal{T}} := \{T \subset \mathcal{T} \mid |T| \le t\}$ be the family of all subsets of $\mathcal{T}$ whose cardinalities are less than or equal to t. Of course, the empty set $\emptyset$ is always contained in $\wp_t^{\mathcal{T}}$.

3.2 A Strong Security Notion

With notations above, we can now discuss security notions for unconditionally secure signature schemes. We start with introducing *exponentially negligible functions* in order to strictly describe a *small error probability* in Requirement 2.

Definition 8 (Exponentially Negligible Function) Let $\epsilon(k)$ be a function defined over the positive integers $k \in \boldsymbol{N}$ that takes non-negative real numbers. Then, $\epsilon(k)$ is called *exponentially negligible* if there exists an integer k_0 and some constant a $(1 < a)$ such that $\epsilon(k) \leq \frac{1}{a^k}$ for all $k \geq k_0$.

Using notations we have introduced, we now formulate the strong security notion in our signature model as follows:

Definition 9 (Strong Security) Let k be a security parameter and $\epsilon(k)$ an exponentially negligible function. For simplicity, we will denote $\epsilon(k)$ by ϵ.

1) For $W \in \mathcal{W}$ such that $V_j, S \notin W$, we define $P_1^{strong}(V_j, W)$ as

$$P_1^{strong}(V_j, W) := \max_{y_W} \max_{M_S=\{(m_S,a_S)\}\in\wp_{\psi}^{\mathcal{M}_k\times\mathcal{A}_k}} \max_{M_{V_j}=\{(m_{V_j},a_{V_j})\}\in\wp_{\psi'-1}^{\mathcal{M}_k\times\mathcal{A}_k}} \max_{M_{V_1},\ldots,M_{V_l},\ldots,M_{V_n}\in\wp_{\psi'}^{\mathcal{M}_k\times\mathcal{A}_k}(l\neq j)} \max_{(m,a)} \Pr(V_j \text{ accepts } (m,a) \mid y_W, M_S, M_{V_j}, M_{V_l}, \{Ver(y_{V_l}, m_{V_l}, a_{V_l}) | (m_{V_l}, a_{V_l}) \in M_{V_l}\} \; (1 \leq l \leq n, l \neq j))$$

where M_S is taken over $\wp_{\psi}^{\mathcal{M}_k\times\mathcal{A}_k}$ such that any element of M_S is a valid signed message; M_{V_j} is taken over $\wp_{\psi'-1}^{\mathcal{M}_k\times\mathcal{A}_k}$ such that $Ver(y_{V_j}, m_{V_j}, a_{V_j}) = false$ for any $(m_{V_j}, a_{V_j}) \in M_{V_j}$; M_{V_l} is taken over $\wp_{\psi'}^{\mathcal{M}_k\times\mathcal{A}_k}$ for $1 \leq l \leq n, l \neq j$; and (m,a) runs over $\mathcal{M}_k \times \mathcal{A}_k$ such that $(m,a) \notin M_S$ and $(m,a) \notin M_{V_j}$. Note that the condition $(m,a) \notin M_S$ means that for any $(m_S, a_S) \in M_S$ either $m \neq m_S$, or $m = m_S$ and $a \neq a_S$ holds. Next we define

$$P_1^{strong} := \max_{V_j,W} P_1^{strong}(V_j, W).$$

2) For $W \in \mathcal{W}$ such that $V_j \notin W$ and $S \in W$, we define $P_2^{strong}(V_j, W)$ as

$$P_2^{strong}(V_j, W) := \max_{x} \max_{y_{W-\{S\}}} \max_{M_{V_j}=\{(m_{V_j},a_{V_j})\}\in\wp_{\psi'-1}^{\mathcal{M}_k\times\mathcal{A}_k}} \max_{M_{V_1},\ldots,M_{V_l},\ldots,M_{V_n}\in\wp_{\psi'}^{\mathcal{M}_k\times\mathcal{A}_k}(l\neq j)} \max_{(m,a)} \Pr(V_j \text{ accepts } (m,a) \mid x, y_{W-\{S\}}, M_{V_j}, M_{V_l}, \{Ver(y_{V_l}, m_{V_l}, a_{V_l}) | (m_{V_l}, a_{V_l}) \in M_{V_l}\} \; (1 \leq l \leq n, l \neq j))$$

where $M_{V_j} = \{(m_{V_j}, a_{V_j})\}$ is taken over $\wp_{\psi'-1}^{\mathcal{M}_k\times\mathcal{A}_k}$ such that $Ver(y_{V_j}, m_{V_j}, a_{V_j}) = false$ for any $(m_{V_j}, a_{V_j}) \in M_{V_j}$; M_{V_l} is taken over $\wp_{\psi'}^{\mathcal{M}_k\times\mathcal{A}_k}$ for $1 \leq l \leq n$, $l \neq j$; and $(m,a) \in \mathcal{M}_k \times \mathcal{A}_k$ runs over invalid signed messages such that $(m,a) \notin M_{V_j}$. We define $P_2^{strong} := \max_{V_j,W} P_2^{strong}(V_j, W)$.

Then, a signature scheme Π is said to be (n,ω,ψ,ψ')*-secure* if

$$\max\{P_1^{strong},\ P_2^{strong}\} \le \epsilon$$

3.3 Relations among Security Notions

One of the purposes in this paper is to clarify which is the strongest among all the security notions that have appeared in unconditionally secure authentication codes and signature schemes. We focus on security notions for the following notable schemes: multireceiver authentication codes (MRA) [9][24], Johansson's scheme [18], Wang and Safavi-Naini's scheme [31] and Hanaoka, Shikata, Zheng and Imai's scheme [15]. Specifically, we analyze a relation among our strong security notion and those of MRA, Johansson's scheme, Wang and Safavi-Naini's scheme, Hanaoka, Shikata, Zheng and Imai's scheme, respectively.

We describe security notions of those schemes as follows. Let Π be a signature scheme (or an authentication code) along with our signature model. Then, Π is said to be $(n,\omega,\psi)^{MRA}$*-secure* if the success probability of all attacks considered in MRA [9][24] is exponentially negligible under the following conditions: there exists at most ω colluders among the users; and the number up to which a signer is allowed to generate signatures is ψ. Similarly, Π is said to be $(n,\omega,\psi)^{HSZI}$*-secure* if the success probability of all attacks considered in Hanaoka, Shikata, Zheng and Imai's scheme [15] is exponentially negligible under the same conditions. Also, we can define $(n,\omega,\psi)^{ext}$*-secure* by slightly modifying security notions of Johansson's scheme [18], and Wang and Safavi-Naini's scheme [31] so as to fit our signature model (the precise definition of $(n,\omega,\psi)^{ext}$*-secure* is described in Appendix).

From the definitions of security notions for the model in Definition 7, an interesting statement can be obtained:

Theorem 3 *The following relations among security notions hold:*

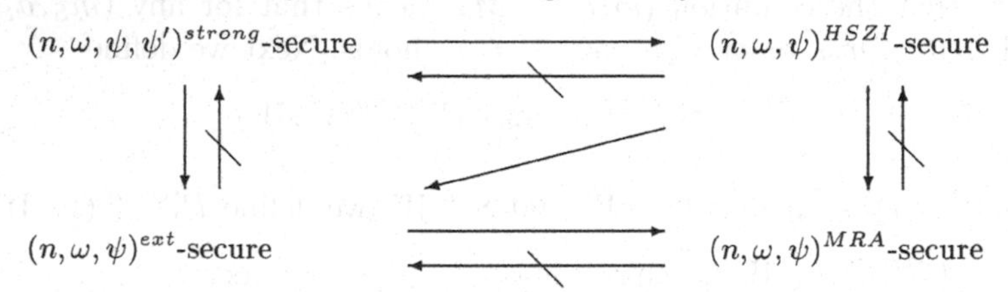

where "X-secure $\longrightarrow$ Y-secure" means that X-secure always implies Y-secure, while "X-secure $\not\longrightarrow$ Y-secure" means that there exists a signature scheme which is X-secure but not Y-secure.

A detailed proof will appear in the full version of this paper.

4 Construction

In this section we propose a construction method for signature schemes which is secure in terms of our strong security notion. We describe the key generation algorithm, *Gen*, signing algorithm, *Sig*, and verification algorithm, *Ver*, using the notations introduced in Section 3.1.

- **Key Generation Algorithm**: The key generation algorithm, *Gen*, which, on input 1^k, picks a k-bit prime power q, constructs a finite field $\boldsymbol{F}_q$ with q elements. It also picks uniformly at random $2n$ elements $\mathbf{v}_1^{(1)}, \mathbf{v}_1^{(2)}, \mathbf{v}_2^{(1)}, \mathbf{v}_2^{(2)}, \ldots, \mathbf{v}_n^{(1)}, \mathbf{v}_n^{(2)}$ in $\boldsymbol{F}_q{}^{\omega+\psi'}$ for verifiers $V_1, V_2, \ldots, V_n$, respectively, and constructs two polynomials $F_d(Y_1, Y_2, \ldots, Y_{\omega+\psi'}, Z)$ $(d=1,2)$ over $\boldsymbol{F}_q$ with $\omega+\psi'+1$ variables $Y_1, Y_2, \ldots, Y_{\omega+\psi'}, Z$ as follows:

$$F_d(Y_1, \ldots, Y_{\omega+\psi'}, Z) = \sum_{i=0}^{\psi} \sum_{j=1}^{\omega+\psi'} a_{ij}^{(d)} Z^i Y_j + \sum_{i=0}^{\psi} a_{i0}^{(d)} Z^i \ (d=1,2),$$

  where the coefficients $a_{ij}^{(d)}$ are chosen uniformly at random from $\boldsymbol{F}_q$. Then, a signing-key for the signer S is $x := (F_1(Y_1, \ldots, Y_{\omega+\psi'}, Z), F_2(Y_1, \ldots, Y_{\omega+\psi'}, Z))$ and a verification-key for the verifier V_i is $y_{V_i} := (\mathbf{v}_i^{(1)}, \mathbf{v}_i^{(2)}, F_1(\mathbf{v}_i^{(1)}, Z), F_2(\mathbf{v}_i^{(2)}, Z))$ for $1 \leq i \leq n$. The algorithm *Gen* returns $(\boldsymbol{F}_q, x, y_{V_1}, y_{V_2}, \ldots, y_{V_n})$.

We consider the case where $\mathcal{M}_k \subset \boldsymbol{F}_q$.

- **Signing Algorithm**: The signing algorithm *Sig* which, on input the signing-key $x = (F_1(Y_1, \ldots, Y_{\omega+\psi'}, Z), F_2(Y_1, \ldots, Y_{\omega+\psi'}, Z))$ and a message m, returns a signature $a := (F_1(Y_1, \ldots, Y_{\omega+\psi'}, m), F_2(Y_1, \ldots, Y_{\omega+\psi'}, m))$.
- **Verification Algorithm**: The verification algorithm *Ver* which, on input (y_{V_i}, m, a), where $a = (F_1(Y_1, \ldots, Y_{\omega+\psi'}, m), F_2(Y_1, \ldots, Y_{\omega+\psi'}, m))$ and $y_{V_i} = (\mathbf{v}_i^{(1)}, \mathbf{v}_i^{(2)}, F_1(\mathbf{v}_i^{(1)}, Z), F_2(\mathbf{v}_i^{(2)}, Z))$, computes evaluation values $e_1^{(d)}, e_2^{(d)}$ $(d=1,2)$ as follows:

$$e_1^{(d)} := F_d(Y_1, \ldots, Y_{\omega+\psi'}, m)|_{(Y_1, \ldots, Y_{\omega+\psi'}) = \mathbf{v}_i^{(d)}}$$
$$e_2^{(d)} := F_d(\mathbf{v}_i^{(d)}, Z)|_{Z=m} \qquad (d=1,2).$$

  Ver then returns *"true"* if $e_1^{(d)} = e_2^{(d)}$ for $d=1,2$, and *"false"* otherwise.

The following theorem proves the security of the above construction in our strong security notion.

Theorem 4 *The above construction results in an (n, ω, ψ, ψ')-secure signature scheme, where ω, ψ, ψ' can be taken in such a way that*

$$0 \leq \omega \leq n, \qquad 0 < \psi < q, \qquad 0 < \psi' \leq q + 1 - \sqrt{q},$$

and the success probability of attacks is less that $1/q$.

Once again a proof for the theorem will be provided in the full version of this paper.

5 Concluding Remarks

In this paper, we have established a sound security notion, which is likely to be the strongest possible, by taking into account the security notion for public-key signature schemes and some desirable requirements for signature schemes in the unconditional security setting. And we have examined relationships among security notions which have appeared in unconditionally secure schemes both for authentication and signature. We have demonstrated that our security notion is the strongest among all the notions proposed so far. An interesting aspect is that our security notion includes that of public-key signature schemes. We have further presented a construction method for unconditionally secure signature schemes which is provable secure in our strong security notion.

Acknowledgement

The authors wish to thank Tatsuaki Okamoto for helpful comments on the previous version. We also thank anonymous referees for their helpful comments.

References

1. M. Abe and T. Okamoto, "A signature scheme with message recovery as secure as discrete logarithm", Advances in Cryptology – ASIACRYPT '99, LNCS 1716, pp. 378–389, Springer, 1999.
2. M. Bellare and P. Rogaway, "The exact security of digital signatures – How to sign with RSA and Rabin", Advances in Cryptology – EUROCRYPT '96, LNCS 1070, Springer, 1996.
3. E. F. Brickell and D. R. Stinson, "Authentication codes with multiple arbiters," Advances in Cryptology – EUROCRYPT '88, LNCS 330, Springer, pp. 51–55, 1988.
4. D. Chaum and H. van Antwerpen, "Undeniable signatures", Advances in Cryptology – CRYPTO '89, Springer, pp. 212–216, 1990.
5. D. Chaum and S. Roijakkers, "Unconditionally secure digital signatures," Advances in Cryptology – CRYPTO'90, LNCS 537, Springer, pp. 206–215, 1990.
6. D. Chaum, E. Heijst and B. Pfitzmann, "Cryptographically strong undeniable signatures, unconditionally secure for the signer," Advances in Cryptology – CRYPTO '91, LNCS 576, Springer, pp. 470–484, 1991.
7. R. Cramer and V. Shoup, "Signature schemes based on the strong RSA assumption", Proc. of the 6th ACM Conference in Computer and Communication Security, 1999.
8. Y. Desmedt and M. Yung, "Arbitrated unconditionally secure authentication can be unconditionally protected against arbiter's attack," Advances in Cryptology – CRYPTO '90, LNCS 537, Springer, pp. 177–188, 1990.
9. Y. Desmedt, Y. Frankel and M. Yung, "Multi-receiver/Multi-sender network security: efficient authenticated multicast/feedback," Proc. of IEEE Infocom'92, pp. 2045–2054, 1992.
10. W. Diffie and M. Hellman, "New directions in cryptography", IEEE Transactions on Information Theory 22, 6, pp. 644–654, 1976.

11. T. ElGamal, "A public key cryptosystem and a signature scheme based on discrete logarithms," IEEE Transactions on Information Theory, 31, 4, pp. 469–472, 1985.
12. R. Gennaro, S. Halevi, and T. Rabin "Secure hash-and-sign signatures without the random oracle", Advances in Cryptology – EUROCRYPT '99, LNCS 1592, pp. 123–139, Springer, 1999.
13. E. N. Gilbert, F. J. MacWilliams and N. J. A. Sloane, "Codes which detect deception," Bell System Technical Journal, 53, pp. 405–425, 1974.
14. S. Goldwasser, S. Micali and R. Rivest, "A digital signature scheme secure against adaptive chosen message attacks", SIAM J. Comput. 17, 2, pp. 281–308, 1988.
15. G. Hanaoka, J. Shikata, Y. Zheng, and H. Imai, "Unconditionally secure digital signature schemes admitting transferability", Advances in Cryptology – ASIACRYPT 2000, LNCS 1976, Springer, pp. 130–142, 2000.
16. G. Hanaoka, J. Shikata, Y. Zheng, and H. Imai, "Efficient and Unconditionally Secure Digital Signatures and a Security Analysis of a Multireceiver Authentication Code", to appear in Proc. of Public Key Cryptography, Springer, 2002.
17. T. Johansson, "Lower bounds on the probability of deception in authentication with arbitration", IEEE Trans. Inform. Theory 40, 5, pp. 1573–1585, 1994.
18. T. Johansson, "Further results on asymmetric authentication schemes," Information and Computation, 151, pp. 100–133, 1999.
19. K. Kurosawa, "New bound on authentication code with arbitration," Advances in Cryptology – CRYPTO '94, LNCS 839, Springer, pp. 140–149, 1994.
20. K. Kurosawa and S. Obana, "Combinatorial bounds for authentication codes with arbitration," Advances in Cryptology – EUROCRYPT '95, LNCS 921, Springer, pp. 289–300, 1995.
21. R. Rivest, A. Shamir and L. Adleman, "A method for obtaining digital signature and public-key cryptosystems," Communication of the ACM, vol.21, no.2, pp. 120–126, 1978.
22. B. Pfitzmann, "Sorting out signature schemes", Proc. of the First ACM Conference on Computer and Communications Security, ACM Press, pp. 74–86, 1993.
23. D. Pointcheval and J. Stern, "Security proofs for signature schemes", Advances in Cryptology – EUROCRYPT '96, LNCS 1070, Springer, 1996.
24. R. Safavi-Naini and H. Wang, "New results on multi-receiver authentication codes," Advances in Cryptology – EUROCRYPT '98, LNCS 1403, pp. 527–541, Springer, 1998.
25. R. Safavi-Naini and H. Wang, "Broadcast authentication in group communication," Advances in Cryptology – ASIACRYPT '99, LNCS 1716, Springer, pp. 399–411, 1999.
26. R. Safavi-Naini and H. Wang, "Multireceiver authentication codes: models, bounds, constructions and extensions," Information and Computation, 151, pp. 148–172, 1999.
27. G. J. Simmons, "Authentication theory/coding theory," Advances in Cryptology – CRYPTO '84, LNCS 196, Springer, pp. 411–431, 1984.
28. G. J. Simmons, "Message authentication with arbitration of transmitter/receiver disputes," Advances in Cryptology – EUROCRYPT '87, Springer, pp. 151–165, 1987.
29. G. J. Simmons, "A Cartesian construction for unconditionally secure authentication codes that permit arbitration," Journal of Cryptology 2, pp. 77–104, 1990.
30. R. Taylor, "Near optimal unconditionally secure authentication," Advances in Cryptology – EUROCRYPT '94, LNCS 950, Springer, pp. 244–253, 1994.
31. Y. Wang and R. Safavi-Naini, "A^3-codes under collusion attacks" Advances in Cryptology – ASIACRYPT '99, LNCS 1716, Springer, pp. 390–398, 1999.

Appendix: A Security Notion for Extended A^2 and A^3-Codes

Johansson's model [18] for a class of broadcast authentication scheme is an extension of that of A^2-codes. Also, Wang and Safavi-Naini's model [31] is an extension of that of A^3-codes. Taking into account security notions of these models, we arrive at the following security notion by modifying their notions so as to fit our signature model. In that sense, the following security notion can also be regarded as that of an extension of A^2 and A^3-codes.

Definition 10 Let k be a security parameter and $V_{arb} \in \mathcal{U} - \{S\}$ an arbiter (or a trusted party).

1. Success probability of impersonation and substitution by verifiers: For $W \in \mathcal{W}$ such that $V_j, V_{arb}, S \notin W$, we define $P^{ext}_{I,S}(V_j, W)$ as

$$P^{ext}_{I,S}(V_j, W) := \max_{y_W} \max_{M \in \wp_{\psi}^{\mathcal{M}_k}, \{(m,a)\}_{m \in M}} \max_{(m',a')} \Pr(V_j \text{ accepts } (m', a') \mid y_W, \{(m,a)\}_{m \in M})$$

   where M is taken over $\wp_{\psi}^{\mathcal{M}_k}$, $\{(m,a)\}_{m \in M}$ is a set of $|M|$ valid signed messages with $m \in M$, and m' is taken over $\mathcal{M}_k$ satisfying $m' \notin M$. Then, $P^{ext}_{I,S}$ is defined as

$$P^{ext}_{I,S} := \max_{V_j, W} P^{ext}_{I,S}(V_j, W)$$

   where V_j is taken over all receivers including V_{arb} and W is taken over $\mathcal{W}$ satisfying $S, V_j, V_{arb} \notin W$.

2. Success probability of attack by colluders including the signer: For $W \in \mathcal{W}$ such that $V_j, V_{arb} \notin W$ and $S \in W$, we define

$$P^{ext}_{signer}(V_j, W) := \max_{x} \max_{y_{W-\{S\}}} \max_{(m,a)} \Pr(V_j \text{ accepts } (m, a) \mid x, y_{W-\{S\}})$$

   where m is taken over $\mathcal{M}_k$ and $a \in \mathcal{A}_k$ is taken such that (m, a) is an invalid signed message, i.e. $a \neq Sig(x, m)$. Then, P^{ext}_{signer} is defined as follows:

$$P^{ext}_{signer} := \max_{V_j, W} P^{ext}_{signer}(V_j, W),$$

   where V_j is taken over all receivers including V_{arb} and W is taken over $\mathcal{W}$ satisfying $V_j, V_{arb} \notin W$ and $S \in W$.

3. Success probability of attack against the sender: For $W \in \mathcal{W}$ such that $S \notin W$, we define

$$P^{ext}_{arbiter\_1}(W) := \max_{y_W} \max_{M \in \wp_{\psi}^{\mathcal{M}_k}, \{(m,a)\}_{m \in M}} \max_{(m',a')} \Pr((m', a') \text{ is a valid signed message generated by } S \mid y_W, \{(m,a)\}_{m \in M})$$

where M is taken over $\wp_{\psi}^{\mathcal{M}_k}$, $\{(m,a)\}_{m\in M}$ is a set of $|M|$ valid signed messages with $m \in M$ and m' is taken over $\mathcal{M}_k$ satisfying $m' \notin M$. Then, $P^{ext}_{arbiter\_1}$ is defined as

$$P^{ext}_{arbiter\_1} := \max_{W} P^{ext}_{arbiter\_1}(W),$$

where W is taken over $\mathcal{W}$ such that $S \notin W$. Here, we note that W runs over $\mathcal{W}$ including the cases $V_{arb} \in W$.

4. Success probability of attack against a verifier by colluders including the arbiter: For $W \in \mathcal{W}$ such that $V_{arb} \in W$ and $S, V_j \notin W$, we define

$$P^{ext}_{arbiter\_2}(V_j, W) := \max_{y_{V_{arb}}}, \max_{y_{W-\{V_{arb}\}}} \max_{M\in\wp_{\psi}^{\mathcal{M}_k},\{(m,a)\}_{m\in M}} \max_{(m',a')}$$
$$\Pr(V_j \text{ accepts } (m',a') \mid y_{V_{arb}}, y_{W-\{V_{arb}\}}, \{(m,a)\}_{m\in M}),$$

where M is taken over $\wp_{\psi}^{\mathcal{M}_k}$, $\{(m,a)\}_{m\in M}$ is a set of $|M|$ valid signed messages with $m \in M$ and m' is taken over $\mathcal{M}_k$ satisfying $m' \notin M$. Then, $P^{ext}_{arbiter\_2}$ is defined as

$$P^{ext}_{arbiter\_2} := \max_{V_j,W} P^{ext}_{arbiter\_2}(V_j, W),$$

where V_j is taken over all receivers except V_{arb}, and W is taken over $\mathcal{W}$ satisfying $V_{arb} \in W$ and $S, V_j \notin W$.

5. Success probability of attack against a verifier by colluders including both the arbiter and the sender: For $W \in \mathcal{W}$ such that $V_{arb}, S \in W$ and $V_j \notin W$, we define

$$P^{ext}_{arbiter\_3}(V_j, W) := \max_{x} \max_{y_{V_{arb}}} \max_{y_{W-\{V_{arb},S\}}} \max_{(m,a)}$$
$$\Pr(V_j \text{ accepts } (m,a) \mid x, y_{V_{arb}}, y_{W-\{V_{arb},S\}}),$$

where (m,a) is taken over $\mathcal{M}_k \times \mathcal{A}_k$ such that (m,a) is not accepted by V_{arb}, i.e. $Ver(m, a, y_{V_{arb}}) = false$. Then, $P^{ext}_{arbiter\_3}$ is defined as

$$P^{ext}_{arbiter\_3} := \max_{V_j,W} P^{ext}_{arbiter\_3}(V_j, W),$$

where V_j is taken over all verifiers except V_{arb}, and W is taken over $\mathcal{W}$ such that $V_{arb}, S \in W$ and $V_j \notin W$.

Let $\epsilon(k)$ be an exponentially negligible function. For simplicity, we denote $\epsilon(k)$ by ϵ. A signature scheme Π along with our signature model is called $(n,\omega,\psi)^{ext}$-*secure* if the following condition is satisfied: under the conditions that there exists at most ω colluders and that the signer is allowed to generate at most ψ signatures, the inequality below holds.

$$\max\{P^{ext}_{I,S}, P^{ext}_{signer}, P^{ext}_{arbiter\_1}, P^{ext}_{arbiter\_2}, P^{ext}_{arbiter\_3}\} \le \epsilon$$

Traitor Tracing with Constant Transmission Rate

Aggelos Kiayias[1] and Moti Yung[2]

[1] Graduate Center, CUNY, NY USA,
akiayias@gc.cuny.edu
[2] CertCo, NY USA,
moti@cs.columbia.edu

Abstract. An important open problem in the area of Traitor Tracing is designing a scheme with constant expansion of the size of keys (users' keys and the encryption key) and of the size of ciphertexts with respect to the size of the plaintext. This problem is known from the introduction of Traitor Tracing by Chor, Fiat and Naor. We refer to such schemes as traitor tracing with constant transmission rate. Here we present a general methodology and two protocol constructions that result in the first two public-key traitor tracing schemes with constant transmission rate in settings where plaintexts can be calibrated to be sufficiently large. Our starting point is the notion of "copyrighted function" which was presented by Naccache, Shamir and Stern. We first solve the open problem of discrete-log-based and public-key-based "copyrighted function." Then, we observe the simple yet crucial relation between (public-key) copyrighted encryption and (public-key) traitor tracing, which we exploit by introducing a generic design paradigm for designing constant transmission rate traitor tracing schemes based on copyrighted encryption functions. Our first scheme achieves the same expansion efficiency as regular ElGamal encryption. The second scheme introduces only a slightly larger (constant) overhead, however, it additionally achieves efficient black-box traitor tracing (against any pirate construction).

1 Introduction

Distributing data securely to a set of subscribers is an important problem in cryptography with a variety of practical applications. A direct solution to this problem is to give to each subscriber a common secret-key. Nevertheless, such a solution is not satisfactory: this enables a subscriber to distribute its secret-key to other parties thus enabling illegal data reception. This situation is identified as piracy in the context of digital content distribution. Preventing piracy via tamper–proof devices is uneconomical and not applicable in many scenarios; additionally software obfuscation has not produced cryptographically strong results that would adequately protect the common secret-key. In light of this, the notion of "traitor-tracing" that originated in [CFN94] suggests a solution to piracy when it is assumed that subscribers' decoders are open and therefore the

L.R. Knudsen (Ed.): EUROCRYPT 2002, LNCS 2332, pp. 450–465, 2002.

secret-keys are accessible. In a traitor-tracing scheme (TTS) each user possesses a different secret-key that allows the reception of the data in a non-ambiguous fashion. The scheme discourages piracy as follows: given a pirate decoder the scheme allows the distributor to recover the identities of some subscribers that collaborated in its construction (henceforth called traitors).

From the time of the primitive's introduction in [CFN94] a series of works [Pfi96,SW98,NP98,KD98,BF99,FT99,GSY99,NP00,SW00,NNL01,KY01b] proposed more efficient/ robust schemes or schemes with advanced capabilities such as revocation and non-repudiation. Two extremely desirable properties of a traitor tracing scheme are (i) Public-key Traitor Tracing where any third party (e.g., any of a number of pay-T.V. stations) is able to send secure messages to the set of subscribers, (ii) Black-Box Traitor Tracing which suggests that the tracing procedure can be accomplished with merely black-box access to the pirate-decoder (something that allows less costly, even remote access tracing).

Traitor Tracing has also its shortcomings: the size of the ciphertexts and keys used by traitor-tracing schemes depends on quantitites such as the number of users and/or the maximum traitor collusion that is expected. Even though progress has been made from the initial scheme of [CFN94] in reducing the "communications overhead" in traitor tracing schemes, so far there has not been a scheme in which the "rate" of the three main efficiency parameters of a traitor tracing scheme: "ciphertext," "encryption-key" and "user-key" size, is constant (where the "rate" of a parameter expresses the ratio of the its size over the size of the plaintexts – which is the security parameter). We will refer to the sum of the rates of the three parameters collectively, as the "transmission rate" of a Traitor Tracing Scheme. The reason we do not concentrate solely on the ciphertext rate, or the "per message" overhead, is that memory costs induced by the size of keys are equally important contributions to the "transmission" costs of a TTS. Minimizing the transmission rate has been, in fact, open since [CFN94] (the issue was reiterated in some stronger form also in [BF99]).

In this work we present the first two constant transmission rate Traitor-Tracing Schemes (for settings where plaintexts can be calibrated to be large enough), thus answering the question that postulated the existence of such schemes in the affirmative. Our results, in comparison to previous schemes, are presented in figure 1 (where the ElGamal scheme which does not provide traitor tracing is given for comparison).

Our methodology starts by investigating the notion of "copyrighted function" proposed by Naccache, Shamir and Stern in [NSS99]. In the copyrighted function setting each member of a set of users possesses a different implementation of a function with the same functionality though, so that an authority is capable of exposing the responsible user(s) if an implementation is used illegally. The techniques presented in [NSS99] applied to one-way (hash) functions and to symmetric encryption (and were implemented based on the RSA function used as a private key function). It was left as an open problem whether it is possible to achieve function copyright based on the Discrete-Logarithm Problem. Here we answer this question in the affirmative. Moreover, we reformulate the setting

	Ciphertext Rate	User-Key Rate	Encryption-Key Rate	Max Traceable Collusion	Black-Box Traitor Tracing
[ElGamal84]	~ 2	~ 1	~ 3	$\times$	$\times$
[CFN94] (Sch.1)	$\mathcal{O}(t^4 \log n)$	$\mathcal{O}(t^2 \log n)$	$\mathcal{O}(t^2 \log n)$	t	$\surd$
[BF99]	$\sim (2t+1)$	$\sim 2t$	$\sim (2t+1)$	t	inefficient (see [KY01a])
TTS Scheme 1	~ 2	~ 1	~ 3	$\Omega(l^c) = t$	?
TTS Scheme 2	~ 3	~ 2	~ 4	$\Omega(l^c) = t$	$\surd$

Fig. 1. A Comparison of our traitor-tracing schemes with previous work. Note that l is a security parameter and c is a constant < 1; the plaintext space in all cases is considered to be $\{0,1\}^l$. The rate of keys/ciphertexts is defined w.r.t. the size of the plaintexts: l.

of [NSS99] in the public-key encryption setting and we then present a simple yet crucial step of our methodology, showing that the goal of copyrighting public-key encryption is equivalent to the construction of a public-key traitor tracing scheme.

The reformulation of the [NSS99]-setting in the public-key context, along with our novel design paradigm for the construction of traitor-tracing schemes based on copyrighted encryption functions and the two concrete copyrighted public-key functions we present, allow us to construct two public-key traitor tracing schemes, which are the first that have constant transmission rate. In our schemes, the distributor has the flexibility of adjusting the size of the plaintexts to accommodate tracing. Such flexibility is always possible in bulk data encryption (or in the public-key setting, bulky transmission of numerous session keys). Our size adjustment method employs collusion-secure codes [BS95], and we further found that in order to support the adjustment while retaining the traitor tracing capability with constant rate, it is crucial to employ the All-or-Nothing-Transform (AONT) [Riv97] prior to encryption (or alternatively employ a threshold assumption similar to the one used in [NP98]). Our first scheme is as efficient, rate-wise, as ElGamal encryption, whereas our second scheme uses slightly extended ciphertexts and keys (by a constant additive factor). However, the second scheme achieves also efficient black-box traitor tracing against any pirate-construction (as in the black-box traitor tracing model of [CFN94]) and not just a single-key traitor as in [BF99]. The scheme relies on ElGamal-like encryption as the [BF99]-scheme and is the first traitor tracing scheme beyond [CFN94] and its variants to achieve black-box traceability.

We note that, interestingly, our schemes employ and combine in a unique way results from all the major contributions in the area: traceability codes introduced in the context of traitor tracing by Chor, Fiat and Naor [CFN94], collusion secure codes defined by Boneh and Shaw [BS95] (in the context of fingerprinting) and the public-key traitor tracing concepts of Kurosawa and Desmedt [KD98], and Boneh and Franklin [BF99].

The intractability assumptions used for the first scheme are the DDH Assumption over the quadratic-residues group modulo a composite and the Quadratic Residuosity assumption, whereas the security of the second scheme is based on DDH-Assumption over a prime order subgroup.

The scope of the design paradigm we propose for the construction of traitor tracing schemes based on copyrighted encryption functions goes beyond the two public-key traitor tracing schemes we propose. It can be readily applied to any basic 2-user copyrighted encryption mechanism yielding a traitor tracing scheme with the same transmission rate as the underlying basic copyrighted encryption function.

2 Preliminaries

Notations. A function $\sigma : \mathbb{N} \to \mathbb{R}$ will be called negligible if for all $c \in \mathbb{N}$ there exists a $l_0 \in \mathbb{N}$ so that for all $l \geq l_0$ it holds that $\sigma(l) < l^{-c}$. Throughout l will denote a security parameter (typically, the plaintext size); all sets of objects that we consider are exponential in size w.r.t. l and contain objects of size polynomial in l. All procedures are polynomial-time in l. If K is a set of objects and f is a procedure that samples an element of K, denote by $k \hookleftarrow_f K$ such element; note that we may occasionally omit f or K from this notation if this is allowed in the context. If K is a set of objects of the same size, let $\mathrm{len}[k \in K]$ denote the size of the objects in K. As stated above $\mathrm{len}[k \in K]$ is polynomial in the security parameter and perhaps it may depend on other factors as well. We use $|x|$ to denote the size of an object x, e.g. $|x| = \lceil \log_2 x \rceil$ if $x \in \mathbb{N}$; also let $[k]$ denote the set $\{1, \ldots, k\}$. If $f(b, v)$ is a function with real values, we write $f(b, v) \sim c$ where $c \in \mathbb{R}$ is a constant if $\lim_{b,v\to\infty} f(b, v) = c$. The notation $a \in_U R$ stands for "a is sampled from R following the uniform distribution."

Next we define the notion of public-key encryption scheme (note that the definition is tailored to our setting).

Definition 1. *A public-kcy encryption scheme is a tuple* $\langle \mathbb{P}, \mathbb{C}, \mathcal{P}, \cup_{pk\in\mathcal{P}} \mathcal{K}_{pk}, G, E, D\rangle$ *so that*
1. $\mathbb{P}$ *and* $\mathbb{C}$ *are the plaintext-space and ciphertext-space respectively. Without loss of generality we assume that the objects in these sets are of the same size.*
2. Key Generation. It holds that: $\langle pk, \kappa\rangle \hookleftarrow_G (\mathcal{P} \times \cup_{pk\in\mathcal{P}} \mathcal{K}_{pk})$ *so that* $\kappa \in \mathcal{K}_{pk}$.
3. Encryption. $E : (\mathcal{P} \times \mathbb{P}) \to \mathbb{C}$ *is a probabilistic poly-time procedure.*
4. Decryption. $D : (\cup_{pk\in\mathcal{P}} \mathcal{K}_{pk} \times \mathbb{C}) \to \mathbb{P}$ *is a deterministic procedure so that* $D(\kappa, E(pk, m)) = m$ *for all* $m \in \mathbb{P}$ *and* $\langle pk, \kappa\rangle \hookleftarrow_G$.
5. Semantic Security (i.e. polynomial indistinguishability): for some $\langle pk, \kappa\rangle \hookleftarrow_G$, *any adversary that given* pk *generates* $m_1, m_2 \in \mathbb{P}$, *when given* $E(m_x)$ *with* $x \in_U \{1, 2\}$ *it can predict* x *with negligible advantage. Note that the definition can be extended to include stronger notions of security such as chosen-ciphertext security or non-malleability.*

2.1 Intractability Assumptions

The security of our schemes is based on the hardness of the Decisional Diffie Hellman (DDH) Problem over a multiplicative cyclic group $\mathbb{G} = \langle g \rangle$:

Definition 2. Decision Diffie Hellman Assumption. *Let $\mathcal{V}$ be the distribution $\{\langle g, g^x, g^y, g^{xy} \rangle \mid x, y < |\mathbb{G}|\}$, and $\mathcal{R}$ be the distribution $\{\langle g, g^x, g^y, g^z \rangle \mid x, y, z < |\mathbb{G}|\}$. The DDH assumption over $\mathbb{G} = \langle g \rangle$ states that any poly-time distinguisher D for the two distributions $\mathcal{V}, \mathcal{R}$ has negligible success probability, i.e. $|\mathbf{Prob}_{X \in \mathcal{V}}[D(X) = 1] - \mathbf{Prob}_{X \in \mathcal{R}}[D(X) = 1]|$ is negligible in $\log |\mathbb{G}|$.*

The DDH assumption has been used in a variety of settings and over many different groups; for an overview and applications the reader is referred to [NR97] [Bon98]. The DDH assumption over a group of prime order is known to be equivalent to the security of ElGamal encryption, see [TY99]. We note here that ElGamal-like encryption with composite modulus has also been used extensively, e.g. [FH96,CG98].

Here we use the DDH assumption (i) over the cyclic subgroup $\mathcal{G}$ of quadratic residues of $\mathbb{Z}_p^*$ of order q, where $p = 2q + 1$ and both p, q are primes; (ii) over the cyclic subgroup $\mathbb{Q}_N$ of quadratic residues of $\mathbb{Z}_N^*$ where $N = pq$ and $p = 2p' + 1, q = 2q' + 1$ with p, q, p', q' all primes. It is believed that DDH over the subgroup of quadratic residues modulo p or modulo N is hard (see [Bon98]). We also utilize the Quadratic Residuosity (QR) Assumption [GM84]:

Definition 3. Quadratic Residuosity Assumption. *If $N = pq$, so that $p = 2p' + 1, q = 2q' + 1$ with p, q, p', q' all primes, any probabilistic algorithm that given $x \in \mathbb{J}_N$ (Jacobi +1 elements) it decides whether $x \in \mathbb{Q}_N$ or $x \in \mathbb{J}_N - \mathbb{Q}_N$, has success probability $1/2 + \epsilon$ where ϵ is negligible in $\log N$.*

2.2 Public-Key Traitor Tracing Schemes

A public-key traitor tracing scheme involves a key-generation (setup) algorithm G, and the corresponding encryption/decryption function as in the public-key encryption setting: an authority uses G to generate $\langle pk, d_1, \ldots, d_n \rangle$ so that each of the d_i "inverts" the public-key pk. Subsequently it publishes pk and privately communicates the key d_i to each user i. From then on users are capable of decrypting messages encrypted using the public-key pk. If a pirate uses t keys given by some users (the traitors) to construct another key for the purpose of implementing an illegal receiver, the authority is able to recover the identity of one of the traitor users given the pirate-key (a procedure called traitor-tracing). Formally,

Definition 4. *An n-user (public-key) traitor-tracing scheme (TTS) is a tuple $\langle \mathbb{P}, \mathbb{C}, \mathcal{P}, \cup_{pk \in \mathcal{P}} (\mathcal{K}_{pk})^n, G, E, D \rangle$ that*
1. Satisfies properties 1,2,3,5 of definition 1.
2. If $\langle pk, d_1, \ldots, d_n \rangle \hookleftarrow_G$ then $D : \cup_{pk \in \mathcal{P}} \mathcal{K}_{pk} \times \mathbb{C} \to \mathbb{P}$ is a deterministic procedure so that $D(d, E(pk, m)) = m$ for all $m \in \mathbb{P}$ and $d \in \{d_1, \ldots, d_n\}$.

3. Tracing. Let $\langle pk, d_1, \ldots, d_n\rangle \hookleftarrow_G$*. There is a procedure* $\mathcal{T}$ *so that: for any adversary* $\mathcal{A}$ *that given* pk *and* $\{d_{i_1}, \ldots, d_{i_t}\}$ *with* $t \leq c$*,* $\mathcal{A}$ *generates some* $d \in \mathcal{K}_{pk}$ *so that* $D(d, E(pk, m)) = m$ *for most* $m \in \mathbb{P}$*,* $\mathcal{T}$ *given* d *is capable of recovering at least one of the indices* i_ℓ*.*
The parameter c *is maximum collusion size allowed by the traitor tracing scheme. We will call such a scheme: an n-user,c-TTS.*

Of course, the pirate may not use directly a certain decryption key d, but instead construct a simulator for the decryption operation that is hard to reverse-engineer and extract its contents. Therefore, it is important for a TTS to allow *black-box traitor tracing*:

3'. Black-Box Tracing. Let $\langle pk, d_1, \ldots, d_n\rangle \hookleftarrow_G$. There is a procedure $\mathcal{T}$ so that for any adversary that given pk and $\{d_{i_1}, \ldots, d_{i_t}\}$ with $t \leq c$ it generates a decryption simulator S so that $S(E(pk, m)) = m$ for almost all $m \in \mathbb{P}$, then $\mathcal{T}$ given oracle access to S is capable of recovering at least one of the indices i_ℓ.

Definition 5. *A n-user,c-TTS with black-box traceability is defined as in definition 4, with item* 3′ *substituting item* 3.

Definition 6. Efficiency Parameters. *The three basic efficiency parameters of traitor tracing schemes are (i) the ciphertext rate* $\frac{\text{len}[c \in \mathbb{C}]}{\text{len}[m \in \mathbb{P}]}$*, (ii) the user-key rate* $\frac{\text{len}[d \in \mathcal{K}_{pk}]}{\text{len}[m \in \mathbb{P}]}$*, and (iii) the encryption-key rate* $\frac{\text{len}[pk \in \mathcal{P}]}{\text{len}[m \in \mathbb{P}]}$*. The transmission rate of the scheme is defined as the sum of the three rates.*

3 Copyrighting a Function

Nacacche, Shamir and Stern [NSS99] introduced a technique for personalizing a certain function f to a set of users. This fingerprinting technique generates a number of personalized copies of f, so that $f_1(x) = \ldots = f_n(x) = f(x)$ for all x. The copies are drawn out of a keyed collection of different versions of f, denoted by $\{f_k\}_{k \in \mathcal{K}}$. It is assumed that there is a "generator" function $F(x, k) = f_k(x)$ for all $x, k \in \mathcal{K}$ that is publicly known and also that $\mathcal{K}$ can be sampled efficiently by some (secret) procedure $\mathcal{G}_\mathcal{K}$. The following definition is from [NSS99], slightly amended:

Definition 7. *A keyed collection* $\{f_k\}_{k \in \mathcal{K}}$ *is called*
(i) c-copyrighted against passive *adversaries* in the strong-sense, *if given c elements of* $\mathcal{K}$ *it is computationally impossible to find another element of* $\mathcal{K}$*.*
(ii) c-copyrighted against passive *adversaries, if there is an analyzer procedure* $\mathcal{T}$ *so that: an adversary given c elements of* $\mathcal{K}$ *constructs another element* κ_0 *of* $\mathcal{K}$*; then,* $\mathcal{T}$ *given* κ_0 *is able to reconstruct at least one of the c elements that were given to the adversary.*
(iii) c-copyrighted against active *adversaries, if there is an analyzer procedure* $\mathcal{N}$ *so that: an adversary given c elements of* $\mathcal{K}$ *produces a simulator* $\mathcal{S}$ *that agrees with* $f_k(x)$ *for almost all inputs* x*, then* $\mathcal{N}$ *with oracle access to* $\mathcal{S}$ *is capable of recovering at least one of the c elements that were given to the adversary.*

In [NSS99] a method was presented that allowed copyrighting a hash function based on RSA-encryption. The basic design paradigm of [NSS99] solved the two-user case first and then the multi-user case was addressed by employing collusion-secure codes [BS95]. Although copyrighting a hash function allows a variety of applications, much greater flexibility is allowed by a method for copyrighting a public-key encryption function. (Note that in [NSS99] a method to copyright the RSA-encryption function was given, but only as a symmetric-encryption function, since no public-components were allowed). In [NSS99] it was left as an open question whether it is possible to achieve a copyright mechanism based on the Discrete-Logarithm Problem. Here we answer this question in the affirmative. Another important question that arises from the work of [NSS99] (who show how to copyright symmetric encryption) is whether it is possible to copyright a public-key encryption function. Next we formalize this notion.

3.1 Copyrighting a Public-Key Function

Definition 8. *A n-key, c-copyrighted Public-Key Encryption Scheme against passive (resp. active) adversaries is a tuple* $\langle \mathbb{P}, \mathbb{C}, \mathcal{P}, \cup_{pk\in\mathcal{P}}\mathcal{K}_{pk}, G_n, E, D\rangle$ *so that*
(i) $\langle pk, d_1, \ldots, d_n\rangle \hookleftarrow_{G_n} \mathcal{P} \times \mathcal{K}_{pk}$ *where* $d_1, \ldots, d_n \in \mathcal{K}_{pk}$.
(ii) $\langle \mathbb{P}, \mathbb{C}, \mathcal{P}, \cup_{pk\in\mathcal{P}}\mathcal{K}_{pk}, G_1, E, D\rangle$ *is a public-key encryption scheme.*
(iii) for any $pk \in \mathcal{P}$, $\{D(\kappa, \cdot) : \mathbb{C} \to \mathbb{P}\}_{\kappa\in\mathcal{K}_{pk}}$ *is c-copyrighted against passive (resp. active) adversaries.*

In the following simple but crucial Lemma we establish the relationship between the above generalization of the [NSS99]-setting and (public-key) traitor tracing:

Lemma 1. *An n-key,c-copyrighted public-key encryption scheme against passive (resp. active) adversaries is equivalent to an n-user,c-TTS (resp. n-user,c-TTS with black-box traceability).*

The Lemma provides a construction methodology for public-key traitor tracing schemes: given an n-key,c-copyrighted public-key encryption scheme the corresponding public-key traitor tracing scheme is the following: the authority uses G to generate a tuple $\langle pk, d_1, \ldots, d_n\rangle$. The key pk is published as the public-key and the decryption-key d_ℓ is given to user ℓ. Any third party can use the encryption algorithm E in combination to pk and send encrypted data to the users that possess the decryption-keys. Traitor tracing is taken care of by the copyright properties of the decryption function: if the security is against passive adversaries, the authority can perform non-black-box traitor tracing. If the copyright security of the decryption function is against active adversaries, the authority can use the analyzer to perform black-box traitor tracing.

4 The Basic Building Block: The Two-User Case

We will consider two alternative settings for copyrighting a public-key encryption function. Following the [NSS99] design paradigm we will consider the 2-

key,1-copyrighted case first. In the following sections we present two 2-key,1-copyrighted public-key encryption schemes. Scheme 1 is more efficient, however scheme 2 allows security against active adversaries.

4.1 Scheme 1

Let $N = pq$ where p, q are two primes so that $p = 2p' + 1, q = 2q' + 1$ with p', q' also prime. The factorization of N is kept secret by the authority. Let $h \in \mathbb{Z}_N^*$ with maximal order, i.e. $\text{ord}(h) = \lambda(N) = 2p'q'$ where $\lambda(N)$ is the Carmichael function, so that $\langle h \rangle = \mathbb{J}_N$ where $\mathbb{J}_N$ is the subgroup of $\mathbb{Z}_N^*$ that contains all elements with Jacobi Symbol $+1$. The element h can be computed easily given the factorization of N as follows: select h_1, h_2 to be generators of the multiplicative groups $\mathbb{Z}_p^*$ and $\mathbb{Z}_q^*$ respectively and compute h by solving the system $h = h_1 \bmod p$ and $h = h_2 \bmod q$ (solvable by the Chinese Remainder Theorem). It follows easily that h_1, h_2 are both quadratic non-residues modulo p, q respectively and as a result $\langle h \rangle = \mathbb{J}_N$. Now let $g =_{\text{df}} h^2 \bmod N$. It is easy to verify that $\langle g \rangle = \mathbb{Q}_N$ (the group of quadratic residues modulo N). Note that $|\mathbb{Q}_n| = p'q'$.

The tuple $\langle N, g, y =_{\text{df}} g^\alpha \bmod N \rangle$ is the public-key of the system, where $\alpha \in_U [p'q']$ with $(\alpha, p'q') = 1$. The set of possible decryption keys is $\mathcal{K}_{pk} =_{\text{df}} \{x \in \mathbb{N} \mid x = \alpha(\text{mod}\phi(N))\}$. The two users are assigned the two (shorter) secret keys of $\mathcal{K}_{pk}$, $\alpha_0 =_{\text{df}} \alpha$ and $\alpha_1 =_{\text{df}} \alpha + \phi(N)$ of $\mathcal{K}_{pk}$. It is immediate that: $g^{\alpha_0} \bmod N = g^{\alpha_1} \bmod N = y$.

Encryption is performed following the ElGamal paradigm ([ElG84]): given a message $M \in \mathbb{J}_N$, the sender computes the tuple $\langle g^k \bmod N, y^k \cdot M \bmod N \rangle$ where $k \in_U [N]$. Note that $\langle g \rangle = \mathbb{Q}_N$ is a group of unknown order for the sender. The decryption procedure is as follows: given $\kappa \in \mathcal{K}_{pk}$, and a ciphertext $\langle A, B \rangle$, the receiver computes the plaintext as follows: $B \cdot (A^{-1})^\kappa \bmod N$. It is easy to verify that the decryption operation inverts encryption. The following lemma shows that the choice of the encryption exponent k from $[N]$ is appropriate:

Lemma 2. *The uniform distribution over $\langle g \rangle$ is statistically indistinguishable from the distribution $\mathcal{D}$ induced over $\langle g \rangle$ by the mapping $k \to g^k \bmod N$ where $k \in_U [N]$.*

Theorem 1. *The public-key encryption function described above is*
(i) Semantically Secure under the DDH Assumption over $\mathbb{Q}_N$ and the QR Assumption in $\mathbb{Q}_N$.
(ii) 1-copyrighted against passive adversaries (in the strong sense): given the public-key pk and a key α_x of $\{\alpha_0, \alpha_1\}$ it is computationally infeasible to construct another key in $\mathcal{K}_{pk}$ under the assumption that factoring N is hard.

Note that the scheme is strictly 1-copyrighted and not 2-copyrighted since if the two users collude it is immediate that they can construct keys in $\mathcal{K}_{pk}$ as follows: given $\alpha_0, \alpha_1 \in \mathcal{K}_{pk}$ it follows that $\alpha_1 - \alpha_0$ equals $\phi(N)$. Subsequently any $\alpha_0 + x(\alpha_1 - \alpha_0)$, where $x \in \mathbb{N}$, is an element of $\mathcal{K}_{pk}$.

Plaintext-Space and Efficiency Parameters. In order to measure efficiency, first we have to specify the plaintext-space: let the plaintext-space for the encryption operation be $\{0,1\}^b$ with $b = |N| - 3$. We have to determine an encoding function $enc : \{0,1\}^b \to \mathbb{J}_N$ that is easily invertible. Given a message $M =_{\mathrm{df}} m_1 m_2 \ldots m_b \in \{0,1\}^b$ let $M' =_{\mathrm{df}} m_1 + 2m_2 + \ldots 2^{b-1} m_b + \frac{N}{4} + 1$. It is easy to see that $\frac{N}{4} < M' < \frac{N}{2}$. Suppose now that $p' = 1(\mathrm{mod} 4)$ and $q' = 3(\mathrm{mod} 4)$. Then, it holds that $(\frac{2}{N}) = -1$ (recall that $N = (2p'+1)(2q'+1)$). Now if $(\frac{M'}{N}) = 1$ the encoding of M is M', else if $(\frac{M'}{N}) = -1$ then the encoding of M is defined as $2 \cdot M'$. This completes the description of *enc*.

The encoding function can be inverted as follows: given $enc(M)$ we compute M' so that $M' =_{\mathrm{df}} enc(M)$ if $enc(M) < N/2$, or $M' =_{\mathrm{df}} enc(M)/2$ if $enc(M) > N/2$. The decoding of $enc(M)$ is the binary representation of $M' - \frac{N}{4} - 1$. The rates of the parameters of the system are illustrated in the figure 2 (recall that $|N| = b + 3$).

Plaintext Space	Ciphertext Rate	User-Key Rate	Public-Key Rate	Max Traceable Collusion
$\{0,1\}$	$\frac{2(b+3)}{b} \sim 2$	$\frac{(b+4)}{b} \sim 1$	$\frac{3(b+3)}{b} \sim 3$	1

Fig. 2. Efficiency Parameters of Scheme 1 (Two-User Setting)

4.2 Scheme 2

Let $\mathcal{G}$ be the group of quadratic residues modulo $p = 2q + 1$ where both p, q, are large primes. It follows that the order of $\mathcal{G}$ is q. Let g be a generator of $\mathcal{G}$. The public-key of the scheme is set to $pk =_{\mathrm{df}} \langle p, f, g, h \rangle$ where $f =_{\mathrm{df}} g^{\alpha}, h =_{\mathrm{df}} g^{\beta}$ and $\alpha, \beta \in_R [q]$. The two users are given two "representations" of α with respect to the "base" g, h, i.e. the authority selects two vectors $\langle d_0, d_0' \rangle, \langle d_1, d_1' \rangle$ over $\mathbb{Z}_q$ so that $d_i + \beta d_i' = \alpha$ for both $i \in \{0,1\}$. The two vectors are chosen so that they are linearly independent over $\mathbb{Z}_q$. Note that the set of all possible keys is $\mathcal{K}_{pk} =_{\mathrm{df}} \{\langle d, d' \rangle \mid d + d'\beta = \alpha(\mathrm{mod} q)\}$.

Encryption is performed as follows: given the public-key $\langle f, g, h \rangle$ and a message $M \in \mathcal{G}$, the encryption of M is $\langle M \cdot f^r \bmod p, g^r \bmod p, h^r \bmod p \rangle$. Decryption works as follows: given one of the two keys $\langle d_i, d_i' \rangle$ and a ciphertext $\langle A, B, C \rangle$ the receiver computes $A(B^{-1})^{d_i}(C^{-1})^{d_i'} \bmod p$. It is easy to verify that the decryption operation inverts encryption.

Theorem 2. *The public-key encryption function described above is*
(i) Semantically Secure under the DDH Assumption over $\mathcal{G}$.
(ii) 1-copyrighted against passive adversaries (in the strong sense): given the public-key information pk and a key $\langle d, d' \rangle \in \mathcal{K}_{pk}$ *it is computationally infeasible to construct another key in* $\mathcal{K}_{pk}$ *under the Discrete-Log assumption over* $\mathcal{G}$.

Note that the scheme is strictly 1-copyrighted and not 2-copyrighted since if the two users collude, they can construct keys in $\mathcal{K}_{pk}$ as follows: given $\langle d_0, d_0' \rangle$ and $\langle d_1, d_1' \rangle$ it holds that $\langle rd_0 + (1-r)d_1, rd_0' + (1-r)d_1' \rangle \in \mathcal{K}_{pk}$ for any $r \in \mathbb{Z}_q$.

Plaintext-Space and Efficiency Parameters. First we specify the plaintext-space: let the plaintext-space for the encryption-operation be $\{0,1\}^b$ with $b = |p| - 2$. We have to determine an easily invertible encoding function $enc : \{0,1\}^b \to \mathcal{G}$. Given $M = m_1 \ldots m_b \in \{0,1\}^b$ let $M' = m_1 + 2m_2 + \ldots 2^{b-1}m_b + 1$. It is easy to verify that $M' \in \{1, \ldots, q\}$. Then, $enc(M) =_{\text{df}} (M')^2 \bmod p$. It is easy to see that $enc(M) \in \mathcal{G}$ for any $M \in \{0,1\}^b$: this is because $\mathcal{G} = \langle g \rangle$ is the subgroup of quadratic residues modulo p. The encoding function enc can be inverted as follows: given $enc(M)$ we compute its two square roots modulo p and let M' be the one that belongs in $\{1, \ldots, q\}$. The decoding of $enc(M)$ is the binary representation of $M' - 1$. The rates of the parameters of the system are illustrated in the figure 3 (recall that $|p| = b + 2$).

Plaintext Space	Ciphertext Rate	User-Key Rate	Public-Key Rate	Max Traceable Collusion
$\{0,1\}^b$	$\frac{3(b+2)}{b} \sim 3$	$\frac{2(b+1)}{b} \sim 2$	$\frac{4(b+2)}{b} \sim 4$	1

Fig. 3. Efficiency Parameters of Scheme 2 (Two-User Setting)

4.3 Scheme 2: Security against Active Adversaries

In this section we establish that scheme 2 is secure against active adversaries.

Theorem 3. *Suppose that there is an adversary $\mathcal{A}$ that:*
(i) Given the public-key information, $\mathcal{A}$ produces a decryption simulator $\mathcal{S}$ that decrypts valid ciphertexts with probability ϵ. Then the Diffie-Hellman Problem is solvable with probability ϵ.
(ii) Given the public-key information pk and a key $\langle d, d' \rangle \in \mathcal{K}_{pk}$, $\mathcal{A}$ produces a simulator $\mathcal{S}$ that decrypts all valid ciphertexts but when given a "randomized" ciphertext of the form $\langle A, g^{r_0}, g^{\alpha r_1} \rangle$ with $r_0, r_1 \in_U [q]$, it outputs a value different than $A/g^{r_0 d + \alpha r_1 d'}$ with probability ϵ. Then the Decision-Diffie-Hellman Problem is decidable with probability ϵ.

Let us now present an analyzer $\mathcal{N}$ that given black-box access to a decryption simulator $\mathcal{S}$ constructed by one of the two users it decides which of the two constructed it:
Description of the Analyzer $\mathcal{N}$: given black-box access to a decryption simulator $\mathcal{S}$, $\mathcal{N}$ selects $a_0, a_1 \in_U \mathbb{Z}_q$ and solves the system $d_0 x + \alpha d'_0 y = a_0$ and $d_0 x + \alpha d'_1 y = a_1$ (note that the system is solvable because of the choice of $\langle d_0, d'_0 \rangle, \langle d_1, d'_1 \rangle$). Then, $\mathcal{N}$ submits to $\mathcal{S}$ the "randomized" ciphertext $\langle A, g^x, (g^\alpha)^y \rangle$. If the output of $\mathcal{S}$ is A/g^{a_0} then $\mathcal{N}$ outputs 0, otherwise, if the simulator's output is A/g^{a_1}, $\mathcal{N}$ outputs 1; finally $\mathcal{N}$ outputs ? if the output of the simulator is not contained in $\{A/g^{a_0}, A/g^{a_1}\}$.

The correctness of $\mathcal{N}$ is guaranteed by theorems 2 and 3. In particular theorem 2(ii) suggests that user 1 cannot incriminate user 2 or use some other key in

$\mathcal{K}_{pk}$; additionally theorem 3(i) suggests that at least one representation should be used by the simulator $\mathcal{S}$; finally theorem 3(ii) suggests that the "randomized" ciphertext used by the analyzer $\mathcal{N}$ cannot be distinguished from regular ciphertexts. Due to theorem 3(ii) the simulator $\mathcal{N}$ will output ? only in the case that both users colluded in the construction of $\mathcal{S}$. This leads to the corollary:

Corollary 1. *Scheme 2 is 1-copyrighted against active adversaries.*

Remark. Scheme 2 can be viewed as a special case of the public-key traitor tracing scheme of [BF99] (for two users). However, the approach we take in extending scheme 2 to capture the multi-user case is different from [BF99].

5 The Multi-user Case

Let $\langle \mathbb{P}, \mathbb{C}, \mathcal{P}, \cup_{pk \in \mathcal{P}} \mathcal{K}_{pk}, G_2, E, D \rangle$ be a 2-key,1-copyrighted (in the strong sense) public-key encryption scheme. In this section, following the [NSS99] design paradigm we compose the two-user case with collusion secure codes. Specifically, we show how to obtain an n-key,c-copyrighted public-key encryption scheme (and thus, by Lemma 1, a public-key traitor tracing scheme) by a parallel combination of independent instantiations of a 2-key,1-copyrighted public-key encryption scheme based on collusion-secure codes. Note that for designing one-way (hash) functions, [NSS99] used nested composition rather than parallel. The parallel approach we choose is crucial for maintaining constant transmission rate.

Key-Generation. Let $\mathcal{C} =_{\text{df}} \{\omega_1, \ldots, \omega_n\}$ be a $\langle n, v \rangle_2$-collusion-secure code over the alphabet $\{0, 1\}$ with v-long codewords, that allows collusions of up to c and has a tracing algorithm that succeeds with probability $1-\epsilon$; collusion secure codes were introduced in [BS95], and further investigated in [SSW00,SW01a,SW01b]. The key-generation procedure, first generates v independent key-instantiations of a 2-user,1-copyrighted scheme:

$$\{\langle pk_i, \kappa_{0,i}, \kappa_{1,i}, E_i, D_i \rangle\}_{i=1}^{v}$$

Without loss of generality we assume that the plaintext-space $\mathbb{P}$ over all instantiations is the same $(= \{0,1\}^b)$ and that $\text{len}[c \in \mathbb{C}_1] = \ldots = \text{len}[c \in \mathbb{C}_1]$. The i-th decryption key of the n-key system is defined as the following sequence $\boldsymbol{\kappa_i} =_{\text{df}} \langle \kappa_{i,\omega_{i,1}}, \ldots, \kappa_{i,\omega_{i,v}} \rangle$ where $\omega_{i,\ell}$ is the ℓ-th bit of the i-th codeword of $\mathcal{C}$. The tuple $\langle pk_1, \ldots, pk_v \rangle$ constitutes the public-key.

Encryption and Decryption. The plaintext space of the n-key system is $\mathbb{P}^v$. A message $\langle M_1, \ldots, M_v \rangle$ is encrypted by the tuple $\langle E_1(pk_1, M_1), \ldots, E_v(pk_v, M_v) \rangle$. Because each user has one key that inverts $E_i(pk_i, \cdot)$ (either $\kappa_{0,i}$ or $\kappa_{1,i}$) for all $i = 1, \ldots, v$ it is possible for any user to invert a ciphertext and compute $\langle M_1, \ldots, M_v \rangle$.

Security Against Passive Adversaries. Suppose $\langle \kappa_1^*, \ldots, \kappa_v^* \rangle$ is a key that was constructed by a coalition of t users s.t. $t \leq c$. Subsequently the tracer constructs a codeword $\omega^* =_{\text{df}} \omega_1^* || \ldots || \omega_v^*$ as follows

$$\omega_i^* =_{\text{df}} 0 \text{ (if } \kappa_i^* = \kappa_{0,i}) \text{ OR } \omega_i^* =_{\text{df}} 1 \text{ (if } \kappa_i^* = \kappa_{1,i}) \text{ OR } \omega_i^* =_{\text{df}} ? \text{ (otherwise)}$$

Because of the fact that each key-instantiation is 1-copyrighted against passive adversaries in the strong sense, if $C =_{\mathrm{df}} \{\omega_{i_1}, \ldots, \omega_{i_t}\}$ is the set of codewords that corresponds to the keys of the coalition of traitor users that constructed $\langle \kappa_1^*, \ldots, \kappa_v^* \rangle$, it holds that $\omega^* \in F(C)$, where $F(C)$ is the *feasible* set of the codewords C (see [BS95]); it follows that if ω^* is given as input to the tracing algorithm of the collusion-secure-code $\mathcal{C}$, and because $|C| \leq c$, we are guaranteed to obtain the identity of one of the traitors with probability $1 - \epsilon$. Note that we assume that a key for all v instantiations is necessary, i.e. partial decryptions of a ciphertext are not useful. We deal with how this can be enforced in more details in section 6 where we describe the two public-key traitor tracing schemes based on this construction.

Security against Active Adversaries. If the underlying 2-key,1-copyrighted public-key encryption scheme is secure against active adversaries then the tracer can construct the codeword ω^* using merely black-box access to the pirate decoder: the tracer constructs a "randomized" ciphertext $\langle C_1, \ldots, C_v \rangle$ where C_i is constructed as dictated by the analyzer procedure $\mathcal{N}$ in the i-th instantiation of the 1-copyrighted public-key scheme. The value ω_i^* is set to be the output of the analyzer for the i-th coordinate (recall that the output of $\mathcal{N}$ is in $\{0, 1, ?\}$). Note that black-box traitor tracing is achieved with merely a single query to the pirate-decoder (plus the time needed for the collusion-secure code's tracer algorithm).

Theorem 4. *Given v-instantiations of a 2-key,1-copyrighted public-key encryption scheme secure against passive (resp. active) adversaries and a $\langle n, v \rangle_2$-collusion secure code secure that allows collusions of up to c, the scheme described above is a n-key,c-copyrighted public-key encryption scheme, and as a result due to Lemma 1 an n-user,c-TTS (resp. n-user,c-TTS of black-box traceability), can be directly obtained.*

Efficiency Parameters. It is easy to see that the derived scheme has the same ciphertext rate, user-key rate and public-key rate as the underlying 2-key,1-copyrighted public-key encryption scheme. This is because the v-fold expansion of these parameters is "cancelled" by the simultaneous v-fold expansion of the plaintext-space.

We remark that the methodology we describe in this section can be used to yield traitor tracing schemes over any type of 2-user 1-copyrighted encryption function (not necessarily public-key).

6 The New Public-Key Traitor Tracing Schemes

The application of the construction of the previous section to the 2-key,1-copyrighted schemes of sections 4.1 and 4.2 together with Lemma 1 yields two public-key traitor tracing schemes. We summarize these results in this self-contained section in the context of traitor tracing. In the following let $\mathcal{C} = \{\omega_1, \ldots, \omega_n\}$ be a collusion secure $\langle n, v \rangle_2$-code over $\{0, 1\}$ with tracing success probability $1 - \epsilon$ against collusions of up to c users.

For convenience we will describe our schemes under the following plausible "threshold" assumption (introduced in [NP98]). The assumption is applicable to many plaintext-space settings. However, by employing All-or-Nothing Transform this assumption is not necessary as illustrated in section 6.3.

Definition 9. Threshold Assumption. *A pirate-decoder that always returns correctly a percentage C of a plaintext of length b where $1-C$ is a non-negligible function in b, is useless.*

6.1 Traitor Tracing Scheme 1

In the following ℓ is interpreted as a value in $\{1,\ldots,v\}$.
Key Generation. The authority selects $N_1,\ldots,N_v$ composites so that $N_\ell = p_\ell q_\ell$ and $p_\ell = 2p'_\ell+1$, $q_\ell = 2q'_\ell+1$ with $p_\ell, p'_\ell, q_\ell, q'_\ell$ all prime. Without loss of generality we assume that $\nu =_{\rm df} |N_1| = \ldots = |N_v|$. The public-key of the system is the set to

$$\langle N_1, g_1, y_1 =_{\rm df} g_1^{\alpha_1} \bmod N_1\rangle, \ldots, \langle N_v, g_v, y_v =_{\rm df} g_v^{\alpha_v} \bmod N_v\rangle$$

where each $\langle g_\ell\rangle = \mathbb{Q}_{N_\ell}$ and $\alpha_\ell \in_U [p'_\ell q'_\ell]$. User i is given as its personal decryption key the tuple $\langle \kappa_{1,\omega_{i,1}},\ldots,\kappa_{v,\omega_{i,v}}\rangle$, where $\kappa_{\ell,x} = \alpha_\ell + x\phi(N_\ell)$ for $x \in \{0,1\}$.
Encryption. Any third party can encrypt a message $\langle M_1,\ldots,M_v\rangle \in \mathbb{Q}_{N_1}\times\ldots\times \mathbb{Q}_{N_v}$ in the following way: $\langle g_1^{r_1} \bmod N_1, y_1^{r_1}\cdot M_1 \bmod N_1,\ldots, g_v^{r_v} \bmod N_v, y_v^{r_v}\cdot M_v \bmod N_v\rangle$ where $r_\ell \in_U [N_\ell]$.
Decryption. Given a ciphertext $\langle A_1,B_1,\ldots,A_v,B_v\rangle$ and a user-key $\langle\kappa_1,\ldots,\kappa_v\rangle$ the decryption is $\langle B_1(A_1^{-1})^{\kappa_1} \bmod N_1,\ldots,B_v(A_v^{-1})^{\kappa_v} \bmod N_v\rangle$.
Traitor Tracing. Suppose that a key $\langle\kappa_1^*,\ldots,\kappa_v^*\rangle$ is constructed by a coalition of $t \le c$ traitors. If all t traitors have a the same key $\kappa_{\ell,x}$ for some $\ell\in\{1,\ldots,v\}$ then because of the fact that the underlying scheme is 1-copyrighted (theorem 1) it is infeasible for them to construct another key to be used for decryption in the ℓ-th coordinate. As a result they have to set $\kappa_\ell^* =_{\rm df} \kappa_{\ell,x}$ (because of the Threshold Assumption: if $\kappa_{\ell,b}$ is missing from the set of keys available to the pirate decoder then it will fail to decrypt a substantial portion of the plaintext). On the other hand if the t traitors have both keys of the ℓ-th coordinate then they may set κ_ℓ^* to either one, or as described in section 4.1 set κ_ℓ^* to some randomized combination of their keys. Now the tracer computes the string $\omega^* =_{\rm df} \omega_1^*\ldots\omega_v^*$, in the following way: $\omega_\ell^* =_{\rm df} x$ if $\kappa_\ell^* = \kappa_{\ell,x}$ where $x\in\{0,1\}$, or $\omega_\ell^* =_{\rm df} ?$ if $\kappa_\ell^* \notin \{\kappa_{\ell,0},\kappa_{\ell,1}\}$. If $C =_{\rm df} \{\omega_{i_1},\ldots,\omega_{i_t}\}$ is the set of codewords that correspond to the traitor keys it is easy to verify that $\omega^*\in F(C)$ (where $F(C)$ is the feasible set of the set of codewords C). The tracer runs the tracing procedure of $\mathcal{C}$ on input ω^*. This will yield with probability $1-\epsilon$ one of the traitors.

The efficiency parameters of the scheme are presented in figure 4.

6.2 Traitor Tracing Scheme 2

Key Generation. The authority selects $p_1,\ldots,p_v$ primes so that $p_\ell = 2q_\ell+1$ with q_ℓ also prime. Without loss of generality we assume that $\nu =_{\rm df} |p_1| = \ldots =$

	Plaintext Space	Ciphertext Expansion Factor	User-Key Expansion Factor	Public-Key Expansion Factor	Max Traceable Collusion with $(1-\epsilon)$-success
TTS 1	$\{0,1\}^{bv}$	$\frac{2v(b+3)}{bv} \sim 2$	$\frac{v(b+4)}{bv} \sim 1$	$\frac{3v(b+3)}{bv} \sim 3$	$\Omega(\sqrt[4]{\frac{v}{\log(n/\epsilon)\log(1/\epsilon)}})$
TTS 2	$\{0,1\}^{bv}$	$\frac{3v(b+2)}{bv} \sim 3$	$\frac{2v(b+1)}{bv} \sim 2$	$\frac{4v(b+2)}{bv} \sim 4$	$\Omega(\sqrt[4]{\frac{v}{\log(n/\epsilon)\log(1/\epsilon)}})$

Fig. 4. Efficiency Parameters of the two Traitor Tracing Schemes, over a $\langle n, v\rangle_2$-collusion secure code of codeword length $v = \mathcal{O}(t^4 \log(n/\epsilon)\log(1/\epsilon))$, where ϵ denotes the error probability of the tracer and t the maximum traitor collusion size ([BS95]). In order to simplify the table we can select $b = v = l^{1/2}$, where l is a security parameter. Note that in order to allow tracing with negligible in n probability of failure the security parameter l (size of plaintexts) should be polylogarithmic in the number of users. This is a plausible condition, satisfied in many settings.

$|p_v|$. The public-key of the system is the set to $\langle p_1, f_1, g_1, h_1\rangle, \ldots, \langle p_v, f_v, g_v, h_v\rangle$ where f_ℓ, g_ℓ, h_ℓ are generators of the q_ℓ-order subgroup $\mathcal{G}_\ell$ of $\mathbb{Z}^*_{p_\ell}$, with known relative discrete-logs for the authority.

Let $\boldsymbol{d}_{\ell,0}$ and $\boldsymbol{d}_{\ell,1}$ be two random, linearly independent representations of f_ℓ w.r.t. g_ℓ, h_ℓ. User i is given as the decryption key the tuple $\langle \boldsymbol{d}_{1,\omega_{i,1}}, \ldots, \boldsymbol{d}_{v,\omega_{i,v}}\rangle$,
Encryption. Any third party can encrypt a message $\langle M_1, \ldots, M_v\rangle \in \mathcal{G}_1 \times \ldots \times \mathcal{G}_v$ in the following way: $\langle M_1 \cdot f_1^{r_1} \bmod p_1, g_1^{r_1} \bmod p_1, h_1^{r_1} \bmod p_1, \ldots, M_v \cdot f_v^{r_v} \bmod p_v, g_v^{r_v} \bmod p_v, h_v^{r_v} \bmod p_v\rangle$ where $r_\ell \in_U [q_\ell]$.
Decryption. Given a ciphertext $\langle A_1, B_1, C_1, \ldots, A_v, B_v, C_v\rangle$ and a user-key $\langle \boldsymbol{d}_1, \ldots, \boldsymbol{d}_v\rangle$ the decryption is computed as follows $\langle A_1 \langle B_1^{-1}, C_1^{-1}\rangle^{\boldsymbol{d}_1} \bmod p_1, \ldots, A_v\langle B_v^{-1}, C_v^{-1}\rangle^{\boldsymbol{d}_v} \bmod p_v\rangle$, where $\langle a, b\rangle^{\langle c,d\rangle} =_{\mathrm{df}} a^c b^d$.
Black-Box Traitor Tracing. Let $\mathcal{S}$ be a pirate-decoder. The tracer prepares the vectors $\langle x_\ell, y_\ell\rangle$ so that they satisfy the system of equations $\langle x, y \log_{g_\ell} h_\ell\rangle \cdot \boldsymbol{d}_{\ell,0} = a_\ell$ and $\langle x, y \log_{g_\ell} h_\ell\rangle \cdot \boldsymbol{d}_{\ell,1} = b_\ell$ where a_ℓ, b_ℓ are random values of $[q_\ell]$. Subsequently it forms the ciphertext $\langle A_1, B_1, C_1, \ldots, A_v, B_v, C_v\rangle$ with A_ℓ chosen randomly over $\mathbb{Z}^*_p$ and $B_\ell =_{\mathrm{df}} g_\ell^{x_\ell} \bmod p_\ell$, $C_\ell =_{\mathrm{df}} h_\ell^{y_\ell} \bmod p_\ell$. The tracer submits this ciphertext to the tracer and observes the decoder's reply $\langle r_1, \ldots, r_v\rangle$. Then it constructs a codeword $\omega^* = \omega_1^* \ldots \omega_v^*$: as follows: If $r_\ell = A_\ell / g_\ell^{a_\ell} (\mathrm{mod} p_\ell)$ then ω_ℓ^* is set to 0; else if $r_\ell = A_\ell / g_\ell^{b_\ell} (\bmod p_\ell)$ then ω_ℓ^* is set to 1. Finally if $r_\ell \notin \{A_\ell / g_i^{a_\ell} \bmod p_\ell, A_\ell / g_\ell^{b_\ell} \bmod p_\ell\}$, ω_ℓ^* is set to ?. It follows from theorem 3 and the threshold assumption that ω^* belongs to the $F(C)$ where $C = \{\omega_{i_1}, \ldots, \omega_{i_t}\}$ is the set of codewords that correspond to the secret-keys assigned to the traitors. Now provided that $t \leq c$ the tracer can recover the identity of one of the traitors by using the tracing algorithm of the code $\mathcal{C}$ with probability $1 - \epsilon$. We note here that a *single* query to the pirate decoder is sufficient for our black-box traitor tracing method.

The efficiency parameters of the scheme are presented in figure 4.

6.3 Obviating the Threshold Assumption

The Threshold assumption was instrumental in the traitor tracing methods of our two schemes since it made it necessary for the pirate decoder to include a key for each of the v components. Nevertheless this can also be enforced by employing an all-or-nothing transform (AONT) [Riv97] (alternatively one can use collusion secure codes under weaker marking assumptions, e.g. [SW01a]).

The two public-key traitor tracing schemes described in sections 6.1 and 6.2 have as plaintext space set of strings $\{0,1\}^{bv}$. Let us formulate the construction of [Riv97] in our setting:

All-or-Nothing Transform. ([Riv97]) Let f be a block-cipher and let K_0 be a publicly known key for f. Given a message $\langle m_1, \ldots, m_{v-1} \rangle \in \{0,1\}^{b(v-1)}$ the sender selects a random key K for f and computes $m'_1, \ldots, m'_v \in \{0,1\}^b$ as follows: $m'_i =_{\text{df}} m_i \oplus f(K, i)$ for $i = 1, \ldots, v-1$; note that $\oplus$ stands for the xor operation. The last block m'_v is computed as $m'_v =_{\text{df}} K \oplus h_1 \oplus \ldots \oplus h_{v-1}$, with $h_i =_{\text{df}} f(K_0, m'_i \oplus i)$ for $i = 1, \ldots, v-1$. The output of the transform is the bitstring $m'_1 \ldots m'_v \in \{0,1\}^{bv}$. It is easy to see that the transform can be inverted by anyone that holds *all* blocks $m'_1, \ldots, m'_v$ as follows: $K = m'_v \oplus f(K_0, m'_1 \oplus 1) \oplus \ldots \oplus f(K_0, m'_{v-1} \oplus (v-1))$ and $m_i = m'_i \oplus f(K, i)$ for $i = 1, \ldots, v-1$.

The concept of AONT has been investigated formally in [CDHKS00]. By employing the above AONT in the encryption and decryption operation of our public-key traitor tracing schemes we enforce the pirate to include a secret-key for each one of the v components and therefore there is no need for the Threshold Assumption. The efficiency loss introduced by the use of the AONT is marginal, and it does not affect the stated expansion factors. Finally, note that another plaintext preprocessing which is possible without much ciphertext expansion is employing one of the preprocessing methods (based on random oracle hash and added randomness) in the case where chosen ciphertext security is required.

References

[Bon98] Dan Boneh, The Decision Diffie-Hellman Problem, In Proceedings of the Third Algorithmic Number Theory Symposium, Lecture Notes in Computer Science, Vol. 1423, Springer-Verlag, pp. 48–63, 1998.

[BF99] Dan Boneh and Matthew Franklin, An Efficient Public-Key Traitor Tracing Scheme, CRYPTO 1999.

[BS95] Dan Boneh and James Shaw, Collusion-Secure Fingerprinting for Digital Data (Extended Abstract), CRYPTO 1995, Springer, pp. 452-465.

[CDHKS00] Ran Canetti, Yevgeniy Dodis, Shai Halevi, Eyal Kushilevitz, and Amit Sahai, Exposure-Resilient Functions and All-or-Nothing Transforms, EUROCRYPT 2000.

[CG98] Dario Catalano, and Rosario Gennaro, New Efficient and Secure Protocols for Verifiable Signature Sharing and Other Applications, CRYPTO 1998.

[CFN94] Benny Chor, Amos Fiat, and Moni Naor, Tracing Traitors, CRYPTO 1994.

[CFNP00] Benny Chor, Amos Fiat, Moni Naor, and Benny Pinkas, Tracing Traitors, IEEE Transactions on Information Theory, Vol. 46, 3, 893-910, 2000.

[ElG84] Taher El Gamal, A Public Key Cryptosystem and a Signature Scheme Based on Discrete Logarithms, CRYPTO 1984.

[FT99] Amos Fiat and T. Tassa, Dynamic Traitor Tracing, CRYPTO 1999.

[FH96] Matthew K. Franklin and Stuart Haber, Joint Encryption and Message-Efficient Secure Computation, Journal of Cryptology 9(4), pp. 217-232, 1996.

[GSY99] Eli Gafni, Jessica Staddon and Yiqun Lisa Yin, Efficient Methods for Integrating Traceability and Broadcast Encryption, CRYPTO 1999.

[GM84] Shafi Goldwasser and Silvio Micali, Probabilistic Encryption, JCSS 28(2): pp. 270-299, 1984.

[KY01a] Aggelos Kiayias and Moti Yung, Self Protecting Pirates and Black-Box Traitor Tracing, CRYPTO 2001.

[KY01b] Aggelos Kiayias and Moti Yung, On Crafty Pirates and Foxy Tracers, Proceedings of the 1st Workshop on Security and Privacy in Digital Rights Management, 2001.

[KD98] K. Kurosawa and Y. Desmedt, Optimum Traitor Tracing and Asymmetric Schemes, Eurocrypt 1998.

[Mil76] G. Miller, Riemann's Hypothesis and Tests for Primality, Journal of Computer and System Sciences, vol. 13, 300–317, 1976.

[NSS99] David Naccache, Adi Shamir, and Julien P. Stern, How to Copyright a Function?, In the Proceedings of Public Key Cryptography 1999, Springer, 188–196.

[NNL01] Dalit Naor, Moni Naor, and Jeffrey B. Lotspiech Revocation and Tracing Schemes for Stateless Receivers, CRYPTO 2001.

[NP98] Moni Naor and Benny Pinkas, Threshold Traitor Tracing, CRYPTO 1998.

[NP00] Moni Naor and Benny Pinkas, Efficient Trace and Revoke Schemes, In the Proceedings of Financial Crypto '2000, Anguilla, February 2000.

[NR97] Moni Naor and Omer Reingold, Number-Theoretic Constructions of Efficient Pseudo-Random Functions, FOCS 1997.

[Pfi96] Birgit Pfitzmann, Trials of Traced Traitors, Information Hiding Workshop, Spring LNCS 1174, pp. 49-63, 1996.

[Riv97] Ron Rivest, All-or-nothing Encryption and the Package Transform, Fast Software Encryption 1997.

[SW00] Reihaneh Safavi-Naini and Yejing Wang, Sequential Traitor Tracing, CRYPTO 2000.

[SW01a] Reihaneh Safavi-Naini and Yejing Wang, Collusion Secure q-ary Fingerprinting for Perceptual Content, Proceedings of the 1st Workshop on Security and Privacy in Digital Rights Management, 2001.

[SW01b] Reihaneh Safavi-Naini and Yejing Wang, New Results on Frameproof Codes and Traceability Schemes, IEEE Transactions on Information Theory, Vol. 47, No. 7, pp. 3029-3033, 2001.

[SSW00] Jessica N. Staddon, Douglas R. Stinson and Ruizhong Wei, Combinatorial Properties of Frameproof and Traceability Codes, Cryptology ePrint Archive, report 2000/004.

[SW98] Douglas R. Stinson and Ruizhong Wei, Combinatorial Properties and Constructions of Traceability Schemes and Frameproof Codes, SIAM J. on Discrete Math, Vol. 11, no. 1, 1998.

[TY99] Yiannis Tsiounis and Moti Yung, On the Security of ElGamal Based Encryption, Public Key Cryptography 1998.

Toward Hierarchical Identity-Based Encryption

Jeremy Horwitz and Ben Lynn

Stanford University, Stanford, CA 94305, USA,
{horwitz|blynn}@cs.stanford.edu

Abstract. We introduce the concept of hierarchical identity-based encryption (HIBE) schemes, give precise definitions of their security and mention some applications. A two-level HIBE (2-HIBE) scheme consists of a root private key generator (PKG), domain PKGs and users, all of which are associated with primitive IDs (PIDs) that are arbitrary strings. A user's public key consists of their PID and their domain's PID (in whole called an address). In a regular IBE (which corresponds to a 1-HIBE) scheme, there is only one PKG that distributes private keys to each user (whose public keys are their PID). In a 2-HIBE, users retrieve their private key from their domain PKG. Domain PKGs can compute the private key of any user in their domain, provided they have previously requested their domain secret key from the root PKG (who possesses a master secret). We can go beyond two levels by adding subdomains, subsubdomains, and so on. We present a two-level system with total collusion resistance at the upper (domain) level and partial collusion resistance at the lower (user) level, which has chosen-ciphertext security in the random-oracle model.

1 Introduction

Shamir asked for an identity-based encryption (IBE) cryptosystem in 1984 [9], but a fully-functional IBE scheme was not found until recent work by Boneh and Franklin [1] and Cocks [4]. Recall that an IBE scheme is a public-key cryptosystem where any arbitrary string is a valid public key. The corresponding private keys must be computed by a trusted third party called the private key generator (PKG) (who possesses a master secret). Users of the system request their private key from the PKG.

We note that the public key infrastructure associated with standard public-key cryptosystems also includes a trusted third party (in the form of a root certificate authority) and allows a hierarchy of certificate authorities [12]: the root certificate authority can issue certificates for other certificate authorities, who in turn can issue certificates for users in their respective domains.

The original system of Boneh and Franklin does not allow for such structure. However, a hierarchy of PKGs is desirable in an IBE system, as it greatly reduces the workload on master server(s) and allows key escrow at several levels. For instance, if the users of the system are employees of corporations, then it is natural to want each corporation to be able to generate the private keys for their employees, so that employees request their keys from their corporation,

L.R. Knudsen (Ed.): EUROCRYPT 2002, LNCS 2332, pp. 466–481, 2002.

rather than the top-level PKG. Only corporations request their domain secret (and only once per corporation) from the top-level PKG. This is the idea behind a hierarchical IBE (HIBE) system. In particular, this is an example of a two-level HIBE (2-HIBE) scheme. (The advantage of an HIBE system over standard PKI is that senders can derive the recipient's public key from their address without an online lookup.)

More precisely, there are three types of entities in a 2-HIBE scheme. There is the root PKG, who possesses a master key. In the upper level, there are domain PKGs, who can request their domain key from the root PKG. Lastly, there are users, who can request private keys from their domain PKG. Each user and each domain has a primitive ID (PID), which is an arbitrary string. (If Alice works for Company.com and her email address is `alice@company.com`, her PID is `alice` and her company's PID is `company.com`.) The public key of a user consists of a tuple of PIDs: the PID of the user and the PID of the user's domain (this public key is also called the user's address) and, as with IBE systems, it is clear that a sender can derive the receiver's public key offline. We can generalize to HIBE schemes with more levels by allowing subdomains, subsubdomains, and so on.

Another application for HIBE systems is generating short-lived keys for portable computing devices. Suppose Alice is planning to embark on a week-long business trip and wants to be able read her encrypted mail while on the road. However, she is also worried that her laptop may be stolen or otherwise compromised, so she does not want to simply copy her private key to the laptop. This dilemma is readily solved with a 2-HIBE system: this time, the upper level consists of people, such as Alice, and the lower level consists of dates, and when an arbitrary user Bob wants to send a message to Alice he uses the tuple of Alice's PID and the PID for the current date as her address. Alice can generate (for example) seven days' worth of keys (from her private key that she has previously requested from a PKG) and transfer these to her laptop. Now if the laptop is compromised, the damage is limited. We note that collusion at the bottom level is not an issue, as Alice will only put a small number of keys on her laptop. This problem can also be solved with a standard (non-hierarchical) IBE scheme by having Alice run her own IBE system [1], but in this case Bob must get Alice's system parameters before he can communicate with her.

In this paper, we give formal security definitions that can model plausible real-life attack scenarios on HIBE systems. In addition to chosen-ciphertext attacks, we must also worry about attacks involving collusion by entities on arbitrary levels. In our example above, for instance, if the domain PKG of one corporation A colludes with employees of another corporation B, they should not be able to decrypt messages of other employees of corporation B (or of any other corporation C, for that matter). In general, an adversary should not be able to decrypt a message encrypted for a particular user in a particular domain (and subdomain, subsubdomain, etc.), even if they have access to the private key of every other user and of every other domain (and subdomain, subsubdomain, etc.), in addition to information obtained from a decryption oracle.

We present a 2-HIBE scheme with total collusion resistance at the upper level and partial collusion resistance at the lower level. (This limitation does not affect its applicability to the above laptop example.) In terms of the corporate setting, even if an arbitrary number of corporations collude, the master secret is safe, but, at the lower level, if more than certain number of employees of a corporation C collude, they can expose C's private key.

Our system requires a bilinear map with certain properties. A suitable map can constructed from the Weil pairing (which is described in [1]). Its performance is sufficiently fast for practical purposes, provided the number of colluding parties allowed in the lower level is not too large. (Its running time and key size involve a term linear in this number.) Additionally, we can employ the same techniques used with the Boneh-Franklin IBE scheme to split secrets across several servers and achieve robustness for free.

2 Definitions

An identity-based encryption scheme (IBE) is specified by four randomized algorithms: Setup, KeyGen (called Extract in [1]), Encrypt, and Decrypt. In brief, Setup generates system parameters that are publicly released and a master key that is given to the PKG only; KeyGen is run by the PKG to generate private keys corresponding to a given primitive ID (PID); Encrypt encrypts a message using a given PID (PIDs are public keys); and Decrypt decrypts a ciphertext given a private key. We shall always take the message space to be $\mathcal{M} = \{0,1\}^m$.

These algorithms must satisfy the standard consistency constraint, namely, when d is the private key generated by algorithm KeyGen when it is given the PID A as the public key, then

$$\forall M \in \mathcal{M} : \mathsf{Decrypt}(\mathsf{params}, A, C, d) = M \ ,$$

where $C = \mathsf{Encrypt}(\mathsf{params}, A, M)$.

An ℓ-HIBE has a family of ℓ key-generation algorithms (KeyGen_i for $1 \le i \le \ell$) instead of just one, and public keys are now ℓ-tuples of PIDs instead of just a single PID.

Definition 1. *A* primitive ID (PID) *is an arbitrary string, i.e., an element of* $\{0,1\}^*$.

Definition 2. *An* address *is an ℓ-tuple of* PID*s.*

An address fully specifies a user's public key.

Definition 3. *A* prefix address (or prefix) *is an i-tuple of* PID*s for some $0 \le i \le \ell$. A prefix address $\langle S_1, \ldots, S_i \rangle$ is said to be a* prefix *of the prefix address $\langle T_1, \ldots, T_j \rangle$ if $i \le j$ and $S_a = T_a$ for $1 \le a \le i$.*

Notice that addresses also happen to be prefix addresses.

Definition 4. *For a non-negative integer ℓ, an* ℓ-level hierarchical identity-based encryption scheme (ℓ-HIBE) *is specified by $\ell + 3$ randomized algorithms:* Setup, KeyGen_i *(for $1 \le i \le \ell$),* Encrypt, *and* Decrypt*:*

Setup: *Input: security parameter $k \in \mathbb{Z}$. Output: system parameters* params *and a master key* mk_ϵ *(which we also call the level-0 key).*
KeyGen_i: *(for $1 \leq i \leq \ell$): Input:* params, $\mathsf{mk}_{\langle S_1,\ldots,S_{i-1}\rangle}$ *(a level-$(i-1)$ key), and an i-tuple of* PIDs *(a prefix address). Output:* $\mathsf{mk}_{\langle S_1,\ldots,S_i\rangle}$ *(a level-i key).*
$\mathsf{Encrypt}$: *Input:* params, *an address, and a message. Output: a ciphertext.*
$\mathsf{Decrypt}$: *Input:* params, *an address, a ciphertext, and a private key* $\mathsf{mk}_{\langle S_1,\ldots,S_\ell\rangle}$. *Output: the corresponding plaintext.*

These algorithms must satisfy the standard consistency constraint, namely, when d is the private key generated by algorithm KeyGen_ℓ *when it is given the address N as the public key, then*

$$\forall M \in \mathcal{M} : \mathsf{Decrypt}(\mathsf{params}, N, C, \mathsf{mk}_{\langle S_1,\ldots,S_\ell\rangle}) = M \ ,$$

where $C = \mathsf{Encrypt}(\mathsf{params}, N, M)$.

Remark 1. For certain values of ℓ, an HIBE is the same as other familiar structures.

- When $\ell = 0$, this definition captures the essence of public-key encryption schemes: the level-0 key corresponds to a private key and the params correspond to the public key (the address is empty when calling $\mathsf{Encrypt}$; each system is associated with only one private key/public key pair).
- When $\ell = 1$, we have a definition of a standard IBE.

2.1 Security

In order to cover realistic attacks, we assume that an attacker may be able obtain private keys at any level except for the master secret, and extend the standard model of chosen-ciphertext security accordingly. We note that if the master secret is compromised, the effects are at least as disastrous as when the root certificate authority is compromised in a public-key cryptosystem. Thus we assume that the precautions taken to guard the master secret are similar to those taken to guard a root certificate authority in real life (e.g., secret splitting, tamper-resistant hardware), rendering it unassailable. Consider the following game played by two parties, an adversary and a challenger:

1. The challenger runs the Setup algorithm (for a given security parameter k) and gives params to the adversary. It does not divulge mk_ϵ.
2. The adversary submits any number of decryption and/or key-generation queries adaptively (i.e., each query may depend on the replies to previous queries). In a decryption query, the adversary sends a ciphertext and an address and is given the corresponding plaintext under the unique key associated with that address (assuming the ciphertext and address are valid). For a key-generation query, the adversary submits any prefix address $\langle S_1, \ldots, S_i\rangle$ (for some $1 \leq i \leq \ell$), and is told the output K_i (where K_j is defined to be $\mathsf{KeyGen}_j(K_{j-1}, \langle S_1, \ldots, S_j\rangle)$ for $0 < j \leq i$ and K_0 is the key returned by Setup).
   In other words, not only can the adversary learn the decryption of any chosen ciphertext, it can also obtain the key corresponding to any prefix address.

3. The adversary then outputs any two plaintexts $M_0, M_1 \in \mathcal{M}$ and any address N on which it wishes to be challenged, subject to the restriction that no prefix of N has been queried in the previous step.
4. The challenger picks $b \in \{0,1\}$ randomly and computes the ciphertext $C = \mathsf{Encrypt}(\mathsf{params}, N, M_b)$. It then sends the challenge C to the adversary.
5. The adversary again issues any number of decryption and/or key-generation queries adaptively, except that it now may not ask for the key corresponding to any prefix of N or for the plaintext corresponding to C under the private key corresponding to N.
6. The adversary outputs $b' \in \{0,1\}$, and wins if $b = b'$.

We call such an adversary an ID-CCA attacker.

Definition 5. *We define an HIBE to be* secure against adaptive chosen-ciphertext attack (ID-CCA) *if no polynomially-bounded adversary has a non-negligible advantage in the above game, that is, for any polynomial f and for any probabilistic polynomial-time algorithm $\mathcal{A}$, $Adv(\mathcal{A}) := \left|\Pr[b = b'] - \frac{1}{2}\right|$ is less than $1/f(k)$. (The probability is over the random bits used by the two parties.)*

Finding even a 2-HIBE that satisfies this security requirement remains an open problem. We describe a 2-HIBE that is secure provided the adversary is limited to n KeyGen_2 queries within its domain (for a given n; unlimited KeyGen_1 queries are allowed). In other words, our system resists arbitrary collusion at the domain level, but resists only limited collusion at the user level.

We will also utilize a weaker notion of security in intermediate steps of our proofs. Consider another game played by two parties, an adversary and a challenger:

1. The challenger runs the Setup algorithm (for a given security parameter k) and gives params to the adversary.
2. The adversary submits some number of key-generation queries adaptively, that is, for each query, the adversary submits any prefix address $\langle S_1, \ldots, S_i \rangle$ (for some $1 \le i \le \ell$), and is told the output K_i (where K_j is defined to be $\mathsf{KeyGen}_j(K_{j-1}, \langle S_1, \ldots, S_j \rangle)$ for $0 < j \le i$ and K_0 is the key returned by Setup).
3. The adversary then outputs any address N on which it wishes to be challenged, subject to the restriction that no prefix of N has been queried in the previous step.
4. The challenger picks a message $M \in \mathcal{M}$ randomly and computes $C = \mathsf{Encrypt}(\mathsf{params}, N, M)$. It sends the challenge C to the adversary.
5. The adversary again issues some number of key-generation queries adaptively, except that it now may not ask for the key corresponding to any prefix of N.
6. The adversary outputs some message $M' \in \mathcal{M}$, and wins if $M = M'$.

We call such an adversary an ID-OWE attacker.

Definition 6. *We define an HIBE to be a* one-way identity-based encryption scheme (ID-OWE) *if no polynomially-bounded adversary has a non-negligible advantage in the above game.*

Both these definitions are generalizations of definitions given by Boneh and Franklin [1].

3 An HIBE Resistant against Domain Collusion

We present a two-level system resistant to collusion at the domain level. The system is based on bilinear forms between two prime-order groups.

3.1 The BDH Assumption

We briefly review definitions given by Boneh and Franklin [1, 2].

Definition 7. *Let G_1, G_2 be groups with prime order q. Then we say a map $e\colon G_1 \times G_1 \to G_2$ is* bilinear *if, for all $g, h \in G_1$ and $a, b \in \mathbb{F}_q$, we have $e(g^a, h^b) = e(g, h)^{ab}$.*

Definition 8. *The* Bilinear-Diffie-Hellman problem *(BDH) for a bilinear function $e\colon G_1 \times G_1 \to G_2$ such that $|G_1| = |G_2| = q$ is prime is defined as follows: given $g, g^a, g^b, g^c \in G_1$, compute $e(g, g)^{abc}$, where g is a generator and a, b, c are randomly chosen from $\mathbb{F}_q$. An algorithm is said to solve the BDH problem with an advantage of ε if*

$$\Pr\left[\mathcal{A}(g, g^a, g^b, g^c) = e(g, g)^{abc}\right] \geq \varepsilon\ .$$

Definition 9. *A randomized algorithm $\mathcal{IG}$ that takes as input a security parameter $k \in \mathbb{Z}$ (in unary) is a* BDH parameter generator *if it runs in time polynomial in k and outputs the description of two groups G_1, G_2 and a bilinear function $e\colon G_1 \times G_1 \to G_2$. We further require that the groups have prime order (which we call q), and denote the output of the algorithm by $(G_1, G_2, e) = \mathcal{IG}(1^k)$.*

Definition 10. *We say that $\mathcal{IG}$* satisfies the BDH assumption *if no probabilistic polynomial-time algorithm $\mathcal{A}$ can solve BDH (for $\mathcal{IG}(1^k)$) with non-negligible advantage.*

For the remainder of the paper we make use of some fixed BDH parameter generator $\mathcal{IG}$ that satisfies the BDH assumption, and use the symbols G_1, G_2, e, q to represent the constituents of its output. Boneh and Franklin [1] also give details on how to implement such a generator (their system also required one), based on the Weil pairing. (In their construction, G_1 is a group of points on a certain elliptic curve and G_2 is a certain subgroup of $\mathbb{F}_{p^2}^{\times}$, for some prime p.)

This assumption was implicitly used by Joux [7] to build a one-round three-party Diffie-Hellman protocol. Other constructions also require the BDH assumption ([8, 10, 11]). Additionally, a bilinear function is needed in a recently described short signature scheme [3].

3.2 A Game Transformation

The BDH assumption is closely tied to the CDH assumption. Recall that the CDH problem asks for g^{ab} given g, g^a, g^b, whereas the goal in the CDH problem is to compute $e(g,g)^{abc} = e(g^{ab}, g^c)$ given g^c in addition to g, g^a, g^b. This similarity between the BDH and CDH assumptions naturally leads to the following transformation on games:

Definition 11. *Using the notation of the previous section, suppose $\mathcal{G}$ is a game where the goal of the adversary is to compute a particular element $g \subset G_1$. Then the* e-transformation of $\mathcal{G}$ *is the same game as $\mathcal{G}$ except now the adversary is also given a random $h \in G_1$ and the adversary's goal is to compute $e(g,h)$.*

We can transform assumptions by applying this transformation to the underlying game. For example, we obtain the BDH assumption (associated with a particular e) when we apply this transformation to the CDH assumption.

It is possible to formulate our assumptions differently: we could have started with assuming that e is a bilinear function such that if a game $\mathcal{G}$ is hard then its e-transformation is also hard. This would simplify our exposition (for example, we need only assume the CDH problem is hard, as that implies that the BDH problem is hard). However, such an assumption is really an abstract description of a class of assumptions, and we prefer the readability gained by relying on a small number of concrete assumptions instead.

Clearly, if an adversary can win a game $\mathcal{G}$, then it can easily win the e-transformation of $\mathcal{G}$. The converse is far from clear.

We shall see that transformed assumptions are required to show that schemes are ID-OWE; without transformation, the assumptions are more natural, but we can only show that an adversary cannot recover a user's private key.

3.3 Linear e-One-Way Functions

We now build up to the definition of a linear e-one-way function, from which one could build an HIBE. We then construct a function that is weaker than e-one-way that will allow for efficiently building a 2-HIBE which is secure against any collusion at the domain level and limited collusion at the user level.

Suppose that we have a function $h\colon G \times X \to G_1$, where G and G_1 are groups, G is of prime order p, X is a set, and $h(g^a, x) = h(g,x)^a$ for all $g \in G$, $x \in X$, $a \in \mathbb{F}_p$.

Definition 12. *The elements $x, x_1, x_2, \ldots, x_n \in X$, $a \in \mathbb{F}_p$, and a generator $g \in G$ are chosen at random. Given x, g, and $\langle x_i, h(g^a, x_i)\rangle$ for $i = 1, 2, \ldots, n$, the problem of computing $h(g^a, x)$ is called the* linear one-way problem *(of size n).*

Definition 13. *We say that h is a* linear one-way function *if no probabilistic polynomial-time (in n and $\log p$) algorithm can solve the linear one-way problem of any size.*

Remark 2. For example, if DDH is hard in $\mathbb{F}_{p^2}^\times$, the Weil pairing is an example of a linear one-way function. More generally, bilinear functions that satisfy BDH give rise to families of linear one-way functions. For example, suppose we have $(G_1, G_2, e) = \mathcal{IG}(1^k)$. Then fix a generator $g \in G_1$ and consider the function $f_g\colon G_1 \to G_2$ defined by $f_g(g_1) := e(g, g_1)$. Now, f_g is one-way, assuming DDH is hard in G_2. To see this, assume that f_g is easy to invert; DDH in G_2 can be solved as follows: given $x, x^a, x^b, x^c \in G_2$ we find their inverses y, y_a, y_b, y_c respectively, and check if $e(y, y_c) = e(y_a, y_b)$. We note that if $\mathcal{IG}$ is constructed as described by Boneh and Franklin, then G_2 is a subgroup of $\mathbb{F}_{p^2}^\times$, a group in which DDH is thought to be hard. (It is also possible to construct elliptic curves where the q-torsion points are contained in $\mathbb{F}_p$ for some large prime q. Inverting the Weil pairing on these curves is equivalent to breaking DDH in $\mathbb{F}_p$.) More generally, this is why the relationship between a game and its e-transform appears to be highly nontrivial: if an algorithm $\mathcal{A}$ could win a game $\mathcal{G}$, given an algorithm $\mathcal{B}$ that wins the e-transform of $\mathcal{G}$, then $\mathcal{A}$ is an algorithm that can invert f_g.

Now suppose that $(G_1, G_2, e) = \mathcal{IG}(1^k)$. Then the e-transformation of the linear one-way problem is called the linear e-one-way problem. (In this problem, we are also given g^r for some random $r \in \mathbb{F}_p$ (in addition to x, g, and $\langle x_i, h(g^a, x_i)\rangle$ for $i = 1, 2, \ldots, n$) and now the goal is to compute $e(h(g^a, x), g^r)$.)

Definition 14. *If no probabilistic polynomial-time algorithm can solve the linear e-one-way problem of any size, then we say that h is a* linear e-one-way function.

If we knew how to construct linear e-one-way functions, we could construct an HIBE scheme as follows:

Setup: Input: $k \in \mathbb{Z}$. Run $\mathcal{IG}(1^k)$ and set (G_1, G_2, e) to be the output. Construct a linear e-one-way function $h\colon G_1 \times \mathbb{F}_q \to G_1$. Choose a random $a \in \mathbb{F}_q$ and a random generator $g \in G_1$. Pick cryptographically-strong hash functions $H_1\colon \{0,1\}^* \to G_1$, $H_2\colon \{0,1\}^* \to \mathbb{F}_q$, and $H_3\colon G_2 \to \{0,1\}^m$ (for some m).
Output: $\mathsf{mk}_\epsilon := a$, and $\mathsf{params} := \langle G_1, G_2, e, g, g^a, H_1, H_2, H_3\rangle$.
KeyGen$_1$: Input: a prefix address $\langle S\rangle$ (the domain name).
Output: $\mathsf{mk}_{\langle S\rangle} := H_1(S)^a \in G_1$.
KeyGen$_2$: Input: an address $\langle S, T\rangle$ (S is the domain PID and T is the user PID).
Let $\mathsf{mk}_{\langle S\rangle} \in G_1$ be the domain key.
Output: $k = \mathsf{mk}_{\langle S,T\rangle} := h(\mathsf{mk}_{\langle S\rangle}, H_2(S \| T)) \in G_2$.
Encrypt: Input: params, $N = \langle S, T\rangle$ (S is the recipient domain's PID and T is the recipient user's PID), and M.
Pick a random $r \in \mathbb{F}_q$.
Output: $C = \langle g^r, M \oplus H_3(s)\rangle$, where $s := e(h(H_1(S), H_2(S \| T)), g)^r$.
Decrypt: Input: params, $N = \langle S, T\rangle$, a ciphertext $C = \langle g^r, V\rangle$, and a user's private key $k := \mathsf{mk}_{\langle S,T\rangle} \in G_1$.
Output: $M = V \oplus H_3(e(k, g^r))$.

It can be shown that this scheme is ID-OWE. By applying the Fujisaki-Okamoto [6] transformation, we obtain a scheme which is ID-CCA. Though finding a linear e-one-way function h remains an open problem, we are able to construct an h such that the linear e-one-way problem for a fixed n is hard, giving rise to a 2-HIBE system that is resistant to (unlimited) domain-level collusion and can tolerate up to n-party user-level collusion. We describe this in the following section. Briefly, we will define h: $G_1^{n+1} \times \mathbb{F}_q \to G_1$ (for some n; q is the prime order of G_1) as $h((g_0, g_1, \ldots, g_n), d) := g_0^{d^0} g_1^{d^1} \cdots g_n^{d^n}$. We then have a linear function h such that, given g and n pairs $\langle x_i, h(g, x_i)\rangle$, it appears hard to determine $\langle x', h(g, x')\rangle$ for any other x'.

3.4 Our Domain-Collusion Resistant Scheme

Let n denote the amount of collusion that we are willing to tolerate at the user level.

Setup: Input: $k \in \mathbb{Z}$. Run $\mathcal{IG}(1^k)$ and set (G_1, G_2, e) to be the output. Choose a random $a \in \mathbb{F}_q$ and a random $g \in G_1$. Pick cryptographically-strong hash functions H_1: $\{0,1\}^* \to G_1^{n+1}$, H_2: $\{0,1\}^* \to \mathbb{F}_q$, and H_3: $G_2 \to \{0,1\}^m$ (where $\mathcal{M} = \{0,1\}^m$ is the message space). For the security proof, we view the hash functions as random oracles.
Output: $\mathsf{mk}_\epsilon := a$ and $\mathsf{params} := \langle G_1, G_2, e, g, g^a, H_1, H_2, H_3\rangle$.

KeyGen$_1$: Input: a prefix address $\langle S\rangle$ (the domain name). Let $\langle g_0, g_1, \ldots, g_n\rangle = H_1(S)$ (so each g_i lies in G_1).
Output: $\mathsf{mk}_{\langle S\rangle} := \langle g_0^a, g_1^a, \ldots, g_n^a\rangle \in G_1^{n+1}$.

KeyGen$_2$: Input: an address $\langle S, T\rangle$ (S is the domain PID and T is the user PID).
Set $d := H_2(S \| T)$ (which is an element of $\mathbb{F}_q$).
Let $\mathsf{mk}_{\langle S\rangle} = \langle g_0^a, g_1^a, \ldots, g_n^a\rangle \in G_1^{n+1}$ be the domain key.
Output: $k = \mathsf{mk}_{\langle S,T\rangle} := \prod_{i=0}^n g_i^{ad^i} \in G_1$.

Encrypt: Input: params, $N = \langle S, T\rangle$ (S is the recipient domain's PID and T is the recipient user's PID), and a message M.
Set $\langle g_0, g_1, \ldots, g_n\rangle := H_1(S)$. Set $d := H_2(S \| T)$.
Pick a random $r \in \mathbb{F}_q$. Then compute $w := e\left(\prod_{i=0}^n g_i^{d^i}, g^a\right)^r \in G_2$.
Output: the ciphertext $\langle g^r, M \oplus H_3(w)\rangle$.

Decrypt: Input: params, $N = \langle S, T\rangle$, a ciphertext $C = \langle U, V\rangle$, and a private key $k = \mathsf{mk}_{\langle S,T\rangle} \in G_1$.
Output: $M = V \oplus H_3(e(k, U))$.

The scheme is consistent because, by the bilinearity of e, we have $e(k, U) = e\left(\prod_{i=0}^n g_i^{d^i}, g^a\right)^r$, when $U = g^r$.

3.5 Proof of Security

Recall that we are restricting the adversary to at most n KeyGen$_2$ queries from the same domain.

Theorem 1. *Suppose $\mathcal{A}$ is an ID-OWE attacker of our cryptosystem with an advantage of ε. Then, if we model H_1, H_2, and H_3 as random oracles, there exists an algorithm $\mathcal{B}$ that can solve the BDH problem with an advantage of $\varepsilon / \left(2(Q_{K_1} + 2Q_{K_2}) Q_{H_1} \binom{Q_{H_2}}{n} \mathbf{e} \right)$, where Q_{K_i} is the total number of* KeyGen$_i$ *queries, Q_{H_i} is the number of H_i queries issued by $\mathcal{A}$, and $\mathbf{e}$ is the base of the natural logarithm.*

Proof. The proof of the theorem is broken into several lemmata. In Lemma 1, we show that an attacker $\mathcal{B}$, whose KeyGen queries are restricted to only KeyGen$_2$ queries from the same domain as the challenge address, is essentially as strong as an arbitrary attacker $\mathcal{A}$. We do so in a manner similar to that used in the analysis of the Boneh-Franklin scheme [1], which is itself partly based on a technique of Coron [5]. In Lemma 2, we define the Bilinear Polynomial Diffie-Hellman (BPDH) game, and give a reduction from the attack by the $\mathcal{B}$ described above to an attack by (an attacker) $\mathcal{C}$ on the BPDH game. Lastly we produce a reduction from an attack by $\mathcal{C}$ on the BPDH game to an attack by $\mathcal{D}$ on the BDH problem. The combination of the three lemmata leads immediately to the theorem. □

Lemma 1. *Suppose there exists an ID-OWE attacker $\mathcal{A}$ with an advantage of ε. Let Q_i be a bound on the number of H_i queries made by $\mathcal{A}$ (for $i = 1, 2$). If we model H_1 as a random oracle, then there exists an ID-OWE attacker $\mathcal{B}$ with an advantage of $\varepsilon / \mathbf{e}Q$, where $Q = Q_1 + 2Q_2$, whose key-generation queries are all* KeyGen$_2$ *queries from the same domain as the challenge domain.*

Proof. After receiving the system parameters, $\mathcal{B}$ passes them on to $\mathcal{A}$. Without loss of generality, we may assume that every key-generation query for a prefix address (domain name) $\langle S \rangle$ or address $\langle S, T \rangle$ has been preceded by an H_1 (domain-level) hash query on the domain PID S. We may also assume that $\mathcal{A}$ issues an H_1 hash query on the challenge domain PID before revealing it.

We will need some auxiliary functions and global variables:

Initially, L is an empty list that will hold information on $\mathcal{B}$'s responses to H_1 queries, and $s_{\text{CHALLENGE}}$ is a string that is set to a special value NULL. Additionally, we will use a unique value REAL (not in G, $\mathbb{F}_q$, etc.) in the proof.

When $\mathcal{A}$ issues an H_2 query for an address $\langle S, T \rangle$ (i.e., a hash query on $S \| T$), $\mathcal{B}$ returns $H_2(S \| T)$.

When $\mathcal{A}$ issues an H_1 query on a domain PID S, $\mathcal{B}$ runs the following algorithm:

1. If L contains a tuple whose first element is S, then
   (a) If L contains $\langle S, r_0, r_1, \ldots, r_n \rangle$, then return $\langle g^{r_0}, g^{r_1}, \ldots, g^{r_n} \rangle$.
   (b) If L contains $\langle S, \text{REAL} \rangle$, then return $H_1(S)$.
2. Otherwise, flip a coin that takes the value 1 with probability p and 0 otherwise (p will be determined later).
   (a) If COIN $= 1$, then pick random $r_0, r_1, \ldots, r_n \in \mathbb{F}_q$. Insert the tuple $\langle S, r_0, r_1, \ldots, r_n \rangle$ into L, and return $\langle g^{r_0}, g^{r_1}, \ldots, g^{r_n} \rangle$.

(b) Otherwise, $\text{COIN} = 0$. In this case, insert $\langle S, \text{REAL} \rangle$ into L and return $H_1(S)$.

Since we are modelling H_1 as a random oracle, $\mathcal{A}$ cannot distinguish between this simulation and the real H_1.

When $\mathcal{A}$ issues a KeyGen_1 query on a prefix address (domain name) $\langle S \rangle$, $\mathcal{B}$ runs the following algorithm:

By assumption, $\mathcal{A}$ has already issued an H_1 query for S.
1. If $\langle S, r_0, r_1, \ldots, r_n \rangle$ is on the list L, return $\langle g^{ar_0}, g^{ar_1}, \ldots, g^{ar_n} \rangle$.
2. Otherwise, $\langle S, \text{REAL} \rangle$ appears on L: output FAILURE and halt.

When $\mathcal{A}$ issues a KeyGen_2 query on an address $\langle S, T \rangle$, $\mathcal{B}$ runs the following algorithm:

Again by assumption, a hash query on S has already been issued. Let $d = H_2(S \| T)$.
1. If L contains $\langle S, r_0, r_1, \ldots, r_n \rangle$, return $\left\langle g^{ar_0 d^0}, g^{ar_1 d^1}, \ldots, g^{ar_n d^n} \right\rangle$.
2. Otherwise, $\langle S, \text{REAL} \rangle \in L$:
   (a) If $s_{\text{CHALLENGE}} = S$, then $\mathcal{B}$ issues the KeyGen_2 query (recall that $\mathcal{B}$ is allowed to do this for the challenge domain).
   (b) If $s_{\text{CHALLENGE}} \neq \text{NULL}$ then $\mathcal{B}$ outputs FAILURE and halts.
   (c) Otherwise, $\mathcal{B}$ sets $s_{\text{CHALLENGE}} := S$ and issues the KeyGen_2 query.

Eventually, $\mathcal{A}$ outputs a challenge address $\langle S, T \rangle$. If $\langle S, \text{REAL} \rangle \notin L$, then output FAILURE. If $\langle S, \text{REAL} \rangle \in L$ and $s_{\text{CHALLENGE}} \neq \text{NULL}$ and $s_{\text{CHALLENGE}} \neq S$, then output FAILURE. Otherwise (when $\langle S, \text{REAL} \rangle \in L$ and ($s_{\text{CHALLENGE}} = \text{NULL}$ or $s_{\text{CHALLENGE}} = S$)), set $s_{\text{CHALLENGE}} := S$.

The next round of queries is handled in the same manner as in the first round.

Finally, $\mathcal{A}$ will output a guess M and halt; then $\mathcal{B}$ outputs M and halts. Clearly, if $\mathcal{A}$ is successful, then so is $\mathcal{B}$.

Recall that Q_1 is the number of KeyGen_1 queries. Then the probability that FAILURE is not output during such a query is at least p^{Q_1} (it is sufficient to have $\text{COIN} = 1$ for each query).

Recall that Q_2 is the number of KeyGen_2 queries. In the worst case, for every KeyGen_2 query on an address $\langle S, T \rangle$, L contains $\langle S, \text{REAL} \rangle$, and, once $s_{\text{CHALLENGE}}$ has been set, any other value for S will cause failure. So the probability that failure is not output during KeyGen_2 queries is bounded from below by $p^{Q_2 - 1}$.

After $\mathcal{A}$ outputs the challenge address, the probability that $\langle S, \text{REAL} \rangle$ is on the list L is $1 - p$, and the probability $s_{\text{CHALLENGE}} = S$ or $s_{\text{CHALLENGE}} = \text{NULL}$ is at least p^{Q_2}. (In the worst case, every KeyGen_2 query is in a different domain, and, trivially, the probability that $s_{\text{CHALLENGE}} = S$ or $s_{\text{CHALLENGE}} = \text{NULL}$ is no less than the probability that $s_{\text{CHALLENGE}}$ remains NULL.)

Let $k = Q_1 + 2Q_2 - 1$. Then $p^k(1-p)$ is a lower bound on the probability that a failure state is not reached. It is minimized when $p = k/(k+1)$, which makes the probability of not reaching a failure state bounded from below by $1/\mathbf{e}(k+1)$.

With $Q = Q_1 + 2Q_2$, we see that $\mathcal{B}$ has an advantage of at least $\varepsilon/\mathbf{e}Q$. $\square$

Definition 15. *The* Computational Polynomial Diffie-Hellman (CPDH) game (of degree n) for a function $H\colon X \to G_1$, *where* X *is a set, is the following game:*

A polynomial $f(x) = c_0 + c_1 x + \cdots + c_n x^n$ with coefficients in $\mathbb{F}_q$ is chosen at random. An element a is chosen at random from $\mathbb{F}_q$.

The attacker is given $g, g^a, g^{c_0}, g^{c_1}, \ldots, g^{c_n}$ and $d \in \mathbb{F}_q$.

Then the attacker picks any $s \in X$, and learns $g^{af(H(s))}$. This step is repeated up to n times. (The attack may be adaptive.)

Lastly, the attacker wins if it can output the value of $g^{af(d)}$.

Remark 3. For $n = 0$, this reduces to the CDH problem. (The adversary is not allowed to make any queries.)

Definition 16. *The* Bilinear Polynomial Diffie-Hellman (BPDH) game (of degree n) for a function $H\colon X \to G_1$ *is the e-transformation of the corresponding CPDH game, i.e., it is the same as the previous game except that the attacker is also given* g^r *for some random* $r \in \mathbb{F}_q$ *and now the attacker's goal is to compute* $e(g, g)^{arf(d)}$.

Remark 4. For $n = 0$ this reduces to the BDH problem.

Lemma 2. *Suppose there exists an ID-OWE attacker* $\mathcal{B}$ *with an advantage of* ε *whose key-generation queries are always* KeyGen_2 *queries of addresses from the same domain as the challenge domain, and furthermore,* $\mathcal{B}$ *makes at most* n *such queries. (*$\mathcal{B}$ *makes* no KeyGen_1 *queries.) Then, there exists an attacker* $\mathcal{C}$ *that can win the BPDH game for* H_2 *with an advantage of* $\varepsilon/(2Q)$*, where* Q *is a bound on the number of* H_1 *queries that* $\mathcal{B}$ *makes.*

Proof. Again we may assume that any key-generation query for an address is preceded by a hash query for that address, and that before the challenge address is output, a hash query for the challenge address will have been issued. We may also assume that each query (for any hash function) is distinct (since previous results can simply be cached). Also, without loss of generality, we may assume that $\mathcal{B}$ makes *exactly* n KeyGen_2 queries.

The algorithm $\mathcal{C}$ is given as input $g, g^a, g^{c_0}, g^{c_1}, \ldots, g^{c_n}, g^r, d$ (using the notation employed in the description of the BPDH game; its goal is to compute $e(g, g)^{arf(d)}$). $\mathcal{C}$ begins by giving $\mathcal{B}$ the system parameters g, g^a.

There is a list L that is used to store H_3 queries and is initially empty.

$\mathcal{C}$ picks a random i between 1 and Q. On the ith H_1 query for S (that $\mathcal{B}$ makes), $\mathcal{C}$ sets $s_{\text{CHALLENGE}} := S$ and returns $\langle g^{c_0}, g^{c_1}, \ldots, g^{c_n}\rangle$. For all other H_1 queries, $\mathcal{C}$ returns $H_1(S)$. Since we are modelling H_1 as a random oracle, the algorithm $\mathcal{B}$ cannot distinguish between this simulation of H_1 and the real H_1.

When $\mathcal{B}$ issues an H_2 query for $\langle S, T\rangle$, $\mathcal{C}$ returns $H_2(S \,\|\, T)$.

When $\mathcal{B}$ issues an H_3 query for $s \in G_2$, $\mathcal{C}$ returns $H_3(s)$, and inserts $\langle s, H_3(s)\rangle$ into L.

When $\mathcal{B}$ issues a KeyGen_2 query on $\langle S, T\rangle$, if $S \neq s_{\text{CHALLENGE}}$, then $\mathcal{C}$ outputs FAILURE and halts. Otherwise, $\mathcal{C}$ issues a query for $g^{af(H_2(S\|T))}$ and returns the result to $\mathcal{B}$.

Eventually, $\mathcal{B}$ gives the challenge address $N = \langle S, T \rangle$ to $\mathcal{C}$. If $S \neq s_{\text{CHALLENGE}}$, then $\mathcal{C}$ outputs FAILURE and halts. Otherwise, $\mathcal{C}$ chooses a random $R \in \{0,1\}^k$, and $\mathcal{C}$ gives the ciphertext $C := \langle g^r, R \rangle$ to $\mathcal{B}$.

The next round of queries is handled in the same manner as in the first round.

Eventually, $\mathcal{B}$ outputs its guess M and halts. Then, the algorithm $\mathcal{C}$ looks for a tuple of the form $\langle s, M \oplus R \rangle$ in L; if it cannot find it, $\mathcal{C}$ outputs FAILURE. If $\mathcal{B}$ is successful (i.e., if $\mathsf{Encrypt}(\mathsf{params}, N, M) = C$), then $M = R \oplus H_3(s)$, where $s = e(g,g)^{arf(H_2(S\|T))}$. If $\mathcal{C}$ finds the tuple in L, $\mathcal{C}$ then knows $n+1$ values of the function $x \mapsto e(g,g)^{arf(x)}$, so $\mathcal{C}$ can compute $e(g,g)^{arf(d)}$ using Lagrange interpolation.

Notice that, since H_3 is a random oracle, if $\langle s, H_3(s) \rangle$ is not found in L, then the decryption of the ciphertext C is independent of the knowledge $\mathcal{B}$ accumulated from its various queries, which means that $\mathcal{B}$ succeeds in this case with probability $1/2^k$.

The probability of success is $(1/Q)(1 - 1/2^k)$ (i must be guessed correctly), which is at least $1/(2Q)$. □

Lemma 3. *Suppose there exists an algorithm $\mathcal{C}$ that can win the BPDH game for a function $H\colon X \to G_1$ with an advantage of ε and suppose H may be modelled as a random oracle. Then there exists an attacker $\mathcal{D}$ that can solve the BDH problem with an advantage of $\varepsilon/\binom{Q}{n}$, where Q is a bound on the number of H queries that $\mathcal{C}$ makes.*

Proof. We may assume that $Q \geq n$, that all H queries are distinct (previous results can be cached), and that a query for $g^{af(H(s))}$ implies that $\mathcal{C}$ has already issued a query for $H(s)$.

The algorithm $\mathcal{D}$ is given g, g^x, g^y, g^z for randomly chosen $x, y, z \in \mathbb{F}_q$ (its goal is to compute $e(g,g)^{xyz}$). Set $y_0 := y$. There is a list L that holds responses to H queries that is initially empty. $\mathcal{D}$ picks random $a_0, a_1, \ldots, a_n, y_1, y_2, \ldots, y_n \in \mathbb{F}_q$. $\mathcal{D}$ then solves the system of equations

$$\begin{aligned} g_0^{a_0^0} \cdot g_1^{a_0^1} \cdots g_n^{a_0^n} &= g^{y_0} \\ g_0^{a_1^0} \cdot g_1^{a_1^1} \cdots g_n^{a_1^n} &= g^{y_1} \\ &\vdots \\ g_0^{a_n^0} \cdot g_1^{a_n^1} \cdots g_n^{a_n^n} &= g^{y_n} \end{aligned}$$

for the g_i. If we define the matrix A by $A_{ij} := a_i^j$, then, with high probability, the a_i are distinct (so A is a Vandermonde matrix), thus guaranteeing a unique solution for the g_i.

Then, $\mathcal{D}$ hands $\mathcal{C}$ the input $\langle g, g^x, g_0, g_1, \ldots, g_n, g^z, a_0 \rangle$.

Let Q be a bound on the number of H queries made by $\mathcal{C}$. $\mathcal{D}$ chooses a random subsequence I of length n from the sequence $(1, 2, \ldots, Q)$.

Let $s_j \in X$ be the jth element on which $\mathcal{C}$ makes an H query. $\mathcal{D}$ answers that query as follows:

If j is the ith element of I, then $\mathcal{D}$ responds with a_i and inserts $\langle s_j, i \rangle$ into L. Otherwise, $j \notin I$ and $\mathcal{D}$ responds with a random number. Since the a_i were chosen at random, the algorithm $\mathcal{C}$ cannot distinguish between our simulation of a random oracle and a real random oracle.

If $\mathcal{C}$ asks for $g^{af(H(s))}$, then, if $\langle s, i\rangle \in L$ for some i, $\mathcal{D}$ replies with g^{y_i}. Otherwise, it outputs FAILURE and halts.

Eventually, $\mathcal{C}$ will output its guess g' for $e(g,g)^{xzf(a_0)}$ and halt; then $\mathcal{D}$ outputs g' and halts.

Let $r_1, r_2, \ldots, r_n \in \mathbb{F}_q$ be such that $g_0 = g^{r_0}, g^{r_1}, \ldots, g_n = g^{r_n}$ (we will never need to explicitly compute the r_i). If $\mathcal{D}$ has not yet failed, and, if $\mathcal{C}$ wins its game, then $\mathcal{C}$ will output $g' = e(g,g)^{xzf(a_0)} = e(g,g)^{xyz}$ (where f is the polynomial $f(u) = r_0 + r_1 u + \cdots + r_n u^n$), which means that $\mathcal{D}$, by outputting $e(g,g)^{xyz}$, will win the BDH game.

Thus, the probability of $\mathcal{D}$ succeeding, given that $\mathcal{C}$ is successful, is at least $1/\binom{Q}{n}$ (it is sufficient for the set I to correspond exactly to the n elements of X for which $\mathcal{C}$ issues $g^{af(H(s))}$ queries). □

Applying the Fujisaki-Okamoto transformation to the cryptosystem yields an ID-CCA 2-HIBE system that tolerates user collusion up to size n; in this system, the Setup algorithm also selects two hash functions H': $\{0,1\}^k \times \{0,1\}^k \to \mathbb{F}_q$ and H'': $\{0,1\}^k \to \{0,1\}^k$, and there is an extra level of hashing during encryption and decryption, as follows (we use the same notation as in the description of the cryptosystem): Encrypt now also picks a random $\sigma \in \{0,1\}^k$, computes $r := H'(\sigma, M)$ (instead of picking a random $r \in \mathbb{F}_q$), and outputs $C := \langle g^r, \sigma \oplus H_3(s), M \oplus H''(\sigma)\rangle$ (as opposed to $\langle g^r, M \oplus H_3(s)\rangle$). Decrypt is modified similarly.

The proof of Lemma 3 involves a reduction that is exponential in n, rendering the system untrustworthy for large n. One could simply assume that the Bilinear Polynomial Diffie-Hellman game is hard to win in order to rectify this, but it would be better to find a polynomial-time reduction from winning the Bilinear Polynomial Diffie-Hellman game to a more natural assumption.

4 Conclusions and Open Problems

We introduced hierarchical identity-based encryption schemes and their related security definitions. We presented a concrete two-level HIBE scheme that is totally collusion-resistant on the upper level and partially collusion-resistant on the lower level, and showed it to be secure under the BDH assumption (in the random-oracle model). An open problem is to construct a two-level HIBE scheme that is totally collusion-resistant on the lower level and at least partially (if not totally) collusion-resistant on the upper level.

One could try to extend our approach to finding a 2-HIBE system to attempt to construct an ℓ-HIBE system for any given ℓ by using a family of $\ell - 1$ linear e-one-way functions such that the output of one is used as the first input of the next. Ideally, there would be a single linear e-one-way function h: $G_1 \times X \to G_1$, and one could apply h recursively to obtain a family of any given size.

Unfortunately, this may not be enough to guarantee that an ID-OWE attacker has negligible advantage. A major hurdle in a proof of security is that it is not clear how to build a simulator that can answer an algorithm's queries for addresses within the same domain but within a different subdomain than the challenge address. Additionally, if we model hash functions as random oracles, then the most obvious approaches will involve security reductions exponential in ℓ (informally, it seems a simulator must guess which hash query to rig at each level).

In any event, there are no known families of linear e-one-way functions of any size (that can accompany a bilinear function satisfying the BDH assumption). However, if we are willing to tolerate n-party collusion at every level, then we may construct a family of functions by extending our scheme. In our system, KeyGen_2 produces one group element from n group elements. This suggests a scheme where a private key at the top level consists of $n^{\ell-1}$ group elements: to generate keys for the next level, sets of n elements are used to produce one group element each; private keys at a given level contain n times the number of group elements in private keys of the level below. At the bottom level a private key consists of a single group element which can be fed into the bilinear function.

Clearly, in this scheme, the key size at the top level increases exponentially with ℓ, so it does not scale well. Even so, for some applications, setting ℓ slightly greater than 2 may be beneficial; e.g., if we take domains, subdomains and users to be corporations, departments and employees, then it may be less likely for n department key generators to collude than for n^2 employees to collude. (It is still possible for n^2 employees to collude and expose the corporation's key, but this requires n employees from each of n different departments.)

References

1. D. Boneh and M. Franklin, "Identity Based Encryption from the Weil Pairing", *Advances in Cryptology: CRYPTO 2001 (LNCS 2139)*, pp. 213–229, 2001.
2. D. Boneh and M. Franklin, "Identity Based Encryption from the Weil Pairing", *Cryptology ePrint Archive*, Report 2001/090, 2001. `http://eprint.iacr.org/2001/090/`
3. D. Boneh, B. Lynn, and H. Shacham, "Short Signatures from the Weil Pairing", *Advances in Cryptology: ASIACRYPT 2001 (LNCS 2248)*, pp. 514–532, 2001.
4. C. Cocks, "An Identity Based Encryption Based on Quadratic Residues", *Cryptography and Coding (LNCS 2260)*, pp. 360–363, 2002.
5. J. Coron, "On the Exact Security of Full Domain Hash", *Advances in Cryptology: CRYPTO 2000 (LNCS 1880)*, pp. 229–235, 2000.
6. E. Fujisaki and T. Okamoto, "Secure Integration of Asymmetric and Symmetric Encryption Schemes", *Advances in Cryptology: CRYPTO '99 (LNCS 1666)*, pp. 537–554, 1999.
7. A. Joux, "A One Round Protocol for Tripartite Diffie-Hellman", *Algorithmic Number Theory : 4th International Symposium, ANTS-IV (LNCS 1838)*, pp. 385–394, 2000.
8. M. Kasahar, K. Ohgishi, and R. Sakai, "Cryptosystems Based on Pairing", *The 2001 Symposium on Cryptography and Information Security*, Oiso, Japan, 2001.
9. A. Shamir, "Identity-Based Cryptosystems and Signature Schemes", *Advances in Cryptology: CRYPTO '84 (LNCS 196)*, pp. 47–53, 1985.

10. E. Verheul, "Evidence That XTR Is More Secure than Supersingular elliptic curve cryptosystems", *Advances in Cryptology: EUROCRYPT 2001 (LNCS 2045)*, pp. 195–210, 2001.
11. E. Verheul, "Self-Blindable Credential Certificates from the Weil Pairing", *Advances in Cryptology: ASIACRYPT 2001 (LNCS 2248)*, pp. 533–551, 2001.
12. ISO/IEC 9594-8, "Information Technology — Open Systems Interconnection — The Directory: Authentication Framework", International Organization for Standardization, Geneva, Switzerland, 1995 (equivalent to ITU-T Recommendation X.509, 1993).

Unconditional Byzantine Agreement and Multi-party Computation Secure against Dishonest Minorities from Scratch

Matthias Fitzi[1], Nicolas Gisin[2], Ueli Maurer[1], and Oliver von Rotz[1]

[1] ETH Zurich, Department of Computer Science,
{fitzi,maurer}@inf.ethz.ch, ovonrotz@ergon.ch
[2] Geneva University, Group of Applied Physics,
Nicolas.Gisin@physics.unige.ch

Abstract. It is well-known that n players, connected only by pairwise secure channels, can achieve unconditional broadcast if and only if the number t of cheaters satisfies $t < n/3$. In this paper, we show that this bound can be improved – at the sole price that the adversary can prevent successful completion of the protocol, but in which case all players will have agreement about this fact. Moreover, a first time slot during which the adversary forgets to cheat can be reliably detected and exploited in order to allow for future broadcasts with $t < n/2$. This even allows for secure multi-party computation with $t < n/2$ after the first detection of such a time slot.

Key words: Byzantine agreement, multi-party computation, unconditional security.

1 Introduction

1.1 Unconditional Broadcast and Multi-party Computation

In this paper we consider a set P of n players. The goal is to achieve broadcast (or multi-party computation, in general), unconditionally secure against an active (Byzantine) threshold adversary that may corrupt up to t of the n players, i.e., the adversary may take full control of the corrupted players and make them deviate from the prescribed protocol in an arbitrary way. *Unconditional security* means that, for some arbitrarily small (but *a priori* fixed) error probability ε, the probability that the protocol fails is at most ε whereas no assumptions are made about the adversary's computational power. As a special case of unconditional security, *perfect security* allows no probability of error ($\varepsilon = 0$).

The goal of broadcast is to have a sender consistently distribute some input value to all players.

Definition 1. *A protocol among n players such that one distinct player s (the sender) holds an input value $x_s \in \mathcal{D}$ (for some finite domain $\mathcal{D}$) and all players eventually decide on an output value in $\mathcal{D}$ is said to achieve* broadcast *(or*

L.R. Knudsen (Ed.): EUROCRYPT 2002, LNCS 2332, pp. 482–501, 2002.

Byzantine agreement) *if the protocol guarantees that all correct players decide on the same output value $y \in \mathcal{D}$, and that $y = x_s$ whenever the sender is correct.*

Some weaker definition of broadcast will be important in the sequel of this paper.

Definition 2. *A protocol among n players such that one distinct player s (the sender) holds an input value $x_s \in \mathcal{D}$ and all players eventually decide on an output value in $\mathcal{D} \cup \{\perp\}$ (with $\perp \notin \mathcal{D}$) is said to achieve* weak broadcast *if the protocol guarantees the following conditions:*

- *If any correct player decides on some value $y \in \mathcal{D}$ then all correct players decide on a value in $\{y, \perp\}$.*[1]
- *If the sender is correct then all correct players decide on $y = x_s$.*

Note that broadcast for any finite domain can be achieved by combining broadcast protocols for a single bit ($\mathcal{D} = \{0,1\}$) [TC84]. Hence, for the constructions in the sequel, we will focus on protocols for binary broadcast.

Broadcast is a special case of the more general problem of secure *multi-party computation (MPC)* where the players want to distributedly evaluate some agreed function on their inputs in a way preserving privacy of the players' inputs and correctness of the computed result.

1.2 Previous Work

Broadcast: For the standard communication model with a complete synchronous network of pairwise authenticated channels, Pease, Shostak, and Lamport [PSL80] proved that perfectly secure broadcast is achievable if and only if less than a third of the players is corrupted: $t < n/3$. This tight bound more generally holds with respect to a network of secure channels and unconditional security, i.e., when even allowing a negligible error probability, as proven by Karlin and Yao [KY]. The first optimally resilient protocol that is efficient was proposed by Dolev et al. [DFF+82]. For the case that broadcast among every subset of three players is possible (in contrast to the standard model with only pairwise communication), Fitzi and Maurer [FM00] proved that (global) broadcast is possible if and only if $t < n/2$. In another line of research, Baum-Waidner, Pfitzmann, and Waidner [BPW91,PW92] proved that broadcast during some precomputation stage allows to later achieve broadcast that tolerates any number of corrupted players ($t < n$), i.e., that the functionality of the prior broadcast can be preserved for any later time.

Multi-party computation: The concept of general multi-party computation (MPC) was introduced by Yao [Yao82] with a first complete solution given by Goldreich, Micali, and Wigderson [GMW87] – though with computational security. Ben-Or, Goldwasser, and Wigderson [BGW88], and, Chaum, Crépeau,

[1] That is, interpreting $\perp$ as "invalid", this condition expresses that no two correct players may decide on valid values that are distinct.

and Damgård [CCD88], proved that, in the standard model with pairwise secure channels, *unconditionally* secure MPC is achievable if and only if $t < n/3$ by giving efficient protocols for the achievable cases. Beaver [Bea89], and independently, Rabin and Ben-Or [RB89] later proved that, when additionally given global broadcast among the players, unconditionally secure MPC is achievable if and only if $t < n/2$ (see also Cramer et al. [CDD+99]). The result in [FM00] hence implies that broadcast among three players (i.e., 2-*cast*) is sufficient in order to achieve MPC for $t < n/2$.

1.3 Contributions

In this paper we investigate how the bound $t < n/3$ for the achievability of broadcast can be improved. Obviously, this requires a modification of the model.

In a first model, additionally to pairwise authenticated communication channels, we assume the existence of an external information source that distributes correlated random variables to the players that correspond to some simple probability distribution. We show that in this model broadcast and MPC are achievable for $t < n/2$.

Our second model assumes only standard communication, namely pairwise secure communication channels, but the goal is to achieve a slightly weaker form of broadcast or MPC, called *detectable broadcast* (or *MPC*), where the adversary can force abortion of the protocol but in which case all correct players have agreement about this fact, i.e., they commonly detect that the protocol was not successful. Besides this, the adversary can neither violate correctness, nor privacy, nor any other condition of the original problem. In other words, the detectable variant of a problem can be seen as the original problem without requiring robustness.

We show that detectable broadcast and MPC are achievable for $t < n/2$ by presenting efficient protocols that are based on the protocols given for the model involving an external information source. Consider, for example, the special case of $n = 3$ and $t = 1$ where broadcast and MPC are not achievable. Our results imply that such a protocol can nevertheless be run in an "optimistic" manner. If no player is corrupted then the protocol satisfies all conditions of the original problem whereas one corrupted player can only make the protocol abort in the worst case. This is strictly more than previously achievable.

Furthermore, a slightly modified version of the protocol achieves that a first time slot where the adversary is not actively cheating can be detected and exploited in order to establish (standard) broadcast for the future. This could be seen as a *lunch-time attack against the adversary*: if once the adversary is absent for a short period[2], the players can secure themselves for future broadcast that will be reliable even when the adversary will be present again. Together with the result of [Bea89,RB89,CDD+99] this more generally allows for future MPC for $t < n/2$ as soon as such a period has been detected.

[2] This could for instance be enforced by rebooting one or more of the servers.

As opposed to the results in the model with an external information source or the results in [Bea89,RB89,CDD+99,FM00,BPW91,PW92], in this model, the bound of $t < n/2$ is achieved *from scratch*, i.e., with no further assumptions on the communication model than pairwise secure channels.

Our "optimistic" model is of particular interest when faults or corruption are expected to be rare but to appear in bursts. Virus infections of servers, for example, only occur from time to time but then it must be expected that many servers get infected at the same time. A first phase where no server is infected can thus be exploited in order to "vaccinate" the system against future infections of any minority of the servers.

2 Summary of Required Previous Results

In this section we briefly summarize previous results that are important for the results derived in this paper.

Proposition 1. *[Bea89,RB89,CDD+99] Consider a set of n players. In the communication model with a complete, synchronous network of pairwise secure channels among the players and global broadcast channels unconditionally secure MPC is (efficiently) achievable if and only if at most $t < n/2$ players are actively corrupted.*

Proposition 2. *[FM00] Consider a set of n players. In the communication model with a complete, synchronous network of pairwise authenticated channels among the players and broadcast among each set of three players (i.e., 2-cast) unconditionally secure global broadcast is (efficiently) achievable if and only if at most $t < n/2$ players are actively corrupted.*

Finally, an unconditionally secure protocol for weak 2-cast can be easily turned into an unconditionally secure protocol for 2-cast. Consider a sender s and two recipients r_0 and r_1. Then the following protocol `Amplify` achieves 2-cast.

Protocol 1: `Amplify` [Precondition: s, r_0 and r_1 have executed weak 2-cast]

1. s decides on his own input to the prior weak 2-cast;
2. r_0, r_1: exchange decision values y_0 and y_1;
3. r_k ($k \in \{0, 1\}$): adopt other recipient's decision value if and only if $y_k = \perp$;
4. r_k ($k \in \{0, 1\}$): **if** $y_k = \perp$ **then** $y_k = 0$ **fi**;

Note that resilience of this 2-cast protocol against one single corrupted player immediately implies resilience against any number of corrupted players since, in the presence of more than one corrupted player, the only condition to be satisfied is that a correct sender decides on his own input value – which is obviously guaranteed.

Proposition 3. *[FM00] Consider a set of* 3 *players. In the model with a complete, synchronous network of pairwise authenticated channels weak* 2*-cast (i.e., weak broadcast) unconditionally secure against one corrupted player implies efficient* 2*-cast unconditionally secure against any number of corrupted players.*

Hence, as follows from the previous propositions, in order to show that global broadcast or MPC are efficiently achievable for $t < n/2$, it is sufficient to show that the simulation of weak 2-cast among any set of three players is possible that is unconditionally secure against one corrupted player. This fact is captured by the following proposition.

Proposition 4. *Consider a set of n players.*

In the model with a complete, synchronous network of pairwise authenticated *channels weak* 2*-cast unconditionally secure against one actively corrupted player implies efficient* broadcast *unconditionally secure against $t < n/2$ actively corrupted players.*

In the model with a complete, synchronous network of pairwise secure *channels weak* 2*-cast unconditionally secure against one actively corrupted player implies efficient* MPC *unconditionally secure against $t < n/2$ actively corrupted players.*

3 Broadcast and MPC with External Information

It can be easily proven that an additional global random source (i.e., a beacon) does not help to improve the classical bound of $t < n/3$ for broadcast (and hence for MPC in general) in the standard model by extending the proofs in [FLM86,FGMO01]. However, by slightly modifying the functionality of such a beacon, as described in the following section, it does. This functionality will be exploited in order to simulate broadcast for $t < n/2$ and hence allowing for general MPC with respect to the same bound.[3]

3.1 The Q-Flip Model

We assume the standard communication model with a complete (fully connected) synchronous network of *pairwise authenticated channels* among the players. But similarly to the model in [FM00] we assume some additional primitive among each triple of players, called *Q-Flip*.

Q-Flip, as described for the three players p_0, p_1, and p_2, is a random generator that (for every invocation), with uniform distribution, generates a random permutation on the elements $\{0,1,2\}$, $(x_0, x_1, x_2) \in \{(0,1,2), (0,2,1), (1,0,2),$

[3] Note that a straightforward solution could be achieved with the help of an external information source that simulates the whole protocol in [BPW91]. However, this would be a rather complex task to be performed by an information source requiring a lot of mathematical structure. In contrast, our solution is based on very simple correlated information.

$(1,2,0),(2,0,1),(2,1,0)\}$, and sends the element x_i ($i \in \{0,1,2\}$) to player p_i. No single player p_i learns more about the permutation than the value x_i which he receives, i.e., a single player does not learn how the remaining two values are assigned to the other players.

The Q-Flip primitive helps to build pairwise one-time pads between the players (see Appendix A) and hence the authenticated channels can be easily turned into *secure channels*. This allows us to assume secure pairwise communication for the rest of this section.

The Q-Flip primitive was originally motivated by quantum entanglement considerations about the Byzantine agreement problem. A detailed description of the quantum physical aspects of our results is given in [FGM01].

3.2 Broadcast for $t < n/2$

Since weak 2-cast secure against one corrupted player implies efficient global broadcast for $t < n/2$ (Proposition 4), it is sufficient to demonstrate the existence of an efficient protocol for weak 2-cast that tolerates one corrupted player. We now describe such a protocol for a sender s and two recipients r_0 and r_1.[4]

Let $x_s \in \{0,1\}$ be the input of sender s, and y_0 and y_1 be the values the players r_0 and r_1 finally decide on (whereas s always implicitly decides on $y_s = x_s$). The primitive Q-Flip is invoked some m times and each player receives a sequence of m elements in $\{0,1,2\}$, i.e., s receives $Q_s = (Q_s[1],\ldots,Q_s[m])$, r_0 receives $Q_0 = (Q_0[1],\ldots,Q_0[m])$, and r_1 receives $Q_1 = (Q_1[1],\ldots,Q_1[m])$, where the triplet $(Q_s[i],Q_0[i],Q_1[i])$ represents the outcome of the i-th invocation of Q-Flip. The protocol now proceeds as follows:

First, s sends to r_0 and r_1 his input bit x_s and the set σ_s of all indices $i \in \{1,\ldots,m\}$ such that s received the complement of x_s for the i-th invocation of Q-Flip (see Figure 1–A):

$$\sigma_s = \{i \in \{1,\ldots,m\} : Q_s[i] = 1 - x_s\} \ . \tag{1}$$

If this is done correctly then σ_s is of large size (i.e., approximately $m/3$, which is important for good statistics on the corresponding Q-Flips) and both recipients r_k ($k \in \{0,1\}$) never received the value $1 - x_s$ for any Q-Flip invocation with respect to σ_s: $\{i \in \sigma_s : Q_k[i] = 1 - x_s\} = \emptyset$.

Let x_k and σ_k ($k \in \{0,1\}$) be the information that (potentially faulty) s actually sent to recipient r_k. The recipients now decide on $y_k = x_k$ if and only if σ_k is of large size and the value $1 - x_k$ was never received with respect to σ_k:

$$y_k := \begin{cases} x_k \ , \text{ if } \ (\sigma_k \text{ large}) \wedge (\{i \in \sigma_k : Q_k[i] = 1 - x_k\} = \emptyset) \\ \perp \ , \text{ else} \ . \end{cases} \tag{2}$$

Now, r_0 sends to r_1 his value y_0 and the set ρ_0 of all indices $i \in \sigma_0$ such that he received x_0 for the corresponding Q-Flip invocation:

$$\rho_0 = \{i \in \sigma_0 : Q_0[i] = x_0\} \ . \tag{3}$$

[4] The protocol descriptions in this paper do not explicitly care about received values that are outside a domain. We implicitly assume that any value received outside some expected domain $\mathcal{D}$ is automatically replaced by an arbitrary value inside $\mathcal{D}$.

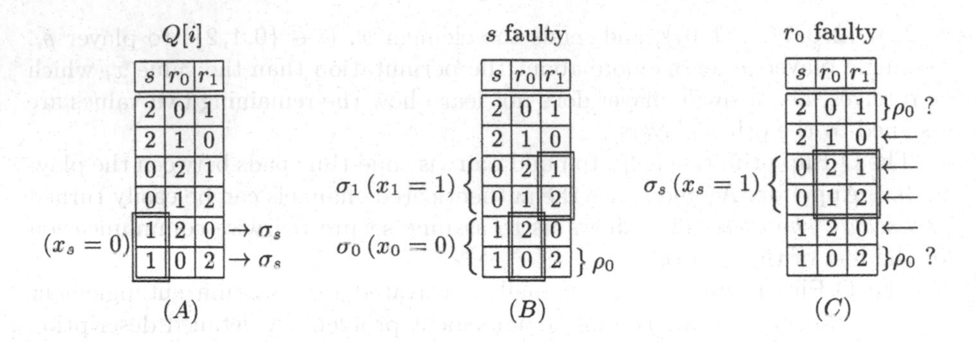

Fig. 1. *(A) Possible outcomes of Q-Flip and selection of σ_s; (B,C) Basic cheating strategies for s and r_0.*

If $y_0 \neq y_1$ but $y_0, y_1 \in \{0,1\}$, y_1 now redecides in the following way:

$$y_1 := \begin{cases} y_0 \text{ , if } (\rho_0 \text{ large}) \wedge (\{i \in \rho_0 \setminus \sigma_1 : Q_1[i] = 2\} \overset{\text{almost}}{=} \rho_0) \\ y_1 \text{ , else .} \end{cases} \quad (4)$$

We give a rough argumentation why none of the players can make the protocol fail.

$\mathbf{r_1}$ faulty: r_1 cannot significantly misbehave since r_1 is silent.

s faulty: In order to make the protocol fail, s must achieve that r_0 and r_1 decide on distinct values $y_0 \neq y_1$ such that $y_0, y_1 \in \{0,1\}$ (Equation (2)) and that r_1 does not redecide on $y_1 = y_0$ according to Equation (4). The only way to achieve this is to select $x_0 \in \{0,1\}$ and $x_1 = 1 - x_0$, and to basically compose large sets σ_0 and σ_1 as shown in Figure 1–B (as shown with respect to $x_0 = 0$).[5] But then, r_0 learns a large set ρ_0 of indices i (Figure 1–B, last row) such that, mainly, $i \notin \sigma_1$ and $Q_1[i] = 2$ which will "convince" r_1 to redecide to $y_1 = y_0$ according to Equation (4).

$\mathbf{r_0}$ faulty: In order to make the protocol fail, r_0 must achieve that r_1 redecides on $y_1 = y_0 \neq x_s$ according to Equation (4). Since correct s sends $\sigma_s = \{i \in \{1,\ldots,m\} : Q_s[i] = x_s\}$ to both players, r_0 cannot come up with a large set ρ_0 of indices i such that most of them satisfy $i \notin \sigma_s$ and $Q_1[i] = 2$ (see Figure 1–C) since r_0 cannot distinguish between the outcomes corresponding to the first and the last row.[6]

The detailed protocol is described by Protocol `Weak 2-Cast`. There, two free protocol parameters are introduced, m_0 ($m_0 = \Omega(m)$; $m_0 < m/6$) for asserting that the sets σ_s and ρ_0 are of sufficiently large cardinality, and λ ($\frac{1}{2} < \lambda < 1$) for

[5] Note that for every selection $i \in \sigma_k$ ($k \in \{0,1\}$) such that $Q_s[i] \neq 1 - x_k$, it holds with a probability of $\frac{1}{2}$ that $Q_k[i] = 1 - x_k$, which makes r_k decide on $\perp$ according to Equation (2).

[6] Note, that r_0 completely learns all instances indicated by an arrow but nothing more about the other instances.

Protocol 2: `Weak 2-Cast` [s to send $x_s \in \{0,1\}$ to r_k $(k \in \{0,1\})$]

0. s, r_0, r_1: invoke primitive Q-Flip m times;
1. $s \overset{\text{send}}{\longrightarrow} r_0, r_1$: x_s, $\sigma_s = \{i \in \{1,\ldots,m\} : Q_s[i] = 1 - x_s\}$; [$r_k$ receives x_k, σ_k]
   s: $y_s := x_s$;
2. r_k: **if** $(|\{i \in \sigma_k : Q_k[i] = x_k\}| \geq m_0) \wedge (\{i \in \sigma_k : Q_k[i] = 1 - x_k\} = \emptyset)$ **then**
   $y_k := x_k$
   else $y_k := \perp$ **fi**;
3. $r_0 \overset{\text{send}}{\longrightarrow} r_1$: y_0, $\rho_0 = \{i \in \sigma_0 : Q_0[i] = y_0\}$; [$r_1$ receives y_{01} and ρ_{01}]
4. r_1: **if** $(\perp \neq y_{01} \neq y_1 \neq \perp) \wedge (|\rho_{01}| \geq m_0) \wedge$
   $(|\{i \in \rho_{01} \setminus \sigma_1 : Q_1[i] = 2\}| \geq \lambda\,|\rho_{01}|)$ **then**
   $y_1 := y_{01}$
   fi;

the test according to Equation (4). Both parameters will be fixed for the final analysis of the protocol, which is given in Appendix C. The analysis yields:

Theorem 1. *In the Q-Flip model, unconditionally secure broadcast and MPC are efficiently achievable for* $t < n/2$.

Proof. The theorem follows from Lemma 9 and Proposition 4. □

The following observation about the given protocol is important for the construction of our protocol for the model in Section 4.

Observation 1: In Protocol `Weak 2-Cast`, player r_1 does not send a single message but only participates in a passive way. Even when turning the protocol into 2-cast (by appending Protocol `Amplify`) the single message sent by r_1 is only considered by r_0 if r_0 has already reliably detected the sender s to be faulty.

4 Detectable Broadcast and MPC for $t < n/2$

We assume the standard communication model with a complete (fully connected) synchronous network of *pairwise secure channels* among the players. Based on the solution for the model in the previous section, we present a protocol for detectable broadcast and MPC among n players that tolerates $t < n/2$ corrupted players.

Definition 3. *A protocol among n players is said to achieve* detectable broadcast *if it satisfies the following conditions:*

- **Correctness:** *If at most $t < n/2$ players are corrupted during the protocol then all correct players commonly accept or commonly reject the protocol. If they accept then the protocol achieves broadcast.*

– **Robustness:** *If no player is corrupted during the protocol then all players accept.*
– **Fairness:** *If any correct player rejects the protocol then the adversary gets no information about the sender's input.*

Note the difference between weak broadcast and detectable broadcast. Weak broadcast only guarantees that no two correct players decide on distinct values inside the domain $\mathcal{D}$. But generally, no player detects whether there are still any correct players that decided on $\perp \notin \mathcal{D}$. In contrast, at the end of detectable broadcast, all correct players agree on whether or not broadcast has been achieved.

Along the lines of [BPW91,PW92], detectable broadcast can be extended to a precomputation protocol that, when once successfully completed, will allow for future broadcast secure against $t < n/2$ corrupted players.

Definition 4. *A protocol among n players is said to achieve* detectable precomputation for broadcast *(or* MPC, *respectively) if it satisfies the following conditions:*

– **Correctness:** *If at most $t < n/2$ players are corrupted during the protocol then all correct players commonly accept or commonly reject the protocol. If they accept then, after the protocol, broadcast (or MPC, respectively) for $t < n/2$ will be achievable.*
– **Robustness:** *If no player is corrupted during the protocol then all players accept.*
– **Independence:** *A correct player's intended input value for any precomputed broadcast (or MPC, respectively) is statistically independent from the information he has sent during the precomputation.*

Independence implies, first, that a correct sender is not required to already know his input for any future protocol during the precomputation, and second, that the adversary gets no information about any future inputs by correct senders.

As opposed to detectable broadcast, the advantage of detectable precomputation for broadcast is that the preparation is separated from the actual execution of the broadcasts, i.e., only the precomputation is "detectable". As soon as the precomputation has been successfully completed standard broadcast with resilience $t < n/2$ is achievable. By applying many detectable precomputations in parallel, as many broadcasts can be precomputed for as will be later required. By using the techniques in [PW92], in order to be able to perform u later broadcasts, the number of broadcasts to be detectably computed for in advance can be made as low as polynomial in n and logarithmic in u.

We now proceed by presenting according protocols for the special case $n = 3$ and $t = 1$. These solutions can then be extended to any n and $t < n/2$ based on Proposition 4.

4.1 Detectable Precomputation for 2-Cast

The idea of our precomputation is to have recipient r_1 (the silent recipient of Protocol `Weak 2-Cast`) take the part of the Q-Flip source to correctly distribute sequences Q, and to base future 2-cast on Protocol `Weak 2-Cast` of Section 3 using Q. Since r_1 is silent during Protocol `Weak 2-Cast`, he will not be able to exploit his complete knowledge about Q in order to mislead anybody else but will be able to completely check consistency of information about Q provided by any other player. In the case that r_1 distributes this information correctly, this setup will allow for some (limited) number of 2-casts by s (and even by r_0 – by having s and r_0 switch their roles).

Clearly, we cannot guarantee that r_1 performs this task correctly. But by cross-checking, s and r_0 will be able to detect any inconsistency in his information that would enable the adversary to make a future 2-cast protocol (based on this information) fail with any reasonable chance.

Finally, any player who has successfully completed this setup procedure will be convinced that the Q-Flip information by r_1 is indeed (mostly) consistent (at least) among all correct players: r_1 since he made up the states himself, and s and r_0 because the cross-check was successful.

However, this does not necessarily imply that the correct players agree on the precomputation outcome. The adversary can still enforce that some correct player successfully completes (i.e., accepts) whereas the other one rejects the outcome.

Finally, in order to achieve agreement on the precomputation outcome, we have s and r_0 "2-cast" whether they accept using instances of the 2-cast protocol the precomputation is actually preparing for but such that enough Q-Flip data remains for one or more 2-casts (i.e., for the later 2-cast(s) we are in fact preparing for). Note that these "2-casts" are not necessarily reliable, but:

Observation 2: (Note that we can assume that $t = 1$, see Proposition 3)
If r_1 distributes inconsistent Q-Flip information then all correct players reject already before the invocation of the two final 2-cast protocols. Hence, if any correct player still accepts just before the invocation of the two final 2-cast protocols then the Q-Flip information is consistent and hence the those protocols achieve 2-cast.

Hence, if either s or r_0 "2-casts" his rejection at the end of the precomputation then all correct players that are not yet rejecting redecide to reject (since they know that the final 2-cast protocols are reliable). On the other hand, if the final 2-cast protocols are unreliable, then all correct players have already decided to reject before, and they simply ignore whatever is communicated during the last round.

We now proceed with a more detailed description of the protocol in three steps. Protocol `VerifSetup` describes how r_1 prepares and distributes Q and how s and r_0 cross-check their information gotten from r_1. Protocol `Conditional 2-Cast` describes an algorithm for 2-cast based on sequences Q as generated by player r_1. Finally, these two protocols are combined in order to obtain our final protocol for detectable precomputation, Protocol `OptPrecomp`.

Verifiable Setup of Q-Flip Information A straightforward way to get a verifiably correct setup Q would be a cut-and-choose manner where r_1 prepares more information than finally required of which s and r_0 would select a random part for testing which then would be discarded whereas the remaining part would be kept. The problem of this approach is that also r_1 (who sets up the states) must learn which part has been discarded – information that would have to be distributed by broadcast, which is not yet available.

As a consequence, in the following protocol, yet some partial information is tested by s and r_0 but is not discarded. It can be shown that the information thus leaked to s and r_0 will not help anyone of them in order to disrupt the final 2-cast based on this setup.

Protocol 3: `VerifSetup` [s, r_0, r_1 to prepare shared Q-Flip Information]

0. All players start with status ACCEPT.
1. r_1 sets up $m = 6\mu$ Q-Flip states each such that each possible outcome occurs exactly μ times; randomly permutes the states, and secretly sends the according values to s and r_0.
2. s and r_0 locally check whether they hold exactly $\frac{m}{3} = 2\mu$ of each value $(0, 1, 2)$. Anybody whose check fails changes to status REJECT.
3. r_0 sends to s his status (ACCEPT or REJECT) and, if he accepts, $\tau \ll m$ random indices in $\{1, \ldots, m\}$ and his according values.
4. s checks whether all those values from r_0 differ from his own values for the according states. If r_0 sent REJECT or if the check fails (i.e., reveals collisions) then s sets his status to REJECT.
5. s tells r_0 his status.
6. If s sent REJECT then r_0 sets his status to REJECT.

Notice that r_1 can undetectedly misbehave by distributing the 6 possible states with non-uniform cardinality, e.g., by repeatedly setting up only the three states determined by $(Q_s[i] = 0, Q_0[i] = 1)$, $(Q_s[i] = 1, Q_0[i] = 2)$, and $(Q_s[i] = 2, Q_0[i] = 0)$ – which makes s and r_0 nevertheless accept according to steps 2 and 4. However, regarding players s and r_0 being correct we shall see that, for the final 2-cast protocol, we only require that each player holds 2μ of each value in $\{0, 1, 2\}$ and that there are (almost) no collisions $Q_s[i] = Q_0[i]$.

Conditional 2-Cast. Conditional 2-cast is a protocol based on one instance of protocol `VerifSetup`. The goal of this protocol is that, given that any correct player accepted the setup protocol, the protocol achieves 2-cast with overwhelming probability. Note that besides its termination we do not require anything for this protocol for the case that all correct players rejected the setup protocol.

The protocol is described by Protocol `Conditional 2-Cast`, which is basically still the same as for the model of Section 3 (i.e., Protocol `Weak 2-Cast` followed by Protocol `Amplify`). We give a short motivation for the changes compared to the former protocol.

- Step 0 drops out since the Q-Flips are provided by Protocol `VerifSetup`.
- Step 1: r_1 has full knowledge of Q – hence only x_s is sent to r_1.
- Step 2: We exploit that exactly 2μ of each value must be prepared (as is required by Protocol `VerifSetup`) in order to enforce good statistical properties of σ_0. Note that the condition $(|\{i \in \sigma_0 : Q_0[i] = x_0\}| \geq m_0)$ must not be checked anymore because faulty r_1 could have undetectedly avoided such states during setup. The only thing to be checked is that s claims no collisions – which is weakened to allow ε_0 collisions because r_1 undetectedly could have set up a small portion of such errors.
- Step 3: r_0 sends his original data from s since r_1 can recalculate everything.

Definition 5. *Let $P_{fail|acc}$ be the probability that conditional 2-cast based on Q (as computed by the corresponding protocol* `VerifSetup`*) does not achieve 2-cast, given that one correct player (s, r_0, or r_1) accepted at the end of protocol* `VerifSetup`.

Protocol 4: `Conditional 2-Cast` [s to distribute $x_s \in \{0,1\}$ to r_k $(k \in \{0,1\})$]

1. $s \xrightarrow{\text{send}} r_0, r_1$: x_s; [r_k $(k \in \{0,1\})$ receives x_k]
   $s \xrightarrow{\text{send}} r_0$: $\sigma_s = \{i \in \{1,\ldots,m\} : Q_s[i] = 1 - x_s\}$; [$r_0$ receives σ_0]
   s: $y_s := x_s$;
2. r_0: **if** $(|\sigma_0| = 2\mu) \wedge (|\{i \in \sigma_0 : Q_0[i] = 1 - x_0\}| < \varepsilon_0)$ **then**
   $y_0 := x_0$
   else $y_0 := \perp$ **fi**;
   r_1: $y_1 := x_1$;
3. $r_0 \xrightarrow{\text{send}} r_1$: x_0, σ_0; [r_1 receives x_{01} and σ_{01}]
4. r_1: determine y_{01} as r_0 determines y_0 according to step 2;
   $\rho_{01} := \{i \in \sigma_{01} : Q_0[i] = y_{01}\}$;
   if $(\perp \neq y_{01} \neq y_1 \neq \perp) \wedge (|\rho_{01}| \geq m_0) \wedge$
   $(|\{i \in \rho_{01} : Q_1[i] = 2\}| \geq \lambda\,|\rho_{01}|)$ **then**
   $y_1 := y_{01}$
   fi;
5. **Amplify**;

It remains to prove that $P_{\text{fail}|\text{acc}}$ can be made negligibly small (with the protocol remaining efficient). Our detailed proof would exceed the space limits of this paper and is hence omitted. We restrict ourselves to an informal (though plausible) argument:

- faulty s or r_0 cannot successfully misbehave since then r_1 is correct and sets up Q correctly; and only a small part of the states is revealed during the precomputation phase.

– r_1 cannot successfully misbehave since Q can only contain a small fraction of incorrect triplets (otherwise the precomputation would have failed), and since r_1 is silent.[7]

Lemma 1. *There exist parameters μ, $\tau < \mu$, ε_0, m_0, and λ for the protocols* `VerifSetup` *and* `Conditional 2-Cast` *such that the following conditions hold:*

- *If Protocol* `Conditional 2-Cast` *with the given parameters is based on a set Q that has been set up with Protocol* `VerifSetup` *with the given parameters which itself has been accepted by at least one correct player then the probability $P_{fail|acc}$ that the protocol does not achieve 2-cast is exponentially small in μ.*
- *Independently of Q, the protocol always terminates.*

Detectable Precomputation for b later 2-casts. The final protocol for detectable precomputation is described in detail by Protocol `OptPrecomp`. It precomputes shared information for b later broadcast instances.

Protocol 5: `OptPrecomp` [s, r_0, r_1 to precompute for b 2-casts]

0. All players start in a status of ACCEPT.
1. Run $b+2$ executions of Protocol `VerifSetup` resulting in sets $S_1, \ldots, S_{b+2}$ of Q-Flip information.
2. s and r_0 decide on ACCEPT if and only if they accepted all $b+2$ executions of Protocol `VerifSetup` in step 1.
3. s and r_0 broadcast their status by applying Protocol `Conditional 2-Cast` with help of the sets S_{b+1} and S_{b+2}.
4. Anybody who received REJECT for one of the 2-casts sets his status to REJECT.

We now show that, whenever the probability $P_{\text{fail}|\text{acc}}$ is negligible, the correct players agree on their decision at the end of the precomputation and that, in case they accept, all b later 2-casts work correctly – both with overwhelming probability.

Lemma 2.

1. *At the end of protocol* `OptPrecomp` *the correct players will agree on their decision (ACCEPT or REJECT) with an error probability of at most $P_1 \leq 2P_{fail|acc}$. Moreover, if all players are correct, they will all agree on ACCEPT with error probability* 0.

[7] More precisely, the single message by r_1 that is sent during the amplification protocol in step 5 is only considered if r_1 is in fact correct.

2. *If the correct players accept the precomputation then, with an overall error probability of at most $P_2 \leq bP_{fail|acc}$, all b later 2-casts work reliably. Moreover, if all players are (and will be) correct, then all 2-casts work with error probability* 0.

Proof.

1. We distinguish two cases.
   (a) *At least one correct player accepts after step 2 of the protocol:* Then each conditional 2-cast in step 3 achieves 2-cast with an error probability of at most $P_{\text{fail}|\text{acc}}$ (see Definition 5). If now all correct players reject at the end of the protocol we are done. On the other hand, suppose that any correct player accepts at the end of the protocol. This player neither sent nor received a REJECT during step 3, and hence, with an error probability of at most $P_1 \leq 2P_{\text{fail}|\text{acc}}$, no other player sent or received a REJECT during step 3, and all correct players accept after step 4.
   (b) *All correct players reject after step 2:* Since no correct player ever changes from REJECT to ACCEPT, all correct players will reject at the end of the protocol.

   Finally, it can be easily seen by inspection that, in the case that all players are correct, all players agree on ACCEPT with error probability 0.
2. Any correct player that accepts at the end of the protocol already accepts after step 2 of the protocol and hence, by the definition of $P_{\text{fail}|\text{acc}}$, the overall error probability is at most $P_2 \leq bP_{\text{fail}|\text{acc}}$. Furthermore, if all players are (and will be) correct, none of the 2-cast protocols can fail. □

The full analysis of Protocol `OptPrecomp` together with Lemma 2 and Proposition 3 yields:

Lemma 3. *In the secure-channels model, efficient detectable precomputation for 2-cast among $n = 3$ players is achievable for any number of actively corrupted players ($t < n$).*

Together with Proposition 1 we get:

Corollary 1. *In the secure-channels model, efficient detectable precomputation for MPC among $n = 3$ players secure against one actively corrupted player ($t = 1$) is achievable.*

4.2 Detectable Precomputation for General n

In order to obtain a detectable precomputation protocol for general n and $t < n/2$, Protocol `OptPrecomp` can be applied in parallel for each triple of players and for every selection of a sender among them (i.e., $3\binom{n}{3}$ parallel protocols). We only need the following, minor changes:

1. The steps 5 of all parallel invocations of protocol `VerifSetup` are merged by having each of the n players tell each other whether or not he accepts with respect to all triples he is involved in, and to have each player reject in step 6 on any single reception of REJECT.

2. Protocol `Conditional 2-Cast` is turned into global conditional broadcast by applying the construction in [FM00].[8]
3. The steps 3 of all parallel invocations of protocol `OptPrecomp` are merged by having each player globally broadcast his status and by having the players decide to reject in step 4 on one single reception of REJECT for any one of the broadcasts.
4. In the final (conditional) broadcast protocol, each player of each a triple is involved as a sender in $2t$ different 2-casts with respect to this triple [FM00]. Furthermore, n instead of 2 conditional broadcasts are executed during step 3 of Protocol `OptPrecomp`. Hence, instead of $b+2$ invocations of `VerifSetup`, $2t(b+n)$ of them are required for each triple of players and for each selection of a sender among them.

The following Theorem immediately follows:

Theorem 2. *In the secure-channels model, efficient detectable precomputation for broadcast and MPC for n players and $t < n/2$ is achievable.*

Furthermore, detectable MPC can even be achieved in a slightly weaker model than with secure channels, namely in a model with classical authenticated channels and quantum channels. Since every player has the possibility to initiate rejection of the precomputation, we can apply quantum key agreement [BB84] in parallel to Protocol `OptPrecomp`. Whenever any quantum channel between two correct players would be eavesdropped they would detect it and just initiate rejection of the whole precomputation.

Theorem 3. *In the model with classical authenticated channels and quantum channels, efficient detectable precomputation for broadcast and MPC for n players and $t < n/2$ is achievable.*

5 Conclusions

To the best of our knowledge we have given the first examples of unconditionally secure protocols for broadcast and MPC in the secure channels model that tolerate $t < n/2$ active player corruptions from scratch, i.e., without any additional assumptions on the communication – at least in a way that the adversary cannot achieve anything more than make all players commonly abort the protocol.

For the case of $n = 3$ players this is strictly more than previously achieved. For the case of general $n > 3$ it is achieved at the price that the adversary may cause non-completion only by corrupting one single player but requiring permanent corruption.

[8] Replacing every invocation of ordinary 2-cast by conditional 2-cast in their protocol immediately yields conditional global broadcast since a player who accepts after step 2 of the merged Protocol `OptPrecomp` knows that all triples involving correct players share correct Q-Flip information.

References

[BB84] C. H. Bennett and G. Brassard. An update on quantum cryptography. In *Proceedings of CRYPTO 84*, volume 196 of *Lecture Notes in Computer Science*, pp. 475–480. Springer-Verlag, 1985, 19–22 Aug. 1984.

[Bea89] D. Beaver. Multiparty protocols tolerating half faulty processors. In *Proceedings of CRYPTO '89*, volume 435 of *Lecture Notes in Computer Science*, pp. 560–572. Springer-Verlag, 1990, 20–24 Aug. 1989.

[BGW88] M. Ben-Or, S. Goldwasser, and A. Wigderson. Completeness theorems for non-cryptographic fault-tolerant distributed computation. In *Proc. 20th ACM Symposium on the Theory of Computing (STOC)*, pp. 1–10, 1988.

[BPW91] B. Baum-Waidner, B. Pfitzmann, and M. Waidner. Unconditional byzantine agreement with good majority. In *Proceedings of STACS '91*, volume 480 of *LNCS*, pp. 285–295, Hamburg, Germany, 14–16 Feb. 1991. Springer.

[CCD88] D. Chaum, C. Crépeau, and I. Damgård. Multiparty unconditionally secure protocols (extended abstract). In *Proc. 20th ACM Symposium on the Theory of Computing (STOC)*, pp. 11–19, 1988.

[CDD+99] R. Cramer, I. Damgård, S. Dziembowski, M. Hirt, and T. Rabin. Efficient multiparty computations secure against an adaptive adversary. In *Proceedings of EUROCRYPT '99*, Lecture Notes in Computer Science, 1999.

[Chv79] V. Chvátal. The tail of the hypergeometric distribution. *Discrete Mathematics*, 25:285–287, 1979.

[DFF+82] D. Dolev, M. J. Fischer, R. Fowler, N. A. Lynch, and H. R. Strong. An efficient algorithm for Byzantine agreement without authentication. *Information and Control*, 52(3):257–274, Mar. 1982.

[FGM01] M. Fitzi, N. Gisin, and U. Maurer. Quantum solution to the byzantine agreement problem. To appear at Physical Review Letters, 87(21). Preliminary version: Quantum Physics, abstract quant-ph/0107127, 2001.

[FGMO01] M. Fitzi, J. A. Garay, U. Maurer, and R. Ostrovsky. Minimal complete primitives for unconditional multi-party computation. In *Proceedings of CRYPTO '01*, Lecture Notes in Computer Science, 2001.

[FLM86] M. J. Fischer, N. A. Lynch, and M. Merritt. Easy impossibility proofs for distributed consensus problems. *Distributed Computing*, 1:26–39, 1986.

[FM00] M. Fitzi and U. Maurer. From partial consistency to global broadcast. In *Proceedings of STOC '00*, pp. 494–503, Portland, Oregon, 2000. ACM.

[GMW87] O. Goldreich, S. Micali, and A. Wigderson. How to play any mental game — a completeness theorem for protocols with honest majority. In *Proceedings of STOC '87*, pp. 218–229, 1987.

[Hoe63] W. Hoeffding. Probability inequalities for sums of bounded random variables. *Journal of the American Statistical Association*, 58(301):13–30, 1963.

[KY] A. Karlin and A. C. Yao. Manuscript.

[PSL80] M. Pease, R. Shostak, and L. Lamport. Reaching agreement in the presence of faults. *Journal of the ACM*, 27(2):228–234, Apr. 1980.

[PW92] B. Pfitzmann and M. Waidner. Unconditional byzantine agreement for any number of faulty processors. In *Proceedings of STACS '92*, volume 577 of *LNCS*, pp. 339–350. Springer, 1992.

[RB89] T. Rabin and M. Ben-Or. Verifiable secret sharing and multiparty protocols with honest majority. In *Proceedings of STOC '89*, pp. 73–85, 1989.

[TC84] R. Turpin and B. A. Coan. Extending binary Byzantine Agreement to multivalued Byzantine Agreement. *Information Processing Letters*, 18(2):73–76, Feb. 1984.

[Yao82] A. C. Yao. Protocols for secure computations. In *Proceedings of FOCS '82*, pp. 160–164, 1982.

A Generation of One-Time Pads in the Q-Flip Model

Alice and Bob can generate a one-time pad (OTP) of length approximately k by using $3k$ Q-Flip invocations shared with an arbitrary third player Charlie and reporting to each other where they got a value different from 2. By this exchange of information Charlie does not get any additional information about the actual outcome of the Q-Flip invocations. Finally, the OTP is formed by those Q-Flip instances where both, Alice and Bob, got either 0 or 1, e.g., by Alice's according bits – which are the complements of Bob's.

B Chernoff and Hoeffding Bounds

For a detailed analysis of our protocols we will apply Chernoff and Hoeffding bounds [Hoe63,Chv79] in order to estimate upper bounds on their error probabilities.

The Chernoff bound gives an upper bound on the probability that of n independent Bernoulli trials the outcome deviates from the expected value by a given fraction.

Let X_i $(1 \leq i \leq n)$ be a sequence of independent random variables with expected value μ. By $\mathcal{C}(\mu, n, \lambda)$ we denote the Chernoff bound as follows

$$\begin{array}{l} \lambda < 1 : \mathcal{C}_{\downarrow}(\mu, n, \lambda) = \text{Prob}(\sum_{i=1}^{n} X_i \leq \lambda\, \mu n) \leq e^{-\frac{\mu n}{2}(1-\lambda)^2} \\ \lambda > 1 : \mathcal{C}_{\uparrow}(\mu, n, \lambda) = \text{Prob}(\sum_{i=1}^{n} X_i \geq \lambda\, \mu n) \leq e^{-\frac{\mu n}{3}(\lambda-1)^2} \end{array} \tag{5}$$

Furthermore, a bound by Hoeffding can be used to estimate tail probabilities of hypergeometric distributions. By the term $\mathcal{H}(N, K, n, k)$ we refer to a setting where N items are given of which K are "good". The experiment consists of selecting n out of the N items at random, and $\mathcal{H}(N, K, n, k)$ denotes the probability that at least k of the n selections are "good". Let $t = \frac{k}{n} - \frac{K}{N}$. The following inequation holds for any t such that $0 \leq t \leq 1 - \frac{K}{N}$ [Chv79],

$$\mathcal{H}(N, K, n, k) = \sum_{i=k}^{n} \binom{K}{i} \binom{N-K}{n-i} \binom{N}{n}^{-1} \leq e^{-2t^2 n} \,. \tag{6}$$

C Broadcast and MPC with an External Information Source: Details

Lemma 4. *Let $\lambda_0 < 1$ and $\lambda_1 > 1$. The probability P_{stats} that of m invocations of Q-Flip any one of the six possible outcomes in $\{(0,1,2), \ldots, (2,1,0)\}$ occurs either less than $m_0 = \lambda_0 \frac{m}{6} < \frac{m}{6}$ or more than $m_1 = \lambda_1 m > \frac{m}{6}$ times satisfies $P_{stats} \leq 6\, max(\mathcal{C}_{\downarrow}(\frac{1}{6}, m, \lambda_0), \mathcal{C}_{\uparrow}(\frac{1}{6}, m, \lambda_1))$.*[9]

[9] See Appendix B for the definition of $\mathcal{C}_{\downarrow}$ and $\mathcal{C}_{\uparrow}$.

Proof. The according probability for each particular outcome, e.g. $(0,1,2)$, can be independently estimated by the Chernoff bound (with random variable X_i representing the i-th Q-Flip). The overall probability is at most as large as the sum of these six probabilities (union bound). □

Lemma 5 (All players correct). *If all players are correct and each possible outcome of Q-Flip appears at least m_0 times, then protocol* `Weak 2-Cast` *achieves weak 2-cast.*

Proof. By the given assumptions we have $\sigma_s = \sigma_0 = \sigma_1$ and $x_s = x_0 = x_1$ after step 1 of the protocol. Since $\sigma_s = \{i \in \{1,\ldots,m\} : Q_s[i] = 1 - x_s\}$, it holds for both recipients r_k that

$$\{i \in \sigma_k : Q_k[i] = 1 - x_k\} = \emptyset \quad \wedge \quad |\{i \in \sigma_k : Q_k[i] = x_k\}| \geq m_0 \ .$$

Hence $y_0 = y_1 = x_s$ after step 3 and hence also at the end of the protocol. □

The following corollary follows from Observation 1 on page 6.

Corollary 2 (r_1 possibly faulty). *If the players s and r_0 are correct and each possible outcome of Q-Flip appears at least m_0 times, then protocol* `Weak 2-Cast` *achieves weak 2-cast.*

It now remains to determine upper bounds on the error probability for the cases that s is faulty (P_s) or that r_0 is faulty (P_{r_0}). The following lemma will be used for the analysis of the former case in the proof of Lemma 7.

Lemma 6. *If each possible outcome of Q-Flip appears at least m_0 times and at most m_1 times and if (faulty) sender s submits $x_0 \in \{0,1\}$ to r_0 and $x_1 = 1 - x_0$ to r_1 and selects k indices $i \in \{1,\ldots,m\}$ such that either*
- *$i \in \sigma_0 \cap \sigma_1$, or*
- *$i \in \sigma_0 \setminus \sigma_1$ such that $Q_s[i] = 2$*

then $y_0 = x_0$ and $y_1 = 1 - x_0$ (i.e., disagreement) holds at the end of the protocol with a probability of at most $(\frac{m_1}{m_0+m_1})^k$.

Proof. If s submits one single index $i \in \sigma_0 \cap \sigma_1$ then either $Q_0[i] = 1 - x_0$ or $Q_1[i] = 1 - x_1$ with a probability of at least $\frac{m_0}{m_0+m_1}$ (since the only information by s is $Q_s[i]$ and hence s risks, with the according probability, to produce a collision on either recipient's side which makes him decide on $\perp$).

On the other hand, if s submits one single index $i \in \sigma_0 \setminus \sigma_1$ such that $Q_s[i] = 2$ (the only possibility in order to achieve that $Q_0[i] = x_0$ and $Q_1[i] \neq 2$) then r_0 decides on $\perp$ with a probability of at least $\frac{m_0}{m_0+m_1}$ (since s risks, with the according probability, to produce a "collision" on r_0's side which makes him decide on $\perp$).

Finally, in order to achieve that $y_0 = x_0$ and $y_1 = 1 - x_0$ holds at the end of the protocol, s must prevent any single collision for all k index selections. This can be achieved with a probability of at most $(\frac{m_1}{m_0+m_1})^k$. □

Lemma 7 (s possibly faulty). *If the players r_0 and r_1 are correct and each possible outcome of Q-Flip appears at least m_0 times and at most m_1 times, then protocol* `Weak 2-Cast` *fails to achieve weak 2-cast with probability at most* $P_s < \left(\frac{m_1}{m_0+m_1}\right)^{(1-\lambda)m_0}$.

Proof. The only way for s to make the protocol fail is to force the recipients to decide on distinct bits, i.e., $y_0 = x_0 = b \in \{0,1\}$ and $y_1 = x_1 = 1 - b$. Hence both recipients must already decide on those values during step 3 of the protocol, which implies $|\rho_{01}| = |\rho_0| \geq m_0$ – since r_0 would set $y_0 = \perp$ otherwise. Furthermore, r_0 must not be able to convince r_1 to redecide on $y_1 = y_{01} = y_0$ during step 5 of the protocol. Since the first two conditions according to step 5 are satisfied, i.e.,

- $\perp \neq y_0 = y_{01} \neq y_1$ (since the recipients hold distinct bits), and
- $|\rho_{01}| \geq m_0$ (see above),

the last condition must be violated, i.e., it must hold that

$$|\{i \in \rho_{01} \setminus \sigma_1 : Q_1[i] = 2\}| < \lambda\, |\rho_{01}| \ .$$

Hence s must find some $\ell > (1-\lambda)\, |\rho_{01}| \geq (1-\lambda)m_0$ indices i such that either

- $i \in \rho_{01} \cap \sigma_1 \quad (\subseteq \sigma_0 \cap \sigma_1)$, *or*
- $i \in \rho_{01} \setminus \sigma_1 \quad \wedge \quad Q_1[i] \neq 2$ (and hence $Q_s[i] = 2$) ,

such that no collision occurs. By Lemma 6 this happens with probability at most $P_s < \left(\frac{m_1}{m_0+m_1}\right)^{(1-\lambda)m_0}$. □

Lemma 8 (r_0 possibly faulty). *If the players s and r_1 are correct and each possible outcome of Q-Flip appears at least m_0 times and at most m_1 times then, for any $\lambda > \frac{m_1}{m_0+m_1}$, protocol* `Weak 2-Cast` *fails to achieve weak 2-cast with a probability of at most $P_{r_0} \leq \mathcal{H}(m, m_1, m_0, \lambda m_0)$.*[10]

Proof. The only way for r_0 to make the protocol fail is to make r_1 adopt $y_1 := y_{01} \neq x_s$ during step 5 of the protocol. Hence the following conditions must hold:

- $(y_{01} = 1 - y_1 = 1 - x_s)$,
- $(|\rho_{01}| \geq m_0)$,
- $(|\{i \in \rho_{01} \setminus \sigma_1 : Q_1[i] = 2\}| \geq \lambda\, |\rho_{01}|)$.

Let $u = |\rho_{01}| \geq m_0$. Since s is correct and hence

$$\{i \in \{1, \ldots, m\} : Q_s[i] = 1 - x_s = y_{01} \ \wedge \ Q_1[i] = 2\} \subseteq \sigma_s = \sigma_1 \ ,$$

[10] See Appendix B for the definition of $\mathcal{H}$.

r_0 must select u indices i such that for λu of them it holds that $i \notin \sigma_s$, and $Q_0[i] = y_{01}$, and $Q_1[i] = 2$. An optimal strategy in order to achieve this is by randomly selecting m_0 indices i such that $Q_0[i] = 1 - x_s$ (corresponding to random selections from the first and the last row in Figure 1–C).

This process corresponds to a hypergeometric distribution with $N = m$, $K = m_1$, and $n = m_0$ (see Section B). The probability for r_0 to succeed is hence given by the tail of this distribution according to $k \geq \lambda m_0$. By Equation (6), for any $\lambda > \frac{m_1}{m_0+m_1}$, this probability can be estimated as

$$P_{r_0} \leq \mathcal{H}(m, m_1, u, \lambda u) \leq \mathcal{H}(m, m_1, m_0, \lambda m_0) \leq e^{-2\left(\lambda - \frac{m_1}{m_0+m_1}\right)^2 m_0} .$$

□

Lemma 9. *For every desired security parameter $k > 0$ there exist parameters m, m_0, and λ such that protocol* `Weak 2-Cast` *has communication and computation complexities polynomial in k and achieves weak 2-cast with an error probability of at most $P_f < e^{-k}$.*

Proof. We let $\lambda_0 = \frac{3}{4}$ and $\lambda_1 = \frac{5}{4}$, and fix the parameterization of the protocol as follows such that there is only one free parameter left, namely m, the number of Q-Flip invocations:

$$\boxed{\begin{array}{lclcl} m_0 & = & \lambda_0 \frac{m}{6} & = & \frac{m}{8} \\ m_1 & = & \lambda_1 \frac{m}{6} & = & \frac{5m}{24} \\ \lambda & = & \lambda_0 & = & \frac{3}{4} \end{array}}$$

Now, as a function of security parameter k, let $m \overset{!}{\geq} 288\,(k+2)$. According to Corollary 2 and Lemmas 4, 7, and 8 we get the following estimations:

$$P_{stats} \leq 6 \max\left(\mathcal{C}_{\downarrow}(\frac{1}{6}, m, \lambda_0), \mathcal{C}_{\uparrow}(\frac{1}{6}, m, \lambda_1)\right) \leq 6e^{-\frac{m}{288}} \tag{7}$$

$$P_{r_1} = 0 \tag{8}$$

$$P_s \leq \left(\frac{m_1}{m_0+m_1}\right)^{(1-\lambda)m_0} = \left(\frac{\lambda_1}{\lambda_0+\lambda_1}\right)^{(1-\lambda)\lambda_0 m} = \left(\frac{8}{5}\right)^{-\frac{m}{32}} < e^{-\frac{m}{69}} \tag{9}$$

$$P_{r_0} \leq e^{-2\left(\lambda - \frac{m_1}{m_0+m_1}\right)^2 m_0} = e^{-2\left(\lambda - \frac{\lambda_1}{\lambda_0+\lambda_1}\right)^2 m_0} = e^{-\frac{3m}{128}} \leq e^{-\frac{m}{43}} \tag{10}$$

Finally, the overall error probability can be estimated by the sum of the probabilities that either the statistics of Q-Flip fail, i.e., that at least one of the six possible outcomes appears less than m_0 or more than m_1 times, or that, given good statistics, a faulty player can nevertheless successfully misbehave:

$$P_f \leq P_{stats} + \max(P_{r_1}, P_s, P_{r_0}) \leq 6e^{-\frac{m}{288}} + e^{-\frac{m}{69}} < 7e^{-\frac{m}{288}} < e^{-\frac{m-576}{288}} \leq e^{-k} .$$

□

Perfectly Secure Message Transmission Revisited
(Extended Abstract)

Yvo Desmedt[1,2] and Yongge Wang[3]*

[1] Computer Science, Florida State University,
Tallahassee, Florida FL 32306-4530, USA,
desmedt@cs.fsu.edu
[2] Dept. of Mathematics, Royal Holloway, University of London, UK
[3] Karthika Technologies Inc.,
ywang@karthika.com

Abstract. Achieving secure communications in networks has been one of the most important problems in information technology. Dolev, Dwork, Waarts, and Yung have studied secure message transmission in one-way or two-way channels. They only consider the case when all channels are two-way or all channels are one-way. Goldreich, Goldwasser, and Linial, Franklin and Yung, Franklin and Wright, and Wang and Desmedt have studied secure communication and secure computation in multi-recipient (multicast) models. In a "multicast channel" (such as Ethernet), one processor can send the same message – simultaneously and privately – to a fixed subset of processors. In this paper, we shall study necessary and sufficient conditions for achieving secure communications against active adversaries in mixed one-way and two-way channels. We also discuss multicast channels and neighbor network channels.

Keywords: network security, privacy, reliability, network connectivity

1 Introduction

If there is a private and authenticated channel between two parties, then secure communication between them is guaranteed. However, in most cases, many parties are only indirectly connected, as elements of an incomplete network of private and authenticated channels. In other words they need to use intermediate or internal nodes. Achieving participants cooperation in the presence of faults is a major problem in distributed networks. Original work on secure distributed computation assumed a complete graph for secure and reliable communication. Dolev, Dwork, Waarts, and Yung [4] were able to reduce the size of the network graph by providing protocols that achieve private and reliable communication without the need for the parties to start with secret keys. The interplay of network connectivity and secure communication has been studied extensively (see, e.g., [1,2,3,4,10]). For example, Dolev [3] and Dolev et al. [4] showed that, in

* Part of the work was done when this author was with Certicom Corp.

L.R. Knudsen (Ed.): EUROCRYPT 2002, LNCS 2332, pp. 502–517, 2002.

the case of k Byzantine faults, reliable communication is achievable only if the system's network is $2k + 1$ connected. They also showed that if all the paths are one way, then $3k + 1$ connectivity is necessary and sufficient for reliable and private communications. However they did not prove any results for the general case when there are certain number of directed paths in one direction and another number of directed paths in the other direction. While undirected graphs correspond naturally to the case of pairwise two-way channels, directed graphs do not correspond to the case of all-one-way or all-two-way channels considered in [4], but to the mixed case where there are some paths in one direction and some paths in the other direction. In this paper, we will initiate the study in this direction by showing what can be done with a general directed graph. Note that this scenario is important in practice, in particular, when the network is not symmetric. For example, a channel from u to v is cheap and a channel from v to u is expensive but not impossible. Another example is that u has access to more resources than v does.

Goldreich, Goldwasser, and Linial [9], Franklin and Yung [7], Franklin and Wright [6], and Wang and Desmedt [15] have studied secure communication and secure computation in *multi-recipient (multicast)* models. In a "multicast channel" (such as Ethernet), one participant can send the same message – simultaneously and privately – to a fixed subset of participants. Franklin and Yung [7] have given a necessary and sufficient condition for individuals to exchange private messages in multicast models in the presence of passive adversaries (passive gossipers). For the case of active Byzantine adversaries, many results have been presented by Franklin and Wright [6], and, Wang and Desmedt [15]. Note that Goldreich, Goldwasser, and Linial [9] have also studied fault-tolerant computation in the public multicast model (which can be thought of as the largest possible multirecipient channels) in the presence of active Byzantine adversaries. Specifically, Goldreich, et al. [9] have made an investigation of general fault-tolerant distributed computation in the full-information model. In the full information model no restrictions are made on the computational power of the faulty parties or the information available to them. (Namely, the faulty players may be infinitely powerful and there are no private channels connecting pairs of honest players). In particular, they present efficient two-party protocols for fault-tolerant computation of any bivariate function.

There are many examples of multicast channels (see, e.g. [6]), such as an Ethernet bus or a token ring. Another example is a shared cryptographic key. By publishing an encrypted message, a participant initiates a multicast to the subset of participants that is able to decrypt it.

We present our model in Section 2. In Sections 3 and 4, we study secure message transmission over directed graphs. Section 5 is devoted to reliable message transmission over hypergraphs, and Section 6 is devoted to secure message transmission over neighbor networks.

2 Model

We will abstract away the concrete network structures and consider directed graphs. A directed graph is a graph $G(V, E)$ where all edges have directions. For a directed graph $G(V, E)$ and two nodes $u, v \in V$,

Throughout this paper, n denotes the number of vertex disjoint paths between two nodes and k denotes the number of faults under the control of the adversary. We write $|S|$ to denote the number of elements in the set S. We write $x \in_R S$ to indicate that x is chosen with respect to the uniform distribution on S. Let $\mathbf{F}$ be a finite field, and let $a, b, c, M \in \mathbf{F}$. We define $\text{auth}(M, a, b) := aM + b$ (following [6,8,13,14]) and $\text{auth}(M, a, b, c) := aM^2 + bM + c$ (following [15]). Note that each authentication key $key = (a, b)$ can be used to authenticate one message M without revealing any information about any component of the authentication key and the each authentication key $key = (a, b, c)$ can be used to authenticate two messages M_1 and M_2 without revealing any information about any component of the authentication key.

Let k and n be two integers such that $0 \le k < n \le 3k+1$. A $(k+1)$-out-of-n secret sharing scheme is a probabilistic function S: $\mathbf{F} \to \mathbf{F}^n$ with the property that for any $m \in \mathbf{F}$ and $(v_1, \dots, v_n) = \text{S}(m)$, no information of m can be inferred from any k entries of $(v_1, \dots, v_n)$, and m can be recovered from any $k+1$ entries of $(v_1, \dots, v_n)$. The set of all possible $(v_1, \dots, v_n)$ is called a code and its elements codewords. We say that a $(k+1)$-out-of-n secret sharing scheme can detect k' errors if given any codeword $(v_1, \dots, v_n)$ and any tuple $(u_1, \dots, u_n)$ over F such that $0 < |\{i : u_i \neq v_i, 1 \le i \le n\}| \le k'$ one can detect that $(u_1, \dots, u_n)$ is not a codeword. If the code is Maximal Distance Separable, then the maximum value of errors that can be detected is $n - k - 1$ [11]. We say that the $(k+1)$-out-of-n secret sharing scheme can correct k' errors if from any $(v_1, \dots, v_n) = \text{S}(m)$ and any tuple $(u_1, \dots, u_n)$ over F with $|\{i : u_i \neq v_i, 1 \le i \le n\}| \le k'$ one can recover the secret m. If the code is Maximal Distance Separable, then the maximum value of errors that allows the recovery of the vector $(v_1, \dots, v_n)$ is $(n-k-1)/2$ [11]. A $(k+1)$-out-of-n Maximal Distance Separable (MDS) secret sharing scheme is a $(k+1)$-out-of-n secret sharing scheme with the property that for any $k' \le (n-k-1)/2$, one can correct k' errors and simultaneously detect $n-k-k'-1$ errors (as follows easily by generalizing [11, p. 10]). Maximal Distance Separable (MDS) secret sharing schemes can be constructed from any MDS codes, for example, from Reed-Solomon code [12].

In a message transmission protocol, the sender A starts with a message M^A drawn from a message space $\mathcal{M}$ with respect to a certain probability distribution. At the end of the protocol, the receiver B outputs a message M^B. We consider a synchronous system in which messages are sent via multicast in rounds. During each round of the protocol, each node receives any messages that were multicast for it at the end of the previous round, flips coins and perform local computations, and then possibly multicasts a message. We will also assume that the message space $\mathcal{M}$ is a subset of a finite field $\mathbf{F}$.

We consider two kinds of adversaries. A passive adversary (or gossiper adversary) is an adversary who can only observe the traffic through k internal nodes.

An active adversary (or Byzantine adversary) is an adversary with unlimited computational power who can control k internal nodes. That is, an active adversary will not only listen to the traffics through the controlled nodes, but also control the message sent by those controlled nodes. Both kinds of adversaries are assumed to know the complete protocol specification, message space, and the complete structure of the graph. In this paper, we will not consider a dynamic adversary who could change the nodes it controls from round to round, instead we will only consider static adversaries. That is, at the start of the protocol, the adversary chooses the k faulty nodes. (An alternative interpretation is that k nodes are static collaborating adversaries.)

For any execution of the protocol, let *adv* be the adversary's view of the entire protocol. We write $adv(M, r)$ to denote the adversary's view when $M^A = M$ and when the sequence of coin flips used by the adversary is r.

Definition 1. *(see Franklin and Wright [6])*

1. *Let $\delta < \frac{1}{2}$. A message transmission protocol is* δ-reliable *if, with probability at least $1 - \delta$, B terminates with $M^B = M^A$. The probability is over the choices of M^A and the coin flips of all nodes.*
2. *A message transmission protocol is* reliable *if it is* 0-reliable.
3. *A message transmission protocol is* ε-private *if, for every two messages M_0, M_1 and every r, $\sum_c |\Pr[adv(M_0, r) = c] - \Pr[adv(M_1, r) = c]| \leq 2\varepsilon$. The probabilities are taken over the coin flips of the honest parties, and the sum is over all possible values of the adversary's view.*
4. *A message transmission protocol is* perfectly private *if it is* 0-private.
5. *A message transmission protocol is* (ε, δ)-secure *if it is ε-private and δ-reliable.*
6. *An (ε, δ)-secure message transmission protocol is* efficient *if its round complexity and bit complexity are polynomial in the size of the network, $\log \frac{1}{\varepsilon}$ (if $\varepsilon > 0$) and $\log \frac{1}{\delta}$ (if $\delta > 0$).*

For two nodes A and B in a directed graph such that there are $2k+1$ node disjoint paths from A to B, there is a straightforward reliable message transmission from A to B against a k-active adversary: A sends the message m to B via all the $2k + 1$ paths, and B recovers the message m by a majority vote.

3 $(0, \delta)$-Secure Message Transmission in Directed Graphs

Our discussion in this section will be concentrated on directed graphs. Dolev, Dwork, Waarts, and Yung [4] addressed the problem of secure message transmissions in a point-to-point network. In particular, they showed that if all channels from A to B are one-way, then $(3k + 1)$-connectivity is necessary and sufficient for (0,0)-secure message transmissions from A to B against a k-active adversary. They also showed that if all channels between A and B are two-way, then $(2k + 1)$-connectivity is necessary and sufficient for (0,0)-secure message transmissions between A and B against a k-active adversary. In this section we assume that there are only $2(k-u)+1$ directed node disjoint paths from A to B, where $1 \leq u \leq k$. We wonder how many directed node disjoint paths from B to A are necessary and sufficient to achieve $(0,\delta)$-secure message transmissions from A to B against a k-active adversary.

Franklin and Wright [6] showed that if there is no channel from B to A, then $2k+1$ channels from A to B is necessary for $(1-\delta)$-reliable (assuming that $\delta < \frac{1}{2}$) message transmission from A to B against a k-active adversary. In the following, we first show that this condition is sufficient also.

Theorem 1. *Let $G(V,E)$ be a directed graph, $A, B \in V$, and $0 < \delta < \frac{1}{2}$. If there is no directed paths from B to A, then the necessary and sufficient condition for $(0,\delta)$-secure message transmission from A to B against a k-active adversary is that there are $2k+1$ directed node disjoint paths from A to B.*

Proof. (Sketch) The necessity was proved in Franklin and Wright [6]. Let $p_1, \ldots, p_{2k+1}$ be the $2k+1$ directed node disjoint paths from A to B. Let $s^A \in \mathbf{F}$ be the secret that A wants to send to B. A constructs $(k+1)$-out-of-$(2k+1)$ secret shares $v = (s_1^A, \ldots, s_{2k+1}^A)$ of s^A. The protocol proceeds from round 1 through round $2k+1$. In round i, A chooses $\{(a_{i,j}^A, b_{i,j}^A) \in_R \mathbf{F}^2 : 1 \leq j \leq 2k+1\}$, sends $(s_i^A, \text{auth}(s_i^A, a_{i,1}^A, b_{i,1}^A), \ldots, \text{auth}(s_i^A, a_{i,2k+1}^A, b_{i,2k+1}^A))$ to B via path p_i, and sends $(a_{i,j}^A, b_{i,j}^A)$ to B via path p_j for each $1 \leq j \leq 2k+1$. In round i, B receives $(s_i^B, c_1^B, \ldots, c_{2k+1}^B)$ via path p_i, and receives $(a_{i,j}^B, b_{i,j}^B)$ via path j for each $1 \leq j \leq 2k+1$. B computes $t = |\{j : c_j^B = \text{auth}(s_i^B, a_{i,j}^B, b_{i,j}^B)\}|$. If $t \geq k+1$, then B decides that s_j^B is a valid share. Otherwise B discards s_j^B. It is easy to check that after the round $2k+1$, with high probability, B will get at least $k+1$ valid shares to s^A. Thus, with high probability, B will recover the secret s^A. In the full version of this paper, we will show that this protocol is a $(0,\delta)$-secure message transmission protocol from A to B. Q.E.D.

Theorem 2. *Let $G(V,E)$ be a directed graph, $A, B \in V$, and $u \geq 1$. If there are $2(k-u)+1 \geq k+1$ directed node disjoint paths from A to B, then a necessary condition for private message transmission from A to B against a k-active adversary is that there are u directed node disjoint paths (these u paths are also node disjoint from the $2(k-u)+1$ paths from A to B) from B to A.*

Proof. First assume that there are less than u directed node disjoint paths from B to A. A strategy that will now be used by the adversary is that controlling $u-1$ nodes to disconnect all directed paths from B to A and controlling $k-u+1$ directed paths from A to B. Thus the adversary could make sure that no feedback message will be sent from B to A. This means that we are left with the same situation as Theorem 1 using $2(k-u)+1$ one-way channels. Since $k-u+1$ of these paths are controlled by the adversary, by Theorem 1, we need $2(k-u+1)+1 > 2(k-u)+1$ directed paths from A to B. This is a contradiction.

Secondly we assume that there are u directed node disjoint paths q_i from B to A, $2(k-u)+1$ paths p_i from A to B, and that p_1 is not node disjoint from q_1. A strategy that will now be used by the adversary is that using one node to control both paths p_1 and q_1, using other $u-1$ nodes to disconnect all directed paths from B to A, controlling $k-u$ other directed paths from A to B. A similar argument as above will show a contradiction. Q.E.D.

In the following we prove a simple sufficient condition.

Theorem 3. *Let $G(V,E)$ be a directed graph, $A, B \in V$. If there are two directed node disjoint paths p_0 and p_1 from A to B, and one directed path q (which is node disjoint from p_1 and p_2) from B to A, then for any $0 < \delta < \frac{1}{2}$, there is a $(0,\delta)$-secure message transmission protocol from A to B against a 1-active adversary.*

Proof. (Sketch) Let $s^A \in \mathbf{F}$ be the secret message that A wants to send to B. In the following we describe the protocol briefly without proof. The details and a generalization will be given in the full version of this paper.

Step 1 A chooses $s_0^A \in_R \mathbf{F}$, $(a_0^A, b_0^A), (a_1^A, b_1^A) \in_R \mathbf{F}^2$, and let $s_1^A = s^A - s_0^A$. For each $i \in \{0,1\}$, A sends $(s_i^A, (a_i^A, b_i^A), \mathrm{auth}(s_i^A, a_{1-i}^A, b_{1-i}^A))$ to B via path p_i.

Step 2 Assumes that B receives $(s_i^B, (a_i^B, b_i^B), c_i^B)$ via path p_i. B checks whether $c_i^B = \mathrm{auth}(s_i^B, a_{1-i}^B, b_{1-i}^B)$. If both equations hold, then B knows that with high probability the adversary was either passive or not on the paths from A to B. B can recover the secret message, sends "OK" to A via the path q, and terminate the protocol. Otherwise, one of equations does not hold and B knows that the adversary was on one of the paths from A to B. In this case, B chooses $(a^B, b^B) \in_R \mathbf{F}^2$, and sends $((a^B, b^B), (s_0^B, (a_0^B, b_0^B), c_0^B), (s_1^B, (a_1^B, b_1^B), c_1^B))$ to A via the path q.

Step 3 If A receives "OK", then A terminates the protocol. Otherwise, from the information A received via path q, A decides which path from A to B is corrupted and recover B's authentication key (a^A, b^A). A sends $(s^A, \mathrm{auth}(s^A, a^A, b^A))$ to B via the uncorrupted path from A to B.

Step 4 B recovers the message and checks that the authenticator is correct.

Q.E.D.

4 (0,0)-Secure Message Transmission in Directed Graphs

In the previous section, we addressed probabilistic reliable message transmission in directed graphs. In this section, we consider reliable message transmission in directed graphs. We first start with necessary conditions.

Theorem 4. *Replacing in Theorem 2: $2(k-u)+1$ by $3(k-u)+1$, provides necessary conditions for $(0,0)$-secure message transmission from A to B.*

Proof. Use an argument as in the proof of Theorem 2, but use the $3k+1$ bound from [4] instead of the $2k+1$ one. Q.E.D.

We will show that if there are $3k+1-u$ paths from A to B and u paths from B to A, then $(0,0)$-secure message transmission from A to B is possible. We first show the simple case for $u=1$.

Theorem 5. *Let $G(V,E)$ be a directed graph, $A,B \in V$. If there are $3k \geq 2k+1$ directed node disjoint paths from A to B and one directed path from B to A (the directed path from B to A is node disjoint from the paths from A to B) then there is a $(0,0)$-secure message transmission protocol from A to B against a k-active adversary.*

Proof. Let $p_1, \ldots, p_{3k}$ be the directed paths from A to B and q be the directed path from B to A. The protocol π proceeds as follows:

Step 1 B sets A_STOP $=0$ and B_STOP $=0$.

Step 2 A chooses a $key^A \in_R \mathbf{F}$ and constructs $(k+1)$-out-of-$3k$ MDS secret shares $v=(s_1^A, ..., s_{3k}^A)$ of key^A. For each $1 \leq i \leq 3k$, A sends s_i to B via the path p_i.

Step 3 Let $v^B=(s_1^B, ..., s_{3k}^B)$ be the shares B receives. If B finds that there are at most $k-1$ errors, B recovers key^B from the shares, sends "stop" to A via the path q, and sets B_STOP $=1$. Otherwise there are k errors. In this case B sends v^B back to A via the path q (note that q is an honest path in this case).

Step 4 A distinguishes the following two cases:
1. A receives $v^A=(s_1^A, ..., s_{3k}^A)$ from the path q. A reliably sends $\mathcal{P}=\{i : s_i^A \neq s_i\}$ to B.
2. A received "stop" or anything else via q. A reliably sends "stop" to B.

Step 5 B distinguishes the following two cases:
1. B reliably receives "stop" from A. B sets A_STOP $=1$.
2. B reliably receives $\mathcal{P}$ from A. If B_STOP $=0$ then B recovers key^B from the shares $\{s_i^B : i \notin \mathcal{P}\}$ (note that $|\{s_i^B : i \notin \mathcal{P}\}|=2k$).

Step 6 A reliably transmits key^A+m^A to B, where m^A is the message to be transmitted.

Step 7 B reliably receives the ciphertext c^B and decrypts the message $m^B = c^B - key^B$.

Note that if B sends v^B to A in Step 3 then k paths from A to B are corrupted and the path q is honest. Thus the adversary will not learn v^B and key. If the adversary controls the path q, then it may change the message "stop" to something else. In this case, A will not be able to identify the corrupted paths from A to B. However, since B has already recovered the key, B will just ignore the next received message. It is straightforward to show that the protocol is $(0,0)$-secure. Q.E.D.

Before proving our main theorem, we describe a variant π' of the protocol π in the proof of Theorem 5. We call B_STOP during the i-th execution of π B_STOP(i) and similar for A_STOP(i). The new protocol π' proceeds as follows:

Step 1 Instead of sending the secret key^A, A first sends $R_1 \in_R \mathbf{F}$ using π.

Step 2 A, B execute Steps 1 and 2 of π for the message R_2 where $R_1 + R_2 = key^A$.

Step 3 If B_STOP(2) = 1 (B_STOP(1) = 1 or 0), then B computes the secret key^B.

Step 4 If B_STOP(1) = 1 and B_STOP(2) = 0, then B and A continue with the rest of π for R_2, and B will be able to compute the secret key^B.

Step 5 If B_STOP(1) = 0 and B_STOP(2) = 0 then A_STOP(2) = 0. In this case, k corrupted paths should have already been identified by both A and B in the second run of π (though A does not know whether it has correctly identified the corrupted paths). A "restarts" the protocol by sending key^A using a $(k+1)$-out-of-$2k$ secret sharing scheme along the $2k$ non-corrupted paths. B excludes the known k bad paths and computes the secret from the secret sharing scheme.

Note that due to the malicious information A received, A may restart the protocol even though B may have already computed the correct secret. In this case, B can just ignore these messages.

Theorem 6. *Let $G(V, E)$ be a directed graph, $A, B \in V$. If there are $3k+1-u \geq 2k+1$ (which implies $k \geq u$) directed node disjoint paths from A to B and u directed paths from B to A (the directed paths from B to A are node disjoint from the paths from A to B) then there is a $(0,0)$-secure message transmission protocol from A to B against a k-active adversary.*

Proof. Let $p_1, \ldots, p_{3k+1-u}$ be the directed paths from A to B, and $q_1, \ldots, q_u$ be the directed paths from B to A. The protocol will be based on the variant protocol π' of Theorem 5. Before we begin, we note that a $(k+1)$-out-of-$(3k+1-u)$ MDS secret sharing scheme can detect k errors and simultaneously correct $k-u$ errors. In the following, we informally describe the protocol. The full protocol will be presented in the full version of this paper.

Step 1 A chooses $R_0 \in_R \mathbf{F}$ and sends R_0 to B via the $3k+1-u$ paths using a $(k+1)$-out-of-$(3k+1-u)$ MDS secret sharing scheme.

Step 2 If B can correct the errors (i.e. there were at most $k-u$ errors in the received shares), B finds R_0. Otherwise B needs help from A (that is, B will send the received shares back to A via all B to A paths). The problems are that:

- B may receive help even if B has never asked. However B can detect this. Therefore B will always work with A on such a protocol.
- A may receive u different versions of "asking for help".

For each of the u paths from B to A, B and A will keep track of the "dishonest" paths from A to B according to the information A received on this path.

Step 3 A now sends R_1 using a $(k+1)$-out-of-$(3k+1-u)$ MDS secret sharing scheme where $key = R_0 + R_1$.

Step 4 If B can correct the errors, B has found the secret. However, B may need to play with A prolonging the protocol due to incorrect paths from B to A. B distinguishes the following two cases:

1. B has not asked help in Step 2. B can ask help now and B will then recover the secret *key*.
2. B has asked help in Step 2. In this case B cannot ask for help again (otherwise the enemy may learn the secret). The protocol needs to be restarted from Step 1 on. We know that in this case there is at least one honest path from B to A. (Indeed, if B asked for help in Step 2, then the number of dishonest paths from A to B is at least $k' \geq k - u + 1$. Assume that all paths from B to A were dishonest then the total number of dishonest parties is $k' + u \geq k + 1$, which is a contradiction.) Since A and B identified (correctly or incorrectly) dishonest parties on the paths from A to B (the version corresponding to the honest B to A path should have correctly identified the dishonest paths), they will only use these paths that were not identified as dishonest. If k' dishonest paths from A to B have been (correctly or incorrectly) identified, a $(k+1)$-out-of-$(3k+1-u-k')$ MDS secret sharing scheme will be used. This MDS secret sharing scheme will only be used for error detection (or message recovery in the case that no error occurs), thus it can be used to detect $3k + 1 - u - k' - k - 1 = 2k - u - k' \geq k - k'$ errors. Due to the fact that this MDS secret sharing scheme cannot detect k errors we need to organize ourselves that B will never use incorrectly identified paths from A to B since otherwise B could compute the incorrect "secret". This is easy to be addressed by having B detect whether a path from B to A is dishonest or not. This is done by having A reliably sends to B what A received via the path q_i from B to A. During each run of the protocol, B will either recover the secret message (when no error occurs) or detect one corrupted path from A to B (A could also detect the corrupted path from A to B according to the information A received on the honest B to A path – though A may not know which path from B to A is honest). Thus the protocol will be restarted at most u times.

After the initial run, B will first use the path q_1 to send the "asking for help" message. Then it will use the path q_2, and then q_3, etc. These steps can be run in parallel. Q.E.D.

Theorem 6 can be strengthened as follows.

Theorem 7. *Let $G(V, E)$ be a directed graph, $A, B \in V$. Assume that there are $3k+1-u \geq 2k+1$ (which implies $k \geq u$) directed node disjoint paths from A to B and u directed paths from B to A. If $3k+1-2u$ paths among these $3k+1-u$ paths from A to B are node disjoint from the u paths from B to A, then there is a $(0,0)$-secure message transmission protocol from A to B against a k-active adversary.*

Proof. The protocol proceeds in the same way as the protocol in the proof of Theorem 6. In addition, at the end of the protocol, A constructs a $(k+1)$-out-of-$(3k+1-2u)$ MDS shares $(s_1, \ldots, s_{3k+1-2u})$ of the secret key^A and sends these shares to B via the $3k+1-2u$ paths which are node disjoint from the paths from the u paths from B to A. If B has determined that all these u paths from B to A have been corrupted, then B will recover the secret key^A from the received shares $(s_1^B, \ldots, s_{3k+1-2u}^B)$ since a $(k+1)$-out-of-$(3k+1-2u)$ MDS secret sharing scheme can be used to detect and correct $k-u$ errors simultaneously. Note that if at least one path from B to A is honest, then B has recovered the secret already and can just ignore this last message. Q.E.D.

We close our discussion on secure message transmission in directed graphs with an application of Theorem 7. Up to now, we have concentrated on the situation that there are more paths from A to B than paths from B to A. The following theorem address the situation that there are more paths from B to A.

Theorem 8. *Let $G(V,E)$ be a directed graph, $A, B \in V$. Assume that there are $k+1$ directed node disjoint paths from A to B and $2k+1$ directed node disjoint paths from B to A. If $k+1$ paths among these $2k+1$ paths from B to A are node disjoint from the $k+1$ paths from A to B, then there is a $(0,\delta)$-secure message transmission protocol from A to B against a k-active adversary.*

Proof. See the full version of this paper. Q.E.D

5 Secure Message Transmissions in Hypergraphs

Hypergraphs have been studied by Franklin and Yung in [7]. A hypergraph H is a pair (V,E) where V is the node set and E is the hyperedge set. Each hyperedge $e \in E$ is a pair (v, v^*) where $v \in V$ and v^* is a subset of V. In a hypergraph, we assume that any message sent by a node v will be received identically by all nodes in v^*, whether or not v is faulty, and all parties outside of v^* learn nothing about the content of the message.

Let $v, u \in V$ be two nodes of the hypergraph $H(V,E)$. We say that there is a "*direct link*" from node v to node u if there exists a hyperedge (v, v^*) such that $u \in v^*$. We say that there is an "*undirected link*" from v to u if there is a directed link from v to u or a directed link from u to v. If there is a directed (undirected) link from v_i to v_{i+1} for every i, $0 \le i < k$, then we say that there is a "*directed path*" ("*undirected path*") from v_0 to v_k. v and u are "*strongly k-connected*" ("*weakly k-connected*") in the hypergraph $H(V,E)$ if for all $S \subset V - \{v,u\}$, $|S| < k$, there remains a directed (undirected) path from v to u after the removal of S and all hyperedges (x, x^*) such that $S \cap (x^* \cup \{x\}) \neq \emptyset$. Franklin and Yung [7] showed that reliable and private communication from v to u is possible against a *k-passive* adversary if and only if v and u are strongly 1-connected and weakly $k+1$-connected. It should be noted that u and v are strongly k-connected does not necessarily mean that v and u are strongly k-connected.

Following Franklin and Yung [7], and, Franklin and Wright [6], we consider multicast as our only communication primitive in this section. A message that is multicast by any node v in a hypergraph is received by all nodes v^* with privacy (that is, nodes not in v^* learn nothing about what was sent) and authentication (that is, nodes in v^* are guaranteed to receive the value that was multicast and to know which node multicast it). We assume that all nodes in the hypergraph know the complete protocol specification and the complete structure of the hypergraph.

Definition 2. *Let $H(V, E)$ be a hypergraph, $A, B \in V$ be distinct nodes of H, and $k \geq 0$. A, B are* k-separable *in H if there is a node set $W \subset V$ with at most k nodes such that any directed path from A to B goes through at least one node in W. We say that W* separates A, B.

Remark. Note that there is no straightforward relationship between strong connectivity and separability in hypergraphs.

Theorem 9. *The nodes A, B of a hypergraph H is not $2k$-separable if and only if there are $2k+1$ directed node disjoint paths from A to B in H.*

Proof. This follows directly from the maximum-flow minimum-cut theorem in classical graph theory. For details, see, e.g., [5]. Q.E.D.

Theorem 10. *A necessary and sufficient condition for reliable message transmission from A to B against a k-active adversary is that A and B are not $2k$-separable in H.*

Proof. First assume that A and B cannot be separated by a $2k$-node set. By Theorem 9, there are $2k+1$ directed node disjoint paths from A to B in H. Thus reliable message transmission from A to B is possible.

Next assume that A, B can be separated by a $2k$-node set W in H. We shall show that reliable message transmission is impossible. Suppose that π is a message transmission protocol from A to B and let $W = W_0 \cup W_1$ be a $2k$-node separation of A and B with W_0 and W_1 each having at most k nodes. Let m_0 be the message that A transmits. The adversary will attempt to maintain a simulation of the possible behavior of A by executing π for message $m_1 \neq m_0$. The strategy of the adversary is to flip a coin and then, depending on the outcome, decide which set of W_0 or W_1 to control. Let W_b be the chosen set. In each execution step of the transmission protocol, the adversary causes each node in W_b to follow the protocol π as if the protocol were transmitting the message m_1. This simulation will succeeds with nonzero probability. Since B does not know whether $b = 0$ or $b = 1$, at the end of the protocol B cannot decide whether A has transmitted m_0 or m_1 if the adversary succeeds. Thus with nonzero probability, the reliability is not achieved. Q.E.D.

Theorem 10 gives a sufficient and necessary condition for achieving reliable message transmission against a k-active adversary over hypergraphs. In the following example, we show that this condition is not sufficient for achieving privacy against a k-active adversary (indeed, even not for a k-passive adversary).

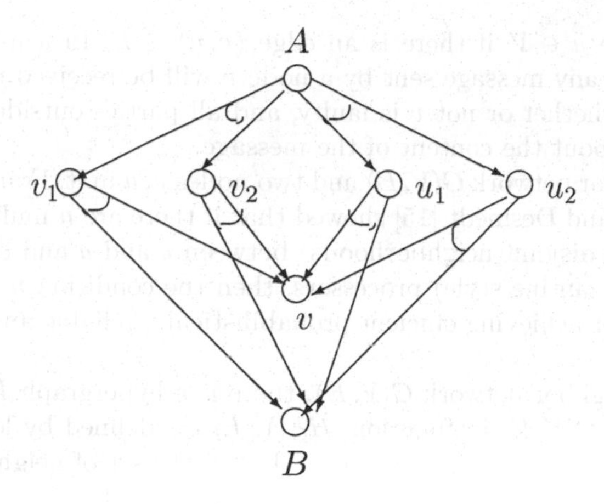

Fig. 1. The hypergraph $H(V, E_h)$ in Example 1

Example 1 *Let $H(V, E_h)$ be the hypergraph in Figure 1 where $V = \{A, B, v_1, v_2, v, u_1, u_2\}$ and $E_h = \{(A, \{v_1, v_2\}), (v_1, \{v, B\}), (v_2, \{v, B\}), (A, \{u_1, u_2\}), (u_1, \{v, B\}), (u_2, \{v, B\})\}$. Then the nodes A and B are not 2-separable in H. Theorem 10 shows that reliable message transmission from A to B is possible against a 1-active adversary. However, the hypergraph H is not weakly 2-connected (the removal of the node v and the removal of the corresponding hyperedges will disconnect A and B). Thus, the result by Franklin and Yung [7] shows that private message transmission from A to B is not possible against a 1-passive adversary.*

Theorem 11. *Let $\delta > 0$ and A and B be two nodes in a hypergraph $H(V, E)$ satisfying the following conditions:*

1. *A and B are not 2k-separable in H.*
2. *B and A are not 2k-separable in H.*
3. *A and B are strongly k-connected in H.*

Then there is a $(0, \delta)$-secure message transmission protocol from A to B against a k-active adversary.

Proof. See the full version of this paper. Q.E.D.

The results in Sections 3 and 4 show that the condition in Theorem 11 is not necessary.

6 Secure Message Transmission over Neighbor Networks

6.1 Definitions

A special case of the hypergraph is the *neighbor networks.* A neighbor network is a graph $G(V, E)$. In a neighbor network, a node $v \in V$ is called a neighbor of another node $u \in V$ if there is an edge $(v, u) \in E$. In a neighbor network, we assume that any message sent by a node v will be received identically by all its neighbors, whether or not v is faulty, and all parties outside of v's neighbor learn nothing about the content of the message.

For a neighbor network $G(V, E)$ and two nodes v, u in it, Franklin and Wright [6], and, Wang and Desmedt [15] showed that if there are n multicast lines (that is, n paths with disjoint neighborhoods) between v and u and there are at most k malicious (Byzantine style) processors, then the condition $n > k$ is necessary and sufficient for achieving efficient probabilistically reliable and perfect private communication.

For each neighbor network $G(V, E)$, there is a hypergraph $H_G(V, E_h)$ which is equivalent to $G(V, E)$ in function. $H_G(V, E_h)$ is defined by letting E_h be the set of hyperedges (v, v^*) where $v \in V$ and v^* is the set of neighbors of v.

Let v and u be two nodes in a neighbor network $G(V, E)$. We have the following definitions:

1. v and u are *k-connected* in $G(V, E)$ if there are k node disjoint paths between v and u in $G(V, E)$.
2. v and u are *weakly k-hyper-connected* in $G(V, E)$ if v and u are weakly k-connected in $H_G(V, E_h)$.
3. v and u are *k-neighbor-connected* in $G(V, E)$ if for any set $V_1 \subseteq V \backslash \{v, u\}$ with $|V_1| < k$, the removal of $neighbor(V_1)$ and all incident edges from $G(V, E)$ does not disconnect v and u, where $neighbor(V_1) = V_1 \cup \{v \in V : \exists u \in V_1(u, v)$ such that $\in E\} \setminus \{v, u\}$.
4. v and u are *weakly (n, k)-connected* if there are n node disjoint paths $p_1, \ldots,$ p_n between v and u and, for any node set $T \subseteq (V \setminus \{v, u\})$ with $|T| \leq k$, there exists an i $(1 \leq i \leq n)$ such that all nodes on p_i have no neighbor in T.

It is easy to check that the following relations hold.

$$\text{weak } (n, k-1)\text{-connectivity } (n \geq k) \Rightarrow k\text{-neighbor-connectivity} \Rightarrow \text{weak}$$
$$k\text{-hyper-connectivity} \Rightarrow k\text{-connectivity}$$

In the following examples, we will show that these implications are strict.

Example 2 *Let $G(V, E)$ be the graph in Figure 2 where $V = \{A, B, C, D\}$ and $E = \{(A, C), (C, B), (A, D), (D, B), (C, D)\}$. Then it is straightforward to check that $G(V, E)$ is 2-connected but not weakly 2-hyper-connected.*

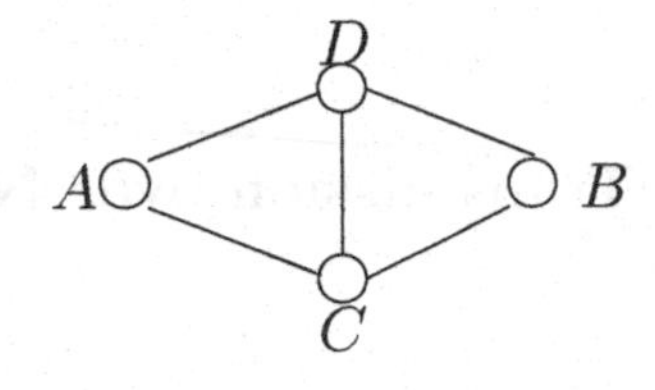

Fig. 2. The graph $G(V, E)$ in Example 2

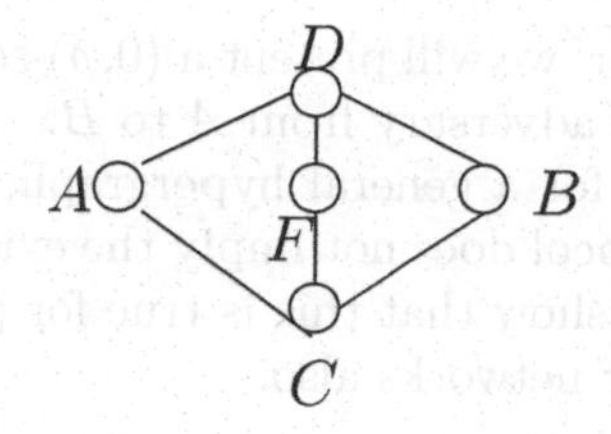

Fig. 3. The graph $G(V, E)$ in Example 3

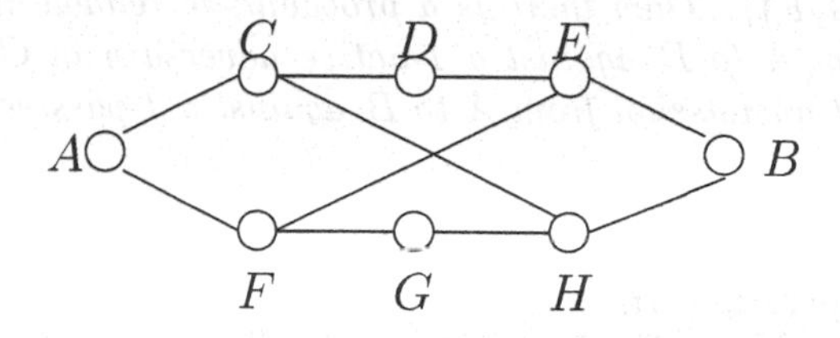

Fig. 4. The graph $G(V, E)$ in Example 4

Example 3 *Let $G(V, E)$ be the graph in Figure 3 where $V = \{A, B, C, D, F\}$ and $E = \{(A, C), (A, D), (C, B), (D, B), (C, F), (F, D)\}$. Then it is straightforward to check that A and B are weakly 2-hyper-connected but not 2-neighbor-connected.*

Example 4 *Let $G(V, E)$ be the graph in Figure 4 where $V = \{A, B, C, D, E, F, G, H\}$ and $E = \{(A, C), (C, D), (D, E)\ (E, B), (A, F), (F, G), (G, H)\ (H, B), (C, H), (E, F)\}$. Then it is straightforward to check that A and B are 2-neighbor-connected but not weakly $(2, 1)$-connected.*

Example 2 shows that k-connectivity does not necessarily imply weak k-hyper-connectivity. Example 3 shows that weak k-hyper-connectivity does not necessarily imply k-neighbor-connectivity. Example 4 shows that k-neighbor connectivity does not necessarily imply weak $(n, k-1)$-connectivity for some $n \geq k$.

6.2 $(0, \delta)$-Secure Message Transmission over Neighbor Networks

Wang and Desmedt [15] have given a sufficient condition for achieving $(0, \delta)$-security message transmission against a k-active adversary over neighbor networks. In this section, we first show that their condition is not necessary.

Theorem 12. *(Wang and Desmedt [15]) If A and B are weakly (n, k)-connected for some $k < n$, then there is an efficient $(0, \delta)$-secure message transmission between A and B.*

The condition in Theorem 12 is not necessary. For example, the neighbor network G in Example 3 is not 2-neighbor-connected, thus not weakly $(2, 1)$-connected. In the full version of this paper, we will present a $(0, \delta)$-secure message transmission protocol against a 1-active adversary from A to B.

Example 1 shows that for a general hypergraph, the existence of a reliable message transmission protocol does not imply the existence of a private message transmission protocol. We show that this is true for probabilistic reliability and perfect privacy in neighbor networks also.

Example 5 *Let $G(V, E)$ be the neighbor network in Figure 5 where $V = \{A, B, C, D, E, F, G\}$ and $E = \{(A, C), (C, D), (D, B), (A, E), (E, F), (F, B), (G, C), (G, D), (G, E), (G, F)\}$. Then there is a probabilistic reliable message transmission protocol from A to B against a 1-active adversary in G. But there is no private message transmission from A to B against a 1-passive (or 1-active) adversary in G.*

Proof. It is straightforward to check that $G(V, E)$ is not weakly 2-hyper-connected. Indeed, in the hypergraph $H_G(V, E_h)$ of $G(V, E)$, the removal of node G and the removal of the corresponding hyperedges will disconnect A and B completely. Thus Franklin and Yung's result in [7] shows that there is no private message transmission protocol against a 1-passive (or 1-active) adversary from A to B. It is also straightforward to check that Franklin and Wright's [6] reliable message transmission protocol against a 1-active adversary works for the two paths (A, C, D, B) and (A, E, F, B). Q.E.D.

Though weak k-hyper-connectivity is a necessary condition for achieving probabilistically reliable and perfectly private message transmission against a $(k-1)$-active adversary, we do not know whether this condition is sufficient. We conjecture that there is no probabilistically reliable and perfectly private message transmission protocol against a 1-active adversary for the weakly 2-hyper-connected neighbor network $G(V, E)$ in Figure 6, where $V = \{A, B, C, D, E, F, G, H\}$ and $E = \{(A, C), (C, D), (D, E), (E, B), (A, F), (F, G), (G, H), (H, B), (D, G)\}$. Note that in order to prove or refute our conjecture, it is sufficient to show whether there is a probabilistically reliable message transmission protocol against a 1-active adversary for the neighbor network. For this specific neighbor network, the trick in our previous protocol could be used to convert any probabilistically reliable message transmission protocol to a probabilistically reliable and perfectly private message transmission protocol against a 1-active adversary.

References

1. M. Ben-Or, S. Goldwasser, and A. Wigderson. Completeness theorems for non-cryptographic fault-tolerant distributed computing. In: *Proc. ACM STOC, '88*, pages 1–10, ACM Press, 1988.
2. D. Chaum, C. Crepeau, and I. Damgard. Multiparty unconditional secure protocols. In: *Proc. ACM STOC '88*, pages 11–19, ACM Press, 1988.
3. D. Dolev. The Byzantine generals strike again. *J. of Algorithms*, **3**:14–30, 1982.
4. D. Dolev, C. Dwork, O. Waarts, and M. Yung. Perfectly secure message transmission. *J. of the ACM*, **40**(1):17–47, 1993.

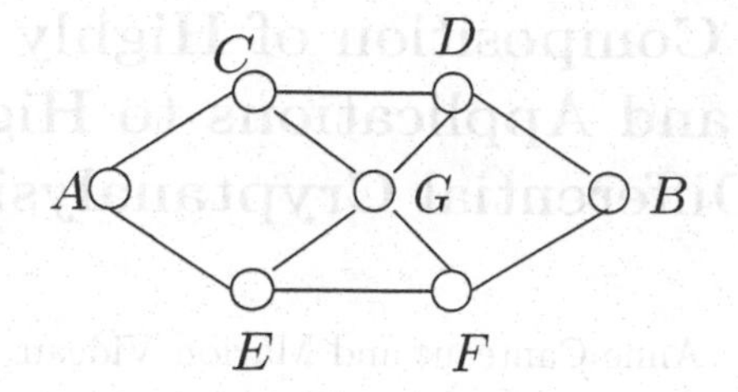

Fig. 5. The graph $G(V, E)$ in Example 5

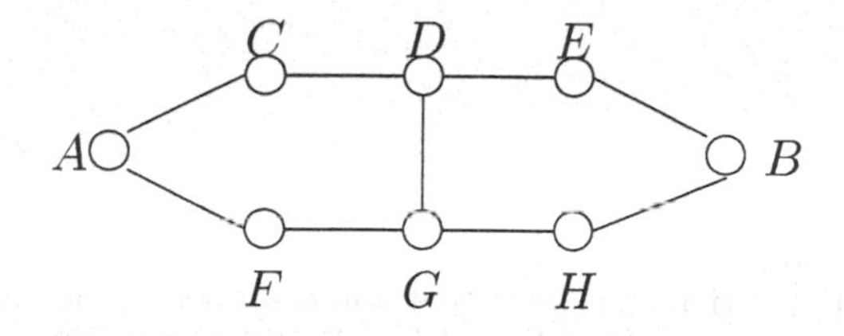

Fig. 6. The graph $G(V, E)$

5. L.R. Ford and D. R. Fulkerson. *Flows in Networks*. Princeton University Press, Princeton, NJ, 1962.
6. M. Franklin and R. Wright. Secure communication in minimal connectivity models. *Journal of Cryptology*, **13**(1):9–30, 2000.
7. M. Franklin and M. Yung. Secure hypergraphs: privacy from partial broadcast. In: *Proc. ACM STOC '95*, pages 36–44, ACM Press, 1995.
8. E. Gilbert, F. MacWilliams, and N. Sloane. Codes which detect deception. *The BELL System Technical Journal*, **53**(3):405–424, 1974.
9. O. Goldreich, S. Goldwasser, and N. Linial. Fault-tolerant computation in the full information model. *SIAM J. Comput.* **27**(2):506–544, 1998.
10. V. Hadzilacos. *Issues of Fault Tolerance in Concurrent Computations*. PhD thesis, Harvard University, Cambridge, MA, 1984.
11. F. J. MacWilliams and N. J. A. Sloane. *The theory of error-correcting codes*. North-Holland Publishing Company, 1978.
12. R. J. McEliece and D. V. Sarwate. On sharing secrets and Reed-Solomon codes. *Comm. ACM*, **24**(9):583–584, September 1981.
13. T. Rabin. Robust sharing of secrets when the dealer is honest or faulty. *J. of the ACM*, **41**(6):1089–1109, 1994.
14. T. Rabin and M. Ben-Or. Verifiable secret sharing and multiparty protocols with honest majority. In: *Proc. ACM STOC '89*, pages 73–85, ACM Press, 1989.
15. Y. Wang and Y. Desmedt. Secure communication in multicast channels: the answer to Franklin and Wright's question. *J. of Cryptology*, **14**(2):121–135, 2001.

Degree of Composition of Highly Nonlinear Functions and Applications to Higher Order Differential Cryptanalysis

Anne Canteaut and Marion Videau

INRIA – projet CODES, B.P. 105 – 78153 Le Chesnay Cedex, France,
{Anne.Canteaut,Marion.Videau}@inria.fr

Abstract. To improve the security of iterated block ciphers, the resistance against linear cryptanalysis has been formulated in terms of provable security which suggests the use of highly nonlinear functions as round functions. Here, we show that some properties of such functions enable to find a new upper bound for the degree of the product of its Boolean components. Such an improvement holds when all values occurring in the Walsh spectrum of the round function are divisible by a high power of 2. This result leads to a higher order differential attack on any 5-round Feistel ciphers using an almost bent substitution function. We also show that the use of such a function is precisely the origin of the weakness of a reduced version of MISTY1 reported in [23, 1].

Keywords. Block ciphers, higher order differential cryptanalysis, Boolean functions, nonlinearity.

1 Introduction

The development of cryptanalysis in the last ten years has led to the definition of some design criteria for block ciphers. These criteria correspond to some mathematical properties of the round function which is used in an iterated block cipher. In particular, the use of a highly nonlinear round function ensures a high resistance to linear attacks [16,17]. The functions with maximal nonlinearity are called almost bent. They only exist for an odd number of variables, but they also guarantee the best resistance to differential cryptanalysis [6]. Such functions are used for instance in the block cipher MISTY [18]. Here, we show that these optimal functions present some particular properties which introduce other weaknesses in the cipher. This vulnerability comes from the fact that all values occurring in the Walsh spectrum of an almost bent function are divisible by a high power of 2. Most highly nonlinear functions of an even number of variables present a similar structure, except the inverse function. Such a spectral property for a round function F leads to an upper bound on the degree of the function $F \circ F$ which grows much slower than $\deg(F)^2$. Therefore, any iterated cipher using an almost bent function may be vulnerable to a higher order differential attack [12,10], even if the round function has a high degree. This weakness leads

L.R. Knudsen (Ed.): EUROCRYPT 2002, LNCS 2332, pp. 518–533, 2002.

to a new design criterion for iterated block ciphers: the Walsh spectrum of the round function should contain at least one value which is not divisible by a higher power of 2. The S-box used in AES is the only known highly nonlinear function which fulfills this requirement.

The paper is organized as follows. Section 2 recalls the main spectral properties of the round function which are involved in differential and linear cryptanalysis. The general principle of a higher order differential attack is described in Section 3. Section 4 then investigates the link between the divisibility of the Walsh coefficients of a function and the degree of the product of its Boolean components. This result leads to a higher order differential attack on any 5-round Feistel cipher using an almost bent substitution function. Finally, we point out that the attack of a reduced version of MISTY1 presented in [23,1] is a direct consequence of the use of almost bent S-boxes. We show that a similar attack can be performed for different block sizes and almost bent S-boxes.

2 Spectral Properties of a Round Function

In an iterated block cipher, the ciphertext is obtained by iteratively applying a round function F to the plaintext. In an r-round iterated cipher, we have

$$x_i = F(x_{i-1}, K_i)$$

where x_0 is the plaintext, x_r is the ciphertext and the r-round keys $(K_1, \ldots, K_r)$ are usually derived from a unique secret key. For any fixed round key K, the round function $F_K : x \mapsto F(x, K)$ is a permutation of the set of n-bit vectors, $\mathbb{F}_2^n$, where n is the block size. The resistance of such cipher to some particular attacks can be quantified by some properties of the round function.

A *Boolean function f of n variables* is a function from $\mathbb{F}_2^n$ into $\mathbb{F}_2$. It can be expressed as a polynomial, called its *algebraic normal form*. The *degree* of f, denoted by $\deg(f)$, is the degree of its algebraic normal form. The following notation will be extensively used in the paper. The usual dot product between two vectors x and y is denoted by $x \cdot y$. For any $\alpha \in \mathbb{F}_2^n$, φ_α is the linear function of n variables: $x \mapsto \alpha \cdot x$.

For any Boolean function f of n variables, we denote by $\mathcal{F}(f)$ the following value related to the Walsh (or Fourier) transform of f:

$$\mathcal{F}(f) = \sum_{x \in \mathbb{F}_2^n} (-1)^{f(x)} = 2^n - 2wt(f) ,$$

where $wt(f)$ is the Hamming weight of f, i.e., the number of $x \in \mathbb{F}_2^n$ such that $f(x) = 1$.

Definition 1. *The* Walsh spectrum *of a Boolean function f of n variables is the multiset*

$$\{\mathcal{F}(f + \varphi_\alpha), \alpha \in \mathbb{F}_2^n\} .$$

The Walsh spectrum *of a vectorial function F from $\mathbb{F}_2^n$ into $\mathbb{F}_2^n$ consists of the Walsh spectra of all Boolean functions $\varphi_\alpha \circ F : x \mapsto \alpha \cdot F(x)$. Therefore, it corresponds to the multiset*

$$\{\mathcal{F}(\varphi_\alpha \circ F + \varphi_\beta),\ \alpha \in \mathbb{F}_2^n \setminus \{0\}, \beta \in \mathbb{F}_2^n\} .$$

A linear attack against a cipher with round function F exploits the existence of a pair (α, β) with $\alpha \neq 0$ such that, for almost all round keys K, the function $x \mapsto \varphi_\alpha \circ F_K(x) + \varphi_\beta(x)$ takes the same value for most values of $x \in \mathbb{F}_2^n$. Therefore, all functions $\varphi_\alpha \circ F_K$ should be far from all affine functions. This requirement is related to the nonlinearity of the functions F_K.

Definition 2. *[21] The* nonlinearity *of a function F from $\mathbb{F}_2^n$ into $\mathbb{F}_2^n$ is the Hamming distance between all $\varphi_\alpha \circ F, \alpha \in \mathbb{F}_2^n$, $\alpha \neq 0$, and the set of affine functions. It is given by*

$$2^{n-1} - \frac{1}{2}\mathcal{L}(F) \quad \text{where} \quad \mathcal{L}(F) = \max_{\alpha \in \mathbb{F}_2^n} \max_{\beta \in \mathbb{F}_2^n} |\mathcal{F}(\varphi_\alpha \circ F + \varphi_\beta)| .$$

Proposition 1. *[6] For any function $F : \mathbb{F}_2^n \to \mathbb{F}_2^n$,*

$$\mathcal{L}(F) \geq 2^{\frac{n+1}{2}} .$$

In case of equality F is called almost bent *(AB).*

This minimum value for $\mathcal{L}(F)$ can only be achieved if n is odd. For even n, some functions with $\mathcal{L}(F) = 2^{\frac{n}{2}+1}$ are known and it is conjectured that this value is the minimum. Note that the Walsh spectrum of a function is invariant under both right and left compositions by a linear permutation of $\mathbb{F}_2^n$.

A particular property of almost bent functions is that their Walsh spectrum is unique.

Proposition 2. *[6] The Walsh spectrum of an almost bent function F from $\mathbb{F}_2^n$ into $\mathbb{F}_2^n$ takes the values 0 and $\pm 2^{\frac{n+1}{2}}$ only.*

This property implies that any almost bent function is *almost perfect nonlinear* [22], i.e., that it ensures the best resistance to differential cryptanalysis. Therefore, the use of an almost bent function as round function (or as substitution function) provides a high resistance to both linear and differential attacks. These functions are used in MISTY [18]. Similarly, AES uses a function of an even number of variables which has the highest known nonlinearity.

3 Higher Order Differential Attacks

Higher order differential cryptanalysis was introduced by Knudsen [12]. As a generalization of differential cryptanalysis, it relies on some properties of higher order derivatives of a vectorial function. In the following, we denote by $\oplus$ the bitwise exclusive-or.

Definition 3. *[14] Let F be a function from $\mathbb{F}_2^n$ into $\mathbb{F}_2^m$. For any $a \in \mathbb{F}_2^n$, the* derivative of F *with respect to a is the function*

$$D_a F(x) = F(x \oplus a) \oplus F(x) .$$

For any k-dimensional subspace V of $\mathbb{F}_2^n$, the k-th derivative of F with respect *to V is the function*

$$D_V F = D_{a_1} D_{a_2} \dots D_{a_k} F ,$$

where $(a_1, \dots, a_k)$ is any basis of V. Moreover, we have for any $x \in \mathbb{F}_2^n$

$$D_V F(x) = \bigoplus_{v \in V} F(x \oplus v) .$$

We now consider an r-round iterated cipher with block size n and round function F. We call *reduced cipher*, the cipher obtained by removing the final round of the original cipher. The reduced cipher corresponds to the function $G = F_{K_{r-1}} \circ \dots \circ F_{K_1}$.

Suppose that there exists a k-dimensional subspace $V \subset \mathbb{F}_2^n$ such that

$$D_V G(x) = c \text{ for all } x \in \mathbb{F}_2^n$$

where c is a constant in $\mathbb{F}_2^n$ which does not depend on the round keys $K_1, \dots, K_{r-1}$. Then, for any round keys the reduced cipher G satisfies

$$\forall x \in \mathbb{F}_2^n, \quad \bigoplus_{v \in V} G(x \oplus v) = c . \qquad (1)$$

This property leads to the following chosen plaintext attack.

1. Select a random plaintext $x_0 \in \mathbb{F}_2^n$ and get the ciphertexts c_v corresponding to all plaintexts $x_0 \oplus v$, $v \in V$.
2. Compute c by applying (1) to the reduced cipher with any round keys (e.g. $K_1, \dots, K_{r-1} = 0$).
3. For each candidate round key k_r, compute

$$\sigma(k_r) = \bigoplus_{x \in V} F_{k_r}^{-1}(c_v) .$$

The key k_r for which $\sigma(k_r) = c$ is the correct last-round key with a high probability. If the attack returns several round keys, it could be repeated for different values of x_0. The running-time of the attack corresponds to 2^{m+k} evaluations of F^{-1} where m is the size of the round key and k is the dimension of V. It requires the knowledge of 2^k chosen plaintexts.

The main problem in this attack is then to find a subspace V satisfying (1) and having the lowest possible dimension. A natural candidate for V arises when the degree of the reduced cipher is known.

Definition 4. *The* degree *of a function F from $\mathbb{F}_2^n$ into $\mathbb{F}_2^n$ is the maximum degree of its Boolean components:* $\deg(F) = \max_{1 \le i \le n} \deg(\varphi_{e_i} \circ F)$ *where $(e_i)_{1 \le i \le n}$ denotes the canonical basis of $\mathbb{F}_2^n$.*

For any F of degree d, we obviously have $D_V F = 0$ for any $(d+1)$-dimensional subspace $V \subset \mathbb{F}_2^n$. Therefore, if the reduced cipher G has degree at most d for all round keys, it is possible to perform a differential attack of order $(d+1)$.

The degree of the round function F provides a trivial upper bound on the degree of the reduced cipher:

$$\deg(G) \leq (\deg(F))^{r-1} \ .$$

This bound was directly used by Jakobsen and Knudsen [10] for breaking a cipher example proposed in [22], whose round function is an almost bent quadratic permutation. Unfortunately, this method can only be used when the degree of the round function is very low. It clearly appears that another approach has to be used when the degree of the round function is strictly greater than $\sqrt{n}$ since $(\deg(F))^{r-1} > n$ for any $r \geq 3$.

4 Divisibility of the Walsh Spectrum and Degree of a Composed Function

In this section, we focus on the degree of a function $F' \circ F$ where F and F' are two mappings from $\mathbb{F}_2^n$ into $\mathbb{F}_2^n$. We show that the trivial bound

$$\deg(F' \circ F) \leq \deg(F') \deg(F)$$

can be improved when the values occurring in the Walsh spectrum of F are divisible by a high power of 2. This situation especially occurs when F is an almost bent function (see Proposition 2).

Definition 5. *The Walsh spectrum of a function F from $\mathbb{F}_2^n$ into $\mathbb{F}_2^m$ is said to be* 2^ℓ-divisible *if all its values are divisible by 2^ℓ. Moreover, it is said* exactly 2^ℓ-divisible *if, additionally, it contains at least one value which is not divisible by $2^{\ell+1}$.*

The divisibility of the values occurring in the Walsh spectrum of a function F provides an upper bound on its degree [15, Page 447]. The following proposition is a direct consequence of [4, Lemma 3].

Proposition 3. *Let F be a function from $\mathbb{F}_2^n$ into $\mathbb{F}_2^m$. If the Walsh spectrum of F is 2^ℓ-divisible, then $\deg(F) \leq n - \ell + 1$.*

The i-th Boolean component of $F' \circ F$ can be expressed as $f'(F_1(x), \ldots, F_n(x))$, where f' is the i-th Boolean component of F' and $(F_1, \ldots, F_n)$ denote the Boolean components of F. Using the algebraic normal form of f', we can write this function as $\sum_J \prod_{j \in J} F_j(x)$ where each product involves at most $\deg(f')$ Boolean components of F. We deduce that the degree of $F' \circ F$ cannot exceed the degree of a product of $\deg(F')$ Boolean components of F.

Now, we focus on the Walsh spectrum of the product of some Boolean functions. We use the following lemma. Its proof can be found in [3].

Lemma 1. *Let $f_1, \dots, f_k$ be k Boolean functions of n variables, with $k > 0$. We have*

$$\mathcal{F}(\sum_{i=1}^{k} f_i) = 2^{n-1}\left[(-1)^k + 1\right] + \sum_{I \subset \{1,\dots,k\}} (-2)^{|I|-1} \mathcal{F}(\prod_{i \in I} f_i) .$$

Moreover, for any nonzero α in $\mathbb{F}_2^n$, we have

$$\mathcal{F}(\sum_{i=1}^{k} f_i + \varphi_\alpha) = \sum_{I \subset \{1,\dots,k\}} (-2)^{|I|-1} \mathcal{F}(\prod_{i \in I} f_i + \varphi_\alpha) .$$

Using the previous relation between the Walsh coefficients of the sum of k Boolean functions and the Walsh coefficients of their product, we obtain:

Theorem 1. *Let $f_1, \dots, f_k$ be k Boolean functions of n variables, with $k > 0$. Suppose that for any subset I of $\{1, \dots, k\}$ we have*

$$\forall \alpha \in \mathbb{F}_2^n, \;\; \mathcal{F}(\sum_{i \in I} f_i + \varphi_\alpha) \equiv 0 \bmod 2^\ell .$$

Then, for any $I \subset \{1, \dots, k\}$ of size at most ℓ, we have

$$\forall \alpha \in \mathbb{F}_2^n, \;\; \mathcal{F}(\prod_{i \in I} f_i + \varphi_\alpha) \equiv 0 \bmod 2^{\ell+1-|I|} . \tag{2}$$

Therefore,

$$\deg(\prod_{i \in I} f_i) \le n - \ell + |I| .$$

Proof. We prove Relation (2) by induction on the size of I. The result obviously holds for $|I| = 1$. We now assume that (2) holds for any I with $|I| \le w$ and we consider a subset $I \subset \{1, \dots, k\}$ of size $w + 1$. From Lemma 1, we have for any $\alpha \in \mathbb{F}_2^n$

$$(-2)^w \mathcal{F}(\prod_{i \in I} f_i + \varphi_\alpha) \equiv \mathcal{F}(\sum_{i \in I} f_i + \varphi_\alpha) - \sum_{\substack{J \subset I \\ J \ne I}} (-2)^{|J|-1} \mathcal{F}(\prod_{j \in J} f_j + \varphi_\alpha) \bmod 2^n .$$

From induction hypothesis, we derive that

$$(-2)^w \mathcal{F}(\prod_{i \in I} f_i + \varphi_\alpha) \equiv \mathcal{F}(\sum_{i \in I} f_i + \varphi_\alpha) \bmod 2^\ell .$$

Therefore, we have

$$\mathcal{F}(\prod_{i \in I} f_i + \varphi_\alpha) \equiv 0 \bmod 2^{\ell - w} .$$

The upper bound on the degree is a direct consequence of (2) and Proposition 3.

By applying the previous theorem to the n Boolean components of a mapping F from $\mathbb{F}_2^n$ into $\mathbb{F}_2^n$, we derive the following corollary.

Corollary 1. *Let F be a function from $\mathbb{F}_2^n$ into $\mathbb{F}_2^n$ such that its Walsh spectrum is 2^ℓ-divisible. Then, the degree of the product of any t Boolean components of F is at most $n - \ell + t$.*

Therefore, for any function F' from $\mathbb{F}_2^n$ into $\mathbb{F}_2^n$, we have

$$\deg(F' \circ F) \leq n - \ell + \deg(F') .$$

When F is an almost bent function, we obtain

$$\deg(F' \circ F) \leq \frac{n-1}{2} + \deg(F') .$$

The result presented in Corollary 1 was already proved for the particular case of *power functions*. Here, we identify $\mathbb{F}_2^n$ with the finite field with 2^n elements, $\mathbb{F}_{2^n}$. In this context, any function F from $\mathbb{F}_2^n$ into $\mathbb{F}_2^n$ can be expressed as a unique univariate polynomial in $\mathbb{F}_{2^n}[X]$, $F(X) = \sum_{u=0}^{2^n-1} a_u X^u$. The degree of F (in the sense of Definition 4) is given by $\deg(F) = \max_{u, a_u \neq 0} w_2(u)$, where $w_2(u)$ denotes the number of ones in the 2-adic expansion of u, $u = \sum_{i=0}^{n-1} u_i 2^i$. The case of power functions is of great interest since all known highly nonlinear mappings are equivalent (up to a linear permutation of $\mathbb{F}_2^n$) to some power functions $x \mapsto x^s$ over $\mathbb{F}_{2^n}$. Now, if we write F' as a univariate polynomial $F'(X) = \sum_{u=0}^{2^n-1} a_u X^u$, we obtain for $F : x \mapsto x^s$ that $F' \circ F(x) = \sum_{u=0}^{2^n-1} a_u X^{us \bmod (2^n-1)}$. Therefore, $\deg(F' \circ F) \leq \max_{u, a_u \neq 0} w_2(us \bmod (2^n - 1))$. This bound is related to the divisibility of the Walsh spectrum of F by the following proposition [2, Coro. 2]. The result is directly derived from McEliece's theorem which provides the weight divisibility of a cyclic code [19]. We refer to [5,2] for the link between cyclic codes and power functions.

Proposition 4. *Let $F : x \mapsto x^s$ be a power function over $\mathbb{F}_{2^n}$. Then, the Walsh spectrum of F is 2^ℓ-divisible if and only if, for any integer u, $1 \leq u \leq 2^n - 1$, we have*

$$w_2(us \bmod (2^n - 1)) \leq n - \ell + w_2(u) .$$

5 Cryptanalysis of 5-Round Feistel Ciphers Using Highly Nonlinear Functions

We now focus on 5-round Feistel ciphers. In a Feistel cipher with block size $2n$, the round function is defined by

$$\begin{array}{rl} F_K\colon \mathbb{F}_2^n \times \mathbb{F}_2^n \to & \mathbb{F}_2^n \times \mathbb{F}_2^n \\ (L, R) \mapsto & (R, L \oplus S_K(R)) \end{array}$$

where S_K is a function from $\mathbb{F}_2^n$ into $\mathbb{F}_2^n$ called the substitution function. In the following, L_i (resp. R_i) denotes the left part (resp. right part) of the output of the i-th round.

In a 5-round Feistel cipher, the right part of the output of the third round, R_3, can be derived from the ciphertext (L_5, R_5) and the last-round key:

$$R_3 = R_5 \oplus S_{K_5}(L_5) \ .$$

Moreover, when we consider any plaintext (x, c_0) whose right part is a given constant c_0, R_3 can be computed from x by only two iterations of the substitution function :

$$R_3(x) = x \oplus c_1 \oplus S_{K_3}(c_0 \oplus S_{K_2}(x \oplus c_1))$$

where x stands for the left half of the plaintext and c_0, c_1 are some constants.

When the Walsh spectrum of the substitution function S_K is 2^ℓ-divisible for all values of K, we can apply Corollary 1. Then, we obtain the following upper bound for the degree of R_3:

$$\deg(R_3) \leq n - \ell + \deg(S) \ .$$

Thus, if we consider the attack described in Section 3, we have exhibited a new attack on the last round key with average running-time of $2^{m+\delta}$, where m is the size of the round key and $\delta = \min(\deg(S)^2 + 1, n - \ell + \deg(S) + 1)$. This attack is feasible as soon as $\delta \leq n$. For example, if S is almost bent, a higher order differential attack can be performed except when $\deg(S) = (n+1)/2$, i.e., when S is an almost bent function of maximum degree.

A similar situation occurs when S is a function of an even number of variables which has the highest known nonlinearity, $\mathcal{L}(S) = 2^{\frac{n}{2}+1}$. All known functions satisfying this property are equivalent (up to a linear permutation of $\mathbb{F}_2^n$) to one of the power functions given in Table 1 (or to one of their inverses) [8]. All optimal functions for n even are such that their Walsh spectra are divisible either by $2^{\frac{n}{2}}$ or by $2^{\frac{n}{2}+1}$, except the inverse function whose Walsh spectrum is exactly 4-divisible. Note that the Walsh spectrum of the inverse function has the smallest possible divisibility for a function whose nonlinearity is even. If the Walsh spectrum of the substitution function S is $2^{\frac{n}{2}+1}$-divisible, then $\deg(S) \leq n/2$. Therefore, the attack is always feasible. When the Walsh spectrum of S is $2^{\frac{n}{2}}$-divisible, the attack can be performed except if $\deg(S) \in \{n/2, n/2+1\}$. These results are summed up in Table 2 (general case).

It is also possible to improve this attack when the round key in the Feistel cipher is inserted by addition, i.e., $S_K(x) = S(x \oplus K)$. In that case, we obtain the following expression for R_3:

$$R_3(x) = x \oplus c_1 \oplus S(c_0 \oplus K_3 \oplus S(x \oplus c_1 \oplus K_2)) \ .$$

Let G be the function defined by $G : x \mapsto S(K_3 \oplus c_0 \oplus S(x \oplus c_1 \oplus K_2))$ and let G' be defined by $G' : x \mapsto S(K_3 \oplus c_0 \oplus S(x))$. Then, we know that $\deg(G') \leq n - \ell + \deg(S)$. The expression of G' shows that the terms containing the constants c_0 or K_3 are the result of the product of at most $(\deg(S) - 1)$ Boolean components of S. Thus, their degree is at most $n - \ell + \deg(S) - 1$. We then deduce that the terms of maximal degree in G' are independent of the constants. In particular we have for any subspace V of dimension $(n - \ell + \deg(S))$:

Table 1. Known power permutations x^s on $\mathbb{F}_{2^n}$, n even, with the highest nonlinearity and exact divisibility of their Walsh spectra

exponents s	condition on n	divisibility	
$2^{n-1}-1$	$n \equiv 0 \bmod 2$	2^2	[13]
2^k+1, with $\gcd(k,n)=2$ and $k<\frac{n}{2}$	$n \equiv 2 \bmod 4$	$2^{\frac{n}{2}+1}$	[9,20]
$2^{2k}-2^k+1$, with $\gcd(k,n)=2, k<\frac{n}{2}$	$n \equiv 2 \bmod 4$	$2^{\frac{n}{2}+1}$	[11]
$2^{\frac{n}{2}}+2^{\frac{n+2}{4}}+1$	$n \equiv 2 \bmod 4$	$2^{\frac{n}{2}+1}$	[7]
$2^{\frac{n}{2}}+2^{\frac{n}{2}-1}+1$	$n \equiv 2 \bmod 4$	$2^{\frac{n}{2}+1}$	[7]
$\sum_{i=0}^{n/2} 2^{ik}$, with $\gcd(k,n)=1, k<\frac{n}{2}$	$n \equiv 0 \bmod 4$	$2^{\frac{n}{2}}$	[8]
$2^{\frac{n}{2}}+2^{\frac{n}{4}}+1$	$n \equiv 4 \bmod 8$	$2^{\frac{n}{2}}$	[8]

Table 2. Higher order differential attack on a 5-round Feistel cipher using a highly nonlinear substitution function S

function S		General case		$S_K(x)=S(x\oplus K)$	
$\mathcal{L}(S)$	div.	differential order	feasibility	differential order	feasibility
$2^{\frac{n+1}{2}}$ n odd	$2^{\frac{n+1}{2}}$	$\deg(S)+\frac{n+1}{2}$	except for $\deg(S)=\frac{n+1}{2}$	$\deg(S)+\frac{n-1}{2}$	always feasible
$2^{\frac{n}{2}+1}$ n even	$2^{\frac{n}{2}+1}$	$\deg(S)+\frac{n}{2}$	always feasible	$\deg(S)+\frac{n}{2}-1$	always feasible
	$2^{\frac{n}{2}}$	$\deg(S)+\frac{n}{2}+1$	except for $\deg(S)\in\{\frac{n}{2},\frac{n}{2}+1\}$	$\deg(S)+\frac{n}{2}$	except for $\deg(S)=\frac{n}{2}+1$

$$\forall a \in \mathbb{F}_2^n,\ \ D_V G'(a) = \bigoplus_{v\in V} G'(a\oplus v) = c$$

where c is independent of any kind of constants. We can see that G is obtained by translating G', so we have:

$$\forall a \in \mathbb{F}_2^n,\quad \bigoplus_{v\in V} G(a\oplus v) = \bigoplus_{v\in V} G'(a\oplus v\oplus c_1\oplus K_2) = D_V G'(a\oplus c_1\oplus K_2) = c.$$

The constant c can be computed, for example, with the null value for all the subkeys. The above attack requires $2^{n-\ell+\deg(S)}$ pairs of plaintexts-ciphertexts and $2^{2n-\ell+\deg(S)}$ evaluations for the function S. It can be performed for any almost bent function S (see Table 2).

6 Higher Order Differential Cryptanalysis on a Generalization of MISTY1

MISTY is a model of block ciphers proposed by Matsui [18] and presented under two forms MISTY1 and MISTY2. MISTY1 is the object of this study. M'1, the version of MISTY1 reduced to 5 rounds without FL functions is provably secure against both differential and linear cryptanalysis. Therefore, the background of the attack is this simplified algorithm. In [23] it is shown that M'1 can be attacked with a 7-th order differential. In [1], the attack is extended to the case where any almost bent power function of degree 3 on $\mathbb{F}_2^7$ is used for the S_7-box.

In this section, we extend the use of this higher order differential attack to a generalization of the algorithm M'1 where the block size becomes $16m$ bits (see Fig. 1). The original value is 64 bits. In this generalization, we show that the weakness of M'1 is due to the use of an almost bent substitution function.

In the following, x_0 and x_1 are the left and right halves of the plaintext. Similarly, (x_{i+1}, x_i) denotes the intermediate value after i rounds.

Notation 1 *Let u be a $16m$ bit word. We denote by u^L, u^R, u^{L_k}, u^{R_k}, respectively the left and right halves of u and the k left and right most bits. The $\|$ symbol stands for the concatenation of two binary words.*

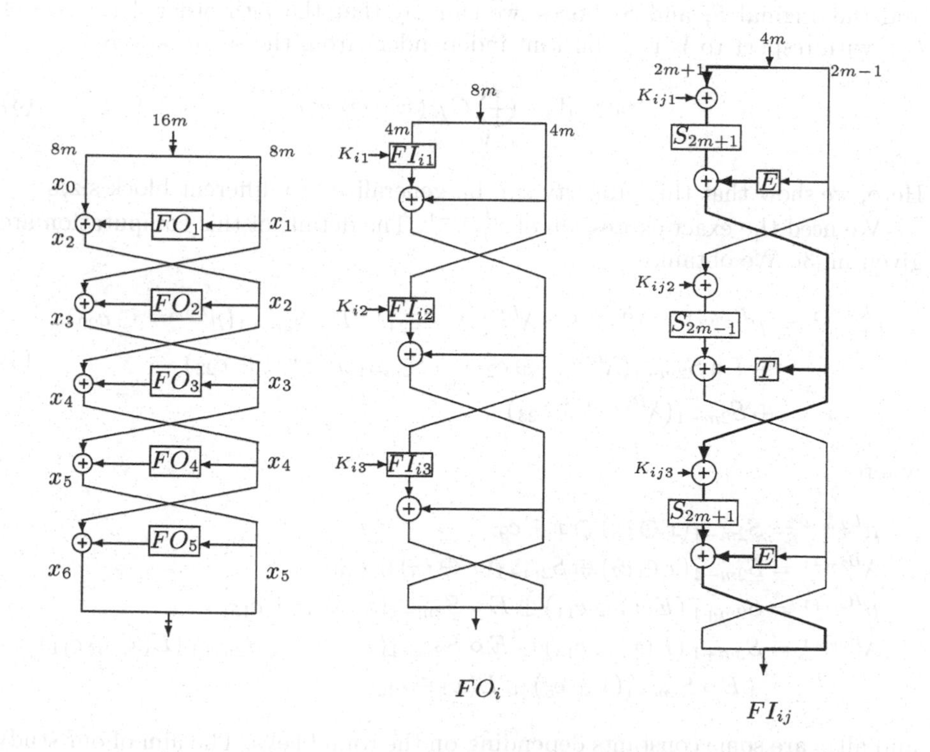

Fig. 1. The 5-round Feistel cipher M'1 with equivalent key schedule

The cipher uses the "zero-extend" function, E, and the 'truncate" function, T, which are respectively defined by:

$$E : \mathbb{F}_2^{2m-1} \rightarrow \mathbb{F}_2^{2m+1}$$
$$(u_1, \cdots, u_{2m-1}) \mapsto (u_1, \cdots, u_{2m-1}, 0, 0) ,$$

$$T : \mathbb{F}_2^{2m+1} \rightarrow \mathbb{F}_2^{2m-1}$$
$$(u_1, \cdots, u_{2m+1}) \mapsto (u_1, \cdots, u_{2m-1}) .$$

The nonlinear part of the cipher consists of two permutations, S_{2m-1} and S_{2m+1}, respectively defined over $\mathbb{F}_{2^{2m-1}}$ and $\mathbb{F}_{2^{2m+1}}$. In the original cipher, we have $S_7(x) = L(x^{81})$ over $\mathbb{F}_{2^7}$ where L is a linear permutation and S_9 is a quadratic almost bent permutation of $\mathbb{F}_{2^9}$.

Let V be the $(2m-1)$-dimensional subspace of plaintexts of $16m$ bits whose form is $(0_{6m+1} \parallel x \parallel 0_{8m})$ where x is in $\mathbb{F}_2^{2m-1}$. Let W denote the subspace $\{(w_0 \parallel 0_{2m-1} \parallel w_1), w_0 \in \mathbb{F}_2^{6m+1}, w_1 \in \mathbb{F}_2^{8m}\}$. We are interested in ciphering plaintexts $P \oplus w$ where $P \in V$ and $w = (w_0 \parallel w_1)$ is a fixed constant in W. We now consider the function G_K defined as follows:

$$G_K : x \mapsto x_4^{L_{2m-1}} .$$

To sum up the higher order differential attack proposed in [23], with $m = 4$ and the original S_7 and S_9 boxes, we can say that the 7-th order derivative of G_K with respect to V is a constant independent from the secret key K:

$$\forall w \in W, \quad \bigoplus_{x \in V} G_K(x \oplus w) = c. \tag{3}$$

Here, we show that this property can be generalized to different block sizes.

We need the exact expression of $x_4^{L_{2m-1}}$. The details of this computation are given in [3]. We obtain:

$$\begin{aligned} x_4^{L_{2m-1}} = \; & \mu^{R_{2m-1}} \oplus \lambda^{R_{2m-1}} \oplus \lambda^{L_{2m-1}} \oplus c_{24} \oplus T \circ S_{2m+1}(\mu^{L_{2m+1}} \oplus c_{20}) \\ & \oplus T \circ S_{2m+1}(\lambda^{L_{2m+1}} \oplus c_{21}) \oplus S_{2m-1}(\mu^{R_{2m-1}} \oplus c_{22}) \\ & \oplus S_{2m-1}(\lambda^{R_{2m-1}} \oplus c_{23}) , \end{aligned} \tag{4}$$

where

$$\begin{aligned} \mu^{L_{2m-1}} &= S_{2m-1}(x \oplus c_5) \oplus x \oplus c_9 \\ \lambda^{L_{2m-1}} &= S_{2m-1}(x \oplus c_7) \oplus S_{2m-1}(x \oplus c_5) \oplus c_{10} \\ \mu^{R_{2m+1}} &= S_{2m+1}(E(x) \oplus c_{11}) \oplus E \circ S_{2m-1}(x \oplus c_5) \oplus c_{15} \\ \lambda^{R_{2m+1}} &= S_{2m+1}(E(x) \oplus c_{13}) \oplus E \circ S_{2m-1}(x \oplus c_7) \oplus S_{2m+1}(E(x) \oplus c_{11}) \\ & \quad \oplus E \circ S_{2m-1}(x \oplus c_5) \oplus E(x) \oplus c_{16} \end{aligned}$$

and all c_i are some constants depending on the round keys. The aim of our study is to determine the degree of the Boolean components of $x_4^{L_{2m-1}}$.

We restrict our study to the case where S_{2m+1} is a quadratic function, as in the original cipher. We suppose that the almost bent permutation S_{2m-1} can be written as $S_{2m-1}(x) = L(x^e)$ where L is a linear permutation. We denote by d the degree of S_{2m-1} and we assume that $2d < 2m-1$, i.e., that the degree of S_{2m-1} differs from the highest possible degree for an almost bent function over $\mathbb{F}_2^{2m-1}$. These conditions obviously imply that we can neglect the terms $T \circ S_{2m+1}(\mu^{L_{2m+1}} \oplus c_{20}) \oplus T \circ S_{2m+1}(\lambda^{L_{2m+1}} \oplus c_{21})$ in (4) for a $(2m-1)$-th order differential.

We denote by $[F]_d$ the terms in the algebraic normal form of F whose degree are at least d. It clearly appears that the terms of degree $2m-1$ in $x_4^{L_{2m-1}}$ correspond to

$$\left[x_4^{L_{2m-1}}\right]_{2m-1} = \left[S_{2m-1}(\mu^{R_{2m-1}} \oplus c_{22})\right]_{2m-1} \oplus \left[S_{2m-1}(\lambda^{R_{2m-1}} \oplus c_{23})\right]_{2m-1} .$$

Terms of Highest Degree in $S_{2m-1}(\lambda^{R_{2m-1}} \oplus c_{23})$

We first consider the terms of highest degree in $S_{2m-1}(\lambda^{R_{2m-1}} \oplus c_{23})$. We make a change of variable, since we consider all $x \in \mathbb{F}_2^{2m-1}$. Then, $[S_{2m-1}(\lambda^{R_{2m-1}} \oplus c_{23})]_{2m-1} = [S_{2m-1}(g(x))]_{2m-1}$ with

$$\begin{aligned} g(x) &= S_{2m-1}(x) \oplus S_{2m-1}(x \oplus c_{28}) \oplus T \circ S_{2m+1}(E(x) \oplus c_{29}) \\ &\quad \oplus T \circ S_{2m+1}(E(x) \oplus c_{30}) \oplus x \oplus c_{31} \\ &= D_{c_{28}} S_{2m-1}(x) \oplus A(x, c_{29}, c_{30}, c_{31}) , \end{aligned}$$

where all terms of A have degree at most 1. Therefore, all terms of $S_{2m-1}(g(x))$ correspond to the product of β_1 components of $D_{c_{28}} S_{2m-1}$ and of β_2 components of $A(x, c_{29}, c_{30}, c_{31})$ where $\beta_1 + \beta_2 = d$. The degree of such a term is then lower than $\beta_1(d-1) + (d-\beta_1)$ as $\deg(D_{c_{28}} S_{2m-1}) \leq d-1$. When $\beta_1 = d$ (and then $\beta_2 = 0$), this term corresponds to a product of derivatives with respect to c_{28}. Hence it has the same value on x and $x \oplus c_{28}$ for all $x \in \mathbb{F}_2^{2m-1}$ and it cannot have degree $2m-1$. Therefore, the degree $2m-1$ can only be obtain for $\beta_1 \leq d-1$. In such cases, the degree admits the upper bound $(d-1)^2 + 1$. It follows that $S_{2m-1}(g(x))$ have degree at most $(2m-2)$ if

$$d < 1 + \sqrt{2m-2} .$$

Note that this condition is satisfied by the original parameters ($m = 4$ and $d = 3$).

Terms of Highest Degree in $S_{2m-1}(\mu^{R_{2m-1}} \oplus c_{22})$

Now, we apply a similar treatment to $S_{2m-1}(\mu^{R_{2m-1}} \oplus c_{22})$, where $\mu^{R_{2m+1}} = S_{2m+1}(E(x) \oplus c_{11}) \oplus E \circ S_{2m-1}(x \oplus c_5) \oplus c_{15}$. We also make a change of variable. Then, $[S_{2m-1}(\mu^{R_{2m-1}} \oplus c_{22})]_{2m-1} = [S_{2m-1}(t(x))]_{2m-1}$ with

$$t(x) = S_{2m-1}(x) \oplus T \circ S_{2m+1}(E(x) \oplus c_{25}) \oplus c_{26} .$$

Moreover, the explicit writing of the almost bent power function $S_{2m-1}(x) = L(x^e)$ leads to:

$$L^{-1}(t(x)) = x^e \oplus Q(x) \oplus A(x, c_{25}, c_{26})$$

where Q contains quadratic terms only and A affine or constant terms (since c_{25} and c_{26} only appear in linear or constant terms). In [1], Babbage and Frisch give the following explanation for the 7-th order differential attack on the original cipher: the only way to obtain a term of degree 7 in $S_{2m-1}(t(x))$ with $d = 3$ is to multiply at least two terms of degree 3 of $L^{-1}(t(x))$ and another term. But, the terms of degree 3 in $L^{-1}(t(x))$ come from the almost bent function S_7, and they observe that the product of any two Boolean components of S_7 has degree at most 5 [1, Fact 2]. This observation is a direct consequence of Corollary 1. Thus, the maximum degree that we can obtain is at most 7.

More generally, all terms in $[S_{2m-1}(t(x))]_{2m-1}$ are the result of the product of β_1 terms from x^e, β_2 terms from $Q(x)$ and β_3 terms from $A(x, c_{25}, c_{26})$, with $\beta_1+\beta_2+\beta_3 = d$. In other terms, we can write them as: $x^{e\lambda_1} \cdot x^{\lambda_2} \cdot x^{\lambda_3} \cdot c$ where λ_1, λ_2 and λ_3 are integers lower than 2^{2m-1} and verifying $w_2(\lambda_1) = \beta_1$, $w_2(\lambda_3) \leq \beta_3$ and $w_2(\lambda_2) \leq 2\beta_2$ as λ_2 is the sum of β_2 integers whose 2-weights equal 2. Such a term depends on a constant only if $\beta_3 \neq 0$. Its degree is then:

$$w_2\left((e\lambda_1 + \lambda_2 + \lambda_3) \bmod \left(2^{2m-1} - 1\right)\right)$$

and the attack could be done as soon as $w_2\left((e\lambda_1 + \lambda_2 + \lambda_3) \bmod \left(2^{2m-1} - 1\right)\right) < 2m - 1$. Now, we derive from Proposition 4

$$\begin{aligned} w_2 &\left((e\lambda_1 + \lambda_2 + \lambda_3) \bmod (2^{2m-1} - 1)\right) \\ &\leq w_2(e\lambda_1 \bmod (2^{2m-1} - 1)) + w_2(\lambda_2) + w_2(\lambda_3) \\ &\leq (m-1) + \beta_1 + 2\beta_2 + \beta_3 \leq (m-1) + d + \beta_2 \end{aligned} \tag{5}$$

as $\beta_1 + \beta_2 + \beta_3 = d$.

Such a term depends on the constants only if $\beta_3 \geq 1$. We then have $\beta_2 \leq d-1$. But the terms including a high value for β_2 ($\beta_2 \geq d-2$) correspond to one of the following particular cases:

- Case $\beta_1 = 0$. Then, we have $\beta_2 + \beta_3 = d$. We deduce that

$$\begin{aligned} w_2((e\lambda_1 + \lambda_2 + \lambda_3) \bmod (2^{2m-1} - 1)) &= w_2(\lambda_2 + \lambda_3 \bmod (2^{2m-1} - 1)) \\ &\leq 2\beta_2 + \beta_3 \leq 2d - \beta_3 \leq 2m - 3 \end{aligned}$$

  since $\beta_3 \geq 1$. Note that this case completely solves the case $\beta_2 = d - 1$.
- Case $\beta_1 = 1$ and $\beta_2 = d - 2$. As $w_2(\lambda_1) = w_2(\lambda_3) = 1$, we have $\lambda_1 = 2^i$ and $\lambda_3 = 2^j$. Therefore,

$$\begin{aligned} w_2((e\lambda_1 + \lambda_2 + \lambda_3) \bmod (2^{2m-1} - 1)) &= w_2(2^i e + \lambda_2 + 2^j \bmod (2^{2m-1} - 1)) \\ &= w_2(e + \lambda_2' + 2^k \bmod (2^{2m-1} - 1)) \\ &\leq w_2(e) + w_2(\lambda_2') + 1 \\ &\leq d + (d-2) + 1 \leq 2m - 3\,. \end{aligned}$$

Both previous situations include the case $\beta_2 \geq d-2$. Now, for any $\beta_2 \leq d-3$, we derive from (5) that

$$w_2((e\lambda_1 + \lambda_2 + \lambda_3) \bmod (2^{2m-1} - 1)) \leq m - 1 + 2d - 3 .$$

This upper bound cannot exceed $(2m-2)$ as soon as

$$d < \frac{m+3}{2} .$$

This study emphasizes that for any block size $16m$, with a S_{2m+1} box of degree 2, the cipher is vulnerable to a higher order cryptanalysis of degree $2m-1$ as soon as the degree d of the almost bent function S_{2m-1} satisfies

$$d < \min(1 + \sqrt{2m-2}, \ \frac{m+3}{2}) .$$

The condition required by the first bound is clearly the most restrictive one, since it does not exploit the almost bent property. For any S_{2m-1} of degree 3, the cipher is vulnerable when $m \geq 4$ and for S_{2m-1} of degree 4 when $m \geq 6$. The attackable degrees are classified in the following table.

m	block size	attackable degrees
3	48	$d \leq 2$
4	64	$d \leq 3$ (original parameters)
5	80	$d \leq 3$
6	96	$d \leq 4$
10	160	$d \leq 5$

Then, our study points out that the property of high divisibility of the Walsh spectrum of the substitution function is at the origin of the vulnerability of such a cipher. This property leads to the following new design criterion: the Walsh spectrum of the substitution function should contain at least one value which is not divisible by a higher power of 2.

References

1. S. Babbage and L. Frisch. On MISTY1 Higher Order Differential Cryptanalysis. In *Proceedings of ICISC 2000*, number 2015 in Lecture Notes in Computer Science, pages 22–36. Springer-Verlag, 2000.
2. A. Canteaut, P. Charpin, and H. Dobbertin. A new characterization of almost bent functions. In *Fast Software Encryption 99*, number 1636 in Lecture Notes in Computer Science, pages 186–200. Springer-Verlag, 1999.
3. A. Canteaut and M. Videau. Weakness of block ciphers using highly nonlinear confusion functions. Research Report 4367, INRIA, February 2002. Available on `http://www.inria.fr/rrrt/rr-4367.html`.

4. C. Carlet. Two new classes of bent functions. In *Advances in Cryptology - EUROCRYPT'93*, number 765 in Lecture Notes in Computer Science, pages 77–101. Springer-Verlag, 1994.
5. C. Carlet, P. Charpin, and V. Zinoviev. Codes, bent functions and permutations suitable for DES-like cryptosystems. *Designs, Codes and Cryptography*, 15:125–156, 1998.
6. F. Chabaud and S. Vaudenay. Links between differential and linear cryptanalysis. In A. De Santis, editor, *Advances in Cryptology - EUROCRYPT'94*, number 950 in Lecture Notes in Computer Science, pages 356–365. Springer-Verlag, 1995.
7. T. Cusick and H. Dobbertin. Some new 3-valued crosscorrelation functions of binary m-sequences. *IEEE Transactions on Information Theory*, 42:1238–1240, 1996.
8. H. Dobbertin. One-to-one highly nonlinear power functions on $GF(2^n)$. *Appl. Algebra Engrg. Comm. Comput.*, 9(2):139–152, 1998.
9. R. Gold. Maximal recursive sequences with 3-valued recursive crosscorrelation functions. *IEEE Transactions on Information Theory*, 14:154–156, 1968.
10. T. Jakobsen and L.R. Knudsen. The interpolation attack on block ciphers. In *Fast Software Encryption 97*, number 1267 in Lecture Notes in Computer Science, pages 28–40. Springer-Verlag, 1997.
11. T. Kasami. The weight enumerators for several classes of subcodes of the second order binary Reed-Muller codes. *Information and Control*, 18:369–394, 1971.
12. L. R. Knudsen. Truncated and higher order differentials. In *Fast Software Encryption - Second International Workshop*, number 1008 in Lecture Notes in Computer Science, pages 196–211. Springer-Verlag, 1995.
13. G. Lachaud and J. Wolfmann. The weights of the orthogonal of the extended quadratic binary Goppa codes. *IEEE Transactions on Information Theory*, 36(3):686–692, 1990.
14. X. Lai. Higher order derivatives and differential cryptanalysis. In *Proc. "Symposium on Communication, Coding and Cryptography", in honor of J. L. Massey on the occasion of his 60'th birthday*, 1994.
15. F.J. MacWilliams and N.J.A. Sloane. *The Theory of Error-Correcting Codes*. North-Holland, 1977.
16. M. Matsui. Linear cryptanalysis method for DES cipher. In *Advances in Cryptology - EUROCRYPT'93*, number 765 in Lecture Notes in Computer Science, pages 386–397. Springer-Verlag, 1993.
17. M. Matsui. The first experimental cryptanalysis of the Data Encryption Standard. In *Advances in Cryptology - CRYPTO'94*, number 839 in Lecture Notes in Computer Science. Springer-Verlag, 1995.
18. M. Matsui. New Block Encryption Algorithm MISTY. In *Fast Software Encryption 97*, number 1267 in Lecture Notes in Computer Science, pages 54–68. Springer-Verlag, 1997.
19. R.J. McEliece. Weight congruence for p-ary cyclic codes. *Discrete Mathematics*, 3:177–192, 1972.
20. K. Nyberg. Differentially uniform mappings for cryptography. In *Advances in Cryptology - EUROCRYPT'93*, number 765 in Lecture Notes in Computer Science, pages 55–64. Springer-Verlag, 1993.
21. K. Nyberg. On the construction of highly nonlinear permutations,. In *Advances in Cryptology - EUROCRYPT'92*, number 658 in Lecture Notes in Computer Science, pages 92–98. Springer-Verlag, 1993.

22. K. Nyberg and L.R. Knudsen. Provable security against differential cryptanalysis. In *Advances in Cryptology - CRYPTO'92*, number 740 in Lecture Notes in Computer Science, pages 566–574. Springer-Verlag, 1993.
23. H. Tanaka, K. Hisamatsu, and T. Kaneko. Strength of MISTY1 without FL function for Higher Order Differential Attack. In *Applied Algebra, Algebraic Algorithms and Error-Correcting Codes*, number 1719 in Lecture Notes in Computer Science, pages 221–230. Springer-Verlag, 1999.

Security Flaws Induced by CBC Padding – Applications to SSL, IPSEC, WTLS ...

Serge Vaudenay

Swiss Federal Institute of Technology (EPFL),
Serge.Vaudenay@epfl.ch

Abstract. In many standards, e.g. SSL/TLS, IPSEC, WTLS, messages are first pre-formatted, then encrypted in CBC mode with a block cipher. Decryption needs to check if the format is valid. Validity of the format is easily leaked from communication protocols in a chosen ciphertext attack since the receiver usually sends an acknowledgment or an error message. This is a side channel.
In this paper we show various ways to perform an efficient side channel attack. We discuss potential applications, extensions to other padding schemes and various ways to fix the problem.

1 Introduction

Variable input length encryption is traditionally constructed from a fixed input length encryption (namely a block cipher) in a special mode of operation. In RFC2040 [2], the RC5-CBC-PAD algorithm is proposed, based on RC5 which enables the encryption of blocks of $b = 8$ words where words are bytes. Encryption of any word sequence with an RC5 secret key K is performed as follows.

1. Pad the word sequence with n words, all being equal to n, such that $1 \leq n \leq b$ and the padded sequence has a length which is a multiple of b.
2. Write the padded word sequence as a block sequence $x_1, \ldots, x_N$ in which each block x_i consists of b words.
3. Encrypt the block sequence in CBC mode with a (either fixed or random or secret) IV with a permutation C defined by RC5 with key K: get

$$y_1 = C(\text{IV} \oplus x_1), \quad y_i = C(y_{i-1} \oplus x_i); i = 2, \ldots, N \tag{1}$$

where $\oplus$ denotes the XOR operation.

The encryption of the message is the block sequence $y_1, \ldots, y_N$.

Although decryption is not clearly defined in RFC2040 [2], it makes sense to assume that the receiver of an encrypted message first decrypts in CBC mode, then checks if the padding is correct and finally removes it. The question is: how must the receiver behave if the padding is not correct? Although the receiver should not tell the sender that the padding is not correct, it is meaningful that non-procession of a decrypted message ultimately leaks this bit of information.

L.R. Knudsen (Ed.): EUROCRYPT 2002, LNCS 2332, pp. 534–545, 2002.

This leads to an attack that uses an oracle for which any block sequence tells if the padding of the corresponding CBC-decrypted sequence is correct according to the above algorithm. The attack works within a complexity of $O(NbW)$ in order to decrypt the message where W is the number of possible words (typically $W = 256$).

A similar attack model was used by Bleichenbacher against PKCS#1 v1.5 [5] and by Manger against PKCS#1 v2.0 [13]. This paper shows that similar attacks are feasible in the symmetric key world.

The paper is organized as follows. We first recall some well known properties and security issues for the CBC mode. We describe several attacks against RC5-CBC-PAD and we introduce the notion of *bomb oracle*. We then discuss extensions to other schemes: ESP, random padding, ... and applications in real life such as SSL, IPSEC, WTLS, SSH2. Next we present some possible fixes which do not actually work like replacing the CBC mode by a double CBC mode, the HCBC mode or other modes which were proposed by the standard process run by NIST. We further propose a fix which does work.

2 CBC Properties

Several security properties of the CBC mode are already known. We think it is useful to recall them in order to remind ourselves of the intrinsic security limits of the CBC mode.

2.1 Efficiency

CBC mode is efficient in practice because we can encrypt or decrypt a stream of infinite length with a constant memory in linear time. Efficiency is comparable to the Electronic Code Book (ECB) mode where each block is encrypted in the same way. The difference between ECB and CBC is a single exclusive or operation. Since the ECB mode is not suitable in most applications because of ciphertext manipulation attacks, and lack of increased message entropy, we prefer to use CBC mode. (See e.g. [14, p. 230] for more details.)

Exhaustive search against CBC mode is related to the length of the secret key. We have yet other bounds related to the block length. First of all, the electronic code book attack has a complexity of W^b. We have other specific attacks related to the intrinsic security of the CBC mode no matter which block cipher is used. These are detailed in following sections.

2.2 Confidentiality Limits

Confidentiality has security flaws. Obviously, when using a fixed IV, one can easily see when two different messages have a common prefix block sequence by just looking at the two ciphertexts.

More generally, when two ciphertext blocks y_i and y_j are equal, one can deduce from Eq. (1) that $y_{i-1} \oplus y_{j-1} = x_i \oplus x_j$.[1] We can then exploit the

[1] This property was notably mentioned in [12, p. 43].

redundancy in the plaintext in order to recover x_i and x_j from $y_{i-1} \oplus y_{j-1}$. This flaw is however quite negligible: since the ciphertext blocks get a distribution which is usually indistinguishable from a uniform distribution, the probability that two b-words blocks out of N are equal is given by the birthday paradox theorem

$$p \approx 1 - e^{-\frac{1}{2}N^2.W^{-b}}$$

where W is the number of possible words. The attack is efficient when N reaches the order of magnitude of $\sqrt{W^b}$. Therefore, for $b = 8$ and $W = 256$, we need about 2^{35} bytes (32GigaBytes) in order to get a probability of success equal to 39% for this attack which leaks information on 16 Bytes only.

2.3 Authentication Limits

The CBC mode can be used to create message authentication codes (MAC). Raw CBC-MAC (i.e. taking the last encrypted block as a MAC) is well known to have security flaws: with the MAC of three messages m_1, m_2, m_3 where m_2 consists of m_1 augmented with an extra block, we can forge the MAC of a fourth message which consists of m_3 augmented with an extra block. This is fixed by re-encrypting the raw CBC-MAC, but this new scheme still has attacks of complexity essentially $\sqrt{W^b}$. (See [15,16,19].)

3 The Attack

Let b be the block length in words, and W be the number of possible words. (We assume that $W \geq b$ and that all integers between 1 and b can unambiguously be encoded into words in order to make the CBC-PAD scheme feasible.)

We say that a block sequence $x_1, x_2, \ldots, x_N$ has a correct padding if the last block x_N ends with a word string of n words equal to n with $n > 0$: 1, or 22, or 333, ... Given a block sequence $y_1, y_2, \ldots, y_N$, we define an oracle $\mathcal{O}$ which yields 1 if the decryption in CBC mode has a correct padding. Decryption is totally defined by a block encryption function C and IV. Oracle $\mathcal{O}$ is thus defined by C and IV.

3.1 Last Word Oracle

For any block y, we want to compute the last word of $C^{-1}(y)$. We call it the "last word oracle".

Let $r_1, \ldots, r_b$ be random words, and let $r = r_1 \ldots r_b$. We forge a fake ciphertext $r|y$ by concatenating the two blocks r and y. If $\mathcal{O}(r|y) = 1$, then $C^{-1}(y) \oplus r$ ends with a valid padding. In this case, the most likely valid padding is the one which ends with 1. This means that the last word of $C^{-1}(y)$ is $r_b \oplus 1$. If $\mathcal{O}(r|y) = 0$, we can try again (by making sure that we pick another r_b: picking the same one twice is not worthwhile).

If we are lucky (with probability W^{-1}), we find the last word with the first try. Otherwise we have to try many r_bs. On average, we have to try $W/2$ values.

Odd cases occur when the valid padding found is not 1. This is easy to detect. The following program eventually halts with the last words of y: one in the typical case, several if we are lucky.

1. pick a few random words $r_1, \ldots, r_b$ and take $i = 0$
2. pick $r = r_1 \ldots r_{b-1}(r_b \oplus i)$
3. if $\mathcal{O}(r|y) = 0$ then increment i and go back to the previous step
4. replace r_b by $r_b \oplus i$
5. for $n = b$ down to 2 do
   (a) take $r = r_1 \ldots r_{b-n}(r_{b-n+1} \oplus 1)r_{b-n+2} \ldots r_b$
   (b) if $\mathcal{O}(r|y) = 0$ then stop and output $(r_{b-n+1} \oplus n) \ldots (r_b \oplus n)$
6. output $r_b \oplus 1$

3.2 Block Decryption Oracle

Now we want to implement an oracle which computes $C^{-1}(y)$ for any y: a "block decryption oracle".

Let $a = a_1 \ldots a_b$ be the word sequence of $C^{-1}(y)$. We can get a_b by using the last word oracle. Assuming that we already managed to get $a_j \ldots a_b$ for some $j \leq b$, the following program gets a_{j-1}, so that we can iterate until we recover the whole sequence.

1. take $r_k = a_k \oplus (b - j + 2)$ for $k = j, \ldots, b$
2. pick $r_1, \ldots, r_{j-1}$ at random and take $i = 0$
3. take $r = r_1 \ldots r_{j-2}(r_{j-1} \oplus i)r_j \ldots r_b$
4. if $\mathcal{O}(r|y) = 0$ then increment i and go back to the previous step
5. output $r_{j-1} \oplus i \oplus (b - j + 2)$

We need $W/2$ trials on average. We can thus recover an additional word within $W/2$ trials. Since there are b words per block, we need $bW/2$ trials on average in order to implement the C^{-1} oracle.

3.3 Decryption Oracle

Now we want to decrypt any message $y_1, \ldots, y_N$ with the help of $\mathcal{O}$. It can be done with $NbW/2$ 2-block oracle calls on average. We just have to call the block decryption oracle on each block y_i and perform the CBC decryption.

One problem remains in the case where IV is secret. Here we cannot decrypt the first block. We can however get the first plaintext block up to an unknown constant. In particular, if two messages are encrypted with the same IV, we can compute the XOR of the two first plaintext blocks.

The attack has a complexity of $O(NbW)$. As an example for $b = 8$ and $W = 256$ we obtain that we can decrypt any N-block ciphertext by making $1024N$ oracle calls on average. The attack is thus extremely efficient.

3.4 Postfix Equality Check Oracle

There are reasons which will be made clear for which we can be interested in *bomb oracles* as defined below. A bomb oracle is an oracle which either gives an answer or *explodes* depending on the input. Of course, the bomb oracle is no longer available after explosion. An attack which uses a bomb oracle fails if the oracle explodes. For instance, we are interested in a bomb oracle $\mathcal{O}'$ which either answers 1 when $\mathcal{O}$ answers 1 or explodes when $\mathcal{O}$ answers 0.

Given a ciphertext $y_1, \ldots, y_N$ and a word sequence $w_1 \ldots w_m$, we want to implement a bomb oracle which checks if $w_1 \ldots w_m$ is a postfix of the decryption of $y_1, \ldots, y_N$ by using $\mathcal{O}'$. Let us first consider that $m \leq b$. We perform the following process.

1. pick a few random words $r_1 \ldots r_{b-m}$
2. take $r_{b-m+k} = w_k \oplus m$ for $k = 1, \ldots, m$
3. send $r|y_N$ to the oracle $\mathcal{O}'$ where $r = r_1 \ldots r_b$
4. if $m = 1$ then
   - take $r'_k = r_k$ for $k = 1, \ldots, b-2, b$ and take $r'_{b-1} = r_{b-1} \oplus 1$

   otherwise
   - take $r'_k = r_k$ for $k = 1, \ldots, b-1$ and take $r'_b = w_m \oplus 1$
5. send $r'|y_N$ to the oracle $\mathcal{O}'$ where $r' = r'_1 \ldots r'_b$
6. output 1

The second oracle call is used in order to eliminate odd cases which are not eliminated by the first one, for instance when $w_m \oplus m \oplus 1$ is a postfix. Obviously, this is a bomb oracle which checks whether $w_1 \ldots w_m$ is a postfix or not.

For $m > b$, we can cut the ciphertext and use the above oracle $\lceil \frac{m}{b} \rceil$ times on each block. As will be noticed, some CBC-PAD variants allow to have paddings longer than b (namely at most $W - 1$), so we can generalize the previous oracle and check postfixes within a single $\mathcal{O}$ oracle call. This will be used against SSL/TLS in Section 5.1.

4 Other Padding Schemes

In Schneier [17, pp. 190–191], a slightly different padding scheme is proposed: only the last word is equal to the padding length, and all other padded words are equal to zero. The padded sequence is thus $00\ldots0n$ instead of $nn\ldots n$. Obviously, a similar attack holds.

IP Encapsulating Security Payload (ESP) [10] uses another slightly different padding: the padded sequence is $1234\ldots n$ instead of $nn\ldots n$. Obviously, a similar attack holds.

Another padding scheme consists of padding with a non blank word then the necessary number of blank words. This is suggested, for instance by NIST [8, App. A] with $W = 2$ (here the blank word is the bit 0). Obviously, a similar attack holds.

One can propose to have the last word equal to the padding length and all other padded words chosen at random (like SSH2). The attack still enables the decryption of the last word of any block. We also have another security flaw: if the same message is encrypted twice, it is unlikely that the last encrypted blocks are equal, but in the case where the padding is of length one. We can thus guess the padding length when the ciphertexts are equal.

5 The Attack in Real Life

Here we discuss various applications. In most of cases, the attack can be (and is) avoided by using appropriate parameters. However, since this is not carefully specified in the standards, our aim is to warn the users about possible bad configurations.

5.1 SSL/TLS

Like in SSL, TLS v1.0 [7] uses the CBC-PAD scheme with $W = 256$ when using block ciphers (default cipher being the RC4 stream cipher though). The only difference is that the padding length is not necessarily less than b but can be longer (but less than $W - 1$) in order to hide the real length of the plaintext. We can thus expect to use a TLS server like the $\mathcal{O}$ oracle.

TLS v1.0 also provides an optional MAC which failed to thwart the attack: when the server figures out that the MAC is wrong, it yields the `bad_record_mac` error. However, the message padding is performed *after* the MAC algorithm, so the MAC does not preclude our attack since it cannot be checked before the padding in the decryption. The situation is a little different in SSL v3.0 since both wrong MAC both invalid padding return the same error. However, the question whether the client can distinguish the two types of error is debatable.

The reason why the attack is not so practical is because the padding format error (the `decryption_failed` error) is a fatal alert and the session must abort. The server thus stops (or "explodes") as soon as the oracle outputs 0. For this reason we consider the bomb oracle $\mathcal{O}'$. We can thus perform the postfix equality check oracle described in Section 3.4. It can be used in order to decrypt by random trial the last word of a block with a probability of success of W^{-1}, the last two words of a block with a probability of success of W^{-2}, ...

Interestingly, TLS wants to hide the real message length itself. We can easily frustrate this feature by implementing a "length equality check bomb oracle" in a very same way: if we want to check whether or not the padding length is equal to n, we take the last ciphertext block y, and we send $r|y$ to the server where the rightmost word of r is set to $n \oplus 1$ and the others are random. Acceptance by $\mathcal{O}'$ means that the right length is n with probability at least $1 - W^{-1}$. Rejection means that n is not the right length for sure.

Since the padding length is between 1 and W, the above oracle may not look so useful. We can still implement another bomb oracle which answers whether or not the padding length is greater than b, i.e. if the length hiding feature of TLS was used: let y_1 and y_2 be the last two ciphertext blocks. We just send $r|y_1|y_2$ with a random block r to $\mathcal{O}'$. Acceptance means that the padding length is at most b with probability at least $1 - W^{-1}$. Rejection means that the padding length is at least $b + 1$ for sure.

5.2 IPSEC

IPSEC [9] can use CBC-PAD. Default padding scheme is similar, as specified in ESP [10]. Standards clearly mention that the padding should be checked, but the standard behavior in the case of invalid padding is quite strange: the server just discards the invalid message and adds a notification in log files for audit and nothing else. This simply means that errors are processed according to non standard rules or by another protocol layer. It is reasonable to assume that the lack of activity of the receiver in this case, or the activity of the auditor, can be converted into one bit of information. So our attack may be applicable.

IPSEC provides an optional authentication mechanism which could protect against our attack, provided that the authentication check is performed *before* the format check of the plaintext. Although used in most of practical applications, this mechanism still has an optional status in IPSEC. As already recommended by Bellovin [4], authentication should be mandatory. Bellovin actually used a side channel which tells the validity of the TCP checksum. His attack was recently extended to the WEP protocol for 802.11 wireless LANs by Borisov et al. [6].

5.3 WTLS

WTLS [1] (which is the SSL variant for WAP) perfectly implements the oracle $\mathcal{O}$ by sending `decryption_failed` warnings in clear. Actually since mobile telephones have a limited power and CPU resources, key establishment protocols with public key cryptography are limited. So we try to limit the number of session initializations and to avoid breaking them. So seldom errors are fatal alerts. Some implementations of WTLS can however limit the tolerance number of errors within the same session, which can limit the efficiency of the attack. This is however non standard.

In the case of mobile telephones (which is the main application of WTLS), WTLS is usually encapsulated in other protocols which may provide their own encryption protocol, for instance GSM. In this case, the extra encryption layer needs to be bypassed by the attacker.

5.4 SSH2

In SSH2, the MAC is optional. When not used, our attack is feasible, but only recovers one word since the padding is mostly random. When used, the MAC is computed on the padded message. Therefore, it is checked before the padding format, which protects against our attack.

6 Fixes Which Do Not Work

6.1 Padding before the Message

One can propose to put the padding in the first block. This only works for CBC modes in which IV is not sent in clear with the ciphertext (otherwise the same attack holds). This also requires to know the total length (modulo b) of the message that we want to encrypt before starting the encryption. When the plaintext is a word stream, this assumption is not usually satisfied. Therefore we believe that this fix is not satisfactory.

6.2 CBCCBC Mode

Another possibility consists of replacing the CBC mode by a double CBC encryption (i.e. by re-encrypting the $y_1, \ldots, y_N$ sequence in CBC mode). We call it the CBCCBC mode.

Unfortunately, a similar attack holds: given y and z we can recover the value of $u = C^{-1}(y) \oplus C^{-1}(y \oplus C^{-1}(z))$ by sending $r|y|z$ trials to the oracle. This is enough in order to decrypt messages: if y is the $(i-1)$th ciphertext block, z is the ith ciphertext block, and if t is the $(i-2)$th ciphertext block, then the ith plaintext block is nothing but $t \oplus u$!

The same attack holds with a triple CBC mode...

6.3 On-Line Ciphers and HCBC Mode

We can look for another mode of operation which "provably" leaks no information. One should however try to keep the advantages of the CBC mode: being able to encrypt a stream without knowing the total length, without having to keep an expanding memory, ... In [3], Bellare et al. presented the notion of on-line cipher. This notion is well adapted for these advantages of the CBC mode.

They also proposed the HCBC mode as a secure on-line cipher against chosen plaintext attacks. The idea consists in replacing Eq. (1) by

$$y_i = C(H(y_{i-1}) \oplus x_i)$$

where H is a XOR-universal hash function which includes part of the secret key. For instance one can propose $H(x) = K_1 x$ in $\mathrm{GF}(W^b)$ where $K_1 \neq 0$ is part of the secret key. (For any fixed a, b, c with $a \neq b$, we have $\Pr[H(a) \oplus H(b) = c] \leq 1/(W^b - 1)$ if K_1 is uniformly distributed, thus H is XOR-universal.)

One problem is that this does not protect against the kind of attack we proposed. For instance we notice that if we get several accepted $r_i|y$ messages with a fixed y, then we deduce that $H(r_i) \oplus x$ ends with a valid padding for an unknown but fixed x. Hence $H(r_i) \oplus H(r_j)$ is likely to end with the word zero. Since this is the last word of $K_1(r_i \oplus r_j)$, we deduce K_1 from several (i, j) pairs. With the knowledge of K_1 we then adapt the attack against the raw CBC. It is even more dramatic here since we indeed recover a part of the secret key.

We outline that with the particular choice of XOR-universal hash function, the claimed security result collapses. Of course, there is no contradiction with the security result since our attack gets extra information from the side channel oracle $\mathcal{O}$, which was not allowed in the security model of [3]: the notion of on-line cipher resistant against chosen plaintext attacks does not capture security against the kind of cryptanalysis that we have proposed.

6.4 Other Modes of Operation

The first stage of the standardization process on modes of operation launched by NIST also contained problematic proposals.[2] Several of the proposals could be generalized as follows. The CBC mode is modified in order to have a XOR before and after the block cipher encryption, depending on all previous ciphertext blocks and all previous plaintext blocks. We replace Eq. (1) by

$$y_i = C(x_i \oplus f_i(x, y)) \oplus g_i(x, y)$$

with public $f_i(x, y)$ and $g_i(x, y)$ functions which only depend on i and all x_j and y_j for $j = 1, \ldots, i-1$. (Note that HCBC is not an example since f_i is not public.)

Assuming that an attacker knows several (x^j, y^j) plaintext-ciphertext pairs written $x^j = x_1^j | \ldots | x_{\ell_j}^j$ and $y^j = y_1^j | \ldots | y_{\ell_j}^j$, and she wants to compute $C^{-1}(y)$ for some given y, she can submit some $y_1^j | \ldots | y_k^j | (y \oplus \delta)$ ciphertexts where $k \leq \ell_j$, $\delta = g_{k+1}(x^j, y^j)$. Acceptance would mean that the block $C^{-1}(y) \oplus f_{k+1}(x^j, y^j)$ ends with a valid padding. Therefore we can decrypt the rightmost word with W samples, two words with W^2 samples, ...

6.5 CBC-PAD with Integrity Check

One can propose to add a cryptographic checkable redundancy code (crypto-CRC) of the whole padded message (like a hashed value) in the plaintext and encrypt

$$\text{message}|\text{padding}|h(\text{message}|\text{padding}).$$

This way, any forged ciphertext will have a negligible probability to be accepted as a valid ciphertext. Basically, attackers are no longer able to forge valid ciphertexts, so the scheme is virtually resistant against chosen ciphertext attacks.

Obviously it is important to pad before hashing: padding after hashing would lead to the a similar attack. The right enciphering sequence is thus

pad, hash, encrypt

Conversely, the right deciphering sequence consists of decrypting, checking the hashed value, then checking the padding value. Invalid hashed value must abort the decipherment.

[2] See `http://csrc.nist.gov/encryption/modes/`

There is still a nice security flaw discovered by David Wagner for this scheme in a subtle attack model.[3] We perform a "semi-chosen plaintext attack": we assume that we can convince the sender to send a message consisting of an unknown x (of known length) concatenated by a chosen postfix, and the goal is to get information on x. We can implement a guess check oracle: if g is a guess for x, we ask the sender to concatenate x with $y|h(g|y)$ where y is such that $g|y$ is a valid padded message. The sender then pads $x|y|h(g|y)$ (with a constant block $bb\ldots b$), appends a message digest, and encrypts the whole sequence in CBC mode. The attacker can then truncate the ciphertext after the $h(g|y)$. If the receiver accepts the truncated message, it means that the guess was right!

7 A Fix Which May Work

The author of this paper first thought that using authentication in the CBC-PAD in order to thwart the attacks was an overkill. Wagner's attack demonstrates that it actually is not. We thus propose to replace the CBC encryption by a scheme which simultaneously provides authentication and confidentiality. As having padding in between authentication and encryption (as is done in TLS) is not a fortunate idea, the authentication-encryption scheme *must* apply on padded plaintexts:

1. take the cleartext
2. pad the message and take the padded message x
3. authenticate and encrypt x
4. transmit the result y

Similarly, the authentication check and decryption *must* be performed *before* the padding check:

1. decrypt and check the authenticity of y
2. take the plaintext x
3. check the padding of x
4. extract the padding and get the cleartext

(Note: we used the subtle difference between "cleartext" and "plaintext" as specified in RFC 2828 [18]: the cleartext is the original message in clear, and the plaintext is the input of the encryption process.)

The question whether authentication must be done before encryption or not is another problem. As an example, Krawczyk [11] recently demonstrated the security of the authenticate-then-CBC-encrypt scheme. We must however be careful about the meaning of this security result: in this proof, attackers are not assumed to have access to side channel oracles like in our model. Therefore the security result may collapse when using an appropriate oracle as for the HCBC mode. Therefore it is not quite clear how this result extends in a model where we can have side channel. We leave this as an open problem.

Despite the lack of formal security result, we believe that this scheme offers the required security.

---

[3] Private communication from David Wagner.

8 Conclusion

We have shown that several popular padding schemes which are used in order to transform block ciphers into variable-input-length encryption schemes introduce an important security flaw. Correctness of the plaintext format is indeed a hard core bit which easily leaks out from the communication protocol.

It confirms that security analysis must not be limited to the block cipher but must rather be considered within the whole environment: as was raised by Bellovin [4] and Borisov et al. [6], we can really have insecure standards which use unbroken cryptographic primitives. This was already well known in the public key cryptography world. We have demonstrated that the situation of symmetric cryptography is virtually the same.

Acknowledgments

I would like to thank Pascal Junod for helpful discussions. I also thank my students from EPFL for having suffered on this attack as an examination topic. After this attack was released at the Rump session of CRYPTO'01, several people provided valuable feedback. I would in particular like to thank Bodo Möller, Alain Hiltgen, Wenbo Mao, Ulrich Kühn, Tom St Denis, David Wagner, and Martin Hirt. I also thank the anonymous referees for their extensive and pertinent comments.

References

1. Wireless Transport Layer Security. Wireless Application Protocol WAP-261-WTLS-20010406-a. Wireless Application Protocol Forum, 2001. `http://www.wapforum.org/`
2. R. Baldwin, R. Rivest. The RC5, RC5-CBC, RC5-CBC-Pad, and RC5-CTS Algorithms RFC 2040, 1996.
3. M. Bellare, A. Boldyreva, L. Knudsen, C Namprempre. Online Ciphers and the Hash-CBC Construction. In *Advances in Cryptology CRYPTO'01*, Santa Barbara, California, U.S.A., Lectures Notes in Computer Science 2139, pp. 292–309, Springer-Verlag, 2001.
4. S. Bellovin. Problem Areas for the IP Security Protocols. In *Proceedings of the 6th Usenix UNIX Security Symposium*, San Jose, California, USENIX, 1996.
5. D. Bleichenbacher. Chosen Ciphertext Attacks Against Protocols Based on the RSA Encryption Standard PKCS#1. In *Advances in Cryptology CRYPTO'98*, Santa Barbara, California, U.S.A., Lectures Notes in Computer Science 1462, pp. 1–12, Springer-Verlag, 1998.
6. N. Borisov, I. Goldberg, D. Wagner. Intercepting Mobile Communications: The Insecurity of 802.11. In *Proceedings of the 7th Annual International Conference on Mobile Computing and Networking*, ACM Press, 2001.
7. T. Dierks, C. Allen. The TLS Protocol Version 1.0. RFC 2246, standard tracks, the Internet Society, 1999.
8. M. Dworkin. Recommendation for Block Cipher Modes of Operation. US Department of Commerce, NIST Special Publication 800-38A, 2001.

9. S. Kent, R. Atkinson. Security Architecture for the Internet Protocol. RFC 2401, standard tracks, the Internet Society, 1998.
10. S. Kent, R. Atkinson. IP Encapsulating Security Payload (ESP). RFC 2406, standard tracks, the Internet Society, 1998.
11. H. Krawczyk. The Order of Encryption and Authentication for Protecting Communications (or: How Secure is SSL?). In *Advances in Cryptology CRYPTO'01*, Santa Barbara, California, U.S.A., Lectures Notes in Computer Science 2139, pp. 310–331, Springer-Verlag, 2001.
12. L.R. Knudsen. *Block Ciphers — Analysis, Design and Applications*, Aarhus University, 1994.
13. J. Manger. A Chosen Ciphertext Attack on RSA Optimal Asymmetric Encryption Padding (OAEP) as Standardized in PKCS#1 v2.0. In *Advances in Cryptology CRYPTO'01*, Santa Barbara, California, U.S.A., Lectures Notes in Computer Science 2139, pp. 230–238, Springer-Verlag, 2001.
14. A.J. Menezes, P.C. van Oorschot, S.A. Vanston. *Handbook of Applied Cryptography*, CRC, 1997.
15. E. Petrank, C. Rackoff. CBC MAC for Real-Time Data Sources. *Journal of Cryptology*, vol. 13, pp. 315–338, 2000.
16. B. Preneel, P. C. van Oorschot. Mdx-MAC and Building Fast MACs from Hash Functions. In *Advances in Cryptology CRYPTO'95*, Santa Barbara, California, U.S.A., Lectures Notes in Computer Science 963, pp. 1–14, Springer-Verlag, 1995.
17. B. Schneier. *Applied Cryptography*, 2nd Edition, John Wiley & Sons, 1996.
18. R. Shirey. Internet Security Glossary. RFC 2828, the Internet Society, 2000.
19. S. Vaudenay. Decorrelation over Infinite Domains: the Encrypted CBC-MAC Case. In *Selected Areas in Cryptography'00*, Waterloo, Ontario, Canada, Lectures Notes in Computer Science 2012, pp. 189–201, Springer-Verlag, 2001. Journal version: *Communications in Information and Systems*, vol. 1, pp. 75–85, 2001.

Author Index

Abdalla, Michel 418
An, Jee Hea 83, 418

Bagini, Vittorio 238
Beaver, Donald 352
Bellare, Mihir 418
Black, John 384
Bresson, Emmanuel 321

Canetti, Ran 337
Canteaut, Anne 518
Chevassut, Olivier 321
Chose, Philippe 209
Clark, John A. 181
Coron, Jean-Sébastien 272
Cramer, Ronald 45

Daemen, Joan 108
Damgård, Ivan 256
Desai, Anand 368
Desmedt, Yvo 502
Dijk, Marten van 149
Dodis, Yevgeniy 65, 83

Fitzi, Matthias 482

Galbraith, Steven D. 29
Gennaro, Rosario 1
Gentry, Craig 299
Gilbert, Henri 288
Gisin, Nicolas 482
Golić, Jovan Dj. 238

Hanaoka, Goichiro 434
Hess, Florian 29
Hevia, Alejandro 368
Horwitz, Jeremy 466

Imai, Hideki 434

Jacob, Jeremy L. 181
Jeong, Sangtae 197
Joux, Antoine 209

Katz, Jonathan 65
Kent, Stephen 398
Kiayias, Aggelos 450
Koprowski, Maciej 256

Krause, Matthias 222
Krawczyk, Hugo 337

Lee, Eonkyung 14
Lee, Sang Jin 14
Lim, Jongin 197
Lynn, Ben 466

Malkin, Tal 400
Matsui, Mitsuru 165
Maurer, Ueli 110, 482
Micciancio, Daniele 1, 400
Miner, Sara 400
Minier, Marine 288
Mitton, Michel 209
Morgari, Guglielmo 238

Nakajima, Junko 165
Namprempre, Chanathip 418

Park, Young-Ho 197
Pointcheval, David 321

Rabin, Tal 83
Rijmen, Vincent 108
Rogaway, Phillip 384
Rotz, Oliver von 482
Russell, Alexander 133

Shikata, Junji 434
Shoup, Victor 45
Smart, Nigel P. 29
Szydlo, Mike 299

Vaudenay, Serge 534
Videau, Marion 518

Wang, Hong 133
Wang, Yongge 502
Woodruff, David P. 149

Xu, Shouhuai 65

Yin, Yiqun Lisa 368
Yung, Moti 65, 450

Zheng, Yuliang 434

Lecture Notes in Computer Science

For information about Vols. 1–2248
please contact your bookseller or Springer-Verlag

Vol. 2248: C. Boyd (Ed.), Advances in Cryptology – ASIACRYPT 2001. Proceedings, 2001. XI, 603 pages. 2001.

Vol. 2249: K. Nagi, Transactional Agents. XVI, 205 pages. 2001.

Vol. 2250: R. Nieuwenhuis, A. Voronkov (Eds.), Logic for Programming, Artificial Intelligence, and Reasoning. Proceedings, 2001. XV, 738 pages. 2001. (Subseries LNAI).

Vol. 2251: Y.Y. Tang, V. Wickerhauser, P.C. Yuen, C.Li (Eds.), Wavelet Analysis and Its Applications. Proceedings, 2001. XIII, 450 pages. 2001.

Vol. 2252: J. Liu, P.C. Yuen, C. Li, J. Ng, T. Ishida (Eds.), Active Media Technology. Proceedings, 2001. XII, 402 pages. 2001.

Vol. 2253: T. Terano, T. Nishida, A. Namatame, S. Tsumoto, Y. Ohsawa, T. Washio (Eds.), New Frontiers in Artificial Intelligence. Proceedings, 2001. XXVII, 553 pages. 2001. (Subseries LNAI).

Vol. 2254: M.R. Little, L. Nigay (Eds.), Engineering for Human-Computer Interaction. Proceedings, 2001. XI, 359 pages. 2001.

Vol. 2255: J. Dean, A. Gravel (Eds.), COTS-Based Software Systems. Proceedings, 2002. XIV, 257 pages. 2002.

Vol. 2256: M. Stumptner, D. Corbett, M. Brooks (Eds.), AI 2001: Advances in Artificial Intelligence. Proceedings, 2001. XII, 666 pages. 2001. (Subseries LNAI).

Vol. 2257: S. Krishnamurthi, C.R. Ramakrishnan (Eds.), Practical Aspects of Declarative Languages. Proceedings, 2002. VIII, 351 pages. 2002.

Vol. 2258: P. Brazdil, A. Jorge (Eds.), Progress in Artificial Intelligence. Proceedings, 2001. XII, 418 pages. 2001. (Subseries LNAI).

Vol. 2259: S. Vaudenay, A.M. Youssef (Eds.), Selected Areas in Cryptography. Proceedings, 2001. XI, 359 pages. 2001.

Vol. 2260: B. Honary (Ed.), Cryptography and Coding. Proceedings, 2001. IX, 416 pages. 2001.

Vol. 2261: F. Naumann, Quality-Driven Query Answering for Integrated Information Systems. X, 166 pages. 2002.

Vol. 2262: P. Müller, Modular Specification and Verification of Object-Oriented Programs. XIV, 292 pages. 2002.

Vol. 2263: T. Clark, J. Warmer (Eds.), Object Modeling with the OCL. VIII, 281 pages. 2002.

Vol. 2264: K. Steinhöfel (Ed.), Stochastic Algorithms: Foundations and Applications. Proceedings, 2001. VIII, 203 pages. 2001.

Vol. 2265: P. Mutzel, M. Jünger, S. Leipert (Eds.), Graph Drawing. Proceedings, 2001. XV, 524 pages. 2002.

Vol. 2266: S. Reich, M.T. Tzagarakis, P.M.E. De Bra (Eds.), Hypermedia: Openness, Structural Awareness, and Adaptivity. Proceedings, 2001. X, 335 pages. 2002.

Vol. 2267: M. Cerioli, G. Reggio (Eds.), Recent Trends in Algebraic Development Techniques. Proceedings, 2001. X, 345 pages. 2001.

Vol. 2268: E.F. Deprettere, J. Teich, S. Vassiliadis (Eds.), Embedded Processor Design Challenges. VIII, 327 pages. 2002.

Vol. 2269: S. Diehl (Ed.), Software Visualization. Proceedings, 2001. VIII, 405 pages. 2002.

Vol. 2270: M. Pflanz, On-line Error Detection and Fast Recover Techniques for Dependable Embedded Processors. XII, 126 pages. 2002.

Vol. 2271: B. Preneel (Ed.), Topics in Cryptology – CT-RSA 2002. Proceedings, 2002. X, 311 pages. 2002.

Vol. 2272: D. Bert, J.P. Bowen, M.C. Henson, K. Robinson (Eds.), ZB 2002: Formal Specification and Development in Z and B. Proceedings, 2002. XII, 535 pages. 2002.

Vol. 2273: A.R. Coden, E.W. Brown, S. Srinivasan (Eds.), Information Retrieval Techniques for Speech Applications. XI, 109 pages. 2002.

Vol. 2274: D. Naccache, P. Paillier (Eds.), Public Key Cryptography. Proceedings, 2002. XI, 385 pages. 2002.

Vol. 2275: N.R. Pal, M. Sugeno (Eds.), Advances in Soft Computing – AFSS 2002. Proceedings, 2002. XVI, 536 pages. 2002. (Subseries LNAI).

Vol. 2276: A. Gelbukh (Ed.), Computational Linguistics and Intelligent Text Processing. Proceedings, 2002. XIII, 444 pages. 2002.

Vol. 2277: P. Callaghan, Z. Luo, J. McKinna, R. Pollack (Eds.), Types for Proofs and Programs. Proceedings, 2000. VIII, 243 pages. 2002.

Vol. 2278: J.A. Foster, E. Lutton, J. Miller, C. Ryan, A.G.B. Tettamanzi (Eds.), Genetic Programming. Proceedings, 2002. XI, 337 pages. 2002.

Vol. 2279: S. Cagnoni, J. Gottlieb, E. Hart, M. Middendorf, G.R. Raidl (Eds.), Applications of Evolutionary Computing. Proceedings, 2002. XIII, 344 pages. 2002.

Vol. 2280: J.P. Katoen, P. Stevens (Eds.), Tools and Algorithms for the Construction and Analysis of Systems. Proceedings, 2002. XIII, 482 pages. 2002.

Vol. 2281: S. Arikawa, A. Shinohara (Eds.), Progress in Discovery Science. XIV, 684 pages. 2002. (Subseries LNAI).

Vol. 2282: D. Ursino, Extraction and Exploitation of Intensional Knowledge from Heterogeneous Information Sources. XXVI, 289 pages. 2002.

Vol. 2283: T. Nipkow, L.C. Paulson, M. Wenzel, Isabelle/HOL. XIII, 218 pages. 2002.

Vol. 2284: T. Eiter, K.-D. Schewe (Eds.), Foundations of Information and Knowledge Systems. Proceedings, 2002. X, 289 pages. 2002.

Vol. 2285: H. Alt, A. Ferreira (Eds.), STACS 2002. Proceedings, 2002. XIV, 660 pages. 2002.

Vol. 2286: S. Rajsbaum (Ed.), LATIN 2002: Theoretical Informatics. Proceedings, 2002. XIII, 630 pages. 2002.

Vol. 2287: C.S. Jensen, K.G. Jeffery, J. Pokorny, Saltenis, E. Bertino, K. Böhm, M. Jarke (Eds.), Advances in Database Technology – EDBT 2002. Proceedings, 2002. XVI, 776 pages. 2002.

Vol. 2288: K. Kim (Ed.), Information Security and Cryptology – ICISC 2001. Proceedings, 2001. XIII, 457 pages. 2002.

Vol. 2289: C.J. Tomlin, M.R. Greenstreet (Eds.), Hybrid Systems: Computation and Control. Proceedings, 2002. XIII, 480 pages. 2002.

Vol. 2291: F. Crestani, M. Girolami, C.J. van Rijsbergen (Eds.), Advances in Information Retrieval. Proceedings, 2002. XIII, 363 pages. 2002.

Vol. 2292: G.B. Khosrovshahi, A. Shokoufandeh, A. Shokrollahi (Eds.), Theoretical Aspects of Computer Science. IX, 221 pages. 2002.

Vol. 2293: J. Renz, Qualitative Spatial Reasoning with Topological Information. XVI, 207 pages. 2002. (Subseries LNAI).

Vol. 2295: W. Kuich, G. Rozenberg, A. Salomaa (Eds.), Developments in Language Theory. Proceedings, 2001. IX, 389 pages. 2002.

Vol. 2296: B. Dunin-Kęplicz, E. Nawarecki (Eds.), From Theory to Practice in Multi-Agent Systems. Proceedings, 2001. IX, 341 pages. 2002. (Subseries LNAI).

Vol. 2297: R. Backhouse, R. Crole, J. Gibbons (Eds.), Algebraic and Coalgebraic Methods in the Mathematics of Program Construction. Proceedings, 2000. XIV, 387 pages. 2002.

Vol. 2299: H. Schmeck, T. Ungerer, L. Wolf (Eds.), Trends in Network and Pervasive Computing – ARCS 2002. Proceedings, 2002. XIV, 287 pages. 2002.

Vol. 2300: W. Brauer, H. Ehrig, J. Karhumäki, A. Salomaa (Eds.), Formal and Natural Computing. XXXVI, 431 pages. 2002.

Vol. 2301: A. Braquelaire, J.-O. Lachaud, A. Vialard (Eds.), Discrete Geometry for Computer Imagery. Proceedings, 2002. XI, 439 pages. 2002.

Vol. 2302: C. Schulte, Programming Constraint Services. XII, 176 pages. 2002. (Subseries LNAI).

Vol. 2303: M. Nielsen, U. Engberg (Eds.), Foundations of Software Science and Computation Structures. Proceedings, 2002. XIII, 435 pages. 2002.

Vol. 2304: R.N. Horspool (Ed.), Compiler Construction. Proceedings, 2002. XI, 343 pages. 2002.

Vol. 2305: D. Le Métayer (Ed.), Programming Languages and Systems. Proceedings, 2002. XII, 331 pages. 2002.

Vol. 2306: R.-D. Kutsche, H. Weber (Eds.), Fundamental Approaches to Software Engineering. Proceedings, 2002. XIII, 341 pages. 2002.

Vol. 2307: C. Zhang, S. Zhang, Association Rule Mining. XII, 238 pages. 2002. (Subseries LNAI).

Vol. 2308: I.P. Vlahavas, C.D. Spyropoulos (Eds.), Methods and Applications of Artificial Intelligence. Proceedings, 2002. XIV, 514 pages. 2002. (Subseries LNAI).

Vol. 2309: A. Armando (Ed.), Frontiers of Combining Systems. Proceedings, 2002. VIII, 255 pages. 2002. (Subseries LNAI).

Vol. 2310: P. Collet, C. Fonlupt, J.-K. Hao, E. Lutton, M. Schoenauer (Eds.), Artificial Evolution. Proceedings, 2001. XI, 375 pages. 2002.

Vol. 2311: D. Bustard, W. Liu, R. Sterritt (Eds.), Soft-Ware 2002: Computing in an Imperfect World. Proceedings, 2002. XI, 359 pages. 2002.

Vol. 2312: T. Arts, M. Mohnen (Eds.), Implementation of Functional Languages. Proceedings, 2001. VII, 187 pages. 2002.

Vol. 2313: C.A. Coello Coello, A. de Albornoz, L.E. Sucar, O.Cairó Battistutti (Eds.), MICAI 2002: Advances in Artificial Intelligence. Proceedings, 2002. XIII, 548 pages. 2002. (Subseries LNAI).

Vol. 2314: S.-K. Chang, Z. Chen, S.-Y. Lee (Eds.), Recent Advances in Visual Information Systems. Proceedings, 2002. XI, 323 pages. 2002.

Vol. 2315: F. Arhab, C. Talcott (Eds.), Coordination Models and Languages. Proceedings, 2002. XI, 406 pages. 2002.

Vol. 2316: J. Domingo-Ferrer (Ed.), Inference Control in Statistical Databases. VIII, 231 pages. 2002.

Vol. 2317: M. Hegarty, B. Meyer, N. Hari Narayanan (Eds.), Diagrammatic Representation and Inference. Proceedings, 2002. XIV, 362 pages. 2002. (Subseries LNAI).

Vol. 2318: D. Bošnački, S. Leue (Eds.), Model Checking Software. Proceedings, 2002. X, 259 pages. 2002.

Vol. 2319: C. Gacek (Ed.), Software Reuse: Methods, Techniques, and Tools. Proceedings, 2002. XI, 353 pages. 2002.

Vol. 2322: V. Mařík, O. Štěpánková, H. Krautwurmová, M. Luck (Eds.), Multi-Agent Systems and Applications II. Proceedings, 2001. XII, 377 pages. 2002. (Subseries LNAI).

Vol. 2324: T. Field, P.G. Harrison, J. Bradley, U. Harder (Eds.), Computer Performance Evaluation. Proceedings, 2002. XI, 349 pages. 2002.

Vol. 2329: P.M.A. Sloot, C.J.K. Tan, J.J. Dongarra, A.G. Hoekstra (Eds.), Computational Science – ICCS 2002. Proceedings, Part I. XLI, 1095 pages. 2002.

Vol. 2330: P.M.A. Sloot, C.J.K. Tan, J.J. Dongarra, A.G. Hoekstra (Eds.), Computational Science – ICCS 2002. Proceedings, Part II. XLI, 1115 pages. 2002.

Vol. 2331: P.M.A. Sloot, C.J.K. Tan, J.J. Dongarra, A.G. Hoekstra (Eds.), Computational Science – ICCS 2002. Proceedings, Part III. XLI, 1227 pages. 2002.

Vol. 2332: L. Knudsen (Ed.), Advances in Cryptology – EUROCRYPT 2002. Proceedings, 2002. XII, 547 pages. 2002.